Drug Discovery

Drug Discovery

Practices, Processes, and Perspectives

Edited by

Jie Jack Li
Bristol-Myers Squibb Company

and

E. J. Corey
Harvard University

A JOHN WILEY & SONS, INC., PUBLICATION

Published by John Wiley & Sons, Inc., Hoboken, New Jersey

Published simultaneously in Canada

For general information on our other products and services or for technical support, please contact our Customer Care Department within the United States at (800) 762-2974, outside the United States at (317) 572-3993 or fax (317) 572-4002.

Wiley also publishes its books in a variety of electronic formats. Some content that appears in print may not be available in electronic formats. For more information about Wiley products, visit our web site at www.wiley.com.

Library of Congress Cataloging-in-Publication Data

Drug discovery : practices, processes, and perspectives / edited by Jie Jack Li, E.J. Corey.
    p. ; cm.
Includes bibliographical references and index.
ISBN 978-0-470-94235-2 (cloth)
I. Li, Jie Jack. II. Corey, E. J.
[DNLM: 1. Drug Discovery—methods. 2. Chemistry, Pharmaceutical—methods. QV 745]

615.1'9—dc23   2012032885

Printed in the United States of America.

10 9 8 7 6 5 4 3 2 1

Dedicated To
Professor Paul Floreancig

Table of Contents

Detailed Table of Contents

Preface

The primary audience of this book is scientists in drug discovery in general and novice medicinal chemists in particular. In academia, most of us are trained as synthetic organic chemists. It takes us many years to learn the trade of drug discovery "on-the-job." This book is to jump-start our understanding of the landscape of drug discovery. Veteran medicinal chemists and process chemists will benefit from this book as well by learning the perspectives of drug discovery authored by experts in different fields of drug discovery.

We are indebted to contributing authors, all of whom are the world's leading experts in drug discovery, for their hard work and care in writing definitive summaries of their respective fields. The pharmaceutical industry is going through an interesting time. We hope that this book will help all scientists in drug discovery appreciate the practices, processes, and perspectives of drug discovery.

Jack Li and E. J. Corey

August 1, 2012

Contributing Authors

Dr. Narendra B. Ambhaikar
Dr. Reddy's Laboratories
CPS
Bollaram Road, Miyapur
Hyderabad-500 049
Andhra Pradesh, India

Dr. Makonen Belema
Medicinal Chemistry
Bristol-Myers Squibb Company
5 Research Parkway
Wallingford, CT 06492, United States

Dr. Ana B. Bueno
DCR&T
Eli Lilly and Company
Avda de la Industria,
28108-Alcobendas
Madrid, Spain

Dr. Ana M. Castaño
DCR&T
Eli Lilly and Company
Avda de la Industria, 30
28108-Alcobendas
Madrid, Spain

Dr. Audrey Chan
Cubist Pharmaceuticals, Inc.
65 Hayden Ave
Lexington, MA 02421, United States

Dr. Christopher W. Cianci
Virology
Bristol-Myers Squibb Company
5 Research Parkway
Wallingford, CT 06492, United States

Dr. Jason Cross
Cubist Pharmaceuticals, Inc.
65 Hayden Ave
Lexington, MA 02421, United States

Dr. Stanley V. D'Andrea
Medicinal Chemistry
Bristol-Myers Squibb Company
5 Research Parkway
Wallingford, CT 06492, United State

Dr. Ira B. Dicker
Virology
Bristol-Myers Squibb Company
5 Research Parkway
Wallingford, CT 06492, United State

Dr. Jeremy J. Edmunds
Medicinal Chemistry
Abbott Bioresearch Center
381 Plantation St.
Worcester, MA, 01605, United States

Dr. Yong He
Cubist Pharmaceuticals, Inc.
65 Hayden Ave
Lexington, MA 02421, United States

Dr. Adam R. Johnson
Biochemical Pharmacology
Genentech Inc.
1 DNA Way
South San Francisco, CA 94080
United States

Dr. Mark R. Krystal
Virology
Bristol-Myers Squibb Company
5 Research Parkway
Wallingford, CT 06492, United States

Dr. Kimberley Lentz
Metabolism and Pharmacokinetics
Bristol-Myers Squibb Company
5 Research Parkway
Wallingford, CT 06492, United States

Dr. Jie Jack Li
Medicinal Chemistry
Bristol-Myers Squibb Company
Route 206 and Province Line Road
Princeton, NJ 08540, United States

Dr. Blaise Lippa
Cubist Pharmaceuticals, Inc.
65 Hayden Ave
Lexington, MA 02421, United States

Dr. John A. Lowe, III
JL3Pharma LLC
28 Coveside Lane
Stonington CT 06378, United States

Dr. Anthony M. Manning
Research & Preclinical Development
Aileron Therapeutics
281 Albany St.
Cambridge, MA 02139, United States

Dr. Nicholas A. Meanwell
Medicinal Chemistry
Bristol-Myers Squibb Company
5 Research Parkway
Wallingford, CT 06492, United States

Dr. Joseph Raybon
Metabolism and Pharmacokinetics
Bristol-Myers Squibb Company
5 Research Parkway
Wallingford, CT 06492, United States

Dr. Angel Rodríguez
Medical Affairs-Diabetes
Eli Lilly and Company
Avda de la Industria, 30
28108-Alcobendas
Madrid, Spain

Dr. M. Dominic Ryan
Cubist Pharmaceuticals, Inc.
65 Hayden Ave
Lexington, MA 02421, United States

Dr. Michael W. Sinz
Metabolism and Pharmacokinetics
Bristol-Myers Squibb Company
5 Research Parkway
Wallingford, CT 06492, United States

Dr. Drago R. Sliskovic
AAPharmasyn, LLC.
3985 Research Park Drive
Ann Arbor, MI 48108, United States

Dr. Kap-Sun Yeung
Medicinal Chemistry
Bristol-Myers Squibb Company
5 Research Parkway
Wallingford, CT 06492, United States

1

History of Drug Discovery

Jie Jack Li

1   Introduction

The history of drug discovery is as ancient as our glorious history of humanity. Shen Nung, *The Divine Farmer*, is fabled to have sampled 365 herbs himself to evaluate their medicinal value as early as 2337 B.C. in China. In 400 B.C. in Greece, Hippocrates, *The Father of Medicine*, decreed the *Hippocratic Oath*, in which a physician is to pledge "I will preserve the purity of my life and my arts". Another Greek Physician, Galen (129–199), influenced 45 generations with his teachings of medicine, transforming medicine from art to science. During the Renaissance, Paracelsus (1493–1541) from Switzerland represented the pinnacle of Western medicine. Since then, a parade of luminaries began to unveil the myth of life. Andreas Vesalius (1514–1564) of Brussels founded the *Science of Anatomy*. William Harvey (1578–1657) of England made one of the greatest discoveries in medicine—*Circulation of Blood*. Dutchman Antonie von Leeuwenhoek (1632–1723) opened our eyes to a whole new world of microbes by inventing the microscope.

The intellectual contributions of these great men established the foundation of modern medicine and heralded the golden age of contemporary drug discovery.

2   Antibacterials

2.1   *Lister and Carbolic Acid*

Although many contributed to the *germ theory*, French chemist Louis Pasteur (1822–1895) transformed medicine with his well-designed experiments and a treatise entitled *Organized Corpuscles Existing in Atmosphere* published in 1855.[1] Pasteur's *tour de force* officially introduced and cemented germ theory within mainstream science. The pasteurization process, heating the food at a specific temperature for a certain time and then cooling it immediately, is still the standard practice. Merely one year later, Joseph Lister (1827–1912) in England successfully applied the germ theory by using carbolic acid (phenol) as an antiseptic during surgery to kill bacteria.

As a surgeon, Lister was appalled at the postsurgical infections, which killed patients with an astonishing 40–60% mortality rate. However, simple and direct application of the pasteurization process would not be acceptable during surgery—after all one could not simply boil the patient in hot water! Inspired by anecdotal success stories of carbolic acid (an ingredient isolated from coal tar, a waste from coal gas production) in deodorizing sewage and in controlling typhoid, Lister introduced carbolic acid as an antiseptic in surgery.[2] It works by solubilizing the phospholipids in cell membranes, thus disrupting the cell membranes. Today, *asepsis* has largely replaced *antisepsis* in the operating rooms.

2.2   Dr. Ehrlich's Magic Bullet

In the 1890s, together with Emil von Behring, Paul Ehrlich (1854–1915) developed a horse serum antitoxin to quell diphtheria. The vaccine saved thousands of children's lives during the 1891 outbreak, for which he won the Nobel Prize in 1908. Since 1910, working with his Japanese associate Sachachio Hata, Ehrlich experimented with numerous chemicals to treat syphilis.[3] They had some success with an arsenic compound atoxyl, which was efficacious but was too toxic. His chemist Alfred Bertheim (1879–1914) at first elucidated the chemical constitution of atoxyl and later synthesized innumerable arsenobenzene compounds including arsphenamine (Ehrlich's 606), which was efficacious with an acceptable safety profile.[4] Ehrlich licensed the drug to Hoeschst, who sold it under the trade name Salvarsan. To find less toxic and more water-soluble ant-syphilitics, Bertheim synthesized neoarsphenamine (Neosalvarsan). Both Salvarsan and Neosalvarsan had a tremendous impact on fighting syphilis, wiping out half of the syphilis infections in Europe in a mere five years (although syphilis was not completely eradicated until the introduction of penicillin in the 1940s).

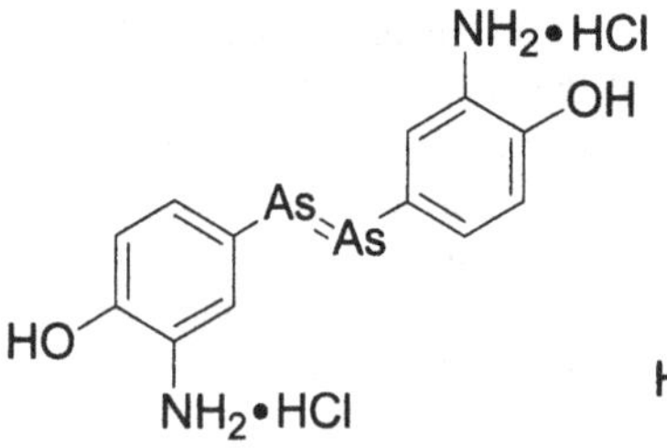

Ehrlich's 606
arsphenamine
(Salvarsan)

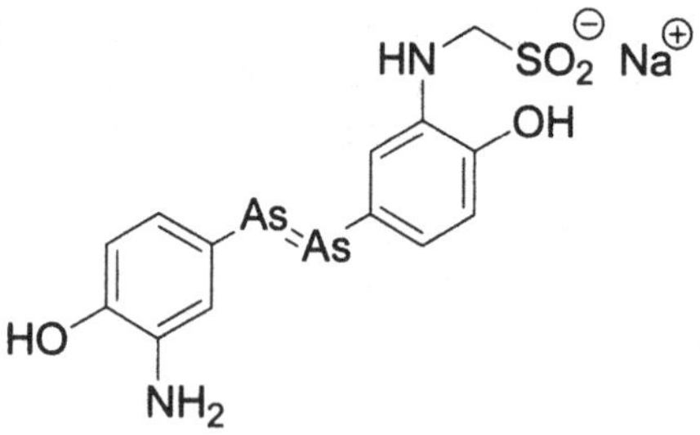

neoarsphenamine
(Neosalvarsan)

Ehrlich was also the first to propose the *side chain theory* and the *receptor theory* to explain how drugs worked. While he is immortalized as *the father of chemotherapy* and with his concept of *magic bullets*, Bertheim, probably the first medicinal chemist in history, is largely forgotten. In addition to his important contributions to the discovery of Salvarsan and Neosalvarsan, Bertheim also published a book *Ein Handbuch der organischen Arsenverbindungen* (*A Handbook of Organic Arsenic Compounds*).

2.3   Domagk and Sulfa Drugs

In 1932, Gerhard Domagk (1895–1964), the head of the Bacteriology Laboratory of I. G. Farbenindustrie Aktiengesellschaft (I. G. Farben), experimented with different dyes available to him in search of antibacterial drugs.[5] Looking for antibacterials from dyes was most likely influenced by Ehrlich's experiences with staining. By injecting dyes in mice infected with

Streptococcus pyogenes bacterium, Domagk discovered that 2′,4′-diaminoazobenzene-4-sulfonamide, later branded as Prontosil, was effective in killing the bacterium without unacceptable toxic effects. The dye was prepared by Josef Klarer (1898–1953), a chemist at the Bayer Company, a branch of I. G. Farben. Later on, another Bayer chemist, Fritz Mietzsch (1896–1958), prepared the salt of Prontosil, enabling liquid formulation that was more amenable for injection. After Domagk's first disclosure in 1935, Prontosil quickly became widely prescribed for streptococcal infections.

In 1935, a French husband and wife team, Professor Jacques Tréfouël and Madame Thérèse Tréfouël, discovered that Prontosil was not active *in vitro*. The real active ingredient for its antibacterial activity is sulfanilamide (mistakenly being called sulfonamide even today), which is generated from *in vivo* metabolism of Prontosil. The mechanism of action (MOA) of sulfanilamides (antimetabolites) is through folate antagonism. Since the structure of sulfanilamide is similar to that of *para*-aminobenzoic acid (PABA), an essential ingredient for cell synthesis, it interrupts bacterial growth.

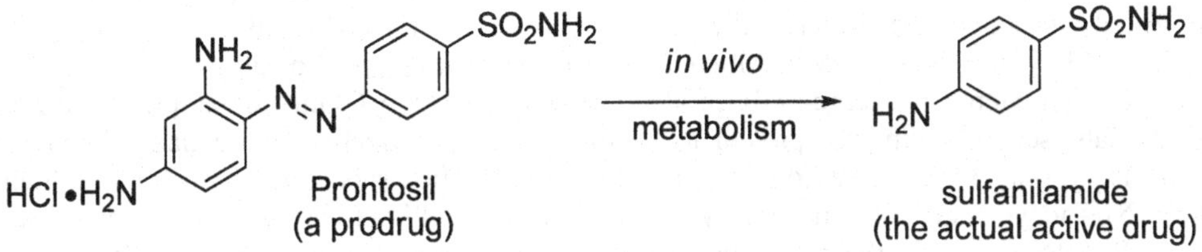

Prontosil
(a prodrug)

in vivo
metabolism

sulfanilamide
(the actual active drug)

Domagk was bestowed the Nobel Prize in 1939. He received numerous letters from patients and doctors, expressing their gratitude for his discovery of Prontosil. In contrast, Klarer, the chemist who first synthesized it, received none.

2.4   *Fleming, Florey, Chain, and Penicillin*

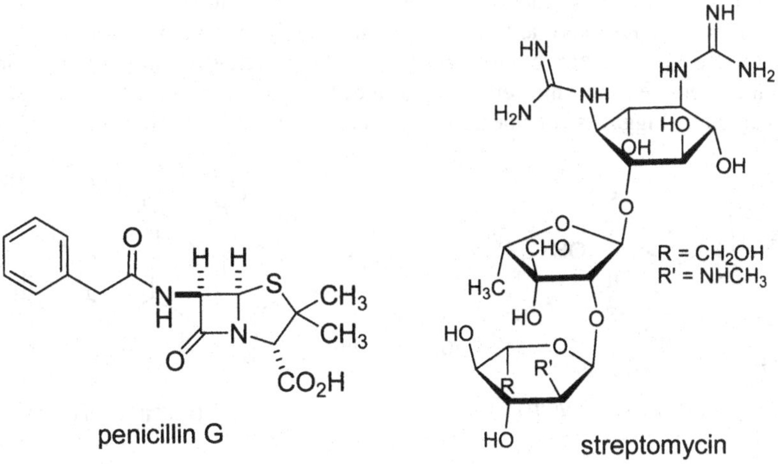

penicillin G

R = CH$_2$OH
R' = NHCH$_3$

streptomycin

Alexander Fleming (1881–1955)[6] actually discovered penicillin in 1928 in England, 4 years before Domagk's Prontosil.   However, more than 15 years elapsed until Howard Florey (1898–1968)[7] and Ernst Chain (1907–1979)[8] isolated enough penicillin and demonstrated its curative effects in both mice and humans. Penicillin quickly replaced Ehrlich's 606 and Domagk's sulfa drugs as the most widely used antibiotic. It works for Gram-positive bacterial infections, including strep and staph infections, pneumonia, gangrene, meningitis, as well as gonorrhea (now, however, a resistant form has emerged) and syphilis. The MOA of penicillin is through inhibition of cell wall synthesis. Because animals, including humans, lack a cell wall, penicillin exerts a bactericidal action selectively on growing or multiplying germs.

2.5   *Waksman, Schatz, and Streptomycin*

Inspired by Fleming's success with penicillin, Selman A. Waksman (1888–1973), a professor of soil microbiology at Rutgers College, began to look for antibiotics in soil in 1939.[9] At first his group isolated a small molecule antibiotic, actinomycin, and then streptothricin. Although both of them killed Gram-negative bacteria, they were so toxic that they also killed test animals.   In October 1943, Waksman's student Albert Schatz isolated streptomycin, an aminosugar.   With assistance from Merck for large-scale production and the Mayo Clinic for animal testing and clinical trials, streptomycin was proven to be both safe and effective in treating tuberculosis. Astonishingly, only 3 years elapsed from its discovery to the first successful treatment of a human patient. Nowadays, it generally takes 12 years and over $1.3 billion to bring a drug to the market.

Streptomycin was the first drug to be effective against Gram-negative bacteria. It was particularly interesting at the time because of its activity against human *tubercle bacillus,* which made it the first specific agent effective in treating tuberculosis. Streptomycin works by inducing the binding of "wrong" tRNA-amino-acid complexes, resulting in synthesis of false protein.

2.6   *Duggar, Conover, and Tetracyclines*

In 1945, 73 year old botanist Benjamin M. Duggar was a consultant for Lederle and led their screening efforts in the hunt for antibiotics.   Coincidentally, a sample from the University of Missouri, where Duggar taught botany 40 years earlier, yielded an antibiotic later named chlortetracycline. Lederle sold chlortetracycline under the brand name of Aureomycin in 1948. Nowadays, Benjamin Duggar is considered the pioneer of tetracycline antibiotics.[10]

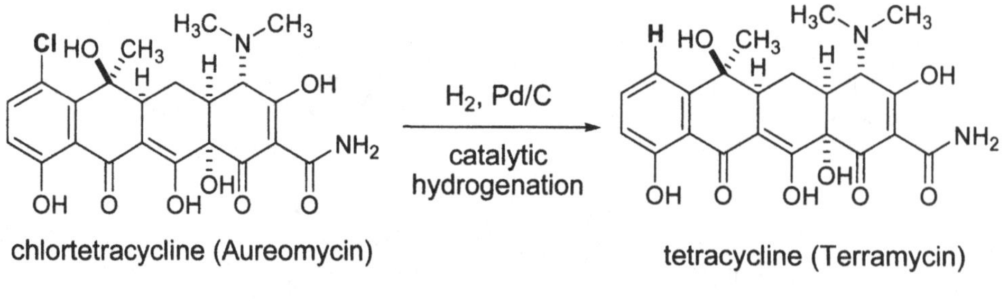

chlortetracycline (Aureomycin)   tetracycline (Terramycin)

In 1949, a yellow powder with strong antibiotic properties was isolated by Pfizer scientists from a soil sample. The soil organism was *Streptomyces rimosus* and the compound was generically known as oxytetracycline. Backtracking revealed that the soil sample was collected at the Terre Haute factory in Indiana owned by Pfizer, which later sold oxytetracycline under the brand name Terramycin.

Later on, Lloyd Conover at Pfizer stunned his colleagues by preparing another powerful antibiotic *chemically* from chlortetracycline. Under carefully controlled catalytic hydrogenation conditions, Conover converted Lederle's chlortetracycline to tetracycline. That was the first example of a semisynthetic compound with antibiotic activities.

Tetracyclines are inhibitors of protein synthesis by inhibiting the binding of tRNA-amino-acid complexes. They are bacteriostatic.

2.7   Quinolones and Zyvox

In 1946, while pursuing better antimalarial drugs by synthesizing chloroquine, George Y. Lesher (1926–1990) at Sterling Winthrop Research Institute at Rensselaer isolated a by-product, nalidixic acid.[11] It was found to be an antibacterial agent during routine screening. But it did not become popular until 1962 when Lesher introduced it into clinical practice for kidney infections. It was also used to treat urinary tract infections because it was excreted via urine in high concentration. Shortly after, the quinolone antibacterial field flourished, rendering thousands of 4-quinolone derivatives as represented by pipemidic acid. Nalidixic acid and pipemidic acid are considered the first-generation quinolone antibacterials. The drawbacks of these drugs were their moderate activity toward susceptible bacteria and poor absorption by the body.

In the early 1980s, fluorinated quinolone (fluoroquinolone) antibacterials were discovered to possess longer half-lives and better oral efficacy than the first-generation quinolones. These so-called second-generation quinolone antibacterials are exemplified by norfloxacin (the first fluoroquinolone discovered in 1980) and ciprofloxacin (Cipro).[12]

The third generation of quinolone antibacterials is still being actively investigated and is urgently needed due to the rapid development of resistance by bacteria toward existing antibacterial drugs. Examples of the third generation of quinolone antibacterials include fleroxacin and tosufloxacin. They are endowed with sufficiently long half-lives to enable a once-daily regimen, along with enhanced activity toward a variety of bacteria.

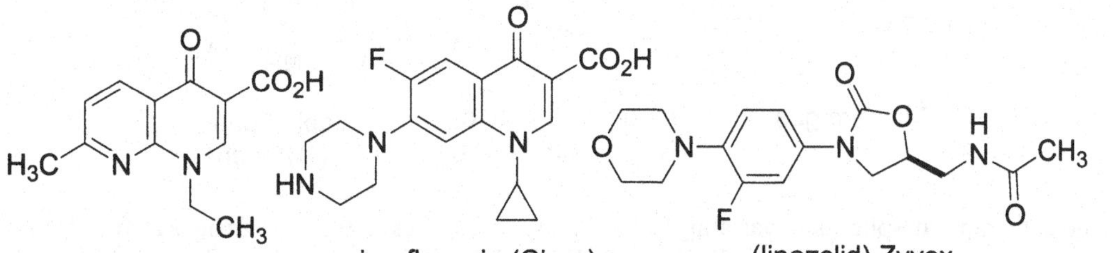

nalidixic acid   ciprofloxacin (Cipro)   (linezolid) Zyvox

The MOA of quinolones is through inhibition of bacterial gyrase (topoisomerase II), thus inhibiting DNA function. Quinolones do not affect human cells because topoisomerase II only exists in bacteria.

Another important category of antibacterial agents is the oxazolidinones as exemplified by linezolid (Zyvox).[13] The genesis of Zyvox began in 1978 when a DuPont patent described some novel oxazolidinones for controlling fungal and bacterial plant pathogens. These compounds and two subsequently optimized drug candidates, DuP 721 and DuP 105, did not become marketed drugs due to their unacceptable toxicities. Steven J. Brickner, a medicinal chemist working at Upjohn, was intrigued by some attributes of this class of compounds after learning of Dupont's exploits in 1987. He immediately began an exploratory oxazolidinone project. Working with two other groups led by Michael R. Barbachyn and Douglas K. Hutchinson, they prepared eperezolid and linezolid and commenced clinical trials for both compounds in 1995. Since linezolid was more advantageous than eperezolid in terms of its pharmacokinetics, linezolid was selected to move forward and won FDA approval in 2000. Pharmacia, which took over Upjohn in 1995, sold it under the trade name Zyvox.

Zyvox works by inhibition of the initial phase of bacterial protein synthesis. It is also a MAO (monoamine oxidase)-B inhibitor but without significant blood pressure liability—bringing an interesting closure to the origin of MAO inhibitors for depression, which came about from the improvement in mood in patients with tuberculosis treated with the original MAO inhibiting agents (see Section 6.2).

3   Cancer Drugs

3.1   *The Origin of Cancer*

Today, the oncogene theory prevails for explaining the origins of cancer.[14,15] The current view is that cancer is a multistep process, characterized by mutations in several oncogenes and the loss of function of *tumor-suppressing genes*. Oncogenes are activated by either an inherited defect or exposure to an outside agent, a carcinogen. However, oncogenes are normally kept inactive by tumor-suppressing genes. In some cases those controlling genes may be mutated, or removed, allowing oncogenes to run rampant. For a tumor to develop, a subject has to lose one tumor-suppressing gene in addition to having two or more oncogenes.

3.2   *Chemotherapy*

Mustard gas

mechlorethamine
(Nitrogen mustard)

cyclophosphamide
(Cytoxan)

Chemotherapy, despite its devastating side effects, saves lives. Yet it originated at first from a chemical weapon, mustard gas. During WWII, the USS *John Harvey* loaded with 100 tons of mustard gas was sunk by Luftwaffe Ju-88 bombers at the Bari Harbor on the Adriatic Sea. Subsequent autopsies of 617 victims revealed that mustard gas destroyed most of their white blood cells,[16] which suggested that it attacked bone marrow preferentially. Since fast division of cells is a hallmark of cancer, mustard-gas-based drugs could be applicable in cancer treatment because they slow the rate of cancer cell division. Nitrogen mustard was invented because mustard gas was impossible to administer due to its volatility. Its mechanism of action is through alkylation of DNA. Subsequently, more elaborate alkylating agents, such as cyclophosphamide (Cytoxan), discovered at Asta–Werke AG in Germany in 1956, have become the staple of chemotherapy.

Cisplatin's chemotherapeutic effect was discovered by Barnett Rosenberg, a physics professor at Michigan State University in 1967.[17] While experimenting with the effects of electric fields on cell growing in culture, he observed that an electric current interfered with cell divisions of *E. coli* bacteria in suspension. Further investigation revealed that the electric current generated by the platinum electrodes resulted in the formation of cationic platinum. Rosenberg tested cisplatin as a drug against intestinal bacteria first and tumors later. Bristol-Myers Squibb (BMS) successfully developed cisplatin and won the FDA's approval in 1978. Cisplatin (Platinol) has become one of the most widely used chemotherapeutics in treating metastatic testicular cancer, ovarian tumors, and bladder cancer. BMS developed a follow-up to cisplatin in collaboration with Johnson Matthey: carboplatin. Sanofi-Aventis licensed oxaliplatin from Switzerland's Debiopharm and developed it for treating colorectal cancer. Approved in 2002 in the U.S. under the name Eloxatin, it is usually administered with 5-fluorouracil and leucovorin in a combination called Folfox. During the past 40 years, a few thousand platinum complexes have been evaluated and about 30 have reached the clinic, but none has surpassed the three original ones which possess superior efficacy and safety profiles.

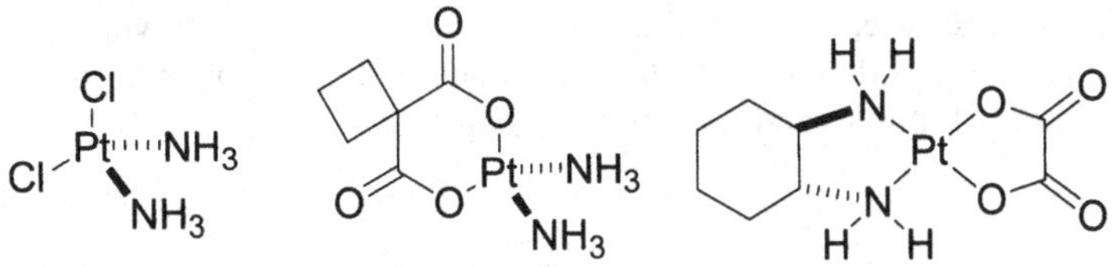

cisplatin (Platinol)   carboplatin (Paraplatin)   oxaliplatin (Eloxatin)

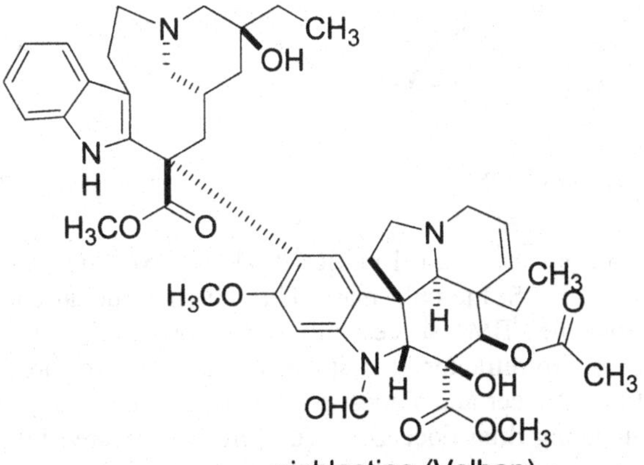

vinblastine (Velban)

In 1958, during their pursuit of diabetes drugs, R. L. Noble and C. T. Beer at the University of Western Ontario in London, Ontario, tested extracts from Madagascar periwinkle (*Vinca rosea*) on rabbits. The animals subsequently developed critically low counts of white blood cells, leaving them with damaged bone marrow and defenseless against bacterial infections. Noble and Beer then tried the plant extract on animals with transplanted tumors because again cancer is characterized by abnormal proliferation of white blood cells. After seeing tumor shrinkage, they further investigated and isolated two important cancer drugs from vinca alkaloids:

vincristine and vinblastine.[18] The vinca alkaloids work by serving as a "spindle poison". They bind to tubulin, one of the key constituents of microtubules, thus preventing the cell from making the spindles it needs to divide. Thanks to vincristine (Oncovin) and vinblastine (Velban), Hodgkin's disease patients now have a 90% chance of survival.

Vinca alkaloids are not the only class of anticancer drugs derived from plants. Taxol is another prominent example. In 1962, Arthur Barclay, under the aegis of the NCI-USDA (National Cancer Institute-United States Department of Agriculture) plant-screening program, traveled to the Gifford Pinchot Forest in Washington State. He collected samples of twigs, leaves, and fruits of a then little known pacific yew tree, *Taxus brevifolia*, and shipped them to the NCI.[19] One of the NCI's contractors was the Wisconsin Alumni Research Foundation, which tested the extracts and found them to be cytotoxic. In 1966, after being rejected by many other laboratories for fear of toxicity, the stem barks found their way to the hands of Monroe E. Wall, chief chemist of the Fractionation and Isolation Laboratory at the Research Triangle Institute in North Carolina. Wall and his colleague, Mansukh C. Wani, isolated the active principle using the "bioactivity-directed fractionation" process. They also elucidated the compound's intricate structure and christened it taxol. The interest in taxol was greatly piqued when Susan Howitz, a professor at the Albert Einstein College of Medicine, discovered that taxol had a completely novel mechanism of action. It exerts its action by stabilizing microtubules (one of the components of the cytoskeleton), resulting in inhibition of mitosis and induction of apoptosis.

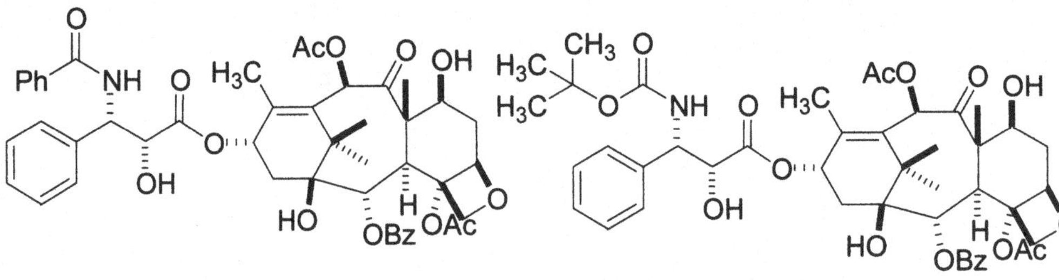

paclitaxel (Taxol)   docetaxel (Taxotere)

The NCI commenced the Phase I clinical trials of taxol in 1984 and Phase II trials in 1987 with positive results. BMS, the only major U.S. pharmaceutical company to have made a bid, was awarded the molecule. BMS successfully carried out the Phase III trials for taxol. The FDA approved taxol for use in refractory ovarian cancer in 1992, for breast cancer in 1994, and later for non-small-cell lung cancer and Kaposi's sarcoma. Meanwhile, French company Rhône-Poulenc marketed a competing drug docetaxel (Taxotere), discovered by French chemist Pierre Potier, who invented docetaxel by a minor modification of taxol (replacing the benzyl amide group on taxol with a *tert*-butoxyl-carbamate group).

3.3   *Hormone Treatment*

The correlation between sex hormones and cancer has been well established. Breast cancer in particular is linked to estrogen abnormality. Tamoxifen, the most frequently prescribed anticancer drug in the world, was initially made as a contraceptive. In the 1960s, Imperial Chemical Industries Ltd. (ICI) chemist Dora M. Richardson discovered ICI-147,741, the *trans*-isomer of triphenylethylene, which would later become tamoxifen. Although it was shown to be an effective contraceptive in rats, it induced ovulation in women, exactly the opposite of what it did to rats. Fortunately, Arthur Walpole along with his endocrinologist colleague Michael J. K. Harper included cancer as an indication in tamoxifen's clinical trials.[20] Tamoxifen was found to be efficacious. It was marketed in the UK as a breast cancer treatment (1973) and as an inducer of ovulation (1975). In 1978, the FDA approved tamoxifen for treatment of estrogen receptor-positive metastatic breast cancer in the US.

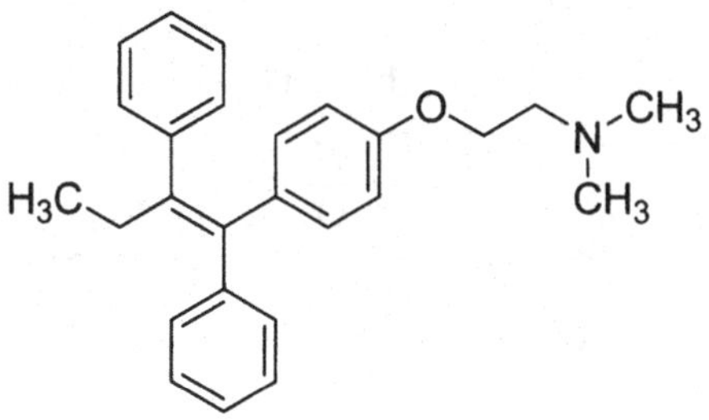

tamoxifen (Nolvadex) 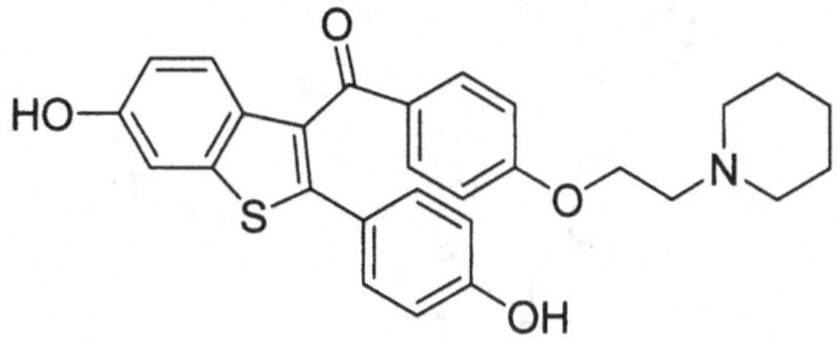   raloxifene (Evista)

Tamoxifen is a *s*elective *e*strogen *r*eceptor *m*odulator (SERM). It modulates the estrogen hormone level by mimicking estrogen. Therefore it is especially effective in treating hormone-receptor-positive breast cancer. A newer SERM, raloxifene (Evista), marketed by Eli Lilly as an osteoporosis drug, renders a 58% reduction of breast cancer.

3.4   *Small-Molecule Protein Kinase Inhibitors*

In contrast to the carpet-bombing approach of old chemotherapy, protein kinase inhibitors, as targeted cancer drugs, are a more effective method for cancer treatment with fewer side effects. Protein kinases are enzymes inside the cell that are capable of donating phosphate groups to target proteins. Protein kinases comprise a family of more than 150 members. They are responsible for signal transduction, turning on and off the switches that control cancer cell growth. Many protein kinases have been implicated in cancer. It has been shown that blocking the functions of protein kinases can stop cancer growth.

Trastuzumab (Herceptin), a bioengineered human monoclonal antibody developed by Genentech, was approved for the treatment of breast cancer in 1998.[21] The success of Herceptin inspired the pursuit of small-molecule protein kinase inhibitors. Using a protein kinase C-α inhibitor as a starting point, Ciba-Geigy chemist Jürg Zimmermann carried out methodical structure-activity relationship (SAR) investigations.[22] From among over 300 analogs emerged imatinib (Gleevec), a selective Bcr-Abl-tyrasine kinase inhibitor. It was launched in 2001 for the treatment of chronic myeloid leukemia (CML). Later study revealed that Gleevec actually blocks a panel of at least 8 protein kinases, including Bcr-Abl, platelet-derived growth factor receptor (PDGFR), and c-kit.

AstraZeneca's gefitinib (Iressa) and OSI's erlotinib (Tarceva), both EGFR inhibitors, entered the market in 2003. Sugen's sunitinib (Sutent) exhibits potent antiangiogenic activity through the inhibition of multiple receptor tyrosine kinases (RTKs). Specifically sunitinib inhibits vascular endothelial growth factor receptors VEGFR1, VEGFR2, and VEGFR3 and platelet-derived growth factor receptors PDGFR-α and PDGFR-β. In addition, sunitinib also targets receptors implicated in tumerogenesis, including fetal liver tyrosine kinase receptor 3 (Flt3) and stem cell factor receptor (c-KIT). Bayer's sorafenib (Nexavar), launched in 2005, is a VEGFR inhibitor. Similar to Gleevec, BMS's dasanitib (Sprycel) and Narvotis's nilotinib (Tasignal) came to market in 2006 and 2007, respectively. They both block the Bcr-Abl kinase, which is largely responsible for causing chronic myeloid leukemia (CML).

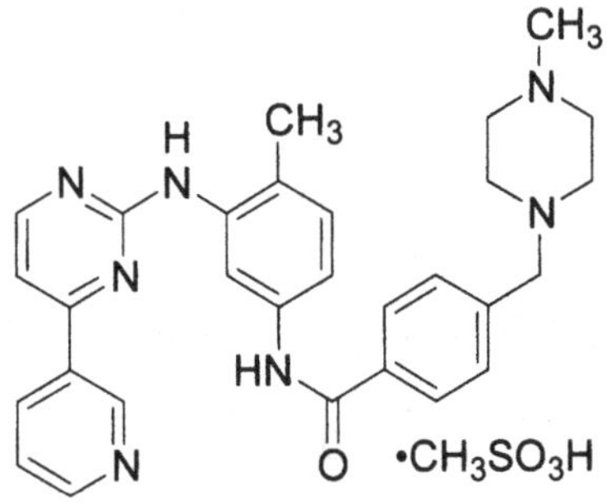

imatinib mesylate (Gleevec)

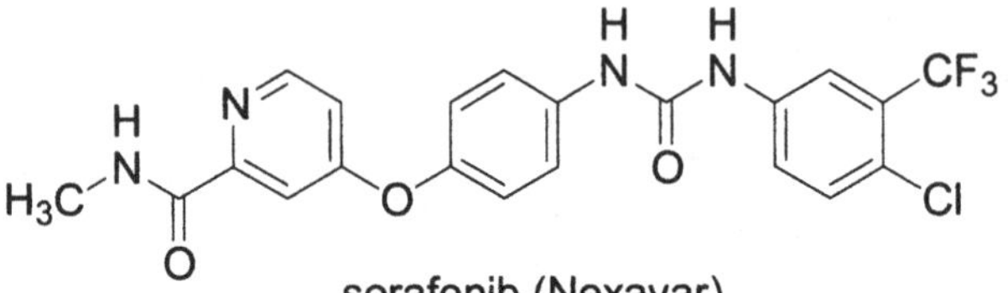

sorafenib (Nexavar)

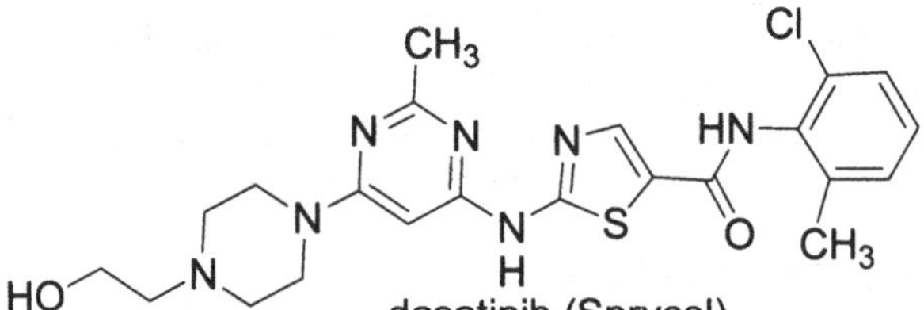

dasatinib (Sprycel)

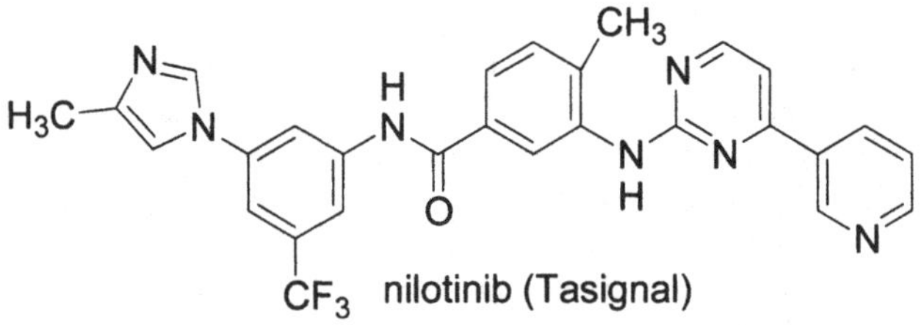

nilotinib (Tasignal)

4   Cardiovascular Drugs

British physician William Harvey's (1578–1657) discovery of the circulation of blood is considered one the greatest discoveries made in physiology.[23] His most celebrated monograph, *Exercitatio Anatomica de Motu Cordis et Sanguinis in Animalibus* (*Anatomical Exercise on the Motion of the Heart and Blood of Animals*) was published in 1628. Over a century later, another Englishman, William Withering, discovered the medicinal use of foxglove in treating cardiac diseases.

4.1   *Withering and Digitalis*

William Withering (1741–1799), a physician in Birmingham, bought from a Gypsy lady the secret formula of her special herbal tea, which showed remarkable results in treating dropsy (similar to today's congestive heart failure).[24] Through careful investigation, Withering correctly concluded that among 20 or so ingredients in the herbal tea, purple foxglove was the active ingredient. Withering spent the next decade exploring the curative effects of digitalis, the major principle. Digitalis, as well as many cardiac glycosides, is toxic when taken in large doses. If the dose is too low, then it is ineffective. Digitalis is only effective when administered at or near toxic dose, so finding the correct dosage is very important. Withering's finding in choosing the precise dosage propelled digitalis to become one of the most valuable cardiac drugs ever discovered.

Digitalis was originally prepared from the powdered leaves of foxglove. Nowadays, digitalis isolated from foxglove leaves is further crystallized to afford cardiac glycosides such as digoxin and digitoxin, which are more easily quantified.

4.2   *Sobrero, Nobel, and Nitroglycerin*

Italian chemist Ascanio Sobrero (1812–1888) first synthesized nitroglycerin in 1847,[25] by nitration of glycerol using a cold mixture of nitric acid and sulfuric acid. When he tasted it, he found it sweet, pungent, and aromatic; but a very minute quantity put upon the tongue produced a violent headache for several hours. That was the first recorded vasodilating effect attributed to a drug.

Alfred Nobel (1833–1896) from Stockholm created dynamite using a porous silica gel to absorb the unstable nitroglycerin. The patented detonator made a fortune for the Nobel family. In explosive production plants, munitions workers often experienced facial flush and severe headache when they returned to work after being away from the factory over the weekend. Further investigation revealed that nitroglycerin was a powerful vasodilator which led to its use as a circulatory system vasodilator.

The MOA of nitroglycerin's treatment of angina was not known until the 1980s. It turns out that soluble guanylated cyclase could be activated by free radicals such as nitric oxide (NO) generated from nitroglycerin. Nitric oxide is a signaling molecule in the cardiovascular system.

In addition to nitroglycerin, many organic nitrates are used to treat angina pectoris. These organic nitrates are typically prepared by nitration of polyols (molecules with many alcohol functionalities). Examples include isosorbide dinitrate (ISDN), pentaerythritol tetranitrate, and erythrityl tetranitrate.

nitroglycerin   merbaphen (Novasurol)

4.3   *Vogl and Diuretics*

Diuretics, also known as *water pills*, and β-blockers are the most prescribed drugs for heart conditions. Serendipity played an important role in the discovery of the first mercurial diuretic.

In 1919, Alfred Vogl, a third-year medical student in Vienna, gave a new mercurial anti-syphilitic, merbaphen (Novasurol), to treat a patient's congenital syphilis.[26] To his astonishment, the patient's urine output was 5–6 times the normal amount. By removing fluid, the pressure on the heart was removed. Mercurial diuretics revolutionized the treatment of severe edema from congestive heart failure and were the primary treatment for this disease until the late 1950s and the emergence of thiazide diuretics.

In 1957, in pursuit of diuretic agents, Merck chemist Frederick C. Novello wanted to make some analogs of an older sulfa drug, dichlorophenamide.[27] Surprisingly, the reaction gave the ring formation product rather than the linear derivatization product. The bicyclic ring formed was a benzothiadiazine derivative, chlorothiazide. Further testing proved it to be a potent diuretic without elevation of bicarbonate excretion, an undesired side effect. Chlorothiazide was the first ever nonmercurial, orally active diuretic drug whose activity was not dependent on carbonic anhydrase inhibition, such as acetazolamide.

chlorothiazide (Diuril)   hydrochlorothiazide (HydroDiuril)

Currently the most frequently prescribed diuretic is hydrochlorothiazide discovered by Ciba scientists led by George deStevens.[28] In 1957, George deStevens became aware of the research of Frederick Novello on the synthesis of disulfonamides in general and chlorothiazide in particular. A simple modification from a double bond on chlorothiazide to a single bond gave him a hydrochlorothiazide that was 10-fold more potent than the prototype. Hydrochlorothiazide was introduced into medical practice in 1959 and within a short time became the drug of choice in the treatment of mild hypertension.

4.4   *Snake Venom and ACE Inhibitors*

Angiotensin converting enzyme (ACE) inhibitors are widely used in treating hypertension, congestive heart failure, and heart attacks. In 1967, John Vane at Oxford University and his colleague Mick Bakhle tested a poisonous venom extract of the Brazilian pit viper *Bothrops jararaca*.[29] It was found to be a potent ACE inhibitor *in vitro*.

Vane suggested to the Squibb Institute that they study snake venom extract for its effects on the cardiovascular system. Biochemist David Cushman and organic chemist Miguel A. Ondetti at Squibb isolated a nonapeptide, teprotide. Using teprotide as a starting point, Cushman and Ondetti curtailed the molecule and replaced its carboxylate group with a thiol (–SH) and achieved a 2,000-fold increase in potency in ACE inhibition. The resulting drug became the first *oral* ACE inhibitor, captopril (Capoten).

captopril (Capoten)　　　　enalapril (Vasotec)　　　　quinapril (Accupril)

To improve on captopril, which suffered some side effects due to the presence of the thiol group, Merck scientists led by Arthur A. Patchett started by replacing the thiol group with the original carboxylate due to thiol's liabilities.[30]　The loss in potency of the carboxylate was compensated by modification of the molecule elsewhere.　They arrived at enalaprilat, which suffered poor oral bioavailability.　They simply converted the acid into its corresponding ethyl ester, creating enalapril, a prodrug of enalaprilat, with excellent oral bioavailability.　One advantage of a prodrug is the delay in onset of action, which can be beneficial for a drug to treat blood pressure.　The longer duration of action also allows a once-daily dosage.　It is also devoid of the side effects associated with the thiol group, including bone marrow growth suppression (due to a decrease in circulating white blood cells), skin rash, and loss of taste.　In 1981, Merck sold enalapril using the brand name Vasotec, which became its first billion-dollar drug in 1988. Another popular ACE inhibitor is quinapril hydrochloride (Accupril), discovered by Parke–Davis.

losartan (Cozarr)　　　　　　valsartan (Diovan)

Since angiotensin II is a potent vasoconstrictor, blocking its action would result in vasodilation.　Dupont–Merck Pharmaceuticals exploited the angiotensin II receptor in the early 1980s.　The fruit of the effort was losartan (Cozaar), which quickly became one of the most important drugs for the treatment of high blood pressure after its launch in 1995.　Other angiotensin II receptor antagonists (also known as *angiotensin receptor blockers*, ARBs) include Novartis's valsartan (Diovan), Sanofi-Synthélabo's irbesartan (Avapro), AstraZeneca's candesartan (Atacand), and Boehringer Ingelheim's telmisartan (Micardis).　They proved to be superior to ACE inhibitors because they did not cause the irritating cough that occurs in a small percentage of patients taking ACE inhibitors.

4.5   Black and Beta-Blockers

As early as 1948, Raymond P. Ahlquist at the Medical College of Georgia speculated that there were two types of adrenergic receptors (adrenoceptors in short), which he termed α-adrenoceptor and β-adrenoceptor.   In 1957, Irwin H. Slater and C. E. Powell at Eli Lilly prepared dichloroisoprenaline (DCI, the *di*chloro analog of *i*soprenaline) and it was later demonstrated to be the first selective β-adrenoreceptor blocking reagent, also known as a β-blocker. However, DCI was not further pursued as a drug because it had a marked undesirable stimulant effect on the heart, an *intrinsic sympathomimetic action* (ISA).

Ahlquist's theory of two adrenergic receptors inspired British pharmacologist James Black to look for drugs with β-receptor blocking properties in a systematic way later known as *rational drug design*.   Before then, drugs were mostly discovered by screening compounds randomly against animal models.

Black famously said, "The most fruitful basis for the discovery of a new drug is to start from an old drug."[31]   At Imperial Chemical Industries (ICI) Pharmaceutical Division, together with chemist James Stephenson, Black led a team to look for β-blockers that were devoid of the stimulant effect on the heart since 1958.   In 1962, Stephenson and his colleagues succeeded in making a β-blocker pronethalol using DCI as the starting point (pronethalol replaced the dichlorobenzene ring with a naphthaline ring). Unfortunately, pronethalol was withdrawn from further development when it was found to cause thymic tumors in mice. In 1964 ICI eventually produced the drug propranolol (Inderal), which possessed a better efficacy and safety profile. Propranolol is now widely used in the management of angina, hypertension, arrhythmia, and migraine headaches. Two additional β-blockers, atenolol (Tenormin) and practolol (Dalzic), were later discovered and marketed by ICI.

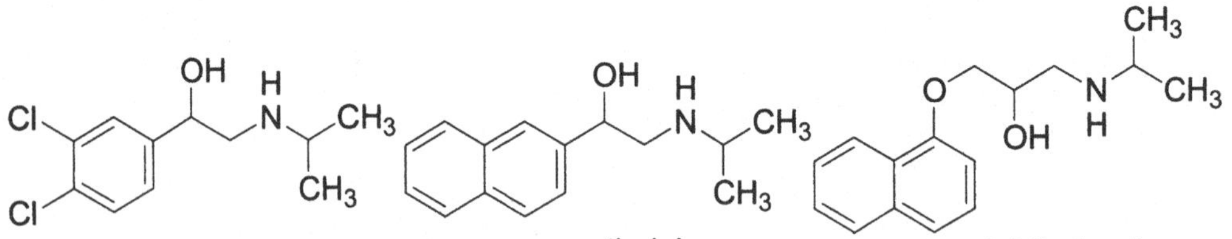

dichloroisoprenaline (DCI)   pronethalol   propranolol (Inderal)

At one point or another, almost all the major pharmaceutical companies had a me-too β-blocker on the market. It brought a windfall to the drug industry.

4.6   Renin Inhibitor

Since renin is extremely specific for angiotensinogen and the first and rate-limiting enzyme of the renin-angiotensin system (RAS), renin inhibition was recognized for decades as an attractive approach for the treatment of hypertension and hypertension-related target organ damage.[32]

Ciba-Geigy was a pioneer in the rennin field in the race to develop orally available renin inhibitors, but the clinical development of aliskiren was in jeopardy following the merger of Ciba-Geigy and Sandoz to form Novartis in 1996 and the successful launch of the antihypertensive angiotensin receptor blocker (ARB) valsartan (Diovan) in 1997. A group of former Ciba-Geigy employees convinced Novartis to out-license the Phase I/II development of aliskiren and formed the biopharmaceutical company Speedel to accomplish this task.[33] Speedel was successful in developing a commercially viable process and demonstrating clinical proof of concept (POC), and Novartis exercised a call-back option in 2002. Following an extensive Phase III development program, aliskiren (Tekturna) has been on the market for treatment of hypertension since 2007.

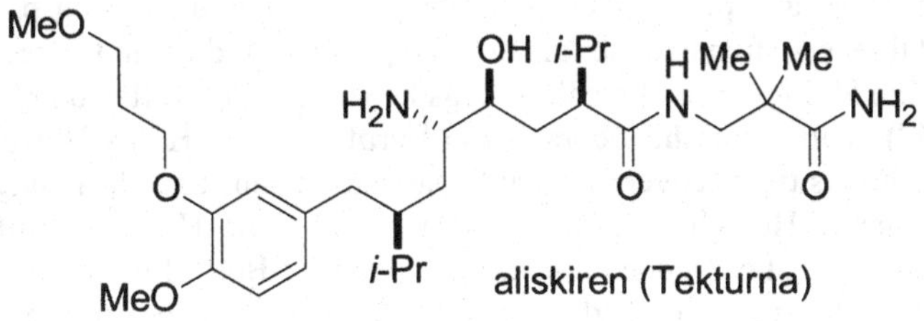

aliskiren (Tekturna)

4.7   Fleckenstein and Calcium Channel Blockers

Calcium channel blockers (CCBs), also known as calcium channel antagonists or calcium entry blockers, are drugs that inhibit the influx of Ca^{2+} ions into cells without affecting inward Na^+ or outward K^+ currents to a significant degree. They are widely used in the treatment of high blood pressure, angina, and rapid heartbeat (tachycardia), including arterial fibrillation.

In 1963, Albrecht Fleckenstein at the University of Freiburg in Germany investigated two newly synthesized coronary vasodilators, prenylamine and verapamil, which had unexplained cardiodepressant side effects.[34] Fleckenstein and his colleagues observed that both compounds exerted a negative inotropic effect caused by calcium. They concluded that this negative inotropism was due to an ability of these drugs to block excitation-induced calcium influx. In 1966 Fleckenstein then coined the term "calcium antagonists" because both drugs mimicked the cardiac effects of simple calcium withdrawal.

In 1969, Professor Kroneberg, the leading pharmacologist of Bayer Company, handed Fleckenstein Bay-a-1040 and Bay-a-7168. Both compounds were strong coronary vasodilators and exerted significant negative inotropic effects on the myocardium. Fleckenstein also found that the mechanism of action of those two drugs appeared to be similar to that of verapamil. Later on Bay-a-1040 and Bay-a-7168 were given generic names nifepidine and niludipine, respectively. Nifepidine (Adalat) and niludipine heralded the beginning of one of the most important classes of calcium antagonists: 1,4-dihydropyridines.[35]

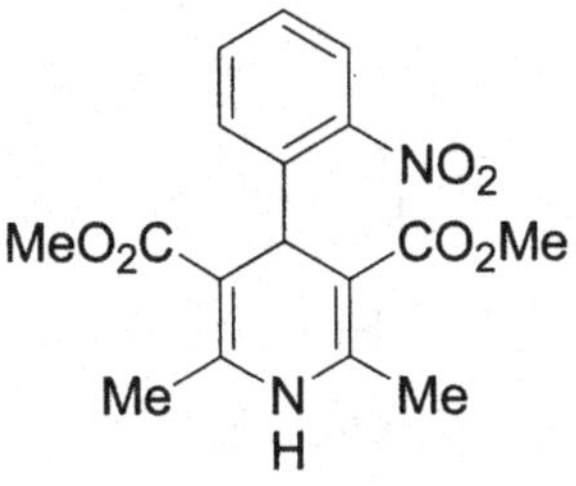

nifedipine (Adalat)

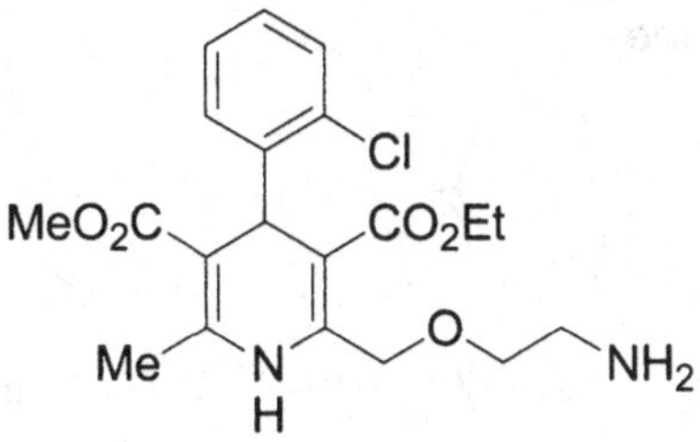

amlodipine (Norvasc)

Bayer's nifepidine is a short-acting calcium channel blocker and thus has to be taken several times a day. On the other hand, Pfizer's amlodipine (Norvasc) has a high bioavailability and a longer half-life in plasma; thus it can be taken once daily.

4.8   Blood Thinners, from Heparin to Plavix

Thrombosis and embolus take place when platelets are overactive, causing promotion of blood coagulation. Blood thinners attenuate the functions of platelets and fibrin to prevent thrombosis or embolus. The first blood thinner in clinical use was heparin, acidic sulfated polysaccharides. Jay McLean (1890–1957), working in the laboratories of Prof. William Henry Howell (1860–1945) at the Johns Hopkins University, discovered the anticoagulation substance from dog's liver isolate in 1917. Two years later, Howell and his student L. Emmett Holt, Jr. further purified the anticoagulation substance and christened it *heparin*.[36] Charles Best of the insulin fame returned at the University of Toronto and embarked on a journey to produce purified heparin since 1929. Working with organic chemists Arthur Charles and David Scott, they prepared the barium and later the sodium salt of heparin, which rendered uniformly consistent composition and potency. That, in turn, enabled Best and Gordon Murray to carry out clinical trials with purified heparin in 1935. Nowadays, small-molecular-weight heparin fractions are used in clinics because only a fraction of the heparin molecule is active in *blocking blood clotting factor Xa* (heparin's MOA).

Heparin, highly sulfated glycosaminoglycan, has to be given intravenously. The first identified oral anticoagulant was warfarin, discovered by Karl Paul Link (1901–1978) at the University of Wisconsin in the 1940s. In 1933, Link began to investigate the cause of cattle death due to internal hemorrhage after ingestion of sweet clovers. By 1939, his group isolated and characterized the active principle as dicumarol. In analogy to hemorrhage of cows, Link sought to look for rodenticide from the dicumarol analogs to kill rats by hemorrhage. One of them, WARF-42, emerged as the best rat poison in 1948.[37] In the early 1950s, clinical trials for warfarin as a blood thinner commenced. It now is the most used oral anticoagulant in history. The MOA of warfarin is inhibition of VKORC1 (vitamin K epoxide reductase complex, subunit 1) and vitamin K epoxide reductase.

Aspirin was synthesized in 1897 by Bayer's Felix Hoffmann as an analgesic. In the 1940s, Lawrence Craven, a family doctor in California, observed aspirin's anticlotting properties. Harvey J. Weiss at Columbia University was one of the first to discover aspirin's antiplatelet effect in 1967. Since 1985, an aspirin a day has been a popular prophylaxis to prevent a second heart attack. Aspirin works by inhibiting prostaglandin synthetase, which explains most of its antiplatelet, antipyretic, and anti-inflammatory properties.

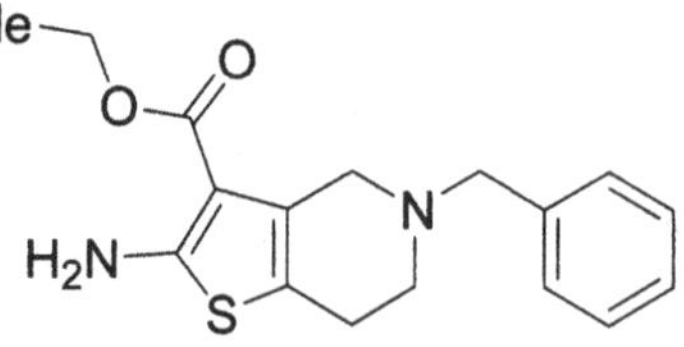

tinoridine (Nonflamin) 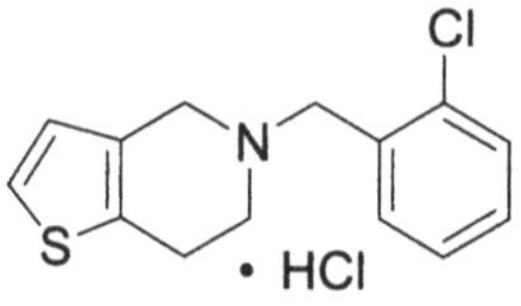   ticlopidine hydrochloride (Ticlid)

In 1972, Jean-Pierre Maffrand of Sanofi made analogs of Yoshitomi's tinoridine (Nonflamin) to find drugs with improved anti-inflammatory properties. But the compounds they made had no anti-inflammatory properties at all. Further scrutiny revealed that they inhibited blood platelet aggregation. Immediately realizing the important roles that platelets played in myocardial infarction and brain ischemia, the team set out to find a platelet aggregation inhibitor superior to aspirin. The fruit of labor was ticlopidine (Ticlid). Unfortunately, a rare but potentially fatal side effect called thrombotic thrombocytopenic purpura (TTP) and several other side effects severely limited ticlopidine's use in patients.[38]

To overcome ticlopidine's shortcomings, Sanofi carried out extensive SAR investigations. PCR 4099, a racemate, was synthesized in 1980. Although more potent and better tolerated than ticlopidine, it caused convulsions in rats, mice, and baboons for certain high dosages. A decision was made to replace PCR 4099 with one of its enantiomers, clopidogrel bisulfate, when clopidogrel was found to have superior safety profile. Sanofi and its American development partner BMS won the FDA approval for clopidogrel (Plavix) in 1997. In 2008, Plavix was the second best-selling drug ever, with sales of $8 billion.[39]

In 2009, Daiichi Sankyo and Eli Lilly began to market a *me-too* of clopidogrel: prasugrel (Effient).

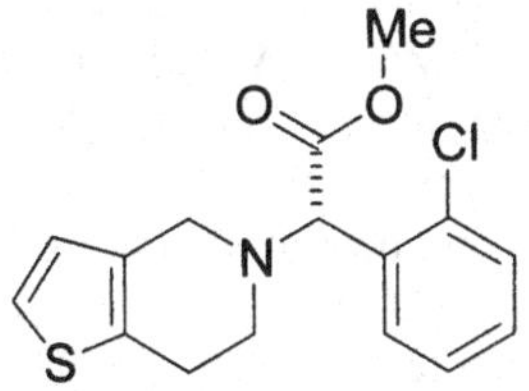

clopidogrel (Plavix) 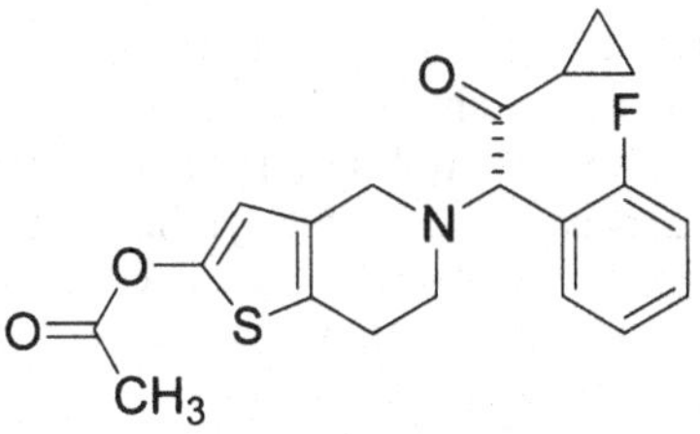   prasugrel (Effient)

5   Cholesterol Drugs

Too much cholesterol, especially the *low-density lipoprotein* (LDL) cholesterol, has been identified as a major *risk factor* for cardiovascular disease by experimental, genetic, and epidemiological evidence.

5.1   *Early Cholesterol Drugs: Niacin and Fibrates*

In the early 1950s, Canadian pathologist Rudolf Altschul at the University of Saskatchewan found that large doses of nicotinic acid decreased serum cholesterol levels in rabbits. Working with Drs. Abram Hoffer and J. D. Stephen at Regina General Hospital, they carried out clinical trials that successfully showed this link. By 1955, nicotinic acid (niacin) in gram doses became the first agent employed for the purpose of lowering cholesterol.[40] The MOA of nicotinic acid is through activation of endothelial lipoprotein lipase as a result of its binding to the *nicotinic acid receptor*, a G-protein-coupled receptor.

In 1954, ICI discovered that clofibrate (Atromid-S) possessed high cholesterol-lowering activity and marketed it in 1958. Parke–Davis's gemfibrozil (Lopid) was the second fibrate on the market. In order to find safer analogs of clofibrate, Parke–Davis screened over 8000 compounds similar to clofibrate using animals.[41] Gemfibrozil was discovered by Paul L. Creger in 1969 and was launched in 1982. Abbott's fenofibrate (Tricor) is also a clofibrate analog. Recently, fibrates were found to be PPARα (*p*eroxisome *p*roliferator-*a*ctivated *r*eceptor-α) agonists, which have been used in treating type II diabetes since the late 1990s.

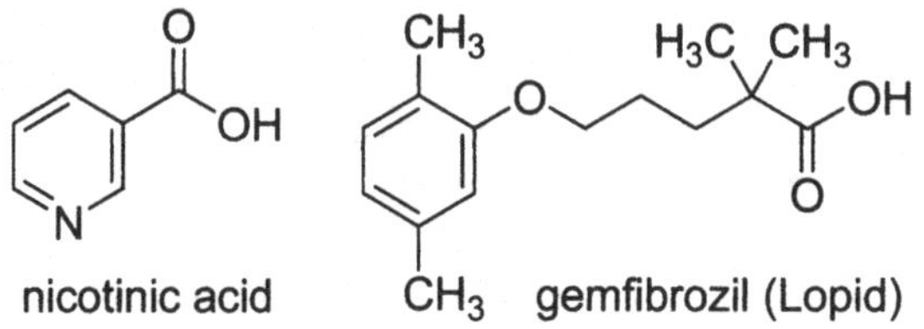

nicotinic acid   gemfibrozil (Lopid)

Recent studies have also shown that risk of drug-drug interactions increases 1400-fold if statins are combined with fibrates. Therefore, fibrates should not be taken with statins (see Section 5.4).

5.2   *Endo, Mevastatin, and Pravachol*

Konrad E. Bloch at Harvard elucidated the biosynthetic pathways of cholesterol in the 1960s. During 1958–1959, it was shown that the conversion of HMG-CoA into mevalonic acid is the major rate-limiting step in cholesterol biosynthesis. Therefore, blocking HMG-CoA reductase would be an ideal approach to stop the synthesis of cholesterol in liver.

About inspired by Fleming's success in identifying penicillin, Akira Endo of Sankyo Pharmaceuticals began in 1971 to screen microbial metabolites for HMG-CoA reductase inhibitory activity. Over a two-year period, Endo's team tested more than 6000 microbial strains for their ability to block lipid synthesis. In August 1973, mevastatin, the first statin, was isolated from the fungus *Penicillium citrium.* Endo's group isolated only 23 mg of crystalline mevastatin by extracting 600 L of the culture filtrate.[42] In 1976, A. G. Brown at Beecham Pharmaceuticals also isolated mevastatin from *Penicillium brevicompactum.* They named it compactin and elucidated its structure by X-ray crystallography, confirming that compactin and mevastatin were indeed the same molecule.[43]

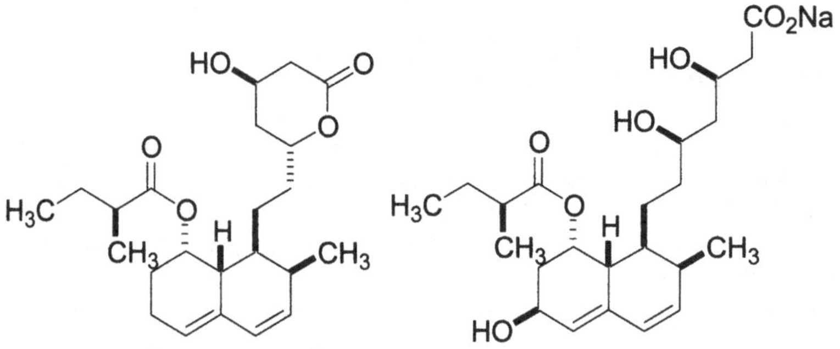

mevastatin = compactin   pravastatin (Pravacol)

In 1976, Endo tested mevastatin in laying hens, dogs, and monkeys and observed efficacy in lowering cholesterol. Akira Yamamoto at Osaka University gave low doses of mevastatin to several of his patients with familial hypercholesterolemia (FH) and saw good results, which encouraged Sankyo to embark on a small clinical trial in early 1979 for patients with severe hypercholesterolemia. But by the mid-1980s, Sankyo halted the trials because long-term safety studies on dogs revealed that some of them started to show intestinal tumors at high doses.[44]

Despite mevastatin's failure to reach the market, Endo's pravastatin, first isolated from the urine of dogs that had been fed mevastatin, succeeded to reach market. Sankyo carried out the clinical trials for pravastatin in 1984. After the successful completion of the trials, Sankyo comarketed pravastatin (Pravachol) with BMS since 1991.

5.3 Merck's Mevacor and Zocor

Under the leadership of Roy Vagelos, Merck embarked on efforts to find HMG-CoA reductase inhibitors in 1978 with Alfred W. Albert as the *Product Champion*. Just a few days after the HMG-CoA project began, they found that the 18th microorganism produced the active principle of the broth, which potently inhibited HMG-CoA reductase. Back-tracking the sample, they identified the microorganism from the culture broth of *Aspergillus terreusi,* a common fungus found around the world. The compound was isolated and characterized and would become lovastatin (Mevacor), which reached the market in 1987 as the first statin. Ironically, nothing with a better profile was ever discovered from screening additional thousands of samples.[45]

Robert L. Smith and his colleagues at Merck synthesized simvastatin, an analog of lovastatin with an extra methyl group on the sidechain. Simvastatin (Zocor), 2.5 times more potent and longer lasting than lovastatin, was launched in 1991.

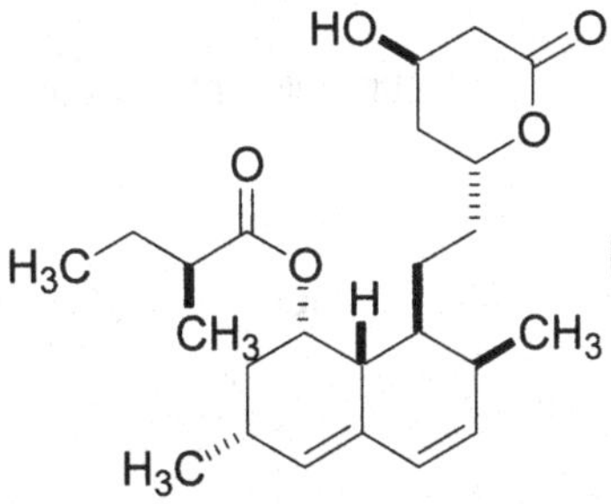

lovastatin (Mevacor) 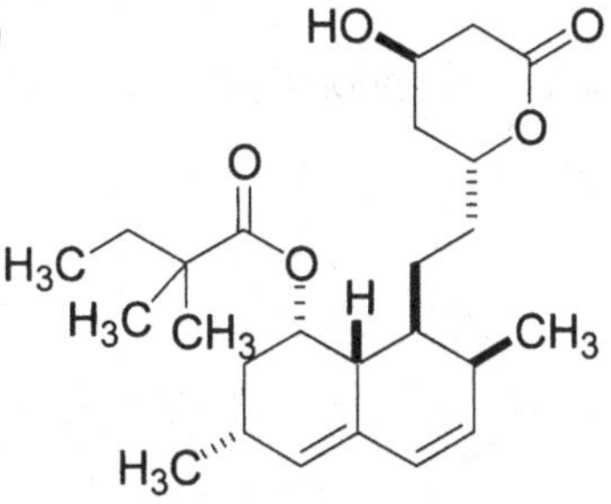   simvastatin (Zocor)

5.4 Lescol, Lipitor, Baycol, and Crestor

In the 1980s, Merck's Alvin K. Willard, Gerald E. Stokker, and their colleagues disclosed their synthetic HMG-CoA reductase inhibitors in which Mevacor's hexahydronaphthalene core structure was replaced with a diphenyl group with four substituents.[46] This demonstrated that it was not necessary to have fermentation products to have the activity against HMG-CoA reductase.

The fourth statin on the U.S. market, Sandoz's fluvastatin sodium (Lescol), was the first synthetic statin. In 1984 Faizulla G. Kathawala at Sandoz replaced mevastatin's hexahydronaphthalene core structure with an indole ring. Unlike mevastatin, Mevacor, and Zocor, Lescol's sidechain was a sodium carboxylate salt instead of a lactone ring.[47]

In the mid-1980s, Bruce D. Roth at Parke–Davis chose the pyrrole ring to replace mevastatin's hexahydronaphthalene core.[48] Initial halogenated pyrroles had similar potency to mevastatin but was toxic to rats. But the pentasubstituted pyrrole with the chiral sidechain (CI-981, atorvastatin) proved both efficacious and with similar toxicology profile to known statins. Although CI-981 was only equipotent to Mevacor in dogs, biologist Roger S. Newton convinced the management to move CI-981 to the clinics. Surprisingly, the Phase I trials showed it had superior efficacy to all known statins. Atorvastatin (Lipitor), the fifth statin on the market since 1997, is also the best in its class, becoming the best-selling drug ever with annual sales of $13 billion in 2008. Lipitor's patent expired on November 31, 2011, which will change the statin landscape significantly as generic atorvastatin floods the market.

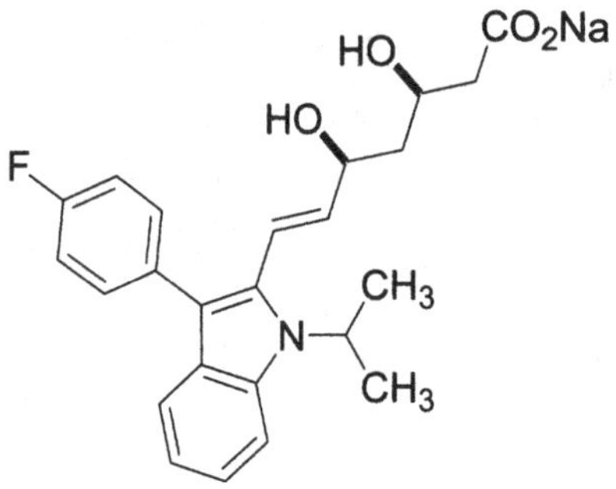

fluvastatin (Lescol)

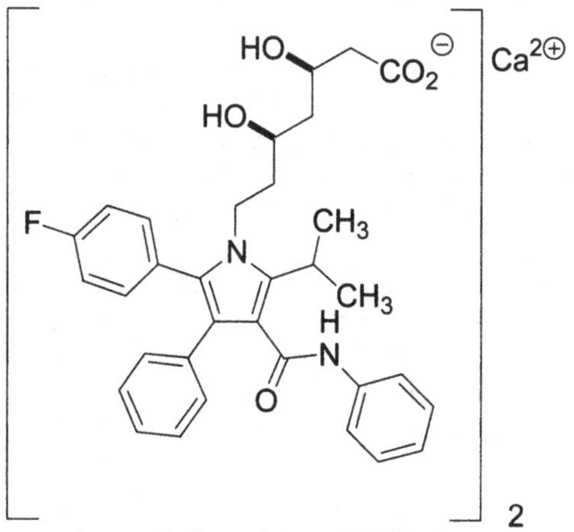

atorvastatin calcium (Lipitor)

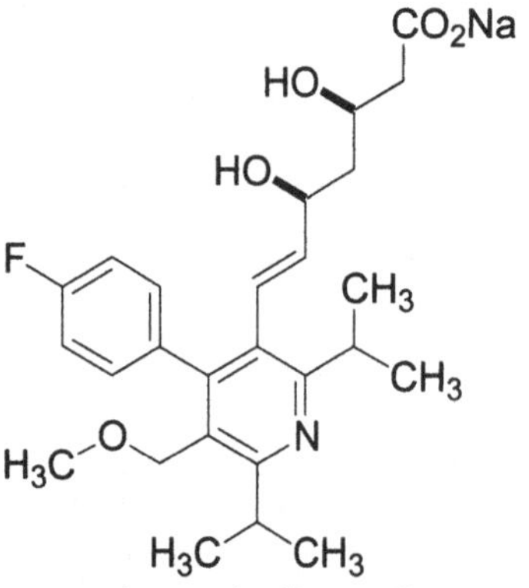

cerivastatin (Baycol)

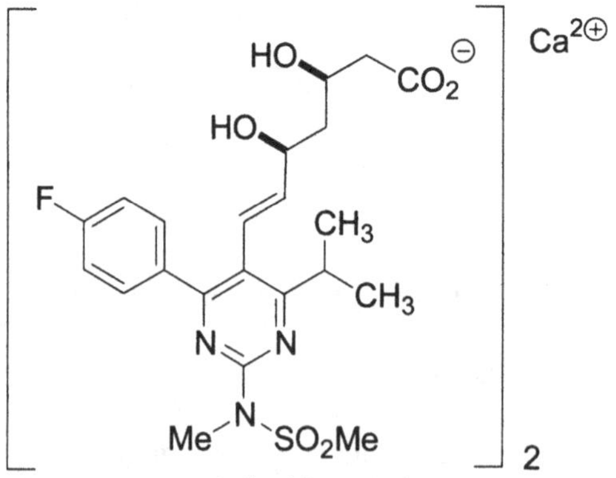

rosuvastatin (Crestor)

Cerivastatin (Baycol) was discovered by Bayer scientists led by medicinal chemist Rolf Angerbauer and biologist Hilmar Bischoff in the late 1980s. It was on the market in 1997 but was withdrawn in 2001 when it was reported that Baycol's serious rhabdomyolysis was about 15–60 times as frequent as with the other statins. A Baycol fatality tended to occur in patients taking higher doses of the drug and in those also taking fibrates, especially gemfibrozil (Lopid).[49] Recent studies have shown that the risks of drug-drug interactions increase 1400 times if statins are combined with fibrates. Both statins and fibrates are metabolized by CYP 3A4 (see Section 5.1).

AstraZeneca licensed rosuvastatin (Crestor), the seventh statin on the market, from Japanese company Shionogi & Co. Ltd. In clinical trials, Crestor lowered levels of LDL by more than 53% and raised levels of HDL by nearly 15%.[50] In 2003, the FDA approved Crestor in 5–40-mg doses, but not the 80-mg dose because of concerns about its risk to the muscles and kidneys.

5.5   *Zetia and Vytorin*

In 1990, Schering-Plough began to work on acyl CoA cholesterol acyltransferase (ACAT) inhibitors to treat atherosclerosis. Because the *in vitro* ACAT inhibition do not correalate *in vivo* animal model efficacy, the medicinal chemists opted to use the onerous route of establishing the SAR guided solely by the *in vivo* animal model. The series of 2-azetidinones afforded ezetimibe (Zetia), a potent inhibitor of cholesterol absorption in several animal models including hamster, dog, and monkey. In humans, ezetimibe reduced LDL cholesterol levels by just 18–20%. Although not as efficacious as statins, it also exhibits desirable effects on triglyceride, apolipoprotein B, and C-reactive protein (CRP) levels.[51]

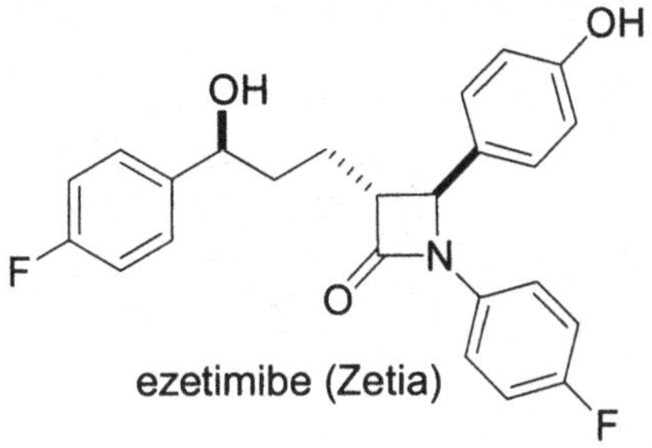

ezetimibe (Zetia)

Because no significant effects on liver enzyme levels have been observed in animals or during the clinical trials of ezetimibe, it is ideal for combination therapies with a statin. Therefore, Merck/Schering-Plough subsequently developed a fixed combination of Zetia and Zocor, which was approved by the FDA in July of 2004. The combined drug is sold under the trade name Vytorin,[52] which quickly became their biggest product. Unfortunately, in 2008, a long-awaited trial showed their cholesterol drug Vytorin failed to slow progression of heart disease better than a cheaper drug, the generic simvastatin (Zocor).[53]   The advantage of Vytorin over generic simvastatin has since been questioned.

6   CNS Drugs

Humans have dabbled with mind-altering medicines for millennia. Alcohol was invented and consumed as early as 5000 B.C. Psychedelic effects of wild mushrooms were recorded in many ancient literatures. Caffeine, nicotine, marijuana, and opium have intertwined with our culture throughout history.

6.1   *Sternbach, Valium, and Minor Tranquilizers*

In 1945, working with chief chemist William Bradley, Frank M. Berger at the Wallace Laboratories investigated a variety of compounds to find an antibacterial for Gram-negative bacteria. One of the drugs, meprobamate (Miltown), synthesized in 1950, caused mice to become paralyzed and their muscles relaxed. It soon proved to be an excellent minor tranquilizer with a preferable safety profile.[54] On the market since 1955, it rapidly achieved outstanding commercial success as an anxiolytic agent.

In 1955, to search new tranquilizers, Leo Henryk Sternbach at Roche prepared 24 benzodiazepines, which he studied for his Ph.D. thesis on azo dyes and dyestuff intermediates. None were active. Just when he was moved to another project, he submitted the last two additional analogs for pharmacologic testing. One of them, chlordiazepoxide (Librium), did well in mice, showing strong hypnotic effects. Subsequently, chlordiazepoxide was tested to be a better tranquilizing agent than Miltown in all species, indicating its phenomenal sedative, muscle-relaxant, and anticonvulsant properties. Librium, on the market in 1960, was followed by diazepam (Valium) three years later. Valium, another benzodiazepine also discovered by Sternbach, is five times more efficacious than Librium.[55] Their MOA is now believed to be through facilitation of synaptic transmission at the γ-aminobutyric acid (GABA) receptor. The benzodiazepine-GABA receptor complex has since been isolated and characterized.

meprobamate (Miltown)   chlordiazepoxide (Librium)   diazepam (Valium)

6.2   *Antidepressants*

Hoffmann-La Roche discovered isoniazid in 1951 for the treatment of tuberculosis. In 1952, Herbert H. Fox and John T. Gibas at Roche prepared many derivatives of isoniazid, including iproniazid (Marsilid). During the clinical trials of iproniazid, three groups, including Nathan S. Kline's simultaneously observed improvement of mood for chronically depressed, hospitalized patients also suffering from tuberculosis. They convinced Roche to put Marsilid on the market to treat depressed patients. After a conference in 1957, an article in the *New York Times* touted Marsilid's miraculous mood elevation effects. Almost overnight, Marsilid found widespread use in treating depression as an off-label prescription. However, some patients who took Marsilid developed jaundice due to liver toxicity. Roche voluntarily withdrew Marsilid from the U.S. market in 1961 with the conviction of quickly finding a safer drug, which never materialized.[56] Iproniazid works by inhibiting monoamine oxidase (MOA, see Section 2.7).

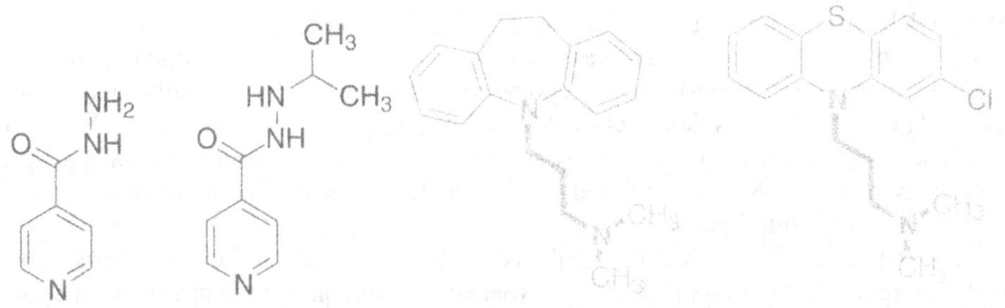

isoniazid   iproniazid (Marsilid)   imipramine (Tofranil)   chlorpromazine (Thorazine)

In 1950, Roland Kuhn tested some of Geigy's antihistamines for possible hypnotic properties, which none of them had.   Five years later, the tranquillizing properties of chlorpromazine (Thorazine) became known and heralded the first highly effective antischizophrenic drug.   Kuhn remembered that some of Geigy's antihistamines he tested produced effects similar to those of chlorpromazine.   He then wrote several letters to Geigy and suggested that further studies of those antihistamines for central nervous system diseases were warranted.   G22355 (imipramine) was synthesized and sent to Kuhn for testing.   Although it did not show any antipsychotic effects on 300 schizophrenics, the clinical trials then moved to test the effects of imipramine on severe depression.   After treating their first three cases, it was already clear that the substance, later known as imipramine, had an antidepressant action.[57]   The emergence of imipramine (Tofranil) in the spring of 1958 ushered in the use of tricyclic antidepressants.   Imipramine works by inhibiting noradrenaline reuptake at the adrenergic endings and serotonin to a lesser extent.

Initially, AB Astra discovered and marketed an inhibitor for serotonin reuptake, zimeldine, the prototype of the *s*elective *s*erotonin *r*euptake *i*nhibitors (SSRIs).   Unfortunately, a rare but serious side effect, Guillain-Barré syndrome, started to surface after zimeldine was approved and administered in a large patient base.   Astra pulled it off the market in the early 1980s.

Eli Lilly was the first to put an SSRI on the market in 1988 with fluoxetine hydrochloride (Prozac).

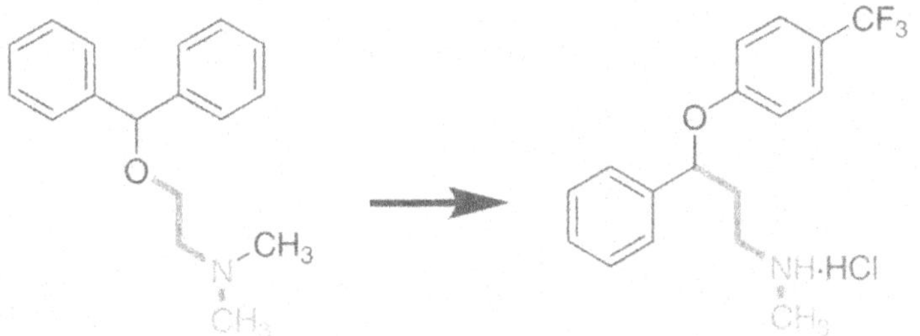

diphenhydramine (Benadryl)   fluoxetine (Prozac)

In collaboration with biologist Robert Rathburn, Lilly's medicinal chemists Bryan Molloy and David Wong looked for antidepressants without acetylcholine modulation, which was the culprit of tricyclic antidepressants' side effects.[58]   Using diphenhydramine (Benadryl) as their starting point, they arrived at the phenoxypropylamine series in 1970, which were devoid of the effect of acetylcholine. One of them, fluoxetine, selectively blocked the removal of serotonin while sparing most other biogenic amines. Unlike some of the earlier compounds, it was relatively inactive in the reserpine-induced hypothermia test in mice. In clinical trials, fluoxetine was found to have a favorable side-effect profile and was much safer in overdose relative to the tricyclics. Fluoxetine (Prozac) rapidly transformed debilitating depression into a manageable disease for many patients. Additional widely known SSRIs are sertraline hydrochloride (Zoloft) and paroxetine hydrochloride (Paxil).

Behaving similarly to SSRIs, venlafaxine (Effexor) is one of the serotonin and norepinephrine reuptake inhibitors (SNRIs), whereas bupropion (Wellbutrin) is one of the norepinephrine and dopamine reuptake inhibitors (NDRIs)—which are also used to help smokers quit.

6.3   *Antipsychotics*

In 1944, Rhone-Poulenc synthesized and marketed an antihistamine, promethazine (Phena).[59] A surgeon in the French Navy, Henri Laborit found that promethazine was superior to other drugs of the time. Rhone-Poulenc sought to enhance promethazine's "side effects" in the central nervous system. Their SAR investigations led to the synthesis of RP-3277 (chlorpromazine) in 1950. Further clinical trials were then carried out under the direction of Jean Delay and Pierre Deniker, two psychiatrists at the L'Hôpital Sainte-Anne de Paris. Under the influence of chlorpromazine, their patients became "disinterested" as well. More importantly, chlorpromazine subdued the hallucinations and delusions of psychotic patients. Chlorpromazine (Thorazine), introduced in 1954 in the US, was the first typical antipsychotic. In the first eight months, over two million patients were administered the drug, contributing to an 80% reduction of the resident population in mental hospitals.

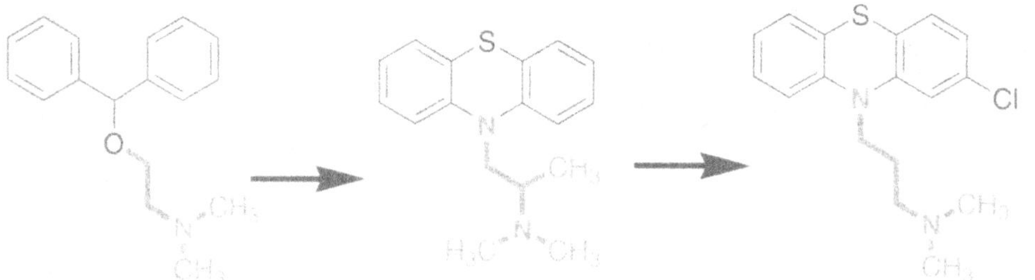

diphenhydramine (Benadryl)   promethazine   chlorpromazine (Thorazine)

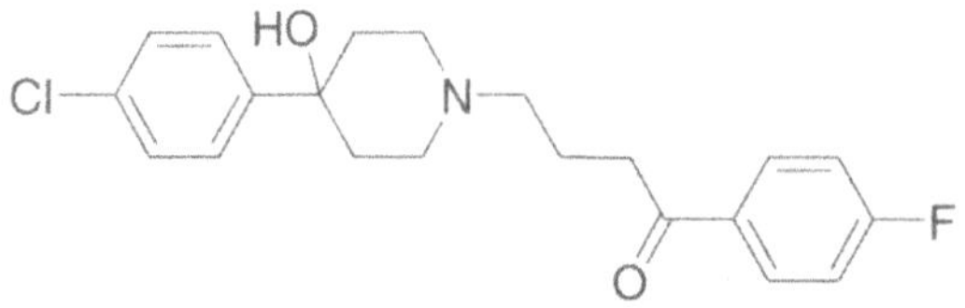

haloperidol (Haldol)

Also a typical antipsychotic haloperidol (Haldol) is 50–100 times more potent than chlorpromazine with fewer side effects such as *extrapyramidal symptoms* (EPS), including Parkinsonian symptoms, akathisia, dyskinesia, and dystonia. It was synthesized in 1958 by Bert

Hermans under the direction of Paul A. J. Janssen.[60] It was many times more potent than chlorpromazine and was both faster and longer acting. Haloperidol was potent orally as well as parenterally. More importantly, it was almost devoid of the antiadrenergic and other autonomic effects of chlorpromazine. Numerous chronic inpatients were able to leave the hospital and live at home thanks to haloperidol. Haloperidol remained one of the most prescribed neuroleptics 40 years after its discovery, until the emergence of atypical antipsychotics.

Clozapine, the first atypical antipsychotic, was developed in 1959 by the small Swiss company Wander AG.[61] During the clinical trials, clozapine showed strong sedating effects and proved to be efficacious for schizophrenia, but it also showed some liver toxicity. Wander received approval to market it in a few European countries in 1971, although liver toxicity limited its widespread use. Clozapine was removed from the market in 1975 due to rare but potentially fatal drug-associated agranulocytosis. Clozapine was reintroduced in 1990 by Sandoz and is now used as a second-line treatment requiring extensive monitoring of the patient's blood cell count.

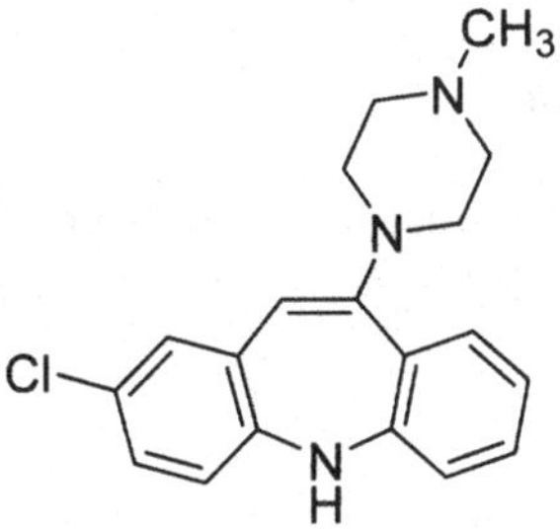

clozapine (Clozaril) 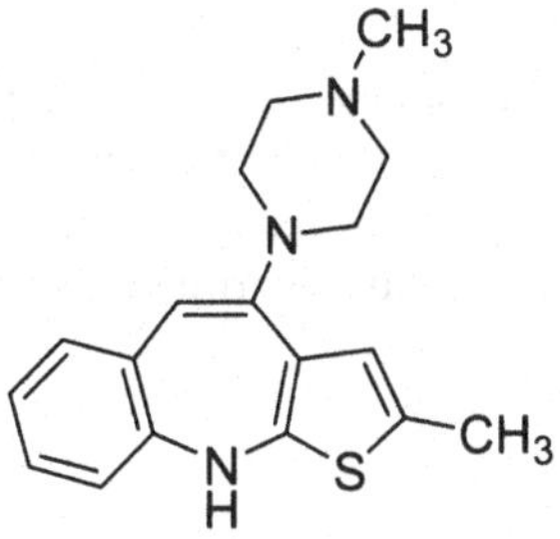   olanzapine (Zyprexa)

The second atypical antipsychotic was risperidone (Risperdal) introduced by Janssen in 1993.[62] Intrigued by the success of a combination therapy of haloperidol and ritanserin (a serotonin antagonist), Janssen initiated a medicinal chemistry program led by chemist Anton Megens. Based on the neuroleptics lenprone and benperidol, they explored a series of benzisoxazole derivatives and ultimately discovered risperidone. It showed a desired combination of very potent serotonin and potent dopamine antagonism.

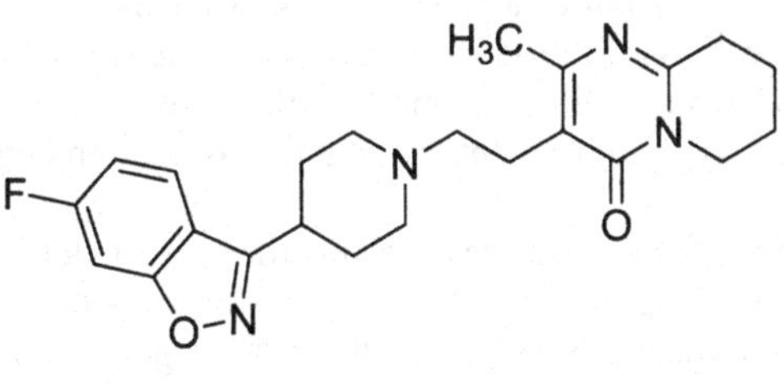

risperidone (Risperdal) 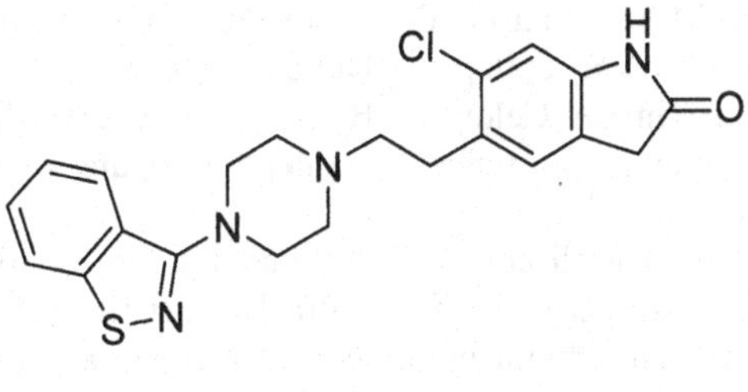   ziprasidone (Geodon)

Additional atypical antipsychotics are Eli Lilly's olanzapine (Zyprexa), AstraZeneca's quetiapine (Seroquel), Pfizer's ziprasidone (Geodon), and Bristol-Myers Squibb's aripiprazole (Abilify). One advantage of Geodon and Abilify is that they have less weight gain side effect.

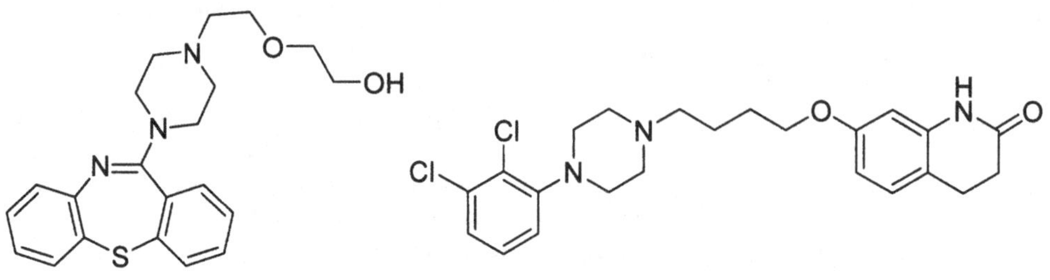

quetiapine (Seroquel)   aripiprazole (Abilify)

6.4   *Drugs for Epilepsy and Bipolar Disorder*

The first drug for epilepsy was phenobarbital (Luminal), discovered by Bayer in 1910. It was the fruit of a larger effort of Bayer's to take advantage of the barbiturates first discovered by Adolf von Baeyer in 1864.

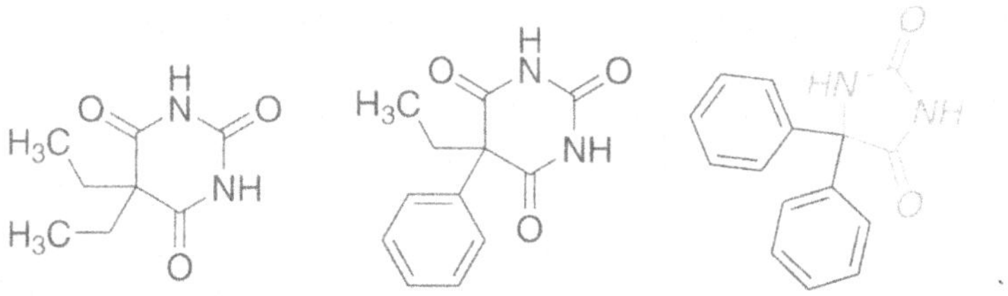

barbital (Veronal)   phenobarbital (Luminal)   phenyltoin (Dilantin)

In 1939, Parke–Davis introduced phenyltoin (Dilantin) as a treatment for epilepsy shortly after its efficacy was discovered by Tracy Jackson Putnam and H. Houston Merritt of Boston City Hospital. In 1954, a chemical and biological research team including L. M. Long, G. M. Chen, and C. A. Miller at Parke–Davis developed phensuximide (Milontin), an anticonvulsant for the control of petit mal epilepsy. Three years later, Parke–Davis introduced another epilepsy drug, methsuximide (Celontin). By then, Parke–Davis's Dilantin, Milontin, and Celotin covered each of the three major types of epileptic seizure (motor symptoms, sensory symptoms, and mental symptoms).[63]

The discovery of lithium for the treatment of manic-depressive disorder was made by John F. Cade at the Bundoora Repatriation Mental Hospital. In 1948, Cade was surprised to notice that lithium urate seemed to have a profound sedative effect on the guinea pigs, which became lethargic after 2 h of latent period. Subsequently, he injected guinea pigs with lithium carbonate and observed the same outcome. He then administered lithium carbonate to 10 of his manic patients and was rewarded with an astonishing triumph: All the patients were cured, and some even went back to work.[64]

Very few people took notice of Cade's pioneering findings except a few Australian psychiatrists. More than 20 years would elapse until 1955 when Danish doctor Mogens Schou reintroduced Cade's discovery to the world and enthusiastically touted the therapeutic power of lithium. Schou's validation and extension of Cade's original observations resulted in wide acceptance of lithium as the gold standard for treating manic patients—it is now the first drug prescribed after a diagnosis of bipolar disorder. Although its mechanism of action is still unknown, it may reverse a major psychotic reaction by boosting the level of serotonin.

7   Anti-inflammatory Drugs

7.1   Cortisone

Rheumatoid arthritis (RA) is a chronic inflammatory disease characterized by pain, swelling, and subsequent destruction of joints. In 1949, Philip S. Hench (1896–1965), a rheumatologist at the Mayo Clinic, worked with his colleague, biochemist Edward C. Kendall (1886–1972), and isolated the active principle that relieved RA symptoms.[65] He named it cortisone to signify the adrenal *cortex* steroid horm*one*. Initially Lewis H. Sarett and Max Tishler at Merck synthesized the first 100 g of cortisone from bile acid.[66] Cortisone became available to every researcher and physician in the United States. Clinical trials showed dramatic effects of cortisone on patients incapacitated by rheumatoid arthritis.

progesterone → (*Rhizopus nigricans*) → 11α-hydroxyprogesterone

In 1949, Chemist Durey H. Peterson and microbiologist H. C. Murray at Upjohn succeeded in converting progesterone to hydroxyprogesterone by introduction of oxygen at carbon-11 by the microorganism *Rhizopus nigricans*.[67] Using 11α-hydroxyprogesterone, Upjohn chemists, led by John A. Hogg, accomplished a practical total synthesis of cortisone in 11 steps, which was commercialized, enabling a commercially viable process for the industrial production of cortisone.

In 1953, Josef Fried and his associate Elizabeth Sabo at the Squibb Institute synthesized 9-fluorocortisol, which not only was much more potent than cortisone but also possessed appreciable mineralocorticoid activity similar to that of cortisone.[68]

7.2   Nonsteroidal Anti-Inflammatory Drugs

Salicylic acid, a component in the bark of willow tree, is aspirin's precursor. The Ebers Papyrus referenced willow's medicinal properties in general and its treatment of rheumatism in pregnant women in particular. In Greece the father of medicine, Hippocrates, recommended using the bark of the willow tree as an analgesic. In 1753, Reverend Edward Stone in England experimented with the extraordinarily bitter willow bark in treating ague (fever from malaria) and intermitting disorders with satisfactory results.

Charles Gerhardt in France and Karl Johann Kraut in Germany synthesized aspirin in 1853 and in 1869, respectively. In 1897, Felix Hoffmann at Bayer prepared a more pure form of aspirin using an improved route. Arthur Eichengrün tested aspirin on himself and did not suffer any heart malady. After aspirin's success in clinics, Heindrich Dreser discovered that aspirin is a prodrug of salicylic acid.[69]

In 1971, John R. Vane (1924–2004) discovered that aspirin works by blocking cyclooxygenase, thus preventing the synthesis of prostaglandins.[70] Another popular analgesic, acetaminophen (e.g., Tylenol), is not an NSAID. It is biologically active as an analgesic possibly through inhibition of cyclooxygenase-3 (COX-3), a subtype of the cyclooxygenase enzyme not identified until just a few years ago.

Charles Winter at Merck, Sharp, and Dohme developed the cotton string granuloma test as a model of inflammatory pain. Using this model, Merck screened about 350 indole compounds and identified indomethacin as a potent anti-inflammatory drug. Indomethacin was initially synthesized by medicinal chemist T. Y. Shen as a plant growth regulator.[71] Indomethacin was introduced in 1964, and is still regarded as a gold standard that combines both anti-inflammatory and analgesic activities.

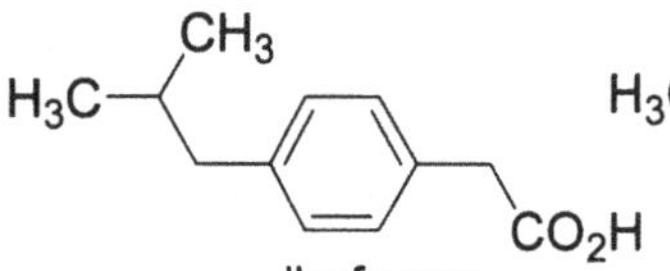

ibufenac 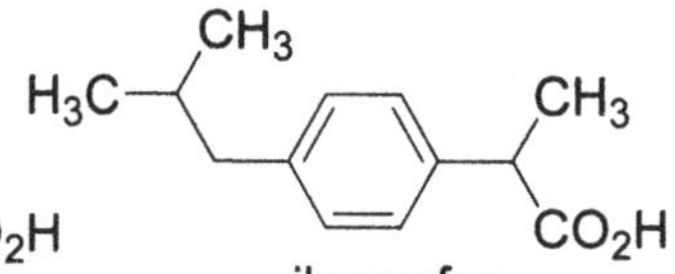   ibuprofen 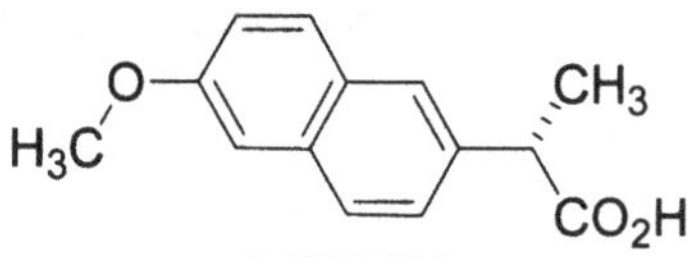   naproxen

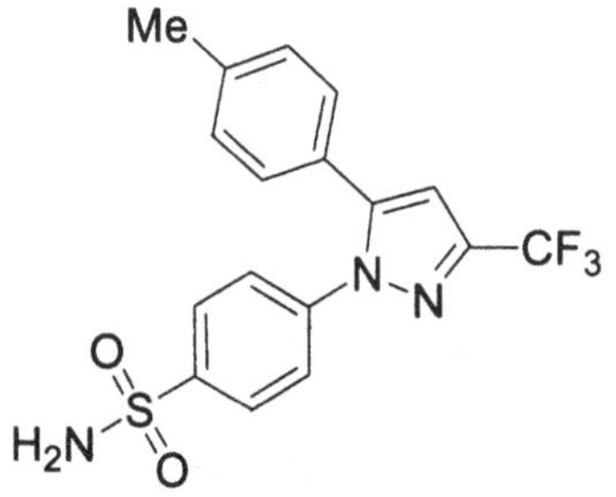

celecoxib (Celebrex) 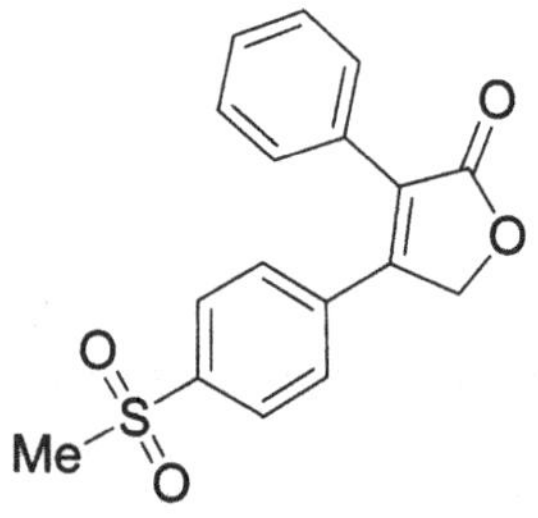   rofecoxib (Vioxx)

At Boots Company in the UK, pharmacologist Stewart S. Adams screened over 600 phenoxylalkanoic acids since 1956 but did not find any analgesics more potent than aspirin. In 1960, chemist John Nicholson synthesized *tert*-butylphenylacetic acid, which was proven to be effective for the treatment of rheumatoid arthritis but caused rash in some patients. A very similar drug, isobutylphenylacetic acid (ibufenac) did not cause a rash but caused liver toxicity in a small number of patients after long-term use. Finally, *isobu*tylphenyl*pro*pionic acid (ibuprofen, Motrin) was found to possess the best safety profile although it was not the most potent one.[72] In the same class, naproxen (Aleve), introduced by Syntex in 1976, was also a propionic acid with similar pharmacology to ibuprofen. Naproxen is twice as potent as ibuprofen and has a much longer half-life in the body.

Under the leadership of Philip Needleman, G. D. Searle came up with the first COX-2 selective inhibitor celecoxib (Celebrex), first synthesized in 1993. Searle began comarketing it with Pfizer since June 1999.[73] Six months later Merck received approval from the FDA to market their version of a COX-2 selective inhibitor, rofecoxib (Vioxx).[74] Both Celebrex and Vioxx quickly became blockbuster drugs for the treatment of osteoarthritis (OA) and rheumatoid arthritis (RA). But Vioxx was withdrawn from the market in 2004 when it was found that it increased the risk of myocardial infarction and stroke in comparison to naproxen.

7.3 *Antiasthmatics*

The first inhaled glucocorticoid, beclomethasone dipropionate (BDP), demonstrated that topical delivery to the lung resulted in reduced systemic side effects (adrenal suppression, oseteoporosis and growth inhibition) typically seen with oral steroid treatments. Unfortunately, glucocorticoids such as BDP have significant bioavailability, and when one considers the surface area of the tracheobronchial mucosa, significant plasma levels and systemic side effects occur at therapeutic doses.

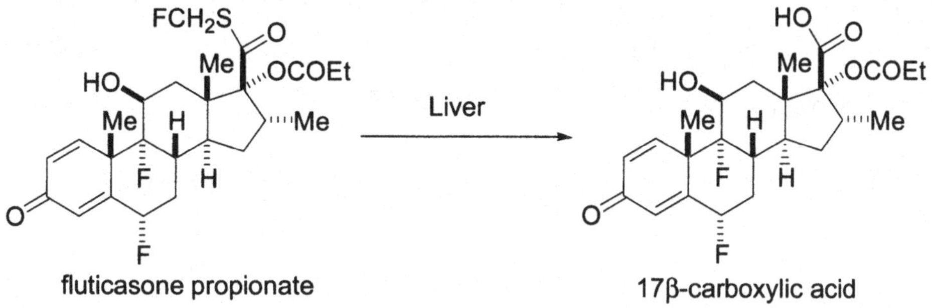

fluticasone propionate → (Liver) → 17β-carboxylic acid

GlaxoSmithKline's fluticasone propionate (Flonase), a prodrug, has a much lower systemic bioavailability. SAR studies led to a series of carbothioates, which were very active *in vivo* when topically applied to rodents but were inactive after oral administration. It was shown that fluticasone propionate underwent first-pass metabolism in the liver to the corresponding inactive 17β-carboxylic acid. Fluticasone is a 100-fold more active than BDP and thousands-fold more active than cortisol, the active form of cortisone. Moreover, Flonase was designed to be pulmonary selective. As a consequence, fluticasone only has 1% oral bioavailability, whereas cortisol has 80% oral bioavailability.

Another asthma drug, salmeterol xinafoate (Serevent),[75] a β-adrenoceptor agonist, works through a different MOA from that of Flonase. It works by dilating the lung's bronchial tubes, which become constricted and make it difficult for asthmatics to breathe. While β-blockers are excellent drugs to lower blood pressure, β-adrenergic agonists are the most prescribed class of drugs for the treatment of asthma. β-Adrenergic agonists are preferred both for the rapid relief of symptoms and for the level of bronchodilation achieved in patients with bronchial asthma. They have now become standard bronchodilators for emergency rooms, and as day-to-day reliever medicine. GlaxoSmithKline combined two asthma drugs with different mechanisms and arrived at Advir, which is comprised of Serevent (a β-adrenergic agonist) and Flonase (a glucocorticoid).

The newest therapy available for the treatment of asthma arises from the recognition of the role of the leukotrienes in the initiation and propagation of airway inflammation. Merck's montelukast sodium (Singulair) is an antagonist of leukotriene receptors.

In 1981, Merck Frosst established biological assays and animal models for modulation of leukotriene receptors in search of a treatment of asthma.[76] To find leukotriene receptor antagonists, they hand screened tens of thousands of compounds from Merck's compound library. They then selected quinolein as their lead compound for their studies on structure-activity relationship (SAR). They arrived at MK-571, which is a 1000-fold more potent than quinolein. The clinical trials in 1989 demonstrated that leukotriene receptor antagonists were effective for treating asthma, thereby confirming the pivotal role of leukotrienes in respiratory disease clinically. Unfortunately, MK-571 caused large increase of liver weight in mice. Merck Frosst scientists discovered that only one enantiomer of MK-571 had the liver side effect. In April 1991, they finally produced montelukast sodium, which possessed desirable attributes, such as high intrinsic potency, good oral bioavailability, and long duration of action for a once daily regimen for asthma.

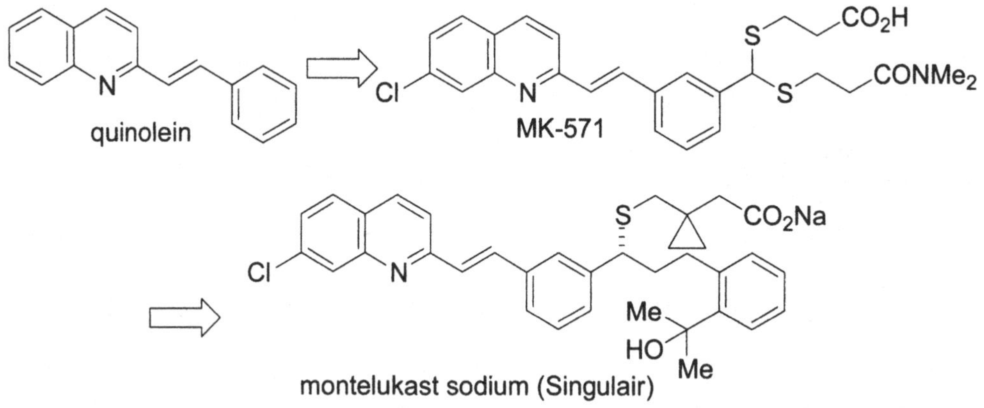

quinolein   MK-571

montelukast sodium (Singulair)

Two additional asthma drugs work via the same MOA as that of Singulair. They are AstraZeneca's zafirlukast (Accolate) and Ono Pharmaceutical's pranlukast (Onon).

8   Antiulcer drugs

8.1   James Black and the Discovery of Tagamet

In 1963, soon after the discovery of propranolol (see Section 4.5), James Black moved from ICI to SK&F and shifted his interests to the effects of histamine. In analogy to β-receptors, Black hypothesized that selectively blocking one of the two histamine receptor would give a drug to treat ulcers. Working with a team of chemists led by C. Robin Ganellin, they did not find a compound that showed any blocking activity in the assays for the first 5 years.[77] From 1968 to 1970, the team, grappling with difficult syntheses, investigated the SAR of the guanyl-histamines. They replaced the strongly basic guanidine group with a thiourea, which in 1970 led to the discovery of burimamide, the first *bona fide* pure H_2 antagonist without agonist effects. Another compound, metiamide, was 10 times more potent than burimamide and healed peptic ulcers in tests on humans, but it caused agranulocytosis. Replacing the thiourea group with cyanoguanidine, they synthesized cimetidine in 1972. It was shown to relieve symptoms and promote healing of lesions in a majority of patients with peptic ulcer disease and did not cause agranulocytosis. Marketed in the US in 1977, cimetidine (Tagamet) became the first blockbuster drug ever in medical history in 1985.

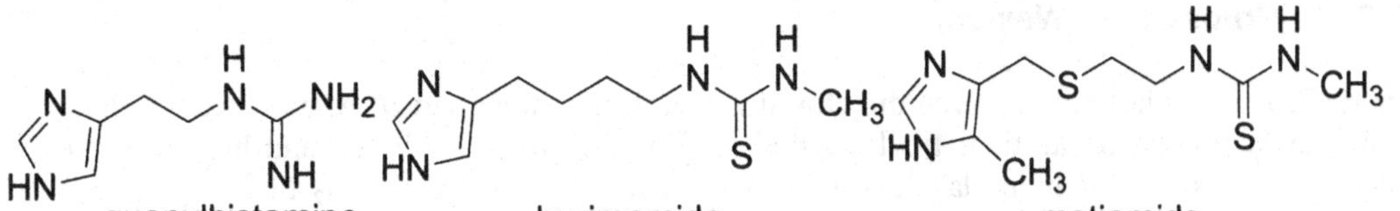

guanylhistamine   burimamide   metiamide

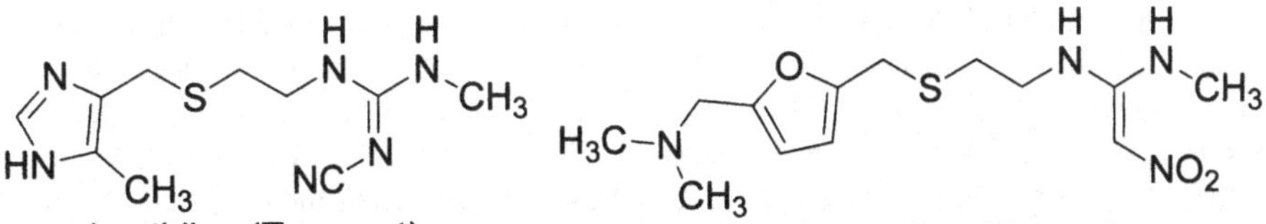

cimetidine (Tagamet)   ranitidine (Zantac)

8.2   *Zantac, Pepcid, and Axid*

In 1972, David Jack and his pharmacologist colleague Roy Thomas Brittain at Allen & Hanburys attended a lecture by James Black, who revealed that burimamide inhibited histamine-induced acid secretion in animals and man. Jack redirected their Peptic Ulcer Project Group to solely focus on the H_2 antagonists using burimamide as the starting point.[78] They struggled for four years in attempts to replace SK&F's imidazole with nonbasic heterocyles without success until chemist John Clitherow carried out a Mannich reaction to install the dimethylaminomethylfuran. At the time, SK&F disclosed that 1,1-diamino-2-nitroethene was a good bioisostere for cyanoguanidine. Allen & Hanburys chemists then attached the nitrovinyl fragment into their own dimethylaminomethylfuran scaffold. The resulting drug AH19065, first prepared in early August 1976, was 10-times more potent in rats than cimetidine and it was also less toxic. AH19065 (ranitidine) was a poorly soluble solid. Its hydrochloride salt (Zantac) was prepared in 1977. Clinical trials subsequently confirmed its efficacy and safety.

In the mid-1970s, a team at Yamanouchi led by Isao Yanagisawa replaced Tagamet's imidazole core with the guanidinothiazople. The guanidinothiazople moiety was initially discovered by ICI in its tiotidine, which was not marketed due to toxic manifestations. Meanwhile, they also replaced Tagamet's cyanoguanidine sidechain with carbamoyl-amidine. The resulting drug was the most potent histamine H_2 antagonist at the time, but it was not stable enough to be manufactured. Replacing the carbamoyl-amidine moiety with an unorthodox sulfamoyl-amidine afforded famotidine (Pepcid). Lilly's nizatidine (Axid), was the fourth histamine H_2 antagonist in the U.S. market.

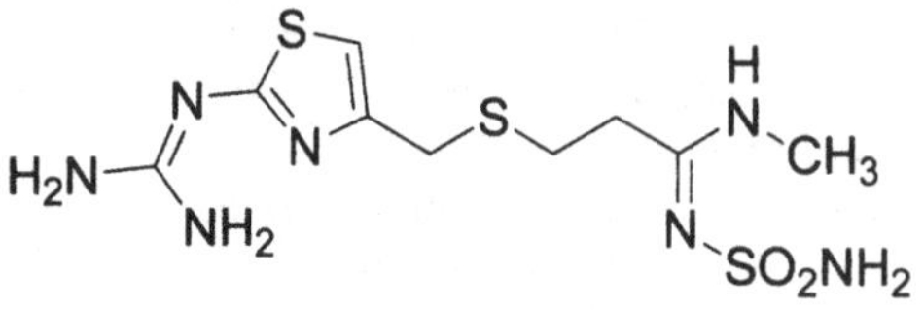

famotidine (Pepcid) 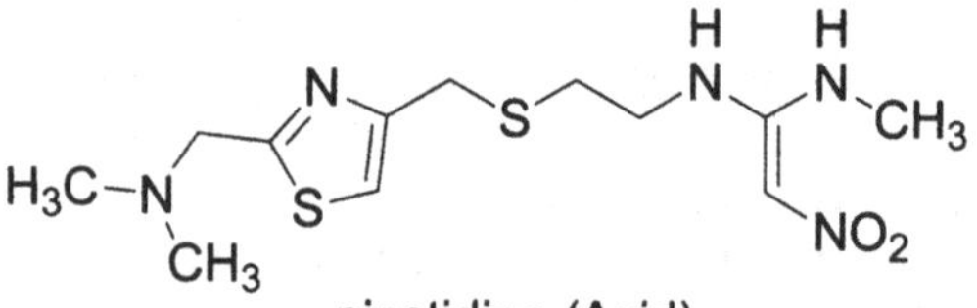   nizatidine (Axid)

8.3   *Prilosec and Nexium*

In 1967, Ivan Östholm, a research director at Astra Hässle (now part of AstraZeneca), initiated a program in the gastrointestinal field[79] called it "Gastrin Project." Black and his colleagues at SK&F demonstrated that imidazole was important for acid control by H_2 histamine receptors. Hässle chemists wisely selected benzimidazole moiety to replace the toxic thioamide portion on Servier's CMN 131. This decision would prove critical. Combining features on CMN 131 and Tagamet gave rise to H 124/26, which was both efficacious and safe. Unfortunately, a Hungarian patent already patented the same compound for the treatment of tuberculosis. The team later isolated timoprazole, an oxidative metabolite isolated from dogs' urine, which was even more active than H 124/26. In 1974, timoprazole was found to cause toxicities in thymus and thyroid by blocking the uptake of iodine to the thyroid gland in rats. In 1976, they arrived at picoprazole, which had antisecretory activity in both rats and dogs but without toxic effect.

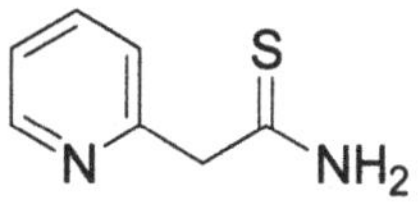

CMN 131 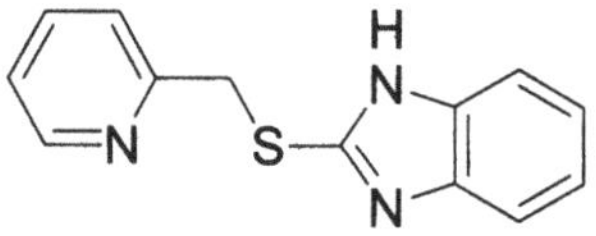   H 124/26 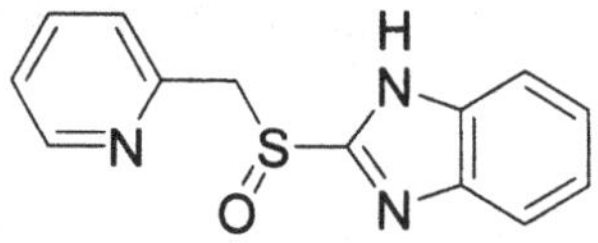   H 83/69 (timoprazole)

In the summer of 1977, Erik Fellenius, a biochemist from Hässle, met George Sachs from Alabama at a conference. The ensuing collaboration revealed that timoprazole's MOA is through inhibition of H^+,K^+-ATPase, the proton pump of the stomach. Sachs and his postdoc Berglindh demonstrated that timoprazole and picoprazole were prodrugs but were converted to the active form after accumulation in the acidic secretory canaliculus of the parietal cells.[80] That was the beginning of proton pump inhibitors (PPIs).

Meanwhile, medicinal chemists kept optimizing the substitutions on the core structure. In January 1979, Ylva Örtengren prepared H168/68,[81] which would eventually become omeprazole (Prilosec). Since 3,5-dimethyl-4-methoxy-pyridine was known to increase the basicity of a pyridine ring, in their enthusiasm to prepare a very potent compound, they went ahead and had Örtengren prepare it without systematically exploring the structure-activity relationship. However, by a stroke of good luck, omeprazole gave higher *in vivo* activity than any other combinations. As a matter of fact, omeprazole was the most powerful inhibitor of stimulated gastric acid secretion in experimental animals at the time. The drug had no sign of serious toxicity in animal models.

Omeprazole (Prilosec) was approved as a treatment for duodenal ulcers and reflux esophagitis in Sweden in 1988 and in the U.S. in 1990.

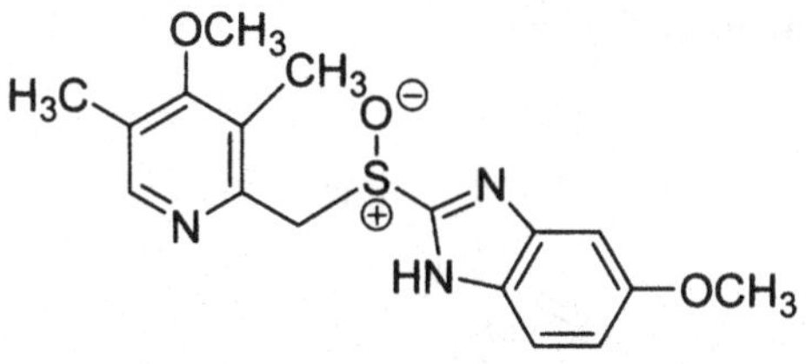

omeprazole (Prilosec) 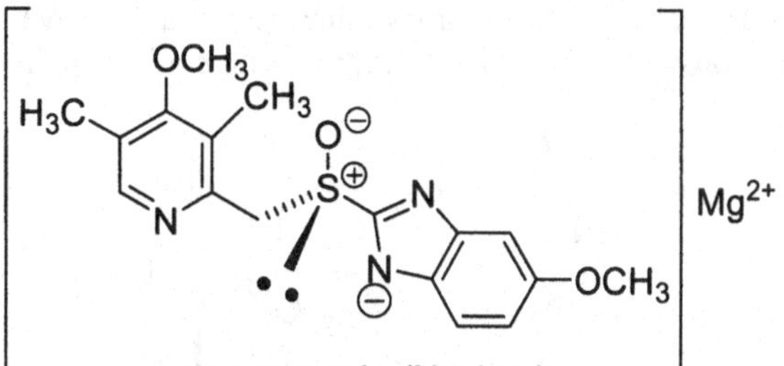   esomeprazole (Nexium)

In 1987, a group led by Gunnel Sundén embarked on a focused program in finding an omeprazole backup with better bioavailability. Among all the compounds in their hands, only one was better than omeprazole. It was the *S*-(−)-enantiomer of omeprazole, esomeprazole (Nexium). When tested in rats, esomeprazole was 4–5 times more bioavailable than the *R*-enantiomer.[82] Another lucky break was that they tested esomeprazole directly in man and saw similar effects to what they observed with rats. Later on, when the two isomers were tested on dogs, no significant difference of efficacy was detected for the two isomers. In this case, rat was a better animal model than dog. If they used dogs for their initial *in vivo* tests, they probably would never have found esomeprazole. Esomeprazole (Nexium) was approved in Sweden in 2000 and in the U.S. in early 2001, just when Prilosec's U.S. patent expired in April.

Lansoprazole (Prevacid) is another proton pump inhibitor sold by TAP Pharmaceuticals, a joint venture of Abbott Laboratories and Takeda. Two other PPIs are pantoprazole (Protonix)[83] and rabeprazole (Aciphex).

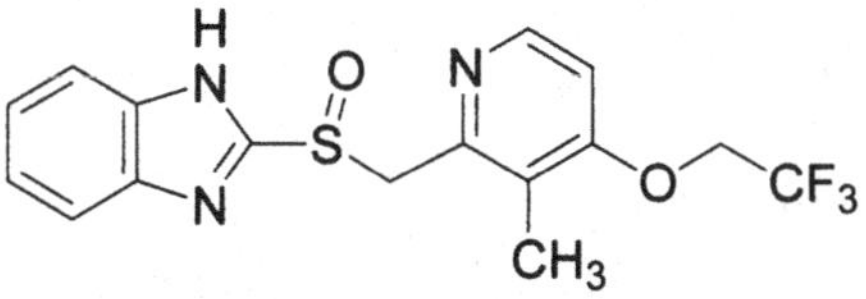

lansoprazole (Prevacid) 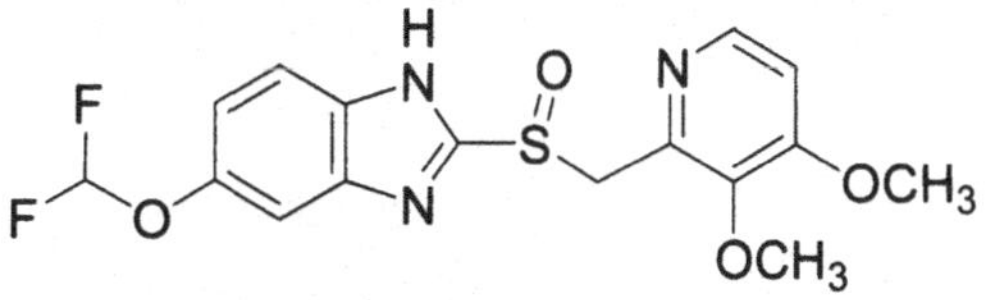   pantoprazole (Protonix)

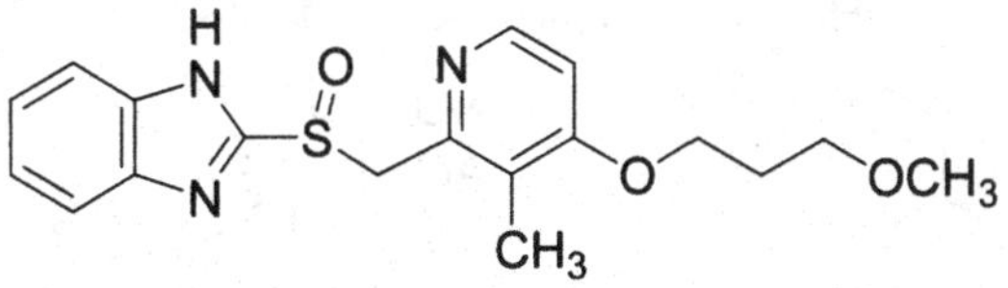

rabeprazole (Aciphex)

9   Antiviral Drugs

The first clinically effective antiviral did not become available in 1962 until 5-iodo-2′-deoxyuridine (IDU) became the first antiviral drug for the treatment of the most common corneal infection of humans: herpes keratitis. It was synthesized by William H. Prusoff at Yale in 1959 as a potential anticancer drug.[84] Since it was not a selective inhibitor of virus replication, it was highly toxic, most noticeably bone marrow suppression. Herbert Kaufman demonstrated IDU could be used for herpes simplex virus (HSV) infection in the eye.[85]   He also found trifluorothymidine (TFT, Viroptic) with similar applications in 1964.

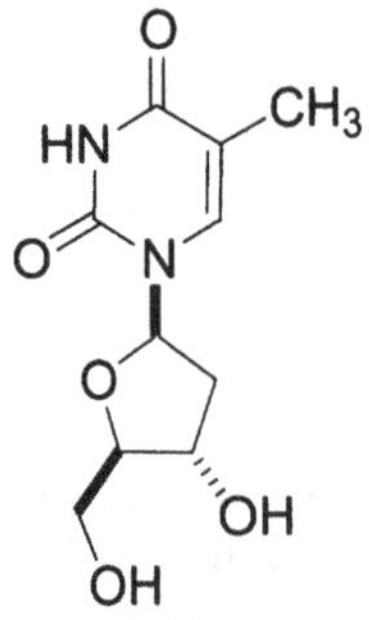

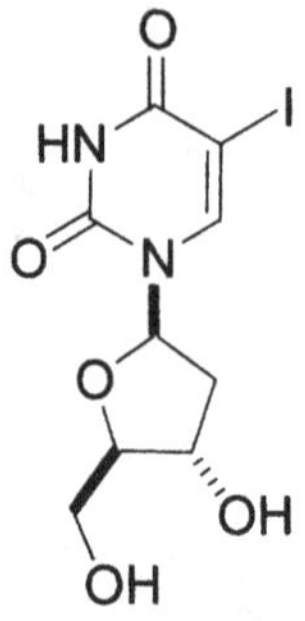

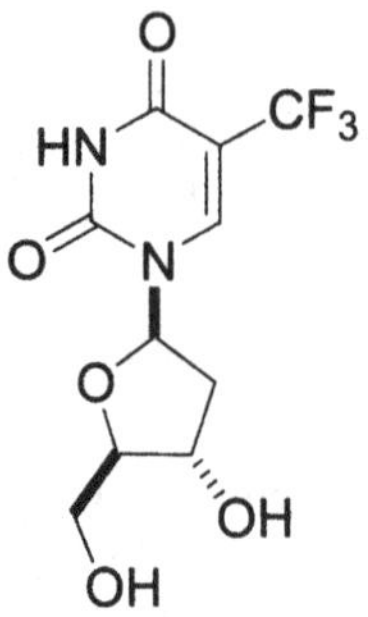

thymidine iododeoxyuridine (IDU) trifluorothymidine (TFT, Viroptic)

9.1   Influenza Drugs

Currently, there are four drugs available for the treatment or prophylaxis of influenza infections: the adamantanes, including amantadine and rimantadine. The adamantanes act as M_2 ion channel inhibitors and interfere with viral un-coating inside the cell.[86]   They are effective only against influenza A and are associated with several toxic effects as well as drug resistance.

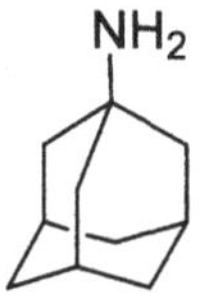

amantadine 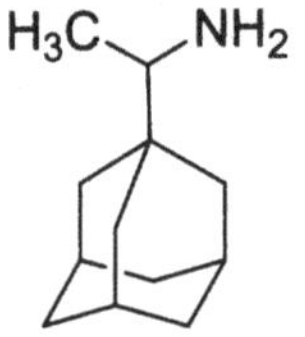   rimantadine

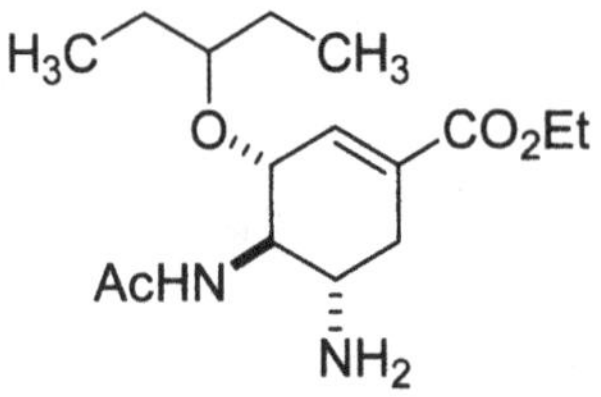

oseltamivir (Tamiflu) 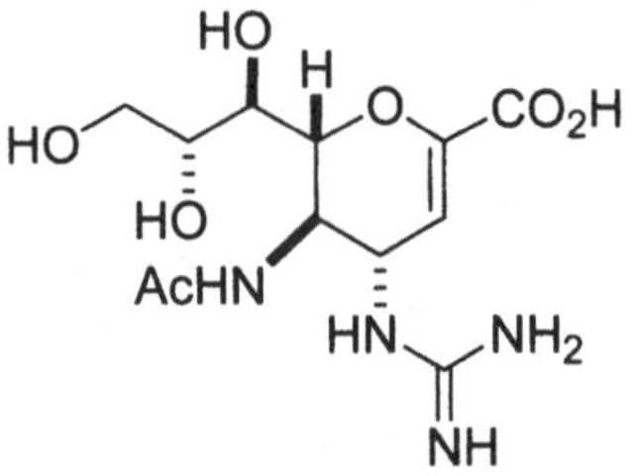   zanamivir (Relenza)

The neuraminidase inhibitors, including oseltamivir (Tamiflu) and zanamivir (Relenza), are newer drugs that have shown greater efficacy for both influenza A and B and are associated with fewer side effects in comparison to the adamantanes. The influenza neuraminidase is one of two major glycoproteins located on the influenza virus membrane envelope. Oseltamivir, the first orally active neuraminidase inhibitor, was discovered by Choung U. Kim and co-workers at

Gilead Sciences in 1995.[87] Gilead and Roche began codeveloping it in 1995 and the FDA approved it in 1999. On the other hand, zanamivir was discovered by Biota Holdings, a small Australian biotechnology concern.[88] In the United States, Biota established an alliance with GlaxoSmithKline for development and marketing of zanamivir .

9.2   HIV Drugs

In 1984, the NCI began to screen antiviral agents as possible treatments for AIDS. In February 1985, one of the compounds, AZT (zidovudine, Retrovir), was found to be active *in vitro*.[89] AZT itself was synthesized by a group led by Jerome Horowitz of the Michigan Cancer Foundation in 1964 as a possible anticancer drug. Burroughs Wellcome acquired the right to AZT and explored the possibility of using it to treat the herpes virus under the guidance of Gertrude Elion, although it did not make it to the market. Wellcome patented AZT as an antiviral drug in 1985 and promptly commenced the clinical trials. The FDA approved the use of AZT in 1987. AZT is a prodrug—it is not active *in vitro*, but its triphosphate is the active agent *in vivo*. Its MOA is the blockade of the HIV reverse transcriptase activity.

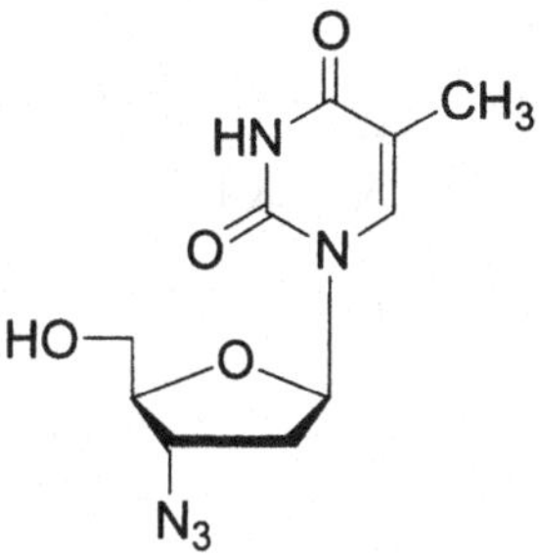

azidothymidine (AZT) 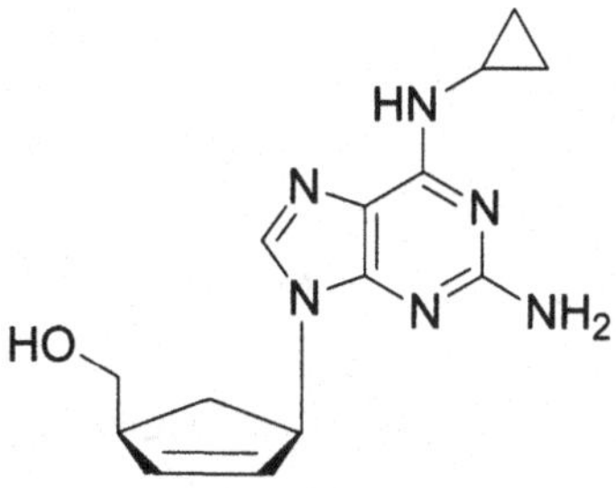   abacavir (Ziagen)

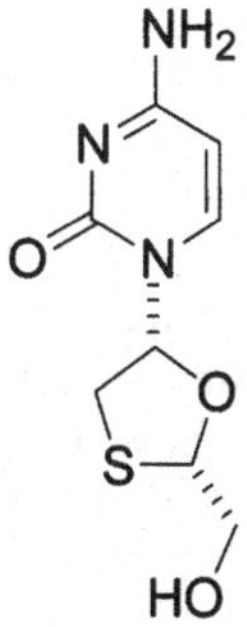

3TC (Epivir) 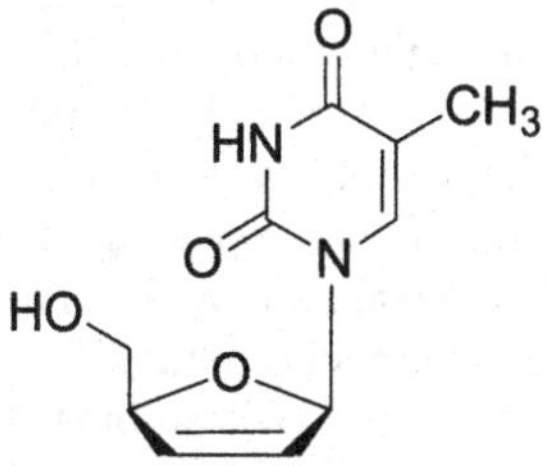   d4t (Zerit) 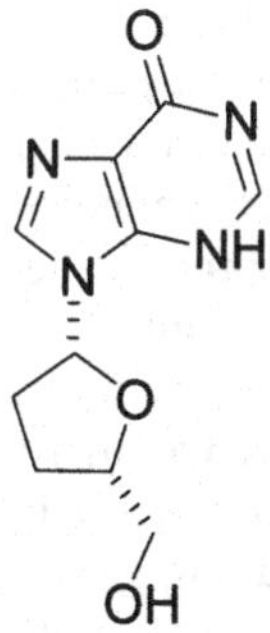   ddI (Videx) 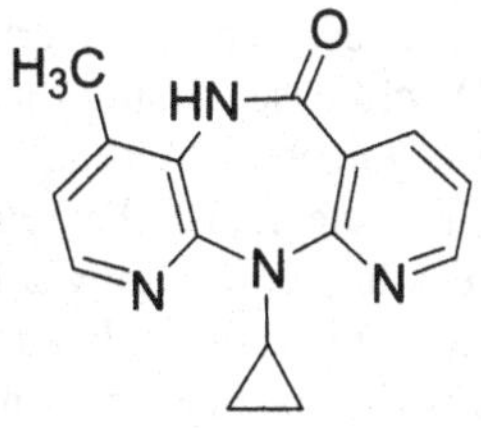   nevirapine (Viramune)

Among the newer reverse transcriptase inhibitors, abacabir (Ziagen) represents a vast improvement over AZT, a nucleotide whose gycosidic core structure is metabolized rapidly.[90] By replacing the oxygen on AZT with a methylene group, carbocyclic nucleoside analogs such as Ziagen are metabolized much slower by the body. Ziagen was developed by GlaxoWellcome (now part of GlaxoSmithKline) using a technology developed by Robert Vince of the University of Minnesota, who licensed the patent to GlaxoWellcome in 1993.

In addition to AZT and Ziagen, many HIV reverse transcriptase inhibitors exist. An organic chemistry professor at Emory University, Dennis Liotta, and his virologist colleague Raymond Schinazi discovered another reverse transcriptase inhibitor, 3TC (lamivudine, Epivir), which allows a once daily regimen.[91] Additional HIV reverse transcriptase inhibitors are BMS's d4T (stavudine, Zerit),[92] 2'-3'-dideoxyinosine (ddI, didanosine, Videx),[93] and Boehringer Ingelheim's nevirapine (Viramune),[94] a nonnucleoside inhibitor.

Saquinavir (Invirase) by Roche was the first HIV protease inhibitor on the U.S. market.[95] In 1986, Roche undertook an ambitious international collaboration to tackle the HIV protease. The chemistry team in Welwyn led by Ian B. Duncan and Sally Redshaw designed some inhibitors using the "transition-state mimic" concept, which was highly successful in producing their potent renin inhibitors. They found that a tripeptide was ideal considering both potency and bioavailability. The Roche team fine tuned the tripeptide, exploring their lead compound systematically by modifying each amino acid residue in turn. In 1991, they arrived at Ro 31-8959 (saquinavir). Roche carried out the clinical trials, led by Keith Bragman, Roche's top European virologist. Saquinavir (Invirase) became the first HIV protease inhibitor for the treatment of AIDS when it was approved by the FDA in December 1995.

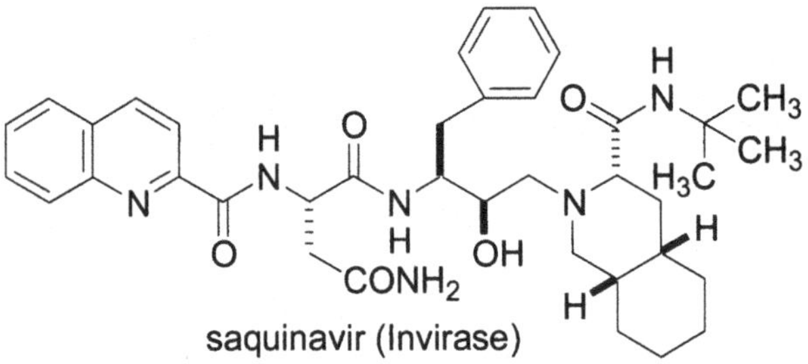

saquinavir (Invirase)

Abbott's team for protease inhibitors was led by an X-ray crystallographer, John Erickson, and a medicinal chemist, Dale Kempf.[96] Instead of screening their renin inhibitors like most drug firms did at the time, they took advantage of Erickson's X-ray crystallography work on the HIV protease, integrating structure-based drug design (SBDD) and traditional medicinal chemistry. They prepared a series of symmetry-based inhibitors to match the C_2-symmetric nature of the HIV protease. Using that approach, they arrived at A-77003, a tetrapeptide. Although A-77003 was potent in binding and cellular assays, it was not bioavailable with extremely high human biliary clearance. By reducing the molecular weight and replacing the existing amino acids with more soluble ones, they achieved an increase in bioavailability. They also identified that the pyridine termini were oxidized into N-oxide by hepatic cytochrome P450. Simply replacing the pyridines with thiazoles and fine tuning gave rise to ritonavir, whose bioavailability was 78% in comparison to 26% for the pyridyl analog. Interestingly, while many other protease inhibitors are metabolized by hepatic cytochrome P450 3A4 (a major isozyme), ritonavir (Norvir) is a potent inhibitor of P450 3A4. As a result, dual protease inhibitor therapy has proven to be a powerful regimen in terms of efficacy and minimizing drug resistance.

Merck began their research on protease inhibitors in 1986 with Irving Sigal as the project champion. Merck's medicinal chemistry team was led by Joel Huff.[97] They initially screened their renin inhibitors for HIV protease inhibition and then carried out rational drug design by taking advantage of the known crystal structure of HIV-1 protease. In 1990 Wayne Thompson arrived at L-689,502, which was active in inhibiting the HIV protease but devoid of renin activity. Unfortunately it was not bioavailable and only effective by injection. Inspired by saquinavir's success, Joseph Vacca successfully incorporated a fragment of saquinavir into L-689,502. Bruce Dorsey, a new hire in 1989 in Vacca's group, and his associate Rhonda Levin succeeded in synthesizing L-735,524 (indinavir). However, in the monotherapy trials HIV developed resistance to indinavir in some patients after six month. Fortunately, it was found that the combination of indinavir and AZT or 3TC was effective in substantially suppressing the virus levels. Merck's studies of combination therapy were the first to prove the efficacy of the cocktail approach and became the standard for the industry. After filing with the FDA in January 1996, indinavir (Crixivan) received approval in March in an accelerated review process.

Other important HIV protease inhibitors include nelfinavir (Viracept) and amprenavir (Agenerase). In 2006, another HIV protease inhibitor darunavir (Prezista) was approved for the treatment of HIV/AIDS patients who are harboring drug-resistant HIV that does not respond to other therapies. The drug was discovered by Arun K. Ghosh at Purdue University, who started the effort strictly as an educational project.[98] They used structure-based drug design, based on X-ray crystallographic analysis of inhibitor-bound HIV protease structures, to optimize inhibitor binding to the backbone. Some older HIV protease inhibitors primarily targeted the protease's amino acid sidechain, which the virus could modify more easily.

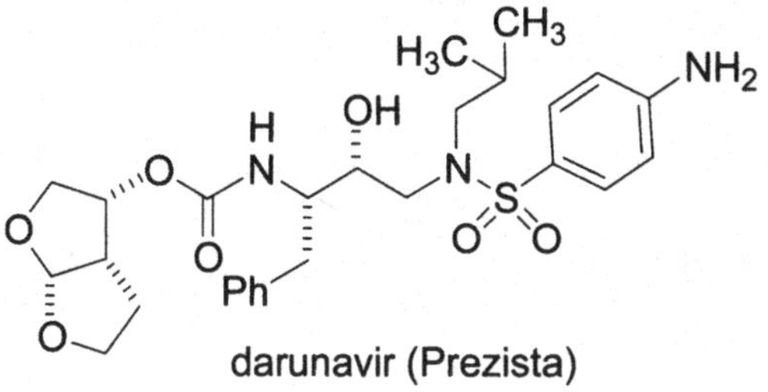

darunavir (Prezista)

In addition to HIV reverse transcriptase inhibitors and HIV protease inhibitors, several other mechanisms yielded successful treatments in the clinics. Merck's raltegravir (Isentress) is the first FDA-approved inhibitor of HIV integrase.[99] Pfizer's maraviroc (Selzentry) is the first-in-class CCR5 antagonist for the treatment of HIV.[100]

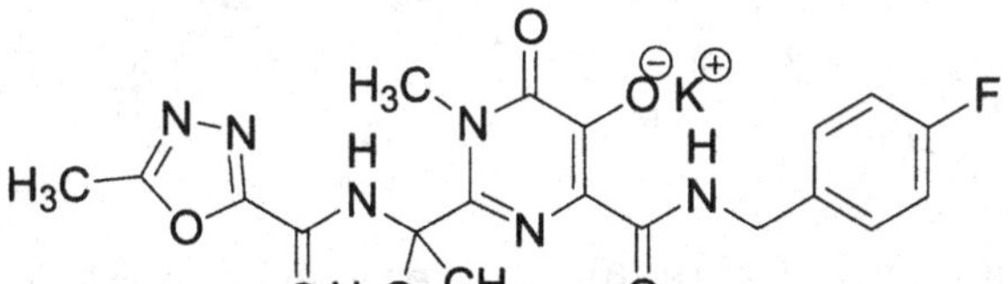

raltegravir (Isentress)

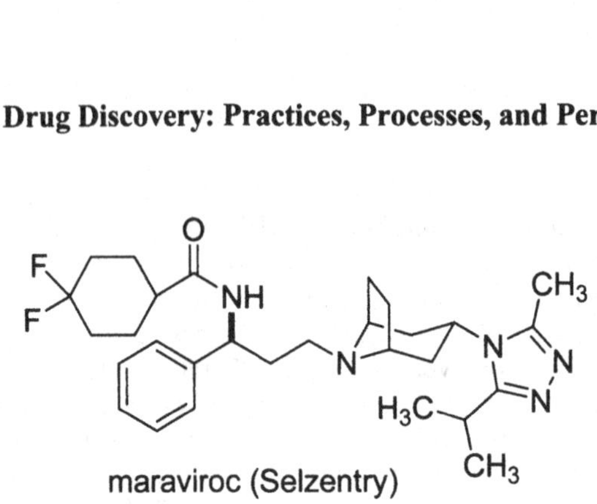

maraviroc (Selzentry)

9.3   *Hepatitis Virus Drugs*

For the treatment of chronic hepatitis B, interferon and five nucleos(t)ides have been approved in many parts of the world.   3TC (lamivudine, Epivir) was the first antiviral drug for HBV. Tenofovir disoproxil fumarate (tenofovir DF) is an orally administered ester prodrug of tenofovir, a nucleotide reverse transcriptase inhibitor that shows potent *in vitro* activity against both hepatitis B virus (HBV) and HIV-1. Current treatments for HBV also include adefovir dipivoxil (Hepsera) and BMS's entecavir (Baraclude).   Interferon, unfortunately, is effective only in a subset of patients It is often poorly tolerated, requires parenteral administration, and is expensive.[101]

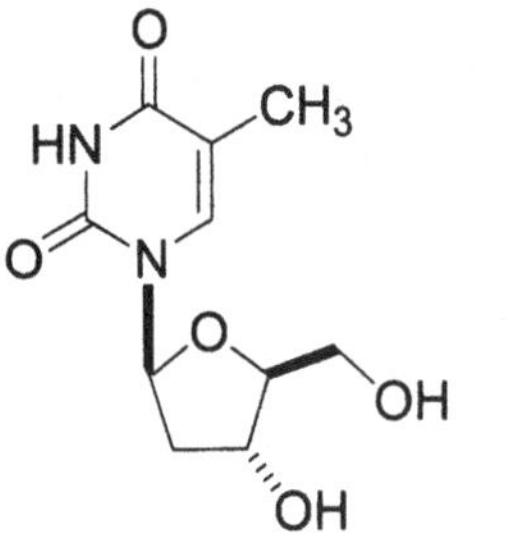

telbivudine (Tyzeka) 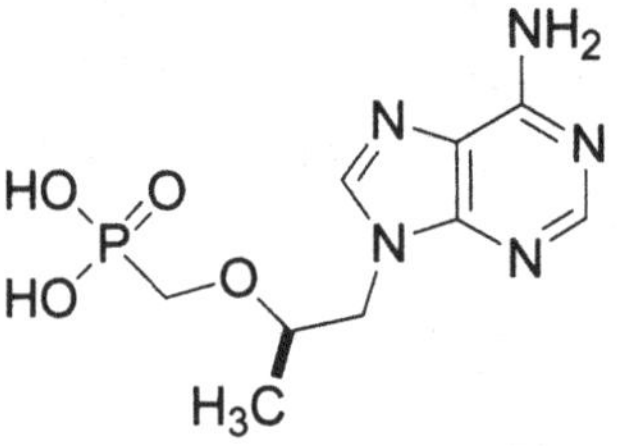   tenofovir (Viread) 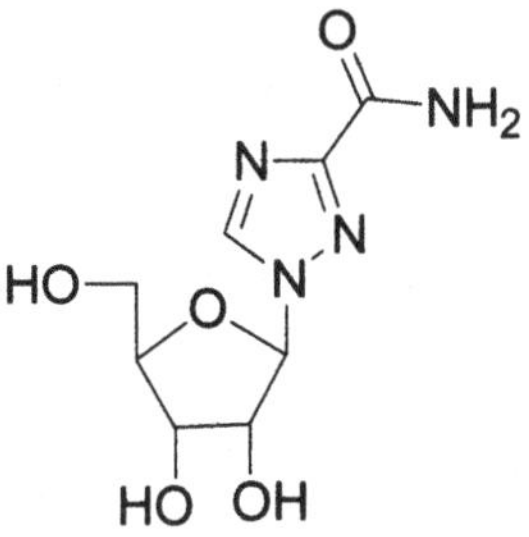   ribavirin (Rebetol)

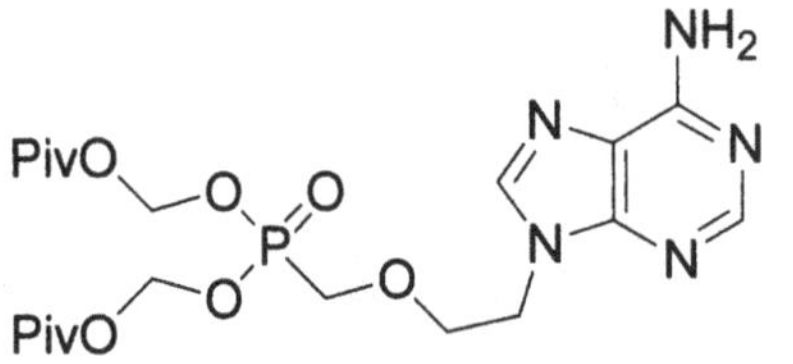

adefovir dipivoxil (Hepsera) 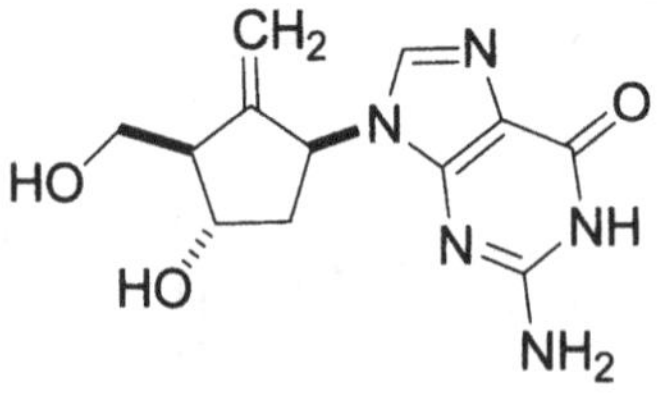   entecavir (Baraclude)

Current therapy for hepatitis C virus (HCV) includes pegylated interferon-α (PEG-Intron) alone or in combination with ribavirin (Rebetol).[102] However, the antiviral activity of interferons is indirect, and ribavirin is a nonspecific agent with inhibitory activity toward some host proteins, a circumstance that has contributed to the considerable effort being expended to identify and develop specifically targeted antiviral therapy for HCV.[103] They include HCV NS2 and NS3/4 protease inhibitors, NS3 helicase inhibitors, NS4B and NS5A, and NS5B replication factor inhibitors, as well as HCV NS5B polymerase inhibitors.

Excitingly, two HCV drugs were approved by the FDA in May 2011 during preparation of the manuscript. Both are HCV NS3–4A serine protease inhibitors. One is boceprevir (Victrelis) by Schering-Plough/Merck and the other is telaprevir (Incivek) by Vertex. These two innovative medicines now avail patients with HCV safe, efficacious, and convenient treatment options.

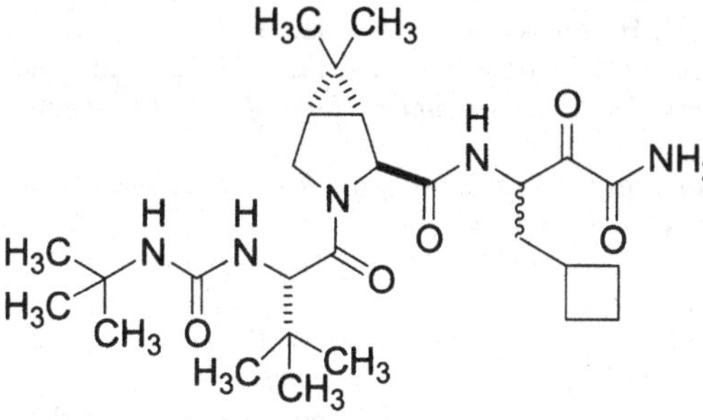

boceprevir (Victrelis)

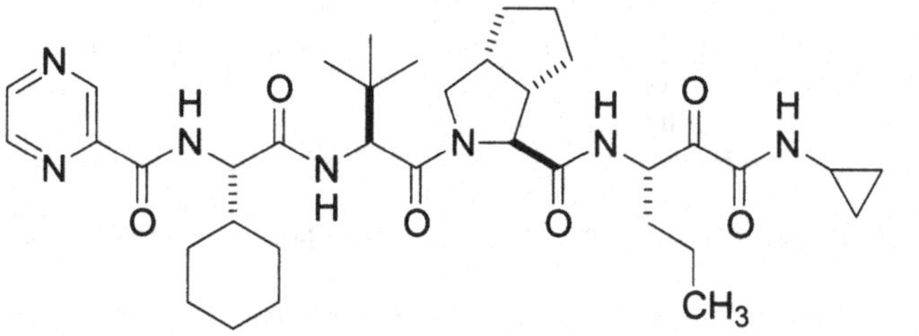

telaprevir (Incivek)

Due to space restraint, I cannot even begin to cover many of the most exciting discoveries going on in the laboratories around the world. History is being written every day in human's crusade to defeat diseases. The ensuing chapters will give an overview of some major aspects in drug discovery.

References

1   Absolon K. B.; Absolon, M. J.; Zientek, R. *Rev. Surg.* **1970,** *27,* 245–258.
2   Bankston, J. *Joseph Lister and the Story of Antiseptics (Uncharted, Unexplored, and Unexplained),* Mitchell Lane Publishers, 2004.
3   Bäumler, E. *Paul Ehrlich: Scientist for Life,* Translated by Grant Edwards, Holmes & Merie: New York, 1984.
4   Riethmiller, S. *Chemotherapy* **2005,** *51,* 234–242.
5   Bankston, J. *Gerhard Domagk and the Discovery of Sulfa,* Mitchell Lane Publishers, Bear, DE, 2003.
6   Otfinoski, S. *Alexander Fleming, Conquering Disease with Penicillin,* Facts on File: New York, NY, 1993.

7   Macfarlance, G. *Howard Florey, the Making of A Great Scientist,* Oxford University Press, 1979.
8   Clark, R. W. *The Life of Ernst Chain, Penicillin and Beyond,* St. Martin's Press, 1985.
9   Gordon, K. *Selman Waksman and the Discovery of Streptomycin,* Mitchell Lane Publishers, Bear, DE, 2003.
10   Rawal, S. Y.; Rawal, Y. B. *West Ind. Med. J.* **2001,** *50,* 105–108.
11   Wentland, M. P. in *Quinolone Antimicrobial Agents,* 2nd ed., Hooper, D. C.; Wolfson, J. S., eds.; Washington D. C., American Society for Microbiology: XII-XIV, **1993.**
12   Hooper, D. C.; Wolfson, J. S., eds. *Quinolone Antimicrobial Agents,* 2nd ed., American Society of Microbiology: Washington, DC, 1993.
13   Batts, D. H.; Kollef, M. H.; Lipsky, B. A.; Nicolau, D. P.; Weigelt, J. A., eds, *Creation of A Novel Class: The Oxazalidinone Antibiotics* Innova Institute for Medical Education: Tampa, FL, 2004.
14   Varmus, H.; Weinberg, R. A. *Genes and Biology of Cancer,* Scientific American Library: New York, 1993.
15   Weinberg. R. A. *Racing to the Beginning of the Road, the Search for the Origin of Cancer,* W. H. Freeman and Company: New York, NY, 1998.
16   Infield, G. B. *Disaster at Bari, the True Story of World War II's Worst Chemical Warfare Disaster!* Ace Books: New York, NY, 1971.
17   Rosenberg, B. in *Cisplatin: Current Status and New Development,* Prestayko, A. W.; Crooke, S. T.; K., eds, Academic, New York, N. Y., **1980**
18   Summer, J. *The Natural History of Medicinal Plants,* Timber Press: Portland, OR, **2000.**
19   Goodman, J.; Walsh, V. *The Story of Taxol: Nature and Politics in the Pursuit of an Anti-Cancer Drug,* Cambridge University Press: Oxford, UK, 2001.
20   Jordan, V. C.   . *Rev. Drug Disc.* **2005,** *2,* 205–213.
21   Bazell, R. *HER-2, the Making of Herceptin, A Revolutionary Treatment for Breast Cancer,* Random House: New York, **1998.**
22   Zimmermann, J. *Chimia,* **2002,** *56,* 428–431.
23   Gregory, A. *Harvey's Heart: The Discovery of Blood Circulation,* Icon Book: Cambridge, UK, **2001.**
24   Roddis, L. H. *William Withering,* Paul B. Hoeber, Inc.: New York, **1936.**
25   Holmes, L. C.; DiCarlo, F. *J. Chem. Ed.* **1971,** *48,* 573–576.
26   Butler, A.; Nicholson, R. *Life, Death and Nitric Oxide,* Royal Society of Chemistry: Cambridge, UK, **2003.**
27   Vogl, A. *Am. Heart J.* **1950,** *39,* 881–883.
28   Beyer, K. H. *Perspect. Biol. Med.* **1977,** *20,* 410–420.
29   deStevens, G. *J. Med. Chem.* **1991,** *34,* 2665–2670.
30   Smith, C. G.; Vane, J. *FASEB J.* **2003,** *17,* 788–789.
31   Raju, T. N. K. *Lancet* **2000,** *355,* 1022.
32   Cee, V. In *Modern Drug Synthesis,* Li, J. J.; Johnson, D. S., eds., Wiley: Hoboken, NJ, 2010, pp 141–158.
33   Watson, C.; Carney, S.; Huxley, A.; Miles, N. *Drug Discov. Today* **2005,** *10,* 881–883.
34   Fleckenstein-Grün, G. *High Blood Pressure* **1994,** *3,* 284–290.
35   Nayler, W. G. *Calcium Antagonists,* Academic Press: London, UK, 1988.
36   Davies, M. K.; Hollman, A. *Heart* **1998,** *80,* 120.
37   Wardrop, D.; Keeling, D. *Br. J. Haematol.* **2008,** *141,* 757–763.
38   Steinhubl, S. R.; Tan, W. A.; Foody, J. M.; Topol, E. J. *JAMA* **1999,** *281,* 806–810.
39   Stein, S. H., U.S.D.J., United States District Court Southern District of New York *Opinion and Order, 02 Civ. 2255 (SHS), Sanofi-Synthelabo; Sanofi-Synthelabo, Inc.; and Bristol-Myers Squibb Sanofi Pharmaceuticals Holding Partnership, Plaintiffs, -against- Apotex Inc. and Apotex Corp.;* June 19, 2007, New York, NY.
40   Carlson, L. A. *J. Intern. Med.* **2005,** *258,* 94–114.
41   Sneader, W. *Drug Discovery: The Evolution of Modern Medicine,* Wiley: New York, **1985.**
42   Endo, A. *Atheroscler. Suppl.*   **4,**   –

43   Brown, A. G.; Smale, T. C.; King, T. J.; Hasenkamp, R.; Thompson, R .H. *J. Chem. Soc. Perkin Trans. 1* **1976,** 1165–1170.

44   Endo, A. *J. Lipid Res.***1992,** *33*, 1569–1582.

45   Vagelos, P. R.; Galambo, L. *Medicine, Science and Merck*, Cambridge University Press: Cambridge, UK, **2004**.

46   Stokker, G. E.; Alberts, A. W.; Anderson, P. S.; Cragoe, E. J., Jr.; Deana, A. A.; Gilfillan, J. L.; Hirshfield, J.; Holtz, W. J.; Hoffman, W. F.; Huff, J. W.; Lee, T. J.; Novello, F. C.; Prugh J. D.; Rooney, C. S. Smith, R. L.; Willard, A. K. *J. Med. Chem.* **1986,** *29*, 170–181.

47   Kathawala, F. G. *Med. Res. Rev.* **1991,** *11*, 121–146.

48   Li, J. J.; Johnson, D. S.; Sliskovic, D. R.; Roth, B. D., *Contemporary Drug Synthesis*, Wiley: Hoboken, NJ, 2004, pp 113–124.

49   Ozdemir, O.; Boran, M.; Gokce, V.; Uzun, Y.; Kocak, B.; Korkmaz, S. *Angiology* **2000,** *51*, 695–697.

50   Rosenson, R. S. *Exp. Rev. Cardiovasc. Ther.* **2003,** *1*, 495–505.

51   Rosenblum, S. B. In *The Art of Drug Synthesis*, Johnson, D. S.; Li, J. J., eds., Wiley: Hoboken, NJ, **2007,** pp 237–254.

52   Toth, P. P.; Davidson, M. H. *Expert Opin. Pharmacother.* **2005,** *6*, 131–139.

53   Davidson, M. H.; Maccubbin, D.; Stepanavage, M.; Strony, J.; Musliner, T. *Am. J. Cardiol.* **2006,** *97*, 223–228.

54   Laties, V. G.; Weiss, B. *J. Chronic Dis.* **1958,** *7*, 500–519.

55   Baenninger, A.; Costae, S.; Jorge A.; Moeller, H.-J.; Rickles, K. *Good Chemistry, The Life and Legacy of Valium Inventor, Leo Sternbach,* McGraw-Hill: New York, **2004**.

56   Kauffman, G. B. *J. Chem. Ed.* **1979,** *56*, 35–36.

57   Ayd, F. J. Jr.; Blackwell, B., eds., *Discoveries in the Biological Psychiatry*, J. B. Lippincott Co.: Philadelphia, PA, **1970**.

58   Molloy, B. B.; Wong, D. T.; Fuller, R. W. *Pharm. News* **1994,** *1*, 6–10.

59   Thuillier, J. *Ten Years That Changed the Face of Mental Illness,* Martin Dunitz: London, **1999**.

60   Granger, B. *Encephale* **1999,** *25*, 59–66.

61   Curzon, G. *Trends Pharm. Sci.,* **1990,** *11*, 61–63.

62   Li, J. J.; Johnson, D. S.; Sliskovic, D. R.; Roth, B. D., *Contemporary Drug Synthesis,* Wiley: Hoboken, NJ, 2004, pp 89–111.

63   Li, J. J. *Triumph of the Heart—the Story of Statins,* Oxford University Press: New York, NY, 2009, pp 74–76.

64   Webb, J. *J. Chem. Ed.* **1976,** *53*, 291–292.

65   Lloyd, M. *Rheumatol.* **2002,** *41*, 582–584.

66   Hirschmann, R. *Steroids* **1992,** *57*, 579–592.

67   Peterson, D. H.; Murray, H. C. *J. Am. Chem. Soc.* **1952,** *74*, 1871.

68   Fried, J. *Steroids* **1992,** *57*, 384–391.

69   Jeffreys, D. *Aspirin: The Remarkable Story of a Wonder Drug,* Bloomsbury: New York, NY. 2004.

70   Vane, J. R. *Nature* **1971,** *231*, 232–235.

71   Shen, T. Y. *Semin. Arthritis Rheum.* **1982,** *12(2, Suppl. 1)*, 89–93.

72   Adam, S. S. *Inflammopharm.* **1999,** *7*, 191–197.

73   Talley, J. J. *Book of Abstracts*, 219th ACS National Meeting, San Francisco, CA, March 26-30, 2000, MEDI-299.

74   Prasit, P.; Wang, Z.; Brideau, C.; Chan, C.-C.; Charleson, S.; Cromlish, W.; et al. *Bioorg. Med. Chem. Lett.* **1999,** *9*, 1773–1778.

75   Sears M. R.; Lotvall, J. *Respir. Med.* **2005,** *99*, 152–170.

76   Young, R. N. *Prog. Med. Chem.* **2001,** *38*, 249–277.

77   Duncan, W. A. M.; Parsons, M. E. *Gastroenterol.* **1980,** *78*, 620–625.

78   Brittain, R. T. *Curr. Clin. Practice Ser.* **1982,** *1,* 5–15.

79   Sjöstrand, S. E.; Olbe, L.; Fellenius, E. In *Proton Pump Inhibitors (Milestones in Drug Therapy)*, Olbe, L., Ed.; Birkhäuser Verlag: Basel, 1999, pp 3–20.
80   Modlin, I. M. *J. Clin. Gastroenterol.* **2006**, *40*, 867–869.
81   Olbe, L.; Carlsson, E.; Lindberg, P. *Nat. Rev. Drug Discov.* **2003**, *2*, 132–139.
82   Lindberg, P.; Carlsson, E. In *Analogue-Based Drug Discovery*, Fischer, J.; Ganellin, C. R., Eds.; Wiley-VCH: Weinheim, Germany, 2006, pp 81–113.
83   Senn-Bilfinger, J.; Sturm E. In *Analogue-Based Drug Discovery*, Fischer, J.; Ganellin, C. R., Eds.; Wiley-VCH: Weinheim, Germany, 2006, pp 114–151.
84   Prusoff, W. H. *Biochim. Biophys. Acta* **1976**, *32*, 295–296.
85   Littler, E. *IDrugs* **2004**, *7*, 1104–1112.
86   Kolocouris, N.; Kolocouris, A.; Foscolos, G. B.; Fytas, G.; Padalko, E.; Neyts, J.; De Clercq, E. *Biomed. Health Res.* **2002**, *55*, 103–115.
87   Lew, W.; Chen, X.; Kim, C. U. *Curr. Med. Chem.* **2000**, *7*, 663–672.
88   Smith, P. W.; Sollis, S. L.; Howes, P. D.; Cherry, P. C.; Starkey, I. D.; Cobley, K. N.; Weston, H.; Scicinski, J.; Merritt, A.; Whittington, A.; et al. *J. Med. Chem.* **1998**, *41*, 787–797.
89   Mehellou, Y.; De Clercq, E. *J. Med. Chem.* **2010**, *53*, 521–538.
90   Vince, R. Personal communications.
91   Borman, S. *Chem. Eng. News* **2007**, *85*, 42–47.
92   Martin, J. C.; Hitchcock, M. J. M.; De Clercq, E.; Prusoff, W. H. *Antiviral Res.* **2010**, *85*, 34–38.
93   de Clercq, E. *Rev. Med. Virol.* **2009**, *19*, 287–299.
94   Adams, J.; Merluzzi, V. J. In *The Search for Antiv. Drugs*, Adams, J.; Merluzzi, V. J., eds., Birkhauser: Boston, 1993, pp 45–70.
95   Duncan, I. B.; Redshaw, S. *Infect. Disease Ther.* **2002**, *25*, 27–47.
96   Kempf, D. J. *Infect. Disease Ther.* **2002**, *25*, 49–64.
97   Dorsey, B. D.; Vacca, J. P. *Infect. Disease Ther.* **2002**, *25*, 65–83.
98   Ghosh, A. K.; Martyr, C. D. in *Modern Drug Synthesis*, Li, J. J.; Johnson, D. S., eds., Wiley: Hoboken, NJ, 2010, pp 29–144.
99   Hunt, J. A. in *Modern Drug Synthesis*, Li, J. J.; Johnson, D. S., eds., Wiley: Hoboken, NJ, 2010, pp 3–15.
100   Price, D. In *Modern Drug Synthesis*, Li, J. J.; Johnson, D. S., eds., Wiley: Hoboken, NJ, 2010, pp 17–28.
101   Zoulim, F.; Locarnini, S. *Gastroenterol.* **2009**, *137*, 1593–1608.
102   De Clercq, E. *J. Med. Chem.* **2010**, *53*, 1438–1450.
103   Meanwell, N. A.; Kadow, J. F.; Scola, P. M. *Ann. Rep. Med. Chem.* **2009**, *44*, 397–440.

2

Target Identification and Validation

Anthony M. Manning

1   Introduction

The discovery and exploitation of new drug targets are a key focus for both academic biomedical research and the pharmaceutical industry. The process of drug target identification and validation has evolved over the last century, and has dramatically changed over the last decade or so. This chapter aims to review our understanding of common drug targets and the technologies that are rapidly identifying and validating new targets.

2   Definition of Drug Targets

The definition of "target" itself is something argued within the pharmaceutical industry and has been evolving in parallel with the evolution of target identification and validation technologies. Generally, the "target" is the naturally existing cellular or molecular structure involved in the pathology of interest on which the drug-in-development is designed to act. Target identification and validation comprise the complex set of experimentation that aims to identify the key molecular drivers of disease and confirm that pharmacological modulation of that target leads to a net clinical benefit in that disease. Given that our current understanding of the molecular drivers of most of the important diseases of mankind is rudimentary, target identification and validation should be recognized as a continuum of knowledge building that occurs throughout the life cycle of a new therapeutic.

Distinctions are often made as to whether a target is "novel," "established" or "validated." These distinctions are typically made by pharmaceutical companies engaged in discovery and development of therapeutics as an attempt to stratify the relative risk that modulation of the target will result in the desired net clinical benefit and therefore deliver a therapeutic agent of value to patients, payers, and the pharmaceutical company itself. Most companies seek to build a portfolio of new drug agents that differ in their risk profiles, with the belief that such a portfolio approach is an appropriate mechanism for managing investment in research and development. In general, "novel" targets are those whose role in disease is speculative and for whom the net clinical benefit of pharmacological modulation remains unclear. The advent of genetic and genomic technologies in the last decade has greatly increased the number of novel targets. It is now possible to interrogate the complete genetic and genomic differences between disease patients and healthy counterparts, and this has identified many hundreds of disease-associated proteins, whose true role as a disease driver and for whom the net result of pharmacological modulation are still speculative. The process of target validation for "novel" targets is complex and lengthy, and begins with understanding the fundamental role of the target in normal physiology, most often in model organisms, and confirmation of the role and mechanism of action in disease, most often utilizing animal models of human disease. As you will read in later sections, there are a number of technologies now available to explore the net clinical benefit of modulation of the activity of novel targets, and these often occur in parallel with the discovery of pharmaceutically relevant modulators of the targets. "Novel" targets are sometimes also those for which there is a substantial amount of information supporting their role in disease but for which a history of successfully identifying therapeutic agents is lacking. For example, the Ras proto-oncogene is well established as a key driver of cancer but is considered a novel target as there historically has been no success in identifying therapeutic agents to modulate this target.[1] It is therefore considered novel due to the novelty of therapeutic modalities targeting Ras.

"Established" targets are those for which there is a good scientific understanding of both how the target functions in normal physiology and how it is involved in human disease, for which prototype pharmaceutical agents are available, but for which the net clinical benefit of pharmacological modulation remains unknown. "Established" targets are associated with a lengthy publication record of multiple laboratories reporting both novel and confirmatory studies supporting this knowledge. The more such information is available, the less investment is assumed to be required to develop a therapeutic directed against the target and the greater the probability of success that such an agent will yield a valuable medicine. "Established" targets also include those that the pharmaceutical industry has had experience mounting drug discovery campaigns against in the past, often using small-molecule, peptide, or protein therapeutic approaches. Such a history provides information on the chemical feasibility of developing a therapeutic against the target and the associated drug discovery tools to support this, including knowledge of structure and function, tools for testing mechanism of action and selectivity, and approaches to design the optimal drug-like properties into a final therapeutic agent. An example of an "established" target would be the phosphoinositide 3-kinase (PI3K). Numerous critical growth factors and cytokines transduce their signals from the cell membrane to the nucleus via protein kinase networks called signal transduction pathways, which have become major targets for anticancer drugs. Significant evidence exists of the key role of PI3K in transducing growth factor signaling in different cancers, and modulation of the target with either gene-silencing technologies or specific kinase inhibitors leads to tumor reductions in mouse models of cancer.[2] PI3K inhibitors therefore constitute a promising class of novel targeted therapies for cancer. Although some of the initial clinical results with these compounds were disappointing, newer agents with more advantageous pharmacokinetic and pharmacodynamic properties have entered clinical trials. Partial responses at drug doses that appear to be tolerated have been seen in patients with advanced leukemia, along with prolonged stable disease in several tumor types.[3] PI3K therefore represents an example of an "established" target: one for which there is a significant amount of biological data and for which small-molecule inhibitors can be identified but for which the net clinical benefit in disease is yet to be established.

"Validated" targets are those for which the role in disease and the net clinical benefit of pharmacological modulation is well understood. Targets of existing medicines can be considered validated, as they have demonstrated clinical efficacy in the disease population and an acceptable safety profile. Therapeutic opportunity often remains for validated targets, as many of the existing medicines either do not fully capture the potential benefit of target modulation or exhibit side effects not directly related to target modulation, but rather as a result of the lack of specificity of the drug for the target. It is often the case that a successful new medicine is quickly followed to market by other agents that can be considered either as "me-too's" or "fast followers." Such agents are often similar or marginally different in their risk/benefit profile but do offer alternatives for patients who either respond to or tolerate agents differently. As a drug class matures, new agents that improve further on the benefit and risk of prior agents emerge. Often agents with better convenience will emerge, such as one tablet per day versus two or one injection per month rather than once a day. These do represent valuable additions to the treatment armamentarium and are recognized as important opportunities by the pharmaceutical industry. For validated targets, the risk of success does not lie so much in whether the target is relevant to disease but rather in whether the therapeutic agent will represent a clinically differentiated medicine with benefit incremental to that of all existing agents. An example of a "validated" target would be cyclooxgenase (COX, also known as prostaglandin synthase or prostaglandin endoperoxide synthetase), an enzyme that is responsible for formation of important biological mediators called prostanoids, including prostaglandins, prostacyclin, and thromboxane. Pharmacological inhibition of COX can provide relief from the symptoms of inflammation and pain. In 1971, Sir John Vane identified the mechanism of action of the anti-inflammatory and analgesic agent acetylsalicylic acid (aspirin) as inhibition of COX.[4] This spurred the development of other inhibitors of this enzyme, leading to the approval of over 10 different nonsteroidal anti-inflammatory drugs (NSAIDs), each with their own relatively different risk/benefit profile. Soon after the discovery of the mechanism of action of NSAIDs, strong indications emerged for alternative forms of COX, but little supporting evidence was found. COX enzyme proved to be difficult to purify and was not sequenced until 1988. In 1992 the human COX-2 enzyme was cloned and its existence, therefore, confirmed.[5] Before the confirmed existence of COX-2, the Dupont company had developed a compound, DuP-697, that was potent in many anti-inflammatory assays but did not have the ulcerogenic effects of NSAIDs. It was quickly noted that DuP-697 was relatively selective for the COX-2 enzyme and hence became the chemical scaffold for synthesis of COX-2 inhibitors. Celecoxib and rofecoxib, the first COX-2 inhibitors to reach the market, were based on DuP-697 and offered evidence to suggest that they spared the gastrointestinal side effects seen with other NSAIDs.[6,7] Alas, such claims were not entirely confirmed, and a higher rate of cardiovascular side effects resulted in the withdrawal of rofecoxib (Vioxx) in 2004. Hence, the COX enzymes are highly validated targets for which multiple therapeutic agents have reached the market. As with all well-validated targets, unique insights into their role in disease and normal physiology have continued to emerge many years after the initial reports of their role in disease and therapeutics.

3   Classification of Currently Utilized Drug Targets

The majority of targets currently selected for drug discovery efforts are proteins, in comparison to lipids or carbohydrates. Numerous studies have reported the number of unique drug targets exploited to date and postulating on the total number of potential drug targets.[8,9] Currently marketed drugs mediate their effects through only a small number of the potential human target proteins, with previously published estimates of the number of current human drug targets ranging from ~200–400, depending on the method used and how the drug targets were defined and categorized. This represents a small fraction of the >30,000 proteins encoded in the human genome.[10]

One of the most comprehensive databases of existing drugs and targets is the publicly available DrugBank database, which drew heavily from earlier data sets and was launched in 2006.[11] The DrugBank database not only has a systematic collection of drug–protein interactions but also contains associations of proteins with consensus genetic annotations, such as Swissprot. The database currently contains information on ~1500 experimental, approved and withdrawn drugs, with up to 107 data fields for each drug that contain information, including current indications, documented drug–target interactions, target protein accession numbers, and pharmacological actions. By analyzing the drugs that were approved by the U.S. Food and Drug Administration during the past three decades and examining the interactions of these drugs with therapeutic targets that are encoded by the human genome, 435 effect-mediating drug targets, which are modulated by 989 unique drugs, have been identified. The most common indication for the drugs in this data set is antihypertensive drugs, followed by antineoplastic and anti-inflammatory agents, and the most common therapeutic targets represent G-protein-coupled receptors and enzymes.

4   Receptors as Drug Targets

4.1   G-Protein-Coupled Receptors

Receptors make up the largest group of currently utilized drug targets: 193 proteins (44% of the human drug targets) are receptors, and 82 (19%) of these are G-protein-coupled receptors (GPCRs). GPCRs are commonly targeted by antihypertensive and antiallergic drugs. By virtue of their large number, widespread distribution, and important roles in cell physiology and biochemistry, GPCRs (also known as seven-transmembrane domain receptors, 7TM receptors, heptahelical receptors, serpentine receptors, and G-protein-linked receptors) play multiple important roles in clinical medicine.[12] GPCRs comprise a large protein family of transmembrane receptors that sense molecules outside the cell and activate internal signal transduction pathways and, ultimately, cellular responses. G-protein-coupled receptors are found only in eukaryotes, including yeast, choanoflagellates, and animals. The ligands that bind and activate these receptors include light-sensitive compounds, odors, pheromones, hormones, and neurotransmitters and vary in size from small molecules to peptides to large proteins. Their expression on the plasma membrane makes GPCRs readily accessible, especially by hydrophilic hormones and drugs, including both agonists and antagonists, and their nonuniformity of expression in different tissues and cell types provides selectivity (in some cases, specificity) in the targeting of these receptors for the activation or blockade of physiological events.

GPCRs are one of the most important drug targets for the pharmaceutical industry. However, existing drugs target only a subset of family members, mainly biogenic amine receptors, so there is enormous potential within the pharmaceutical industry to exploit the remaining family members, including the >100 orphan receptors for which no existing ligands have so far been identified.[13] The exact size of the GPCR superfamily is unknown, but nearly 800 different human genes (or ≈4% of the entire protein-coding genome) have been predicted from genome sequence analysis. Although numerous classification schemes have been proposed, the superfamily is classically divided into three main classes (A, B, and C) with no detectable shared sequence homology between classes. The largest class by far is Class A, which accounts for nearly 85% of the GPCR genes. Of Class A GPCRs, over half of these are predicted to encode olfactory receptors while the remaining receptors have known endogenous ligands or are classified as orphan receptors. Despite the lack of sequence homology between classes, all GPCRs share a common structure and mechanism of signal transduction.

4.2   Ligand-Gated Ion Channels

Ligand-gated ion channels, which are the second largest receptor target class, are most commonly targeted by hypnotic drugs and sedatives. Ion channels are pore-forming proteins that help establish and control the voltage gradient across the plasma membrane of cells by allowing the flow of ions down their electrochemical gradient. They are present on all membranes of cells (i.e., the plasma membrane) and intracellular organelles (e.g., nucleus, mitochondria, endoplasmic reticulum, golgi apparatus).[14] Such "multisubunit" assemblies usually involve a circular arrangement of identical or homologous proteins closely packed around a water-filled pore through the plane of the membrane or lipid bilayer. There are over 300 types of ion channels in a living cell. Ion channels may be classified by the nature of their gating, the species of ions passing through those gates, the number of gates (pores), and localization of proteins. Further heterogeneity of ion channels arises when channels with different constitutive subunits give rise to a specific kind of current. Because ion channels underlie the nerve impulse and because "transmitter-activated" channels mediate conduction across the synapses, channels are especially prominent components of the nervous system. Absence or mutation of one or more of the contributing types of channel subunits can result in loss of function and, potentially, underlie neurological diseases. Ion channels are important therapeutic targets in a range of indications, including arrhythmia, hypertension, local anesthesia, pain, stroke, epilepsy, depression, bipolar disorder, COPD, autoimmune disorders and diabetes.[15]

Early ion channel drug discovery used classical pharmacological approaches in which profiling in animal models designed to simulate human disease states was used to optimize compound activity, even if the nature of the molecular target was unclear. Serendipity, insight, and brute force effort drove these drug discovery efforts and resulted in a number of notable successes, including successful therapies and discovery of research tools that have been invaluable in mapping out signaling pathways, purifying channel proteins, and characterizing gating mechanisms, all of which have sustained the present era of ion channel drug discovery. With the advent of a more complete understanding of cellular physiology and identification of the molecular components that constitute individual channel types and that regulate their function, researchers are now focusing on a molecular-based strategy to identify drugs targeting this protein class.[16]

4.3   Receptor Tyrosine Kinases

The third largest receptor target class, receptor tyrosine kinases, has been targeted frequently by anticancer drugs.[17] The kinase activities of these receptors are classified as distinct from intracellular nonreceptor kinases, such as ABL, the active site of which is the target of imatinib and which are classified as enzymes.[18] Receptor tyrosine kinases are the high-affinity cell surface receptors for many polypeptide growth factors, cytokines, and hormones. Of the 90 unique tyrosine kinase genes identified in the human genome, 58 encode receptor tyrosine kinase proteins. Receptor tyrosine kinases have been shown not only to be key regulators of normal cellular processes but also to have a critical role in the development and progression of many types of cancer. Growth factors, such as epidermal growth factor (EGF), activate the EGFR and the classical RAS-dependent mitogen-activated protein kinase (MAPK) cascade, which is required for the proliferation of some cells and the differentiation of others. However, the uncontrolled activation of this pathway is now known to cause cancer. This can occur as a result of the overexpression of EGFR or mutation to constitutively active forms. Interest in receptor tyrosine kinases as drug targets emerged based on data demonstrating that they are overexpressed in many cancers, and that modulation of their kinase activities results in inhibition of cell proliferation and tumor regressions in experimental models. These preclinical results were validated in multiple clinical trials demonstrating both delay in progression and prolonged overall survival of cancer patients. A number of EGFR inhibitors have been approved for the treatment of cancer.[19]

4.4   Nuclear Receptors

Nuclear receptors act as ligand-inducible transcription factors by directly interacting as monomers, homodimers, or heterodimers with DNA response elements of target genes as well as by "crosstalking" to other signaling pathways.[20] Nuclear receptors play a pivotal role in homeostatic metabolism via the controls of activities and processes at distal sites in the body. Signaling molecules, in some cases non-protein small molecules, traverse the body and ultimately relay their chemically encoded information to a nuclear receptor at the target tissue. The nuclear hormone receptor (NHR) is a classic example of a nuclear receptor for such small-molecule chemical messengers. The NHR is well adapted for this type of function because it not only specifically binds the small molecule but also is capable of relaying or transducing a complex set of signals carried along by the properties of the ligand. The nuclear hormone receptor superfamily includes receptors for thyroid and steroid hormones, retinoids and vitamin D, as well as different "orphan" receptors of unknown ligand. Ligands for some of these receptors have been recently identified, showing that products of lipid metabolism such as fatty acids, prostaglandins, or cholesterol derivatives can regulate gene expression by binding to nuclear receptors. The NHR family comprises at least 48 known members that are involved in the regulation of diverse metabolic functions.[21] NHRs have a rich and long-standing history in drug discovery. This can be attributed to features inherent to this class of targets. NHRs have been designed by nature to selectively bind "drug like" small molecules, and a diverse set of biologically important functions can be regulated through a single ligand-activated receptor. The NHR field remains an area of intense drug discovery research with most of the current effort directed toward the improvement of current NHR drugs or screening currently unexploited NHRs.[22]

5   Enzymes as Drug Targets

Enzymes are the second largest group of drug targets in the human genome, with 124 target-encoding genes, comprising 29% of all human drug targets. Hydrolases are the most common class of enzymatic drug targets, comprising 42% of all human enzyme drug targets. They are followed by oxidoreductases and transferases, which comprise 27% and 19% of all human enzyme drug targets, respectively. In addition, the majority (78%) of enzyme targets are soluble proteins, not membrane-associated proteins. Anti-inflammatory treatments most frequently act on enzyme targets; for example, the most studied anti-inflammatory target pathway—the eicosanoid metabolic pathway—is modulated via the enzyme targets cyclooxygenase 1 and cyclooxygenase 2, which belong to the oxidoreductase family and are targeted by acetylsalicylic acid. Other common enzyme targets for drugs include DNA polymerases, angiotensin-converting enzyme, and the monoamine oxidases.

Protein phosphorylation regulates most aspects of cell life, whereas abnormal phosphorylation is a cause or consequence of disease. A growing interest in developing orally active protein kinase inhibitors has culminated in the approval of several agents for the treatment of cancer.[23] Cyclosporin and rapamycin were the first compounds to be approved as drugs that exert their effects by inhibiting a particular protein phosphatase or protein kinase.[24,25] However, these compounds were not developed as a result of this knowledge, and their clinical efficacy was known before their mechanism of action was elucidated. The human genome contains at least 500 protein kinase genes, and up to 30% of all human proteins may be modified by kinase activity.[26] Protein kinases have now become the second most important group of drug targets, after G-protein-coupled receptors.

6   Transporter Proteins as Drug Targets

Transporters are membrane proteins that control the influx of essential nutrients and ions and the efflux of cellular waste, environmental toxins, and other xenobiotics. Consistent with their critical roles in cellular homeostasis, approximately 2000 genes in the human genome code for transporters or transporter-related proteins.[27] The functions of membrane transporters, may be facilitative (not requiring energy) or active (requiring energy). Transporter proteins are the third most common class of targets, with 67 target-encoding genes that comprise 15% of all human drug targets. This category encompasses voltage-gated channels, active transporters and solute carriers among other transporter proteins and is commonly targeted by antihypertensive drugs, diuretics, anaesthetics and antiarrhythmic drugs. Voltage-gated channels, which are often targeted by anaesthetics and antiarrhythmic drugs, are the most common type of transporter drug target.

More than 20 members have been identified in the neurotransmitter transporter family. These include the cell surface reuptake mechanisms for monoamine and amino acid neurotransmitters and vesicular transporter mechanisms involved in neurotransmitter storage. The norepinephrine and serotonin re-uptake transporters are key targets for antidepressant drugs.[28] Clinically effective antidepressants include those with selectivity for either norepinephrine or serotonin uptake and compounds with mixed actions. The dopamine transporter plays a key role in mediating the actions of cocaine and the amphetamines and in conferring selectivity on dopamine neurotoxins.[29]

7   Modern Technologies Employed in Target Identification and Validation

The advent of molecular biology and, in particular, of genomic sciences is having a deep impact on drug discovery in general and target identification and validation specifically. Recombinant proteins and monoclonal antibodies, generated using the tools of molecular biology, have greatly enriched our therapeutic armamentarium. Genome sciences, combined with the computational power of bioinformatics, are facilitating the dissection of the genetic basis of multifactorial diseases and the identification of the most suitable points of attack for future medicines. The role of functional genomics in modern drug discovery is to prioritize these targets and to translate that knowledge into rational and reliable drug discovery. Approaches including human genetics, RNA expression profiling, proteomics, antisense and RNA interference, and model organisms are powering modern target identification and validation and are spurring the development of new therapeutic modalities that extend beyond small-molecule chemicals and biologic-derived protein therapeutics to attack important disease-causing targets.

7.1   Human Genetics

Identifying genes associated with human diseases is a powerful mechanism for defining potential therapeutic targets. The first human disease whose genetic basis was understood was sickle cell anemia, a blood disorder characterized by red blood cells that assume an abnormal, rigid, sickle shape. Pauling and colleagues established in 1949 that sickle cell anemia is a molecular disease in which affected individuals have a different form of hemoglobin in their blood.[30] This paper helped establish that genes control not just the presence or absence of enzymes (as genetics had shown in the early 1940s) but also the specific structure of protein molecules. The sickle cell gene defect represents a single-nucleotide polymorphism (SNP) (A to T) of the β-globin gene, which results in glutamic acid being substituted by valine at position 6. This minor amino acid change results in a protein that polymerizes at low oxygen concentrations, changing the shape of red blood cells and causing them to lodge in small blood vessels as well as break into pieces that can interrupt healthy blood flow. These problems result in anemia and episodes of tissue ischemia and damage.

Presently, more than 5000 diseases have been identified that are caused by mutations in single genes.[31] Many of these mutations are associated with either a gain or loss of function of the protein encoded by that gene. As such, this genetic knowledge defines precisely the molecular target for developing therapeutic agents that can either antagonize an over active protein or replace a missing activity. This genetic knowledge, along with shorter development time and costs afforded by the Orphan Diseases Act in the US, has spurred the development of a growing number of therapeutics for the treatment of rare genetic diseases.[32] For example, there are more than 40 monogenic genetic diseases classified as lysosomal storage disorders (LSDs), each resulting from an inherited genetic defect that causes an enzymatic deficiency or malfunction, resulting in accumulation of substrate in cell lysosomes.[33] Most LSDs can present across a continuum of clinical severity, but they are all progressive in nature and may cause multisystemic, irreversible damage that can be seriously debilitating and even life threatening in severe phenotypes. An understanding of the genetic defect in LSDs has led to the development of a number of biotechnology-derived enzyme replacement therapies, including Cerezyme (imiglucerase) to treat Gaucher disease, Fabrazyme (agalsidase beta) to treat Fabry disease, and Myozyme (alglucosidase alfa) to treat Pompe disease, an inherited, progressive, and often fatal LSD.[34]

In contrast to monogenic diseases such as Gaucher or Fabry, common human diseases such as diabetes, cardiovascular, and autoimmune diseases are the result of a complex mixture of multiple genetic and environmental interactions that accumulate over the lifetime of an individual. The complexity of these common diseases is magnified by the fact that patients present with clinical disease at varying stages of pathogenesis, making it difficult to treat the underlying causes with a more general approach. Notwithstanding, protein and nucleotide sequence changes associated with human disease have accumulated over the last decade and a large body of literature has appeared on human disease-associated mutations, normal sequence variation, and alterations that acquire pathological significance when combined with other deleterious alleles or second-site mutations. With this information compiled into organized databases, it is now possible to conduct large-scale, comprehensive analyses of human disease genes. The information of genes and mutations causing disease is stored in several databases, such as OMIM, Homophila, LocusLink, The Human Gene Mutation Database, and Genecards.[35] These data acquire additional discriminatory power with the availability of multiple genome sequences from model organisms in which pathophysiological features common to human disease can be interrogated experimentally. This contributes substantially to our understanding of the genetic basis of human disease.

Although more than 99% of human DNA sequences are the same across the human population, polymorphic variations in DNA sequence could have a major impact on how humans respond to disease; to environmental stresses such as bacteria, viruses, toxins, and chemicals; and to drugs and other therapies. Thus, genetics and polymorphisms have a significant impact in target discovery. Single-nucleotide polymorphisms, or SNPs, are DNA sequence variations that occur when a single nucleotide in the genome sequence is altered. SNPs are more common than other types of polymorphisms and occur at a frequency of approximately 1 in 1000 base pairs throughout the genome. Some of these differences may alter gene products in ways that alter protein expression or function and thereby may confer susceptibility or resistance to a disease and contribute to the severity or progression of disease. Several large-scale approaches to defining genetic polymorphisms associated with disease have been developed over the last decade. Genome-wide association studies (GWAS), also known as whole-genome association studies (WGAS), are an examination of many common genetic variants in different individuals to see if any variant is associated with a trait. GWAS typically focus on associations between SNPs and traits like major diseases. These studies normally compare the DNA of two groups of participants: people with the disease (cases) and similar people without (controls). Each person gives a sample of DNA, from which millions of genetic variants are read using SNP arrays. If one type of the variant (one allele) is more frequent in people with the disease, the SNP is said to be "associated" with the disease. The associated SNPs are then considered to mark a region of the human genome which influences the risk of disease. In contrast to methods which specifically test one or a few genetic regions, GWAS investigates the entire genome. The first GWAS was published in 2005 and investigated patients with age-related macular degeneration.[36] It found two SNPs which had significantly altered allele frequency when compared with healthy controls. As of 2011, hundreds or thousands of individuals have been tested, over 1,200 human GWAS have examined over 200 diseases and traits, and almost 4000 SNP associations have been found.[37] GWAS identify SNPs and other variants in DNA which are associated with a disease but cannot on their own specify which genes are causal. Further studies are required to localize the specific polymorphisms and any genes affected. To date only ~5% of the disease-causing nonsynonymous mutations hitherto identified have been reported to have a direct effect on catalytic or ligand-binding properties of the proteins studied. For this reason, GWAS should be considered a useful tool for regional mapping of disease-associated genome alterations but has limited resolution to identify the actual molecular basis of gene defects driving disease.

Several notable examples of genetic definition of human disease leading to the approval of novel therapies have demonstrated the impact of this approach. The target for the CCR5 antagonist, maraviroc, was identified based on human genetic data showing that a naturally occurring genetic variation (del 32 polymorphism) in CCR5 reduces its ability to act as a co-receptor for the HIV virus, effectively protecting individuals carrying two copies of this polymorphism from HIV infection.[38] Additionally, human genetic studies on those subjects without a functional CCR5 receptor provide insights into safety of this approach. Imatinib, an inhibitor of the BCR/Abl oncogene, was developed after cytogenetic studies showed that 95% of patients with chronic myelogenous leukemia (CML) also had a 9/22 chromosomal translocation that resulted in the fusion of BCR and Abl genes and the constitutive activation of Abl.[39] This genetic research provided the confidence to proceed with an inhibitor to the BCR/Abl in CML patients and resulted in the rapid development of a highly efficacious therapeutic.

More recently, an alternative technological approach to GWAS has evolved to enable the complete sequence analysis of an individual's DNA. High-throughput, short-read DNA sequencers have revolutionized the field of genomics and have accelerated the pace of target discovery. The new technologies simultaneously read millions of short, 50- to 200-nucleotide DNA sequences from a pool of randomized genomic fragments in a single experiment. The process, which would have taken months with older technologies, is finished in a few days. Prior to the completion of the Human Genome Project, these short reads would have been difficult to interpret, as their genomic origin would have been unknown. However, the human reference sequence is now used to computationally map these sequences to the genome and to identify polymorphisms and novel mutations in a patient's DNA. The ability to sequence an individual's entire genome is now a feasible enterprise at a cost and speed that was unthinkable even 5 years ago. One of the first reports utilizing whole-genome sequencing in human disease was the analysis of an acute myeloid leukemia genome and its matched normal counterpart, obtained from the patient's skin.[40] The investigators identified 10 genes with acquired mutations; 2 were previously described mutations thought to contribute to tumor progression, and 8 were novel mutations present in virtually all tumor cells at presentation and relapse, whose function is not yet known. Whole-genome sequencing is an unbiased method for discovering initiating mutations in cancer genomes and for identifying novel genes that may respond to targeted therapies.

The basic examples discussed above show that when target selection is based on human disease relevance at the functional genetic level, the likelihood of developing a successful therapeutic is increased. In these cases, the relative risk that a therapeutic agent that modulates this target will not provide clinical benefit is reduced. These examples now serve as the basis for using human genetics to build disease relevance in the pursuit of therapeutics for more common and complex diseases.

7.2   Gene Family Mining

Another approach to identify targets for drug discovery is to search the human genome for additional members of families that are already known to contain drug targets. The advantages of identifying extra members of such "druggable" gene families are twofold: the potential identification of novel drug targets and the generation of knowledge of additional members of a gene family that can be used to help design drugs that are selective only for the target under consideration.

A good example of such an approach is the GPCR family, which comprises the largest group of previously drugged targets. The exact size of the GPCR superfamily is unknown but nearly 800 different human genes (or ≈4% of the entire protein-coding genome) have been predicted from genome sequence analysis. A large majority of human-derived GPCRs still remain promising drug targets, and thus a key goal of target discovery is to identify both the GPCR and its ligand(s) and to determine if they participate in the pathogenesis of human disease. With the unprecedented accumulation of genomic information, databases and bioinformatics have become essential tools to guide GPCR target discovery. The GPCRDB and IUPHAR receptor database (IUPHAR-RD) are representatives of widely used public databases covering GPCRs.[41] A major obstacle in novel GPCR discovery is the identification of natural ligands. GLIDA (GPCR-Ligand

DAtabase) is a public GPCR-related Chemical Genomics database designed to simultaneously mine biological information on GPCRs and chemical information on their ligands.[42] It provides various analytical data regarding GPCR–ligand correlations by incorporating bioinformatics and chemoinformatics techniques, and thus it is useful for GPCR-related target discovery. Even though newly identified members of the GPCR superfamily represent novel drug targets, much work remains to define the role of the GPCR in normal physiology and disease pathology. Attempts to characterize orphan GPCR receptors and determine whether they are indeed novel potential drug targets rely on a process often described as "reverse pharmacology."[43] This strategy uses the orphan receptor as a "hook" to capture its own ligand from cells, and the ligand is then used to explore the biological and pathophysiological role of the receptor. Drug discovery efforts are initiated simultaneously with biological characterization with the goal of generating agonistic and antagonistic ligand mimetics that can further define the function of the receptor. The laborious process of validating the therapeutic potential of a novel target is believed to have more chance of success when the target belongs to a family of proteins that has previously yielded drugs.

7.3 Gene Expression Profiling in Target Discovery

One of the highest-throughput methods available today for identifying genomic differences between disease and healthy individuals is RNA profiling (also referred to as gene expression profiling). There are several methods available for RNA profiling, but most of these consist of attaching DNA probe sequences to glass, labeling the RNA, and hybridizing this to the DNA array. Analysis of the expression level of more than 10,000 genes can be routinely determined in a single hybridization from very small starting genetic material. Although the phenotype of a cell is largely determined by the expressed proteins and their interactions with each other and the environment, gene expression profiling currently offers greater throughput and generally greater coverage of the genome and can provide data from smaller samples. However, although many differences at the gene expression level are reflected in differences in the protein level, there is not always a good correlation, and RNA levels cannot reflect protein modifications or interactions. Nevertheless, for new target discovery, the most common applications compare RNA levels in diseased tissues with those from normal tissues, either from humans or animal models or in tissues or cells that were drug treated. The initial data set from an RNA-profiling experiment typically has 200,000–400,000 data points and there are various tools available for handling these large data sets. The outcome of the application of any or all of these tools is a list of genes, typically tens to hundreds of genes that are probably modulated in disease and which therefore merit further characterization as potential drug targets.

Gene expression profiling has been extensively applied to the analysis of tumor tissues. Large-scale gene expression profiling of normal and cancerous tissue can aid the differentiation of tumors with similar morphological appearance, predict patient outcome independently of conventional prognostic factors, and select for response or resistance to specific anticancer therapies. Key studies of leukemia and lymphoma have examined a number of hypotheses concerning gene expression profiling. Leukemias with rearrangements of the mixed-lineage leukemia (MLL) gene are characterized by having a particularly poor outcome with current therapies. Microarray analysis determined that MLL was an entity distinct from ALL and AML.[44] Within the classification set of genes, FLT3 was highly expressed in MLL, and as a receptor tyrosine kinase, represents an attractive novel target for rationale drug development. Stam and colleagues profiled infant and childhood ALL and noted increased expression of hENT1, the nucleoside transporter used by the deoxycytidine analog Ara-C, in MLL tumor cells. Increased expression of this nucleoside transporter correlated with sensitivity to Ara-C, suggesting a potential explanation for the sensitivity of these tumors to Ara-C treatment and a potential therapeutic strategy.[45]

Human autoimmune diseases are also well suited for the application of gene expression profiling. Sampling of blood cells and target tissues has already revealed many important pathways contributing to this spectrum of disorders, and many commonalities are emerging. For instance, clinically distinct diseases such as systemic lupus erythematosus, Sjögren's syndrome, dermatomyositis, and psoriasis all show evidence for dysregulation of the type I interferon pathway.[46] In one of the earliest microarray studies in pediatric cases of systemic lupus erythematosus (SLE), it was demonstrated that all but one of the pediatric patients exhibited up-regulation of type 1 interferon (IFN) inducible genes, and the only patient lacking this signature had been in remission for over 2 years.[47] In addition, it was found that treating SLE patients with high dose IV steroids, which are used to control disease flares, results in the silencing of the IFN signature. A Phase 1 trial to evaluate the safety, pharmacokinetics, and immunogenicity of anti-IFNα monoclonal antibody therapy in adult SLE patients was recently conducted.[48] The antibody elicited a specific and dose-dependent inhibition of overexpression of type I IFN-inducible genes in both whole-blood and skin lesions from SLE patients at both the transcript and protein levels. As expected, overexpression of BLyS/BAFF, a type I IFN-inducible gene, also decreased with treatment. Thus, this first trial supports the proposed central role of type I IFN in human SLE. These data suggest that autoimmune diseases may eventually be categorized at the level of gene expression and have identified novel targets for therapeutic development, most notable of which is interferon-α blockade.

Gene expression profiling has now become a standard approach to the identification and validation of new molecular targets for therapeutic intervention in human disease. Gene expression profiling is also being used to improve the drug discovery process itself and to characterize clinical development candidates. This technology is being applied to investigate the mechanism of action and to determine on-target versus off-target effects during the lead optimization and development candidate process. The comparison of gene expression changes induced by the test therapeutic with that produced by knockout or knockdown of the target—increasingly by the use of RNA interference—is proving to be exceptionally valuable for defining the selectivity of the prototype pharmaceutical agent. Furthermore, there are already several examples of the use of microarrays to determine global genome expression changes that are induced in patients by drug treatment. The application of gene expression profiling to target identification and validation has led to advances in our understanding of disease pathogenesis and to novel drug targets for therapeutic intervention. It will remain a powerful tool in the future.

7.4   Proteomics

The genome determines its potential for gene and protein expression but does not specify which proteins are expressed in the various types of cell in an organism or individual, to what level they are expressed, or the extent of their posttranslational modifications. Proteomics is used to determine differential protein expression, posttranslational modifications, and alternative splicing and processed products. Two-dimensional gel electrophoresis is often used to fractionate the numerous proteins from a cell or tissue and to identify differentially expressed or modified proteins. This is followed by mass spectrometry to identify the individual protein spots of interest from the gels. This approach has lower throughput than gene expression profiling, but the resulting differentially expressed or modified proteins identified by proteomics represent the actual potential drug target or disease-associated molecules. Because of the complexity of protein expression, proteomic analyses studies often simplify the sample being studied by examining only a fraction of the proteome, thereby focusing the study to address more specific enquiries. Generally, this is achieved by subcellular fractionation or by affinity methods such as so-called activity-based probes for binding substrates to a specific class of proteases. A number of higher throughput and more sensitive proteomic profiling technologies are emerging and will greatly enable the application of this tool to target identification and validation.[49]

One example of the application of proteomics to target discovery is the analysis of plasma membrane proteins in cancer.[50] Plasma membrane proteins that are exposed on the cell surface have important biological functions, such as signaling into and out of the cell, transporting ions and nutrients, and mediating cell-cell and cell-matrix interactions. The expression level of many of the plasma membrane proteins involved in these key functions is altered on cancer cells, and these proteins may also be subject to post translational modification, such as altered phosphorylation and glycosylation. Additional protein alterations on cancer cells confer metastatic capacities. A number of cell surface proteins have already been successfully targeted in cancer using protein biologic drugs, such as human antibodies, and these have enhanced survival of several groups of cancer patients.[51] The combination of novel analytical approaches and sub-cellular fractionation procedures has made it possible to study the plasma membrane proteome in more detail and provide both greater classification of tumor types and drug responses and a number of novel drug targets. Using two isogenic cell lines with opposite metastatic capabilities in nude mice and an optimized cell surface membrane protein purification, Lund *et al.* identified 16 cell surface proteins as potential markers of the ability of breast cancer cells to form distant metastases.[52] These proteins represent novel targets for therapeutic modulation of breast cancer metastases. Many similar studies have been conducted in different tumor types, predominantly in cell lines where optimal protein labeling and purification techniques can be applied.[53]

Proteomics can also be used to identify protein–protein interactions and intracellular signaling networks associated with disease. Protein-protein interactions are important for nearly all biological processes, and it is known that aberrant protein-protein interactions can lead to human disease and cancer. Recent evidence has suggested that protein interaction interfaces describe a new class of attractive targets for drug development. Perreau *et al.* recently reported on a protein-protein interaction network controlled by beta-amyloid (Aβ), the primary constituent of the amyloid plaque and thought to be the causal "toxic moiety" of Alzheimer's disease.[54] Despite much research focused on demonstrating a causative role for both Aβ and its parent protein, amyloid precursor protein (APP), in Alzheimer's disease, the interactions of APP and its cleavage products that contribute to cellular dysfunction have not been fully elucidated. These investigators curated all published work characterizing both amyloid precursor protein (APP) and Aβ interactions, determined by the techniques described above, to create a protein interaction network of APP and its proteolytic cleavage products, with annotation, where possible, to the level of APP binding domain and isoform. This is the first time that an interactome has been refined to the domain level, essential for the interpretation of APP due to the presence of multiple isoforms and processed fragments. Gene ontology and network analysis were used to identify potentially novel functional relationships among interacting proteins. Such interactome information has yielded multiple novel targets of APP that may be causative in the cellular dysfunction leading to Alzheimer's disease and may themselves represent targets for future drug discovery. In a further evolution of interactome analysis, Nordstrom *et al.*[55] generated mice that harbor a mouse prion protein promoter-driven cDNA encoding human APP-695 fused to a *C*-terminal affinity tag. Using this tag, they prepared mild detergent lysates from transgenic mouse brain cortical membrane preparations and isolated a number of previously identified APP-interacting proteins. In addition to these factors, mass spectrometric analysis revealed the presence of NEEP21 as a novel interacting protein. The authors demonstrated that NEEP21, a neuronal-specific endosomal protein that regulates protein trafficking within cells, profoundly affects the processing of APP and Aβ production. NEEP21 may itself represent a target for drug therapy, to slow the accumulation of processed APP and therefore the onset and progression of Alzheimer's disease.

Full characterization of protein interaction networks of protein complexes and their dynamics in response to various cellular cues will provide essential information for us to understand how protein complexes work together in cells to maintain cell viability and normal homeostasis. Affinity purification coupled with quantitative mass spectrometry has become the primary method for studying *in vivo* protein interactions of protein complexes and whole-organism proteomes.[56] Recent developments in sample preparation and affinity purification strategies allow the capture, identification, and quantification of protein interactions of protein complexes that are stable, dynamic, transient, and/or weak. Current efforts have mainly focused on generating reliable, reproducible, and high confidence protein interaction data sets for functional characterization. The availability of increasing amounts of information on protein interactions in eukaryotic systems and new bioinformatics tools allow functional analysis of quantitative protein interaction data to unravel the biological significance of the identified protein interactions. High-throughput interaction analysis has also resulted in the mapping of networks in yeast, worms, fruit fly, and mammals.

The Human Protein Reference Database (HPRD) is a rich resource of experimentally proven features of human proteins.[57] Protein information in HPRD includes protein-protein interactions, post translational modifications, enzyme/substrate relationships, disease associations, tissue expression, and subcellular localization of human proteins. Although protein–protein interaction data from HPRD have been widely used by the scientific community, its phosphoproteome data has not been exploited to its full potential. HPRD is one of the largest documentations of human phosphoproteins in the public domain. Currently, phosphorylation data in HPRD comprise 95,016 phosphosites mapped on to 13,041 proteins. Additionally, enzyme-substrate reactions responsible for 5930 phosphorylation events were also documented. Significant improvements in technologies and high-throughput platforms in biomedical investigations led to an exponential increase of biological data and phosphoproteomic data in recent years.

As technologies for high-throughput protein characterization improve, so too will the impact of proteomic investigation on the identification and validation of drug targets.

7.5   *Oligonucleotides*

A key step in the process of target validation requires the identification of a pharmacological agent capable of modulating the target in a manner somewhat similar to that anticipated for the final therapeutic agent. This can be either a prototype therapeutic agent or one of several research tool compounds that are capable of modulating the target. Antisense oligonucleotides have a rich history both as target validation tools and as therapeutic agents. The potential of single-stranded antisense oligonucleotides to knock down the expression of a targeted gene has been known for 25 years and has been widely exploited for research and target validation.[58] The use of oligonucleotides in the target and drug discovery process is based on the premise that they can be rapidly and specifically used to simulate the biological and pharmacological effects of target inhibition by therapeutic agents in cellular assays, in animal models of disease, and even in humans. Antisense inhibition of target gene expression is initiated by Watson–Crick binding of the antisense oligonucleotide to its target mRNA. Normal translation of the mRNA is subsequently prevented by one of several mechanisms, including induced degradation of the message, interference with the splicing process or a physical blocking of the translational machinery.

Antisense oligonucleotides are also being researched as therapeutics to treat cancers (including lung cancer, colorectal carcinoma, pancreatic carcinoma, malignant glioma, and malignant melanoma), cardiovascular disease, diabetes, amyotrophic lateral sclerosis (ALS), Duchenne muscular dystrophy, and diseases such as asthma and arthritis. Most potential therapies have not yet produced significant clinical results. A single antisense oligonucleotide drug has made it to the market; Vitravene (fomivirsen) was approved in 1998 by the US Food and Drug Administration for the treatment of cytomegalovirus (CMV) retinitis.[59] Vitravene blocks translation of viral mRNA by binding to the complementary sequence of the mRNA transcribed from the coding segment of a key CMV gene. However, due to inherent instability of oligonucleotides, the utility of oligonucleotides as therapeutics has been somewhat limited. In recent years, major advances have been achieved by the development of novel, chemically modified nucleotides with improved properties, such as enhanced nuclease resistance, high target affinity, and low toxicity.[60] Peptide nucleic acids and locked nucleic acids are two of the most promising examples of this class of new building blocks. In peptide nucleic acids, the deoxyribose phosphate backbone is replaced by polyamide linkages. Locked nucleic acids are characterized by a methylene bridge, which connects the oxygen at the 2'-position of the ribose with the carbon at the 4'-position. Locked nucleic acid gapmers with four to five modified monomers at each end and a central stretch of eight to ten unmodified DNA nucleotides in the center were significantly more stable than unmodified ONs and had an AS potency exceeding that of isosequential phosphorothioates or 2'-O-methyl RNA gapmers.[61]

Antisense oligonucleotides have been valuable tools in target validation, allowing the early assessment of target inhibition in cell culture, animal models, and even human clinical trials. In 1978 Zamecnik and Stephenson were the first to report that a specific 13-mer oligonucleotide can act as hybridization competitor to 35S RNA and inhibit Rous sarcoma viral RNA replication in cell culture.[62] Since then many hundreds of targets have been explored via the use of specific antisense oligonucleotides to suppress their expression. Although in vitro cellular data can help in target validation, determining gene function *in vivo* is much more valuable and might be predictive of human disease. In animals, the most widely used antisense molecules are those that work through an RNase H-mediated mechanism. Systemically dosed phosphorothioate-containing oligonucleotides distribute to a wide variety of tissues, including liver, kidney, bone, and adipose tissue. The distribution has been shown to strongly correlate with a reduction in target gene expression in the tissues accumulating the highest levels of antisense oligonucleotide.[63]

Double-stranded RNA reagents that operate by the RNA interference (RNAi) mechanism, such as small interfering RNA (siRNA) and the vector-driven expression of short hairpin RNA (shRNA), have recently supplanted antisense oligonucleotides as optimal gene knockdown reagents. They induce an enzyme-driven degradation of mRNA by a ribonuclease complex known as the RNA-induced silencing complex (RISC). In 2006, Andrew Fire and Craig C. Mello shared the Nobel Prize in Physiology or Medicine for their work on RNAi interference in the nematode worm *C. elegans*.[64] Two types of small ribonucleic acid (RNA) molecules—

microRNA (miRNA) and small interfering RNA (siRNA)—are central to RNA interference. The sequence specificity of RNAi can be utilized experimentally to silence specific genes by the transfection of siRNAs into mammalian cells. This technology has been expanded into RNAi libraries encompassing reagents that target a wide range of transcripts, allowing the role of multiple genes in a cellular process to be assessed in an unbiased fashion. RNAi screens have been used to identify genes important for cancer cell phenotypes, including cell viability. An important difference between the mechanistic action of siRNAs and antisense oligonucleotide is that these reagents are reported to operate in the cytoplasm only. Intracellular expression of shRNAs that function by the RNAi mechanism are reported to show long-term target suppression. A further advantage of RNAi is the possibility of delivering the reagents by viral vectors into cell types that have typically been difficult to transfect. For example, Hahn and colleagues screened a small panel of colorectal tumor cell lines with an RNA interference library to identify CDK8, a gene that not only controlled tumor cell viability but also modulated WNT signaling, an oncogenic pathway commonly active in colorectal cancer.[65] By integrating these screen data with the genetic profiles of colorectal adenocarcinomas, they demonstrated that the CDK8 gene was also amplified in a significant proportion of colorectal tumors, suggesting that it could be a promising drug target. Furthermore, the CDK8 gene copy alteration could also serve as a biomarker with which to select patients for treatment with a CDK8 targeting agent, once developed.

The process of target validation utilizes the integration of a wide variety of disparate data types, such as gene expression profiles, immunohistochemical profiles, metabolic profiles, and forms of functional analysis. With the availability of next-generation sequencing that offers the rapid dissection of genome and transcriptome sequences, it is anticipated that a large number of potential therapeutic targets will be identified. Oligonucleotides serve as a valuable tool in understanding the physiological and pathophysiological role of these targets.

7.6   Model Organisms

Widespread conservation of DNA and protein sequences, gene function, and signaling pathways across diverse organisms has provided the rationale for the use of model organisms as 'surrogates' for human patients. It has long been assumed that experimental results and knowledge obtained using these model systems can be applied to understanding equivalent processes, pathways, and mechanisms in the human situation. Model organisms used in target identification and validation can range from nonhuman primates to yeast. Rodents are the most common type of mammal employed in models of human disease, and extensive research has been conducted using rats, mice, gerbils, guinea pigs, and hamsters. Among rodents, the majority of studies, especially those involving disease, have employed mice, not only because their genomes are so similar to that of humans and can be genetically engineered, but also because of their availability, ease of handling, high reproductive rates, and relatively low cost of use. Other common experimental organisms include fruit flies, zebra fish, and baker's yeast.

Model organisms can be used for identifying pathways and networks regulating key physiological processes and can also be used to model the pathophysiology of human disease. A useful resource for cataloging animal models of disease is provided by the National Institute of Health's National Center for Research Resources as part of its support for researchers. The resource, LAMHDI (an initiative to Link Animal Models to Human Disease), is designed to accelerate the research process by providing biomedical researchers with a simple, comprehensive Web-based resource to find the best animal models for their research.[66]

Baker's yeast (*Saccharomyces cerevisiae*) is a single-cell organism used in the bread-making industry. As a simple eukaryote *S. cerevisiae* have a nucleus containing chromosomes which divide in a similar manner to human cells, and share many other basic biological properties with humans. The yeast genome is just over 12 million base pairs in length and contains about 6000 genes. About 20% of human disease genes have counterparts in yeast, suggesting that such diseases may result from the disruption of basic cellular processes, such as DNA repair, cell division, or the control of gene expression. The yeast genome was completed in 1996 and other projects have been initiated to determine the functions of all 6000 genes.[67] Yeast have been extensively exploited to investigate functional relationships involving these genes and to test new drugs. Approximately one third of yeast genome open reading frames (ORFs) had no known function four years after their discovery. The goal of the Saccharomyces Genome Deletion Project was to generate as complete a set as possible of yeast deletion strains with the overall goal of assigning function to the ORFs through phenotypic analysis of the mutants.[68] To date, the Saccharomyces Genome Deletion Project Consortium disrupted 90% of the yeast genome. More than 20,000 strains of mutant yeast have been generated and made available for academic research worldwide. Four different mutant collections were generated: haploids of both mating types, homozygous diploids for nonessential genes, and heterozygous diploids, which contain the essential and nonessential ORFs. Results to date revealed that 18% of the genes are essential for growth on rich glucose media and approximately 15% of the homozygous diploid disruptions cause slow growth in this type of media. An early application of the yeast genome deletion set was in the investigation of diseases associated with mitochondrial dysregulation.[69] To date, 102 heritable human disorders have been attributed to defects in nuclear-encoded mitochondrial proteins in humans. Many mitochondrial diseases remain unexplained, however, in part because only 40–60% of the presumed 700–1000 proteins involved in mitochondrial function and biogenesis have been identified. Steinmetz *et al.* conducted a systematic functional screen using the pool of yeast deletion mutants to identify mitochondrial proteins. Three million measurements of strain fitness identified 466 genes whose deletions impaired mitochondrial respiration, of which 265 were new. When human orthologs for these genes were identified and linked to heritable diseases using genomic map positions, 24 novel disease candidate genes were assigned to reported disease intervals. Such analysis can be valuable in identifying potential disease-related genes and represent a powerful approach to target identification based on phenotypic analysis.[70]

Another large project involves utilizing the yeast two-hybrid system and mass spectrometry to catalogue all the different protein interactions that occur in yeast cells. Interacting proteins often function in conserved complexes or pathways, suggesting that a pathway found in yeast might therefore exist in humans, and characterizing the interacting proteins in yeast might help to identify the corresponding proteins in humans. The identification of interacting proteins is useful because they may provide alternative drug targets. For example, the product of a human disease gene might be an unsuitable drug target, perhaps due to extensive polymorphism. However, interacting proteins in the same pathway might show less variability and would be better targets for drug development. The yeast proteome and its interactome (that is, the sum of all protein interactions) are the best studied of all organisms. Currently there are about 3000 verified protein interactions and several thousand non-verified interactions known in yeast.[71] Independent studies estimated that there may be more than 30,000 interactions in yeast, although most estimates rather suggest 15,000–25,000. It remains unknown how many of these interactions are really essential. The average yeast protein appears to have about 5 interactions, but this number may represent an overestimate because many proteins of yet unknown function exhibit fewer interactions. Nevertheless, most proteins can be connected in a huge network of interactions. The protein interaction network of yeast is highly dynamic although there are 1500 or more proteins involved in several hundred stable complexes. The dynamics and regulation of protein interaction networks, e.g., by protein modifications, are only now being explored.

Two higher eukaryotic model organisms that are increasingly employed in target identification and validation are the fruitfly *Drosophila melanogaster* and the zebrafish *Danio rerio*. All share essential features that are necessary for a good genetic model system, including ease of culture, short generation time (two weeks and three months, respectively), and the production of large numbers of progeny. Equally important is the availability and continual improvement of methods for experimental and genetic manipulation; these methods allow researchers to carry out genome-scale genetic screens and analyze the identified genes and pathways at a level of sophistication that is not typically practical in other organisms. Using *Drosophila*, researchers have developed models for many complicated pathologies, including the complex, multistep processes of oncogenic transformation and metastasis.[72] Additionally, researchers have generated transgenic *Drosophila* lines that overexpress mutated proteins causing flies to undergo neurodegeneration and serving as models for disorders such as Alzheimer's disease and Huntington's disease.[73] Because the zebrafish has most of the same organs found in mammals, it is a much more useful model than *Drosophila* and *C. elegans*. Most human genes have homologues in zebrafish, and the functional domains of proteins, such as ATP-binding domains of kinases, are almost 100% identical between homologous genes, although the similarity over the entire protein is only about 60%.[74] Because protein function largely resides in functional domains where drugs often bind, the zebrafish is a highly valid model for studying gene function and drug effects in humans. Apoptotic processes in zebrafish and mammals are similar, and zebrafish homologues of most mammalian apoptosis-related genes have been identified.[75] Screening for apoptosis inducers can be performed by looking for their effects in the zebrafish embryo. Apoptosis is an important mechanism for morphogenesis and homeostasis, and abnormalities in apoptosis are involved in many other diseases in addition to cancers, such as neurodegenerative diseases. Using forward genetic screens, researchers have identified many zebrafish mutants that display abnormal apoptosis, which can serve as models for anti-apoptotic drug screening.[76]

The mouse is the most frequently used organism in which to model complex human disease. Over 1000 mutant mouse strains exist, and most of these mutants are models for inherited genetic diseases.[77] Models for human genetic disease have been made by mutating the same gene in mice that is responsible for the human condition for about 100 genes, and in most cases, these models replicate many of the corresponding human disease phenotypes. These diseases include several types of cancer, heart disease, hypertension, metabolic and hormonal disorders, diabetes, obesity, osteoporosis, glaucoma, skin pigmentation diseases, blindness, deafness, neurodegenerative disorders (such as Huntington's or Alzheimer's disease), psychiatric disturbances (including anxiety and depression), and birth defects (such as cleft palate and anencephaly).[78] Other techniques for engineering the mouse genome, including knock-in, conditional knockout, and transgenics, have made it possible to create specific gene sequence alterations and manipulate the levels and patterns of target gene expression. Using these techniques, researchers can generate specific disease models to validate targets as therapeutic intervention points and screen drug candidates. For example, researchers have generated many mouse models for Huntington's disease by introducing different versions of human Huntington protein fragments carrying expanded poly-glutamine repeats.[79] These transgenic or knock-in mice showed phenotypes characteristic of Huntington's disease patients, including the formation of neuronal inclusion bodies and apoptosis in certain brain regions. Such a demonstration of induction of disease by insertion of the novel target represents powerful validation of the role of that target in disease pathogenesis.

One drawback to using transgenic, knock-in, and knockout mice to study human diseases is that many human disorders occur late in life, and when genes are altered in the mouse to model such diseases, the mutations can profoundly affect development and manifest their biological effect early in life. These effects have precluded using mouse transgenics to study targets associated with many human adult-onset diseases. New gene deletion technology has made it possible to generate mutations in specific tissues and at different stages of development in the mouse, including adulthood. To do this, mice with two different types of genetic alterations are needed: one that contains a conditional vector, which is like an "on switch" for the mutation, and one that contains specific sites (called loxP) inserted on either side of a whole gene, or part of a gene, that encodes a certain component of a protein that will be deleted.[80] A conditional vector for the gene is made by inserting recognition sequences for the bacterial Cre recombinase (loxP sites) using homologous recombination in ES cells. The vector contains a drug-resistant marker gene that allows only the targeted ES cells to survive when exposed to the drug. Thus, the mutant ES cells can be selected and injected into the host mouse embryo, which is implanted into a foster mother. The resulting offspring are chimeras and have multiple populations of genetically distinct cells. Chimeric offspring are then crossed, and the resulting generation of offspring has the recombinase effector gene. The mice containing the Cre recombinase under the control of tissue-specific or inducible regulatory elements are crossed to the mice with the desired loxP sites. When Cre is expressed, recombination occurs at the loxP sites, which delete the intervening sequences, and the resulting mutation is induced in specific regions and times. These conditional mutant models are becoming increasingly popular, and international initiatives have been created to accommodate the demand.[81]

In addition to the gene-specific methods for altering the mouse genome to reflect human genomic variations associated with disease, a number of indirect approaches have been developed to randomly mutate the mouse genome and to screen for phenotypes that are similar to human diseases. Thus, instead of being driven by the disease mutation, these methods are driven by the disease phenotype and seek to identify the genes involved in the pathogenesis of that phenotype. Two of the most effective ways to generate random genetic mutations are by exposing organisms to X-rays or to the chemical N-ethyl-N-nitrosourea (ENU). X-rays often cause large deletion and translocation mutations that involve multiple genes, whereas ENU treatment is linked to mutations within single genes, such as point mutations. ENU can produce mutations with many different types of effects, such as loss and gain of function. These types of mutagenesis approaches have also been successfully employed in other model organisms such as the zebrafish.

Animal models have greatly improved our understanding of the cause and progression of human genetic diseases and have proven to be a useful tool for discovering targets for therapeutic drugs. Nonetheless, despite promising results with certain preclinical treatments in animal models, the same treatments do not always translate to human clinical trials. Most available animal models re-create only some aspects of the particular disease in humans and must therefore be used cautiously in the identification and validation of targets and in drug discovery. More and more, animal models of human disease are viewed more as a model for specific molecular mechanisms associated with disease and a tool in understanding the physiological consequences of modulation of that mechanism. Nonetheless, model organisms are of immense value in target identification and validation and with improvements in the development of appropriate models for human disease, will be even more valuable in the future.

8   Impact of Therapeutic Modalities on the Selection Drug Targets

A major influence on the selection of targets for drug discovery is the ability to identify a specific therapeutic agent that can modulate that target effectively and specifically. Even if a molecular target is identified and validated to be a key mediator of the pathogenesis of disease, it cannot be considered an attractive drug discovery target unless it has features that enable the identification of a selective and potent pharmacological agent that can modulate the target effectively.

Small-molecule drugs have dominated therapeutic target selection for decades. The majority of existing drugs are small molecules, and the key rules for small-molecule drug discovery and optimization are well understood. However, even though small-molecules are relatively effective in modulating G-protein-coupled receptors, ion channels, kinases, and proteases, they are relatively ineffective at blocking protein–protein interactions that are important for disease.

With the emergence of genetic and genomic technologies in the 1980s and on, recombinant proteins and therapeutic antibodies have become important modalities for the development of therapeutics. The types of drug targets that can be addressed by these agents, including extracellular receptors, soluble mediators such as cytokines, and growth factors, and the ability to replace protein functions with recombinant protein therapeutics have greatly expanded the treatment options for patients. Extracellular protein-protein interactions are ideal targets for protein therapeutics. Since the first therapeutic antibody for a cancer indication was approved by the FDA for marketing in 1997, antibodies have become increasingly important in the fight against cancer, autoimmune, and infectious diseases. To date, more than 22 unconjugated and 5 conjugated monoclonal antibodies have been approved for these indications. This number will rapidly escalate as there are over 300 monoclonal antibodies reported to be in in pre- and clinical development. Antibodies provide excellent affinity and specificity of target recognition. Furthermore, because of their relatively large size, *in vivo* stability, and ability to be sequestered from the blood by interacting with FcRn receptors on endothelial cells, they can have extended pharmacokinetic half-lives (days to weeks).[82] There are a number of drawbacks for monoclonal antibodies as therapeutics with perhaps the most significant being the inability to access intracellular targets.

Even though synthetic peptides are a proven class of medicines, with more than 40 marketed drugs and about 300 molecules in clinical trials, current peptide therapeutics are also limited to modulating extracellular therapeutic targets (e.g., receptors), and only as a result of significant transformation of their chemical structures have second-generation peptidomimetic drugs been advanced for specific intracellular therapeutic targets.

A number of emerging therapeutic modalities have greatly expanded the potential number of drug targets.   These include oligonucleotide therapeutics such as antisense, small-interfering RNA (siRNA), and micro RNA (miRNA), and technologies for modulating protein-protein interactions such as Stapled Peptides. As previously noted, oligonucleotide therapeutics have recent significant focus and investment from the pharmaceutical industry, but due to fundamental challenges in stability and drug delivery, these modalities have yet to made a large impact on patient care.

Investigation of the molecular basis of human disease has revealed that numerous intracellular and extracellular protein–protein interactions serve as critical control points in disease mechanisms, including cellular survival/death, signal transduction, and gene regulation. Such protein–protein interactions have largely eluded small-molecule drug targeting strategies, and intracellular protein-protein interactions have eluded the development of protein therapeutics such as monoclonal antibodies. Protein-protein interactions are frequently mediated by the α-helix structures of proteins, and it is predicted that about 1500–3000 α-helical protein–protein interaction of therapeutic relevance exist.[83]

Stapled Peptides are peptides capable of forming stable alpha helical structure as a result of "hydrocarbon stapling".[84] This chemical stabilization of the α-helix retains the secondary structure of the α-helix that is critical for high-affinity target binding, prevents degradation by proteolytic enzymes, and facilitates uptake into cells. As such, Stapled Peptides exhibit excellent pharmacokinetics and the ability to modulate intracellular protein–protein interactions important for disease. Stapled Peptides targeting key drug targets associated with cancer, including p53,[85] Ras,[86] Bcl2 family members,[87] and Notch,[88] have been reported. These targets are considered "undruggable" by existing modalities. As such, Stapled Peptides represent an important emerging therapeutic modality for the development of new medicines.

During the process of target identification and validation, it remains critical to select targets for drug discovery based on their "druggability." The emergence of novel modalities for modulating the activity of key drivers of disease will expand greatly the number of therapeutic targets available for drug discovery efforts.

9   The Future

This chapter has outlined both the current targets that are highly utilized for drug discovery and development and the technologies that are exponentially expanding the number and range of targets for future efforts. Coupled with the development of novel therapeutic modalities that can expand the number of targets of potential therapeutic value, it is clear that we are at the beginning of a renaissance in our understanding and treatment of disease. This is indeed an exciting time to be starting a career in drug discovery.

References

1.   Gysin, S.; Salt, M.; Young, A.; McCormick, F. *Genes Cancer* **2011**, *2*, 359–372.

2.   Courtney, K. D.; Corcoran, R. B.; Engelman, J. A. *J. Clin. Oncol.* **2010**, *28*, 1075–1083.

3.   Castillo, J. J.; Furman, M.; Winer, E. S. *Expert Opin. Investig. Drugs* **2012**, *21*, 15–22.

4.   Ferreira, S. H.; Moncada, S.; Vane, J. R. *Nature* **1971**, *231*, 237–239.

5.   Hla, T.; Neilson, K. *Proc. Natl. Acad. Sci. USA.* **1992**, *89*, 7384–7388.

6.   Penning, T. D.; Talley, J. J.; Bertenshaw, S. R.; Carter, J. S.; et al. *J. Med. Chem.* **1997**, *40*, 1347–1365.

7.   Prasit, P.; Wang, Z.; Brideau, C.; Chan, C.-C.; et al. *Bioorg. Med. Chem. Lett.* **1999**, *9*, 1773–1778.

8.   Drews, J.; Ryser, S. *Nat. Biotechnol.* **1997**, *15*, 1350.

9.   Hopkins, A. L.; Groom, C. R. *Nat. Rev. Drug Discov.* **2002**, *1*, 727–730.

10.   Venter, J. C.; Adams, M. C.; Myers, E. W.; Li, P.; et al. *Science* **2001**, *291*, 1304–1351.

11.   Rask-Andersen, M.; Almen, M. S.; Schioth, H. B. *Nat. Rev. Drug Discov.* **2011**, *10*, 579–590.

12.   Insel, P. A.; Tang, C.-M.; Hahntow, I.; Michel, M. C. *Biochim. Biophys. Acta* **2007**, *1768*, 994–1005.

13.   Chung, S.; Funakoshi, T.; Civelli, O. *Br. J. Pharmacol.* **2008**, *153*, S339–S346.
14.   Nasiripourdori, A.; Taly, V.; Grutter, T.; Taly, A. *Toxins* **2011**, *3*, 260–293.
15.   Clare, J. J. *Discov. Med.* **2010**, *9*, 253–260
16.   Kaczorowski, G.; McManus, O. B.; Priest, B. T.; Garcia, M. L. *J. Gen. Physiol.* **2008**, *131*, 399–405
17.   Zwick, E.; Bange, J.; Ullrich, A. *Trends Mol. Med.* **2002**, *8*, 17–23
18.   Druker, B. J.; Lydon, N. B. *J. Clin. Invest.* **2000**, *105*, 3–7.
19.   Sequist, L. V.; Lynch, T. J. *Ann. Rev. Med.* **2008**, *59*, 429–442.
20.   Aranda, A.; Pascual, A. *Physiol. Rev.* **2001**, *81*,1269–304.
21.   Zhang, Z.; Burch, P. E.; Cooney, A. J.; Lanz, R. B.; et al. *Genome Res.* **2004**, *14*, 580–590.
22.   Moore, J. T.; Collins, J. L.; Pearce, K. H. *Chem. Med. Chem.* **2006**, *1*, 504–523.
23.   Zhang, J.; Yang, P. L.; Gray, N. S. *Nat. Rev. Cancer* **2009**, *9*, 28–39.
24.   Handschumacher, R. E.; Harding, M. W.; Rice, J.; Drugge, R. J.; Speicher, D. W. *Science* **1984**, *226*, 544–547.
25.   Brown, E. J.; Albers, M. W.; Shin, T. B.; Ichikawa, K.; et al. *Nature* **1994**, *369*, 756–758.
26.   Manning, G.; Whyte, D.B.; Martinez, R.; Hunter, T.; Sudarsanam, S. *Science* **2002**, *298*, 1912–1934.
27.   Yee, S. W.; Chen, L.; Giacomini, K. M. *Pharmacogenomics* **2010**, *11*, 475–479.
28.   Iversen, L. *Mol. Psychiatry* **2000**, *5*, 357–362.
29.   Beaulieu, J.-M.; Gainetdinov, R. R. *Pharmacol. Rev.* **2011**, *63*, 182–217.
30.   Pauling, L.; Harvey, A.; Itano, S. J.; Singer, I.; Wells, C. *Science* **1949**, *110*, 543–548.
31.   Brinkman, R. R.; Dub, M.-P.; Rouleau, G. A.; Orr, A. C.; Samuels, M. E. *Nat. Rev. Genet.* **2006**, *7*, 249–260
32.   Melnikova, I. *Nat. Rev. Drug Discov.* **2012**, *11*, 267–268.
33.   Meikle, P. J.; Hopwood, J. J.; Clague, A. E.; Carey, W. F. *JAMA* **1999**, *281*, 249–254.
34.   Rohrbach, M.; Clarke, J. T. *Drugs* **2007**, *67*, 2697–716.
35.   Frodsham, A. J.; Higgins, J. P. T. *BMC Med. Res. Methodol.* **2007**, *7*, 31.
36.   Klein, R. J.; Zeiss, C.; Chew, E. Y.; Tsai, J. Y.; et al. *Science* **2005**, *308*, 385–389.
37.   Johnson, A. D.; O'Donnell, C. J. *BMC Med. Genet.* **2009**, *10*, 6.
38.   Samson, M.; Libert, F.; Doranz, B. J.; Rucker, J.; et al. *Nature* **1996**, *382(6593)*, 722–725
39.   Capdeville, R.; Buchdunger, E.; Zimmermann, J.; Matter, A. *Nat. Rev. Drug Discov.* **2002**, *1*, 493–502.
40.   Ley, T. J.; Mardis, E. R.; Ding, L.; Fulton, R.; et al. *Nature* **2008**, *456*, 66–72.
41.   Horn, F.; Bettler, E.; Oliveira, L.; Campagne, F.; et al. *Nucleic Acids Res.* **2003**, *31*, 294–297.
42.   Okuno, Y.; Tamon, A.; Yabuuchi, H.; Niijima S.; et al. *Nucleic Acids Res.* **2008**, *36*, 907–912.
43.   Lappano, R.; Maggiolini, M. *Nat. Rev. Drug Discov.* **2011**, *10*, 47–60.
44.   Gabrovska, P. N.; Smith, R. A.; Haupt, L. M.; Griffiths, L. R. *Open Breast Cancer J.* **2010**, *2*, 46–59
45.   Ronald, W.; Stam, R. W.; den Boer, M. L.; Meijerink, J. P.; et al. *Blood*, **2003**, *101*, 1270–1276.
46.   Chaussabel, D.; Pascaul, V.; Bancherau, J. *BMC Biol.* **2010**, *8*, 84–92.
47.   Bennett, L.; Palucka, A. K.; Arce, E.; Cantrell, V.; et al. *J. Exp. Med.* **2003**, *197*, 711–723.
48.   Yao, Y.; Richman, L.; Higgs, B. W.; Morehouse, C. A.; et al. *Arthritis Rheum.* **2009**, *60*,1785–1796.
49.   Rotilio, D.; Della Corte, A.; D'Imperio, M.; et al. *Thromb. Res.* **2012** *129*, 257–262.
50   Leth-Larsen, R.; Lund, R. R.; Ditzel, H. J. *Mol. Cell Proteomics.* **2010**, *9*, 1369–1382.
51.   Scott, A. M.; Wolchok, J. D.; Old, L. J. *Nat. Rev. Cancer*, **2012**, *12*, 278–287.
52.   Lund, R.; Leth-Larsen, R.; Jensen, O. N.; Ditzel, H. J. *J. Proteome Res.* **2009**, 8, 3078–3090.
53.   Moseley, F.; Bicknell, K. A.; Marber, M. S.; Brooks, G. *J. Pharm. Pharmacol.* **2007**, *59*, 609–628.
54.   Perreau, V. M.; Orchard, S.; Adlard, P. A.; Bellingham, S. A.; et al. *Proteomics* **2010**, *10*, 2377–2395.
55.   Norstrom, E. M.; Zhang, C.; Tanzi, R.; Sisodia, S. S. *Neurosci.* **2010**, *30*, 15677–15685 .
56.   Sardiu, M. E,; Washburn, M. P. *J. Biol. Chem.* **2011**, *286*, 23645–23651.

57. Peri, S.; Navarro, J. D.; Kristiansen, T. Z.; Amanchy, R.; et al. *Nucleic Acids Res.* **2004**, *32*, D497–501.
58. Rayburn, R. R.; Zhang, R. *Drug Discov. Today*, **2008**, 13, 513–521.
59. Bennett, C. F.; Crowsert, L. M. *Curr. Opin. Mol. Ther.* **1999** *1*, 359–371.
60. Orr, R. M. *Curr. Opin. Mol. Ther.* **2001**, *3*, 288–294.
61. Zamecnik, P. C.; Stephenson, M. L. *Proc. Natl. Acad. Sci. USA*, **1978**, *75*, 280–284.
62. Levin, A. A. *Biochim. Biophys. Acta* **1999**, *1489*, 69–84.
63. Fire, A.; Xu, S.; Montgomery, M.; Kostas, S.; Driver, S.; Mello, C. C. *Nature* **1998**, *391*, 806–811.
64. Kurreck, J. *Eur. J. Biochem.* **2003**, 270, 1628–1644.
65. Firestein, R.; et al. *Nature* **2008**, *455*, 547–551.
66. www.lamhdi.org, accessed June 1, 2012.
67. Goffeau, A.; et al. *Science* **1996**, *274*, 546–567.
68. Winzeler, E. A.; et al. *Science* **1999**, *285*, 901–906.
69. Steinmetz, L. M.; Scharfe, C.; Deutschbauer, A. M.; et al. *Nat. Genet.* **2002**, *31*, 400–404.
70. Yu, H.; Braun, P.; Yildirim, M. A.; et al. *Science* **2008**, *322*, 104–110.
71. Goll, J.; Uetz, P. *Genome Biol.* **2006**, *7*, 223.
72. Pagliarini, R. A.; Xu, T. *Science* **2003**, *302*, 1227–1231.
73. Shulman, J. M.; et al. *Curr. Opin. Neurol.* **2003**, *16*, 443–449.
74. Langheinrich, U. *Bioessays* **2003**, *25*, 904–912.
75. Inohara, N.; Nunez, G. *Cell Death Different.* **2000**, *7*, 509–510.
76. Abdelilah, S.; et al. *Development* **1996**, *123*, 217–227.
77. Hardouin, N.; Nagy, A. *Clin. Genet.* **2000**, *57*, 237–244.
78. Rosenthal, N.; Brown, S. *Nature Cell Biol.* **2007**, *9*, 993–999.
79. Bates, G. P.; Hockly, E. *Curr. Opin. Neurol.* **2003**, *16*, 465–470.
80. Bedell, M. A.; Largaespada, D. A.; Jenkins, N. A.; Copeland, N. G. *Genes Dev.* **1997**, 11, 11–43.
81. http://www.knockoutmouse.org/aboutkomp, accessed May 1, 2012.
82. Daugherty, A. L.; Mrsny, R. J. *Adv. Drug Deliv. Rev.* **2006**, *58*, 686–706.
83. Jochim, A. L.; Arora P. S. *ACS Chem. Biol.* **2010**, *5*, 919–923.
84. Walensky, L. D.; Kung, A. L.; Escher, I.; Malia, T. J.; et al. *Science* **2004**, *305*, 1466–1470.
85. Bernal, F.; Wade, M.; Godes, M.; Davis, T. N.; et al. *Cancer Cell* **2010**, *18*, 411–422.
86. Patgiri, A.; Yadav, K. K.; Arora, P. S.; Bar-Sagi, D. *Nat. Chem. Biol.* **2011**, *7*, 585–587.
87. LaBelle, J. L.; Katz, S. G.; Bird, G. H.; Gavathiotis, E.; et al. *J. Clin. Invest.* **2012**, *122*, 2018–2031.
88. Moellering, R. E.; Cornejo, M.; Davis, T. N.; Del Bianco, C.; et al. *Nature* **2009**, *462*, 182–188.

3

In Vitro and *in Vivo* Assays

Adam R. Johnson

1 Introduction

Congratulations! As a medicinal chemist, you have chosen a challenging, rewarding career during which you will have the opportunity to dramatically impact humankind by improving, extending, and saving the lives of patients. Your excellent academic training thus far, however, may not have prepared you for many of the practical challenges of drug discovery that you will encounter once your inventions exit the fume hood.

Your molecules must survive a gauntlet of biochemical, cellular, pharmacokinetic, and animal tests before they can be deemed good enough to be dosed in human clinical trials. A practical reality of careers in medicinal chemistry is that it is extremely rare to be the one to synthesize a molecule that completes the journey from *in vitro* and *in vivo* assays to patients and is ultimately approved as a medicine that can bring relief to the world.

This chapter is designed to introduce medicinal chemists to preclinical drug discovery assays and to equip you to identify and overcome some of the practical challenges your molecules will encounter. We will probe the basic types of *in vitro* (biochemical and cellular) and *in vivo* (animal pharmacology and pharmacokinetics) assays that are routinely employed in the preclinical "testing funnel" of a small-molecule drug discovery project. You will see what challenges exist along this cascade of assays and what you can do to increase your chances of synthesizing a molecule that will one day help patients.

2 The Testing Funnel

The flowchart on the next page shows a simplified, generic testing funnel for a small-molecule drug project, consisting of the preclinical assays in which compounds will be tested. Compounds synthesized by medicinal chemists generally must first show potency in a target-driven biochemical assay, such as a receptor binding or enzyme activity assay. The *in vitro* biochemical assay defines how potently a molecule acts on its protein target in a closed, equilibrium system. Biochemical assays are generally high throughput, automated with pipetting robotics, and placed at the beginning of the assay cascade.

In parallel with biochemical target testing, compounds can be tested in a battery of *in vitro* pharmacokinetics assays that quantify their ADME properties (absorption, distribution, metabolism, and excretion). Such assays define whether compounds have, for example, good membrane permeability and low metabolism by oxidative enzymes such as the cytochromes P450 (CYP) and conjugating enzymes such as the glucuronosyltransferases, sulfotransferases, and acetyltransferases. Metabolism of your precious molecules by CYPs, or intrinsic clearance, can be quantified in liver microsomal preparations isolated from human, rat, and mouse. It is important to minimize the intrinsic clearance of your molecules by CYPs so their concentrations can remain high *in vivo*.

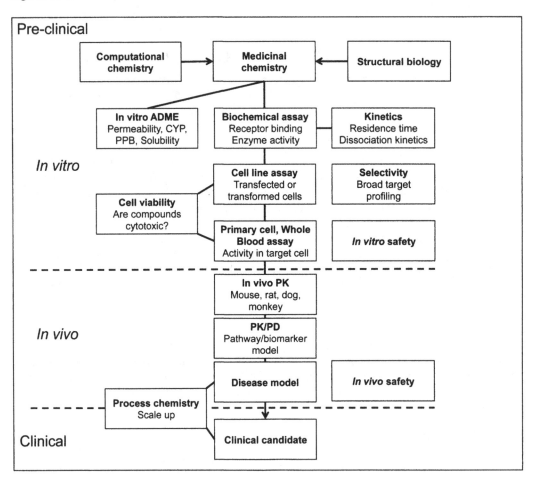

An *in vitro* plasma protein-binding (PPB) assay provides a readout that will be used later on in the cascade after compounds are tested in animals. In a PPB assay, compounds are incubated with serum albumin protein from human, rat, or mouse and then subjected to equilibrium dialysis to define the amount of compound that is bound to plasma protein. The percent of compound that is bound to plasma proteins in this closed system will be used to calculate the concentration of free, unbound drug in the plasma after *in vivo* dosing.

Note, however, as emphasized by Smith *et al,*[1] that PPB is typically not a parameter that medicinal chemists should aim to minimize with hopes of increasing drug exposure and improving efficacy *in vivo*. Rather, the PPB value is a number that is used only to estimate the free, unbound drug concentration required *in vivo* to inhibit the biological readout either in a pharmacodynamics (PD) assay or in an animal model of disease (*vide infra*). Smith clearly explains that reductions in PPB alone will simply result in the free compound being more available to be metabolized (by CYPs, *etc*.) and no net increase in free drug will result. Efforts to design and synthesize analogs with lower PPB will not likely result in higher free drug concentrations *in vivo* but will instead simply redirect the molecule to be metabolized by CYPs and cleared.[1] A more productive path for the medicinal chemist is to aim to decrease the intrinsic clearance of your putative drug molecules by CYPs. Reducing PPB alone, without reducing metabolism, will not be a fruitful strategy.

If, however, your molecules are extremely highly bound (e.g., 99.9% or more bound) to plasma proteins, it will be technically difficult to accurately quantify their percent binding. In such cases, of course, it would be a wise goal to reduce the protein binding in order to enable accurate measurements of the percent bound.

After showing some ability to inhibit (or activate) the disease target in the biochemical assay, the next hurdle is a cell-based assay. Often the cell assay uses a "cell line" that has been engineered or mutated or is somehow different from a native cell that has not been manipulated. For example, the engineered cell line may overexpress the target protein and could also overexpress a modified form of the direct cellular substrate of the target. To get a robust signal, cells can be transiently transfected with plasmid DNA that will enable a downstream readout of the signaling pathway. This type of gene transcription "reporter" readout can be used for targets that are upstream in a pathway leading to activation of transcription factors such as NF-κB (nuclear factor for site B in κ enhancer) or STAT (single transducer and activator of transcription). However, one must realize that although such engineered or mutant cell lines are extremely useful in filtering out molecules that cannot cross cell membranes, these types of cell assays can still be somewhat artificial with respect to the underlying cell and pathway biology. The cells used are often not the same type as the true target cell. They may be engineered by incorporation of nonnative DNA to overexpress proteins or use reporter sequences that are different from the native sequence, and the signal may not be dependent on a native signaling stimulus. Nevertheless, engineered cell assays are widely used and can be very effective to define structure-activity relationships (SAR) within chemical series. If you run into problems translating results from an engineered cell assay to the next level assay, however, you may want to test your compounds in an assay using primary cells to see if they provide a more predictive assay.

Primary cells are cells that are freshly isolated from blood or tissues and have not been further engineered. Primary cells generally express the native types and levels of receptors and cellular proteins. Thus, the agent used to induce a signal can be the same stimulus that induces the disease pathology. In this respect, data from primary cell assays may translate better to results from testing in animal models of disease. In addition, the downstream readout in a primary cell assay can be the same cytokine, phosphorylated protein, oncogene, or inflammatory gene as is involved in the disease. This increases one's ability to translate from potency in cells to efficacy in *in vivo* studies.

If your molecule survives this suite of cell assays, your compound will then face the unpredictable *in vivo* world. The first *in vivo* test of a molecule is to define its pharmacokinetics (PK), or time-concentration relationship. This will define whether it can be dosed orally and intravenously and achieve a measurable level in the blood or tissue of interest and define the time course or kinetics of this process.

Once the PK of the molecule is known, you can test whether and how potently the compound can cause the desired biological effect *in vivo*, *i.e.*, its pharmacodynamics (PD). This is often termed a "PK/PD assay." At this stage of *in vivo* testing, the aim is to determine whether the dosed drug can access and modulate the target. Unlike the closed system of a biochemical or cell assay, where the drug concentration remains constant, *in vivo* testing is an open system in which the concentration of drug peaks after dosing and absorption, then decreases over time due to distribution, metabolism, and excretion. Drug molecules must reach the tissue or cell that expresses the target, and there achieve a concentration of free drug (not bound to plasma proteins or associated with membranes) that is high enough to bind to and modulate the target receptor or enzyme. It is useful to monitor *in vivo* the same signaling pathway, gene expression, or other readout as was monitored in the *in vitro* cell assay. For example, in the development of JAK1 (Janus kinase 1) inhibitors for arthritis, the JAK1-dependent phosphorylation of transcription factor STAT3 was monitored *in vivo* following stimulation by the inflammatory cytokine, interleukin-6 (IL-6).[2] Achieving *in vivo* modulation of the same biological pathway that is thought to be involved in the human disease is an important milestone in the life of a small-molecule drug discovery project.

The next challenge will be to show that your molecule can block, slow, or alleviate the pathology in an animal model of disease. Such efficacy models can use a single dose of drug or can be multi-day experiments that will require you to consider the drug levels in the animal after multiple doses. In an acute study, demonstrating efficacy can be as simple as showing the drug can reduce the level of blood glucose, reduce blood pressure, or reduce total plasma cholesterol. In a more complex animal model, one might need to show that your molecule can reduce tumor volume, improve cognitive function in order to navigate a maze, or reduce the degradation of cartilage that occurs in rheumatoid or osteoarthritis.[3]

Blocking pathology in an animal model, however, does not always translate to efficacy in treating human disease. Animal models often do not closely mimic human pathology. Although some of the symptoms and gross pathology may resemble the human conditions, the underlying molecular basis or cause of the pathology in human disease can differ significantly from the molecular pathology in the animal model.[4,5] While not covered in this chapter, safety is as important as efficacy. Your molecules will need to be safe when dosed to animals or they will never progress to human dosing. Assuming your molecule is efficacious in the animal disease model, and it is shown to be safe in multiple animal species from mouse and rat, to dog and/or monkey, then your compound may be ready to be named a clinical candidate and will be on its way to phase 1 human clinical trials. Before you can synthesize that clinical candidate, however, you will need to start the journey with some "lead matter."

Traditionally, medicinal chemists have been charged with selecting the lead matter to prosecute. Aside from judicious target selection, the choice of lead chemical matter is one of the most important decisions a project faces. If the lead matter has poor physicochemical properties such as low solubility, high log*P*, or high polar surface area, it will be extremely challenging to develop molecules that are efficient modulators of the target *in vitro*, potent modulators of the target biology in cells, and can be dosed at reasonably low doses to animals and show robust biological efficacy. Lead matter or ideas for lead matter can be obtained from previous successful projects, from patents, from the scientific literature, by an in-license deal, from biophysical or activity-based fragment screens, or from high-throughput screens of broad or focused libraries of small molecules.

3   *In Vitro* Assays

3.1   High-Throughput Screening as a Source of Lead Matter

Projects often exploit high throughput screening (HTS) in biochemical or cell-based assay format to identify active molecules ("hits") from several thousands to millions of compounds in a corporate chemical screening library. Often, the compounds in a pharmaceutical company's library have been assembled from several sources such as previous internal medicinal chemistry projects, purchased combinatorial or singly synthesized compound libraries, natural product libraries, as well as purchased private collections. The purities of some of the combinatorial compound libraries generated in the early days of combinatorial chemistry may be less pure than more recently synthesized libraries. Catalysts, solvents, or other reactants or side products could often be present in a sample generated by combinatorial chemical synthesis. Low-level but highly reactive chemical catalysts or reactants present in test samples can result in false positives that require significant resource to deconvolute.

In addition, various sources of commercially available lead matter often have varying degrees of freedom to operate associated with them. A chemical library from vendor X may have been purchased by many pharmaceutical companies and as such it may have a potentially lower likelihood of generating a novel molecule for a particular target, whereas the compounds synthesized by internal medicinal chemists for previous projects may be more novel and potentially more likely to lead to compounds that are outside existing patent space. In addition, the chemists who generated the internal chemical matter are an excellent source of information regarding the benefits and liabilities of the chemical matter and may help save time and effort by directing the new team to the most fruitful directions.

Shown below is a general flowchart for a HTS campaign. In this example, compounds are tested once at a single concentration if the library is large (i.e., on the order of 1 million compounds or more). Screening of smaller libraries (e.g., 100,000-500,000 compounds) increases flexibility. It may be possible to test each compound at multiple concentrations or even in a full titration during the primary HTS run. Readers who are interested in alternative approaches to HTS assays can find online an excellent technical guide that has been compiled by academic and industrial scientists.[6]

After the initial single-point test, compounds with activity that is significantly outside the noise of the assay (generally > 3 standard deviations of the mean of the uninhibited control wells) are selected for re-testing in a full titration. This confirms their activity in the original HTS assay and allows one to roughly rank the potency of the primary hits. Generally, the IC_{50} tests are carried out using the same liquid stock of compound as was used in the initial screen. Thus, at this IC_{50} stage 1 is only testing the reproducibility of the HTS assay and whether it will confirm the initial identification of the compounds as active hits. At this stage it is also critical to determine whether a hit is simply interfering with the assay signal detection or readout due to, for instance, the compound being highly fluorescent or having a strong absorption spectrum or causing some other interference artifact.[7]

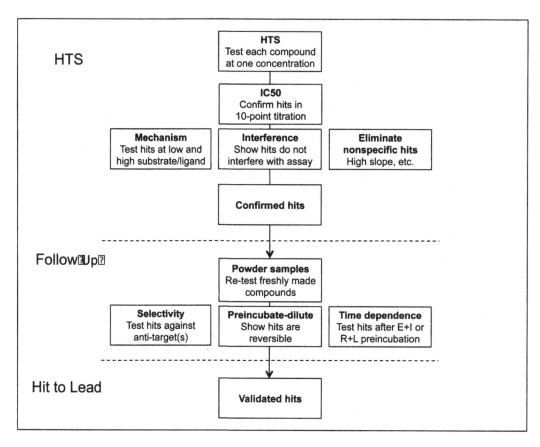

In addition, some early probing of mechanism of inhibition can be initiated. One approach is to also run the IC_{50} tests at a higher (or lower) substrate or ligand concentration than used in the HTS to see whether the IC_{50} values shift to lower (or higher) potency. Such shifts can indicate competitive (or uncompetitive) binding or inhibition. If the potency is reduced when the substrate or ligand concentration is increased, this is consistent with competitive inhibition. Generally, competitive behavior means that the hits are indeed binding to a unique site on the target and are specifically preventing substrate or ligand binding. If the potency of the compound does not change when the ligand or substrate is varied, this could be an exciting finding. You may have identified some noncompetitive hits that may interact at a potentially novel allosteric binding site.[3] On the other hand, it could mean that these hits are undesirable, problematic nonspecific inhibitors that do not bind to a unique site on the target. Apparent noncompetitive inhibitors could simply be covalent modifiers of your target protein. This is great if they are highly specific for binding and reacting only with your protein. On the other hand, covalent modifiers could be a significant risk if they are not highly selective for binding to your target but are broadly reactive with proteins, DNA, or other biomolecules. At this stage, one often has several hundred to several thousand hits of potential interest and it is not feasible to clarify the reversibility of inhibition by every hit. In addition, one may not have significant amounts of compound samples available to run such studies at this point in the cascade.

Once the hits have been *confirmed* (e.g., are reproducible) from the original sample source, they must be *validated*. This is achieved through the testing of freshly prepared samples. Corporate compound libraries often have samples of frozen liquid stocks as well as powder stocks available for such follow up studies. If no such samples are available, any interesting hits will have to be re-purchased or re-synthesized, and these activities often are very time-consuming and expensive. Additional detailed mechanism of action studies can be carried out in the follow up phase, after the confirmed hits have been validated.

The most important responsibility of the HTS team, aside from developing and implementing a robust assay, is to ensure that their list of hits contains few if any nonspecific ("promiscuous") inhibitors, as these false positives will fail to progress during hit-to-lead chemistry efforts. Synthesizing analogs of compounds that are not true ligands/modulators of the target will be a time consuming and costly effort for both the medicinal chemist and the team. Fortunately, much work over the past decade in both academic and industrial laboratories has resulted in a wider appreciation and deeper understanding of the significant presence of nonspecific inhibitors in screening libraries. In addition, numerous assay strategies have been published that can be applied to reduce the number of nonspecific hits that are found in HTS assays, and to separate nonspecific hits from real hits.[7-11] Assay additives such as a low concentration of a nonionic detergent or addition of protein such as bovine gamma globulins can clarify whether a hit is real or is a nonspecific false positive.[8,9] By applying these techniques, fewer nonspecific inhibitors will be present in hit lists that are presented to medicinal chemistry teams. As will be discussed later in this chapter, it is also important to carefully inspect the IC_{50} curves and scrutinize any compounds that have high slopes or which have inhibition curves that otherwise look odd or unusual.

3.2   The IC_{50} as a Measure of Inhibitory Potency

The most common measurement of inhibitor potency in the medicinal chemistry literature is the IC_{50}, or the concentration of inhibitor that results in 50% inhibition (half-maximal inhibition) of the enzyme activity or ligand binding. With the automation available today, it is technically easy to run titrations of compounds across several orders of magnitude of concentration. Challenges can arise when a poorly behaved compound or a less than robust assay is involved. In addition, IC_{50} values can change if the conditions of the assay vary. If another laboratory reports an IC_{50} it may not be comparable to yours unless the assay conditions are the same. This limits one's ability to compare IC_{50} values between different types of assays and across laboratories.

Many parameters can affect the measured IC_{50} of a compound: increased or decreased substrate or ligand concentration; a change in the concentration of target protein (in the case of tight-binding inhibition); a change in pH, salt concentration, or temperature; the presence of detergent, its type and concentration; the presence of proteins such as serum albumin; the construct of the protein including differences in phosphorylation and/or activation state for kinases or metabolic enzymes; the concentration of solvent (dimethyl sulfoxide, DMSO); the presence, type, and concentration of reducing agents; whether the signal is truly captured under steady state conditions; and whether the inhibitor is time dependent or slow-binding. One cannot compare IC_{50} values across assays or between laboratories for the same target unless the salient parameters of the assay are identical. As will be discussed later, a more standardized measure of affinity is the inhibition constant (K_i) or dissociation constant (K_d) for an inhibitor or ligand. Nevertheless, with this advance warning, we will inspect some IC_{50} plots.

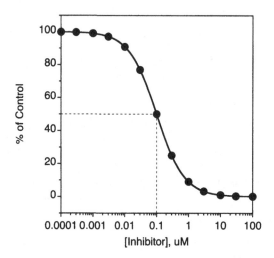

Shown above is a textbook inhibition IC_{50} curve. Characteristics of a high-quality IC_{50} curve are: (1) there are numerous points defining the curve, with at least two points per log unit of concentration; (2) the fitted line approaches 100% of control at low inhibitor concentration and 0% control at high inhibitor concentration; (3) there are several data points at concentrations both above and below the IC_{50} (in this example, IC_{50} is 0.1 μM); (4) there are several points on the linear middle region of the curve; (5) the slope at the midpoint is −1; and (6) the fitted line passes through all of the data points. The data above are fit to the equation:

$$\% \text{ of Control} = \text{max} \, / \, 1 + ([I]/IC_{50})^{\text{slope}}$$

where max = 100%. If you want to plot % Inhibition instead of % of control, or if you have a receptor assay where you are monitoring activation, you can plot binding or activity data in terms of % Inhibition (or Activation) by fitting to the equation

$$\% \text{ Inhibition (or \% Activation)} = \text{max} \, / \, 1 + (IC_{50}/[I])^{\text{slope}}$$

which describes an increasing sigmoidal function.

Note, however, that not all inhibition curves will appear as nice as the textbook curve above. When inhibition curves deviate from this example, it is prudent to investigate the causes as they may relate to poor behavior of the compound or a lack of robustness in the assay, both of which should be addressed. As a rule of thumb, you can recognize a good data set because the data will pass from 90% Control to 10% Control over a concentration range of approximately 2 log units if the slope is −1. As the medicinal chemist who is designing, synthesizing, purifying, and characterizing the small-molecule test compounds, you should feel free to meet with your biochemist and inspect the inhibition curves together. Clear communication is especially important in the event that the structural modifications you installed did not lead to the desired improvements in potency, selectivity, or affinity.

As an example, shown below is an inhibition curve for a compound with very low solubility. The inhibition reaches a rough plateau due to a reduction in the concentration of soluble compound in the assay well at high inhibitor concentration. In this case, one can still generate an IC_{50} (= 0.16 μM in this case) by omitting the data that clearly deviate from a good inhibition curve. But the omission of data points and the poor behavior of the compound in the assay must be noted in your database (e.g., "potential poor solubility, 1–100 μM points omitted"). In addition, the slope is much lower than ideal (–0.7 in this curve), and this fact should also be captured in the database as it speaks to both the performance of the compound and the quality of the assay results.

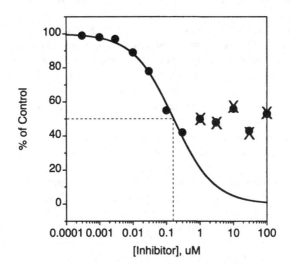

3.3   Solubility

One of the biggest challenges of *in vitro* assays is dealing with compounds that have low aqueous solubility that crash out of solution when they are added to the assay buffer or culture medium.[12,13] Low solubility compounds can be fully soluble at 10 mM in DMSO, but they can show <1 μM solubility when diluted in the aqueous buffers and nutrient media used in biochemical and cell-based assays. In addition, the way that your biochemist dilutes the concentrated inhibitor sample for the IC_{50} titrations can adversely affect the resultant potency measured in the assay. For example, if the 10 mM inhibitor stock in DMSO is first diluted directly into an aqueous buffer or cell culture medium, compounds with low solubility can immediately precipitate out of solution. Making a serial dilution of the compound at this point, after it has already precipitated or aggregated in the buffer or medium, generally results in stock solutions that do not contain the desired concentration of inhibitor.

A better dilution approach is to prepare the serial dilution of the 10 mM compound stock completely in DMSO. Low-solubility compounds will not likely fall out of solution in DMSO solvent. At this point, some will make an intermediate dilution of the inhibitor, for example, 10- to 100-fold into buffer or medium. The result for low-solubility compounds again is that they can come out of solution once they are exposed to the aqueous solution. To make matters worse, these intermediate dilutions of compound in buffer or medium are sometimes allowed to sit on the bench for a few hours before being used. The longer that compounds are in aqueous solution, the more likely it is that the low-solubility compounds will progressively precipitate out of solution or aggregate with other inhibitor molecules. The best approach is to transfer a small aliquot of the serially diluted compound in DMSO directly into the assay plate or cell culture, ensure that it mixes well with the aqueous buffer or medium and then initiate the experiment immediately. In this way, you can still learn about the biochemical or cell activity of a poorly soluble compound, and you will not lose a pharmacophore or a template of inhibitor molecules. A poorly soluble molecule that is a *bona fide* inhibitor of the target can be modified chemically to improve the solubility. A molecule representing a template or a pharmacophore that is not detected as active because the molecule was not actually in solution (and therefore was not truly tested in the assay) is lost forever.

The storage of DMSO-solvated samples of pharmaceutical compounds has been a topic of much investigation over the past decade. Scientists from academic and industrial compound management groups have carried out experiments to define the optimal storage conditions for their chemical library compounds.[13–17] Compounds were solvated in DMSO and then stored at ambient temperature or frozen in the presence and absence of humidity and oxygen. After various periods of time and after different numbers of freeze/thaw cycles, the presence and concentration of the compound in solution were monitored.

Several observations were made. (1) The main reason compound samples lost integrity was because the compound precipitated out of solution. The concentration of sample (e.g., 10 mM vs. 20 mM) impacted the stability of the sample solution. More concentrated samples were less stable, especially if the compounds had low solubility. (2) For samples that were initially of high purity, chemical degradation was not a major problem. (3) Exposure to oxygen did not cause degradation or loss of compound from the stored solutions. (4) Absorption of water into the DMSO solvent was observed, and the presence of water may increase the hydrolytic degradation of compounds. (5) Freezing and thawing the samples up to 11 times was generally not a problem for compounds with good solubility, but for compounds with low solubility repeated freeze/thaws were problematic. (6) There was no difference in compound solution stability or integrity between samples stored in glass or polypropylene containers.

The main recommendations coming from these studies are that for short-term storage (i.e., weeks) DMSO-solvated samples can be kept at ambient temperature, as this preserves the integrity of low-solubility compound solutions better than subjecting them to freeze/thaw cycles. For long-term storage, DMSO-solvated compounds should be stored frozen under low-humidity conditions. While the samples can be frozen/thawed up to 11 times without significant loss of the sample, it is preferable to avoid repeated freeze/thaw cycles, especially if dealing with low-solubility compounds.

Thus, an important goal of the medicinal chemist should be to synthesize soluble compounds that can be assayed with fidelity, will generate reliable data in biochemical, cell, and *in vivo* assays, and for which solutions can be preserved long term in DMSO-solvated form as this will extend the life of the sample and reduce the need for resyntheses.

3.4   Nonspecific (Promiscuous) Inhibitors

Another challenge that you may face with biochemical and cell assays is what is termed nonspecific, or promiscuous, inhibition. This phenomenon is frequently caused by compound aggregation, during which compounds with high hydrophobic and/or aromatic character can interact to form a large aggregate that then can bind and partially denature the enzymes and receptors used in the biochemical assays.[7–10] Adsorption of the target protein to a large molecular aggregate can impair or inactivate the target protein, resulting in reduction in the assay signal that is not due to specific, stoichiometric binding of the small-molecule to the target. Some experimental approaches have been published to both identify promiscuous inhibitors from screening data sets (i.e., through the use of surface plasmon resonance) and avoid identifying them (i.e., by including an aggregate-disrupting additive such as a nonionic detergent or a carrier protein such as bovine gamma globulins).[8,9,11] The most efficient path for the project team, and specifically for the medicinal chemist, is to design a robust screening assay that will not identify promiscuous inhibitors in the first place.

Nevertheless, if your HTS assay does identify promiscuous inhibitors, there are ways to recognize them and differentiate them from the desired target-based inhibitors. Promiscuous inhibitors frequently display weak (micromolar) potency, inhibition curves that have a high slope, and potency that increases with preincubation. Shown below at left is the inhibition curve for a nonspecific inhibitor, measured with no preincubation of enzyme with inhibitor. The compound has weak activity and the slope of the curve is high (–2.0). As a result, the curve does not have many data points defining the linear portion and, thus, the accuracy of the IC_{50} = 5.9 μM is low. Nonspecific aggregating inhibitors often show time dependent inhibition.[8,10] Shown at right, below is the inhibition curve for the same nonspecific inhibitor following a 15-min. preincubation of enzyme with inhibitor. The potency is slightly higher (IC_{50} = 3.1 μM) than without preincubation. Note, too, that the slope is also higher (–3.7) and the accuracy of the IC_{50} will be even more tenuous now because there is an even less well-defined curve of inhibition. Such promiscuous inhibitors can be found in HTS libraries that have been populated with molecules purchased from commercial sources and which have poor physicochemical properties and are rich in molecules having multiple aromatic rings. Interestingly, nonspecific inhibition can also occur within templates of *bona fide* inhibitor lead matter as structural modifications are made to the molecules.[18]

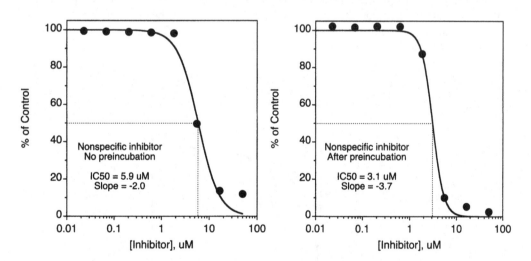

Promiscuous inhibitors can cause high false-positive hit rate in high throughput biochemical screens and can also cause discrepancies in structure–activity relationships for series of true stoichiometric lead matter. Because promiscuous compounds are not binding to a discrete site on the target, attempts to improve potency through synthetic medicinal chemistry will lead only to frustration.

3.5   *Mechanism of Inhibition Studies*

In addition to confirming the validity of your chemical matter, it is informative to define the kinetic mechanism of inhibition as competitive, noncompetitive, or uncompetitive with substrate. Before delving into the kinetics of inhibitor binding, it is useful to understand some basics of enzyme-catalyzed reactions. Two excellent sources of additional details on enzyme kinetics and inhibitor analysis are the texts of Segel and Copeland.[19,20] Please consult those references if you want additional information regarding evaluation of enzyme inhibitors in drug discovery.

Shown below is the scheme for an enzyme-catalyzed reaction. Binding of substrate S to enzyme E occurs with an associated substrate binding affinity, K_s. The rate constants for substrate binding (k_1) and dissociation (k_2) define $K_s = k_2/k_1$. The rate-limiting chemical step of the reaction is described by the rate constant, k_{cat}:

$$E + S \underset{k_2}{\overset{k_1}{\rightleftharpoons}} ES \xrightarrow{k_{cat}} E + P$$

The Michaelis constant, or K_m, relates to the kinetic constants in this scheme as

$$K_m = (k_2 + k_{cat})/k_1$$

The figure below shows a textbook hyperbolic plot (Michaelis-Menten plot) of enzyme activity vs. substrate concentration for the simple enzyme-catalyzed scheme described above. The initial velocity of the reaction, v_o, is related to the substrate concentration [S], the maximal velocity V_{max}, and K_m as described by the Michaelis-Menten equation for a rectangular hyperbola:

$$v_o = V_{max} [S] / (K_m + [S])$$

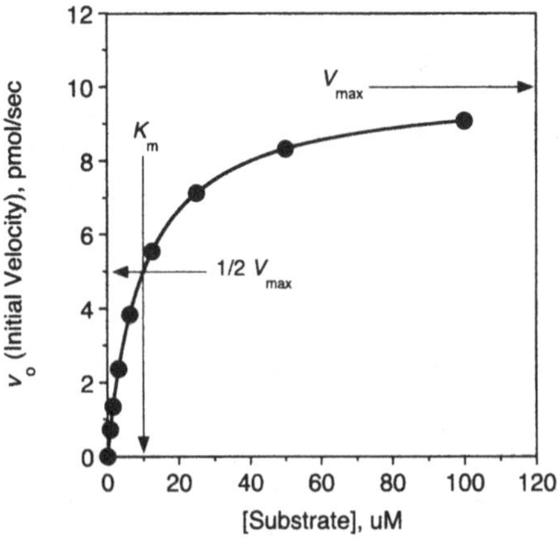

You can see in the plot that as the substrate concentration is increased the initial velocity approaches V_{max}. As illustrated graphically in this plot, K_m is the substrate concentration that gives reaction velocity that is equal to ½ V_{max}. In this example, V_{max} = 10 pmol/s and K_m = 10 μM.

A common assumption often made is that the K_m is equal to the substrate binding affinity. Inspection of the equation for K_m above will show that this assumption is sometimes true, but not always. If $k_{cat} << k_2$, the K_m will approach the substrate binding affinity, K_s. If, however, k_{cat} is not $<< k_2$, then the K_m will not be equal to the substrate binding affinity, but will default to the more complex relationship, $K_m = (k_2 + k_{cat})/k_1$, with the result being that $K_m > K_s$. If you assumed that $K_m = K_s$, you could run into problems as you try to understand the kinetic mechanisms of your inhibitors by using steady state kinetics assays, and try to draw conclusions about what your inhibitors are doing that induce changes in V_{max} or K_m. Although not covered here, when the enzyme catalyzes a multi-substrate reaction (such as a kinase), the relationships of rate constants and K_m are much more complicated.[19] Thus, for multi-substrate enzymes one must be even more cautious in making any assumptions about the relationships between substrate binding affinity and K_m.

Now that we have seen how the uninhibited enzyme-catalyzed reaction proceeds and how one determines kinetic constants, we will look at how to determine the mechanism of inhibition of the inhibitor molecules. Previously we presented the reaction scheme for the uninhibited enzyme. The scheme below introduces equilibria for inhibitor binding. This scheme shows the original enzyme-substrate equilibrium, but now it includes the equilibria involved for an enzyme that can reversibly bind substrate and inhibitor, I.

In this scheme, K_i represents the kinetic inhibition constant for E + I ⇌ EI, the interaction of a competitive inhibitor with the free enzyme. On the other hand, an uncompetitive inhibitor binds only to the ES complex (or to some form of enzyme other than free E) with affinity of αK_i (α < 1), according to the ES + I ⇌ ESI equilibrium. The α value modifies the substrate binding K_s and inhibitor binding K_i terms in cases when the binding affinity of substrate is changed by the presence of bound inhibitor, and vice versa. Note that competitive inhibitors do not bind to ES ($\alpha = \infty$), whereas uncompetitive inhibitors do not bind to free E. Finally, noncompetitive inhibitors bind to both free E and the ES complex with equal affinity ($\alpha = 1$). Noncompetitive inhibitors bind at a site that is distinct from the active site where substrate(s) binds. With this basic understanding of inhibitor binding equilibria, let's look at the relationships between K_i values and the commonly measured IC_{50}s.

An equation that is commonly used in drug discovery for this purpose is the Cheng–Prusoff equation. However, this is a misnomer. Cheng and Prusoff actually derived and presented several equations that describe the relationships between IC_{50} and K_i values for different kinetic mechanisms of inhibition.[21] Three of the equations presented in their seminal paper relate K_i and IC_{50} values for a single-substrate enzyme as follows:

Competitive inhibitor: $K_i = IC_{50} / (1 + [S]/K_m)$

Noncompetitive inhibitor: $K_i = IC_{50}$

Uncompetitive inhibitor: $K_i = IC_{50} / (1 + K_m/[S])$

Because most small-molecule drugs are competitive inhibitors of their targets and drug discovery projects have traditionally focused on competitive inhibitors, the Cheng-Prusoff equation that is most familiar to drug discovery researchers is the one that describes the relationship between IC_{50} and K_i for a competitive inhibitor.

      Cheng and Prusoff presented six more equations in their paper covering more complicated multi-substrate enzyme situations. The interested reader is encouraged to consult that reference, as well as an enzymologist, for further insight.[21] In addition, please refer to Cheng and Prusoff's paper to appreciate all of the assumptions behind use of these equations. With this basic understanding of how IC_{50} relates to K_i for the simple kinetic mechanisms of inhibition, we can introduce a high throughput (shortcut) and a lower throughput (classical) method to define the kinetic mechanism of inhibition.

      A shortcut approach to assess the mechanism of inhibition is to measure IC_{50} values at different substrate or ligand concentrations. This approach is useful during HTS when you have several hundred or thousand hits to evaluate. The figure below depicts how IC_{50} changes as the substrate concentration varies for competitive, noncompetitive, or uncompetitive inhibitors.

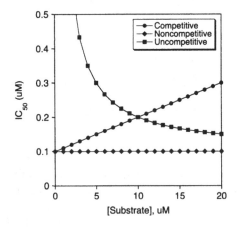

In this example, the inhibitor has target affinity $K_i = 0.1$ µM towards an enzyme that has a substrate $K_m = 10$ µM. As illustrated in this figure (and by rearranging the Cheng-Prusoff equations above to solve for IC_{50}), a competitive inhibitor will be less potent (IC_{50} increases) as the substrate or ligand concentration is increased: $IC_{50} = K_i (1 + [S]/K_m)$. The potency of noncompetitive inhibitors will be unaffected by changes in substrate concentration: $IC_{50} = K_i$. Uncompetitive inhibitors have low potency at low substrate levels but will become more potent as the substrate or ligand concentration increases: $IC_{50} = K_i (1 + K_m/[S])$. Assuming you have a reasonably precise biochemical assay, measuring IC_{50} values at low and high substrate concentrations may be sensitive enough to bin compounds into different mechanistic classes. The IC_{50} changes you observe will be characteristic of the kinetic mechanism of inhibition.

The classical approach to define kinetic mechanism of inhibition is to carry out a steady state kinetics of inhibition experiment wherein one monitors the enzyme activity while varying the concentrations of substrate and inhibitor. Basically, you determine K_m for the substrate at different inhibitor concentrations. As a general rule, the substrate concentration in this type of experiment should range from $0.25 \times K_m$ to $5 \times K_m$. (For an uncompetitive inhibitor, you may need to test even lower substrate concentrations, as will be explained later). The concentrations of inhibitor tested in this type of study do not need to span several orders of magnitude as they do in IC_{50} experiments. Rather, the inhibitor concentrations used should be high enough to achieve >10% inhibition, but low enough to have <90% inhibition (i.e., not so high as to abolish the assay signal).

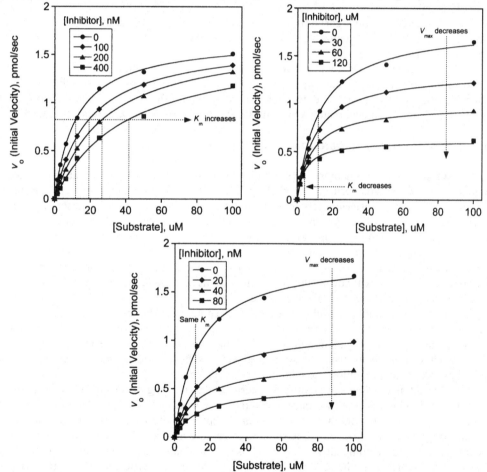

Competitive inhibitor: K_i = 157 nM, K_m = 12.1 µM at I = 0 up to 40.1 µM at I = 400 nM
Uncompetitive inhibitor: K_i = 64 µM, K_m = 12.2 µM at I = 0 down to 4.5 µM at I = 120 µM
Noncompetitive inhibitor: K_i = 31 nM, K_m = 12.9 µM to 14.1 µM from I = 0 to 80 nM

Shown on the previous page are some real steady state kinetics of inhibition data collected for a competitive inhibitor (left), an uncompetitive inhibitor (right), and a noncompetitive inhibitor (lower middle plot). The Michaelis constants (K_m) and inhibition constants (K_i) were determined by global fitting of the inhibition data to the model equations for the respective mechanism of inhibition.[19] As evident in the upper left plot, the competitive inhibitor increases the apparent K_m for the substrate (K_m shifts to the right as the inhibitor concentration increases), but does not change the V_{max}. All of the curves approach the same V_{max} as the substrate concentration increases toward saturation. These are the hallmarks of competitive inhibition. As shown in the upper right plot, the uncompetitive inhibitor causes a decrease in both the V_{max} and the K_m (the K_m moves to the left as inhibitor concentration is increased). Uncompetitive inhibition is common for multi-substrate enzymes and extremely rare for single substrate enzymes.[19,20] Finally, in the bottom plot the noncompetitive inhibitor decreases V_{max} but has no effect on the K_m.

With the robust curve-fitting algorithms available today, there is no longer a need to plot this type of inhibition assay data in double-reciprocal (Lineweaver–Burk) style in order to graphically determine the kinetic mechanism.

The shortcut IC_{50} shift and the more rigorous steady state kinetics of inhibition method are two straightforward approaches that are useful to define the mechanism of inhibition of your chemical matter. These assays are valuable not only to profile your HTS hits, but also to characterize the advanced lead matter as you make structural changes to your molecular templates.

3.6   How to Identify Allosteric Inhibitors

Now that we know how a noncompetitive inhibitor performs in mechanism of inhibition assays, it is a good time to discuss the term "allosteric," since these terms are often used interchangeably but imprecisely. The term allosteric refers to a binding site or location. It does not describe a kinetic mechanism of inhibition like the term noncompetitive does. Allosteric modulators bind to a site that is distinct from (non-overlapping) the active site of the target. One must be careful not to assume all allosteric inhibitors are kinetically noncompetitive. An allosteric inhibitor could be noncompetitive, uncompetitive or even competitive with substrate. Since noncompetitive and uncompetitive inhibitors cannot occupy the same site as the substrate or ligand, they must by nature be binding at a different, or allosteric, site.

Because noncompetitive inhibitors do not lose potency as the concentration of substrate is raised, they are extremely attractive as therapeutic agents.[3] There is interest in being able to selectively identify allosteric inhibitors in HTS campaigns. However, there is a perception that special conditions (such as using extremely elevated substrate or ligand concentrations) are required in order to identify allosteric inhibitors in an HTS assay.

The truth is that one can identify allosteric inhibitors using normal assay conditions, and there is no need to increase the substrate concentration in order to "favor" allosteric inhibitors. As shown in the example table below as well as in an in-depth statistical analysis,[22] a concentration of substrate or ligand equal to K_m or K_d is suitable to capture all classes of inhibitors, competitive, noncompetitive, and uncompetitive. Any allosteric inhibitors that exist in your screening collection will be present in the HTS hit list, assuming they are of sufficient potency to have been detected. For a practical example, assume you will run an HTS at 5 µM test compound and you set your hit cutoff to 50% Inhibition. Consider two inhibitors that you may identify, one with $K_i = 0.5$ µM and the other with $K_i = 5$ µM. What types of inhibitor mechanisms will you be able to identify? The table below shows how much inhibition will be measured at different substrate concentrations above and below the K_m. The % Inhibition values in the table were calculated using the Cheng–Prusoff equations for competitive, uncompetitive, and noncompetitive inhibitors, and the equation: % inhibition = $100 / 1 + (IC_{50}/[I])^{slope}$.

It is clear from this table that noncompetitive inhibitors will show equal inhibition at all substrate concentrations. The ability to detect competitive and uncompetitive inhibitors improves as the substrate or ligand concentration approaches K_m. Thus, running the HTS assay at a substrate concentration close to the K_m, you maximize your ability to identify all true hits, regardless of their kinetic mechanism of inhibition. By running the screen at an excessively high substrate concentration, competitive inhibitors will inhibit less and your ability to detect them will decrease. It is usually preferable to capture all types of real inhibitors in the initial screening assay. You can always pare down your list of chemical matter after you have all the hits in hand and have characterized their mechanisms of inhibition and determined their binding sites through structural studies such as NMR or X-ray crystallography. If you miss detecting an inhibitor in the initial HTS assay, that inhibitor is gone forever.

Mechanism	Inhibitor K_i, uM	% Inhibition at various substrate concentrations relative to K_m				
		0.1 x K_m	0.5 x K_m	1 x K_m	2 x K_m	10 x K_m
Competitive	0.5	90	87	83	77	48
	5	48	40	33	25	8
Noncompetitive	0.5	91	91	91	91	91
	5	50	50	50	50	50
Uncompetitive	0.5	48	77	83	87	90
	5	8	25	33	40	48

3.7 Selectivity

With this background, we apply a few of the principles that have been presented thus far to explore an issue that arises often in kinase inhibitor projects: selectivity. It is generally taught that one should define selectivity of a compound based on its K_d or K_i values for the target vs. the off-target(s).[20,23] Selectivity is more accurately defined in cell-based or *in vivo* assays. By definition, K_d and K_i values represent an intrinsic affinity when the ATP substrate concentration = 0. A selectivity comparison based on K_d or K_i values ignores the fact that different kinases have different affinities for and reactivities with ATP and that the cellular ATP concentration is not 0. For ATP-competitive kinase inhibitors, an assessment of selectivity based on K_i values will generally underestimate the real selectivity that will be evident in cells, where the inhibitors will have to compete with high levels of ATP for target and non-target binding. Using K_i values is fine to assess target or off-target SAR, but using K_i values to define selectivity between or among different enzymes with different kinetic properties will not be highly predictive.

To illustrate this concept, consider a kinase inhibitor that has K_i = 10 nM with its target kinase X and the same K_i with an off-target kinase Y that we wish to avoid. Is this inhibitor too non-selective to progress?   Based on its K_i value for the target and off-target biochemical selectivity is 1×. This compound may be deemed too non-selective for further testing. One might expect this inhibitor to inhibit both kinase X and kinase Y in cells and *in vivo* with equal potency. However, this was not found to be the case.

Kinase	ATP K_m (µM)	Inhibitor K_i (nM)	Biochemical Selectivity	Cellular IC$_{50}$ (nM)*	Cellular Selectivity	Inhibition in cell by 100 nM Inhibitor**
Target X	500	10	1×	30	34×	77%
Off-target Y	10	10		1010		9%

*IC$_{50}$ = K_i (1 + [S] / K_m), and [S] = 1000 µM ATP in cells
**% Inhibition = 100/(1 + IC$_{50}$/ [I])

As shown in the table above, when tested in cell assays at 10 × K_i (100 nM), this inhibitor inhibited kinase X by 77%, but inhibited kinase Y only 9%. Why? It turns out that these kinases have vastly different K_m values for ATP (500 µM for target X vs. 10 µM for off-target Y). Biochemical selectivity is low (1×), but cellular selectivity is high (34×). The difference in ATP K_m values made it easier to inhibit the kinase with the higher K_m (kinase X) and more difficult to inhibit the lower K_m off-target kinase Y. An efficient approach to overcome these limitations of *in vitro* biochemical assays could be to run all your biochemical kinase assays (for the primary target as well as off-target kinases) using, for example, at least 1 mM ATP. The resultant IC$_{50}$ or K_i^{app} values you generate will better reflect the potencies achieved in cells, and your SAR data may more efficiently prioritize your best inhibitors. Knight and Shokat have written an excellent review covering features of selective kinase inhibitors. The interested reader will learn more from that article.[23]

3.8   *Tight Binding Inhibition*

As your synthetic ideas come to fruition and the potency of your molecules improves, there may come a point when the biochemical IC$_{50}$ assay reaches its limit of sensitivity. When the K_i of an inhibitor is in the same range of concentration as (or lower than) the concentration of active enzyme or receptor used in the assay, this is considered tight-binding inhibition. One of the assumptions of enzyme inhibition kinetics assays is that the inhibitor concentration, [I], is in great excess over the enzyme concentration, [E], so that the free inhibitor concentration approximates the total inhibitor concentration. Tight-binding inhibition situations no longer conform to this assumption. With tight-binding inhibitors, the Activity vs. Inhibitor data no longer fit to a textbook IC$_{50}$ curve: the slope will be high, and the IC$_{50}$ values of high potency inhibitors will reach a limiting value of ½[E], even if they truly have higher potency or affinity. Under these circumstances, the Morrison equation for tight binding inhibition is needed in order to accurately quantify the potency and affinity of the inhibitors.

The Morrison equation allows determination of an apparent K_i (K_i^{app}) and total enzyme concentration ([E]$_T$) from inhibition data.[20,24,25] This K_i^{app} takes into account the potential for tight binding and depletion of inhibitor. In the Morrison equation below, v_i and v_o are the inhibited and uninhibited initial velocity, respectively, while [I]$_T$ represents the total inhibitor concentration.

$$\text{Fractional activity} = \frac{v_i}{v_o} = 1 - \frac{([E]_T + [I]_T + K_i^{app}) - \sqrt{([E]_T + [I]_T + K_i^{app})^2 - 4[E]_T[I]_T}}{2[E]_T}$$

When tight-binding is evident, the K_i^{app} value is more precise than a simple IC_{50}. The plots below show successively improved ways of fitting tight binding inhibition data. The first plot (below, left) shows a fit to an IC_{50} model, where the top is fixed at 100% Activity, the bottom is fit at 0% Activity, and the slope is (unreasonably) fixed at –1.0 for the purpose of this example. The resultant IC_{50} = 3.3 nM (R^2 = 0.986). Although the R^2 value is high, the points do not fit well to the curve. In addition, the slope of the data appears much higher than the –1.0 slope expected for an ideal binding interaction. In the second plot below (right), the data are again fit to an IC_{50} model, but now the slope has been allowed to vary. In this case, the data fit much better to the line, the IC_{50} = 3.4 nM and the slope = –1.5 (R^2 = 0.998). While this R^2 indicates a better fit, the IC_{50} still is inaccurate. Why? It turns out that in this experiment the enzyme concentration used was 3 nM, which is close to the apparent IC_{50}. Thus, at low inhibitor concentrations, the assumption that [I] >> [E] no longer holds. This case is a classical tight binding inhibition situation, and it requires application of the Morrison equation.

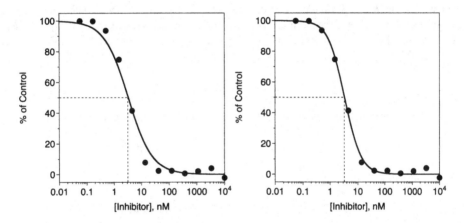

To apply the Morrison equation (in the form as written above) to the data that are currently plotted in terms of % of Control, one must first convert the data to Fractional Activity by dividing % of Control by 100. Fitting the Fractional Activity data to the Morrison equation gives the inhibition curve below, which provides an excellent fit of the data to the fit line, and generates K_i^{app} = 0.83 nM (R^2 = 0.998). This K_i^{app} is a much more robust value than the initial, poorly fitted IC_{50} = 3.3 nM from the first plot.

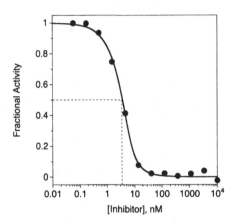

Now if you know the type of inhibitor this is (competitive, uncompetitive, noncompetitive), you can convert the K_i^{app} value to a more precise true K_i by taking into account the concentration of substrate used in the assay and the K_m for that substrate. In the example above, the assay contained 5 μM substrate, the $K_m = 3.48$ μM, and the inhibitor is competitive with substrate. Using the kinetic equations for various mechanistic types of inhibitors,[19] we can convert K_i^{app} to K_i. The relationship shown below for a competitive inhibitor allows one to convert $K_i^{app} = 0.83$ nM to the true $K_i = 0.34$ nM.

$$K_i = K_i^{app} / (1 + [S]/K_m) \qquad \text{Competitive}$$

In the event that your inhibitor was uncompetitive with substrate, the conversion would be:

$$K_i = K_i^{app} / (1 + K_m/[S]) \qquad \text{Uncompetitive}$$

While for a noncompetitive inhibitor:

$$K_i = K_i^{app} \qquad \text{Noncompetitive}$$

Although the quadratic form of the Morrison equation may appear complicated, it is not difficult to implement use of this equation in routine assay reporting. For the data analysis of inhibition curves, the Morrison equation can be coded into the software so that the inhibition data will be fitted to generate a K_i^{app} value. Or the Morrison equation could be further modified to account for the mechanism of inhibition, the substrate K_m, and the concentration of substrate used in the assay to generate true K_i values. Finally, although the Morrison equation accounts for the tight binding scenario, you do not need to have tight binding inhibitors to apply it. The equation is completely flexible and is able to precisely define K_i values for non-tight binding inhibitors as well.

3.9   *Slow-binding Inhibition, Reversibility, and Residence Time*

To this point, we have dealt only with inhibitors that bind rapidly to the target in a single fast step. As shown in the scheme below, however, this simple mechanism may be incomplete. Many enzymes and receptors bind inhibitory or stimulatory ligands in a multistep mechanism. The initial binding event is fast, but step(s) subsequent to the initial binding event are slow relative to the initial binding step and slow relative to the speed of substrate binding and enzyme catalysis.[26,27]

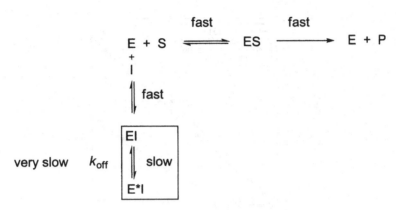

The slow conversion of the EI form of the enzyme to a lower energy state, E*I, may be due to a conformational change in the enzyme, the inhibitor, or both.[26-28] Slow-binding inhibitors have a rate constant k_{off} for the reversion of the E*I complex that is typically very slow.[26-28] This can make the lifetime of the E*I complex very long, and it gives the inhibitor a long residence time bound to the target. The residence time is defined as:

$$\text{Residence time} = 1/k_{off}$$

Drugs with long residence times should have an advantage *in vivo* over drugs that display fast binding and dissociation kinetics with their target. In the open system of *in vivo* biology, a drug that dissociates rapidly from its target will lose efficacy once the drug is removed from the target cell or tissue by blood flow or metabolism. A drug with a long residence time, on the other hand, will remain bound to its target long after the circulatory system washes away the unbound drug from the target tissue. Drugs with long residence time can remain bound and effectively inhibit their target even when the concentration of the drug in the plasma is well below the cellular IC_{50} of the drug. This allows for lower drug doses to be used. Some examples of drugs with long residence times are (residence time): the D2 dopamine receptor antagonist haloperidol (58 min);[30] allopurinol, a xanthine oxidase inhibitor (433 min);[31] the epidermal growth factor receptor kinase inhibitor lapatinib (7.2 h);[32] the HIV-1 protease inhibitor darunavir (>14 d);[33] finasteride, a steroid 5α-reductase inhibitor (>43 d);[34] and acyclovir triphosphate, an inhibitor or *Herpes simplex* DNA polymerase (60 days).[35] Many more examples of drugs with long residence times have been compiled in review articles on the subject.[26-29]

How would you know if your inhibitor has a long residence time? One way to quantify the residence time of a slowly dissociating inhibitor is with a preincubation-dilution (jump dilution) experiment. In this type of study, the target enzyme or receptor is first incubated with inhibitor at a concentration of $10 \times IC_{50}$, generally for an hour or more so that the system can approach equilibrium. A control preincubation sample contains enzyme plus vehicle but no inhibitor. Then, the enzyme samples are rapidly diluted 100-fold into an assay mixture and the activity in the diluted samples is monitored over time. In the activity assay the concentration of inhibitor is $0.1 \times IC_{50}$ and little inhibition (9%) should occur. The plot below shows kinetic traces for the formation of product over time upon dilution of enzyme samples preincubated with different types of inhibitors.

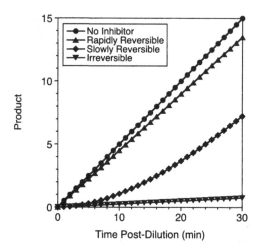

As can be seen in the figure, the control enzyme incubated with no inhibitor immediately catalyzes formation of product, as does enzyme that had been incubated with a rapidly reversible inhibitor, whereas an irreversible inhibitor (e.g., a covalent inhibitor or an extremely slowly reversible noncovalent inhibitor) will show very little product formation over time. A slowly reversible inhibitor will show a lag in product formation in this type of jump dilution experiment. The time course data for recovery of activity for the slowly reversible inhibitor sample can be fit to an exponential equation that generates an observed rate constant (k_{obs}) for the recovery of activity.[28] This k_{obs} corresponds to the rate constant k_{off} for reversion of E*I to re-form EI. Inhibitor cannot dissociate directly from the E*I complex. However, once E*I reverts back into the EI form, the inhibitor is able to dissociate. Free E can then bind substrate and catalyze its intended reaction. Using jump dilution studies one can quantify the residence time of inhibitors and define what structural modifications engender slow dissociating capability.

To have a long residence time, an inhibitor does not need to have extremely high affinity for the target. For long residence time, however, the inhibitor must dissociate slowly from the target. The kinetic constants that drive inhibitor affinity are the association (k_{on}) and dissociation (k_{off}) rate constants, where $K_d = k_{off}/k_{on}$. These kinetic values can vary widely within a chemical series. In the "kinetic map" illustrated below, k_{off} and k_{on} have been plotted for a number of analogs within a few chemical series of kinase inhibitors. The diagonal lines denote the K_d, ranging from 1 to 10,000 nM, as labeled. Compounds that lie along the same diagonal have the same binding affinity, but can have different kinetics of association and dissociation. An important parameter that drives efficacy in an open system (such as in an *in vivo* study) is the residence time, which was defined as $1/k_{off}$. Let us compare residence times for a few inhibitors that have equal affinity.

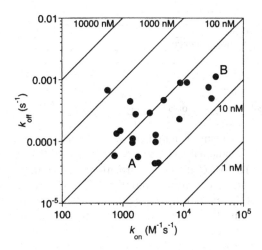

Compounds A and B lie along the same diagonal and therefore have the same affinity, K_d = 30 nM. However, these two molecules have greatly different dissociation kinetic rate constants and, therefore, residence times (A = 416 min, B = 15 min). In an *in vitro* binding or activity assay where the inhibitor concentration is fixed and the system is at equilibrium, these compounds would show the same potency or affinity. However, when tested in an open system where the concentration of drug is not held constant (e.g., due to blood flow, absorption, metabolism, clearance), the compound that remains bound to the target longer—the inhibitor with the longest residence time—will retain efficacy longer than a compound that dissociates rapidly and is effectively washed away. The molecular basis of long residence time drug–target interactions is only now starting to be systematically studied and understood. While the kinetics of slow-binding, long residence time inhibitors has long been appreciated, the conformational adaptation of the target protein with the long-residence-time ligands is more recently being addressed.[26,28]

3.10   Mechanism, Kinetics, and Potency Comparisons

Before moving on to cell-based assays, we will consider review how thorough one must be in order to accurately assess potency values. Consider an enzyme inhibitor that is reported in the medicinal chemistry literature to have IC_{50} = 5.2 nM. Your team has also been designing inhibitors of this target and your most potent compound so far has IC_{50} = 3.5 nM. You believe you are on par with or slightly ahead of the competition. But is your compound any better than the published compound? We will apply the fundamentals presented in this chapter and find out.

You synthesize the literature inhibitor so the biochemist can profile it in a battery of *in vitro* assays. Reading the Methods provided in the medicinal chemistry article you learn that the published IC_{50} was determined in an assay where the active enzyme concentration was 5 nM, so the reported IC_{50} = 5.2 nM may underestimate its true potency. As seen in the table below, applying the Morrison equation to account for tight binding would generate a K_i^{app} = 0.78 nM. This is more potent than the published IC_{50}, and better than the IC_{50} of your best inhibitor. Your biochemist tests this inhibitor in a higher sensitivity assay (using 0.025 nM enzyme) and generates K_i^{app} = 0.78 nM. As mentioned previously, an IC_{50} provides limited mechanistic insight.

Potency estimate	Inhibition value (nM)
IC_{50} from competitor's published SAR paper	5.2
K_i^{app} taking into account tight-binding situation	0.78
K_i if noncompetitive	0.78
K_i if uncompetitive	0.55
K_i competitive	0.22
True K_i^* due to slow-binding, competitive inhibition	**0.04**
K_i of your non-time-dependent, non-tight-binding competitive inhibitor	1

If we clarify the kinetic mechanism of inhibition, we may get a different picture of the inhibitor. The Methods section of the medicinal chemistry paper indicated that the substrate was used at 25 µM in the assay and $K_m = 10$ µM. Using this information and the mechanistic equations presented in Section 3.8 of this chapter, you can convert the K_i^{app} to a true K_i value depending on the mechanism of inhibition. If the literature compound was noncompetitive with substrate (this is not common), it would have a true $K_i = 0.78$ nM. If instead it were an uncompetitive inhibitor (even less common) would have a true $K_i = 0.55$ nM. Your biochemist reports that this inhibitor is competitive with substrate so the $K_i^{app} = 0.78$ nM changes to a true $K_i = 0.22$ nM. Knowing the mechanism of inhibition helps to clarify the inhibitor potency.

Until now, we have assumed this inhibitor binds the target and achieves equilibrium rapidly. However, your biochemist assesses the effect of preincubation on the potency and determines that the inhibitor is actually a time-dependent, slow-binding inhibitor. After a long preincubation with enzyme, the inhibitor potency shifts from $K_i = 0.22$ nM to $K_i^* = 0.04$ nM. The K_i^* represents the true equilibrium affinity of the literature compound, and it is much better than the reported $IC_{50} = 5.2$ nM. Finally, in a jump dilution experiment your biochemist determines that the literature inhibitor has an extremely slow dissociation rate constant, $k_{off} = 0.000004$ s^{-1}. Thus, the residence time of the inhibitor is 4,167 min, or 69 h, which equals 2.9 days. Assuming there is no potential liability due to on-target toxicity by inhibiting the target, this literature inhibitor has a very desirable molecular mechanism of action.

By contrast, your biochemist has determined that your best inhibitor with $IC_{50} = 3.5$ nM is competitive with substrate, so it has $K_i = 1$ nM [$K_i = IC_{50} / (1 + [S]/K_m)$]. It is not tight-binding or time-dependent. Thus, your best inhibitor, kinetically and with respect to ultimate potency, performs very differently from the inhibitor reported in the literature.

To end this section, it is fitting to provide some context regarding the ideal precision of biochemical assays. In the hands of a careful operator, the precision of a biochemical assay can be as low as 1.5–2-fold, but the assay precision needs to be carefully understood before a chemist can determine if biochemical potency differences of 2–3-fold represent significant structure-activity relationships.

3.11   Cell Assays

In the current era of target-based and structure-aided drug design, imparting biochemical potency is often straightforward. Compounds with excellent biochemical potency, however, may not effectively get into cells and some hydrophobic or amphiphilic molecules may partition into cell or organelle membranes and be less accessible to bind to their intended target. As you fully optimize biochemical potency and understand mechanism of action of your compounds at the molecular and kinetic level, the next important hurdle is to achieve cellular activity. For the initial cell assay, it is important to monitor as proximal a readout as is possible. For example, a cell assay where one monitors the direct phosphorylation of a native cellular substrate by a target kinase would be an ideal first tier assay. Later on, it will be valuable to assess the ability of your molecules to modulate the biological pathway(s) of interest by measuring more downstream functional readouts that occur subsequent to engagement or activation of the target of interest. If the target is an intracellular protein, the cell line assays clarify whether the inhibitor can cross the cell membrane and functionally inhibit or activate the intracellular (cytosolic or nuclear) target.

In small-molecule drug discovery projects, such cell-based assays often use a cell line, or cells that have been transfected with DNA (by viral or lipid-mediated DNA delivery) to express the target protein and even its substrate, or has been transformed in some way that alters its pathway and cell biology relative to a naïve, non-transformed cell.

The fact that such cell line assays use somewhat engineered cell biology is a concern, but the information one gains in showing that compounds can get into the cell is nonetheless useful. These cell assays are generally high throughput and can be automated to handle testing IC_{50}s for 50–100 compounds per run against one or more cell assays. Cell lines can be kept in culture for many passages. In other words, cells are allowed to grow to the point where they almost fully cover the surface of the culture plate and are near confluence, then are harvested, diluted with fresh medium and seeded (passed) into new culture plates at a lower cell density, and allowed to grow again to near confluence. Depending on the cell line and its genetic stability, the assay signal (window) may remain stable over many passages, it may remain stable for some period of time and then abruptly decrease, or an initially robust assay window might steadily decrease as the passage number increases. As the assay window decreases, however, the variability of the IC_{50} values from the assay may increase. Thus, cell passage number is an important criterion that is monitored by biochemists to assure reproducible SAR data.

In addition to cell lines, assays that use primary cells are a valuable option for relevant cell biology readouts. Unlike secondary culture cell line assays where the cells have been passed over many passages, a primary culture cell assay is one in which the cells have never been sub-cultured or transformed. Primary cells are freshly isolated from the tissue of interest or from donated whole blood, and thus have not been kept in culture or passed like the cell lines. In this sense, it is believed that primary cells should behave more like native cells do *in vivo*. The pathway biology, protein expression, and responses to small-molecule inhibitors may more closely mimic what happens *in vivo*. However, as primary cells from different donors have significant genetic, immunological, and activation differences, donor-to-donor variability is a common challenge of primary cell assays. In addition, depending on the type of assay and the stimulus used to activate the cells, cells from the same donor may be more or less responsive when donated on a different day. One common source of primary cells is peripheral blood mononuclear cells, or PBMCs. A PBMC preparation is isolated from whole blood and contains lymphocytes such as T cells, B cells, monocytes, and macrophages but does not contain red blood cells. Primary cells can also be isolated from tissues such as liver (hepatocytes), skin (keratinocytes), and kidney (proximal tubule epithelial cells). Primary cells are generally more challenging to work with since they are not immortal, and the assays and data are generally lower throughput.

In characterizing the cellular potency of a compound, it is vital to distinguish between on-target cellular readouts and non-selective effects on cell health. For any cell assay where the small-molecule treatment is 8–12 hours or longer, cell viability must be assessed to confirm that the small-molecule treatment is not simply decreasing signal by adversely affecting the health of the cell and grossly perturbing cell physiology. Common cell viability readouts used are measurements of cell ATP levels (by an ATP-dependent luciferase assay) or the cellular reducing power in NAD(P)H levels (via redox active tetrazolium dyes). When the cell assay is fairly short (less than 6–8 hours), it is possible that a compound that does cause adverse effects on the cell health may nonetheless not disrupt the ATP or NAD(P)H levels, and more sensitive readouts of cell viability would be needed. Chemistry teams often prosecute several different series of chemical matter in parallel. The different series may affect cell health in different ways. Using a cell viability assay that only measures one readout may be insufficient to detect the potentially different effects on cell health of multiple series of chemical matter.

Cell assay technologies today are versatile enough to easily monitor more than just a single parameter. The application of multi-readout high content imaging cell health assays is becoming more common.[36,37] High-content imaging is a method wherein fluorescently-labeled dyes or probes are used with a laser-based imaging system to simultaneously measure multiple cellular readouts such as cell number, nuclear size, lysosomal mass, phosphorylation and nuclear translocation of transcription factors, and even cell cycle regulation.[36,37] Effects of small molecules on these various readouts can be monitored with full IC_{50}s in relatively high-throughput format.

As compounds advance from biochemical to cellular assays, they often experience decreases in potency. These biochemical-to-cell potency shifts can vary from only 5- to 10-fold for proximal readouts to several hundred-fold for more downstream readouts, or when a molecule is highly protein bound. In the specific case where the target is a kinase, biochemical-to-cell shifts for ATP-competitive inhibitors can be large when the kinase has a low K_m for ATP, whereas shifts are lower when the kinase has a high K_m for ATP.[23] These shifts are a result of the cellular concentration of ATP being high, in the 1-10 mM range,[38] whereas biochemical SAR assays are often carried out at much lower ATP concentrations. For projects whose target protein exists in cells in the blood, a whole blood assay can be devised to gauge the potency of compounds in the presence of full serum and the attendant serum proteins, and in the context of the other potentially interacting cells. Human serum contains 34-54 mg/mL albumin and 0.6-1.2 mg/mL α_1-acid glycoprotein, as well as a number of other proteins and factors.[39,40] However, it is generally accepted that of the numerous protein components in blood, small drug molecules bind mainly to albumin and α_1-acid glycoprotein. In a whole blood *in vitro* assay which contains serum proteins, compounds can appear to be much less potent than when tested in an *in vitro* cell culture assay which typically contains only 10% serum. While plasma protein binding is not generally a parameter worth modulating by chemical modifications, as explained previously by Smith,[1] the effects of serum proteins on cellular potency are nevertheless particularly important for estimating efficacious doses that will need to be administered *in vivo*.

The higher complexity of cellular and whole blood cell assays (relative to biochemical assays that generally use purified proteins) can erode assay precision, and potency differences of two- to four-fold in cellular assays may no longer be considered significant. One can usually get a good idea of the precision of a cellular assay by assessing the variability in potency of the positive control compound used in the assay, after it has been tested over many assay runs. Once the cellular or whole blood assay variability is understood, one can then more accurately sort compounds based on their cellular potency values.

4   *In Vivo* Assays

4.1   *Pharmacokinetics and Pharmacodynamics (PK and PD)*

Now that your compounds have passed the *in vitro* assay hurdles, testing in animals is needed to prioritize which molecules possess the safety and efficacy needed to progress to human testing. In the pharmacokinetics (PK) testing, you will learn *what the body does to the drug*. Assuming the intended route of administration is an oral drug, the compound is separately dosed orally and intravenously to mice and/or rats and the absorption, distribution, metabolism, and excretion (ADME) is evaluated by measuring the concentration of drug in the plasma as a function of time after dose administration. Other routes of administration, such as for an inhaled drug or topically dosed dermatological drug will present their own challenges and technical hurdles.

The other goal of *in vivo* testing is to evaluate the pharmacodynamics (PD), or *what the drug does to the body*. In a hypothetical example of such an experiment, mice are dosed with compound before activating a biological pathway by injection of an inflammatory cytokine into the peritoneum of the animal. At various times after the cytokine stimulus, blood, plasma, and/or tissue samples are obtained and the level of drug in the plasma and the biological pathway readout are quantified. The figure below (left) shows the unbound (free) plasma drug concentrations over time for an orally dosed drug along with the biological effect on the cytokine level in serum. The biological pathway is inhibited maximally when the drug concentration is highest. As the drug concentration decreases over time, the pathway inhibition decreases as the drug concentration falls below the cellular IC_{50}. (In this example, the cellular assay is a short, acute readout and the cell incubation medium does not contain serum. The cellular IC_{50} does not need to be corrected for protein binding of the drug). This is the situation for a traditional rapidly dissociating drug.

On the other hand, recall that a drug with a long residence time should show sustained inhibition of the target and pathway. In the figure below (right) are shown data for a drug that has identical PK properties as the drug in the left panel, but the drug in the right panel has a long residence time on its target. The PK of the long-residence-time drug is identical to that of the rapidly dissociating drug. However, in the case of the long-residence-time drug, its slow dissociation from the target enables the biological pathway to be inhibited long after the plasma levels of the drug fall below the cellular IC_{50} of the compound. Slow-binding inhibition has been recognized and documented for several decades. However, the systematic search for and characterization of drugs having long residence times has only recently been rekindled. Examples of drugs with long residence times in the current pharmacopeia have been highlighted and proclaimed as a route to better efficacy and a greater likelihood of clinical success.[26–29]

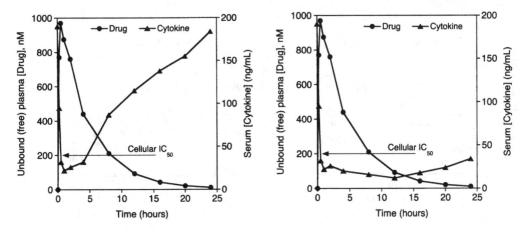

Note that in these theoretical PK/PD time course examples the free or unbound drug concentration is used, rather than the total drug concentration. Drug that is bound to serum proteins or sequestered in membranes is not able to bind to the target protein and exert any pharmacological effect.

An alternative, equally valid, approach to defining PK/PD relationships is to determine your cell potency using a whole blood assay format (which contains all the serum proteins that are present in blood). In the PK/PD vs. Time plot, you plot the *in vivo* drug levels in terms of Total plasma drug concentrations rather than unbound concentrations. Either approach is viable to build your understanding of PK/PD, but the approach using total drug levels is a bit more flexible because it does not require generating PPB data for every compound tested *in vivo*. An excellent review rigorously covering the subject of plasma protein binding in drug discovery has recently appeared.[1]

Another approach to study PK/PD relationships is to test a dose response of the test compound, to see whether the PD shows a clear dependence on drug dose. Below is a data set from a hypothetical *in vivo* animal PK/PD model dose response study. In this study, there are 5 animals per treatment group, including a group of naïve animals that are not induced by any external stimulus. The remaining groups all are induced by some treatment to the animals, potentially by dosing an inflammatory cytokine or some other stimulatory biological or biochemical agent. In addition, all groups receive the drug vehicle, either alone (as in the naïve and 0 mg/kg groups) or with the test compound added (in the 3, 10, 30, 100, and 200 mg/kg groups). One could also characterize the time-dependence of the drug PK and PD. Or, as in this study, withdraw blood at a single predetermined time after the stimulus is applied and determine the levels of plasma drug and serum cytokine. The PK data (left, below) show that in this *in vivo* model the concentration of drug correlates linearly with the dose given. There is low variability in the drug concentrations among the animals within a group (right, below). In the PD measurements, on the other hand, there is a somewhat more variable response to the drug among animals within a group. Such variability is not unexpected and is often the case when assessing biological responses within a population of animals. Nevertheless, the different dose groups of animals show a measurable PD response that increases with dose. This successful study defines the plasma concentration needed to provide significant inhibition of the PD readout. In addition, the dotted line in the PK plot shows that the free plasma drug concentration for the 30, 100, and 200 mg/kg dose groups surpasses the IC_{50} of the drug as defined in the cell-based assay.

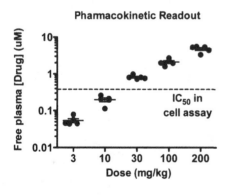

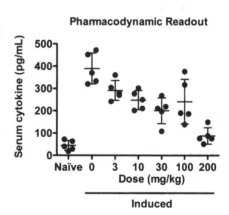

In contrast to the fairly simple and high throughput biochemical and cell-based assays, experiments using living animals, and the inherent challenges of animal variability and dosing, further increase assay variability. Because animal studies generally have high cost and low throughput, it becomes more critical to carefully consider in advance the number of animals used and sample replicates to be processed. In addition, the study design must ensure that the number of animals per treatment group will be sufficient to enable the application of statistical analyses to help interpret the study data and to prioritize your lead molecules properly based on the *in vivo* data.

4.2 Animal Models of Disease

Animal models of human disease are used to build confidence that your drug molecules will have a therapeutic effect in humans. However, efficacy of a molecule in a rodent model does not always translate to efficacy in humans. Animal models of human disease are often accelerated versions of pathology that mimic the human disease condition in some respect(s). However, there are certainly examples of animal models that do not closely mimic human disease on a molecular or etiological level.[4,5]

In general, in a disease model the animals are dosed with drug in either a preventive (prophylactic) schedule, where the drug is given before the onset or initiation of the pathological symptoms, or in a therapeutic mode, where the drug is dosed starting some time after the disease pathology has been initiated or observed. Efficacy following therapeutic dosing in animal models is a much more challenging hurdle to overcome than is efficacy achieved through prophylactic dosing. Because each animal model is typically developed to act as a surrogate of a certain phase of a disease, it is impossible in this chapter to describe any of them in completeness.

Nevertheless, it is important to realize that small-molecule drugs that advance to clinical trials may display unexpected pharmacological effects *in vivo*. When this occurs, it is imperative to dig deeper into the underlying biology, often at the molecular and pathway biological level. Careful investigations can often clarify the cause(s) of the discrepancy between the original scientific hypothesis for targeting the pathway and the *in vivo* outcome.

In order to fully validate your therapeutic target, it is important to connect the dots from *in vitro* to *in vivo* assays and then into human clinical trials. If you can demonstrate *in vivo* that your drug molecule can bind to and modulate the molecular target and that this translates into inhibition of the biological pathway that is thought to be related to the disease symptoms, pathology, and outcome, you will have a better chance of being able to convince your drug development organization to advance your molecule into human clinical trials.

4.3 Scaling Up

Your initial synthesis of inhibitor molecules may only be on a small scale of around 10-30 mg of material. This amount of compound (assuming MW 500) is sufficient to prepare 2-6 mL of a 10-mM stock of compound. This should supply all of the biochemical and cell-based critical path assays, the *in vitro* ADME testing, and perhaps even some protein X-ray crystallography studies. In order to support *in vivo* dosing for PK measurements in mice or rats, a modest scale-up on the order of 100-1000 mg may be needed, and may also support a small-scale PK/PD study as well. When your molecule progresses to testing in a disease model, you will conduct what may be a challenging large-scale multi-step synthesis and purification of your lead in order to generate anywhere from 1 g to 50 g of your molecule. The total amount needed will depend on the animal size (dog > rat > mouse) and study duration (1-month collagen-induced arthritis study > 1-week multiple sclerosis model study > 1-day neurobehavioral study). For *in vivo* safety studies, where the aim is to dose high levels in order to elicit and observe the toxicity of the molecule, synthesis of hundreds of grams of compound may be necessary. *In vivo* efficacy in additional disease models, safety studies in other animal species, and formulation studies to select the medicinal product may also be warranted once your molecule reaches this stage in the project. Before your precious molecule surpasses its *in vivo* preclinical efficacy and safety hurdles, you should engage process chemists. Large-scale synthetic routes will be needed to scale up the kilograms of API (active pharmaceutical ingredient) needed to support the clinical studies and beyond. It is wise to begin discussing synthetic routes with process chemists as early in the lead evaluation process as is practical.

5   Outlook

In small-molecule drug discovery, it has been said that chemical matter trumps biological rationale. Of course, this does not mean that an in-depth understanding of disease biology is unimportant or unnecessary. What this means is that a small-molecule drug discovery organization can do nothing to modulate even the most compelling therapeutic target unless it has chemical lead matter available to interrogate that target. A deep and rigorous understanding of biological rationale is obviously essential for delivery of valuable medicines. While not covered here, this can include *in vivo* biological animal studies, gene knock-outs and knock-ins, conditional knock-outs, genome-wide association studies, and diagnostic strategies to understand patient stratification. But in the end, it is the quality and functional activity of the chemical matter that will ensure success.

   In this chapter, we have reviewed the basics of *in vitro* and *in vivo* assays used in pre-clinical drug discovery. For a successful career in drug discovery: (1) communicate openly and collaborate with the biochemists, cell biologists, and pharmacologists on your projects to enable you to get the most precise and relevant mechanistic, kinetic, cellular, and *in vivo* efficacy data you need to drive your compound design; (2) make soluble molecules so that their activities can be accurately measured; and (3) use a rigorous testing cascade to fully optimize the physicochemical properties as well as the *in vitro* and *in vivo* activities of your compounds to ensure your designs will yield the next medicine to help patients.

References

1   Smith, D. A.; Di, L.; Kerns, E. H. The Effect of Plasma Protein Binding on *In Vivo* Efficacy: Misconceptions in Drug Discovery, *Nat. Rev. Drug Discov.* **2010**, *9*, 929–939.

2   Zak, M.; Mendonca, R.; Balazs, M.; *et al.* Discovery and Optimization of *C*-2 Methyl Imidazopyrrolopyridines as Potent and Orally Bioavailable JAK1 Inhibitors with Selectivity over JAK2, *J. Med. Chem.* **2012**, *55*, 6176–6193.

3   Johnson, A. R.; Pavlovsky, A. G.; Ortwine, D. F.; Prior, F.; Man, C.-F.; Bornemeier, D. A.; Banotai, C. A.; Mueller, W. T.; McConnell, P.; Yan, C.; Baragi, V.; Lesch, C.; Roark, W. H.; Wilson, M.; Datta, K.; Guzman, R.; Han, H.-K.; Dyer, R. D. Discovery and Characterization of a Novel Inhibitor of Matrix Metalloprotease-13 That Reduces Cartilage Damage *in vivo* without Joint Fibroplasia Side Effects, *J. Biol. Chem.* **2007**, *282*, 27781–27791.

4   Barve, R. A.; Minnerly, J. C.; Weiss, D. J.; Meyer, D. M.; Aguiar, D. J.; Sullivan, P. M.; Weinrich, S. L.; Head, R. D. Transcriptional Profiling And Pathway Analysis Of Monosodium Iodoacetate-Induced Experimental Osteoarthritis In Rats: Relevance To Human Disease, *Osteoarthritis and Cartilage* **2007**, *15*, 1190–1198.

5   Jimenez-Gomez, J.; Osentoski, A.; Woods, J. H. Pharmacological Evaluation Of The Adequacy Of Marble Burying As An Animal Model of Compulsion and/or Anxiety, *Behavioural Pharmacology* **2011**, *22*, 711–713.

6   Sittampalam, G. S.; Weidmer, J.; Auld, S.; et al., eds. *Assay Guidance Manual [Internet]*. Bethesda, MD: Eli Lilly & Company and the National Center for Advancing Translational Sciences; **2004-** ; Available from: http://www.ncbi.nlm.nih.gov/books/NBK53196/.

7   Thorne, N.; Auld, D. S.; Inglese, J. Apparent Activity In High-Throughput Screening: Origins Of Compound-Dependent Assay Interference, *Current Opin. Chem. Biol.* **2010**, *14*, 1–10.

8   Shoichet, B. K. Screening in a spirit haunted world, *Drug Discov. Today* **2006**, *11*, 607–615.

9   Liu, Y.; Beresini, M. H.; Johnson, A.; Mintzer, R.; Shah, K.; Clark, K.; Schmidt, S.; Lewis, C.; Liimatta, M.; Elliott, L. O.; Gustafson, A.; Heise, C. E. Case Studies of Minimizing Nonspecific Inhibitors in HTS Campaigns That Use Assay-Ready Plates, *J. Biomol. Screening* **2012**, *17*, 225–236.

10   McGovern, S. L.; Caselli, E.; Grigorieff, N.; Shoichet, B. K. A Common Mechanism Underlying Promiscuous Inhibitors from Virtual and High-Throughput Screening, *J. Med. Chem.* **2002**, *45*, 1712–1722.

11   Giannetti, A. M.; Koch, B. D.; Browner, M. F. Surface Plasmon Resonance Based Assay for the Detection and Characterization of Promiscuous Inhibitors, *J. Med. Chem.* **2008**, *51*, 574–580.

12   Di, L.; Kerns, E. H. Biological Assay Challenges From Compound Solubility: Strategies For Bioassay Optimization, *Drug Discov. Today* **2006**, *11*, 446–451.

13   Waybright, T. J.; Britt, J. R.; McCloud, T. G. Overcoming Problems of Compound Storage in DMSO: Solvent and Process Alternatives, *J. Biomol. Screening* **2009**, *14*, 708–715.

14   Cheng, X.; Hochlowski, J.; Tang, H.; Hepp, D.; Beckner, C.; Kantor, S.; Schmitt, R. Studies on Repository Compound Stability in DMSO under Various Conditions, *J. Biomol. Screening* **2003**, *8*, 292–304.

15   Ilouga, P. E.; Winkler, D.; Kirchhoff, C.; Schierholz, B.; Wölcke, J. Investigation of 3 Industry-Wide Applied Storage Conditions for Compound Libraries, *J. Biomol. Screening* **2007**, *12*, 21–32.

16   Kozikowski, B. A.; Burt, T. M.; Tirey, D. A.; Williams, L. E.; Kuzmak, B. R.; Stanton, D. T.; Morand, K. L.; Nelson, S. L. The Effect of Room-Temperature Storage on the Stability of Compounds in DMSO, *J. Biomol. Screening* **2003**, *8*, 205–209.

17   Kozikowski, B. A.; Burt, T. M.; Tirey, D. A.; Williams, L. E.; Kuzmak, B. R.; Stanton, D. T.; Morand, K. L.; Nelson, S. L. The Effect of Freeze/Thaw Cycles on the Stability of Compounds in DMSO, *J. Biomol. Screening* **2003**, *8*, 210–215.

18   Ferreira, R. S.; Bryant, C.; Ang, K. K. H.; McKerrow, J. H.; Shoichet, B. K.; Renslo, A. R. Divergent Modes of Enzyme Inhibition in a Homologous Structure-Activity Series, *J. Med. Chem.* **2009**, *52*, 5005-5008.

19   Segel, I. H. *Enzyme Kinetics*, Wiley: New York, **1975**.

20   Copeland, R. A. *Evaluation of Enzyme Inhibitors in Drug Discovery: A Guide for Medicinal Chemists and Pharmacologists*, Wiley, Hoboken, NJ, **2005**.

21   Cheng, Y.-C.; Prusoff, W. H. Relationship Between the Inhibition Constant (K_I) and the Concentration of Inhibitor Which Causes 50 Per Cent Inhibition (I_{50}) of an Enzymatic Reaction, *Biochem. Pharmacol.* **1973**, *22*, 3099–3108.

22   Buxser, S.; Vroegop, S. Calculating the Probability of Detection For Inhibitors In Enzymatic Or Binding Reactions In High-Throughput Screening, *Anal. Biochem.* **2005**, *340*, 1–13.

23   Knight, Z. A.; Shokat, K. M. Features of Selective Kinase Inhibitors, *Chem. Biol.* **2005**, *12*, 621–637.

24   Morrison, J. F. Kinetics of the Reversible Inhibition of Enzyme-Catalyzed Reactions by Tight-Binding Inhibitors, *Biochim. Biophys. Acta* **1969**, *185*, 269–286.

25   Williams, J. W.; Morrison, J. F. The Kinetics of Reversible Tight-Binding Inhibition, *Methods Enzymol.* **1979**, *63*, 437–467.

26   Morrison, J. F.; Walsh, C. T. The Behavior and Significance of Slow-Binding Enzyme Inhibitors, in *Advances in Enzymology and Related Areas of Molecular Biology*, Meister, A. ed. Wiley: New York, **1988**, *61*, 201–301.

27   Swinney, D. C. Biochemical Mechanisms of Drug Action: What Does It Take For Success? *Nat. Rev. Drug Discov.* **2004**, *3*, 801–808.

28   Copeland, R. A. Conformational Adaptation In Drug-Target Interactions And Residence Time, *Future Med. Chem.* **2011**, *3*, 1491–1501.

29   Tummino, P. J; Copeland, R. A. Residence Time of Receptor-Ligand Complexes and Its Effect on Biological Function, *Biochemitry* **2008**, *47*, 5481–5492.

30   Kapur, S.; Seeman, P. Antipsychotic Agents Differ in How Fast They Come Off The Dopamine D_2 Receptors. Implications for , *J. Psychiatry Neurosci.* **2000**, *25*, 161-166.

31   Massey, V.; Komai, H.; Palmer, G.; Elion, G. On the Mechanism of Inactivation of Xanthine Oxidase by Allopurinol and Other Pyrazolo[3,4-*d*]pyrimidines, *J. Biol. Chem.* **1970**, *245*, 2837-2844.

32   Wood, E. R.; Truesdale, A. T.; McDonald, O. B.; Yuan, D.; Hassell, A.; Dickerson, S. H.; Ellis, B.; Pennisi, C.; Horne, E.; Lackey, K.; Alligood, K. J.; Rusnak, D. W.; Gilmer, T. M.; Shewchuk, L. A Unique Structure for Epidermal Growth Factor Receptor Bound to GW572016 (Lapatinib): Relationships among Protein Conformation, Inhibitor Off-Rate, and Receptor Activity in Tumor Cells, *Cancer Res.* **2004**, *64*, 6652-6659.

33   Dierynck, I.; De Wit, M.; Gustin, E.; Keuleers, I.; Vandersmissen, J.; Hallenberger, S.; Hertogs, K. Binding Kinetics of Darunavir To Human Immunodeficiency Virus Type 1 Protease Explain The Potent Antiviral Activity And High Genetic Barrier, *J. Virol.* **2007**, *81*, 13845-13851.

34   Bull, H. G. G.-C. M.; Andersson, S.; Baginsky, W. F.; Chan, H. K.; Ellsworth, D. E.; Miller, R. R.; Stearns, R. A.; Bakshi, R. K.; Rasmusson, G. H.; Tolman, R. L.; Myers, R. W.; Kozarich, J. W.; Harris, G. S. Mechanism-Based Inhibition of Human Steroid 5α-Reductase by Finasteride: Enzyme-Catalyzed Formation of NADP-Dihydrofinasteride, a Potent Bisubstrate Analog Inhibitor, *J. Am. Chem. Soc.* **1996**, *118*, 2359-2365.

35   Furman, P. A.; St. Clair, M. H.; Spector, T. Acyclovir Triphosphate Is a Suicide Inactivator of the Herpes Simplex Virus DNA Polymerase, *J. Biol. Chem.* **1984**, *259*, 9575-9579.

36   Zanella, F.; Lorens, J. B.; Link, W. High Content Screening: Seeing Is Believing, *Trends Biotechnol.* **2010**, *28*, 237–245.

37   Lu, S.; Jessen, B.; Strock, C.; Will, Y. The Contribution of Physicochemical Properties To Multiple *In Vitro* Cytotoxicity Endpoints, *Toxicology in vitro* **2012**, *26*, 613–620.

38   Beis, I.; Newsholme, E. A. The Contents of Adenosine Nucleotides, Phosphagens And Some Glycolytic Intermediates In Resting Muscles From Vertebrates And Invertebrates, *Biochem. J.* **1975**, *1*, 23-32.

39   http://www.nlm.nih.gov/medlineplus/ency/article/003480.htm, Albumin – serum: MedlinePlus Medical Encyclopedia, accessed 2012-10-08.

40   Colombo, S.; Buclin, T.; Décosterd, L. A.; Telenti, A.; Furrer, H.; Lee, B. L.; Biollaz, J.; Eap, C. B. Orsomucoid (Alpha1-Acid Glycoprotein) Plasma Concentration And Genetic Variants: Effects On Human Immunodeficiency Virus Protease Inhibitor Clearance And Cellular Accumulation, *Clin. Pharmacol. Ther.* **2006**, *80*, 307-318.

4

Drug Metabolism and Pharmacokinetics
In Drug Discovery

Kimberley Lentz, Joseph Raybon, and Michael W. Sinz

1 Introduction

Evaluation and integration of druglike properties are essential requirements of the drug discovery process. Druglike properties are the intrinsic physicochemical properties of a molecule that impart their effects on the absorption, metabolism, distribution, and elimination (ADME) of a potential new drug candidate. Prior to the 1990s, drug discovery was the realm of chemistry and biology, which were charged with synthesizing and testing molecules for activity and then progressing them into drug development. It was not until the drug development stage that a single molecule was tested for druglike properties, most commonly with *in vivo* animal models, such as mice, rats, dogs, and monkeys. Often these drugs would fail at this stage or later in clinical development due to poor ADME properties that could not be circumvented during the development process. During the 1990s, *in vitro* tools and LC-MS became more commercially available and pharmaceutical companies began testing larger numbers of compounds earlier. By the year 2000, with the advent of molecular biology, robotics, liquid handlers, and high-throughput screening (HTS) assays, ADME discovery screening became essential and commonplace in the pharmaceutical industry. It was at this time that chemistry, biology, and ADME scientists began working in collaboration to bring forward potent compounds with druglike properties. As a result, drug failures due to ADME issues significantly decreased. Winston Churchill once said *"Those that fail to learn from history, are doomed to repeat it."* The same can be said if we choose to ignore the physico-chemical and ADME properties of a molecule and focus solely on optimizing biological potency. Extremely potent compounds rarely make good drugs and often a compromise needs to be made by balancing potency with ADME properties.

     Common properties of molecules that can have a significant impact on ADME fall into two general categories: (1) structural features or physicochemical properties and (2) ADME-toxicology properties. Often the former has an influence on the latter as a molecule's characteristics can have an impact on ADME-toxicology. For example, solubility or LogP can affect drug absorption and tissue distribution or the presence of functional groups that can undergo reactive biotransformation which may lead to toxicity or drug-drug interactions (DDIs). Physicochemical and structural features which influence druglike properties include hydrogen bonding (donors/acceptors), polar surface area, LogP, molecular weight, solubility, and reactive functional groups or functional groups that can form reactive centers as a result of metabolism (e.g., Michael acceptors or formation of aromatic amines, respectively).

Solubility is a significant issue when it comes to limiting drug-like properties and prosecuting a speedy and successful drug discovery campaign. Solubility can affect two key aspects of drug discovery: reliability of *in vitro* assay results and drug absorption. Drug candidates are often tested *in vitro* at concentrations as high as 10–40 μM and compounds with poor solubility will often precipitate at or below these concentrations. Precipitation of drug in these *in vitro* assays can lead to misinterpretation of the results, such as false negatives, as well aswreak havoc with structure-activity relationships (SAR) which are often incorrect or misleading. The addition of higher concentrations of organic solvents (e.g., DMSO) or protein can often help increase the solubility of such compounds, but often at the expense of inactivating or inhibiting the enzymes in the assay hence invalidating the results. Poorly soluble drugs also have poor dissolution properties leading to low and variable drug exposure which also leads to variable efficacy in preclinical and human efficacy studies. Variable exposure and efficacy due to poor solubility are drug properties to be avoided as these drugs are less likely to be prescribed and will have lower patient compliance. Often the introduction of an ionizable group or atom, such as a basic nitrogen, can improve solubility sufficiently to avoid these issues and still maintain good potency. If more dramatic increases in solubility are necessary, then consideration of a prodrug with a highly soluble functional group (e.g., phosphate) should be considered.

The concept of predicting good drug-like properties from physico-chemical assessments came to the forefront with Lipinski's Rule of Five which employs four parameters (all multiples of five) that have been shown to lead to good drug-like properties for solubility and permeability.[1] The rule of five states that poor absorption or permeation is more likely when there are more than 5 H-bond donors, 10 H-bond acceptors, the molecular weight is greater than 500, and the calculated LogP is greater than 5. Since this highlighted publication, many researchers have developed similar rules using a variety of other factors which are often represented visually for ease of translation and interpretation. Figure 1 illustrates one such prediction of bioavailability using a radar plot with points of: LogP, molecular weight, polar surface area, water solubility, and number of rotatable bonds.[2] Compounds which fall inside the shaded area conform to drug-like properties while those that fall outside the shaded region exceed the limits of drug-like properties.

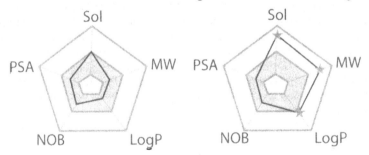

Figure 1. Radar plots of two compounds using various drug-like parameters necessary for good bioavailability [molecular weight (MW), polar surface area (PSA), water solubility (WS), and number of rotatable bonds (NOB)]. The compound on the left falls within the acceptable parameter ranges, while the compound on the right could have issues with solubility, molecular weight, and/or LogP.

Common ADME properties that influence drug-like properties include such characteristics as permeability, protein or tissue binding, drug concentration at the target organ or enzyme site, pathways of elimination (metabolic, biliary, urinary), enzymes involved in metabolism, interactions with uptake or efflux transporters, half-life, bioavailability, and the potential for drug-drug interactions (enzyme inhibition or induction). Many of these properties can be evaluated in a drug discovery setting through *in vitro* screening (solubility, permeability, DDI potential, transporter involvement, enzymes involved in metabolism, plasma protein binding, etc.) while others require *in vivo* testing in preclinical animal models (defining elimination pathways, drug distribution, clearance, half-life, bioavailability, etc.). Whereas the ultimate goal is to develop new and better drugs to treat human diseases, the challenge becomes extrapolating these human *in vitro* and animal *in vivo* data into predictions of human outcome or drug performance. The remaining sections on drug metabolism and pharmacokinetics will describe in greater detail the essential elements of these disciplines and how they relate to good or poor ADME drug-like properties, as well as the predictability of extrapolating preclinical *in vitro* and *in vivo* data to humans.

2   Drug Metabolism

Drug metabolism is a defense mechanism to remove foreign compounds or xenobiotics that have made their way into the body. Metabolism is considered a detoxification pathway by which the molecules are chemically modified to make them more polar thereby facilitating elimination from the body and renders them pharmacologically inactive. This is the general rule with xenobiotics, however not all biotransformation reactions yield more polar or inactive metabolites. For example, *N*-methylation tends to make compounds less soluble and more difficult to eliminate and the glucuronidation of morphine results in a biologically active conjugated morphine metabolite more potent than the parent drug.[3] Although these tend to be the exceptions and not the rule, they should not be forgotten in an overall assessment of possible metabolism scenarios. Avoiding unwanted toxicity is a given in all situations, and although active metabolites are generally unwanted they can sometimes provide additional pharmacological activity in certain situations that may be beneficial. For example, a metabolite which is formed at the site of action or an active metabolite formed from a rapidly metabolized parent drug (e.g., prodrug). An example is Lipitor (atorvastatin) which is metabolized to two aromatic hydroxylated metabolites in the liver: both metabolites are pharmacologically active and formed in the liver which is the target organ for cholesterol lowering.

Understanding the metabolism of newly developed drugs is also important because of the central role metabolism plays in determining the developability of new drug candidates. For example, the rate of drug metabolism in part determines the bioavailability, clearance, and half-life of a drug and combining these parameters defines how much and how often a drug needs to be administered. Both these factors (dose and frequency) can be significant hurdles to new drugs especially if either is too high making the drug unmanageable to develop or market.

2.1   Drug Metabolizing Enzymes

Drug metabolism occurs everywhere within the body to differing extents; however the greatest metabolism occurs in those organs involved in elimination and those to which the drug is first exposed. Therefore the most common drug metabolizing organs are: liver, GI tract, and kidney, as well as the plasma/blood compartment (which is involved in hydrolytic and reductive reactions). Other organs with lower metabolism capacity are brain, skin, lung, and nasal mucosa.[4]

     Metabolism reactions are broadly categorized into Phase I and Phase II reactions, whereby Phase I reactions are generally considered functionalization reactions (oxidation, reduction, or hydrolysis reactions) and Phase II reactions are considered conjugation reactions. The family of Phase I enzymes are comprised of the cytochrome P450 (CYP) enzymes, the flavin monooxygenases, the molybdenum hydroxylases (xanthine oxidase and aldehyde oxidase), the monoamine oxidases (MAO-A and MAO-B), dihydrodiol dehydrogenase, and various alcohol and aldehyde dehydrogenases. The family of Phase II enzymes includes glucuronosyltransferases, sulfotransferases, methyl transferases, actetyltransferases, glutathione transferases, and amino acid transferases.[5] Out of all these metabolic processes the two most predominant types of metabolism involved in the biotransformation of drugs tend to be oxidative metabolism by CYP enzymes and glucuronidation. Figure 2a and 2b illustrates the sites and functional groups that undergo metabolism by Phase I enzymes and Figure 3a and 3b illustrates the variety of Phase II conjugative reactions.

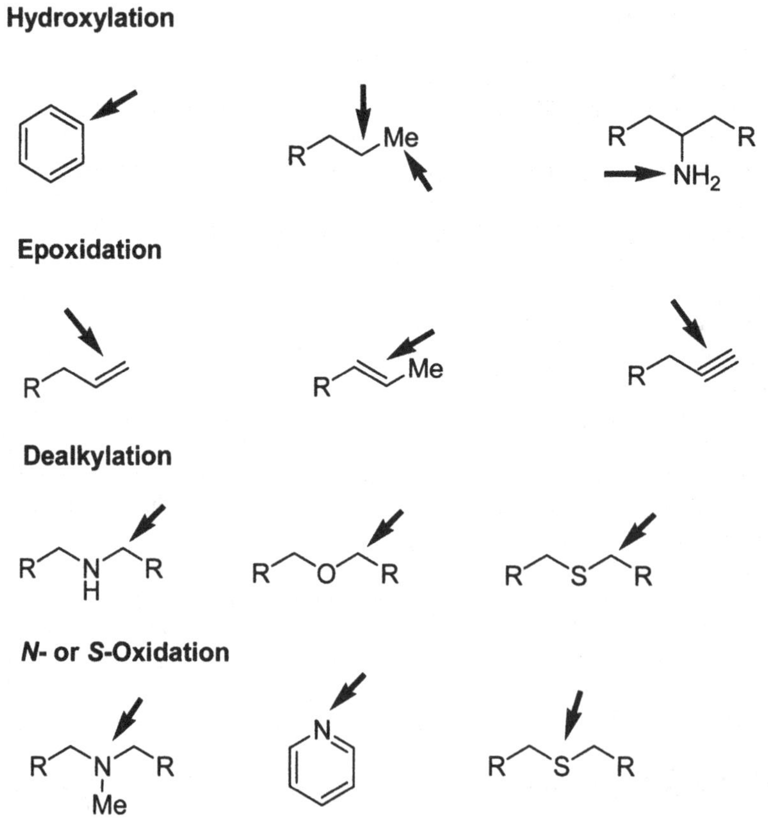

Hydroxylation

Epoxidation

Dealkylation

N- or S-Oxidation

Figure 2a. Examples of Phase I Oxidative, Reductive, and Hydrolytic Metabolic Reactions

Oxidation
(alcohols and ketones)

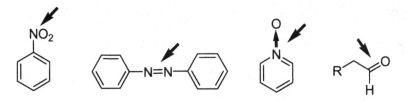

Reductions
(ketones, double bonds, nitro and azo compounds, sulfoxides and
N-oxides, disulfides, quinone, dehalogenation)

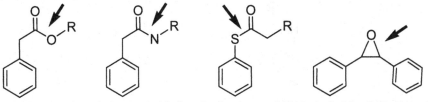

Hydrolytic Reactions
(esters, amides, thioesters, epoxides, and peptides)

Figure 2b. Examples of Phase I Oxidative, Reductive, and Hydrolytic Metabolic Reactions

Glucuronidation
(primary, secondary, and tertiary alcohols; phenols; hydroxylamines;
aromatic and aliphatic carboxyls; carbamic acids, amino and sulfhydryl groups)

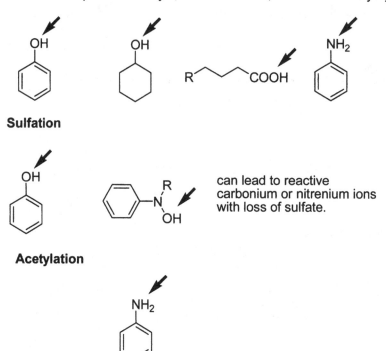

Sulfation

can lead to reactive
carbonium or nitrenium ions
with loss of sulfate.

Acetylation

Methylation
(primary, secondary, and tertiary amines;
aromatic amines; phenols; and aromatic sulfhydryl groups)

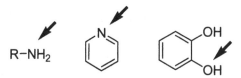

Figure 3a. Examples of Phase II Conjugative Metabolic Reactions

Glutathione Conjugation
(A, Displacement to an electron-withdrawing group such as halogen or nitro group;
B, Attack of an arene oxide intermediate; or addition to an electron-deficient
double bond)

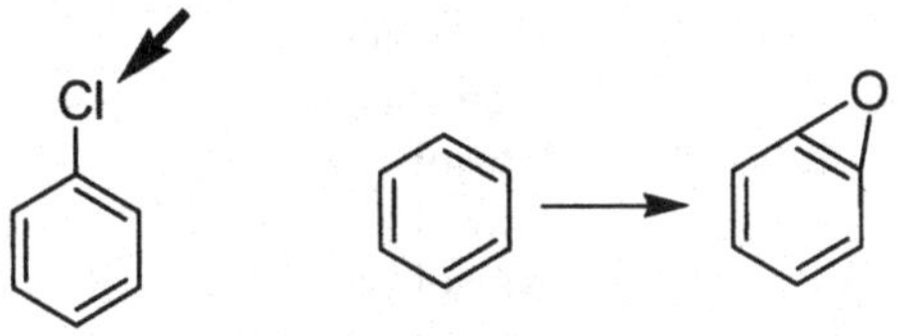

Amino Acid Conjugation
(Amino acids involved in conjugation include glycine, glutamine, and taurine)

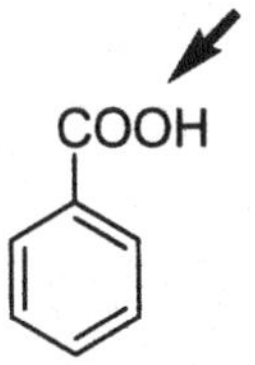

Figure 3b. Examples of Phase II Conjugative Metabolic Reactions

2.2   *In vitro and In vivo Models to Evaluate Drug Metabolism*

Although metabolism occurs throughout the body in many organs, the *in vitro* models employed in drug discovery and preclinical development tend to be single enzymes, subcellular fractions, or cellular models derived from liver tissue. The single enzyme system is a useful tool to study enzymatic activity in a controlled environment which can be easily manipulated. This model is highly amenable to high throughput screening assays as the enzymes can be stored at –80 °C for long periods of time and taken out for use on a routine basis. They are also durable and remain active even with manipulation by liquid handlers and robotic platforms. The most commonly employed single enzyme family is the cytochrome P450 (CYP) family of isozymes which have been used to evaluate: metabolic stability, drug-drug interaction potential (inhibition), reaction phenotyping, the effect of allelic variants on metabolism or elimination, the potential to undergo bioactivation or covalent binding, and as bioreactors to generate larger quantities of metabolites for identification purposes.[6-9]   Many of the other individually expressed drug metabolizing enzymes (e.g., glucuronyltransferases, sulfotransferases, flavin monooxygenases, and *N*-acetyltransferases) can also be used for these purposes; however they are used less often since these enzymes play a less significant role in drug metabolism than the CYPs.

By use of tissue homogenization and differential centrifugation it is possible to obtain various subcellular liver fractions, such as S-9, cytosol, and microsomes. The S-9 fraction is essentially a liver homogenate without nuclei, mitochondria, or cell membranes and contains all the enzymes found in the cytosolic and microsomal fractions. Further centrifugation of the S-9 fraction yields the cytosol and microsomal fractions each with different sets of drug metabolizing enzymes, Table 1.

Table 1. Subcellular Location of Drug Metabolizing Enzymes

Reaction	Enzymes	Subcellular Location or Tissue
Phase I		
Oxidation	cytochrome P450	microsomes
	flavin monooxygenase	microsomes
	aldehyde oxidase	cytosol
	xanthine oxidase	cytosol
	alchohol dehydrogenase	cytosol
	aldehyde dehydrogenase	cytosol
	monoamine oxidase	mitochondria
	dihydrodiol dehydrogenase	cytosol
Hydrolysis	esterases/amidases	microsomes, cytosol, plasma
	epoxide hydrolase	microsomes and cytosol
Reduction	sulfur reductase	cytosol
	quinone reductase	microsomes and cytosol
	reductive dehalogenation	microsomes
	azo/nitro reductase	microsomes, cytosol, gut microflora
	carbonyl reductase	microsomes, cytosol, blood
Phase II		
Conjugation	glucuronosyltransferase	microsomes
	sulfotransferase	cytosol
	methyltransferase	microsomes and cytosol
	N-acetyltransferase	cytosol
	glutathione conjugation	microsomes & cytosol
	amino acid conjugation	microsomes

These subcellular fractions can be stored at −80 °C for long periods of time and as such serve as a convenient source of enzyme for drug metabolism studies. Due to their ready availability, subcellular fractions are used in a variety of drug metabolism studies, such as metabolic stability, enzyme inhibition, reaction phenotyping, bioactivation and bioreactor studies, and metabolite identification.[6,9–11] The microsomal fraction containing the smooth endoplasmic reticulum is the most commonly employed *in vitro* model because most drugs are metabolized by the CYPs or glucuronidation enzymes both of which are found in the microsomal fraction. The cytosolic fraction is used less frequently, however when the major enzyme involved in metabolism of a drug is located in the cytosol, this model becomes essential (e.g., sulfation or aldehyde oxidase). The S9 fraction has utility when complex metabolic pathways involving multiple metabolic reactions need to be studied simultaneously.

Primary animal or human hepatocytes isolated from liver are the predominate cell-based model to evaluate drug metabolism.[4] Like the aforementioned subcellular fractions, hepatocytes can be freshly isolated or cryopreserved and stored for long periods of time which makes them highly useful for routine assays to assess metabolic stability, metabolite identification, and induction. Depending on the study type, the hepatocytes can be used as a cell suspension (generally for 2–3 h) or in cell culture (1–5 days).[4] Hepatocytes in suspension represent a fully functional unit of drug metabolism from the liver containing *in vivo*-like concentrations of enzymes and cofactors and the ability to perform complex and multi-step metabolic reactions. Therefore, hepatocytes in suspension are commonly used to determine metabolic stability or predict human metabolic clearance, and compare metabolic pathways or metabolites across multiple species.[4,12–15] Hepatocytes in culture are predominately used to assess enzyme induction (transcriptional activation) and the potential for drug-drug interactions and less often to evaluate metabolic clearance due to the changing expression levels of drug metabolizing enzymes in culture.[16,17]

   In vivo metabolism and disposition of drug and metabolites in preclinical animal species is equally important in understanding the overall potential of a new drug to succeed. The most commonly sampled biological matrices from *in vivo* drug metabolism studies include plasma, urine and bile, as well as feces in some situations. In such studies, these matrices can be used for metabolite identification and quantitation purposes. Plasma and urine sampling from *in vivo* animal studies is rather straightforward, however bile collection requires bile duct cannulated animals which is common in rats, but more challenging in mice, dogs, and monkeys. Many drugs are eliminated in bile as parent drug and/or metabolites and without the use of bile duct cannulated animals it would be difficult to assess the complete picture of a drug's elimination, metabolic pathways, or extent of metabolism.

2.3   *Metabolic Stability*

In drug discovery, metabolic stability is measured by quantitating the disappearance of parent drug from either liver microsomes or hepatocytes across species. These estimations of metabolic clearance are then compared to actual *in vivo* clearance values from preclinical animal pharmacokinetic studies. If there is a good correlation between the *in vitro* prediction and the total body clearance determined *in vivo*, then the *in vitro* metabolic system (microsomes and/or hepatocytes) can be used for screening larger numbers of discovery candidates and developing SAR. It is extremely important to develop this *in vitro–in vivo* correlation early in a program to provide a level of confidence that the *in vitro* metabolic stability screen is providing useful information to guide further synthesis of improved analogs. If the hepatic metabolic clearance is the predominant elimination pathway *in vivo*, then there is generally a good correlation, however if hepatic metabolic clearance is not the predominant elimination pathway then there may be a poor correlation and the screen should not be used to guide SAR. Elimination pathways that can confound this correlation include: significant extrahepatic metabolism or elimination by non-metabolic pathways, such as direct biliary or renal elimination of parent drug.

   When metabolic stability is an issue there are several different approaches commonly employed to increase metabolic stability (i.e., lower *in vivo* clearance). Approaches often used when Phase I metabolic reactions are an issue include: 1) reducing the lipophilicity of the drug; 2) blocking a site of hydroxylation by replacing the hydrogen(s) with fluorine(s); 3) blocking a site of metabolism through cyclization; 4) eliminating or replacing a functional group with an isostere less susceptible to metabolism; or 5) changing the chirality near or at the site of metabolism. When Phase II conjugation is the predominant metabolic pathway, then the following modifications can help improve metabolic stability: 1) mask the site of conjugation; 2) remove the site of conjugation; 3) introduce electron withdrawing groups near the site of conjugation; or 4) create steric hindrance near the site of conjugation.[18]

   As part of, or separate from the metabolic stability experiments in microsomes or hepatocytes, the appropriate use of metabolite identification in drug discovery can provide valuable information.   Metabolite identification is important in drug discovery in order to understand the pharmacological or toxicological properties of the metabolites.   As discussed previously, metabolism is a detoxification process that often produces inactive metabolites. However, this is not always the situation and it is important to understand if metabolites have the potential to be toxic or active and compare these metabolic profiles across species including human, both *in vitro* and *in vivo*, in order to assess the potential pharmacological or toxicological consequences.

2.4   *Reaction Phenotyping*

Pharmaceutical companies routinely determine the major enzymes involved in the metabolism of new drugs *in vitro* to predict potential drug interactions that may occur *in vivo*, such as those due to metabolism by polymorphic drug metabolizing enzymes   (e.g., CYP2D6 or *N*-acetyltransferases) or co-administration with potent inhibitors or inducers of CYP450 enzymes.[6] Reaction phenotyping involves the use of recombinant enzymes, liver microsomes, human hepatocytes and clinical studies with polymorphic populations. There are three *in vitro* techniques that can address reaction phenotyping and many researchers will utilize at least two of them if not all three.[5] The first technique utilizes recombinantly expressed single enzymes. In this situation, the drug is incubated with individual enzymes and turnover of parent drug or formation of metabolite(s) can be measured. Those enzymes which demonstrate turnover are considered to be involved in the metabolism of the compound and through appropriate scaling of these turnover numbers the major enzymes involved in metabolism can be elucidated.   A second common approach utilizes human liver microsomes incubated with selective chemical inhibitors or inhibitory antibodies for individual CYP450 enzymes.   These chemical inhibitors or inhibitory antibodies selectively knock-out or eliminate the activity of a single CYP450 enzyme and when compared to control (no inhibitor) incubations, the relative contribution of each CYP450 can be determined.   The third *in vitro* technique termed 'correlation analysis' correlates the rate of substrate turnover to the rate of turnover of an isoform selective probe substrate. For example, the rate of metabolism (drug candidate) correlates with the CYP3A4-mediated formation of 6-β hydroxytestosterone, which is only mediated by CYP3A4. If the rates of metabolism demonstrate a good correlation, then it is presumed that CYP3A4 is involved in the metabolism of the compound. This is then done with selective probe reactions for each of the other CYP enzymes whereby good correlations indicate enzymes involved in metabolism while poor correlations indicate enzymes not involved in metabolism of the compound.

2.5   *Drug-Drug Interactions*

There are two principle types of DDIs: inhibition or induction of enzyme activity.   Enzyme inhibition leads to an increase in drug plasma exposure due to a lack of drug metabolism (or elimination) whereas enzyme induction leads to a decrease in drug exposure due to enhanced metabolism (or elimination).   These two types of DDIs occur through distinctly different mechanisms which will be described in greater detail along with the model systems employed to predict human drug-drug interactions for each.

2.5.1 Enzyme Inhibition

Enzyme inhibition can be categorized into two types, reversible and irreversible inhibition. Competitive reversible inhibition is the most common type of inhibition and DDI found with drug metabolizing enzymes. Simply put, competitive inhibition is the competition of a probe substrate and the drug discovery candidate for the same active site on the enzyme. Reversible CYP inhibition is routinely evaluated in early drug discovery by means of recombinantly expressed CYP enzymes and fluorescent probes both of which are amenable to high throughput screening. Assays such as this are useful in assessing CYP inhibition potential (IC_{50} determination) or rank ordering compounds or chemotypes and for conducting SAR to eliminate a CYP inhibition liability. In order to determine a more accurate DDI potential, labs will perform similar inhibition studies with human liver microsomes (IC_{50} or K_i determinations).[19] By determining the ratio of the inhibition parameter (IC_{50} or K_i) and the efficacious plasma concentration [I], a general idea of the likelihood of a DDI can be determined as follows: [I] / K_i > 1, implies a drug interaction is likely; 1.0 > [I] / K_i > 0.1, implies a drug interaction is possible; [I] / K_i < 0.1, implies a drug interaction is unlikely.

The second type of inhibitory drug interaction known as mechanism-based or metabolism-based inhibition (MBI) is a result of inactivation of a drug metabolizing enzyme by a metabolite or metabolic intermediate that forms an adduct with the enzyme or forms a metabolite-intermediate-complex (MI-complex) with the enzyme. Certain functional groups on drug molecules have a propensity to undergo reactive metabolite formation and cause irreversible inhibition. These functional groups include: acetylenes, alkenes, organosulfo-compounds, arlyamines, tertiary amines, cyclopropyl amines, hydrazines, furans, thiophenes, dihaloalkanes, and methylene dioxyphenyl-containing compounds.[20] Whenever such structures are present in a molecule, rigorous testing should be conducted to evaluate the potential for reactive metabolite formation that may lead to irreversible inhibition or covalent binding to other macromolecules such as proteins or nucleic acids.

Drugs that cause irreversible inhibition distinguish themselves from ordinary reversible inhibition by several characteristics: 1) they may display a delayed onset of inhibition; 2) their inhibition properties are typically greater than what would be expected from ordinary reversible inhibition parameters (IC_{50} or K_i); and 3) the inhibition effects persist after the inhibitor has been eliminated due to the fact that enzyme has been destroyed and new enzyme must be synthesized to regain basal enzyme levels. The mechanistic differences between reversible and irreversible inhibition require additional experimental methods to appropriately characterize the mechanism and potential drug-drug interaction from irreversible inhibition.[21] Examples of irreversible inhibition (along with experimental methods) can be found in the references for furafylline and lapatinib.[22,23]

2.5.2 Enzyme Induction

The most common mechanism of CYP induction is transcriptional gene activation. For drug metabolizing enzymes, transcriptional activation is mediated by transcription factors, such as the aromatic hydrocarbon receptor (AhR), constitutive androstane receptor (CAR), and pregnane X receptor (PXR), also known as the steroid X receptor (SXR).[24,25] The general concept for nuclear receptor signaling, as exemplified by PXR, follows that in the absence of a ligand (drug), the nuclear receptor is inactive. Drug binding to the ligand-binding domain (LBD) of the nuclear receptor induces conformational changes that lead to activation of the receptor and subsequent transcriptional activation of target genes. PXR-derived nuclear hormone receptor models (ligand binding and cell-based PXR transactivation assays) are the most common high throughput assays to evaluate enzyme induction due to the simplicity of the assay and the importance of PXR target genes, such as CYP3A4, CYP2B6, and CYP2Cs in precipitating drug interactions.[26] Although cell-based transcription activation assays exist for CAR (CYP2B6) and AhR (CYP1A) these are used less frequently as screening assays and more often to address mechanistic questions.

Beyond the receptor-based screens, cultured primary human hepatocytes are the most widely accepted model for assessing the potential of drug candidates to induce human CYP enzymes, primarily CYP1A2, CYP2B6, and CYP3A4. The hepatocyte model is able to best predict the outcome of enzyme induction and drug interactions due to the ability to simultaneously capture multiple nuclear receptor-mediated pathways rather than a single pathway as with the transcriptional activation models above. Primary human hepatocyte culture systems have been shown to effectively model human *in vivo* induction responses and are recognized by regulatory agencies as an effective tool for assessing induction potential.[27,28] The enzyme induction data from *in vitro* methods are known to correlate well with clinical observations, provided the *in vitro* experiments are performed at pharmacologically relevant concentrations of drug (i.e., efficacious drug concentration).

3   Pharmacokinetic Fundamentals

Pharmacokinetics is the study of the processes involved with the overall absorption, distribution, metabolism and excretion of drugs or drug products. In its simplest definition, pharmacokinetics is what the body does to the drug once it is administered.

The route in which drugs are administered will have a profound impact on how much and how fast a drug reaches the blood or systemic circulation. Some routes involve direct access to the systemic circulation (intravenous or IV), while other routes require that the drug cross a physiological membrane such as the intestine (oral or PO), skin (topical), lung (inhalation) or by diffusion from muscle (intramuscular or IM), peritoneum (intraperitoneal or IP) or under the skin (subcutaneous or SC). Once drug reaches the blood, it is available to rapidly distribute between blood cells and plasma proteins, as well as other tissues. Drugs which are more lipid soluble (high LogP) can undergo extravascular distribution more readily and accumulate into the intracellular fluid of tissues. When a drug distributes into the liver, it is available to be metabolized. However, in addition to drug metabolism, some drugs can be directly excreted by organs such as the kidney (into the urine) and the liver (through bile ducts into the gall bladder). The overall pathway of a compound from administration to elimination is shown in Figure 4.

Pharmacokinetic (PK) studies are quantitative in nature and provide a snapshot of the time course of drug in the body. For PK studies in drug discovery, drug is first dissolved in a pharmaceutically acceptable vehicle (which can be aqueous based or non-aqueous based) and administered to a non-clinical species usually via IV and PO. The blood sampling times and intervals between samples should be adequate to define the peak plasma concentration of drug (C_{max}) and the extent of drug absorption (AUC). The sampling time points should also be sufficiently comprehensive and extend long enough to capture complete elimination of the drug (> 5–7 elimination half-lives) beyond the time at which peak plasma concentration levels are measured.

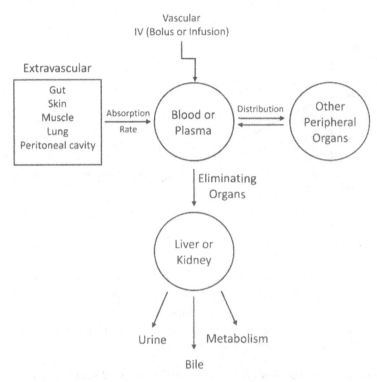

Figure 4. Schematic representation of common ADME pathways for a hypothetical drug

Plasma is obtained from blood samples by centrifugation, then extracted and quantitated for the concentration of drug at each time-point using LC/MS/MS. Plasma concentration vs. time profiles are plotted and data are used to calculate several PK parameters which will be discussed in further detail. Based on the shape of the PK profile, knowledge of various ADME processes can be inferred. Persistence of drug levels in the blood or plasma, coupled with knowledge of the drug's pharmacologic activity, can help determine the dose and or dosing frequency (e.g., once/day; twice/day, etc.). Moreover, PK data can be used in the optimization of new drug candidates.

Figure 5 shows a hypothetical PK profile for a drug administered both IV and PO. When compound is administered IV, it is available to rapidly distribute throughout the body, without dependence on the process of drug absorption. Once in the systemic circulation, it can be distributed or eliminated. Each of these processes can be described by mathematical expressions comprised of different rate constants, volume and mass terms. It is the interplay of these variables that determine the plasma concentration at any given time-point. When compound is administered orally, it must first be absorbed prior to reaching the systemic circulation. A prerequisite for good oral absorption requires adequate solubility and permeability across the gastrointestinal mucosa. Drug concentrations will start off low and increase with time until reaching a concentration maximum (C_{max}). Beyond C_{max}, drug levels will typically decrease. The terminal concentration at the end of the dosing interval is termed minimal concentration (C_{min}). For rapidly absorbed compounds, the period after C_{max} reflects its distribution and elimination phase.

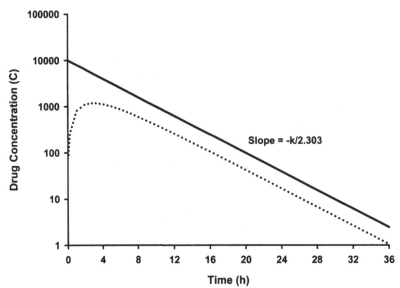

Figure 5. Blood/plasma concentration versus time profiles for a hypothetical drug after either intravenous (solid line) or oral administration (dotted line)

3.1   Noncompartmental Analysis (NCA) of Pharmacokinetic Data

Several PK parameters can be derived from the concentration vs. time profile for a given drug through the use of non-compartmental analysis (NCA). This technique is a model independent approach allowing for resolution of the primary PK parameters of interest. This method has also been termed SHAM analysis which represents the determination of the following metrics 1) $\underline{S}$lope for approximation of the elimination rate (k_{el}) and half-life ($T_{1/2}$); 2) $\underline{H}$eight for assessing the peak drug concentration (C_{max}) and its corresponding time (T_{max}); 3) $\underline{A}$rea for quantitation of the area under the PK curve (AUC) giving a measure of overall exposure and; 4) $\underline{M}$oment which is the area under the curve from the plot of plasma concentration × time vs. time (commonly referred as the area under the first moment curve, AUMC) allowing for quantitation of parameters like mean residence time (MRT) and various distribution volumes which will be discussed later in this section. The areas under the curve (AUC and AUMC) are calculated by a summation of the total area of a series of trapezoids drawn under the curve termed as the trapezoidal rule, Figure 6.

The $T_{1/2}$ describes how long it takes to reduce plasma drug levels by 50%. When plotting plasma concentration vs. time on a semi-logarithmic plot, $T_{1/2}$ will be estimated using the terminal phase of the PK profile as it is a first order process.

$$T_{\frac{1}{2}} = \frac{0.693}{k_{el}}$$

Equation 1

where k_{el} is the elimination rate constant and can be calculated from the slope of the terminal linear portion of the PK curve.

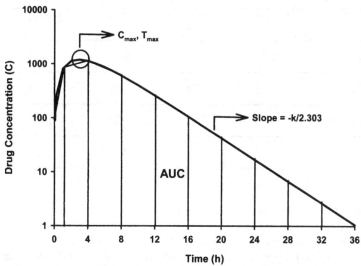

Figure 6. Example of primary PK parameter estimations using non-compartmental analysis based upon observed blood/plasma concentration versus time data for a hypothetical drug after oral administration.

When plasma drug levels are analyzed, a concentration of drug is determined. In order to understand the amount of drug in the body, plasma concentrations of drug need to be normalized by a factor known as the apparent volume of distribution (V_D). The V_D is the proportionality constant that relates the total amount of drug in the body to the measured concentrations in the biological matrix that was sampled (e.g., plasma). For a drug administered IV, the amount of drug in the body immediately after administration is the actual dose administered. The V_D is calculated by the following relationship,

$$Dose = C_0 \times V_D$$
Equation 2

where Dose is the actual dose administered and C_0 is the plasma concentration of drug extrapolated back to time = 0.

The actual value of the volume of distribution does not have a physiologic meaning in terms of understanding exactly which fluids or tissues the drug occupies, thus the term "apparent" volume of distribution is typically used. There is however the finite limit that volume of distribution cannot be any smaller than the physiological plasma or blood volume. For example, for drug A, if plasma drug levels after IV administration are low, the V_D will be large. Conversely for drug B, if plasma drug levels are high, V_D can be quite small, assuming dose is the same for both compounds. The magnitude of the V_D can be an indication of the amount of drug outside the blood. Thus, the higher the apparent V_D, the greater the concentration of drug distributed in peripheral tissues or other extravascular compartments. The physiologic volumes of several compartments are shown in Table 2, and are useful references for placing the calculated V_D into physiological context.

Table 2. Physiologic Volumes of Body Fluids across Nonclinical Species and Human

	Mouse (0.02 kg)	Rat (0.25kg)	Dog (10 kg)	Monkey (5 kg)	Human (70 kg)
Total Body Water (mL)	14.5	167	6036	3465	42000
Intracellular Fluid (mL)	-	92.8	3276	2425	23800
Extracellular Fluid (mL)	-	74.2	2760	1040	18200
Plasma Volume (mL)	1.0	7.8	515	224	3000

References[29,30]

Clearance (Cl) of a drug is dependent on the ability of organs such as the kidney or liver to either metabolize or excrete the drug. In units of volume/time, Cl represents the volume of blood cleared of drug through one of these organs in a given time period. When examining a PK profile, a measure of total body clearance can be calculated. Total body clearance is the cumulative sum of each individual clearance pathway that contributes to total drug elimination. Total body clearance can be calculated after an IV dose by the following equation:

$Cl = Dose/AUC$ Equation 3

Determination of total body Cl is more complicated after extravascular administration since not all drugs are completely bioavailable, meaning only a fraction of the dose administered to the site of absorption (e.g., gut, muscle, skin, etc.) will actually enter the systemic circulation. Thus, when Cl is estimated from an oral PK profile, it is actually a function of the bioavailability (%F) or the fraction of the dose that reaches systemic circulation, and therefore is expressed as, Cl/F and is calculated as follows:

$Cl = F \times Dose/AUC$ Equation 4

The %F after extravascular administration can be calculated by comparing the drug exposure achieved to that obtained following IV administration. It is assumed that the bioavailability of an IV dose is 1; thus, the %F for a compound administered PO is:

$$\%F = \frac{AUC_{PO} \times Dose_{IV}}{AUC_{IV} \times Dose_{PO}} \times 100$$

Equation 5

The mean residence time (MRT) is the mean time the drug remains in circulation, or the average time the drug molecules of a given dose reside in the body. This can only be quantitated following IV administration because it becomes confounded by absorption time upon extravascular administration. For a drug with monoexponential kinetics following IV administration the MRT is:

$MRT = AUMC/AUC$ Equation 6

The half- life and elimination rate constant can also be expressed as a function of the MRT as follows:

$$T_{\frac{1}{2}} = \frac{0.693}{MRT}$$

Equation 7

$$k_{el} = \frac{1}{MRT}$$

Equation 8

Further, the apparent volume of distribution at steady state (Vss) can be calculated as follows:

$V_{ss} = Cl \times MRT$ Equation 9

Substituting with Equations 3 and 6 yields:

$V_{ss} = (Dose \times AUMC)/AUC^2$ Equation 10

The volume of distribution at steady state (V_{ss}) differs from the V_D in that it is an approximation of this value at the perceived steady state plasma concentration, whereas V_D is more commonly referred to as the initial volume of distribution. It is critical to understand that estimation of all these basic PK parameters is dependent on the accurate determination of a drug's AUC. Therefore, the sampling frequency and intervals in the design of the PK study are very important in accurately determining the drug's true AUC in the plasma. NCAmay fail to adequately capture any non-linearity in drug exposure as a function of dose and does not yield specific rate constants for processes involved in the distribution of drug to various tissues or compartments.

3.2   Compartmental Analysis of Pharmacokinetic Data

As stated previously, PK analysis involves the utilization of mathematical expressions which can be used to describe the rate of change in drug concentration in the body as a function of time. This can be done by describing the body as a series of compartments to allow simplification of the system. The one-compartment model is the simplest of all compartmental-based models, wherein the entire body is treated as a single homogenous entity for PK analysis of blood/plasma concentration vs. time data. It assumes drug distributes rapidly throughout the entire body, and is eliminated by a first-order process (k_{el}) after IV administration. Plasma concentrations of a drug exhibiting one-compartment PK are linear with respect to time when plotted on a semilog axis and can be explained by the following monoexponential equation:

$C = C_0 \times e^{-k_{el} \times t}$ Equation 11

where C_0 is the initial concentration at time = 0 and the k_{el} represents the first-order rate of elimination, Figure 5.

Following oral administration, the additional process of drug absorption must be considered. Plasma drug concentrations will be influenced by both the rate of absorption (k_a) and elimination (k_{el}), as well as bioavailability and can be explained using the Batemen Function:

$$C = \frac{(k_a \times F \times Dose)}{V_D \times (k_a - k_{el})} \times (e^{-kel \times t} - e^{-ka \times t})$$

Equation 12

While other techniques exist for solution of the model parameters shown above, nonlinear regression analysis is the most accurate means. Several commercial software packages are available for this purpose (Kinetica, Phoenix/WinNonlin, PK Solutions, ADAPT). The ability to obtain accurate parameter estimates from existing PK data affords researchers the opportunity to simulate drug exposure under new conditions (i.e., dose, route, dose interval, altered clearance) or integrate drug exposure and pharmacology data which will be discussed in subsequent sections.

Often times, drugs administered via the IV route show two or more distinct linear phases when plotted on a semi-log axis. These drugs are said to follow multi-compartmental distribution. In these instances, drug is rapidly distributing to various peripheral tissues. After the drug reaches equilibrium with these tissues, the plasma concentration time profile will then reflect first-order elimination of the drug from the body. These types of profiles require even more complex compartmental models to fit the observed data and are beyond the scope of this work.

Compartmental modeling can be disadvantageous when there is lack of a meaningful physiological basis for the derived PK parameters or when there is lack of rigorous criteria to determine the true number of compartments required to describe drug disposition. As a result, they are highly sensitive to sampling frequency or sampling site. In these cases, when the simplest approach for characterization fails, most PK parameters can be estimated using NCA; however, if a drug exhibits non-linearity with dose, compartmental approaches may be better.

4   Pharmacokinetic Studies in Support of Drug Optimization

Advances in technology and progress in molecular biology have led to an increase in the numbers of new pharmacological targets. Coupled with high throughput screening and automation of combinatorial synthesis, it is now possible for drug discovery programs to be faced with multiple lead series requiring optimization for a given target. Overall, potency at the target of interest remains imperative to the discovery of a clinical candidate, but multiple properties related to the potential of these molecules to become drug products for convenient dosing in patients must remain in the forefront of a successful discovery effort. In addition to numerous *in vitro* screens such as metabolic stability, CYP inhibition, protein binding, aqueous solubility and chemical stability, it is often necessary to determine how molecular changes influence overall *in vivo* pharmacokinetics, as this process incorporates the totality of multiple processes involved with drug absorption, distribution, metabolism and elimination.

Animal studies can be placed into two basic types: pharmacokinetic screens or full pharmacokinetic studies. In PK screening studies, fairly un-optimized compounds are investigated for general trends in PK, and parameter estimates are often compared to the data generated by *in vitro* screens to establish correlations. This approach allows chemistry the confidence to conclude whether higher throughput assays such as metabolic stability truly reflect *in vivo* clearance. Once such a correlation can be established, SAR efforts can be led largely by *in vitro* data to gain efficiency. In the example just provided, the PK screen would only involve IV dosing, as Cl is best calculated from IV data. Other programs may face the challenge of oral exposure and in these cases, oral screens are more appropriate. Often the dose is fixed across a series of compounds and C_{max}, T_{max} and AUC are compared. Whether the appropriate screen is IV or PO, the number of animals per study can vary in a screening mode. Typical PK studies may contain N = 3 or higher, but screens can be dosed in as little as N = 1, bearing in mind the risk that a false negative may eliminate a compound from further program progression. This is the most likely design if the species most closely linked to human is a large animal, such as dog or monkey. When the weight of the preclinical species increases, so does the amount of compound required to dose that species. It may not be practical to prepare large quantities of prototype molecules for an N = 3 design in dog, thus the N = 1 approach may be more convenient.

When rudimentary PK for a large number of leads is desired, cassette dosing may be an option. In a cassette approach, two of more molecules are combined in the same dosing solution and administered to the animal. Blood samples are monitored for the presence of each drug simultaneously allowing more drugs to be studied per animal. Extreme care must be taken when designing the drug cassette. Drugs with equal or very close MWs should not be combined as they will be difficult to resolve by LC/MS/MS analysis. Another drawback is that the total dose to the animal is usually fixed (e.g., 10 mg/kg), thus requiring the dose of each component of the cassette to be lowered significantly (e.g., a four-drug cassette implies a dose of 2.5 mg/kg for each drug). If the drug has a high clearance or poor absorption, lowering dose may be a disadvantage to obtaining quantifiable plasma drug exposures. In addition, some understanding of each compound's drug-drug interaction potential should be considered. If one component of the cassette is a potent CYP inhibitor or if saturation of elimination processes occurs, it may provide false positives in terms of overall exposure. Any "hits" or promising compounds from cassette studies should always be followed up in discrete animal studies to confirm the PK conclusions. Despite these potential downsides, this approach does allow significantly more compounds to be studied *in vivo* in a shorter time frame.

Full pharmacokinetic studies are most appropriate for completely characterizing the disposition of a drug. They are often conducted across species (mouse, rat, dog and monkey) for comparative purposes and often entail dosing by both the IV and oral routes. Numerous considerations impact the selection of dose, the blood sampling schedule and study design; therefore, a flexible approach should always be permitted. As stated previously, IV dosing allows for the determination of CL and Vss, which have a significant impact on overall plasma drug levels. One or more oral doses allow for the determination of oral bioavailability as well as an assessment of the rate of drug absorption. The optimal blood sampling scheme for these studies depends upon the anticipated profile of the drug. For a high clearance compound (as indicated by a short $T_{1/2}$ in an *in vitro* metabolic stability assay), it may be more efficient to cluster several blood samples over a shorter duration, say times 0–8 h, rather than to have blood samples more periodically dispersed over a longer duration. In a program whose target requires rapid onset after oral dosing, it would make more sense to dose orally and include additional time points immediately after dosing through 2 h. The total amount of blood that can be removed over the course of a PK study is generally no more than 10% of the animal's circulating blood volume (as prescribed by the Institutional Animal Care and Use Committee). Thus, the total number of samples per study may remain fixed, but the sampling interval is flexible.

Table 3. Compound Amounts Required to Conduct Typical Pharmacokinetic Characterization Studies Across Species

Species	Weight (kg)	Dose (mg/kg)	Approximate Amount of Compound Required* (mg)
Mouse	0.02	1 (IV)	1
		10 (PO)	2–3
Rat	0.25	1 (IV)	1–2
		10 (PO)	12–13
Dog	10	1 (IV)	40
		10 (PO)	400
Monkey	5	1 (IV)	20
		10 (PO)	200

*Compound approximations for rat, dog and monkey assume an N = 3 design with additional compound to allow for loss to oral gavage tubes and IV syringe pumps. Due to sampling in mice, most PK studies are a composite designs (N = 3 mice/timepoint) which usually requires 9 total mice to perform the study. Compound amount will increase with weight of animal.

The more involved the PK design, the more compound will be required to perform it. Table 3 summarizes the compound requirement to conduct a full PK study IV (1 mg/kg) and PO (10 mg/kg) across species. These are generally good starting points for PK studies on lead compounds.

Other considerations for the type of *in vivo* study design may include whether the compound has already been shown to produce a robust pharmacodynamic effect or is more of a tool compound to understand the influence of SAR on a parameter such as drug clearance or bioavailability. For potent compounds with demonstrated *in vivo* efficacy, that are of high interest to a program, the investment in more cumbersome PK studies is probably warranted.

5   Absorption and Permeability

Drug absorption involves the ability of dissolved drug to cross biological membranes such as the small intestine. Good absorption is essential for drugs whose intended route of administration is via the oral route. The oral bioavailability of drug can be related by the following:

$$F = F_a \times F_g \times F_h \qquad\qquad\qquad \text{Equation 13}$$

where F_a is the fraction of drug absorbed and F_g and F_h are the fraction escaping gut and hepatic metabolism, respectively. Thus, strategies to limit metabolism are clearly important, but good bioavailability cannot be achieved in the absence of good absorption. Two very fundamental factors that impact both the rate and extent of a drug's absorption are permeability and solubility.

In vitro, techniques exist for studying the absorption potential of drug candidates. Some early methods involved the isolation of intestinal segments from animals which were stretched in Ussing Chambers to allow for the addition of drug on one side, with measurement of drug appearance on the other. These methods were cumbersome and highly variable as they were dependent on the quality and reproducibility associated with the tissue isolation. In the early

1990s, a vast improvement to the field of absorption models occurred with the introduction of an immortalized colorectal carcinoma cell line, Caco-2.[31] Caco-2 cells can be plated on a semi-permeable membrane and will grow to confluency and differentiate in cell culture over a 21-day period, developing morphology consistent with that of enterocytes in the small intestine. For studies designed to assess drug absorption, drug is dissolved in buffer and added to the top of the Transwell® (apical side) containing the confluent cell monolayer. Samples are then taken from the bottom (basolateral side) and analyzed for appearance of drug. The Permeability coefficient (P_c) of a drug can be calculated according to the following equation:

$$P_c = \frac{dA/dt}{S \times C_0}$$

Equation 14

where dA/dt is the flux of drug across the monolayer (nmole/sec), S = surface area of the cell monolayer (cm^2), C_0 is the initial concentration (μM) in the donor compartment.

Permeability across the small intestine is achieved either transcellularly or paracellularly. Transcellular transport includes the active uptake or passive diffusion of drug through the intestinal epithelial cells themselves into blood circulation. Paracellular transport is the process where drug is transported around the epithelial cells through the tight junctions into the blood. Typically, only small very polar solutes are transported in this manner. Most lipophilic drugs will be absorbed via transcellular permeability. The link between drug permeability and the fraction absorbed in humans was first established by Artusson and Karlsson.[32] A sigmoidal relationship between permeability and human absorption was established which further enabled the comparison of P_c data for unknown compounds to that of known standards which represent good absorption (i.e., metoprolol) or poor absorption (i.e., ranitidine) reference agents.

In addition to providing an assessment of a compound's passive permeability as mentioned previously, it is possible to use the Caco-2 model to study active efflux. Efflux is the active process by which drugs are secreted by transporters from the enterocyte back into the intestinal lumen, thus limiting their ability to reach the systemic circulation. For studies designed to assess drug efflux, the experiment initiates with drug in the basolateral chamber and measures the appearance of drug in the apical side. By performing permeability studies in Caco-2 monolayers in opposing directions, efflux substrate properties can be concluded if the secretory transport (basolateral-to-apical permeability) is significantly larger than the absorptive one (apical-to-basolateral permeability). Caco-2 cells express a number of efflux transporters which include MDR1 (P-glycoprotein or P-gp), BCRP (Breast Cancer Resistance Protein), and MDR2 and it is possible for compounds to be substrates of one or multiple efflux transporters. Studies using co-incubations of drug and specific chemical inhibitors of these efflux systems can help tease out which specific transporters are involved. Additional discussion of the role of these transporters in drug disposition will be covered in Section 6.

Caco-2 has now become an industry standard for the measurement of permeability. It is relatively high throughput, allows for the ranking of compounds at very early stages in drug discovery, and can help triage compounds for additional *in vivo* testing. The disadvantage is that the model is cell-based and drugs must be dissolved in buffer to enable the transport experiments. In addition, drugs with poor aqueous solubility or cytotoxic may yield erroneous results. This can be especially limiting for compounds with low aqueous solubility, as the use of higher concentrations of solvent can compromise cellular integrity and confound interpretation of the data.

For drugs with poor solubility, it may be possible to dissolve them in higher concentrations of co-solvent and still assess absorption potential *in vitro* using the Parallel Artificial Membrane Permeability Assay (PAMPA). The PAMPA model consists of a lecithin-based lipid system which forms a membrane in a sandwich plate much like that used in the Caco-2 assay. The lipid closely resembles *in vivo* membrane composition and permeability is measured and compared to known reference agents known to be passively absorbed in humans. In contrast to Caco-2, there are no tight junctions or active influx or efflux transport mechanisms. Thus, PAMPA only examines the passive, intrinsic membrane permeability rate of compounds.

Once a general idea of a compound's permeability and aqueous solubility can be obtained, it is possible to make some preliminary predictions of the compound's anticipated PK behavior. The Biopharmaceutics Classification System (BCS) bins a compound's propensity for solubility and permeability into one of four categories.[33] A compound is considered to have good solubility if the highest intended dose strength can be dissolved in 250 mL (approximate volume of the human stomach) over the pH range 1 to 6.5. A compound is considered to have good permeability if its *in vitro* permeability coefficient exceeds that of a suitable reference agent (i.e., metoprolol) known to have > 90% fraction absorbed in man or has been shown to have > 90% absorbed in a human intestinal perfusion study. A schematic of this system is shown in Figure 7A. This binning system provides four categories. Within each category, some generalizations can be made. Class I (high solubility, high permeability) includes compounds which generally have good absorption in man. They are sufficiently soluble to allow for drug to dissolve *in vivo* and can efficiently cross the GI mucosa. Class II (low solubility, high permeability) contains compounds that retain the high permeability characteristics of Class I compounds, but suffer from low aqueous solubility which could impact the amount of drug able to be dissolved in the GI lumen prior to the absorption process. These compounds are generally more lipophilic in nature. Class III (high solubility, low permeability) drugs are very soluble (often very polar molecules), but lack sufficient lipophilicity for good absorption properties. Those compounds in Class IV (low solubility, low permeability) suffer from both poor aqueous solubility and poor permeability. Often times it is not clear which parameter is the rate-limiting step to be observed poor oral exposure with Class IV compounds. It is also difficult to find large numbers of marketed drugs in the BCS Class IV category, which is probably due to the fact that these types of compounds are subject to higher attrition in the drug development cycle due to poor human PK.

Solubility and permeability are two very important *in vitro* parameters for predicting oral exposure, but clearly the process is more complex as described throughout this chapter. Metabolism and drug transporters (either influx or efflux) can also influence drug disposition. This concept was addressed by Wu and Benet in an extension of the original BCS Classification System entitled the Biopharmaceutics Drug Disposition Classification System (BDDCS).[34] In the BDDCS, classification based on solubility is defined the same as in the BCS; however, classification based on permeability is replaced by classifying drugs for extent of metabolism. The purpose of the BDDCS was to enable the prediction of drug disposition for new drug candidates and ascribe the potential for compounds in each class to be subject to certain drug-drug interactions. Figure 7B, shows the breakdown of classes within the BDDCS. For drugs in Class I (high solubility, high metabolism) or Class II (low solubility, high metabolism), the major route of drug elimination is via Phase I and Phase II metabolism. For drugs in Class III (high solubility, low metabolism) and Class IV (low solubility, low metabolism), the major route of drug elimination is typically through the kidney or bile as excretion of intact drug. The BDDCS was also later extended to describe the role of drug transporters on the compounds in each BDDCS class.[35]

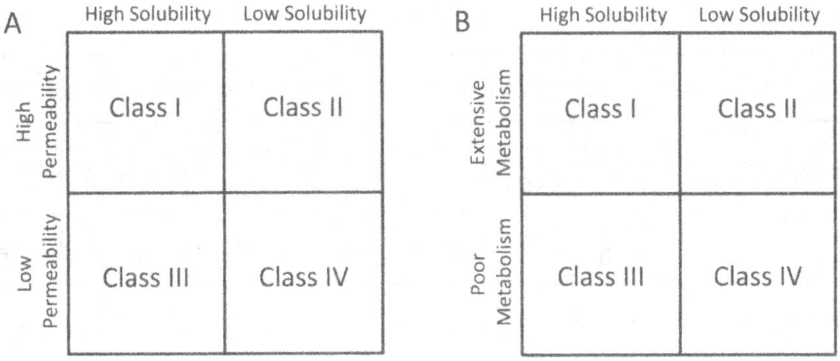

A | High Solubility | Low Solubility
B | High Solubility | Low Solubility

A. High Permeability: Class I (High Solubility), Class II (Low Solubility). Low Permeability: Class III (High Solubility), Class IV (Low Solubility).

B. Extensive Metabolism: Class I (High Solubility), Class II (Low Solubility). Poor Metabolism: Class III (High Solubility), Class IV (Low Solubility).

Figure 7. (A) The Biopharmaceutics Classification System (BCS) and (B) The Biopharmaceutics Drug Disposition Classification System (BDDCS)

6   Drug Transporters

Understanding the involvement of drug transporters *in vivo* remains one of the most challenging phenomena governing drug distribution, elimination, and interaction of drugs. Significant strides in molecular biology have enabled the cloning and characterization of multiple influx and efflux transporters *in vitro* (cell lines, vesicles, etc. many of which are now commercially available). However, translating data that suggests a compound is a substrate and/or inhibitor of a certain drug transporter in the context of *in vivo* PK can be more cumbersome. In some cases, genetically modified knock-out (KO) animals exist which allow the direct comparison of the PK in a wild-type animal to that of its KO phenotype, but these studies are limited largely to mouse only. Table 4 highlights some of the more relevant influx and efflux transporters located in the intestine, liver, kidney and blood brain barrier.[36]

In the kidney, transporters in the Organic Cation Transporter (OCT) and Organic Anion Transporter (OAT) family are responsible for the transport of hydrophilic, low molecular weight cations or anions, respectively, from the proximal tubules out into the urine. Lipophilic drugs with a net positive or negative charge tend to be inhibitors of these transporters and have the potential to cause drug interactions when co-administered with OCT or OAT substrates (e.g., cimetidine and probenacid). Some of the clinically meaningful drug interactions via this mechanism are the increase in acyclovir exposure (> 40%) when administered with probenecid or the increase in metformin AUC (> 50%) when co-dosed with cimetidine.[37,38]

Within the hepatocytes of the liver, the OATP superfamily plays a large role in the transport of amphiphilic organic compounds. The substrate specificity is broad and includes endogenous substances such as bile acids, bilirubin, anionic peptides and steroids, as well as a significant number of drugs such as the statins, fexofenadine, telmisartan, valsartan and olmesartan. Drug interactions attributed to the OATP family have been highlighted for OAT1B1 and OAT1B3 such as the increase in pravastatin exposure (> 800% increase in AUC) when coadministered with cyclosporin (an OATP inhibitor) or the increase in glyburide exposure (> 100% increase in AUC) after an acute dose of rifampicin.[39,40]

Table 4. Transport Proteins in Various Tissues in Humans

Tissue	Transporter(s)	Role in Drug Disposition
Intestine	PEPT1/2, OATP, MCTI, ASBT	Drug absorption
	P-gp, MRP2, BCRP	Efflux into intestinal lumen
Liver	OATP, OCT family, NTCP	Uptake into hepatocytes
	P-gp, BCRP, BSEP, MRP2, MATE1	Efflux into bile
	OAT7, OSTα, MRP3,4,5	Efflux from hepatocyte
Kidney	OAT family, OCT2	Uptake into proximal tubules
	MRP2/4, MATE, P-gp	Excretion into urine
	OAT4, URATI, PEPT1/2, OCTN1/2	Reabsorption from urine
Blood Brain Barrier	OATP1A2 and OAT2B1	Uptake into the brain
	P-gp, BCRP, MRP4/5	Efflux from brain to blood

At both the small intestine and the blood brain barrier (BBB), key transporters involved in limiting drug entry are P-gp and BCRProtein. Cell lines expressing these transporters are available for *in vitro* work, and typically involve the study of drug transport in both the absorptive and secretary direction, such that an efflux ratio can be determined (see Section 5). In addition, KO phenotypes for both P-gp and BCRP are commercially available which allow for the study of these transporters in the mouse. Both P-gp and BCRP are located in the liver and kidney as well, where they serve to efflux drug into the bile or urine, thus enhancing drug elimination. Most notable drug interactions involving P-gp have been observed with the P-gp substrate digoxin, where significant increases in digoxin exposure (> 60%) were observed when P-gp inhibitors such as ritonavir and ranolazine were co-administered.[41,42]

Until specific inhibitors for each of these transporters are available (most tend to cross react with other transporters or drug metabolizing enzymes), the design of specific human drug interaction studies with respect to transporters is no simple task. However, extending the BDDCS to include the potential for generalized transporter involvement is possible. For BDDCS Class I compounds, transporter effects in the intestine and the liver are minimal and likely not considered clinically relevant. Recall that drugs in this class tend to be highly soluble and subject to extensive metabolism. Conversely, those in BDDCS Class II (low solubility, high metabolism) may include compounds not only with increased metabolism, but also with a greater likelihood for active efflux in the gut since their poor solubility often makes it difficult to obtain sufficient solubility to saturate this active process. BDDCS Class III compounds are often absorptive transporter substrates. These compounds are highly soluble, but can exhibit both low passive permeability and CYP-mediated metabolism *in vitro*. Those that are able to enter the systemic circulation may depend on active uptake processes. As described for BCS Class IV compounds, BDDCS Class IV compounds can contain a mixed bag of absorptive and efflux transporter involvement, but the fact remains that this class of compounds are often plagued with multiple development issues with respect to achieving and maintaining adequate PK.

7   Protein Binding

The presence of proteins, lipids and lipoproteins in the blood and in biological tissues/cellular membranes creates the potential for binding interactions to occur with small molecules. These interactions can vary in affinity and magnitude based upon the physico-chemical properties of a given molecule and may have implications with regards to its PK, toxicity, and/or efficacy. The focus of this section will be to describe the basic principles of these binding interactions and how changes in protein binding may affect the clearance and volume of distribution of a compound, thereby resulting in changes in drug exposure which could contribute to alterations in toxicity-efficacy margins. For additional information on this topic please refer to the following references.[43–46]

7.1   *Protein Binding Theory*

Consider the following law of mass-action in equation 15:

$$F + P \; \underset{k_2}{\overset{k_1}{\rightleftharpoons}} \; B$$

Equation 15

where unbound drug (F) and free protein (P) can interact to form protein-drug complex (B). At equilibrium, the ratio of the second-order association rate (k_1) and first-order rate of disassociation (k_2) is equal to the association rate constant, K_A. Conversely, the inverse is the K_D or dissociation rate constant. If we assume that:

$$P = P_t + B$$

Equation 16

where P_t is the total protein concentration, the association rate constant, K_A, can be expressed as follows:

$$K_A = \frac{B}{(P_t - B) \times F}$$

Equation 17

and rearrangement of Equation 17 to solve for B and substitution of K_D for K_A, yields the following relationship:

$$B = \frac{P_t \times F}{K_D \times F}$$

Equation 18

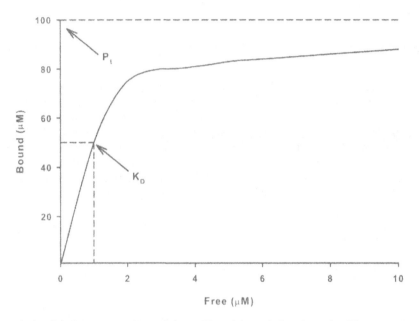

Figure 8. Relationship between unbound drug (F) and bound drug/protein (B) concentration for a hypothetical drug, where P_t is the total number of binding sites and K_D is the concentration of free drug at which bound drug is equal to 50% of P_t

A graphical representation of the relationship between B vs F is illustrated in Figure 8 which takes the form of a typical capacity plot where the parameters governing the binding interaction, namely the K_D (affinity) and P_t (total capacity), in Equation 18 could be resolved by simple nonlinear regression. Given the relationship described in Equation 18, one can clearly see that any impact on either the total protein concentration or on drug dissociation could cause alterations in free drug concentrations. Two of the primary circulating plasma proteins responsible for drug binding are albumin and alpha-1-acid glycoprotein (AAG). While there is a limited amount of information related to alterations in albumin concentration, AAG concentrations have been shown to have significant variations under certain physiologic or pathologic conditions.

Assuming the drug in question preferentially associates with AAG, then the potential impact of disease state on free drug exposures may need to be placed in appropriate context.

7.2 Role of f_u in Clearance and Volume of Distribution

As discussed in Section 3, the two primary PK parameters that regulate drug exposure are Cl and V_D; therefore the discussion on the impact of protein binding on PK will be restricted to these two parameters. Moreover, the discussion will also be limited to issues related to oral dosing as this is typically the desired route of administration from an industrial perspective. The relationship Cl has with respect to average drug exposure (i.e., AUC of the concentration vs. time curve) is given in Equation 19.

$$AUC = \frac{F \times f_u \times Dose}{Cl}$$

Equation 19

where AUC describes the total drug exposure and F represents the oral bioavailability and f_u is the unbound fraction. Clearance has the potential to have a profound impact on the total drug exposure, (i.e., an increase in Cl will reduce exposure and a decrease in Cl will increase exposure). However, plasma protein binding and more importantly, drug free fractions (f_u), will not always have an effect on Cl. If the compound of interest undergoes elimination primarily by hepatic metabolism, then its Cl can be described according to the well-stirred model[47] (Equation 20) where drug is assumed to distribute instantaneously and homogeneously throughout the liver with equivalent unbound inflow and outflow concentrations.

$$Cl_{hepatic\ blood} = \frac{Q_{hepatic} \times f_u \times Cl_{u,int}}{Q_{hepatic} + f_u \times Cl_{u,int}}$$

Equation 20

where $Q_{hepatic}$ is the hepatic blood flow, f_u is the fraction unbound and $Cl_{u,int}$ is the unbound intrinsic clearance. For compounds that are low clearance or low extraction, $Q_{hepatic} \gg f_u \times Cl_{u,int}$ and Equation 20 is simplified to:

$$Cl_{hepatic\ blood} = f_u \times Cl_{u,int}$$

Equation 21

In this case, compounds that undergo low extraction by the liver (low intrinsic clearance) will be more affected by changes in f_u. In this case, increases in unbound fraction (lower protein binding) will result in an increase in hepatic clearance. Conversely, compounds with a decrease in free fraction (higher protein binding) will display a lower hepatic clearance.

However, when we consider the other extreme with compounds undergoing higher clearance/high extraction with $f_u \times Cl_{u,int} \gg Q_{hepatic}$, then Equation 20 reduces to:

$$Cl_{hepatic\ blood} = Q_{hepatic}$$

Equation 22

Therefore, compounds that are already under rapid clearance will be unaffected by changes in protein binding fractions. Based upon these various relationships it can be stated that low extraction compounds are capacity limited and dependent on protein binding, while high extraction compounds are flow-limited and Cl is independent of protein binding.

For compounds that undergo renal clearance via glomerular filtration in the kidney (GFR), renal Cl can be defined as:

$$Cl_{renal} = f_u \times GFR \hspace{3cm} \text{Equation 23}$$

which is dependent upon protein binding. However, if active/passive reabsorptive and active secretory processes are regulated by renal transporters (Section 5) then the scenario becomes more complicated and the Cl_{renal} will also become dependent upon transporter expression levels and activity. Whereas the goal of most discovery paradigms are to optimize total body clearance (Cl_T) the interplay with protein binding can be important.

Described in Section 3, the V_D is the proportionality constant that relates the total amount of drug in the body to the measured concentrations in the biological matrix that was sampled (e.g., plasma). The $V_{D,plasma}$ can then be calculated according to the Gillette Equation[48–50] assuming that drug is equally distributed between plasma and blood cells:

$$V_{D,plasma} = V_{plasma} + \frac{f_{u,P}}{f_{u,T}} \times V_T$$

$$\text{Equation 24}$$

where, V_{plasma} is the physiological plasma volume (~3 L; see Table 2), $f_{u,P}$ and $f_{u,T}$ are the free fractions in plasma and tissue, respectively, and V_T is the extravascular volume to which the drug is distributed. For acids, the degree of V_D is typically smaller (i.e., less extravascular distribution) while bases and neutrals usually display higher distribution volumes. The impact on V_D can be assessed readily by assuming two extremes and focusing the discussion to the plasma protein binding. If the free fraction in plasma is very low (i.e. highly protein bound), then $V_{plasma} \gg (f_{u,P}/f_{u,T}) \times V_T$ and thus $V_D \approx V_{plasma}$ and therefore, the effect of protein binding will have a minimal effect on the observed V_D. However, when the plasma free fraction is high (minimal protein binding) then $(f_{u,P}/f_{u,T}) \times V_T \gg V_{plasma}$ and therefore, the apparent V_D will change as a function of protein binding.

7.3   *Experimental Techniques*

To date, the two most common approaches for assessing free fraction either *in vitro* or *in vivo* are equilibrium dialysis or ultrafiltration. In the former, the biological matrix of choice (e.g., plasma, serum, tissue homogenate) containing compound either following *in vivo* sampling or simply by spiking into blank matrix is dialyzed against blank physiological buffer across a semi-permeable membrane (e.g., methylcellulose) until free concentrations on the matrix side and buffer side are at equilibrium. Much of the industry has adopted this as a gold standard and high-throughput techniques have been developed in 96-well formats. The caveats associated with this approach are: 1) long incubation times required for equilibration (8–24 h); 2) sample stability/degradation and 3) ability to accurately quantify low free fractions due to the effect of drug dilution.[51]

Ultrafiltration is of most use for samples obtained following drug administration *in vivo*. Basically, plasma/serum/blood are placed in a sample collection tube that is separated by a semi-permeable membrane analogous to that used in equilibrium dialysis. Samples are then centrifuged and the ultrafiltrate containing free drug is collected into a reservoir at the bottom of the tube. A variety of systems are available with differing membrane composition varying in collection volumes. The obvious caveats associated with this approach are: 1) nonspecific binding of drug to the membrane; 2) low ultrafiltrate volume; 3) difficulty in controlling temperature, and 4) concentrating effect in the ultrafiltrate.[52]

7.4   *Optimization of f_u in Drug Discovery*

In recently published datasets of clinically successful drugs, a majority of compounds have been reported to be greater than 95% protein bound.[53] This brings to light an important consideration on the utility of optimizing potential drug candidates for protein binding and whether there is any added benefit. Liu et al. recently evaluated the utility of protein binding optimization in a discovery setting using a theoretical understanding of the impact that f_u plays on unbound drug concentrations and analysis of experimental data. The results from their analysis is that while free fraction does play a role in PK and PD properties, after oral administration the hepatic intrinsic clearance more strongly governs the unbound plasma concentrations *in vivo*. Therefore, while free fraction is an important characteristic of a compound, compounds should not be triaged because of a high degree of protein binding. The ideal optimization strategy in drug discovery should not look to address free fraction independently, but focus on drug potency *in vitro* and *in vivo* as well as important PK characteristics like clearance. Moreover, as will be discussed at the end of this chapter, the nominal value of the dose appears to be the major determinant in a drug's clinical success and is more of a function of drug potency and PK and less driven by the degree of protein binding.

8   **Pharmacokinetics and Pharmacodynamics**

As stated previously, PK is the study of the time course of the disposition of drug and drug-related components in plasma. An understanding of these principles allows for description of the temporal patterns using mathematical expressions to better understand the relationship between dose and drug concentration in bodily fluids, as well as the rates that govern ADME characteristics of a molecule. Additionally, it enables researchers to make predictions/extrapolations under varying scenarios to aid in dose selection, optimization of dosing-regimens, or selection of alternative routes of administration.

Pharmacodynamics (PD) is the study of the relationship between drug concentrations in bodily fluids and the magnitude and duration of its pharmacologic effect. Similarly, mathematical expressions incorporating an understanding of the mechanism of action and other relevant physiological factors can be used to describe the temporal patterns in drug response. The combination of PK–PD integrates a quantitative measure of a drug's ADME and pharmacological properties which can help provide insight into the early stages of target validation and lead characterization/optimization. This not only aids in compound selection, but also prediction of a drug's PK and its effect in humans. Gabrielsson et al recently published a comprehensive 'PK–PD Guide' for medicinal chemists which serves as an overview on how PK–PD should be integrated in the discovery process and highlights relevant concepts and acceptable practices.[54]

Concepts related to PK andprotein binding have been discussed previously, thus subsequent discussions will be focused on the interplay between PK and PD. While it would seem implicit that optimization of an *in vitro* measure of binding potency (k_d) or half-maximal inhibitory/stimulatory concentration (IC_{50} or SC_{50}) would be an ideal SAR strategy, not considering the time drug is available for interaction with the target (i.e., CL or $T_{1/2}$) and magnitude of the concentrations at the target site (f_u and V_D) would shed little light on the onset, duration, and intensity of the desired effect *in vivo*. Therefore, it would be prudent to optimize these variables in parallel by utilizing PK-PD characterization of *in vivo* pharmacology thus helping identify structural features that could be conserved or modified.

8.1   *Reversible Direct Effects*

The potential impact of reversible direct effects can be illustrated by the simple scenario where drug in the plasma and target site are in equilibrium and the magnitude of a given reversible effect is directly proportional to the target site concentration at a given time point. Equation 25 and Figure 9A describe the log-linear relationship between the intensity of the effect (E) and drug concentration:

$$E = E_0 + m \times logC$$
Equation 25

where m is the slope and E_0 is the intercept. For reasons of simplicity, assume that the drug was administered by rapid IV injection and displays monoexponential PK.

As illustrated in Figure 5, the drug concentration versus time can be described by the following equation:

$$C = C_0 \times e^{-k \times t}$$
Equation 26

where Co is the initial concentration at t=0 and k represents the first-order rate of elimination. By taking the natural logarithm of both sides:

$$ln\ C = ln\ C_0 - k \times t$$
Equation 27

and then converting the logarithm in Equation 27 to the base of 10(log):

$$log\ C\ = log\ C_0 \times \frac{k \times t}{2.303}$$
Equation 28

Combining this relationship with Equation 25 and rearranging:

$$E = E_0 - \frac{k \times m}{2.303} \times t$$
Equation 29

An illustration of the effect vs. time (or PD) profile for this example based upon Equation 29 and the drug's PK (Figure 5) is shown in Figure 9B. It can be clearly seen that the time course of the response is in equilibrium with the time course of drug concentration in plasma (i.e., same peak and trough of response and exposure).

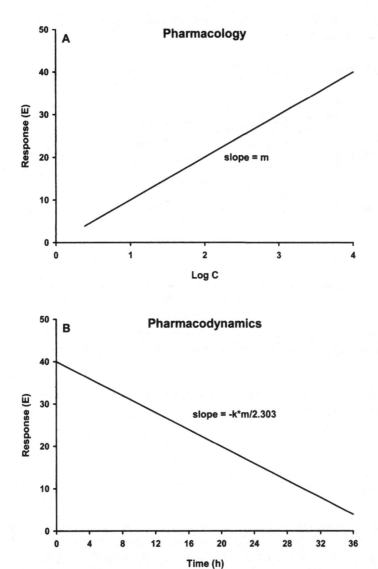

Figure 9. Prototypical relationship between (A) response vs drug concentration and (B) response vs time for a hypothetical drug which exerts a reversible linear direct effect

Given that the intensity of the effect is directly proportional to the rate constant for elimination, one can appreciate the importance in understanding and optimizing the drugs clearance and half-life in order to control not only the intensity of the effect, but also the duration. An example of the impact of changing the half-life of a drug (rate of elimination) is illustrated in Figure 10, where an increase in the half-life from 3 to 10 h results in a slower decay in the time course of the pharmacological response, thus maintaining an elevated PD response over a given dosing interval. This type of analyses would allow researchers the ability to understand how differences in dose and/or dosing regimen would impact efficacy and perhaps screen for molecules with more desirable characteristics.

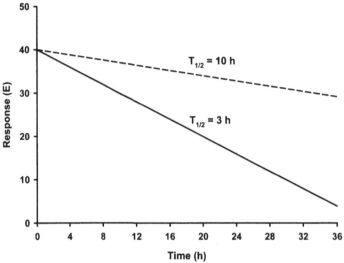

Figure 10. Effect of increasing a drug's blood/plasma half-life ($T_{1/2}$) on its pharmacodynamic relationship (response vs. time) for a hypothetical drug which exerts a reversible linear direct effect

In addition to simple direct effects behaving both linearly or log-linearly, drugs can also behave in a nonlinear fashion and these exposure-response relationships are best described using the Hill equation (Equation 30 and Figure 11)

$$E = \frac{E_{max} \times C^Y}{EC_{50}{}^Y + C^Y}$$

Equation 30

where E_{max} is the maximal effect/capacity parameter of the process, EC_{50} represents the sensitivity constant of the drug which is a measure of its intrinsic potency for the target and γ is the Hill coefficient which is a measure of a drugs cooperativity describing the steepness of the exposure-response relationship.

     A variety of mathematical expressions have been derived to accommodate the various mechanisms that drugs exert their effects. These can be agonistic or antagonistic resulting in either stimulation or suppression of various physiological or pathological effects.

     In all of these scenarios, regardless of the mechanism, drug concentrations are directly related to the given effect. It should also be clear that a drug's pharmacologic activity is not only dictated by its intrinsic potency, but that the driving function for the PD response lies in its PK. Thus, a major determinant of the time course and magnitude of a given response requires that PK be optimized in parallel with the parameters governing target engagement.

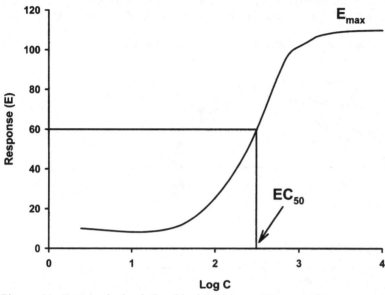

Figure 11. Prototypical relationship between response vs drug concentration for a hypothetical drug which acts in a nonlinear direct and reversible manner, where E_{max} is the maximal effect and EC_{50} is the drug concentration at which the response is equal to 50% of E_{max}

8.2 Time-Delayed Effects

The aforementioned examples pertain to processes that occur in parallel (i.e., time of peak exposure coincides with peak response). However, more complex scenarios arise when there are temporal delays between plasma drug concentrations and the observed drug effect. In these cases, a typical graphical representation of exposure versus effect, as shown previously in Figure 9A and Figure 11, would demonstrate a hysteresis loop. In these situations, there would be an equivalent magnitude of response occurring at two different drug concentrations, as illustrated in Figure 12. This phenomenon could be due to a variety of reasons, such as equilibrium delay between the plasma and target compartments, active metabolites, drugs that exert their effects indirectly on the rates of production and/or degradation of the response variable, and where tolerance or desensitization of the effect occurs in the absence of changes in the PK. For a more detailed mathematical explanation of these additional complexities refer to the various reviews and texts that have been provided at the end of Section 9.

In the context of delayed equilibration, characterization of the lag time between peak plasma concentration and peak effect could be performed using a biophase or effect-compartment model (also known as the link model).[55] This model is based upon the principles of Fick's Law of Diffusion and is described by the following equation

$$\frac{dC_e}{dt} = k_{eo} \times (C_p - C_e)$$

Equation 31

where, C_e is the concentration in the hypothetical effect-compartment, C_p is the drug concentration in the plasma and k_{eo} is the first-order distribution rate from the plasma to the effect compartment.

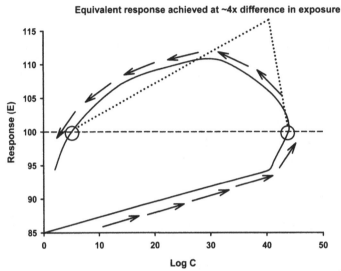

Figure 12. The effect of temporal delays between peak concentration and peak response on the response vs concentration relationship for a hypothetical drug resulting in counter-clockwise hysteresis

This assumes that drug in the 'biophase' does not contribute to the PK in the plasma and that the equations governing the pharmacologic response, such as those described in Equation 25 and 30, would be tied to C_e as opposed to C_p. This approach allows for approximation of the 'hypothetical' effect site concentrations by estimation of the distribution rate, k_{eo}, to account for the time delay. This will ultimately collapse the hysteresis loop when plotting the anticipated effect-site concentration versus pharmacological response and will also allow for resolution of pertinent PD parameters, such as E_{max} and EC_{50}.

Interestingly, this model could also be used to account for the delayed onset of effect when the pharmacologic response is a function of active metabolite exposures. While this would serve as a gross over-simplification of the system and the observed *in vivo* potency could only be tied to plasma exposures of the parent drug, it would provide some useful information in the drug discovery setting when there is little to no understanding of the metabolite PK or potency of the metabolite(s).

In other situations, the temporal disconnect between exposure and effect could be due to factors other than distributional delays or delays in the formation of active metabolites, such as when the compound acts indirectly on the rates of production and/or degradation of the response variable. In these cases, utilization of the link/biophase model would not allow for accurate resolution of the PD parameters that govern the response and as a result could not be extrapolated to other dose levels. In light of these complex scenarios, a series of PD models were developed to aid in the characterization of these reversible effects and were appropriately termed physiologic indirect response models, also known as turnover models. A schematic representation of these models is shown in Figure 13. Depending on whether the compound exhibits inhibition or stimulation of the response and whether it involves the input or output of the response variable, one of the four basic models can be chosen based upon an understanding of the mechanism of action and the target (e.g., enzyme, receptor, mediator, etc.). Models I and IV describe the inhibition of the zero-order production and stimulation of the first-order degradation of the response, respectively. In either case, the observed PD profile would exhibit a reduction in the response measure as a function of time with a gradual decay to pre-dose conditions as drug is eliminated. Models II and III describe the stimulation of the zero-order production and inhibition of the first-order degradation of the response, respectively. In these cases, the observed PD profile would exhibit an increase or build-up of the response measure as a function of time with a gradual decay to pre-dose conditions as drug is eliminated. So an understanding of the actual mechanism of action of a drug is crucial during the model selection process as it allows for incorporation of physiological context to the predictions.

Family of Indirect Response Models

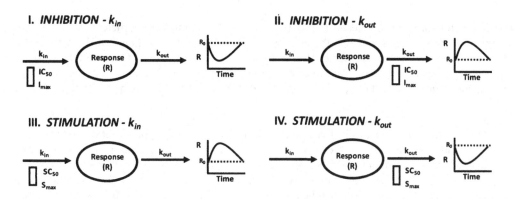

Figure 13. Four basic indirect response models describing the production (k_{in}) and dissipation (k_{out}) factors governing the response variable (R) where drug concentrations can either result in inhibition (I_{max}; IC_{50}) or stimulation (S_{max}; SC_{50}) of these processes. Adapted from Jusko and Sharma.[65,66]

While there remain other situations where effects can be delayed relative to exposure, such as slow on/off rates with receptors, receptor desensitization, inhibition/stimulation of a precursor upstream to the response, or delays due to signal transduction, the reader is advised to consult the various references provided at the end of Section 9 for a more detailed description of these more complex situations.

9   Predicting Human Pharmacokinetics

Another consideration in the discovery environment is the need to extrapolate PK from nonclinical species to predict the PK in humans. Because of species differences in the expression levels and activities in metabolic enzymes and transporters, often times it is difficult to know what species or scaling approach would best predict human PK. In light of this, researchers tend to use a combination of approaches to arrive at a selected human dose. The most common of these approaches used in the industry is allometric scaling of clearance and apparent volume of distribution. The primary assumption of allometry is that physiological variables such as clearance, volume of distribution, blood flow, heart rate or other biochemical processes from lower to higher species are related to body weight or surface area. By plotting the physiologic variable (e.g. Cl, V_{ss}) on the y-axis vs. the body weight for a range of species on a log-log plot, a linear relationship can be achieved. Performing linear regression using the following power equation (Equation 32) allows for the estimation of an unknown variable for a given species by plugging in the desired body weight (e.g., human BW = 70 kg).[56]

$$y = a \times W^b$$ Equation 32

where y is the variable of interest (e.g. Cl, V_{ss}), a is the allometric coefficient, W is the body weight of a given species and b is the allometric exponent. Figure 14 gives an illustration of an allometry plot of Cl across species for a hypothetical compound. Other common scaling approaches utilize *in vitro* data from liver microsomes, hepatocytes and plasma protein binding in conjunction with *in vivo* data to provide a greater degree of confidence in the prediction of human Cl.[47,57,58]

Once a prediction of clearance and volume are obtained, approximations of absorption properties and oral bioavailability can be made based upon a compendium of preclinical results obtained for a given compound. With a desired target plasma exposure for efficacy in mind (e.g., C_{max}, C_{min} or AUC), a dose to achieve the requisite target can be derived. Utilization of the anticipated human PK model can then be used to forecast the human PD profile. Based upon these estimations, researchers can put these findings into proper context with an understanding of the pharmaceutics limitations (i.e., can the dose actually be delivered), possible DDI's and toxicity margins to determine whether progression of the compound is feasible, practical and safe.

For additional details related to basic PK and PD theory and PK-PD please refer to the following references.[44,59-61]

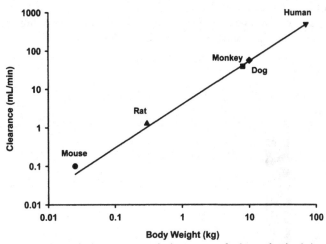

Figure 14. Allometric relationship between total clearance of a hypothetical drug and body weight

9.1 *Importance of Dose*

Throughout this chapter, factors which govern the plasma exposure of a drug have been discussed (Cl, V_D, %F). Additionally, the topic of pharmacodynamics reviewed that it is possible to gain knowledge regarding the *in vivo* potency of a drug against the target of interest. Remember, this *in vivo* potency needs to be decoupled from the *in vitro* potency data that is often collected early in a discovery screening tier. In the latter, exquisite potency may be achievable because drug resides in a well, 100% available to the target or enzyme of interest. *In vivo* potency has the benefit of incorporating multiple factors that may limit the drug from getting to the site of action (ADME). By integrating PK-PD relationships, it is possible to establish the plasma drug levels that are required for a drug to yield robust efficacy. Combining principles of PK-PD and factors such as aqueous solubility and route of administration (concepts which pertain to drug delivery), it becomes apparent that a successful drug candidate needs to retain sufficient *in vivo* potency such that the dose required for efficacy is able to be delivered in a reasonably sized drug product.

For example, a compound with low systemic clearance and high oral bioavailability may still not be successful if maintenance of extremely high plasma drug concentration is required for efficacy (i.e., anti-infectives, as a class of drugs, are often plagued with this challenge). In such cases, the only way to drive efficacy is to raise dose. When dose is high, the risk of a number of undesirable effects (i.e., toxicity, drug-drug interaction, inability to dose escalate, cost of goods, formulation challenges) increases. Thus, exquisite potency *in vivo* will always enable dose to trend in the right direction. With sufficient potency, it may in fact be tolerable to advance compounds with less than perfect pharmacokinetics. The important point to remember is that drug candidate optimization is a balancing act.

In order to illustrate this concept, an analysis of the dose of the top 200 prescribed drugs of 2010 by sales is shown in Figure 15.[62] This analysis looked only at the highest available dose strength for single agent oral drug products (N = 98). It is clear from this analysis, that in general, successful oral drug products are typically those whose oral dose is < 100 mg (about 62%). Casting the net a bit further, it becomes apparent that a total of 86% of the oral drug products have doses less than 400 mg. There are only 14 products with high oral dose strengths ≥ 500 mg. Of those, six are anti-infectives, two are pharmaceutical grade nutritionals (niacin and omega-3 fatty acids), one is an oral anti-cancer agent, one is an anti-rejection drug, and three are for indications that could be considered localized intestinal targets like colitis or binders of dietary bile acids and phosphates.

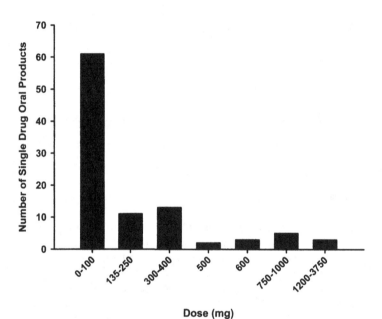

Dose (mg)

Figure 15. Highest available dose strength for single agent oral drugs products on the top 200 prescribed drug list of 2010

There is however one drug in the ≥ 500 mg category worthy of discussion. Keppra (levetiracetam) is an anti-epileptic (CNS target) with the highest available oral dose strength of 1000 mg. Levetiracetam is rapidly and almost completely absorbed and is not appreciably metabolized by CYPs. The bioavailability of the tablet formulation is nearly identical to when drug is administered as solution, indicating rapid dissolution of the tablet formulation. Overall, the PK would be considered excellent. So why is the dose so high? Anti-epileptics are required to cross the BBB, thus any brain penetration issues could limit the drug's ability to be available at the site of action, even if the plasma PK was reasonable. However, a microdialysis study in rat revealed that levitiracetam readily crossed the blood brain barrier and that half-life in the brain was actually longer than serum.[63] In other words, both the PK and tissue distribution at the site of action is excellent, however the human dose remains high.

The answer may be that levitiracetam appears unique in its mechanism of action among the class of antiepileptic drugs and that the exact mechanism of how it works is not well understood. In animal studies, levetiracetam was administered at extremely high doses (up to 80 mg/kg) to demonstrate reductions in taurine concentrations in both the frontal cortex and hippocampus.[63] The binding affinity to that site (1–5 μM) correlates with efficacy with some (not all) *in vivo* epilepsy models.[64] Thus, by conventional standards, this drug is not appreciably potent. Had the PK not been as outstanding and the potential for drug-interactions low, this type of scenario could have resulted in the potential for failed clinical trials, at least as far as PK is concerned. This outlier further strengthens the point that *in vivo* potency is not the only determinant of a successful drug candidate.

10 Summary

Appropriate animal and human *in vitro* liabiltiy screening (solubility, permeability, transporter involvment, metabolic stability, inhibition or induction potential, etc.) in combination with metabolism and pharmacokinetic evaluations *in vivo* (preclinical animals species) will lead to compounds moving into drug development with greater probablities for success in the clinic. Also, a good discovery effort will integrate ADME into its candidate optimization process and balance *in vivo* potency with PK, such that typical oral drug doses for efficacy remain < 500 mg once a day (ideally ~ 100 mg). At these lower doses and exposures, patient compliance increases and limitations such as solubility, toxicity, off-target pharmacology, and drug-drug interactons are minimized or eliminated.

References

(1)   Lipinski, C. A.; Lombardo, F.; Dominy, B. W.; Feeney, P. J. *Adv. Drug Deliv. Rev.* **2001**, *46*, 3.

(2)   Ritchie, T. J.; Ertl, P.; Lewis, R. *Drug Disc. Today* **2011**, *16*, 65.

(3)   Trecant, C.; Dlubala, A.; George, P.; Pichat, P.; Ripoche, I.; Troin, Y. *Eur. J. Med. Chem.* **2011**, *46*, 4035.

(4)   Pearson, P.; Wienkers, L. *Handbook of Drug Metabolism*; 2 ed.; Informa Healthcare: New York, 2009.

(5)   Parkinson, A. In *Casarett & Doull's Toxicology: The Basic Science of Poisons*; Sixth ed.; Klaassen, C., Ed.; McGraw-Hill: New York, **2001**, p 133.

(6)   Zhang, H.; Davis, C. D.; Sinz, M. W.; Rodrigues, A. D. *Expert Opin. Drug Met.* **2007**, *3*, 667.

(7)   Tang, W.; Wang, R. W.; Lu, A. Y. *Curr. Drug Metab.* **2005**, *6*, 503.

(8)   Nicoli, R.; Bartolini, M.; Rudaz, S.; Andrisano, V.; Veuthey, J. L. *J. Chromatogr.* **2008**, *1206*, 2.

(9)   Carlson, T. J.; Fisher, M. B. *Comb. Chem. High T. Scr.* **2008**, *11*, 258.

(10)   Obach, R. S. *Curr. Top. Med. Chem.* **2011**, *11*, 334.

(11)   Mohutsky, M.; Wrighton, S.; Ring, B. In *Handbook of Drug Metabolism*; 2 ed.; Pearson, P., Wienkers, L., Eds.; Informa Healthcare: New York, **2009**, p 445.

(12)   Mao, J.; Mohutsky, M. A.; Harrelson, J. P.; Wrighton, S. A.; Hall, S. D. *Drug Metab. Dispos.* **2011**.

(13)   Lave, T.; Coassolo, P.; Reigner, B. *Clin. Pharmacokinet.* **1999**, *36*, 211.

(14)   Foster, J. A.; Houston, J. B.; Hallifax, D. *Xenobiotica* **2011**, *41*, 124.

(15)   Barter, Z. E.; Bayliss, M. K.; Beaune, P. H.; Boobis, A. R.; Carlile, D. J.; Edwards, R. J.; Houston, J. B.; Lake, B. G.; Lipscomb, J. C.; Pelkonen, O. R.; Tucker, G. T.; Rostami-Hodjegan, A. *Curr. Drug Metab.* **2007**, *8*, 33.

(16)   Sinz, M.; Wallace, G.; Sahi, J. *AAPS J.* **2008**, *10*, 391.

(17)   Chu, V.; Einolf, H. J.; Evers, R.; Kumar, G.; Moore, D.; Ripp, S.; Silva, J.; Sinha, V.; Sinz, M.; Skerjanec, A. *Drug Metab. Dispos.* **2009**, *37*, 1339.

(18)   Kearns, E.; Li, D. In *Drug-like Properties: Concepts, Structure Design and Methods*; Kearns, E., Li, D., Eds.; Elsevier: Amsterdam, **2008**, p 137.

(19)   Obach, R. S. *Curr. Opin. Drug Di. De.* **2009**, *12*, 81.

(20)   Stepan, A. F.; Walker, D. P.; Bauman, J.; Price, D. A.; Baillie, T. A.; Kalgutkar, A. S.; Aleo, M. D. *Chem. Res. Toxicol.* **2011**, *24*, 1345.

(21)   Grimm, S. W.; Einolf, H. J.; Hall, S. D.; He, K.; Lim, H. K.; Ling, K. H.; Lu, C.; Nomeir, A. A.; Seibert, E.; Skordos, K. W.; Tonn, G. R.; Van Horn, R.; Wang, R. W.; Wong, Y. N.; Yang, T. J.; Obach, R. S. *Drug Metab. Dispos.* **2009**, *37*, 1355.

(22)   Kunze, K. L.; Trager, W. F. *Chem. Res. Toxicol.* **1993**, *6*, 649.

(23)   Takakusa, H.; Wahlin, M. D.; Zhao, C.; Hanson, K. L.; New, L. S.; Chan, E. C.; Nelson, S. D. *Drug Metab. Dispos.* **2011**, *39*, 1022.

(24)   Tirona, R. G.; Kim, R. B. *J. Pharm. Sci.* **2005**, *94*, 1169.

(25)   Wang, H.; LeCluyse, E. L. *Clin. Pharmacokinet.* **2003**, *42*, 1331.

(26)   Sinz, M.; Kim, S.; Zhu, Z.; Chen, T.; Anthony, M.; Dickinson, K.; Rodrigues, A. D. *Curr. Drug Metab.* **2006**, *7*, 375.

(27)   Hewitt, N. J.; Lechon, M. J.; Houston, J. B.; Hallifax, D.; Brown, H. S.; Maurel, P.; Kenna, J. G.; Gustavsson, L.; Lohmann, C.; Skonberg, C.; Guillouzo, A.; Tuschl, G.; Li, A. P.; LeCluyse, E.; Groothuis, G. M.; Hengstler, J. G. *Drug Metab. Rev.* **2007**, *39*, 159.

(28)   FDA Draft Guideline: Drug interaction studies - study design, data analysis, and implications for dosing and labeling. http://www.fda.gov/cder/guidance/index.htm, FDA **2006**.
       (29)   Davies, B.; Morris, T. *Pharm. Res.* **1993**, *10*, 1093.

(30)   Altman, P. *Blood and Other Body Fluids*; Federation of American Societies for Experimental Biology: Washington DC, **1961**.

(31)   Hidalgo, I. J.; Raub, T. J.; Borchardt, R. T. *Gastroenterology* **1989**, *96*, 736.

(32)   Artursson, P.; Karlsson, J. *Biochem. Biophys. Res. Commun.* **1991**, *175*, 880.

(33)   Amidon, G. L.; Lennernas, H.; Shah, V. P.; Crison, J. R. *Pharm. Res.* **1995**, *12*, 413.

(34)   Wu, C. Y.; Benet, L. Z. *Pharm. Res.* **2005**, *22*, 11.

(35)   Shugarts, S.; Benet, L. Z. *Pharm. Res.* **2009**, *26*, 2039.

(36)   Giacomini, K. M.; Huang, S. M.; Tweedie, D. J.; Benet, L. Z.; Brouwer, K. L.; Chu, X.; Dahlin, A.; Evers, R.; Fischer, V.; Hillgren, K. M.; Hoffmaster, K. A.; Ishikawa, T.; Keppler, D.; Kim, R. B.; Lee, C. A.; Niemi, M.; Polli, J. W.; Sugiyama, Y.; Swaan, P. W.; Ware, J. A.; Wright, S. H.; Yee, S. W.; Zamek-Gliszczynski, M. J.; Zhang, L. *Nat. Rev. Drug Discov.* **2010**, *9*, 215.

(37)   Laskin, O. L.; de Miranda, P.; King, D. H.; Page, D. A.; Longstreth, J. A.; Rocco, L.; Lietman, P. S. *Antimicrob. Agents Chemother.* **1982**, *21*, 804.

(38)   Somogyi, A.; Stockley, C.; Keal, J.; Rolan, P.; Bochner, F. *Br. J. Clin. Pharmacol.* **1987**, *23*, 545.

(39)   Hedman, M.; Neuvonen, P. J.; Neuvonen, M.; Holmberg, C.; Antikainen, M. *Clin Pharmacol. Ther.* **2004**, *75*, 101.

(40)   Zheng, H. X.; Huang, Y.; Frassetto, L. A.; Benet, L. Z. *Clin. Pharmacol. Ther.* **2009**, *85*, 78.

(41)   Ding, R.; Tayrouz, Y.; Riedel, K. D.; Burhenne, J.; Weiss, J.; Mikus, G.; Haefeli, W. E. *Clin. Pharmacol. Ther.* **2004**, *76*, 73.

(42)   Jerling, M. *Clin. Pharmacokinet.* **2006**, *45*, 469.

(43)   Benet, L. Z.; Hoener, B. A. *Clin. Pharmacol. Ther.* **2002**, *71*, 115.

(44)   Gabrielsson, J.; Dolgos, H.; Gillberg, P. G.; Bredberg, U.; Benthem, B.; Duker, G. *Drug Disc. Today* **2009**, *14*, 358.

(45)   Schmidt, S.; Gonzalez, D.; Derendorf, H. *J. Pharm. Sci.* **2010**, *99*, 1107.

(46)   Smith, D. A.; Di, L.; Kerns, E. H. *Nat. Rev. Drug Discov.* **2010**, *9*, 929.

(47)   Riley, R. J.; McGinnity, D. F.; Austin, R. P. *Drug Metab. Dispos.* **2005**, *33*, 1304.

(48)   Gibaldi, M.; McNamara, P. J. *Eur. J. Clin. Pharmacol.* **1978**, *13*, 373.

(49)   Gillette, J. R. *Ann. N Y Acad. Sci.* **1971**, *179*, 43.

(50)   Wilkinson, G. R.; Shand, D. G. *Clin. Pharmacol.Ther.* **1975**, *18*, 377.

(51)   Banker, M. J.; Clark, T. H.; Williams, J. A. *J. Pharm. Sci.* **2003**, *92*, 967.

(52)   Whitlam, J. B.; Brown, K. F. *J. Pharm. Sci.* **1981**, *70*, 146.

(53)   Liu, X.; Chen, C.; Hop, C. E. *Curr. Top. Med. Chem.* **2011**, *11*, 450.

(54)   Gabrielsson, J.; Fjellstrom, O.; Ulander, J.; Rowley, M.; Van Der Graaf, P. H. *Curr. Top. Med. Chem.* **2011**, *11*, 404.

(55)   Sheiner, L. B.; Stanski, D. R.; Vozeh, S.; Miller, R. D.; Ham, J. *Clin. Pharmacol. Ther.* **1979**, *25*, 358.

(56) Mordenti, J. *J. Pharm. Sci.* **1986,** *75*, 1028.

(57) Huang, C.; Zheng, M.; Yang, Z.; Rodrigues, A. D.; Marathe, P. *Pharm. Res.* **2008**, *25*, 713.

(58) Lin, J. H. *Drug Metab. Dispos.* **1995,** *23*, 1008.

(59) Gabrilsson, J.; Weiner, D. *Pharmacokinetics and Pharmacodynamic Data Analysis: Concepts and Applications*; 3 ed.; Swedish :Pharmaceutical Press: Stockholm, Sweden, **2000**.

(60) Levy, G.; Gibaldi, M. *Pharmacokinetics*; Springer Verlag: Berlin, **1975**; Vol. 28.

(61) Shargel, L.; Yu, A. *Applied Biopharmaceutics and Pharmacokinetics*; 4 ed.; Appleton and Lange: Stamford, CT, **1999**.

(62) Bartholow, M. In *Pharmacy Times* **2010**; http://www.pharmacytimes.com/publications/issue/2011/May2011/Top-200-Drugs-of-2010, accessed Nov. 2011.

(63) Tong, X.; Patsalos, P. N. *Br. J.Pharmacol.* **2001,** *133*, 867.

(64) Surges, R.; Volynski, K. E.; Walker, M. C. *Ther. Adv. Neurol. Disord.* **2008**, *1*, 13.

(65) Jusko, W. J.; Ko, H. C. *Clin. Pharmacol. Ther.* **1994,** *56*, 406.

(66) Sharma, A.; Jusko, W. J. *Br. J. Clin. Pharmacol.* **1998,** *45*, 229.

5

Cardiovascular Drugs

Drago R. Sliskovic

1   Introduction

Cardiovascular diseases are a heterogeneous group of disorders of the heart and blood vessels. The World Health Organization International Classification of Disease, 9th revision; (ICD-9) identifies over 60 separate cardiovascular diseases [including cerebrovascular disease (a disease of the blood vessels supplying the brain)] in eleven separate groupings consisting of the following diseases: hypercholesterolemia (high blood cholesterol); rheumatic heart disease; hypertensive disease; ischemic heart disease; pulmonary circulatory diseases; other forms of heart disease, including heart failure, cardiac dysrrhythmias, and cardiomyopathies; cerebrovascular disease; disease of arteries, arterioles, and capillaries, including atherosclerosis and peripheral artery disease; and diseases of veins and lymphatics. The most prevalent classes are diseases of the arteries (including atherosclerosis and peripheral artery disease), hypercholesterolemia, heart failure, and hypertensive disease.[1] Heart failure is the inability of the heart to supply sufficient blood flow to meet the body's needs. There are many causes of heart failure but the main ones, for the purposes of this review, are ischemic heart disease (reduced blood flow to heart muscle) and hypertension (elevated systemic arterial blood pressure). Other causes include acute myocardial infarctions, cardiomyopathies and valvular disease. The main cause of ischemic heart disease is atherosclerosis. This disease arises from a deposition of lipids and extracellular matrix resulting in the formation of plaques in the intima of large and medium-sized arteries. These plaques are associated with a reduction in the arterial lumen and a predisposition to thrombosis whose clinical manifestations include myocardial infarction, unstable angina pectoris, deep-vein thrombosis, pulmonary embolism, and stroke. Independent risk factors for atherosclerosis include hypercholesterolemia, hypertension, diabetes mellitus, and lifestyle factors such as smoking, sedentary lifestyle, and type A personality.

According to the World Health Organization (WHO), an estimated 17.1 million people died from cardiovascular disease in 2004, representing 29% of all global deaths.[2] In 2006 it was estimated that 81,100,000 people in the USA have one or more forms of cardiovascular disease and 831,272 people (34.3% of all deaths) died from the disease. Of these 425,425 died from coronary heart disease caused by atherosclerosis, the leading cause of death in the USA.[3] The economic burden of these diseases is staggering with health care costs estimated at $50–150 billion per year worldwide.[4]

2   Early History of Coronary Heart Disease (CHD)

P. R. Lichtlen,[5] in his paper:[11] History of Coronary Heart Disease, identified Leonardo da Vinci (1452–1519) as an early investigator of the cardiovascular system when he described the coronary arteries as "embedded in greasy material." This was one of the first observations of atherosclerosis, the causative disease involved in the development of CHD. This finding was reiterated by Friedrich Hoffmann (1660–1742) of the University of Halle, who wrote that the origins of CHD lie in "the reduced passage of blood within the coronary arteries." None of these conclusions would have been possible without the seminal work of William Harvey, who in 1628 thoroughly described the circulatory system. The development of the stethoscope by R. T. H Laennec (1781–1826) led to significant progress in the treatment of cardiovascular disease in the 18th and 19th centuries. Since heart disease patients often presented with angina pectoris (chest pain as a result of myocardial ischemia), physicians during this period utilized various treatments to relieve anginal pain. In addition, patients with end-stage heart failure, also presented with symptoms such as pulmonary congestion, peripheral edema, fatigue and breathlessness. All of these symptoms are associated with low cardiac output (CO) and the treatments of the 18th and 19th centuries were employed to increase the contractility of heart muscle and make it beat harder. It was not until the 20th century that an understanding of the etiology of this disease developed. In 1912, James B. Herrick (1861–1954) concluded that a gradual narrowing of coronary vessels was a possible cause of heart disease. Herrick is also famous for creating the term "heart attack." The first coronary heart catheterization was performed in 1929 by Werner Forssmann. Mason Sones (1918–1985) developed selective coronary angiography by injecting small amounts of contrast medium directly into the coronary arteries. This landmark technique became a routine procedure in the Cleveland Clinic in 1959. The treatment of coronary artery disease changed in 1967, when Rene Favaloro of the Cleveland Clinic performed the first coronary bypass surgery. This open-heart technique involved removing a vein from another part of a patient's body and using it to connect the aorta and coronary arteries, creating a "bypass" around blocked, coronary arteries. The first coronary angioplasty was performed in 1977 by Andreas Gruentzig at a Zurich hospital. During the procedure, a catheter (tiny tube) is inserted into a coronary artery that is largely blocked and a tiny balloon is inflated, compressing the plaque against the walls of the artery. In 1986, tiny metal tubes called stents were developed and utilized to prop open the artery and maintain coronary blood flow. Jacques Puel and Ulrich Sigwart, working at a French hospital, inserted the first stents in a human coronary artery.[5]

In parallel with these surgical interventions, there was an ongoing development of pharmacological treatments. As previously mentioned, early treatments aimed to simply relieve symptoms such as angina pectoris and to make the failing heart beat harder; however, it was not until the mid-20th century, when specific molecular targets were identified for specific risk factors such as hypertension and hypercholesterolemia, that effective treatments for CHD became available.

The drugs described in this review target cardiovascular risk factors such as hypercholesterolemia and hypertension; some of these same drugs are used to treat secondary manifestations, such as angina, and heart failure. The anticoagulant and antiplatelet drugs prevent clot formation and coagulation and are used extensively in patients at risk for acute myocardial infarction (MI) and stroke. Antifibrinolytic agents are used to destroy clots that have already formed within arteries and veins.

3 Lipid-Lowering Agents

Atherosclerosis is an insidious disease of the arteries which can begin in the first decade of life with the deposition of lipids within the arterial intima. The resulting fatty streaks, whose most prominent component is the cholesteryl ester-enriched macrophage foam cell, eventually progress to form more complicated, fibrous lesions which can obstruct blood flow, creating turbulence which often results in laceration of the atheroma and emboli formation. The resulting MI is the leading cause of death in the industrialized nations.

Early hypotheses of the pathogenesis of this disease by Virchow[6] were confirmed by the landmark studies of N. N. Anitschkow, who fed rabbits egg yolks and produced a high level of cholesterol in the blood as well as atherosclerosis in the aorta and coronary arteries.[7] Since these experiments were performed, there have been many clinical studies which have established hypercholesterolemia as a definite risk factor for the development of atherosclerosis in man. The landmark Framingham Heart Study provided solid and unarguable evidence that hypercholesterolemia (and hypertension and smoking) was a definitive risk factor for the development of CHD. This study was carried out by the National Heart Institute in the small town of Framingham, Massachusetts, beginning in 1950 and continuing to this day. Measurements were made of most of the relevant risk factors known at the time.[8] Over the years it was shown that the risk factors were at least additive. Later studies identified additional risk factors, including diabetes, obesity, low high-density lipoproteins (HDL), lack of exercise, family history of CHD, and others. It is now generally accepted that atherosclerosis is initiated or aggravated by a variety of environmental and genetic factors. The Expert Panel of the National Cholesterol Education Program (NCEP) in the USA has recognized 10 such risk factors, including smoking, hypertension, severe obesity, gender, and diabetes mellitus.[9] However, among these risk factors, only an elevated serum total or, more specifically, low-density lipoprotein cholesterol (LDL-C) level has been shown to be an unequivocal independent risk factor for increased morbidity due to MI.[10] Hypercholesterolemia is defined as a total serum cholesterol level $\geq$ 240 mg/dL and in 2010 it had a prevalence in the US population of 16.2%. Thus, the lowering of plasma cholesterol by dietary or pharmacological intervention has become an accepted preventative measure against CHD. However, it was not until the landmark results of the Lipid Research Clinic Coronary Primary Prevention Trial (LRC-CPPT) became available in 1984 that a widespread acceptance was established among clinicians about the desirability of treating a single risk factor, asymptomatic hypercholesterolemia, and its efficacy in reducing coronary risk.[11] This study demonstrated that long-term treatment (seven years) with the bile acid sequestrant cholestyramine (**1**) resulted in a 12.5% decrease in LDL-C, which was associated with a 19% decrease in the incidence of MI and CHD death. The Helsinki Heart Study showed that treatment with gemfibrozil (**2**) led to an 11% decrease in LDL-C and an 11% increase in HDL-cholesterol (HDL-C); these changes were associated with a 34% decrease in the CHD end points of MI and/or deaths.[12] The Cholesterol Lowering Atherosclerosis Study (CLAS) demonstrated that aggressive therapy with niacin (**3**) and another bile acid sequestrant, cholestipol (**4**), resulted in a 43% reduction in LDL-C and a 37% elevation in HDL-C. Repeat angiography demonstrated significant reductions in both the progression and regression of atherosclerotic lesions.[13] However, in none of these trials was a statistically significant difference in overall mortality demonstrated.

The drugs involved in these trials were state of the art at the time; however, they all exhibited some undesirable side effects, had poorly understood mechanisms of action, and were limited in their efficacy.

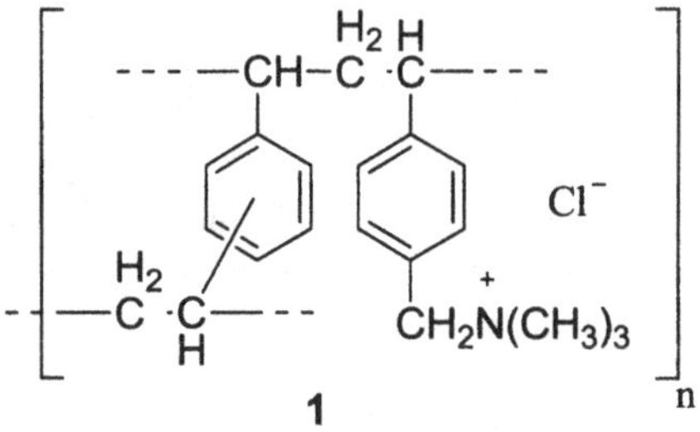

3.1   *Nicotinic Acid Derivatives*

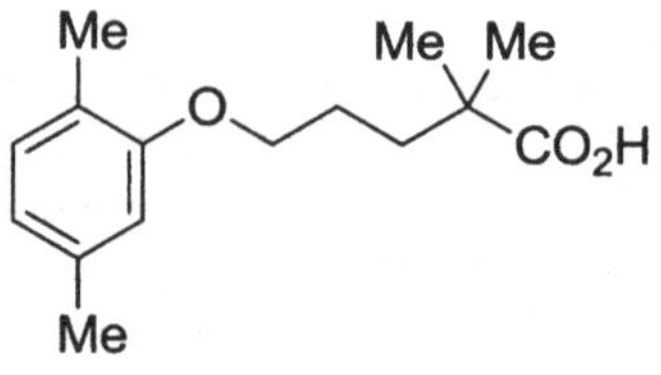

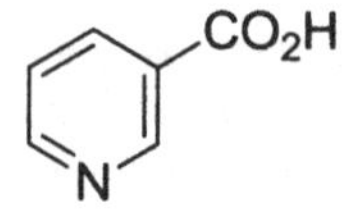

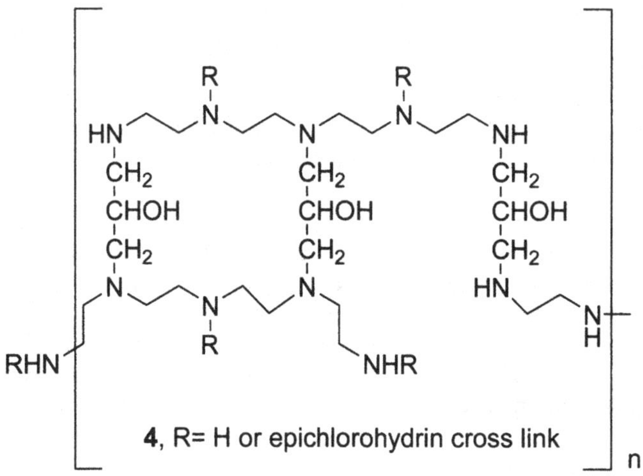

Since its first synthesis in 1897 and the subsequent discovery, in 1937, of its role as a vitamin (vitamin B_3), nicotinic acid (**3**, niacin) and its biological actions has fascinated scientists for well over a century. In milligram doses, it is used to treat niacin deficiency and its associated disease, pellagra. However, in 1955, research interest really accelerated when Rudolf Altschul at the University of Saskatchewan reported that nicotinic acid, this time in gram doses, lowered plasma cholesterol in both normal and hypercholesterolemic patients.[14] Subsequent studies revealed that it also lowered VLDL (very low density lipoprotein), and triglycerides (TGs) as well as elevating HDL.[15] The precise mechanisms contributing to the lipid-lowering effects were obscure for many years; however, the inhibition of lipolysis (lipase-mediated conversion of esterified fatty acids to free fatty acids and subsequent release into the circulation) in adipose tissue has emerged as the most viable mechanism to explain the observed hypolipidemic effects.[16] It is only relatively recently that a greater understanding of the precise cellular mechanism has evolved with the discovery of the nicotinic acid receptor that is highly expressed in adipose tissue and to which nicotinic acid is a high-affinity ligand, although the endogenous ligand is unknown. It is thought that activation of this receptor gives rise to an inhibitory G-protein signal, which reduces adipocyte cAMP concentrations and thus inhibits lipolysis.[17]

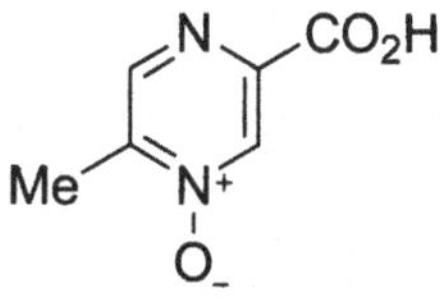

For many years this was the only drug available for lowering plasma cholesterol. However, it was quickly recognized that at these multigram doses niacin exhibited some unpleasant side effects, the most common ones being an intense flushing sensation on the face and upper part of the body and an increase in uric acid that predisposes the patient to the development of gout. The flushing was subsequently found to be due to the niacin-mediated effect of stimulating the biosynthesis of prostaglandin D_2, a potent vasodilator acting at the prostaglandin D_2 receptor, increasing blood flow and thus leading to flushes.[16] Efforts continue to this day to discover novel nicotinic acid analogs and/or different formulations to lessen the severity of these side effects. One analog available in Europe and Asia, acipimox (5, Olbetam), has been shown to exhibit less flushing than niacin although it was not as efficacious.[18] Niaspan is a prolonged-release formulation of niacin that exhibits reduced flushing when compared to immediate-release niacin; it was initially approved by the FDA in 1997. It is administered only once a day at bedtime, in contrast to immediate-release niacin, which must be administered 3–4 times daily.[19] One of the more novel approaches to dealing with this problem was the development of Merck's Cordaptive, a combination of niacin and laropiprant, a prostaglandin D_2 receptor antagonist, that was intended to decrease the vasodilation associated with niacin therapy and thus lessen the flushing.[20] However, in April 2008, the FDA rejected this drug when the beneficial effect on flushing was not as great as was expected and was comparable to those patients receiving just aspirin. However, in that same year it was approved as Tredaptive in the EU. Interestingly, combinations of various statins and niacin have been approved; Advicor (lovastatin/niacin) in 2001 and Simcor (simvastatin/niacin) in 2008.

3.2   Bile Acid Sequestrants (BAS)

It has been known for many years that the bile acids secreted into the intestine are very efficiently reabsorbed and returned to the liver. If these bile acids were then trapped, using a substance that bound bile acids, in the intestine and excreted in the feces, it would require an increase in cholesterol biosynthesis, and subsequent new bile acid synthesis, to maintain a steady state. This was later confirmed when it was shown that as the bile acid pool becomes depleted, cholesterol 7α-hydroxylase, the rate-limiting enzyme in the production of bile acids from cholesterol, is upregulated, which increases the conversion of cholesterol into bile acids. This increased demand for cholesterol in hepatocytes results in an increase in the transcription and activity of HMG-CoA reductase (catalyzes the rate-limiting step in cholesterol biosynthesis) and a concomitant increase in the number of hepatic LDL receptors. These effects result in an increased clearance of LDL-C from the blood, resulting in decreased serum LDL-C levels.[21] In 1953, Siperstein et al. showed that ferric chloride, which precipitates bile acids in vitro, lowered plasma cholesterol levels in cholesterol-fed cockerels when fed a high-cholesterol diet.[22] In 1960, Tennent et al. at Merck Sharp and Dohme Research Laboratories showed that high-molecular-weight, nonabsorbable polymers containing a quaternary ammonium anion bound bile acids strongly and lowered plasma cholesterol levels in normocholesterolemic cockerels and dogs.[23] One of these polymers, cholestyramine (3, MK-135, Questran), is a solid and was shown to be relatively nontoxic when dosed for over a year. Cholestyramine was first used in clinical practice in 1961 to treat a patient with xanthomatous biliary cirrhosis where cholesterol lowering was observed in addition to a resolution of the associated pruritus (due to retained serum bile acids).[24] In 1963, it was also used to treat the pruritus of obstructive jaundice patients and some plasma cholesterol lowering was observed.[25] In 1964, cholestyramine was shown to lower plasma cholesterol in two siblings with familial hypercholesterolemia.[26] After efforts were initiated to improve palatability, cholestyramine was approved in 1973 for the treatment of hypercholesterolemia in patients resistant to dietary modifications. Since these drugs are not orally absorbed, they produce minimal side effects. Constipation, bloating, and abdominal discomfort are the most frequent patient complaints. Colestipol (**4**, Colestid), a high-molecular-weight insoluble copolymer of tetraethylenepentamine and epichlorohydrin, was discovered in 1970 and approved in 1977.[27] In 1994 a more convenient pill form of this resin was approved. The development of new resins continues in an effort to develop low -dose resins with a lower incidence of side effects and more efficient binding of bile acids. Colesevelam hydrochloride (Welchol), approved in 2000, has a unique water-absorbing hydrogel formulation that is claimed to lessen the abdominal side effects associated with this class of drugs.[28] Colestimide (Cholebine) was approved for usage in Japan in 1999 and it is claimed to be more potent than cholestyramine.[29] These agents have also been shown to lower blood glucose in type 2 diabetes mellitus patients, although the mechanism of action is poorly understood.[30]

3.3   *Fibric Acids*

Fibric acids (and fibrates, their ester analogs) continue to be generally effective in lowering elevated plasma TGs by about 40–50% and LDL-C by about 10–15% and is one of the few compound classes that actually elevate HDL-C (by about 10–20%). The fibrates emerged from a screening program run by ICI in the mid-1950s. In 1962, Thorp and Wareing of ICI described the hypercholesterolemic activity of clofibrate (**6**, Atromid-S) in rats fed a normal diet.[31] In humans, clofibrate was quite effective, decreasing plasma cholesterol by about 20%; however in the WHO (World Health Organization) clofibrate trial it was shown that, despite having a favorable effect on cardiac events, clofibrate, for reasons as yet unknown, actually increased overall mortality.[32,33] The relative success of clofibrate stimulated a flourish of research into the discovery of more potent and safer analogs. After screening about 8700 compounds, researchers at Parke Davis synthesized gemfibrozil (**2**, Lopid) in 1968. Launched in 1982, this compound was very effective at lowering plasma TGs and VLDL and moderately effective at lowering plasma LDL and elevating plasma HDL-C levels.[34] In 1975, the French company Fournier discovered fenofibrate (**7**, and its corresponding acid, fenofibric acid, **8**).[35] It received marketing approval in the USA in 1994 and then established a marketing partnership with Abbott to market the drug as Tricor (approved in 2001). In 2008, fenofibric acid (Trilipix) was approved as the first and only fibrate to be used in combination with a statin. Other fibrates such as bezafibrate (**9**, Bezalip) and ciprofibrate (**10**, Lipanor) are available in Europe and Asia. For many years the exact mechanism of action of these agents was poorly understood but was thought to include: a) induction of lipoprotein lipolysis through increased lipoprotein lipase (LPL) activity; b) induction of free fatty acid uptake and a reduction of hepatic TG production; c) formation of an LDL particle with an increased affinity for the LDL-receptor; d) reduction in cholesteryl ester and TG exchange between VLDL and HDL; and e) increase in HDL production. However, the exact molecular target that mediates these changes was unknown for many years.[36] It was known that fibrates induce peroxisome proliferation by inducing the transcription of genes involved in peroxisomal β-oxidation and is mediated by transcription factors called peroxisome proliferator-activated receptors (PPARs). It was subsequently found that fibrates are agonists of PPARα. PPARα agonism, among a whole host of other things, stimulates the expression of the major HDL-apolipoproteins (A-I, A-II), decreases the expression of hepatic apoC-III transcription, which enhances the clearance of TG-rich lipoproteins, and stimulates the expression of LPL.[37] In general, the fibrates are well tolerated although combination therapy with statins has led to reports of kidney failure occurring. Statin/gemfibrozil combinations are particularly problematic whereas statin/fenofibrate combinations have much less associated risk. This difference in fibrate safety has been attributed to pharmacokinetic interactions. Future developments with fibrates are aimed at developing tissue-selective agents with targeted gene-selective activities, i.e. the so-called SPPARMs (selective PPAR modulators). This concept is supported by the demonstration of differential PPARα-dependent effects of gemfibrozil and fenofibrate on hepatic apoA-I expression. Whereas both of these agents elevate HDL in humans, only fenofibrate elevates apoA-I levels, whereas gemfibrozil has little or no effect. *In vitro* profiling has shown that fenofibrate acts as a full agonist of PPARα, whereas gemfibrozil acts as a partial agonist.[38]

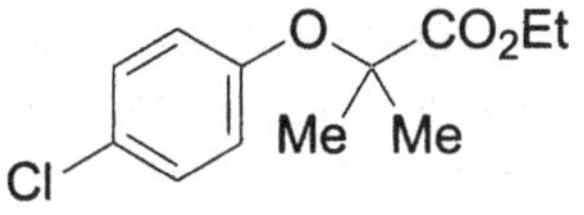

6

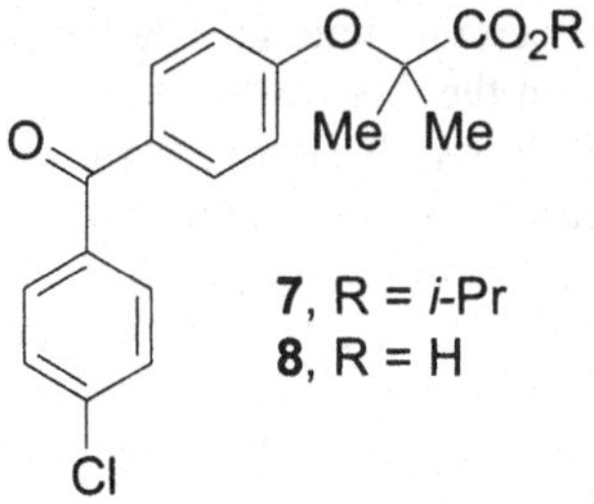

7, R = *i*-Pr
8, R = H

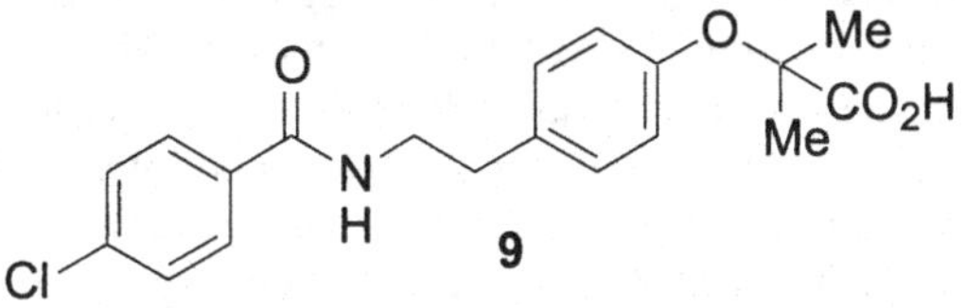

9

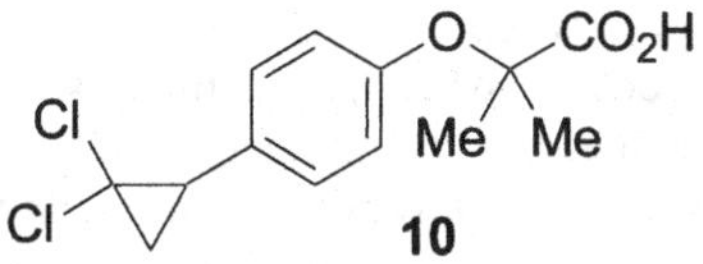

10

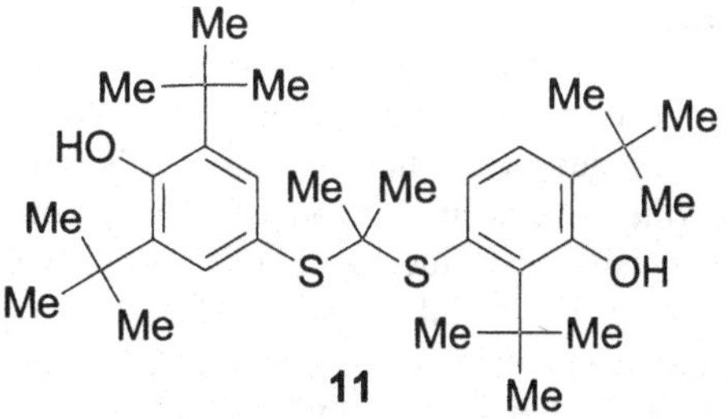

11

3.4   *Probucol*

The hypocholesterolemic activity of probucol (11, Lorelco) was first demonstrated in animals by Barnhart *et al.* in 1970.[39] Efficacy in humans was demonstrated in the period 1969–1971. Despite its modest effects on lowering lipids in humans, it was approved by the FDA in 1977. What made this compound very interesting was the fact that it had a dramatic ability to promote the regression of cutaneous and tendinous xanthoma (cholesterol rich depositions just under the skin) in patients, independent of its cholesterol-lowering effect.[40] Originally synthesized as an antioxidant by the Consolidation Coal Company, it was subsequently shown that, *in vitro*, probucol inhibits the oxidative modification of LDL and inhibits foam cell formation. However, it also significantly decreases HDL-C in humans as well as causing an elongation of the QT interval on ECG.[41]

Although studies continue to this day to elucidate the mechanism of action of this interesting drug, it was withdrawn from the U.S. market in 1995 in the face of the growing dominance of the statins. It is still marketed in Japan.

As can be seen, the physician did not have very many effective options for lowering plasma cholesterol in patients. The advent of the statins changed all of that.

3.5   *HMG-CoA Reductase Inhibitors*

Cholesterol is an essential component of all eukaryotic cells. It is utilized as a structural component in the membranes of cells as well as in the production of steroid hormones, bile acids, lipoproteins, and certain vitamins. Cholesterol within the body originates from two sources, either by absorption from the diet (exogenous source), which accounts for 300–500 mg/day in humans, or by endogenous biosynthesis within the tissues of the body (700–900 mg/day).[42]

Cells possess an enzymatic pathway to synthesize the cholesterol they need. In a series of enzyme catalyzed reactions, acetyl-CoA is transformed into cholesterol. These reactions take place both in the cytosolic compartment and on specific membranes.[43] The rate-limiting step in this sequence is the conversion of hydroxymethylglutaryl-CoA (HMG-CoA) into mevalonic acid and it is catalyzed by the enzyme HMG-CoA reductase (HMGR).[44] Inhibition of any of the steps in this pathway would, in theory, lower the amount of cholesterol produced by the body. In order to maintain cholesterol homeostasis within the cell, negative-feedback mechanisms, involving cholesterol, suppress the activities of both HMG-CoA synthase (HMGS) and HMGR.[45] The rate of cholesterol synthesis within the liver is decreased dramatically when animals or humans are fed a high-cholesterol diet. The opposite occurs on a low-cholesterol diet when endogenous biosynthesis is increased to compensate for the low levels of cholesterol obtained from the diet. This regulatory mechanism is mediated by the transcription factors, SREBPs (sterol regulatory element-binding proteins 1 and 2), and their ability to sense the levels of intracellular cholesterol and turn on/off the synthesis of key enzymes in the biosynthetic pathway.[46] When the cell is replete with cholesterol, the activities of both HMGS and HMGR decrease by more than 90% and the cells produce only the small amounts of mevalonate needed for the production of dolichol and ubiquinone.[44] In addition, when cellular sterols rise, the LDL receptor (LDL-R) gene is repressed, avoiding further overaccumulation of cholesterol. Conversely, when intracellular sterol levels drop, the cell responds by increasing cholesterol synthetic rates and the number of LDL-R's. Thus, the mechanism by which inhibition of HMGR lowers plasma LDL levels is by upregulating LDL-R activity and increasing the clearance of plasma LDL.[47]

Curran and Azarnoff outlined the desirable characteristics of an inhibitor of cholesterol biosynthesis. They identified the preferred site of action to be post-acetoacetate, since before that, crucial intermediates are required for other cellular processes, and presqualene, because it was thought that steroid precursors of cholesterol were themselves atherogenic.[48] The earliest known inhibitor was shown to be 2-phenylbutyric acid, which was shown to inhibit the formation of acetyl-CoA.[49] Even though this compound was shown to be hypocholesterolemic in rats,[50] it had no effect in humans.[51] Other early inhibitors were based on the observation that premenopausal women are less susceptible to atherosclerosis than men.[52] This was thought to be due to the protective effects of estrogen, so a search for a nonestrogenic estrogen was undertaken. Structure–activity studies on chlorotrianisene (**12**), a synthetic estrogen, yielded triparanol (MER-29, (**13**)).[53] This compound was shown to be hypocholesterolemic in rats[54] and was eventually approved for human use in 1960.[55] However, in 1962 it was withdrawn from use due to the development of a number of side effects, including cataract formation,[56] alopecia, and icthyosis.[57] It was later shown that the compound inhibited the biosynthetic pathway at the penultimate step, i.e.; the conversion of desmosterol to cholesterol, and that desmosterol accumulated in the lenses of patients.[58]

It was recognized early in the development of HMGR inhibitors that either substrate or product analogs could act as competitive inhibitors. 3-Hydroxy-3-methylglutaric acid has been shown to lower plasma cholesterol in man and shown to inhibit HMGR; however, additional mechanisms of action could not be ruled out.[59]

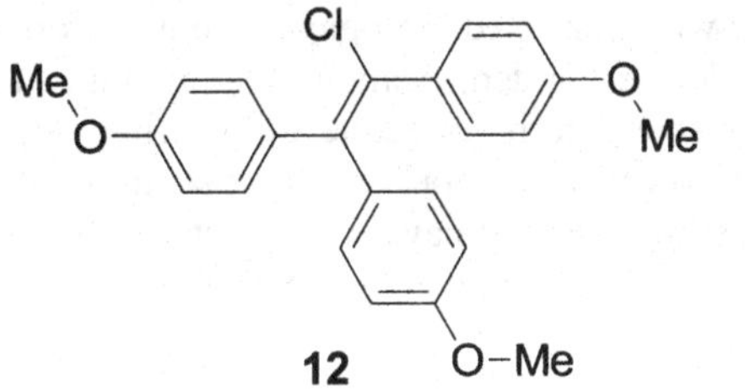

12

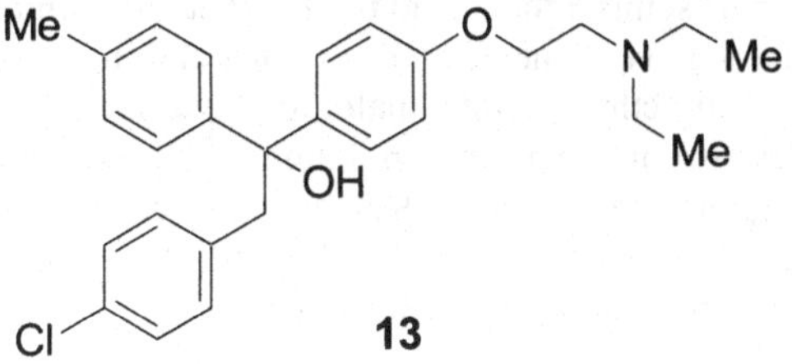

13

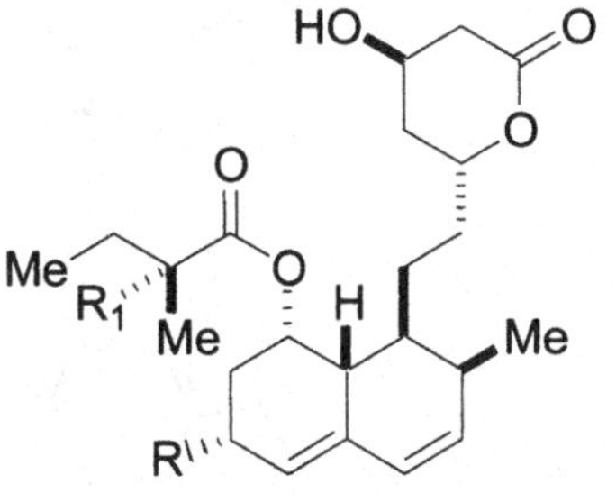

14 R = H, R$_1$ = H
15, R = CH$_3$, R$_1$ = H
17, R = R$_1$ = CH$_3$

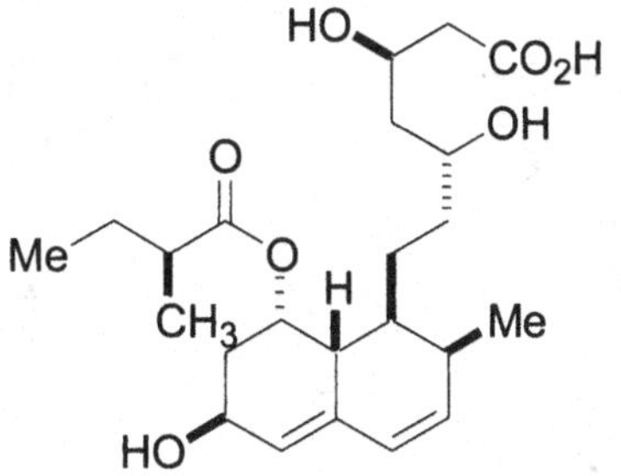

16

The seminal finding in the quest for inhibitors of HMGR was the discovery of the fungal metabolites mevastatin (**14**) and lovastatin (**15**). In 1973, Endo and co-workers, working at Sankyo in Japan, discovered mevastatin (**14**, ML-236B) in the culture broth of the fungus *Penicillium citrinum*.[60] It was shown to be identical to a compound (called compactin) isolated by Brown *et al.* in 1976 from cultures of *Penicillium brevicompactum*.[61] The active form of these compounds is the ring-opened dihydroxy acid form. Inhibition of HMGR by these compounds is reversible and competitive with respect to HMG-CoA and noncompetitive with respect to NADPH. These discoveries initiated a new search for more potent inhibitors. In 1979, Endo discovered a closely related analog of mevastatin, called monacolin K, from cultures of the fungus *Monascus ruber*.[62] The same compound (**15**, lovastatin, originally called mevinolin) was isolated, independently by workers at Merck, from *Aspegillus terreus* just three months earlier.[63] Several active compounds have been derived from either compactin (**14**) or lovastatin (**15**) by microbial conversion. The most noteworthy of these compounds is pravastatin (**16**, CS-514), the 6β-hydroxy open acid form of compactin. This compound was originally identified, by workers at Sankyo in 1983, by microbial transformation of compactin using *Norcardia autotrophica*.[64] It was also identified as a urinary metabolite of compactin in the dog.[65]

The ability of these compounds to inhibit cholesterol biosynthesis has been evaluated *in vitro* and *in vivo*.[62] Compactin (14), lovastatin (15), and pravastatin (16) have all been shown to be effective hypocholesterolemic agents in humans. Both lovastatin (Mevacor, approved in 1987) and pravastatin (Pravachol, approved in 1991) are marketed around the world for the treatment of hypercholesterolemia. In the mid-1980s, compactin was discontinued in clinical trials due to toxicity in dogs.[66]

The discovery of the fungal metabolites has led to a number of SAR studies designed to identify the pharmacophore necessary for HMGR inhibition and to use this information in the design of semisynthetic and totally synthetic inhibitors with improved biological activity. Workers at Merck have pioneered these efforts and their studies led to the identification of simvastatin (17), the 2,2-dimethylbutyrate analog of lovastatin.[67] It was shown to be effective in lowering plasma cholesterol in dogs and in humans.[68] Simvastatin (Zocor) was approved for use in humans throughout the world in 1991 and was used in the landmark Scandinavian Simvastatin Survival Study (4S) trial that established the utility of this class of agents in positively affecting morbidity and mortality in patients with CHD.[69]

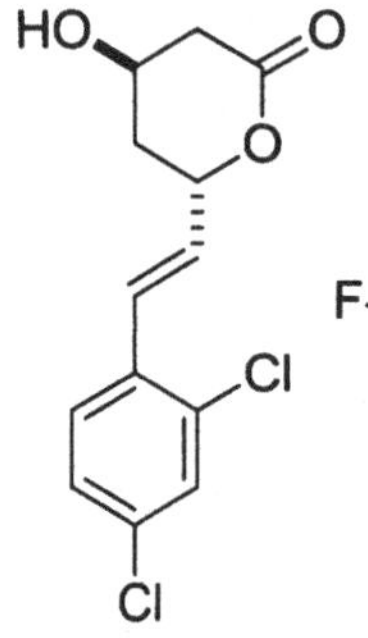

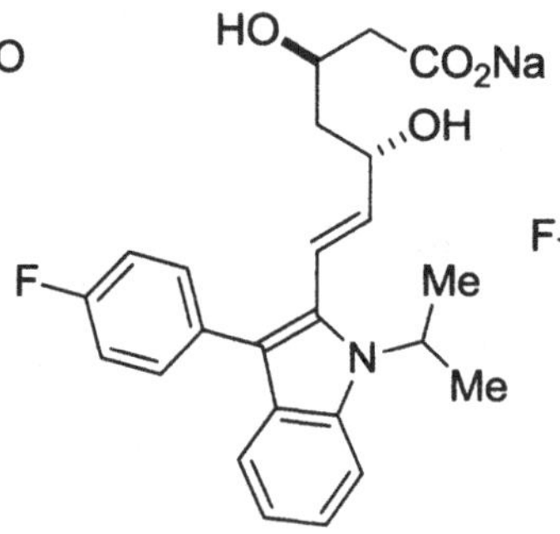

Because of the success of the fungal metabolite HMGR inhibitors at lowering plasma cholesterol and LDL-C in animals and man, major efforts to elucidate the key structural features needed for potent inhibition were initiated at several pharmaceutical and academic research institutions. The first reports of totally synthetic inhibitors, incorporating the mevalonate moiety found in the naturally occurring fungal metabolites, originated from the Sankyo[70] and Merck laboratories.[71] The work from Merck proved seminal. Utilizing a 2,4-dichlorophenyl group as a surrogate for the hexahydronaphthalene ring system of the fungal metabolites, it was demonstrated that all of the biological activity was contained in the dextrorotatory mevalonolactone side-chain isomer (+)-18,[71] which was subsequently demonstrated to correspond to the specific 4R, 6R stereochemistry found in both HMG-CoA and the fungal metabolites.[72] With the key topological features for potent inhibition defined, it was left to others to conclude that the actual nature of the centerpiece was relatively unimportant and served only as a template to hold the lactone and a lipophilic moiety in the proper spatial arrangement and provide sufficient lipophilicity for tight binding to the enzyme.[73] Because of this observation and the seemingly essential nature of the mevalonolactone moiety, most of the effort in the synthetic inhibitors has focused on variations of the central template. This has led to a host of inhibitors containing a wide variety of templates.

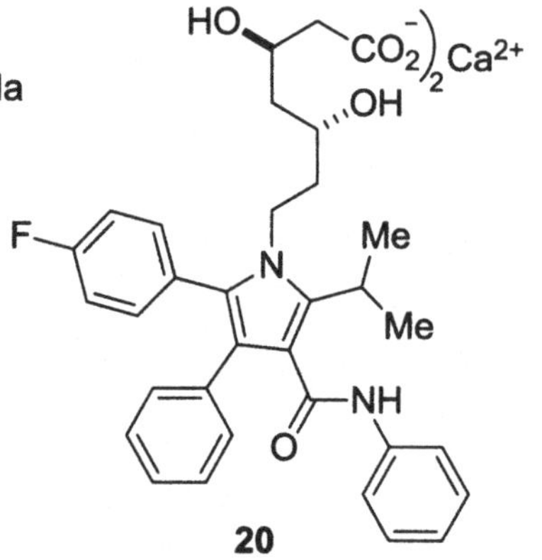

The first totally synthetic HMGR inhibitor approved for use in the United States in 1994 was fluvastatin (**19**, Lescol, Sandoz).[74]

    The stepwise development of the pyrrole series is illustrative of the evolution of synthetic inhibitors. The early SAR in this series involved the definition of the optimal substituents in the 2- and 5-positions of a pyrrole appended to the mevalonolactone by a saturated 2 carbon spacer at the pyrrole 1-position.[73] In this study, it was found that the 2-position was optimally substituted by a 4-F-phenyl group. At the 5-position, it was found that potency increased with increasing substituent size, with a maximum at isopropyl. An examination of substitution at the 3- and 4-positions revealed that, although introduction of a bulky substituent into either the 3 or 4-position led to an improvement in potency *in vitro*, potency equivalent to the fungal metabolites was obtained only when both positions were substituted by bulky lipophilic groups.[75] The best activity was found in compound **20** (atorvastatin, CI-981, Lipitor,), originally synthesized in 1988 by workers at Parke-Davis in Ann Arbor, Michigan. This compound was shown to be very effective in clinical trials and superior to the existing statins on the market. Atorvastatin received FDA approval in December 1996 and was launched in early 1997. Lipitor went on to become the best selling drug in history with first-year annual sales of $1 billion climbing to $12.9 billion in 2007.

Four months after the Lipitor launch, the FDA approved cerivastatin (**21**, Baycol). This was the sixth statin to reach the market and was extremely potent (initially approved at doses of 0.2 and 0.3 mg/day) when compared to other statins, whose doses typically fall in the range 10–80 mg/day.[76] However, upon reaching the market, and thus reaching more patients, the frequency of a serious side effect, rhabdomyolysis, increased.[77] Rhabdomyolysis is a disorder in which myocytes in muscle tissue break down, which overwhelms the kidneys with cellular debris and ultimately leads to kidney failure. This side effect was noted very rarely with the other five statins on the market. Eventually deaths were reported among patients on Baycol and the drug's manufacturer, Bayer, voluntarily withdrew the drug in 2001. The adverse publicity surrounding the Baycol withdrawal made it harder for new statins to be approved. Rosuvastatin (**22**, Crestor) was the seventh statin to reach the market in 2003. It was originally synthesized by workers at Shionogi & Co Ltd. and licensed to Astra Zeneca.[78] This compound acquired the sobriquet of "superstation" since it is probably the most efficacious statin on the market with LDL-C reductions of up to 63% on a single daily dose of 40 mg/kg. In the wake of the Baycol withdrawal the FDA delayed its approval decision for over a year until the FDA thoroughly addressed concerns over its safety profile. The mechanism of statin-induced myalgia is thought to involve, at least in part, inhibition of HMGR in nonhepatic tissues (especially muscle). Thus it was thought that the development of myalgia could be reduced by targeting statins to hepatic tissues and limiting peripheral exposure.[79] It was already known that the hepatoselectivity of a statin is related to its degree of lipophilicity.[80] Thus, in general, lipophilic statins often achieve higher levels of exposure in peripheral tissues compared with hydrophilic statins, which tend to be more hepatoselective. Hepatoselectivity with statins of different lipophilicities has been demonstrated in primary rat hepatocytes and myocytes (muscle cells). A comparison of the potency of these compounds to inhibit cellular cholesterol biosynthesis in these cell types provides a measure of hepatoselectivity. It was shown that lipophilic statins, such as Baycol, were nonselective, whereas hydrophilic statins such as Pravachol and Crestor were more hepatoselective. It is thought that lipophilic statins passively and nonselectively diffuse into both hepatic and nonhepatic cells, while the less permeable hydrophilic statins rely on active transport mechanisms for cellular penetration.[79] It has been shown that both Pravachol and Crestor are substrates for active transport via the organic anion transporter (OAT), which are principally expressed in hepatocytes. The active uptake of the hydrophilic statins by the liver reduces drug concentrations in the plasma and peripheral tissues, thus reducing the potential for muscle-related side effects.[81] Research continues to this day in the quest to identify hepatoselective statins.[79]

It has generally been assumed that the reduction in plasma cholesterol achieved by statin therapy is the predominant, if not the only mechanism, underlying their beneficial effects on CAD. However, this assumption is increasingly being challenged and a subgroup analysis of large clinical trials has suggested that the beneficial effects of statins may extend to mechanisms beyond cholesterol reduction.[82] Meta-analyses of cholesterol-lowering trials suggest that the risk of a coronary event in individuals treated with a statin is significantly lower than in individuals treated with other agents despite a comparable reduction in cholesterol levels in both groups.[83] Further evidence is provided by angiographic trials, which have demonstrated clinical improvements with statins that far exceed changes in the size of atherosclerotic lesions.[84] The definitive mechanisms by which the statins exert these additional effects are still a matter for conjecture.

3.6 *Cholesterol Absorption Inhibitors*

As previously mentioned, the cholesterol carried in LDL originates from de novo synthesis within the tissues of the body and from the diet. As has been shown, the statins were effective agents at blocking the de novo synthesis of cholesterol within the tissues of the body. In the early 1980s a concerted effort began to identify agents that would block the absorption of dietary cholesterol. One of the initial targets of this research was the enzyme acyl-CoA:cholesterol O-acyltransferase (ACAT). This is an intracellular enzyme that is located within the endoplasmic reticulum of cells. It catalyzes the esterification of cholesterol and promotes the storage of cholesteryl esters within the cytoplasm of the cell. In the intestinal enterocyte, the cholesterol absorbed during intestinal transport undergoes ACAT-mediated esterification before incorporation into large lipoprotein particles called chylomicrons. In the liver, ACAT-mediated esterification of cholesterol is involved in the production and release of VLDL. ACAT also plays an important role in foam cell formation and atherosclerosis by participating in the accumulation of cholesteryl esters in macrophages. The esterification activity has been attributed to two enzymes, cholesterol ester hydrolase (CEH) and ACAT. Over the years, the bulk of the evidence has pointed to ACAT as being responsible for intracellular cholesterol esterification. Inhibition of this enzyme has been a key target for the pharmaceutical industry and academia.[85] Many inhibitors have been discovered and many have shown profound cholesterol lowering in animal models however; the cholesterol lowering observed in humans has been disappointing. One ACAT inhibitor, avasimibe (**23**), initially discovered by workers at Parke–Davis in 1993,[86] reduced plasma TGs and VLDL cholesterol by up to 30%.[87]Avasimibe failed to reduce plaque volume in atherosclerotic patients after two years of dosing.[88] Pactimibe (**24**) also failed to show a reduction in plaque volume and secondary plaque-related end points in an 18-month study (ACTIVATE trial) of patients with atherosclerosis.[89] In the CAPTIVATE trial, pactimibe once again failed to live up to its earlier promise. In this study it failed to prevent atherosclerotic progression in patients with familial hypercholesterolemia and also caused an actual increase in cardiovascular events.[90] The most important discovery to date in the cholesterol absorption field was the discovery of ezetimibe (**25**), a potent inhibitor of cholesterol absorption that has been approved worldwide for the treatment of hypercholesterolemia (Zetia, Schering Plough). This compound was originally identified as an ACAT inhibitor[91]; however, its underwhelming potency against this target always left doubts that this was its true mechanism of action. It had been thought that the process of cholesterol absorption is mediated by a specific cholesterol transporter. To identify this transporter, researchers at Schering Plough searched human and rodent expressed sequence tag (EST) databases for protein sequences that are highly expressed in the intestine and that contain characteristics of transporters (transmembrane domain, an extracellular signal peptide, N-linked glycosylation sites, and most importantly a sterol sensing domain). They identified a single rodent protein with these features and discovered that it is homologous to human Niemann-Pick C1 Like 1 protein (NPC1L1, also known as NPC3).[92] To test whether NPC1L1 is required for cholesterol absorption *in vivo*, they created a knockout mouse and showed that cholesterol absorption was almost completely abolished in these mice and the animals were insensitive to ezetimibe. The development of a binding assay, using enterocyte brush border membranes (BBMs), unequivocally established NPC1L1 as the direct target of ezetimibe.[93]

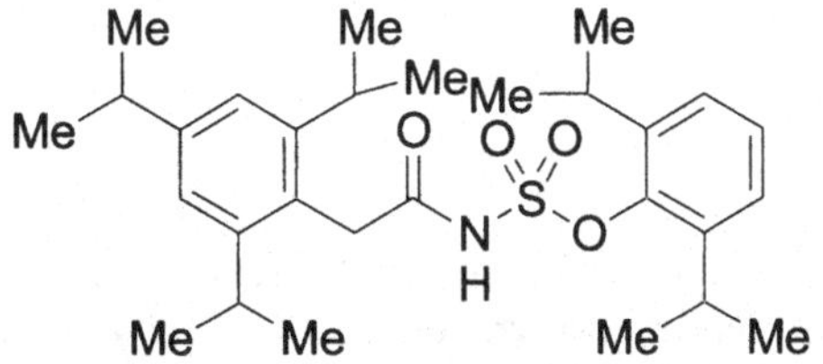

23

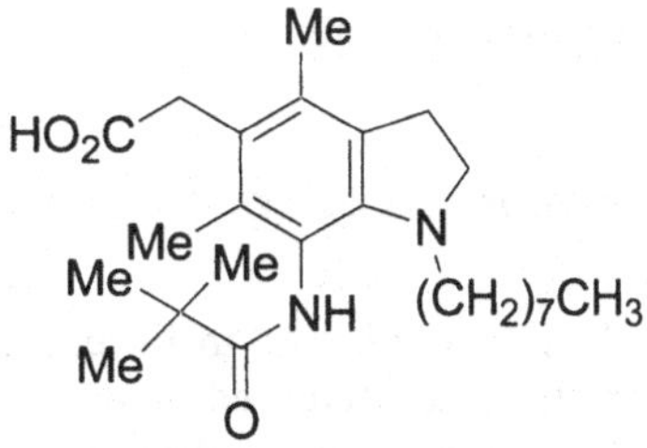

24 hemisulfate

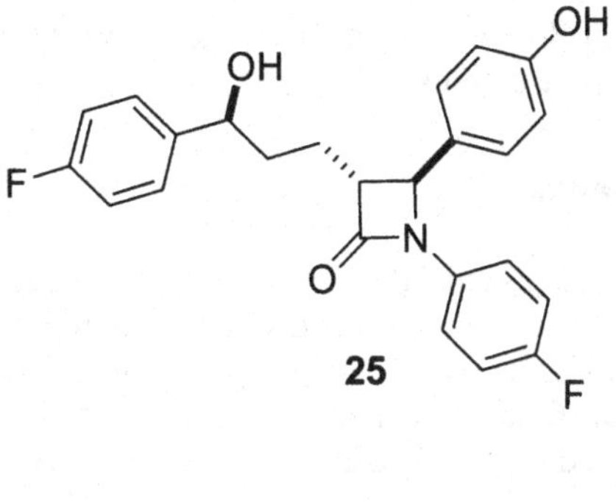

25

Depriving the body of its exogenous source of dietary cholesterol causes a feedback mechanism to compensate by upregulating endogenous biosynthesis. To overcome the expected compensatory increase in cholesterol biosynthesis, resulting from the decrease in both biliary and dietary cholesterol to the liver from the intestine, ezetimibe was combined with simvastatin, a potent HMGR inhibitor, and marketed jointly by Schering Plough and Merck as Vytorin. An effective ad campaign showing how this combination lowered plasma cholesterol by attacking both endogenous and exogenous sources of cholesterol propelled the sales of this combination to $1.9 billion in the US in 2007. However, the publication in January 2008 of the results of the ENHANCE (Ezetimibe and Simvastatin Hypercholesterolemia Enhance Atherosclerosis Regression) trial has raised serious questions about the ultimate efficacy (reducing atherosclerotic plaque and demonstrating clinical benefit) for this modality. This was a two year study comparing daily therapy with 80 mg simvastatin plus either placebo or 10 mg ezetimibe on the average change in carotid intima thickness in patients with familial hypercholesterolemia. Surprisingly, it concluded that this combined therapy did not result in a significant difference in changes in intima-media thickness as compared with simvastatin alone, despite decreases in LDL-C and C-reactive protein.[94] In addition, the results of the SEAS (Simvastatin and Ezetimibe in Aortic Stenosis) trial showed a potential increase in cancer associated with the use of these drugs in combination. There are numerous caveats associated with both of these trials; however, there is no doubt that enthusiasm for this combination has dropped precipitously.[95]

In 1993 the ACAT1 gene was cloned and was quickly followed by the discovery of ACAT2, thus confirming the presence of two ACAT isoforms. This has led to some interesting findings. ACAT1 KO mice showed an accumulation of toxic levels of unesterified cholesterol and mice with macrophages engineered to lack ACAT1 had increased atherosclerotic lesions in a mouse model of atherosclerosis. It was concluded that ACAT1 inhibition (as exhibited by the known ACAT inhibitors) was not desirable. However, mice lacking ACAT2 had a restricted capacity to absorb cholesterol and were protected from atherosclerosis in mouse models of atherosclerosis. Thus, ACAT2 specific inhibition would seem to be desirable and is currently the focus of ongoing research.[96]

3.7 Triglyceride-Lowering Agents

The major class of drugs used to lower plasma triglycerides are the fibrates. Over-The-Counter (OTC) fish oil has also been used to lower triglycerides. Lovaza, originally approved in 2004, is a highly purified fish oil capsule developed by GlaxoSmithKline to lower very high triglyceride levels ($\geq$ 500 mg/dL). The mechanism of action of fish oils is not completely understood. Potential mechanisms of action include inhibition of acyl-CoA:1,2-diacylglycerol acyltransferase, increased mitochondrial and peroxisomal β-oxidation in the liver, decreased lipogenesis in the liver, and increased plasma lipoprotein lipase activity.[97]

3.8 HDL-Elevating Agents

Epidemiological studies have identified HDL-C as a strong, independent, inverse predictor of CHD risk.[98] In the Framingham Heart Study, HDL-C was a more potent risk factor for CHD than LDL-C, total cholesterol, or plasma TGs.[98] An analysis of four large studies has indicated that each 1 mg/dL increase in HDL-C is associated with a 2–3% decrease in the risk of developing CHD.[99] These findings have led to the formal endorsement of HDL-C as an independent risk factor for CHD. So the decision to treat patients with low HDL-C is based on one assumption that an elevation in HDL-C translates into a direct health benefit to the patient. This assumption is increasingly being called into question. Evidence from large-scale randomized clinical trials that elevating HDL-C reduces risk is sparse.[100] The most compelling results, to date, are from the Veterans Affairs High-Density Lipoprotein Cholesterol Intervention Trial (VA-HIT)[101] and the Helsinki Heart Study (HSS).[12] In VA-HIT, CHD patients randomized to gemfibrozil experienced an 11% increase in HDL-C, a 35% decrease in plasma TGs, and no change in LDL-C. Major coronary events were reduced by 22% compared with placebo.[101] The question before the investigators in VA-HIT was how much of the 22% reduction in events could be explained by individual increases in HDL-C. A subsequent regression analysis revealed that only 23% of the relative reduction in risk of coronary events could be explained by the individual one-year changes in HDL-C. In other words, 77% of the benefit of gemfibrozil was unexplained and may be attributable to other mechanisms other than HDL-C elevation. A number of other studies have also questioned the HDL-C hypothesis. In the Bezafibrate Infarction Prevention (BIP) trial, it was shown that, despite an 18% increase in HDL-C, there was only a 9% reduction in coronary events.[102] In the Heart and Estrogen/Progestin Replacement Study (HERS), hormone treatment raised HDL-C by 10% but there was no reduction in the occurrence of coronary events.[103]

In contrast to LDL, a well known pro-atherogenic particle, it is widely believed, based in large part on animal studies, that HDL can inhibit atherogenesis. The exact mechanism by which HDL exerts its protective effect is largely unknown, but the most popular mechanistic explanation has been that HDL facilitates the uptake of peripheral cholesterol and its return to the liver for excretion in the bile and feces. This was a concept first introduced by Glomset in 1968[104] and subsequently termed "reverse cholesterol transport (RCT)". It was later suggested that RCT could be a protective mechanism against atherosclerosis[105] and Miller and Miller suggested that HDL might protect against atherosclerosis by promoting RCT.[106]

This antiatherogenic function of HDL is generally attributable to apo A-I, the principal protein of HDL. The biosynthesis of HDL is complex and involves the synthesis and secretion of Apo AI, followed by the largely extracellular acquisition of lipids (phospholipids and cholesterol) and the assembly and generation of the mature HDL particle. Apo AI is synthesized in the hepatocyte and enterocyte and secreted into the plasma in a lipid-poor form. This newly secreted apo AI rapidly acquires free cholesterol and phospholipids from tissues via the hepatic and intestinal adenosine triphosphate (ATP)-binding cassette AI (ABCAI) transporter to from discoidal HDL particles.[107] The genetic absence of ABCAI forms the molecular basis of Tangier disease, which is associated with extremely low levels of HDL-C and apo AI.[108] The discoidal HDL particles interact with lecithin:cholesterol acyltransferase (LCAT), an enzyme that catalyzes the transfer of a fatty acid from a phospholipid to free cholesterol and which converts a proportion of free cholesterol into cholesterol esters that migrate into the hydrophobic core in a process that converts the disc into a mature, spherical HDL particle.[107] LCAT deficiency in humans and in mice causes markedly reduced levels of HDL-C and one form of LCAT genetic deficiency causes Fish Eye disease in humans.[109] The nascent particles also acquire more lipid from other peripheral tissues and from lipoproteins. When TG-rich lipoproteins undergo hydrolysis of the TG core, surface phospholipids are shed and acquired by HDL. Thus, the activity of lipoprotein lipase is inversely associated with HDL-C levels.[110] Lipoprotein-derived phospholipids are transferred to HDL by phospholipid transfer protein (PLTP).[111] Free cholesterol may be effluxed to mature HDL particles by passive diffusion or by a receptor-mediated pathway including SR-B1 or the ATP-binding cassette transporter G1 (ABCG1).[112]

The major pathway of HDL catabolism is uptake by the liver and the best understood mechanism is that mediated by the hepatic scavenger receptor class B1 (SR-B1). SR-B1 promotes hepatic uptake of HDL-C without mediating the degradation of HDL apolipoproteins in a process known as selective uptake. Studies in hepatocytes suggest that SR-B1 mediates the internalization of whole HDL particles, with subsequent removal of cholesterol and resecretion of smaller cholesterol-depleted HDL particles.[113]

In humans, there is clearly at least one alternative pathway by which HDL-C is metabolized and ultimately transported to the liver, a pathway mediated by the cholesterol ester transfer protein (CETP). CETP transfers TGs from apo B-containing lipoproteins in exchange for HDL-CE, thus resulting in CE depletion and TG enrichment of HDL. The resulting TG-enriched HDL particles are more readily hydrolyzed by hepatic lipase to generate smaller HDL particles, which are more effective in promoting reverse cholesterol transport. While the impact of CETP-mediated lipid transfer on atherogenesis remains controversial, many studies in both animals and humans support the hypothesis that inhibiting CETP activity is beneficial. Rodents naturally lack CETP, and when engineered to express it, they experience substantial reductions in HDL-C levels. A human CETP deficiency is known and results in an extremely high HDL-C level.[107]

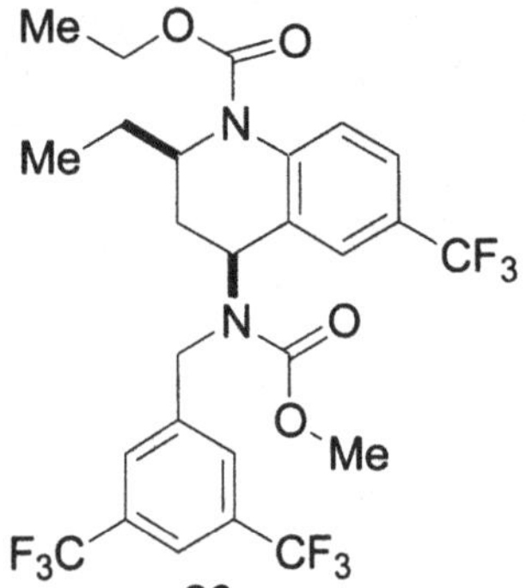

26 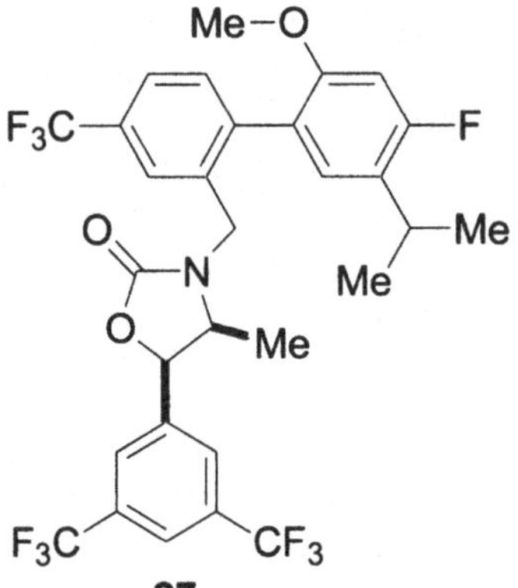   **27**

The efficacy of CETP inhibition on reducing cardiovascular events was further questioned in 2007 when the final results of the ILLUMINATE (Investigation of Lipid Level Management to Understand Its Impact in Atherosclerotic Events) trial were released. This study was designed to test the hypothesis that inhibiting cholesterol ester transfer protein (CETP) with a CETP inhibitor, torcetrapib (**26**), would protect against cardiovascular disease in humans. This trial involved 15,000 subjects, with CHD or CHD risk equivalence, in seven countries over a period of 4.5 years. As predicted, treatment with torcetrapib increased HDL-C by 72% and decreased LDL-C by 25% over and above the changes induced by atorvastatin.[114] However, on December 2, 2006, after a median follow-up of 550 days, the trial was terminated because there was a statistically significant excess of deaths and CV events in the group treated with torcetrapib. It was also found that patients dosed with torcetrapib had increases in blood pressure, serum bicarbonate, serum sodium, and serum aldosterone as well as decreases in serum potassium. These changes are consistent with the activation of the renin-angiotensin-aldosterone system. It has been speculated that the adverse clinical outcome associated with the use of torcetrapib may have been the consequence of off-target pharmacology. It was subsequently shown that torcetrapib induced the synthesis of both aldosterone and cortisol in adrenal cortical cells. The question of whether this was a mechanism-based toxicity or compound-specific toxicity was definitively answered with the publication of the DEFINE (Determining the Efficacy and Tolerability of CETP Inhibition with Anacetrapib) study results. This 18-month study evaluated the efficacy and safety profile of anacetrapib (**27**), a CETP inhibitor structurally dissimilar from torcetrapib, in people with CHD or at high risk for CHD. Anacetrapib was shown to have robust effects on lowering LDL (39.8%) and elevating HDL (138.1%). It did not result in the adverse cardiovascular effects observed with torcetrapib; it did not alter blood pressure, electrolyte levels, or serum aldosterone levels.[115] This finding has reignited the enthusiasm for CETP inhibition as a viable means of attaining HDL elevation.

This preceding discussion on RCT and its attendant receptors, transporters, proteins and enzymes indicates how important this particular function of HDL has become in the search for new strategies to elevate HDL-C levels in humans. Upregulation of hepatic and/or intestinal ABCA1, upregulation of PLTP, upregulation of LCAT, upregulation of SRB1 and CETP inhibition are now considered viable drug discovery approaches to the modulation of HDL-C levels in humans.[116]

However, in addition to its role in RCT, HDL also exhibits other properties that may explain how HDL either prevents or reduces CHD. HDL has been shown to possess antioxidant, anti-inflammatory, and antithrombotic activities and going forward future investigators may have to devote more attention to the effects of their potential treatments on the actual functional properties (e.g., antiinflammatory, antioxidant, antithrombotic) of HDL and not necessarily on the absolute levels of HDL-C. New therapies may improve the function of HDL without necessarily increasing the steady-state level of HDL-C.[117]

4   Antihypertensive Agents

Hypertension is a chronic medical condition in which the systemic arterial pressure is elevated [diastolic (resting) blood pressure $\geq$ 90 mm Hg, systolic blood pressure $\geq$ 140 mm Hg, or a combination of both]. There are two major classifications of hypertension. One, "essential hypertension," is a disorder in which no definitive cause of the disease can be identified; the other is "secondary hypertension" where a definitive cause can be identified. The most common form is "essential hypertension", which accounts for nearly 90% of all cases. This suggests that clearly defined causes underlying the disease can only be identified in a relatively small percentage of patients. In addition, this suggests that the management of the disorder may require different approaches since the basis for the elevated pressure may differ among individuals. This explains the large number of agents of differing mechanisms available to treat this disease.[118]

Hypertension is present in 92% of all patients presenting with any cardiovascular disease and the American Heart Association estimates that nearly 80 million individuals in the USA have high blood pressure, as previously defined. Hypertension also contributes to nearly 50% of all cardiovascular diseases either independently or as a comorbid factor. In the USA, hypertension is diagnosed in nearly 30% of the population, however, only 68% were taking antihypertensive medications, and of those patients, only 44% had their disease adequately controlled by that medication. Untreated hypertension is a major risk factor for sudden death due to MI, stroke, CHF, and renal failure.[119]

Hypertension is a complicated disease and research into understanding the mechanisms that contribute to the development of this disease is ongoing. In the heart the left ventricle pumps blood into the systemic circulation via the aorta, and a pressure is generated. The mean arterial pressure (MAP) is the product of cardiac output (CO) and the total peripheral resistance (TPR). Therefore, to maintain blood pressure in the vasculature, at physiologically relevant levels, the factors that influence both CO and TPR need to be tightly controlled. These factors can be under neural, hormonal or local control and the drugs that have been developed to modulate these factors form the basis of all antihypertensive therapies. In addition to the vasculature, the heart and the kidney indirectly influence blood pressure. For example, total blood volume is regulated by renal function, an elevation in blood volume increases CO, and the diuretic antihypertensive drugs enhance sodium and water removal by the kidneys and thereby decrease blood volume and CO. Cardiac output is the product of stroke volume (SV) and the number of beats per minute (heart rate); thus the β-blockers exert their antihypertensive effect by decreasing heart rate and thus decreasing CO. Changes in stroke volume (and CO) can be accomplished by changes in the contractility (inotropy) of the ventricular muscle fibers; the calcium channel blockers (CCBs) have been employed to decrease the contractility (and thus the CO) in hypertensive patients. The most important mechanism for changing TPR involves changes in vessel lumen diameter. All arterial and venous vessels under basal conditions exhibit some degree of smooth muscle contraction that determines the diameter and hence tone of the vessel. Vascular tone is determined by many different competing vasoconstrictor and vasodilator influences acting on the blood vessel. These influences can be separated into extrinsic factors that originate from outside of the organ or tissue in which the blood vessel is located, and intrinsic factors that originate from the vessel itself or the surrounding tissue. The primary function of extrinsic factors is to regulate arterial blood pressure by altering systemic vascular resistance, whereas intrinsic mechanisms are important for local blood flow regulation within an organ. In general, extrinsic factors such as sympathetic nerves and circulating angiotensin II increase vascular tone (i.e., cause vasoconstriction); however, some circulating factors (e.g., atrial natriuretic peptide) decrease vascular tone. Intrinsic factors include endothelial factors such as nitric oxide and endothelin, which can either decrease or increase tone, respectively; local hormones/chemical substances (e.g., arachidonic acid metabolites, histamine, and bradykinin can either increase or decrease tone. Drugs affecting TPR include vasodilators such as CCBs, alpha blockers, ACE inhibitors, angiotensin receptor blockers and direct-acting vasodilators. As can be seen, the regulation of blood pressure involves a number of neural (autonomic) and, humoral (hormonal) factors. Neural mechanisms involve the autonomic nervous system and in particular, two subdivisions: the sympathetic nervous system (SNS) and the parasympathetic nervous system (PNS). The SNS stimulates the heart and constricts blood vessels resulting in a rise in MAP. The PNS depresses cardiac function and dilates selected vascular beds. The most important humoral mechanisms involved in hypertension include the renin-angiotensin-aldosterone system (RAAS), circulating catecholamines (epinephrine and norepinephrine), vasopressin, and endothelin. All of these systems directly or indirectly alter cardiac function, vascular function, and arterial pressure.[120]

The factors determining MAPs and the mechanisms regulating those factors have been elucidated over the past one hundred years. Early antihypertensive therapies lowered blood pressure without an understanding of the mechanisms involved, it was not until the mid-20th century that the elucidation of these mechanisms became clearer and exact molecular targets, suitable for intervention, were identified. At this point the field exploded and the golden era of antihypertensive therapy began.

There are very early writings as far back as 2600 BC that describe the use of leeches and venesection (bleeding) to reduce the blood volume in patients with "hard pulse disease."[121]

However, it was not until the publication of William Harvey's (1578–1657) book "De Motu Cordis" ('On the Motion of The Heart and Blood') in 1628 that an understanding of hypertension became possible. The first blood pressure measurement was made by the Reverend Stephen Hales (1677–1761) in 1733 using a glass tube inserted into the artery of a horse. In 1856, Faivre performed the first invasive measurements of blood pressure in humans; however, at the same time noninvasive methods were being developed. In 1855 Karl Vierordt introduced the sphygmograph, a device with an inflatable cuff that obliterated the arterial pulse. This apparatus was further refined by the work of Etienne-Jules Marey in 1860, von Basch in 1881, and Riva-Rocci in 1896. At this point in time only the systolic blood pressure was measured since the importance of diastolic pressure had not been established. In 1905, Nikolai Korotkoff, a young Russian surgeon, used an inflatable cuff and stethoscope to identify the sound of blood rushing through the arteries that corresponded to the diastolic and systolic blood pressures, thus establishing the method, used to this day, in the routine measurement of blood pressure.[122] Initial descriptions of hypertensive disease appeared in the works of Thomas Young in 1808 and Richard Bright in 1836.[121] F. A. Mahomed, using a primitive sphygmograph, first described hypertension in patients without renal disease in 1872.[123] Disability and death from untreated hypertension and its complications were rampant prior to 1950. Deaths from stroke were high; hypertension was the major cause of CHF; hypertensive emergencies were frequent; malignant hypertension was rapidly and uniformly fatal; and hypertension was a major risk factor underlying deaths from MI. The Framingham Heart Study established that hypertension was a major risk factor for the development of CHD.[8] However, prior to 1950, there were no effective antihypertensive therapies available. Prior to effective pharmacological treatments, treatment consisted of sodium restriction (established in 1904), the surgical destruction of the sympathetic nerve trunk (sympathectomy, first performed in 1923), and the intravenous infusion of bacterial products (pyrogen therapy) to induce fever, which was based on the observation that fever lowers blood pressure.[121,124]

4.1 Early Vasodilators

The earliest antihypertensive medications were all vasodilators with a mechanism of action closely related to that observed with organic nitrates such as nitroglycerine. In 1900, sodium thiocyanate (NaSCN) was the first chemical to be used in the treatment of hypertension by Truepel and Edinger. This is a vasodilator that acted on vascular smooth muscle cells; however, its usage was limited due to its many side effects.[121] A closely related compound, sodium nitroprusside ($Na_2[Fe(CN)_5NO]\cdot2H_2O$), was also recognized as a potent vasodilator and in 1929 its value as a hypotensive agent in humans was demonstrated for the first time. I. H. Page introduced this compound as Nipride in 1945. In the mid-1950s this drug became the standard treatment for treating hypertensive emergencies and is still the drug of choice to this day. Because it is unstable to light, it must be administered as a continuous IV infusion with a maximal hypotensive effect occurring within two minutes. Side effects are due to excessive vasodilation and the toxic accumulation of cyanide leading to lactic acidosis.[125] In 1949, Ciba Geigy patented hydralazine (28, Apresoline) and it was introduced as one of the first orally active antihypertensive agents in the USA.[126] Its mechanism of action is not entirely clear, but it does have multiple direct effects on vascular smooth muscle, including the stimulation of the formation of the potent vasodilator nitric oxide (NO). Because of its short half-life, it cannot be used for the management of chronic hypertension, but it still finds use today in treating acute hypertensive emergencies, hypertension in pregnancy due to preeclampsia, and pulmonary hypertension. During the development of the thiazide diuretics in the mid-1950s, many companies embarked on efforts to discover close analogs. In 1961, workers at Schering Plough described a new series of antihypertensive agents structurally related to the thiazide diuretics that do not cause diuresis (because of the absence of the second sulfonamide group) but exhibited a potent vasodilatory action due to the activation of potassium channels in arterial smooth muscle cells. One of these compounds, diazoxide (**29**, Hyperstat IV), is still used today as an IV treatment for hypertensive emergencies.[127]

4.2   *Autonomic Nervous System (Sympatholytic) Approaches to Hypertension*

4.2.1   *Adrenergic Neuron Blockers*

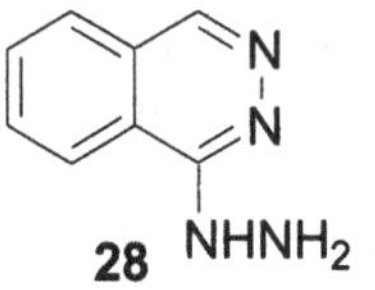

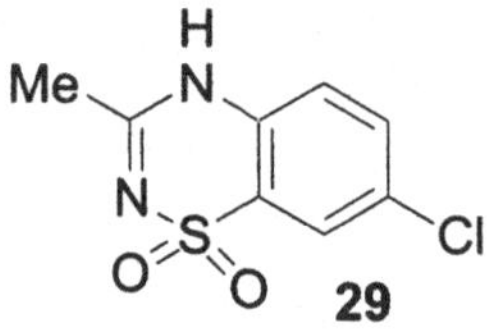

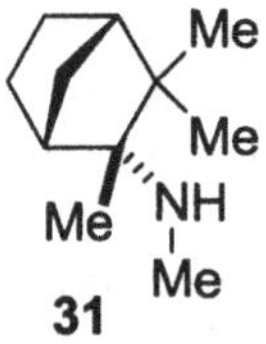

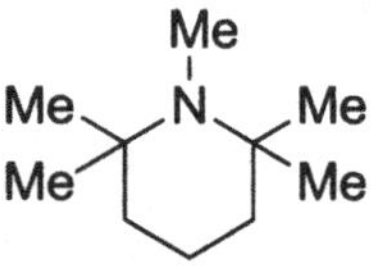

As mentioned previously, a sympathectomy was a drastic way of surgically treating hypertension.[121] The size of blood vessels is controlled by the sympathetic nervous system, and in 1946 it was shown that the same effect, i.e., hypotensive effect mediated by vasodilation caused by blocking the neurotransmission of electrical signals, mediated by the neurotransmitter norepinephrine, could be achieved by a chemical substance, tetraethylammonium chloride [(Et)$_4$N$^+$Br$^-$].[124] An improved version of this compound, hexamethonium (**30**), was available for use by 1951, but these early compounds suffered from poor bioavailability and a lack of specificity for sympathetic ganglia, often hitting parasympathetic ganglia leading to unpleasant side effects.[128] In 1955, workers at Merck discovered mecamylamine (**31**), a compound with much improved bioavailability,[129] and in 1958 both ICI and May and Baker announced the discovery of pempidine (**32**), a much simpler compound with a better side-effect profile.[130] However, these compounds still blocked parasympathetic ganglia.[131] The breakthrough in overcoming this problem was the identification of a compound, xylocholine (**33**), that blocked nerve impulses in adrenergic nerves at dose levels that did not block parasympathetic ganglia. However, this compound had cholinergic activity, and this initiated a search for similar compounds that had no cholinergic activity. In 1957, workers at Ciba discovered as series of compounds that eventually yielded guanethidine (**34**).[132] It was later found that this compound is transported into the presynaptic nerve by the norepinephrine transporter (NET) where it replaces NE in the transmitter vesicles, leading to a depletion of NE stores in the nerve endings. Despite being supplanted by more effective treatments, guanethidine and some closely related analogs [e.g. guanadrel (**35**)] are still used in some parts of the world. Reserpine (**36**), an alkaloid isolated from *Rauwolfia serpentina* (Indian snakeroot), was introduced in 1953 as both an antihypertensive agent and a tranquillizer.[133] Reserpine acts by blocking the vesicular monoamine transporter (VMAT) which transports neurotransmitters from the cytoplasm of the presynaptic nerve into vesicles for subsequent release into the synaptic cleft. Reserpine is still an option as a treatment, but it is seldom prescribed today due to a number of CNS-related side effects.

Effective hypertension treatments really became possible with the emerging understanding of the "adrenoreceptors." These receptors are activated by norepinephrine where it binds reversibly to the receptor and causes a conformational change in the receptor that translates into a biochemical event in the cell. The cells bearing these receptors are called effector cells since they produce the effect seen by sympathetic stimulation. This stimulation of the sympathetic nervous system causes what is known as the "fight-or-flight" responses, including increasing the rate and force of heart contraction, increasing blood pressure, shifting blood flow to skeletal muscles, and dilating bronchioles and pupils. The field took an enormous step forward in 1948 when Ahlquist subclassified these receptors into both α- and β-adrenoreceptors according to their responses to different adrenergic receptor agonists.[134] Over the years a further subclassification yielded two types of α-adrenoreceptors, the α1 and α2 subtypes, and three types of β-adrenoreceptors, the β1, β2, and β3 subtypes. Each receptor is coupled through a G-protein to different effector mechanisms. The clinical utility of these adrenoreceptors became apparent when one considers the effector responses of only a few organs and tissue types innervated by the SNS. For hypertension, one of the most relevant subtypes is the α1-receptor present on smooth muscle in the peripheral vasculature where stimulation causes constriction and a concomitant rise in blood pressure. Adrenergic stimulation of the heart, primarily through the β1-receptor, causes an increase in the rate and force of contraction; thus antagonism of this receptor would decrease heart rate and the force of contraction and thus lower blood pressure. The α2-receptor resides on the outer membrane of the nerve terminus where it serves as a sensor of the amount of NE present in the synapse; as more NE is released, the α2-receptor is stimulated and causes an inhibition of further NE release.

4.2.2   Centrally Acting Sympatholytics

α2 adrenergic agonists

The oldest centrally acting antihypertensive drug is methyldopa (**37**, Aldomet). This drug was originally synthesized by workers at Merck, Sharp, and Dohme as a NE biosynthesis inhibitor[135] and its hypotensive activity was discovered in 1960 by Sidney Udenfriend and colleagues at the NIH in Maryland.[136] It was originally thought that this compound caused vasodilation of blood vessels by locally depleting NE stores; however, it was subsequently shown that this compound actually worked by acting as an agonist of the presynaptic α2-receptor. Methyldopa is actually a prodrug that is converted to α-methylnorepinephrine by dopamine-β-hydroxylase. It decreases total peripheral resistance with little change in CO or heart rate; however, its use has declined due to CNS-mediated adverse events. It is occasionally used in pregnancy-induced hypertension.

The major group of centrally acting antihypertensive agents is the imidazolines and closely related compounds. The discovery of this class of drugs goes back to 1939 when scientists at Ciba in Switzerland combined structural elements of NE, the phenylethylamine moiety, with the imidazole present in histamine.[137] They produced tolazoline (**38**), an α-adrenoreceptor antagonist. Imidazolines were shown to be vasodilators and subsequent compounds were found to be useful in the relief of nasal congestion. In an attempt to find new nasal decongestants, H. Stähle of Boehringer-Ingelheim synthesized compound St155 in 1962.[138] This compound, now named clonidine (**39**, Catapres), lowered blood pressure by acting as a centrally acting α-adrenergic receptor agonist which was more selective for the α2-subtype than the α1-subtype. Clonidine fools the brain into believing that NE levels are higher than they really are and it causes the brain to reduce its signals to the adrenal medulla, which inhibits norepinephrine levels in the blood. The result is a lowered heart rate and blood pressure. Clonidine usage has now been supplanted by newer therapies; its use is limited by CNS-mediated side effects. Additional SAR studies on the imidazolines showed that the imidazoline ring was not necessary for activity. This work resulted in the discovery of ring-opened analogs of clonidine such as guanfacine (**40**, Tenex)[139] and guanabenz (**41**, Wytensin).[140] Guanfacine is more selective for the α2-receptor than clonidine, but the mechanism of action and adverse event profile for both guanfacine and guanabenz are similar to those seen with clonidine. Although it has long been considered that the α2-receptor is the primary target for these agents, another potential target, the imidazoline (I₁) receptor, has emerged relatively recently. The concept of an imidazoline receptor was first proposed by Bousquet in 1984 based on the *in vivo* action of clonidine and clonidine analogs that act as antagonists rather than agonists at the α2-receptor.[141] Two selective I_1 receptor agonists, rilmenidine (**42**, Hyperium) and moxonidine (**43**, Physiotens), have been approved in Europe but not in the US. Compared to clonidine, the side-effect profile of these two agents is much improved.[142]

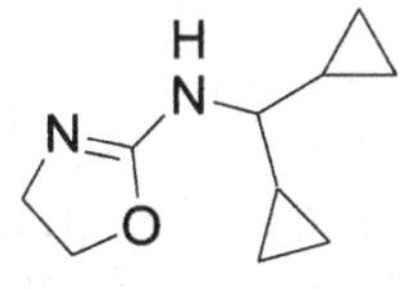

42 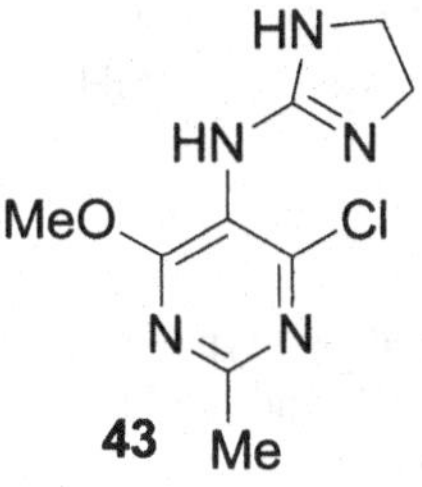   **43** Me

4.2.3   Peripherally Acting Sympatholytics

α1 adrenergic antagonists

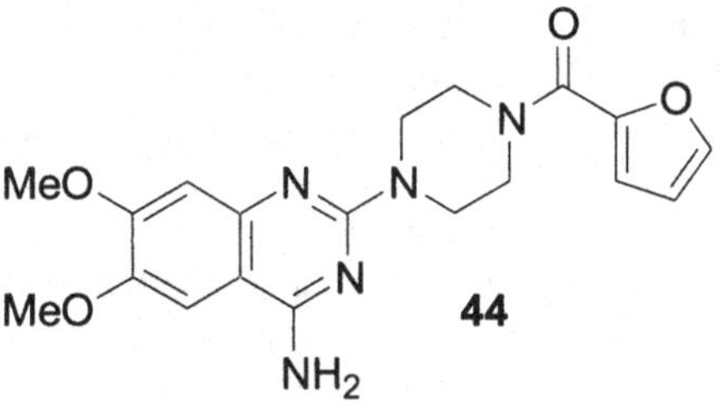

44 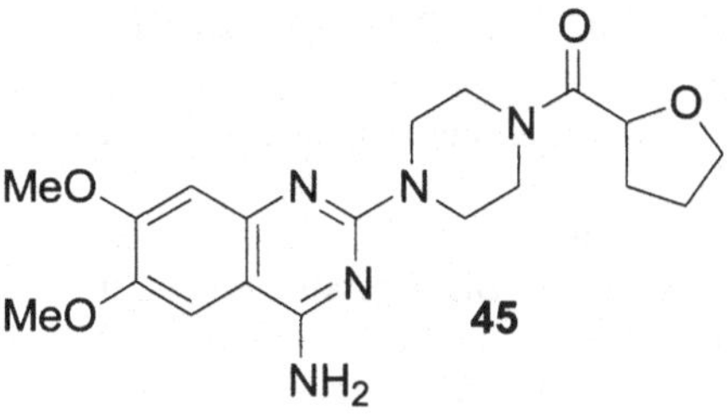   **45**

Blockade of α1-adrenergic receptors inhibits vasoconstriction induced by endogenous catecholamines; peripheral vasodilation may occur in both arteries and veins. The result is a fall in blood pressure because of decreased peripheral resistance. This hypotensive effect is opposed by baroreceptor reflexes that cause an increase in heart rate and CO as well as fluid retention. The most clinically used class of α1-adrenergic receptor antagonists are the quinazolines. The prototypical quinazoline is prazosin (**44**). The quinazolines were first synthesized by workers at Miles laboratories, in Indiana, in the early 1960s. Researchers at Pfizer began to make quinazoline analogs, which resulted in the discovery of prazosin (Minipress).[143] Prazosin does not increase heart rate and has little α2-receptor blocking activity. It is effective for the initial management of hypertension; however, since it was shown, in the ALLHAT (Antihypertensive and Lipid-Lowering Treatment to Prevent Heart Attack Trial) trial, that these compounds may precipitate heart failure, it is primarily used in conjunction with other antihypertensive agents.[144] The major adverse event associated with these agents is the "first-dose" effect, which is the development of orthostatic hypotension after the first few doses. This can be managed by slowly escalating the dose and taking it at bedtime. Two other structurally related compounds, tetrazosin (**45**, Hytrin) and doxazosin (**46**, Cardura), have found increased clinical utility in the treatment of benign prostatic hyperplasia (BPH).[145]

46

β-Adrenergic antagonists

In 1967, Lands *et al.* described two types of β-adrenergic receptors, β1 and β2.[146] A β3-adrenergic receptor has also been cloned[147] and evidence exists for the possible existence of a β4-adrenergic receptor.[148] β1-Receptors occur predominantly in the heart, whereas β2-receptors are mainly found in the lungs, liver, kidney, and uterus. All β-adrenergic antagonists (β-blockers) counter the effects of catecholamines on the cardiovascular tissues. When catecholamines such as norepinephrine bind to the receptor, Gs protein couples the activated receptor to adenyl cyclase and generates cAMP, which then activates protein kinase A (PKA), which phosphorylates the membrane calcium channel and increases calcium entry into the cytosol. This calcium loading accounts for the positive inotropic effect (increased force of contractions). The relaxation of cardiac muscle (lusitropic effect) is mediated by PKA phosphorylation of both troponin I and phospholamban. Increased pacemaker current (I_f) in the sinus node leads to an increased cardiac rate (chronotropic effect).

The discovery of the β1-adrenergic receptor antagonist, propranolol (**49**), in the early 1960s by Sir James Black revolutionized the treatment of angina pectoris and hypertension and is considered to be one of the most important contributions to clinical medicine and pharmacology of the 20th century. Black's entry into cardiovascular medicine stemmed from his initial interest in the treatment of angina pectoris. At this time in the mid-1950s, the major therapy for the treatment of angina was the use of nitrate vasodilators, which increased blood supply and therefore the amount of oxygen to the heart. This treatment also caused facial flushing and headache. Black's seminal insight was to propose a novel treatment for angina that, instead of increasing the supply of oxygen *to* the heart, would work by reducing the demand for oxygen *from* the heart. Black was aware that myocardial oxygen consumption was determined by both systemic arterial pressure and heart rate, but reducing arterial pressure was dangerous and could lead to MI. Black thus turned his attention to heart rate, which was determined largely by the autonomic nervous system. With knowledge of the Ahlquist paper on adrenergic receptor subtypes, Black developed an approach that had the goal of antagonizing the actions of the sympathetic hormones norepinephrine and epinephrine on the heart. Thus the search for compounds (β-blockers) that block the receptors responsible for determining heart rate was initiated in 1956.[149,150] In 1958, Powell and Slater at ICI published a paper describing the possible use of dichloroisoprenaline (**47**) as a bronchodilator.[151] Black later showed that this compound was the first selective β-receptor antagonist. A vigorous medicinal chemistry program by scientists at ICI led to the discovery of pronethalol (**48**, Alderlin) in 1960. This compound was a potent β-blocker that did not antagonize the α-receptor and was shown to be effective in humans at decreasing heart rate and producing increased exercise tolerance in patients with angina. Although approved in 1963, this compound did not enjoy widespread clinical use because it was shown to produce thymic tumors in mice. It was also shown to have intrinsic sympatomimetic activity (ISA), i.e. not only antagonizing the β-receptor but also acting as a partial agonist. It was discarded in favor of another compound, propranolol (**49**, Inderal), that was shown to be safer than pronethalol.[149,152] Clinical trials were started in 1964 and propranolol was launched in 1965 for the treatment of angina.[153] Evidence quickly mounted that propranolol reduced both mortality and morbidity in angina sufferers.[154] In 1964, Prichard and Gillam published the first paper describing the use of propranolol as an antihypertensive drug,[155] and in 1969 it was approved for that indication in Europe and in 1973 it received approval as an antihypertensive in the USA. For his work on β-blockers (and H_2 receptor antagonists), James Black received the 1988 Nobel Prize for Physiology and Medicine.

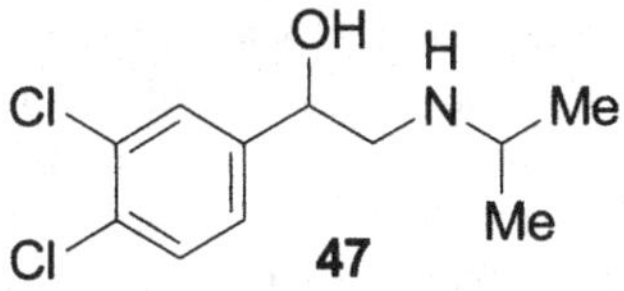

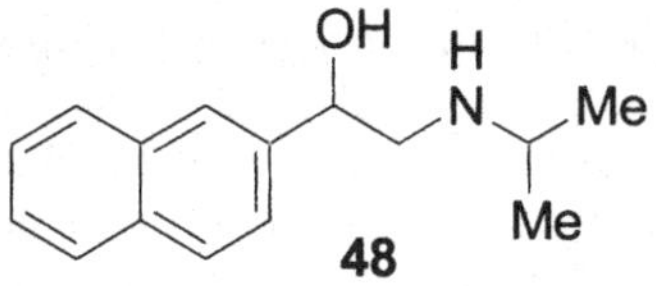

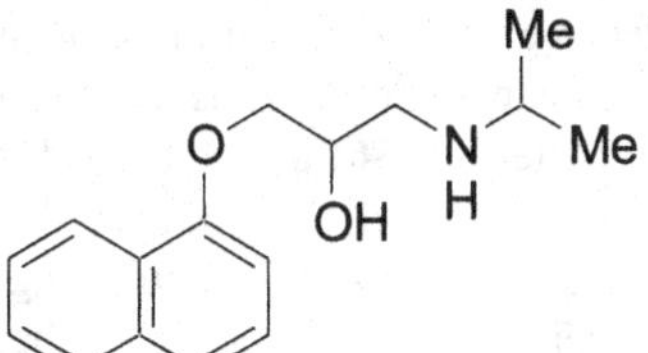

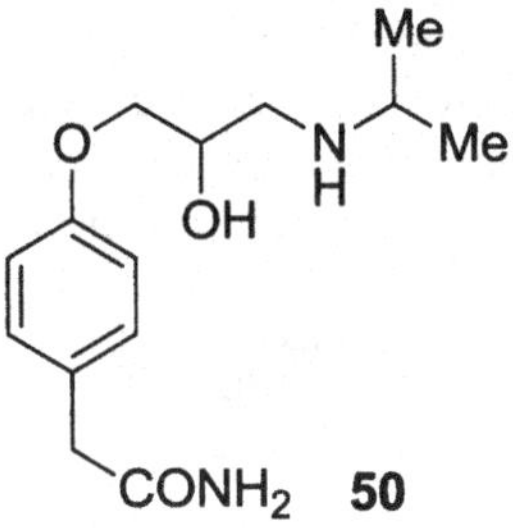

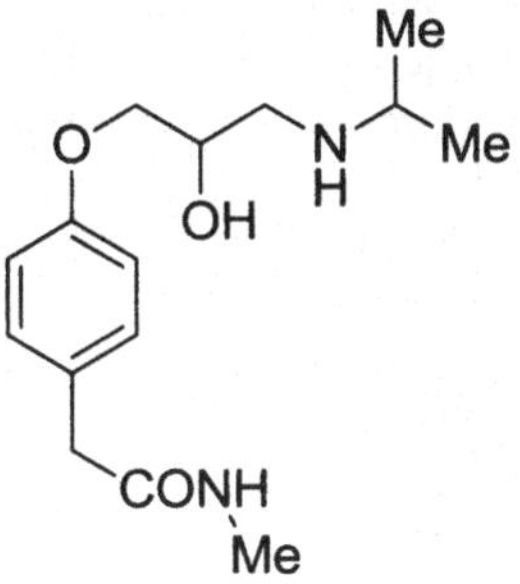

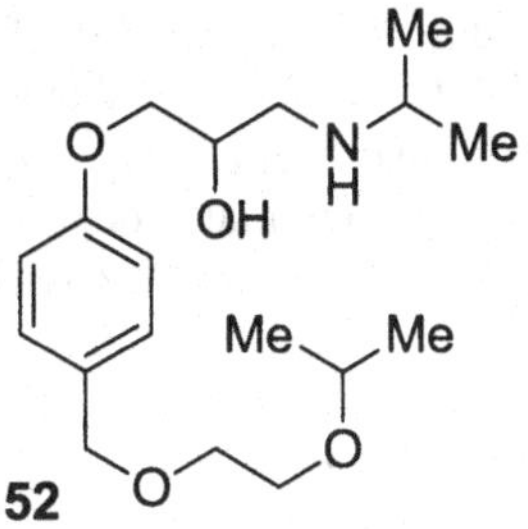

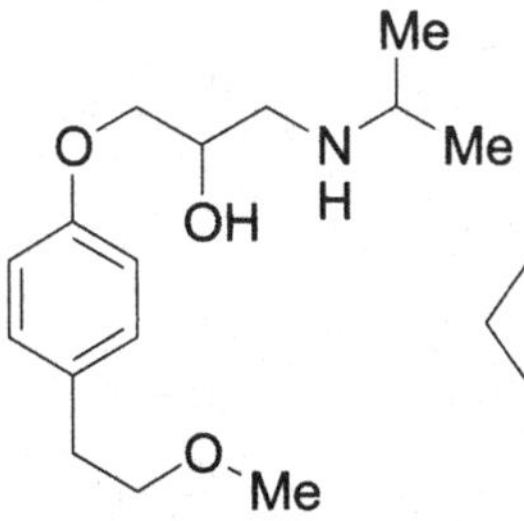

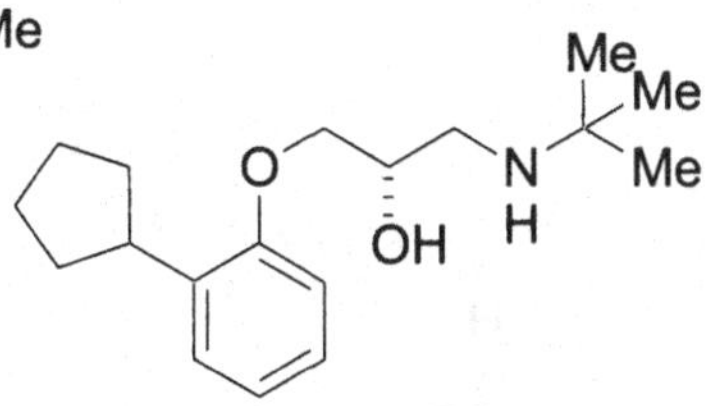

Propranolol was subsequently shown to cause bronchoconstriction in patients with asthma. This was due to antagonism of the β2-receptor in bronchial muscle and thus began the search for β1-selective antagonists. In 1966, workers at ICI turned their attention to practolol (**50**, Eraldin) as the first β1-selective antagonist. Unlike propranolol, this compound was significantly less lipophilic and thus did not penetrate the CNS to the degree that propranolol does and thus does not cause CNS side effects. Practolol was launched in 1970, but in 1975, it was withdrawn from the market because it was found to cause blindness in some patients. This was later found to be a compound-specific toxicity called oculomucocutaneous syndrome that manifested itself by keratoconjunctivitis, sclerosing peritonitis, and pleurisy. ICI quickly replaced this compound with another hydrophilic compound, atenolol (**51**, Tenormin), which was launched in 1976 primarily as an antihypertensive. This compound was β1-selective and devoid of CNS side effects due to its hydrophilicity. This "ideal" compound became one of the best selling heart drugs, and within ten years, Tenormin and its related products generated worldwide sales of about $1 billion.[149] Other β1-selective antagonists followed and include bisoprolol (**52**) and metoprolol (**53**). It is important to note that none of these compounds are cardiospecific and at high doses they can still adversely affect asthma and peripheral vascular disease patients. They can also increase insulin resistance and hasten the development of hyperglycemia. Additionally, they have been shown to be less effective in lowering blood pressure in hypertensive African-American patients compared to a Caucasian patient. Nonselective β-blockers such as propranolol, penbutolol (**54**), pindolol (**55**), and carteolol (**56**) are contraindicated in patients with asthma. β-blockers with ISA activity, such as penbutolol (**54**), pindolol (**55**), and carteolol (**56**), cause less cardiodepression and resting bradycardia as well as displaying neutral effects on lipid and glucose metabolism.[156,157] Newer β-blockers, the so-called 3rd generation, have additional attributes. Carvedilol (**57**, Coreg) and labetolol (**58**, Normodyne) are mixed α/β-blockers that also exhibit direct vasodilator activity. Vasodilation via α-blockade lowers peripheral vascular resistance to maintain CO, thus preventing bradycardia.[158,159] Nebivolol (**59**, Nebilet) produces vasodilation by increasing nitric oxide bioavailability in the vascular endothelium and thus has a more beneficial hemodynamic profile than traditional β-blockers.[160] The third-generation β-blockers are associated with fewer adverse reactions and potentially beneficial effects on the arterial vasculature. They also have beneficial effects on insulin sensitivity as well as on atherogenic risks factors and endothelial function.

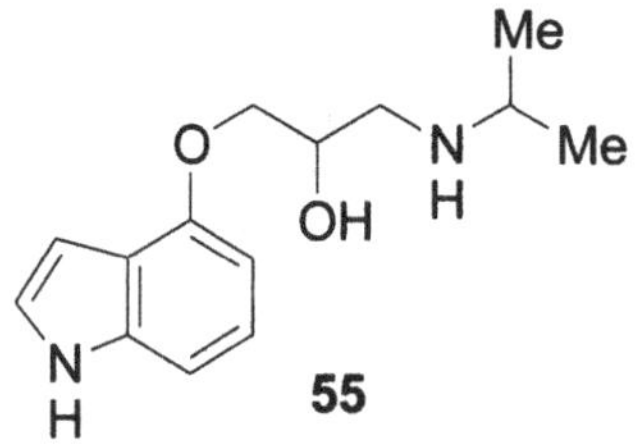

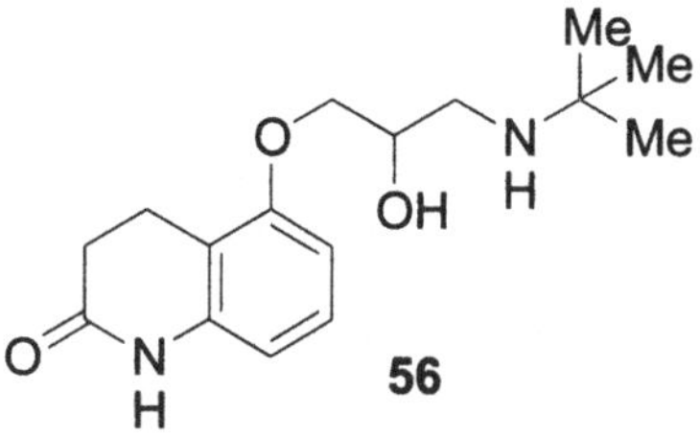

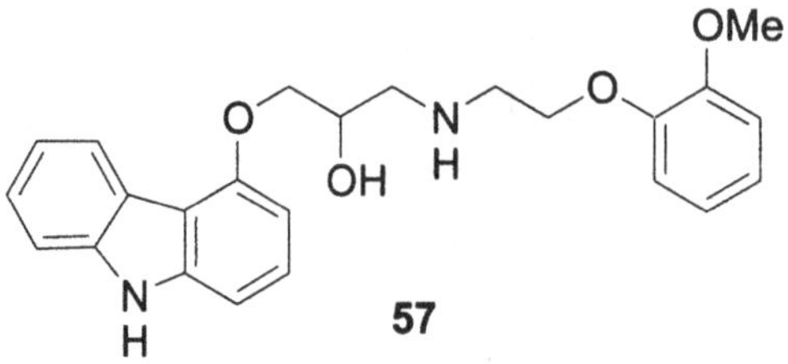

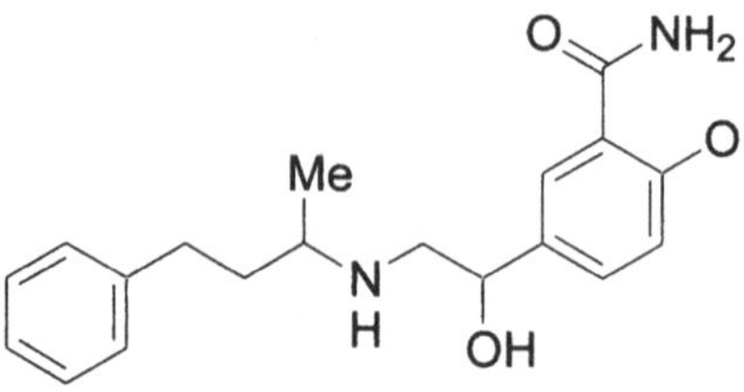

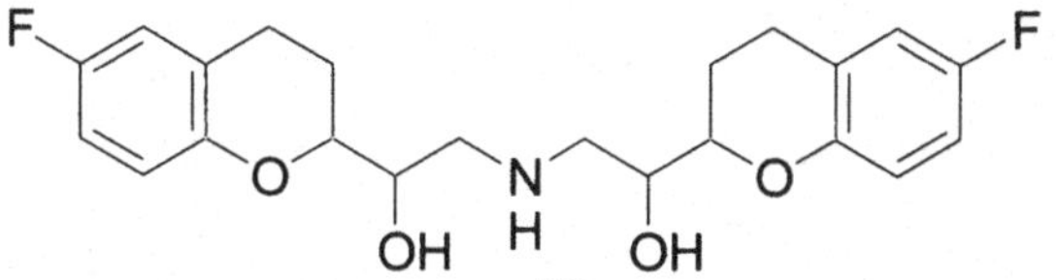

59

In recent years, controversy has surrounded the use of β-blockers as initial therapy in hypertension. In 1992, a large outcomes trial in elderly British patients (propranolol versus a diuretic) found little benefit of β-blockers against stroke and none against coronary events.[161] In 1998, a systematic review of trials showed that in the elderly β-blockers gave worst outcomes than did the diuretics.[162] β-Blockers were also found to increase new-onset diabetes, and in the resulting analysis it was concluded that β-blockers should not remain the first choice in the treatment of hypertension because of a failure to provide adequate protection against cardiovascular disease. The mechanism proposed was that central aortic systolic pressure decreased less when compared to brachial pressure.[163] It remains to be seen whether the newer agents, carvedilol and nebivolol, have better metabolic and hemodynamic profiles.

4.3   Diuretics

The importance of the kidney in blood pressure control and the pathophysiology of hypertension have been known for many years. The basic premise underlying the mechanism of action of a diuretic is to decrease CO and peripheral resistance, the major determinants of arterial pressure. They do this by acting on the kidney and increasing urine flow and the rate of excretion of sodium ion (natriuresis) and chloride ion. Since sodium chloride is the major determinant of extracellular fluid volume and total blood volume is regulated by renal function, an elevation in blood volume increases CO. Diuretics exert their hypotensive effect by reducing blood volume and CO, although initially there is a slight increase in peripheral resistance quickly followed by a decrease in vascular resistance.

Compounds that increase urine flow rate have been known for centuries. Many of the early diuretics were herbal remedies. Probably the best known were the xanthine alkaloids present in tea, cocoa, and coffee. The diuretic action of caffeine was described in 1886 by Bronne at the University of Strasburg. Both theobromine and theophylline were also shown to be potent diuretics.[164] In 1919, Vogl in Vienna, Austria, discovered the diuretic action of mercurial antisyphalitic drugs.[165] The first effective mercurial diuretic was merbaphen (**60**, Novasurol); however due to its renal toxicity, in 1924, this compound was supplanted by mersalyl (**61**) as the first clinically useful diuretic.[166] Since these drugs could only be administered intravenously, a search began for an orally administered, nontoxic diuretic.

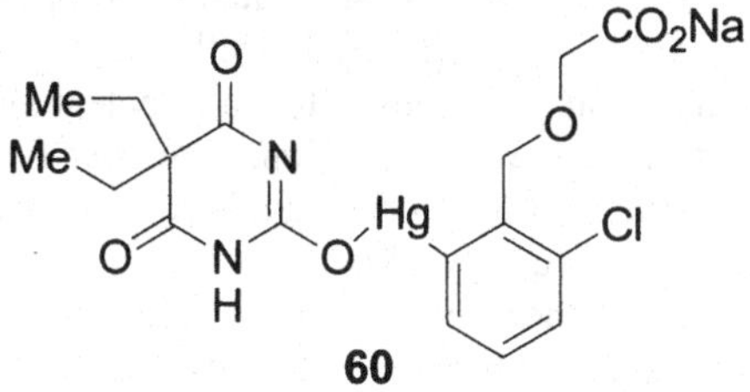

60 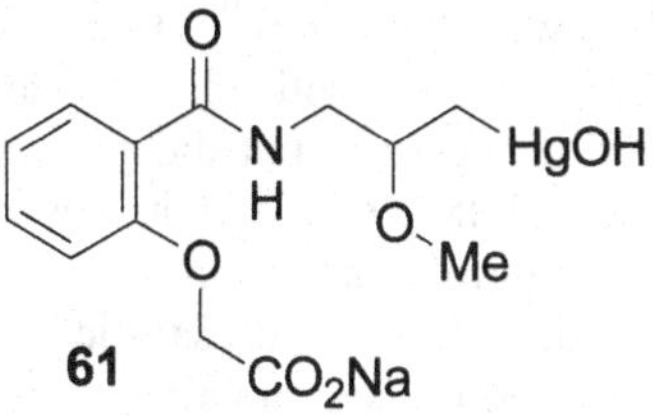   **61**

4.3.1 Thiazide Diuretics

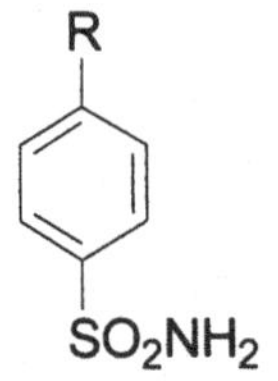

62, R = NH$_2$
64, R = CO$_2$H

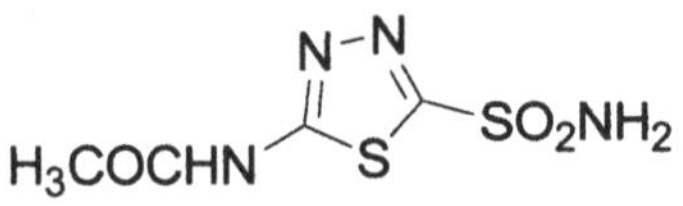

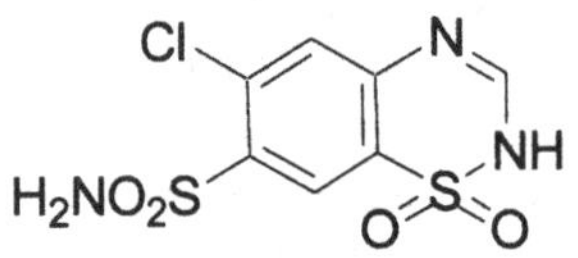

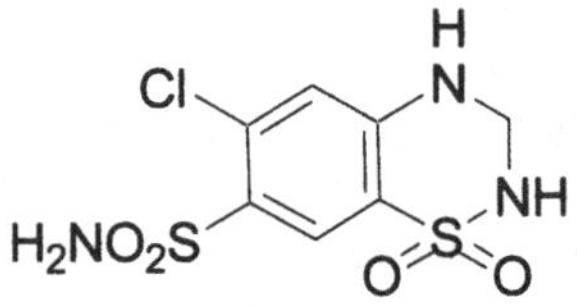

Soon after its introduction as a chemotherapeutic agent, sulfanilamide (**62**) was shown to cause an alkaline diuresis in patients who had received a large dose.[167] In 1942, Hober suggested that this could be accounted for by an increased excretion of sodium bicarbonate caused by the inhibition of a family of enzymes called the carbonic anhydrases.[168] These enzymes catalyze the rapid interconversion of carbon dioxide and water to bicarbonate and protons to maintain acid-base balance in blood and other tissues and to help transport carbon dioxide out of tissues. Inhibition of this enzyme leads to a lesser exchange of hydrogen ions for sodium ions in the distal portion of the renal tubule. The sodium ions, bicarbonate, and associated water molecules are then excreted and a diuretic effect is noted. The diuretic effect was demonstrated clinically in 1949, when Schwartz administered high-dose sulfanilamide to heart failure patients, however, this approach was quickly abandoned due to toxicity.[169] Researchers at Lederle Laboratories followed up on these findings by synthesizing a number of heterocyclic sulfanilamide analogs. One compound, acetazolamide (**63**, Diamox), was found to be a very potent inhibitor of the carbonic anhydrases, and it was introduced clinically as the first orally administered diuretic in 1952.[170] Its use was limited because of the systemic acidosis it produced; however, it is still used to reduce the rate of aqueous humor formation and intraocular pressure in glaucoma patients. At the time of Schwartz's work with sulfanilamide, Beyer of Sharp and Dohme began a search for carbonic anhydrase inhibitors that acted in the proximal end of the renal tubule and therefore increased the secretion of chloride ion in the form of sodium chloride. This was thought to be an effective antihypertensive therapy since it was known (since 1904!) that low-salt diets were an effective means of controlling high blood pressure. The first carbonic anhydrase inhibitor they found to increase chloride secretion was 4-sulfonamidobenzoic acid (**64**), although it still increased bicarbonate secretion, indicating a lack of specificity in the kidney.[171] Continued SAR investigation by Novello and Sprague led to the identification of a benzothiadiazine derivative that was shown to be a potent diuretic and increased chloride secretion but did not increase bicarbonate excretion. The first reports of its discovery appeared in 1957, and it was given the name chlorothiazide (**65**, Saluric). It was introduced into clinical practice in 1958. It was subsequently found that the dihydro derivative hydrochlorothiazide (**66**, Hydrosaluric) was ten times more potent and this compound subsequently entered clinical practice in 1959 and quickly became a mainstay in the treatment of hypertension.[172] For the first time it was possible to reduce blood pressure to normal levels for long periods of time with few side effects. Much later it was found that the major mechanism of action of the thiazide diuretics was not inhibition of carbonic anhydrase but the inhibition of the Na^+Cl^- symporter located in the distal convoluted tubule. These drugs compete with the chloride binding site of the Na^+Cl^- symporter and inhibit the reabsorption of sodium and chloride ions.[173] Other structurally distinct inhibitors of this symporter have been discovered and include the quinazolinones [as exemplified by metalazone (**67**, Mykrox) and quinethazone (**68**, Hydromox)], phthalimidines [as exemplified by chlorthalidone (**69**, Hygroton)], and indolines [as exemplified by indapamide (**70**, Lozol)].

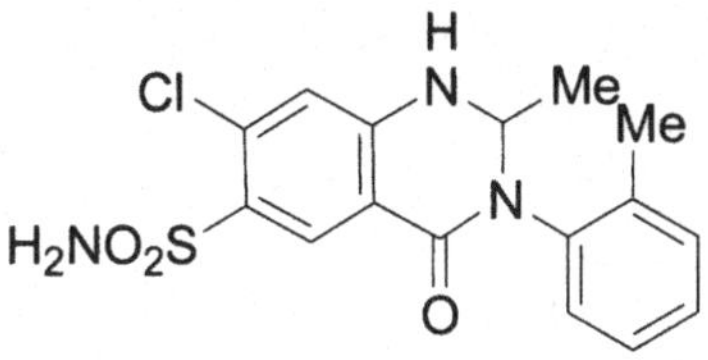

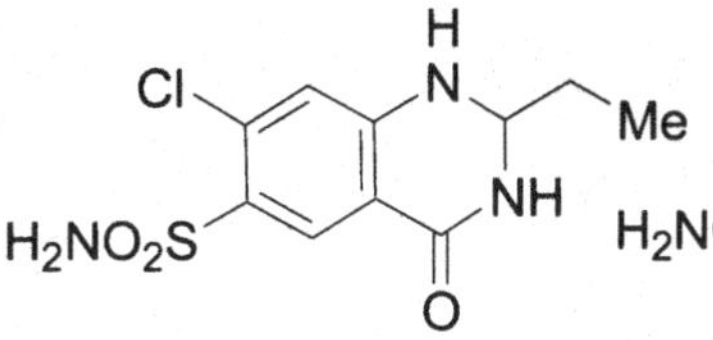

It is well known that approximately 30–40% of individual in the USA are salt sensitive in that they respond to a high-sodium intake with concomitant increases in blood pressure. Sodium restriction decreases blood pressure, and many of these individuals respond readily to thiazide diuretic therapy. In 1964, the Veterans Cooperative Study established for the first time that antihypertensive intervention with a thiazide diuretic, reserpine and hydralazine, in patients with moderate to severe hypertension led to remarkable decreases in the incidence of death, strokes, and other cardiovascular events in hypertensive patients.[174]

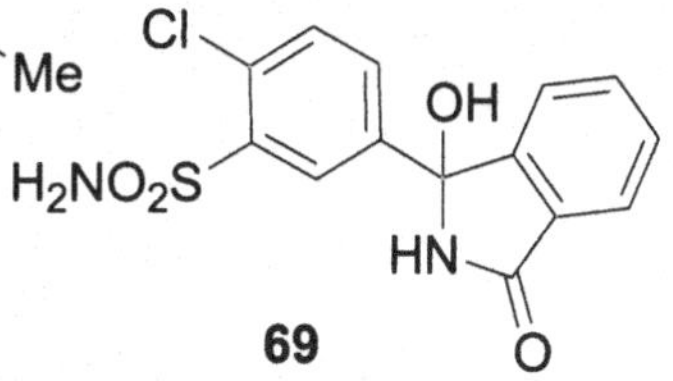

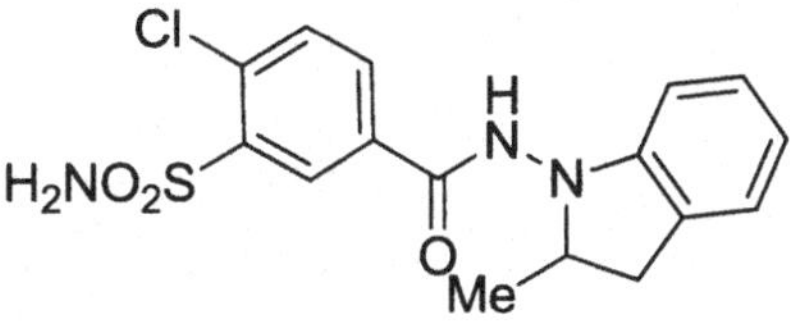

4.3.2 High-Ceiling or Loop Diuretics

High-ceiling diuretics get their name from the fact that they produce a peak diuresis much greater than that observed with the thiazide diuretics. Their main site of action is on the thick ascending limb of the loop of Henle (hence the name loop diuretics), where they inhibit the luminal $Na^+/K^+/2Cl^-$ symporter.[175] The prototypical loop diuretic is furosemide (**71**, Lasix). This compound was derived from dichlorphenamide (**72**), an early carbonic anhydrase inhibitor, by workers at Hoechst and was introduced into clinical practice in 1962.[176] The major use of loop diuretics is in the treatment of acute pulmonary edema, and chronic CHF and reducing the edema associated with nephritic syndrome.

4.3.3   Potassium-Sparing Diuretics

These compounds are usually employed for their potassium-sparing actions to offset the effects of other diuretics that increase potassium excretion. They act as sodium channel blockers by blocking luminal sodium channels in the distal tubule of the kidney and preventing the reabsorption of sodium ions and blocking the secretion of potassium ions. The main drugs in this class are triamterene (**73**, Dyrenium) and amiloride (**74**, Midamor). Triamterene is a pteridine and amioloride is an aminopyrazine structurally related to triamterene as a ring-opened analog.[177] Both of these drugs are seldom used as sole agents in the treatment of edema or hypertension; their major utility is in combination with other diuretics.

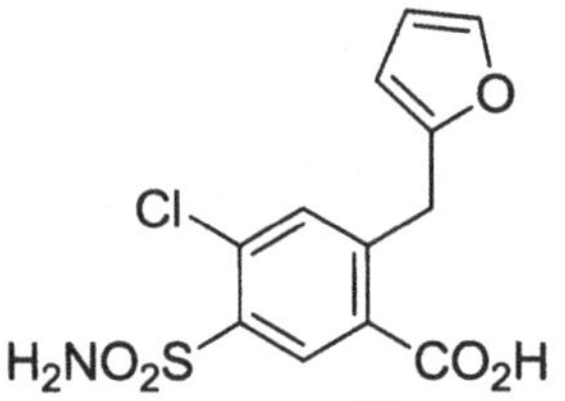

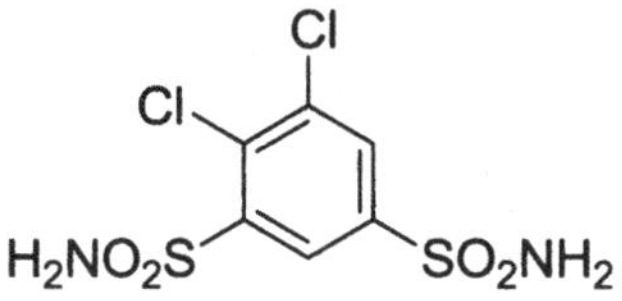

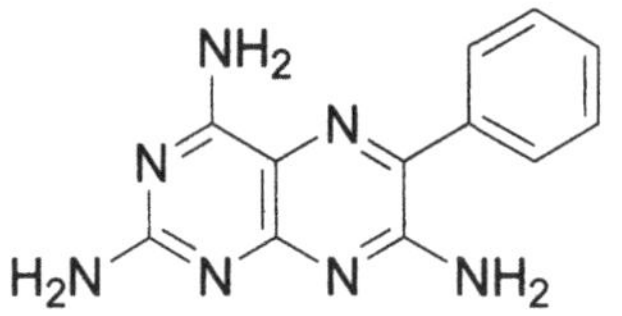

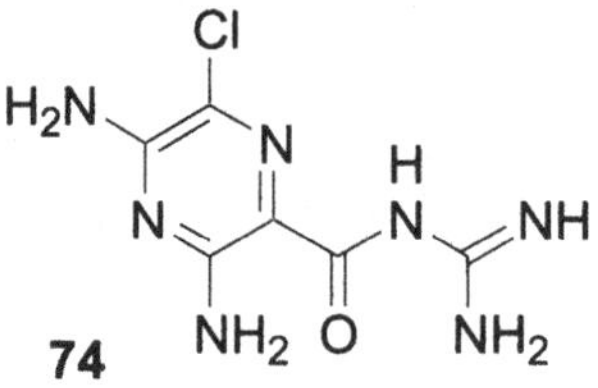

4.3.4   Aldosterone Antagonists

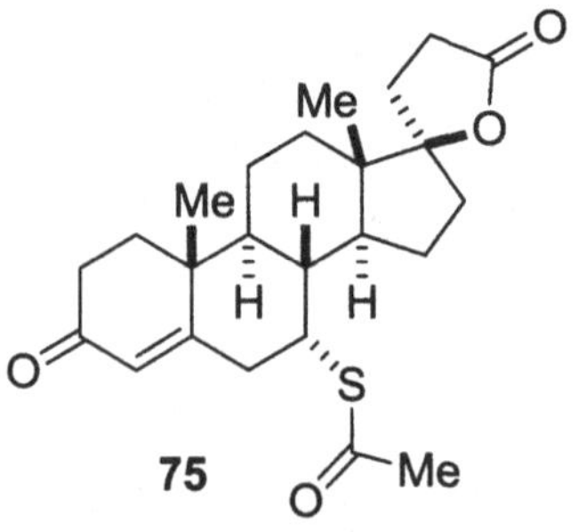

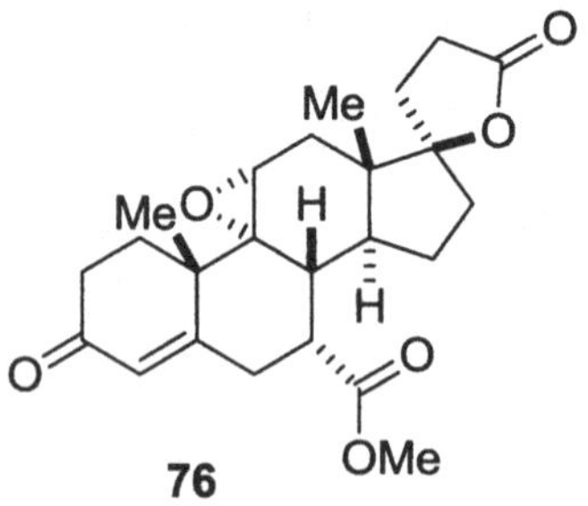

Aldosterone is a potent mineralocorticoid secreted by the adrenal cortex that promotes salt and water retention and potassium and hydrogen ion excretion. It binds to a cytoplasmic mineralocorticoid receptor found in the epithelial cells in the late distal tubule and collecting duct. Spironolactone (**75**, Aldactone), originally synthesized in 1959 by workers at G. D. Searle in Chicago,[178] is a competitive antagonist of this receptor and competitively inhibits aldosterone binding, thus leading to decreased reabsorption of sodium and chloride ions and associated water as well as exhibiting a K^+-sparing effect. These compounds, like other K^+-sparing diuretics, are often coadministered with thiazide or loop diuretics in the treatment of edema and hypertension.[179] In 2002, eplerenone (**76**, Inspra) was approved by the FDA. This compound is closely related to spironolactone and is reported to be more selective for the mineralocorticoid receptor.[180] It is indicated for the reduction of risk of cardiovascular death in patients with CHF and left ventricular dysfunction within 3–14 days of an MI in combination with standard therapies.

4.4   Calcium Channel Blockers

It has been known for over a hundred years that myocardial contractility depends on the extracellular concentration of calcium ions. Calcium acts as a cellular messenger linking external excitation with a cellular response. Increased cytosolic concentrations of calcium ion results in the binding of calcium ion to a regulatory protein, either troponin C in cardiac and skeletal muscle or calmodulin in vascular smooth muscle cell, with the result of producing a positive inotropic effect (increased myocardial contractility). In the absence of calcium ion, the heart stops beating. The cardiac action potential, driven by an inward flux of sodium ions, remains virtually unaltered in the absence of calcium. The lack of mechanical activity with nearly unchanged electrical activity is called electromechanical uncoupling. Both vascular tone and contraction are primarily determined by the availability of calcium from both extracellular and intracellular sources.[181]

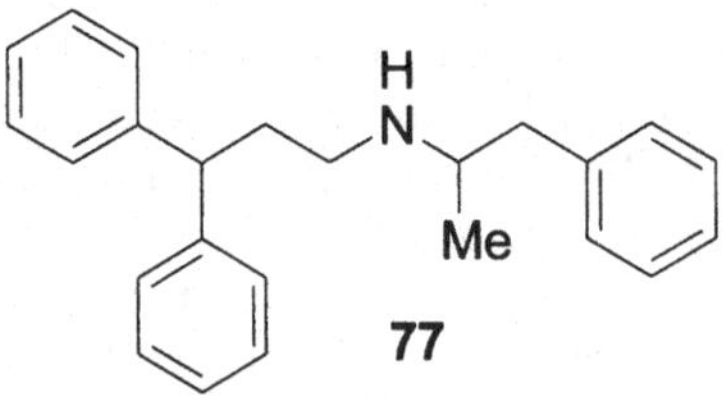

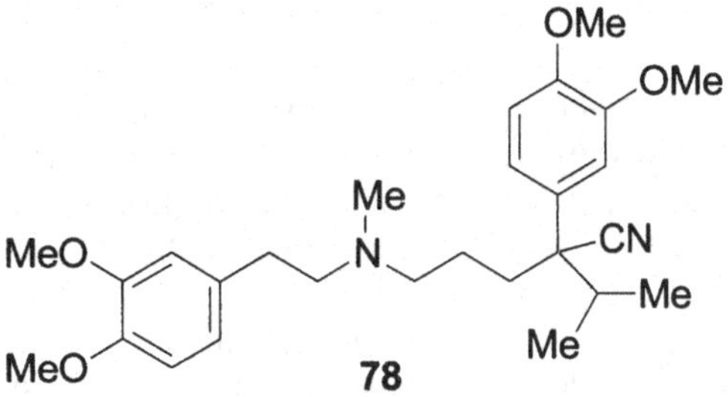

        The genesis of calcium channel blockers as a treatment of hypertension began in 1960 with the publication of a paper by E. Lindner at Hoechst AG on the effects of prenylamine (**77**, Segontin), a potential new antianginal agent. This compound produced coronary vasodilatation but produced additional negative inotropic and chronotropic effects not seen with classical coronary vasodilators.[182] Similar effects were also noted in 1962 for verapamil (**78**). In 1963, both prenylamine and verapamil were given to Albrecht Fleckenstein at the University of Freiburg. He performed electrophysiological experiments in isolated cardiac muscle on both compounds and found that they both exhibited a negative inotropic effect with no changes in electrical activity. Because these effects on cardiac muscle were the same as calcium withdrawal, they were called calcium antagonists. The effects of prenylamine and verapamil were negated by calcium ions, epinephrine, and norepinephrine.[183] Because of the antagonistic effects of epinephrine and norepinephrine, it was thought that the calcium antagonists were β-blockers; however, in 1967, Fleckenstein suggested that the negative inotropic effect of prenylamine and verapamil resulted from inhibition of excitation-contraction coupling and that the mechanism of action involved the reduction in movement of calcium ions into the cardiac cell.[184]

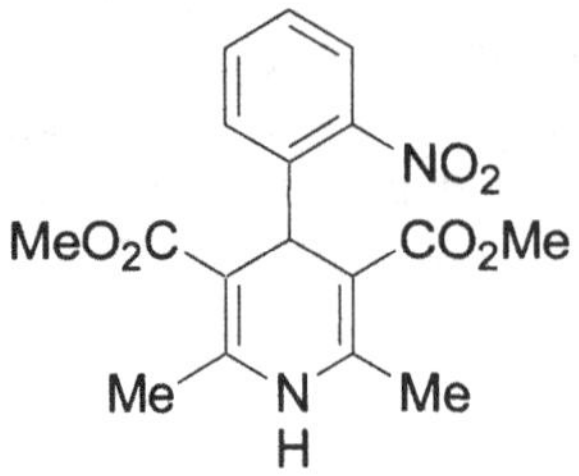

79

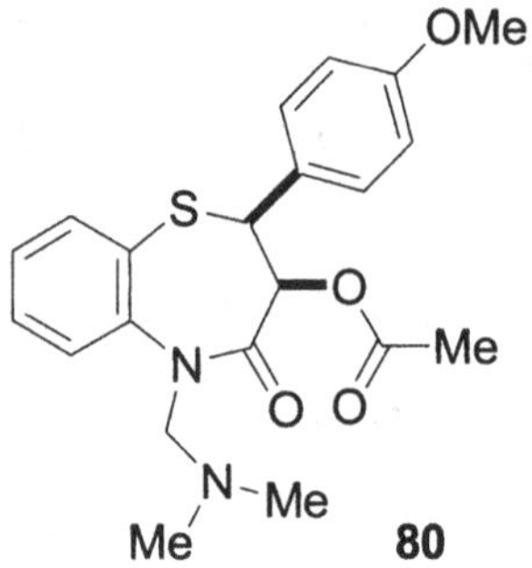

80

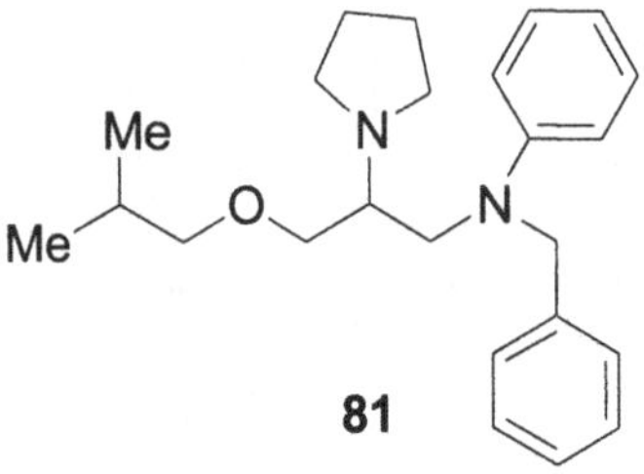

81

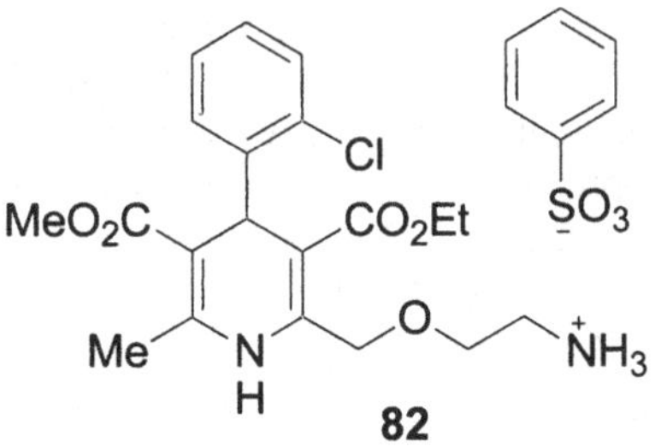

82

In 1969, Kroneberg of the Bayer Company asked Fleckenstein to evaluate a compound, Bay-a-1040 (**79**). He found it to have a similar mechanism of action to verapamil. The first description of Bay-a-1040, later named nifedipine (Adalat), was in 1971 by Bossert and Vater of the Bayer Company.[185] It was subsequently shown in 1972 by M. Kohlhardt of the University of Freiburg, using the voltage-clamp method, that the calcium antagonists block the inward movement of calcium ions through the slow calcium channel, later called the L-type calcium channel.[186] To date, it is recognized that there are six functional subclasses of potential-dependent calcium channels; T, L, N, P, Q, and R. They differ in location and function, however, the L (long-lasting, large) channel is the most extensively studied since it is the site of action of the currently available calcium channel blockers (CCBs). All of the L-type CCBs inhibit the flow of calcium ions into the myocardial cell, where they inhibit excitation-contraction coupling, resulting in vasodilation and a fall in blood pressure and antianginal effects and no undesirable myocardial depression. There are four main classes of CCBs currently employed for the treatment of hypertension, angina, arrhythmias and CHF: the phenylalkylamines, the benzothiazepines, the dihydropyridines and the diaminopropanols. They are differentiated by their pharmacokinetic and pharmacodynamic effects. The first-generation drugs in each class are verapamil (**78**, phenylalkylamine originally approved by the FDA in 1981), diltiazem (**80**, benzothiazepine originally approved in 1982), and nifedipine (**79**, 1,4-dihydropyrine originally approved in 1981). Early dihydropyridines such as nifedipine are characterized by a rapid onset of action and short duration of action due to short half-lives and thus require twice-daily dosing.[187] The use of nifedipine and related calcium channel blockers was curtailed after the publication, in 1995, of a meta-analysis that showed that nifedipine increased mortality in patients with CAD.[188] Bepridil (**81**,Vascor), the lone example of a diaminopropanol, is unique from the above-mentioned agents in that its actions are not solely based on its ability to block L-type calcium channels. It also blocks fast sodium channels as well as receptor-operated calcium channels and is thus considered nonselective. These additional mechanisms are responsible for bepridil's ability to inhibit cardiac conduction, slow AV nodal conduction, slow heart rate, and prolong the QT interval. Although approved by the FDA in 1990, bepridil was withdrawn voluntarily from the US market in 2004 because of its ability to induce the ventricular arrhythmia, torsade de pointes. The second-generation compounds includes the extended- or slow-release (SR or ER) variants of the main first-generation 1,4-dihydropyrines and includes nifedipine XL, felodipine ER, and nicardipine XL. In addition, diltiazem CD and verapamil SR variants have been produced. All of these variants have a more gradual onset of action and a longer half-life. The third generation of calcium channel blockers includes amlodipine (**82**, Norvasc), originally approved in 1990, with an intrinsically longer half-life than the first-generation agents. As a class, the calcium channel blockers have manageable side effects which are mainly due to excessive vasodilation resulting in edema, flushing, hypotension, nasal congestion, headache, and dizziness. Additional side effects such as palpitations, chest pains, and tachycardia arise from sympathetic responses to the vasodilatory effect.[187]

4.5   *The Renin-Angiotensin-Aldosterone System (RAAS)*

The RAAS is one of the most powerful regulators of blood pressure and electrolyte fluid/homestasis. The systematic study of this system and the targeted synthesis of compounds that inhibit at various points in the pathway have led to a number of antihypertensive treatments characterized by their specificity and efficacy. Inhibitors of this system are the newest therapies available and sales for these agents are counted in billions of dollars annually.

The RAAS is a proteolytic cascade in which renin, an aspartyl protease secreted by the kidney in response to a reduction in renal blood flow, blood pressure, or sodium concentration, cleaves angiotensinogen, an α_2-globulin secreted by the liver, to an inactive decapeptide called angiotensin I. Angiotensin I is then cleaved by angiotensin converting enzyme (ACE), a relatively nonspecific metalloprotease, to produce the octapeptide angiotensin II. Angiotensin II then binds to the angiotensin II receptors AT_1 and AT_2 located in the kidney, heart, vascular smooth-muscle cells, brain, adrenal glands, platelets, adipocytes, and the placenta. This results in aldosterone secretion and sodium retention by stimulation of the angiotensin receptors present on the adrenal cortex, resulting in elevated blood pressure due to vasoconstriction. Interestingly, the nonspecificity of ACE allows it to participate in another important system, the kallikrein-kinin system (KKS). Bradykinin is a nonapeptide which acts locally to produce pain, cause vasodilation, increase vascular permeability, and stimulate prostaglandin synthesis. Similar to angiotensin II, bradykinin is produced by the proteolytic cleavage of a precursor peptide. Cleavage of kininogen by the protease kallikrein produces a decapeptide known as kallidin (or lysyl-bradykinin). Cleavage of the N-terminal lysine by aminopeptidase produces bradykinin. The cleavage of bradykinin by ACE produces several inactive peptides. So, in addition to producing a potent vasoconstrictor (angiotensin II), ACE also inactivates a potent vasodilator (bradykinin).

Thus, research efforts over the last several decades have been directed towards developing drugs capable of suppressing RAAS by the inhibition of renin release, by blocking the formation of angiotensin II by inhibition of ACE, and by antagonism of angiotensin II at the AT_1 receptor.[189]

The history of the discovery of RAAS began in 1898 with the studies of Tigerstedt and Bergman at the Karolinska institute in Finland. They analyzed the effects of rabbit renal extracts on arterial pressure and found that it contained a pressor compound they named renin. In addition, they showed that renal vein blood increased blood pressure when injected into nephrectomized animals.[190] Based on these observations, they explained that the association between renal disease and cardiac hypertrophy, first noted by Richard Bright in 1836, was due to the release by the kidney of a potent vasoconstrictive substance. In the following years, many attempts were made to develop an experimental model of arterial hypertension by manipulating renal function. It was not until 1934 that Goldblatt and colleagues showed it was possible to produce persistent hypertension in dogs by constricting the renal arteries.[191] With this model in hand, a vigorous search for the mechanisms responsible for the increase in blood pressure began. Two research groups, working independently, simultaneously reached similar conclusions on the matter. One group was under the leadership of Bernardo Houssay at the Physiology Institute of the School of Medicine in Buenos Aries, Argentina, and the other group was under the leadership of Irvine Page at the Eli Lilly Research Laboratories in Indianapolis. In 1940, Braun Menedez and his colleagues in Argentina and Page and Helmer in the USA reported that renin was an enzyme that acted on a plasma protein substrate to catalyze the formation of the actual pressor substance, an octapeptide called hypertensin by the former group and angiotonin by the latter. These names persisted until 1961 when the pressor substance was renamed angiotensin and the plasma substrate was called angiotensinogen.[192] Another seminal finding was the identification of ACE. In the period 1954–1956, Skeggs and colleagues reported that renin liberates angiotensin I, which is converted by a factor in horse plasma to angiotensin II, in the presence of chloride ion. They named this factor angiotensin converting enzyme (ACE).[193] The synthesis of the angiotensin II was accomplished for the first time in 1957 by the group led by Page,[194] then in Cleveland and then by Schwyzer in Basle, Switzerland.[195] Erdos and colleagues at the University of Illinois, Chicago in 1967 independently identified an enzyme they named kininase II that inactivated bradykinin. They subsequently showed that kininase II and ACE were the same enzyme and thus provided a link between the RAAS and KKS.[196] This characterization of ACE prompted the search for inhibitors of this enzyme as a potential new treatment for hypertension.

4.5.1   Angiotensin-Converting Enzyme (ACE) Inhibitors.

83

The initial validation of ACE as a therapeutic target originated with a study by Ferreira in 1965 that showed that an extract produced by the Brazilian pit viper *Bothrops jararaca* potentiated the action of bradykinin.[197] This extract was originally termed bradykinin potentiating factor (BPF) and was later shown to be a mixture of peptides containing 5-13 amino acids.[198] These peptides were subsequently shown to be potent and specific inhibitors of ACE. One of these peptides, teprotide (**83**, SQ 20,881),[199] was shown to have the greatest *in vivo* potency in inhibiting ACE, and clinical trials established its efficacy as an antihypertensive;[200] however, it also displayed the typical shortcomings of peptidic drugs in general, a lack of oral activity.

84   **85**, R = H
                87, R = Et   **86**

Thus, began the search for an orally active ACE inhibitor which ushered in the beginning of rational drug design. SAR studies indicated that the optimal C-terminal inhibitory sequence was Phe-Ala-Pro and that the venom peptides were substrate analogs that bound to substrate-binding sites in the ACE active site. In 1973, Byers and Wolfenden had described the synthesis of benzylsuccinic acid inhibitors of carboxypeptidase A (CPA), an enzyme thought to be mechanistically related to ACE.[201] Cushman and Ondetti recognized that part of the binding affinity of benzylsuccinic acid to CPA originated with the coordination of the active-site zinc atom by the carboxyl group, and thus they hypothesized that a similar succinyl derivative would inhibit ACE if its structure was analogous to the dipeptide product of ACE activity. On the basis of the Phe-Ala-Pro sequence derived from the BPF peptides, they synthesized methylsuccinyl-Pro and indeed found it to be a specific inhibitor of ACE.[202] Another major breakthrough in potency came with the replacement of the carboxyl group with a much more potent zinc binding motif, the sulfhydryl group. This compound, later called captopril (**84**), was 10-20 more potent than teprotide in inhibiting the contractile and vasopressor responses to angiotensin I. Captopril (Capoten) became the first inhibitor in clinical use (first approved in 1981).[203] The sulfhydryl group was not only responsible for the potency of captopril but was also linked to a number of side effects, such as skin rashes and taste disturbances. In 1980, Patchett and colleagues at Merck described their efforts to discover nonsulfhydryl ACE inhibitors. In their design, they reinstalled the carboxyl group in place of the sulfhydryl group and utilized additional binding sites not utilized by captopril, the S_1 binding site and a hydrogen binding site for the amide nitrogen (of the substrate) scissile bond. SAR studies were performed on a series of *N*-carboxyalkyl dipeptides of the general formula $R\text{-}CHCO_2H\text{-}A_1\text{-}A_2$. The optimal group occupying the S_1 pocket was found to be benzylmethylene, whereas $A_1\text{-}A_2$ was Ala-Pro or Lys-Pro leading to enalaprilat (**85**) and lisinopril (**86**, Prinivil, introduced in 1987), respectively.[204] Enalaprilat was subsequently shown to have poor oral bioavailability; however, the ethyl ester prodrug enalapril (**87**, Vasotec, introduced in 1985) had superior oral bioavailability.[205] The design of captopril, enalaprilat, and lisinopril was later extended by others, and a total of approximately 17 ACE inhibitors have been approved for clinical use. The differences among these compounds are due to different functionalities binding the active-site zinc and the S_2' pocket.[189] Fosinoprilat (**88**) contains a phosphinic acid moiety that also binds zinc; however, this compound was too hydrophilic and exhibited poor oral bioavailability. The prodrug fosinopril (**89**, Monopril, introduced in 1991) contains an acyloxyalkyl group which is more lipophilic and exhibits better oral bioavailability.[206]

88, R = H

89, R =

ACE inhibitors are first-line therapy for hypertension, CHF, left ventricular systolic dysfunction, MI, and diabetic nephropathy. The landmark HOPE (Heart Outcomes Prevention Evaluation) trial established the value of ACE inhibitor therapy in preventing events in patients with established cardiovascular disease.[207] Side effects can either be mechanism related or traced to specific functional groups within the molecule (e.g., the captopril sulfhydryl group causing rash and taste disturbances). Hypotension results from an enhanced pharmacological effect and the hyperkalemia observed results from a decrease in aldosterone secretion secondary to a decrease in angiotensin II production. The most prevalent side effect is cough, which arises from a lack of selectivity for ACE inhibition.[208] Since ACE inhibitors also prevent the breakdown of bradykinin and since bradykinin stimulates prostaglandin synthesis, it is thought that the cough results from increased levels of both bradykinin and prostaglandins. Angioedema can also occur in rare instances.

4.5.2   Angiotensin II Receptor Antagonists (or Blockers) ARBs

The phenomenal success of the use of ACE inhibitors for the treatment of hypertension prompted many researchers to look for alternative ways to interfere with the RAAS cascade. In addition, inhibiting ACE was not without its drawbacks. As mentioned previously, ACE is a nonspecific protease which is also responsible for the degradation of bradykinin; in fact inhibition of ACE, and the resulting potentiation of bradykinin, has been implicated as the root cause of the dry cough noted in the 5–10% of the population treated with ACE inhibitors.[208] Inhibition of the terminal step of the RAAS cascade, i.e.; angiotensin II receptor blockade, offers a highly specific approach to inhibition of the RAAS system regardless of the source of angiotensin II. In fact, the search for ARBs began in the early 1970s with the discovery of peptides that were analogs of the natural agonist, angiotensin II. The prototypical compound arising from these studies was the octapeptide saralasin (Sar-Arg-Val-Tyr-Val-His-Pro-Ala). This peptide suffered from poor oral bioavailability and short duration of action.[209] In addition, it exhibited significant agonist activity. The search for nonpeptidic antagonists of the angiotensin II receptor was not fully realized until 1982 when Furukawa and co-workers at Takeda Chemical Industries disclosed a series of 1-benzylimidazole-5-acetic acid derivatives.[210] Two compounds from this series, S-8307 and S-8308 (**90**), were studied by the DuPont group and were found to specifically block the receptor, and although they were shown to be relatively weak antagonists, they were shown to be devoid of agonist activity.[211,212] Molecular modeling studies studied the overlap of angiotensin II with S-8308 and established a number of common structural features. Improvements in lipid solubility by the addition of additional lipophilic groups, and replacement of the carboxylic acid by bioisosteres such as tetrazole led to the disclosure of losartan (**91**, DuP 753, Cozaar) in 1990. This compound was shown to be a potent and selective antagonist of the angiotensin II receptor as well as exhibiting antihypertensive effects in a number of animal models.[213] This compound was launched in the U.S. market in 1995. Working independently, researchers at Smith Kline Beecham discovered a compound, SK&F 108566 (**92**, eprosartan, Teveten, launched in 1997), using a different molecular model than that employed in the discovery of losartan. This was the first nonbiphenyltetrazole antagonist.[214] Since the approval of losartan and eprosartan, several other derivative compounds have been approved for clinical use. All are potent antagonists with bioavailability that ranges from 15% for eprosartan mesylate to as high as 70% for irbesarten (**93**). Valsartan (**94**, Diovan, launched in 1997) is a nonheterocyclic analog where the imidazole group of losartan has been replaced with an acylated amino acid. Irbesartan is longer acting than both losartan and valsartan and has the highest oral bioavailability than any other antagonist. Candesartan cilexetil (**95**, Atacand, launched in 1998) and olmersartan medoxomil (**96**, Benicar, launched in 2002) are both potent antagonists that utilize a prodrug to overcome poor oral absorption. Telmisartan (**97**, Micardis, launched in 1999) is another nonbiphenyltetrazole that has the longest elimination half-life (24 h) of all the antagonists.[189] The Losartan Intervention for Endpoint Reduction in Hypertension (LIFE) trial was the first randomized, controlled-outcome trial to show that a particular antihypertensive drug class conferred benefits beyond blood pressure reduction and was more effective than other classes in preventing cardiovascular events and mortality.[215] Overall, this class of drugs is well tolerated and has a side-effect profile comparable with placebo.

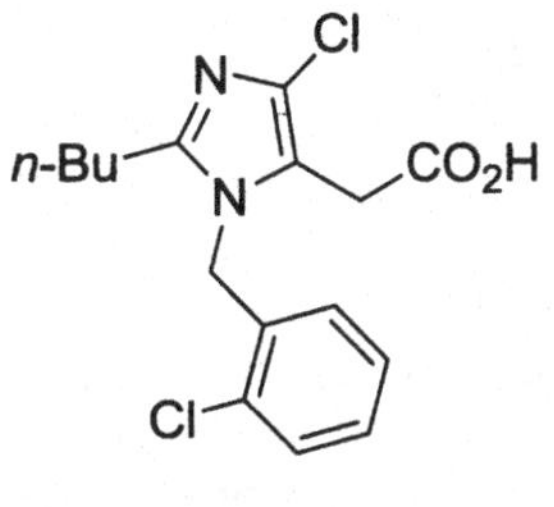

90

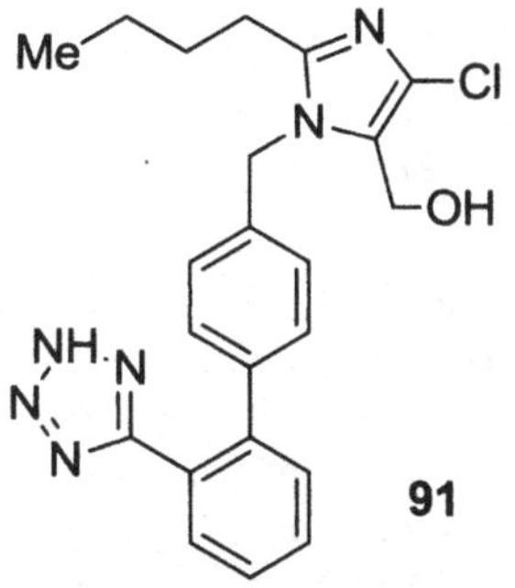

91

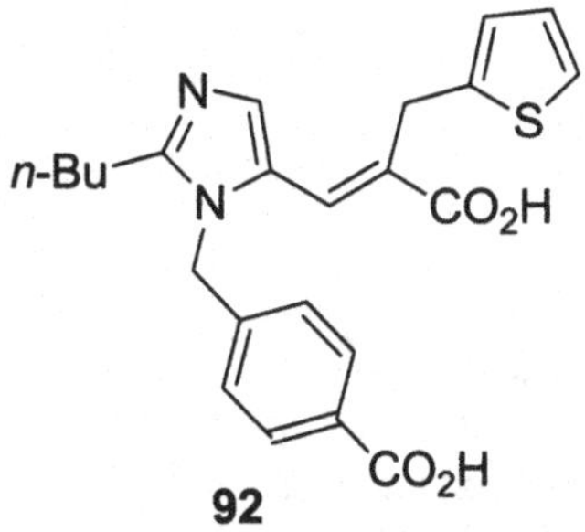

92

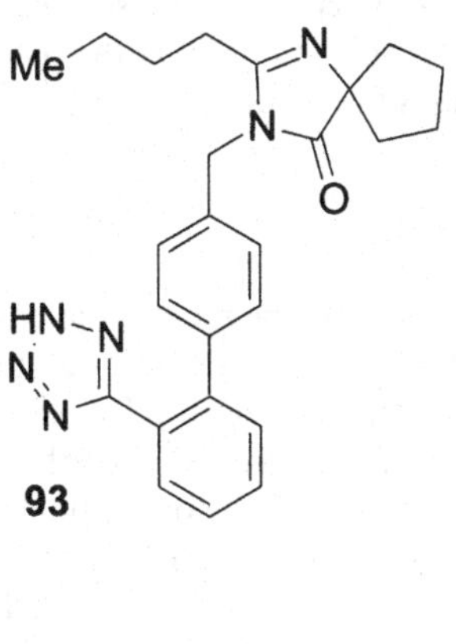

93

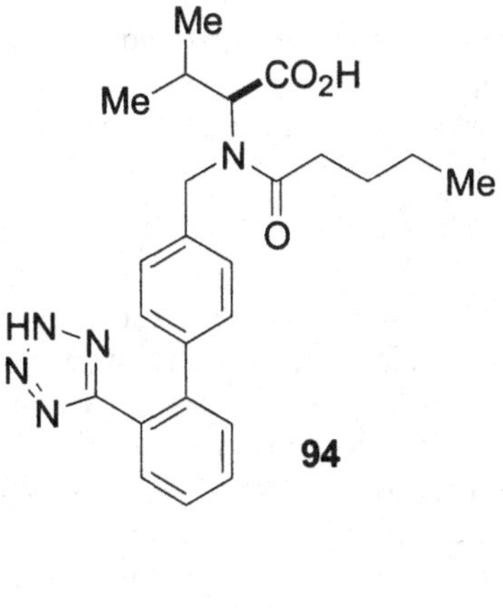

94

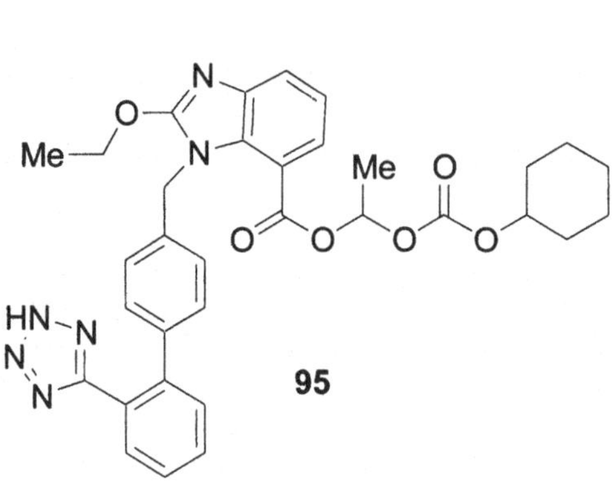

95

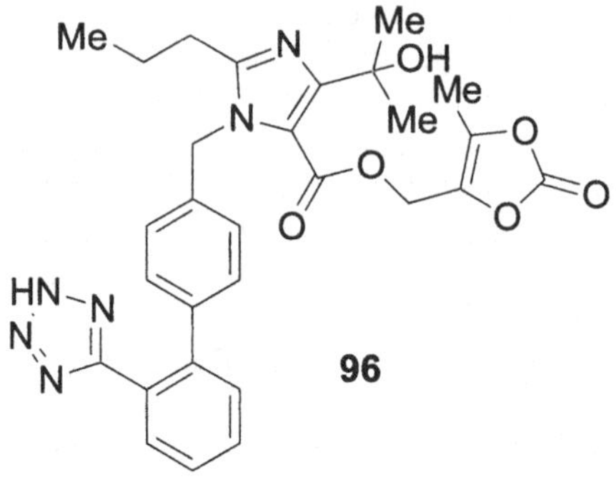

96

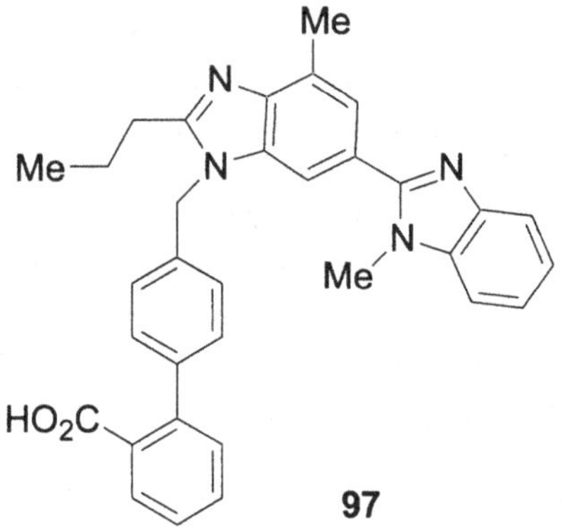

97

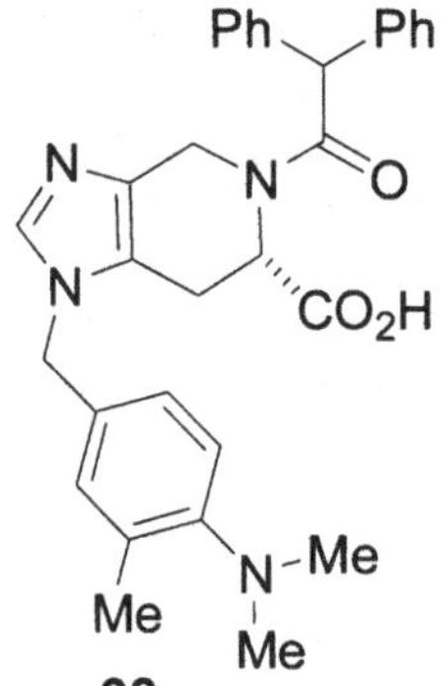

98

The angiotensin II receptor exists in at least two subtypes, type 1 (AT_1) and type 2 (AT_2). The distribution of these receptor subtypes vary from species to species and from tissue to tissue within the same species. The AT_1 receptors are located in most tissues and mediate the cardiovascular, renal, and CNS effects of angiotensin II. Losartan and the other angiotensin II receptor antagonists all show selectivity for this subtype and prevent or reverse all the known effects of angiotensin II. The role of the AT_2 receptor remains controversial. Effects mediated by the AT_2 receptor may include inhibition of cell growth, fetal tissue development, and differentiation processes. The first nonpeptide AT_2-selective ligands were a series of tetrahydroimidazopyridines [e.g., PD-123,319 (**98**)] first disclosed in 1991 that have proven to be invaluable tools for determining the possible roles for the AT_2 receptor.[216]

4.5.3   Renin Inhibitors

The most logical target in the RAAS has long been considered to be rennin, the enzyme at the top of the enzymatic cascade. Since the advent of ACE inhibitors and ARBs it has been shown that there is a powerful counter regulatory mechanism that is activated during RAAS blockade and is responsible for the flat dose-response relationship observed with the use of these inhibitors. Therapeutic interventions that either reduce angiotensin II (by ACEi) or attenuate AT_1 signaling (by ARBs) ultimately result in enhanced release of renin, thus establishing the fact that renin levels are controlled by the amount of angiotensin II present in the plasma and the tissues. This renders renin a highly attractive target for therapeutic intervention since renin inhibition is the only way to reduce plasma renin activity. Renin inhibitors bind to the active site of renin, an aspartic protease, with high specificity for its endogenous substrate angiotensinogen. While it has been realized that renin inhibition could be a highly effective therapeutic modality for the treatment of hypertension, the design of small nonpeptidic inhibitors has been a considerable challenge for nearly three decades.[217] It was not until the advent of structure-based drug design (SBDD), using X-ray crystallography and computer-assisted molecular modeling, in the 1980s that small, nonpeptidic renin inhibitors became a reality. The development of direct renin inhibitors tracks nicely with the rise and utilization of SBDD methods in medicinal chemistry. The earliest inhibitors of renin began to appear in the mid-1970s and were simple peptide analogs of angiotensinogen. These compounds were not very potent and suffered from the usual liabilities ascribed to peptidic drugs.[218]

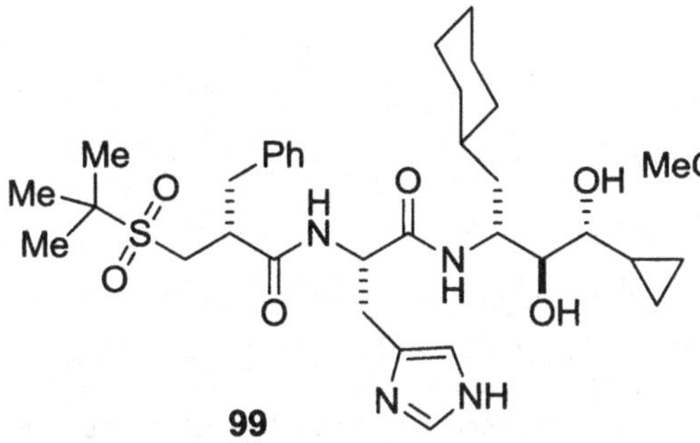

99

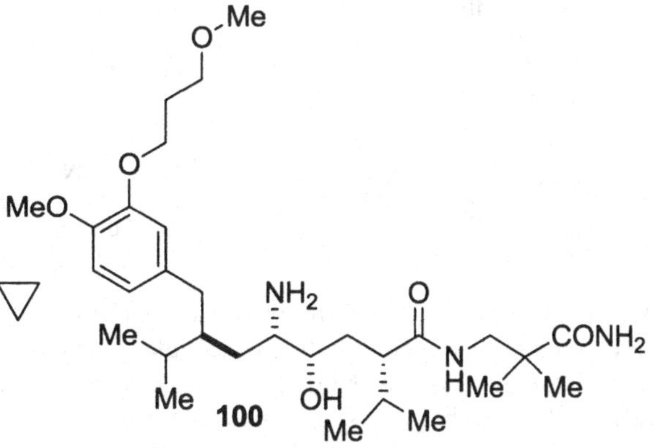

100

The next generation of inhibitors was peptidomimetic analogs, such as remikiren (**99**) that mimicked the putatative transition state of the enzyme-catalyzed amide bond hydrolysis as nonhydrolyzable replacements of the scissile cleavage site.[219] These compounds were considerably more potent, and when administered by a parenteral route, lowered blood pressure in both animals and humans. Further improvements led to longer durations of action and oral activity in humans, albeit at high doses. In the mid-1990s pharmaceutical companies began high-throughput screening (HTS) of large corporate collections of compounds. In 1999, researchers at Roche described the discovery by HTS of the first nonpeptide small molecule renin inhibitor.[220] Utilizing SBDD methods, Novartis discovered aliskiren (**100**, Tekturna, approved by the FDA in 2007). This compound interacts with several binding pockets in distinct regions around the active site of renin. It also binds to a previously unrecognized subpocket S3^{sp} of renin that extends from the S3-binding site. Binding to this sub-pocket is essential for the high potency of aliskiren. Clinical trials in normal volunteers were initiated in 2000 and a dose-dependent decrease in plasma renin activity, angiotensin I, and angiotensin II levels were shown across all doses. The first Phase I/II clinical trials in hypertensive patients were performed by Speedel, which licensed the compound from Novartis. In these, and subsequent Phase III clinical trials, aliskiren was shown to be a safe and effective antihypertensive medication. The frequency of adverse events was lower than in patients treated with placebo.[221] In addition, the RAAS is strongly implicated in the development of hypertension-related end-organ damage (e.g., CHF and end-stage renal disease). Organ protection in transgenic animal models has been observed with both ACEi and ARBs.[222] Aliskiren also shows protection in these same animal models.[223] In healthy human subjects, a dose-dependent increase in renal plasma flow is observed following a single dose of aliskiren and the authors of this study conclude that aliskiren induced a renal vasodilation greater than what has been previously seen with ACEi or ARBs.[224]

4.6   *Endothelin Antagonists*

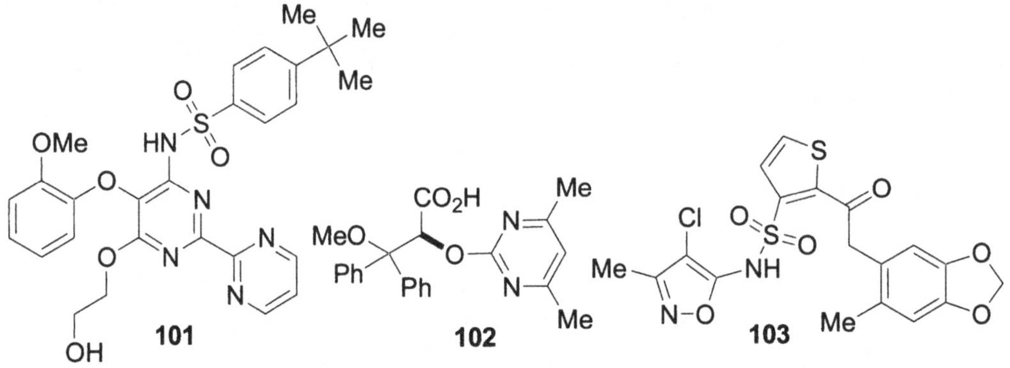

101   **102**   **103**

Although, in its simplest form, the endothelium is just the lining layer of the arterial wall, it also plays a major role in contributing to vascular tone control. It has been shown to produce the potent vasodilator nitric oxide (NO). It has also been shown to release endothelin-1 (ET-1), a potent vasoconstrictor that plays an important role in maintaining vascular tone by activating the endothelin-A (ET$_A$) receptor. ET-1 is released from vascular endothelial cells in response to a variety of factors, including angiotensin II, catecholamine, growth factors, free radicals, and mechanical stress. ET-1 is synthesized de novo as a pre-pro-protein (pre-pro-ET-1) that undergoes stepwise cleavage in the cytoplasm to form big ET-1, which is further cleaved to form ET-1 by endothelin-converting enzyme (ECE). In the vasculature, ET-1 exhibits autocrine and paracrine effects on the vascular endothelium and neighboring smooth-muscle cells. These effects are mediated by two receptor subtypes, ET$_A$ and ET$_B$ receptors. Activation of these receptors results in vasoconstriction and protein kinase C-mediated cell proliferation.[225] Elevated ET-1 plasma concentrations have been reported in patients with essential hypertension.[226] Because of the differential distribution of ET receptors in different vascular beds, the potential application of ET receptor antagonists has focused on pathological pulmonary vasoconstriction and pulmonary hypertension. Pulmonary arterial hypertension can develop independently of systemic hypertension and can result in life-threatening right-heart failure.[227] Thus began the search for fast-acting pulmonary artery vasodilators. The first orally active endothelin receptor antagonist, Ro 46-2005 (**101**, bosentan, Tracleer), was first reported in 1993. It was originally identified during a screening campaign to identify novel hypoglycemic agents. This screen identified this class of pyrimidinyl sulfonamides as weak inhibitors of ET-1 binding. Subsequent structural modifications identified Ro-46-2005 as a potent ET$_A$ receptor antagonist devoid of hypoglycemic activity.[228] It was shown to be efficacious in a number of animal models.[229] Clinical trials established its utility in the treatment of pulmonary hypertension, and it was subsequently approved by the FDA in 2001. Ambrisentan (**102**, Letairis, approved in 2007) is also a pyrimidinyl analog that is selective for the ET_A receptor. Sitaxentan sodium (**103**, Thelin), a closely related thienyl sulfonamide, was approved for use in the EU, Canada, and Australia; however, in 2010, Pfizer voluntarily removed it from the market due to reports of fatal hepatotoxicity. Subsequently all patients receiving ET receptor antagonists now require monthly monitoring of liver function.

5   Antithrombotic Drugs

We have previously described agents that treat some of the most important risk factors (hypercholesterolemia and hypertension) in the development of CAD. These risk factors contribute to the initial development of a site of injury in the vessel wall. The subsequent formation of a clot (or thrombus) and all of its negative sequelae are the single largest cause of mortality and morbidity in the Western world. Cardiac conditions involving clot formation include acute MI, valvular heart disease, unstable angina, and atrial fibrillation. Thrombotic conditions in the vasculature involve free-floating clots called emboli and include venous thromboembolism and peripheral vascular disease (PVD). The most significant conditions are pulmonary embolism in the lung and cerebrovascular strokes in the brain. The complex process of clot formation is called coagulation. It is an important component of hemostasis, a natural process in which a damaged blood vessel wall is initially covered by platelets and then a fibrin mesh to produce a clot that stops the bleeding and initiates the repair of the damaged vessel. When this process is deranged, there can be an increased risk of bleeding or an increased formation of obstructive clots (thrombosis). The pathophysiology of arterial thrombosis differs from that of venous thrombosis, and this is evident from the different ways that they are treated. In broad terms, arterial thrombosis is treated with drugs that target platelets, and venous thrombosis is treated with drugs that target the proteins of the coagulation cascade.

The primary event leading to arterial thrombosis is the rupture of an atherosclerotic plaque in the artery wall. The thrombi that form at ruptured plaques are rich in platelets that are disc-shaped cells that circulate in the blood and rapidly form a hemostatic plug at sites of vascular injury. They are recruited to these sites through the interaction of specific platelet cell surface receptors with collagen and von Willebrand factor (vWF). After this adhesion the receptor-mediated binding of additional platelets (platelet aggregation) then results in the rapid growth of the thrombus. This also results in platelet activation and the release of stored granules. These granules contain ADP, serotonin, platelet-activating factor (PAF), vWF, and other substances which, in turn, activate additional platelets. A major pathway of activation involves the cleavage and consequently, the activation of the platelet receptor PAR1 (protease-activated receptor 1, also known as the thrombin receptor) by the protease thrombin, which is activated by the blood coagulation cascade. Platelet-to-platelet aggregation is mediated by fibrinogen binding to the activated integrin αIIbβ3 receptor on the platelet surface. This completes the process of primary hemostasis. The coagulation cascade, which mediates secondary hemostasis, is the sequential process by which coagulation factors present in the blood interact and are activated, ultimately generating fibrin, the main component of the thrombus.

The coagulation cascade can be subdivided into three pathways: the extrinsic pathway (tissue factor and factor VIIa), which is the primary activator of the cascade; the intrinsic pathway (factor XIIa, factor XIa, factor IXa, and factor VIIIa), which amplifies the cascade; and the common pathway (factor Xa, factor Va, and thrombin), which generates thrombin and fibrin. The pathways are a series of reactions in which an inactive enzyme precursor, typically a serine protease (zymogen), and its glycoprotein cofactor are activated to become active components that then catalyze the next reaction in the cascade, ultimately resulting in cross-linked fibrin. The main role of the extrinsic pathway is to generate. a "thrombin burst," a process in which thrombin is released instantaneously. Following damage to the blood vessel, factor VII is activated and catalyzes the conversion of factor X to factor Xa. The intrinsic pathway involves the sequential activation of factor XII, factor XI, and factor IX. Factor IXa initiates the activation of factor X to factor Xa. The underlying purpose of the intrinsic pathway is the maintenance of homeostasis while the extrinsic pathway is activated by trauma. The intrinsic and extrinsic pathways come together in the common pathway with the conversion of factor X to its activated form, factor Xa. The final steps in the cascade involve the conversion of prothrombin (factor II) to thrombin (factor IIa) by factor Xa. Thrombin in turn catalyzes the conversion of fibrinogen to soluble fibrin, which then becomes insoluble fibrin through the action of factor XIIIa, a transamidase, which catalyzes the formation of cross-links in distinct fibrin molecules to form cross-linked fibrin aggregates, the main structural component of a clot. The principal inhibitor of thrombin in normal blood circulation is antithrombin III.[230,231]

Many of the theories on the clotting of blood and components of the coagulation cascade were identified in the late 19th century to mid-20th century. Platelets were identified in 1865 and their function was elucidated by Giulio Bizzozero in 1882.[232] Fibrin was first described by Johannes Muller (1801–1858) and fibrinogen was described by Rudolf Virchow (1821–1902). The recognition that the coagulation process was an enzymatic cascade was simultaneously reported in 1964 by MaFarlane in the UK[233] and by Davie and Ratnoff in the USA.[234]

In the case of acute thrombotic events, such as MI, drugs that reduce the growth of a thrombus can be administered and the main target of these drugs is platelets. Antiplatelet drugs are also used prophylactically to reduce the incidence of arterial thrombosis in patients with cardiovascular disease. The primary targets of antiplatelet therapy are molecules involved in platelet activation and aggregation, and these will be discussed in subsequent sections. Another treatment for acute thrombotic events is the degradation of fibrin by using clot-busting drugs such as tissue plasminogen activator and streptokinase.

5.1   Antiplatelet Agents

5.1.1   Aspirin

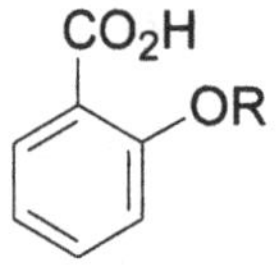

CO_2H

OR   **104**, R = COMe
             105, R = H

Aspirin (**104**, acetylsalicylic acid, ASA) is the oldest and most commonly used antithrombotic (and antiplatelet) agent used today. In addition to its use in treating thrombotic disorders, this so-called "wonder drug" is also the gold standard for antipyretic, analgesic, and anti-inflammatory drugs.

The first recorded descriptions of the therapeutic benefits of willow bark and other plant sources (e.g., wintergreen plant, poplar) of salicylates were made by Hippocrates about 2400 years ago. The modern history of salicylates began around 1757 when the Reverend Edward Stone of Chipping-Norton in Oxfordshire, England, began a search for a substitute for cinchona bark.

Cinchona bark (Jesuit's bark) was a source for the preeminent antipyretic drug quinine; however, the cinchona tree was only indigenous to South America. In 1630 cinchona bark was found to be successful in treating malaria, so securing a readily available cheap substitute in Europe was greatly needed. In 1757, Stone first tasted the bark of the white willow tree (*Salix alba vulgaris*) and was impressed by its severe bitterness, being somewhat reminiscent of Jesuit's bark. In 1763, after six years of clinical trials, Stone reported his findings on the use of willow bark in patients suffering from inflammatory disorders and fevers. Over the next century steady progress was made in the isolation and purification of salicylates from natural sources. The earliest attempt at isolating the active agent from a natural source was by Fontana and Brugnatelli in Italy between 1826 and 1829. They isolated salicin, a glycoside, from willow bark. Piria, in Paris, isolated salicyclic acid (**105**) from salicin in 1838. In the next fifty years, the synthetic production of salicylates went into high gear. Interestingly, Charles Gerhardt first synthesized acetylsalicylic acid (ASA) in 1853; however, the significance of this compound would not become evident for another 45 years! In the latter half of the 19th century, there was a growing desire to find salicylates with a much better side effect profile. This quest was championed by workers at the Bayer Company in Germany where chemist Felix Hoffmann rediscovered the work of Gerhardt and developed an improved synthetic pathway for ASA. He subsequently tested the drug on himself and his father who was suffering from crippling arthritis. Heinrich Dreser at the University of Bonn, Germany, promoted the use of ASA in a paper published in 1899 and is credited with coining the new trade name, Aspirin, in January 1899. The drug was introduced commercially by the Bayer Company five months later. Aspirin use soared due to its antipyretic, analgesic and anti-inflammatory effects. Interest in its antithrombotic effects did not awaken until 1940 when Karl Link (the discoverer of dicumarol and warfarin) showed that dicumarol spontaneously breaks down into salicyclic acid, and that, after experimenting on himself, he found that ASA was a weak anticoagulant. He considered this a dangerous side effect and it was not until 1948 that Gibson first proposed that ASA be tried as a treatment for vascular diseases. In that same year, L. L Craven, a physician in Glendale, California, began treating all his older male patients with ASA to prevent MI.[235] At the time of these clinical discoveries, the mechanism of action of ASA was unknown until 1967 when Weiss demonstrated that the MOA involved platelet dysfunction.[236] However, it was not until 1971 that John Vane's historic report showed that ASA suppressed the synthesis of prostaglandins and thromboxanes by inhibiting an enzyme, later identified as cyclooxygenase.[237,238] Vane received the Nobel Prize in Medicine for this work in 1982. Low-dose, long-term use of ASA irreversibly blocks the synthesis of thromboxane A2, a substance that has been shown to activate platelets and increase platelet aggregation. Since the 1970s, ASA has been shown to have benefit in both primary and secondary prophylaxis of MI, MI in the acute phase, unstable angina, and transient ischemic attacks. In high-risk patients, aspirin reduces vascular death by ~15% and nonfatal vascular death by ~30%, as evidenced by a meta-analysis of over 100 randomized trials. Aspirin may also be of benefit in the primary prevention of cardiovascular events, but the effect is more modest. Its biggest drawback is that it increases the risk of gastrointestinal bleeding. By irreversibly inhibiting cyclooxygenase, it also prevents the production of certain prostaglandins that protect the stomach mucosa from damage by hydrochloric acid.[239]

Despite the wealth of data that support the use of aspirin, it remains a suboptimal antiplatelet agent. There is a substantial cohort of patients who display clinical "aspirin resistance" and experience an ischemic event despite taking aspirin. The underlying cause of this "resistance" is still in question, but its incidence has been reported to be in the range 5–65% depending on the assay and other variables.[240]

5.1.2   *ADP Receptor (P2Y$_{12}$) Antagonists*

106   **107**   **108**

Platelets require adenosine diphosphate (ADP) to bind to other platelets and to facilitate additional platelet adhesion and aggregation. The platelet response to ADP is mediated by a family of nucleotide receptors called P2 receptors. After platelet activation, released ADP promotes platelet activation via G-protein-linked P2Y$_1$ and P2Y$_{12}$ purinergic receptors leading to further platelet activation, aggregation, and other platelet functions such as platelet shape change, secretion, and development of procoagulant and proinflammatory activities. The result of ADP signaling through the P2Y$_1$ receptor is calcium ion mobilization, a change in platelet shape, and rapidly reversible platelet aggregation. ADP signaling through the P2Y$_{12}$ receptor results in the amplification of stable platelet aggregation and secretion.[241]

The first ADP receptor antagonist was the thienopyridine ticlopidine (**106**, Ticlid). This compound was initially discovered in 1972 by Jean-Pierre Maffrand of Sanofi in a search for novel anti-inflammatory agents. Although this compound had no anti-inflammatory activity, it was shown to inhibit platelet aggregation.[242] This compound is metabolized by cytochrome P450 in the liver, and it is the resulting active metabolite that irreversibly antagonizes the P2Y$_{12}$ receptor.[243] Subsequently, there has developed a large body of data supporting the efficacy of ticlopidine in conditions such as unstable angina, coronary artery bypass surgery, peripheral artery bypass surgery, and cerebrovascular disease. Ticlopidine is given orally twice a day and was the first P2Y12 antagonist approved in the US in 1991. It has been shown that ticlopidine can cause neutropenia (decrease in white blood cell count) and rarely thrombotic thrombocytopenic purpura (TTP), a much more serious disorder.[244] To overcome these shortcomings, Sanofi carried out extensive SAR studies and identified a racemic compound, PCR 4099, in 1980 that was shown to be more potent and better tolerated than ticlopidine. It was eventually decided to develop a single enantiomer, clopidogrel (**107**, Plavix). This compound is also a prodrug that is given once daily and has a much better side-effect profile than ticlopidine. It was developed by Sanofi and a development partner, Bristol Myers Squibb, and won FDA approval in 1997. This drug has become a true blockbuster and in 2010 was the third best selling drug in the US with sales approaching $5 billion. Clopidogrel represents a major advance in antiplatelet therapy, and the clinical benefit of adding clopidogrel to aspirin therapy has been demonstrated by large multicenter, randomized controlled trials.[245] For example, the CURE (Clopidogrel in Unstable Angina to Prevent Recurrent Events) study showed that patients with unstable angina or non-ST segment elevation MI experienced a 20% relative risk reduction if they were randomized to clopidogrel plus aspirin instead of placebo plus aspirin.[246] This study and others demonstrated the benefit of dual antiplatelet therapy where, although it is clear that clopidogrel is more effective than aspirin alone, dual antiplatelet inhibition is more than additive in preventing thrombus formation. Challenges still remain with this therapeutic modality: There is a significant patient population that is hyporesponsive to clopidogrel therapy; clopidogrel has a relatively slow onset of action compared to aspirin; and finally there is the ongoing incidence of ischemic events, including stent thrombosis, in patients treated with clopidogrel and aspirin.[241,247] The search for improved ADP receptor antagonists continues. Prasugrel (**108**, Effient, Lilly/Daiichi Sankyo) is also a thienopyridine prodrug; however, metabolism to its active metabolite is much more efficient than that of clopidogrel because it is metabolized by an esterase and is less dependent on cytochrome P450 enzymes. Clinical studies showed that a loading dose of 60 mg prasugrel results in a more rapid, potent, and consistent inhibition of platelet function than the standard clopidogrel loading dose of 300 mg. Prasugrel was approved by the FDA in 2009.[248] Ticagrelor (**109**, Brilinta, Astra Zeneca) is given orally twice a day and received EU approval in 2010 and FDA approval in July 2011. It is not a thienopyridine but a cyclopentyl triazolopyrimidine nucleoside that is a direct and reversible ADP receptor antagonist. It acts at a different binding site on the $P2Y_{12}$ receptor than ADP and is thus regarded as an allosteric antagonist. In addition, it does not require metabolic activation. Like prasugrel, it acts more rapidly and is a more potent inhibitor of platelets than clopidogrel. Another nucleoside, cangrelor (**110**), is also a direct-acting, reversible $P2Y_{12}$ antagonist. Unlike the other $P2Y_{12}$ antagonists, it is administered intravenously, and its effects are rapidly reversed after the end of the infusion. However, two Phase III trials were stopped early due to a lack of efficacy.[241]

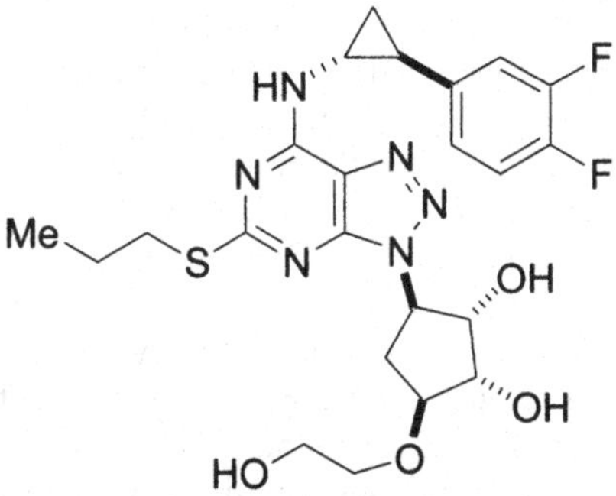

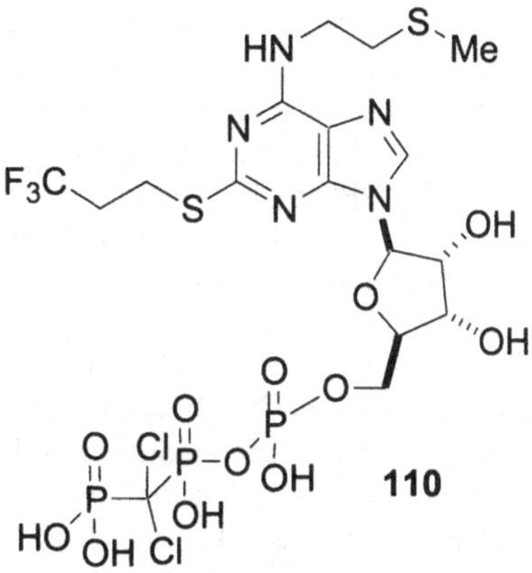

5.1.3 Glycoprotein IIb/IIIa Antagonists

The final obligatory step in platelet aggregation is the binding of fibrinogen to the glycoprotein IIb/IIIa receptor, an integrin found in high concentration on the platelet membrane. Antagonism of this receptor was thus seen as an attractive antiplatelet therapeutic modality.[241,247] The evolution of this new therapy happened over many decades and began with a new look at an old disease. Glanzmann thrombasthenia (GT) is an autosomal recessive disorder which manifests itself as a lifelong moderate to severe bleeding tendency. It was first described in 1918, but it was not until the 1960s when it was demonstrated that patients' platelets failed to aggregate at all in response to a variety of agonists such as ADP.[249] One important clue to both the pathogenesis of GT and the mechanism by which normal platelets aggregate was the observation made independently by

Zucker[250] and Jackson[251] that GT patients are severely deficient in fibrinogen. In the mid-1970s, Nurden and Caen[252] and Phillips[253] independently identified deficiencies of two different platelet membrane glycoproteins in several GT patients. These glycoproteins were designated glycoprotein (GP) IIb and GPIIIa. It took until the late 1970s to establish that the binding of fibrinogen to the platelet surface is necessary for normal platelet aggregation to occur.[254] The GPIIb/IIIa receptor became the prime candidate as the platelet "fibrinogen" receptor and studies in 1980 confirmed that, in GT patients, the GPIIb/IIIa receptor was defective.[255] At this time the technology for preparing monoclonal antibodies using murine hybridomas was becoming available and Barry Coller at the State University of New York at Stony Brook produced a monoclonal antibody (10E5) that could abolish the platelet aggregation of normal platelets, block platelet-fibrinogen interactions, and inhibit clot retraction.[256] Coller produced another antibody, 7E3, which reacted with dog platelets in well-characterized dog models of arterial thrombosis.[257] Subsequent *in vivo* studies confirmed that this monoclonal antibody was a potent platelet aggregation inhibitor and may be useful in treating patients with unstable angina, transient ischemic attacks, and MI.[258] In 1986, 7E3 was licensed to Centocor. Because of concern about the immunogenicity of the murine antibody in human subjects, chimeric molecules were constructed which consisted of mouse-derived variable regions linked to human-derived constant regions. One of these molecules, mouse/human chimeric c7E3 Fab (abciximab, ReoPro), entered clinical development and was eventually approved in late 1994. It is administered intravenously and is used in patients undergoing high-risk angioplasty and atherectomy.[259] As with all the antiplatelet agents, its major side effect is bleeding.

In the early 1980s, a striking discovery was made concerning the mechanism by which fibrinogen binds to the GPIIb/IIIa receptor. It was found that a three-amino-acid sequence, arginine–glycine–aspartic acid (RGD), in fibrinogen was crucial for binding to the GPIIb/IIIa receptor.[260] It was subsequently shown that vWF, fibronectin, vitronectin, and thrombospondin could all bind to the GPIIb/IIIa receptor and that they all contained the RGD sequence.[261] It was at this time that the first nonantibody antagonists began appearing. The seminal finding that prompted the search for nonantibody antagonists was the discovery of trigramin, a small protein containing 72 amino acids identified and purified from the venom of the Indian green tree viper, which was shown to be a potent GPIIb/IIIa receptor antagonist. It was also shown to contain the RGD sequence that allowed it to bind to the GPIIb/IIIa receptor.[262] Many other peptide antagonists of the GPIIb/IIIa receptor were identified in the venom of other viper and pit viper species. These antagonists were termed "disintegrins" and they were, in general, nonspecific antagonists of RGD-dependent receptors such as GPIIb/IIIa, $\alpha_v\beta_3$ (vitronectin), and $\alpha5\beta1$ (fibronectin). Most of the disintegrins contain the RGD sequence; however, barbourin, isolated from the pygmy rattlesnake (*Sistrurus m. barbour*), contains the KGD inhibitory sequence and is highly specific for the GPIIb/IIIa receptor with no reactivity for the closely related vitronectin receptor. The conservative Lys-for-Arg substitution appears to be the sole structural feature that is responsible for this specificity.[263] This specificity information provided the impetus for researchers at COR Therapeutics in San Francisco to design cyclic peptide mimics of barbourin. They found that small, conformationally constrained peptides containing the KGD sequence were potent and specific antagonists of the GPIIb/IIIa receptor versus other closely related integrins. This work, first described in 1993, eventually yielded a cyclic heptapeptide, C68-22 (**111**, eptifibatide, Integrilin).[264] This compound was shown to be an effective antithrombotic agent in a number of animal models of thrombosis when dosed intravenously. Like abciximab, it finds its greatest utility in the setting of percutaneous coronary interventions and acute coronary syndromes. Once again, bleeding is the major side effect to be considered. Eptifibatide was approved by the FDA in 1998.

In a quest to discover a nonpeptidic GPIIb/IIIa receptor antagonist, workers at Merck screened a subset of their compound library containing amine and carboxylic acid functionalities that had a distance separation of 10–20 Å (approximating the distance between the basic guanidine and the acidic carboxylate of the RGD sequence). Initially disclosed in 1992, the initial hit weakly inhibited ADP-mediated platelet aggregation and was of comparable potency to the RGDS tetrapeptide sequence. Optimization led to the discovery of L-700,462 (**112**, tirofiban, Aggrastat), a tyrosine derivative of high potency.[265] The inclusion of the sulfonamide group was critical in attaining the high potency observed, and it is now known that this group binds to a region of the GPIIb/IIIa receptor which is not exploited by the cyclic peptides. Although it is a small molecule, this compound is not orally bioavailable, and it has a relatively short half-life in animals and humans, properties that make it suitable as a parenteral antithrombotic agent. It was approved by the FDA in 1998.

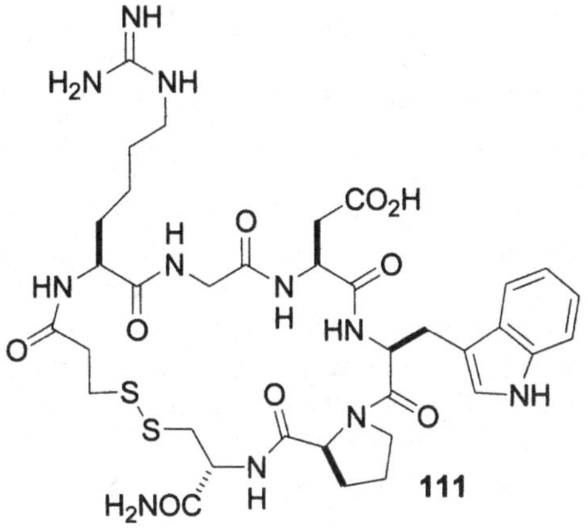

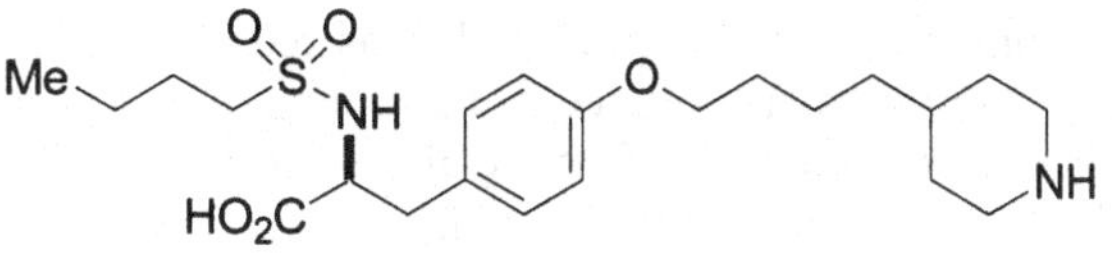

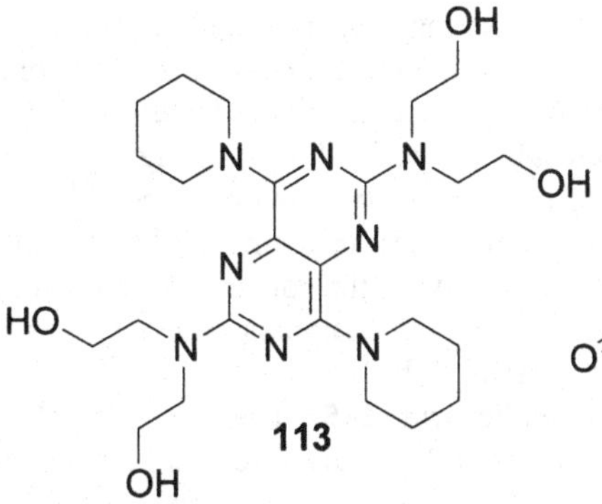

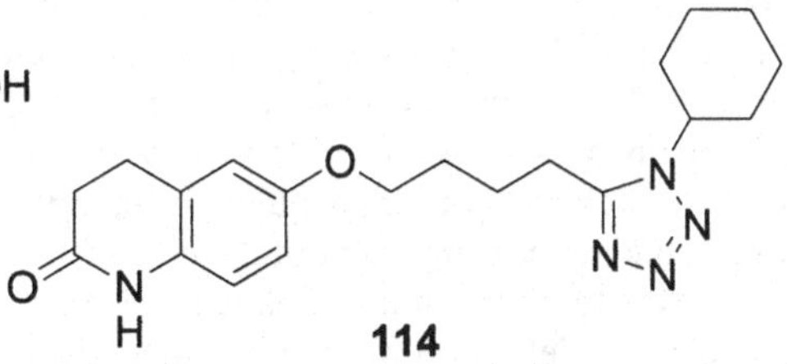

Numerous clinical trials have established the benefits of IV GPIIb/IIIa receptor antagonists in the treatment of coronary artery disease. However, large-scale placebo-controlled randomized trials of oral GPIIb/IIIa receptor antagonists have failed to provide commensurate reductions in ischemic end points despite potent inhibition of platelet aggregation. In fact, Phase

III trials of oral antagonists all found that these agents caused an excess of major bleeding without any significant benefit. A meta-analysis of four major trials failed to find a benefit for these agents (xemilofiban, orbofiban, sibrafiban) and actually found an increase in the rate of all-cause mortality with these agents. The reasons for this discrepancy between IV and oral therapies remain a subject of much conjecture.[266] Except in high-risk cases, the clinical use of GPIIb/IIIa receptor antagonists has decreased in favor of the increased usage of ADP receptor antagonists.

5.1.4   Phosphodiesterase Inhibitors

Dipyridamole (**113**, Persantine) was shown to have antiplatelet activity and has been approved as a coronary vasodilator since 1961. The antiplatelet effects have been reported to be due to several mechanisms of action, including inhibition of phosphodiesterase (PDE) enzymes and the blockade of adenosine uptake, both of which result in increased intraplatelet cAMP levels, and consequent blockade of the platelet response to ADP. In 1999 the FDA approved Aggrenox, a modified-release formulation of dipyradamole and low-dose aspirin for stroke prevention.[241] Cilostazol (**114**, Pletal) is an oral-selective PDE3 inhibitor which also has both vasodilatory and antiplatelet activity. It was approved by the FDA in 1999 for the treatment of intermittent claudication. Side effects (headaches, GI symptoms, and skin rash) cause ~15% of patients to discontinue the drug.[241]

5.2   **Anticoagulant Agents**

Deep-vein thrombosis and pulmonary embolism are the third leading cause of cardiovascular-associated death after MI and stroke. Thrombi that form in the veins are rich in fibrin and trapped red blood cells and are referred to as red clots (as opposed to platelet-rich thrombi that form in arteries, which are referred to as white clots). Anticoagulants are used to treat a wide variety of chronic conditions that involve arterial or venous thrombosis, including the prevention of venous thromboembolism and the long-term prevention of ischemic stroke in patients with atrial fibrillation. The most commonly available anticoagulant drugs work by inhibiting steps in the coagulation cascade.[230,231]

Bloodletting has long been thought to have merit in the treatment of a variety of disorders. One of the most common methods of bloodletting was the application of leeches (*Hirudo medicinalis*) to the skin of the patient. The ability of the leech to absorb a large amount of the patient's blood was due to the action of the leeches saliva and its ability to keep the blood flowing. In 1884, J. B. Haycraft recognized that leech saliva contained an anticoagulant factor. At the turn of the 20th century a dry powdered extract of leeches' heads was introduced[267] and given the name hirudin by Jacoby in 1904. In 1957, it was determined that hirudin directly inhibited thrombin, the key enzyme in the coagulation cascade that catalyzes the conversion of fibrinogen to soluble fibrin, the major component of the clot.[268] Its use as an anticoagulant clinically was limited since the best way to deliver it to the patient was still via the leech!

Another widely used injectable anticoagulant is heparin (**115**). The discovery of heparin can be traced back to the late 1800s.[235] In 1880, Schmidt-Mulheim disclosed that an endogenous anticoagulant of hepatic origin was released after dogs were injected with peptone, an enzymatic digest of animal protein. In 1905, Morawitz noted that extraction of tissues by organic solvents gave procoagulants, whereas aqueous extracts of the residue after organic solvent extraction showed anticoagulant activity. In 1912, William Howell of Johns Hopkins isolated a procoagulant substance he called cephalin, from dog brain. In 1916, Jay McLean, a premedical student in

Howell's laboratory, found that aged cephalin and another phosphatide, heparphosphatide extracted from dog liver, had lost their procoagulant activity and became anticoagulant. McLean announced his discovery of "antithrombin" on February 19, 1916, at the University of Pennsylvania. In 1918, Howell and Holt renamed heparphosphatide heparin and confirmed its solubility in organic solvents. In 1922, Howell reported the discovery of another heparin that was also extracted from dog liver, but this was an aqueous extract. Further purification of this new heparin by Howell and Holt between 1922 and 1928 revealed that it was a sulfated carbohydrate of high aqueous solubility. In 1933, pure crystalline heparin was produced by a Toronto group led by C. H. Best (of insulin isolation fame), and in 1935, Hedenius and Wilander injected purified heparin into human volunteers successfully with no apparent side effects. This marked the beginning of the clinical era of heparin, and it became the standard parenteral treatment for a variety of venous and arterial thrombotic disorders. Low-molecular-weight-heparins (LMWHs, MW < 8000 Da) have greater efficacy and are associated with less bleeding when compared to unfractionated heparin (MW= 5000-40000 Da). Synthetic preparations of the LMWHs appeared in the 1970s and clinical trials followed in the 1980s. The first approved by the FDA in 1993 was enoxaparin sodium (Lovenox). The original insights into heparin's mechanism of action occurred in 1939 when Brinkhous *et al.* reported that heparin did not prevent clotting of isolated thrombin and fibrinogen; there was a serum cofactor whose activation was necessary for heparin's action.[269] This cofactor was identified as antithrombin III by Abildgaard in 1968, and shortly thereafter the exact molecular mechanisms of heparin's mechanism of action were elucidated. In short, heparin binds antithrombin III, induces a conformational change, and irreversibly binds to and inhibits the active site of thrombin. Thus, heparin is regarded as an indirect [Antithrombin (AT)-dependent] thrombin inhibitor.[270]

115

116

The discovery of oral anticoagulants began in the early 20th century when farmers in the northern prairie states of North America began planting hardy sweet clover plants (*Melilotus alba, M. officinalis*) for use as an animal feed in the harsh climate.[235] However, within two decades, a new disease, sweet clover disease, decimated cattle herds, causing fatal spontaneous bleeding. In 1922, F. W. Schofield, a veterinarian in Alberta, Canada, showed that this disease was caused by eating hay prepared from spoiled sweet clover.[271] L. M. Roderick, another veterinarian of North Dakota, found that ther hemorrhagic factor present in the spoiled sweet clover acted by reducing the activity of prothrombin, a precursor of thrombin. The next major step in unraveling this mystery occurred in 1933 when Karl Link, at the Agricultural Experiment Station at the University of Wisconsin, was tasked with the job of finding a strain of sweet clover free of coumarin (**116**), a substance known to be responsible for its bitter taste and sweet smell when freshly cut. However, in the wake of an encounter with a distraught farmer, Link dedicated himself to the isolation of the hemorrhagic factor. In June 1939, Link's associate, Harold Campbell, isolated crystals of the hemorrhagic agent and then another two colleagues, Charles Huebner and Mark Strahman, solved the hemorrhagic agent's structure as 3,3′-methylenebis-(4-hydroxycoumarin) (117) and synthesized it in 1940.[272] This compound received the name dicumarol, which would serve in the USA as both generic and trade name. Interestingly, at this time, Link showed that vitamin K reversed the actions of either spoiled sweet clover or dicumarol. O. Meyer at Wisconsin General Hospital was the first to document its anticoagulant effects in humans in February 1941; however, the first published report came in June 1941 from the Mayo group headed by E. V. Allen.[273] The use of dicumarol in the treatment of acute MI became widespread in the forties and fifties, but in the sixties and seventies there was growing concern over the incidence of hemorrhage in patients receiving this therapy and it quickly fell out of favor. Meanwhile, Link developed an unusual interest in the history of rodent control. He was determined to develop the ideal rat poison. He found dicumarol weak and unreliable as a rodenticide, but then he searched through some compounds synthesized by Ikawa from 1942 through 1944.[274] Two compounds, 42 and 63, emerged as potent and reliable; however, compound 42 was chosen for further development and given the name warfarin (118, Coumadin) (from the first letters of the Wisconsin Alumni Research Foundation, with the suffix *arin*). This was launched in 1948 as the "ideal" rat poison.[275] S. Shapiro of New York[276] and O. Meyer[277] at the University of Wisconsin tested warfarin in human volunteers in 1953 and reported it to be far superior to dicumarol. It was launched commercially in 1954, and interestingly, when President Dwight Eisenhower suffered an MI that year, he was treated with warfarin. Warfarin became the standard treatment for all venous or arterial thrombotic conditions requiring long-term treatment. The precise mechanism of action of warfarin was not elucidated until 1974 when Stenflo *et al.* described the posttranslational carboxylation of vitamin K-dependent clotting factors.[278] In 1978, Whitlon *et al.* and Bell independently found that warfarin acts by inhibiting the enzyme vitamin K epoxide reductase (VKOR), the C1 subunit (VKORC1) being the specific target.[279] Despite careful monitoring, the incidence of major bleeding is about 1–3% of warfarin-treated patients per year.

117   118

5.2.1 Direct Thrombin Inhibitors

It has been shown that drugs that inhibit thrombin can bind to three possible domains on thrombin: the active site and two exosites. Exosite 1 is the binding site for substrates such as fibrin and orientates the appropriate peptide bonds in the active site. Exosite 2 is the heparin-binding domain. From this understanding, there have emerged two different classes of direct thrombin inhibitors. Bivalent direct thrombin inhibitors (DTIs) inhibit thrombin at both the active site and exosite 1, whereas univalent DTIs bind only to the active site. Hirudin is a bivalent DTI whose isolation from natural sources is difficult, and this has prompted the development of recombinant forms such as lepirudin (Refludan) and desirudin (Revasc) and synthetic congeners such as bivalirudin (Angiomax).[280]

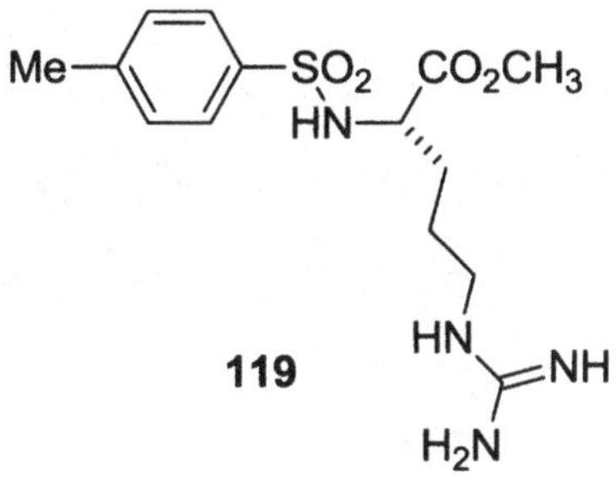

119

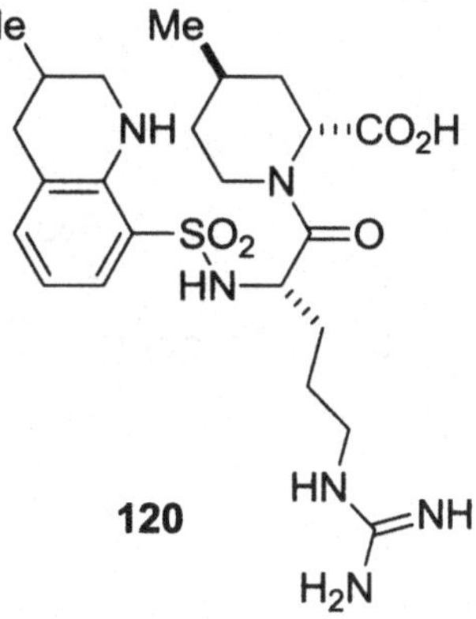

120

Thrombin is a trypsin like serine protease that catalyzes the conversion of fibrinogen to fibrin by cleaving the peptide bond between arginine and glycine in a fibrinogen sequence Gly-Val-Arg-Gly-Pro-Arg. It has been known for many years that synthetic arginine esters such as N^{α}-tosyl-L-arginine methyl ester (**119**, TAME) are hydrolyzed by thrombin and inhibit thrombin by binding to the active site noncovalently.[281] However, TAME is also hydrolyzed by trypsin and other trypsin like proteases. In 1980, Okamoto and colleagues at Mitsubishi Chemical Industries reported on their efforts to develop potent and specific noncovalent inhibitors of thrombin by systematically modifying the TAME molecule.[282] Their work, initially reported in 1975,[283] focused on replacement of the ester group of TAME with tertiary amides and optimization of the arylsulfonamide moiety. One compound, no. 805, was shown to be a potent, specific, and fully reversible thrombin inhibitor.[284] In several animal models, the antithrombotic effect was similar to heparin. This compound, argatroban (**120**, Acova), was shown to be effective in clinical trials and received its first approval in Japan in 1990 for treating peripheral vascular disorders. It was approved by the FDA in June 2000 as an anticoagulant in patients with heparin-induced thrombocytopenia and in 2002 as an anticoagulant in patients with or at risk of heparin-induced thrombocytopenia while undergoing percutaneous coronary intervention. It is administered intravenously.[285] Many other companies invested considerable resources into the development of similar compounds. The publication of the X-ray crystal structure of human thrombin to four active site directed thrombin inhibitors greatly aided the design and synthesis of new compounds.[286] One highly desirable feature of any new compound developed was that it be orally bioavailable. One compound, melagatran, was shown to be a potent inhibitor that was efficacious in animal models.[287] It entered the clinic as an IV therapy; however, a double prodrug of melagatran, ximelagatran (**121**), was the first orally active thrombin inhibitor. The carboxylic acid group of melagatran was esterified and the amidine was converted to amidoxime. Ximelagatran (Exanta) was first approved in France in 2003 for the prevention of venous thromboembolic events in patients undergoing major elective orthopedic surgery. In 2006, Astra Zeneca announced it was voluntarily withdrawing ximelagatran due to elevations in liver enzymes leading to liver damage. Despite this disappointment, another double prodrug, dabigatran etexilate (**122**, Pradaxa), recently received FDA approval in October 2010, after receiving EU approval in March, 2008, thus making it the first oral anticoagulant since warfarin to reach the market in the USA.[288]

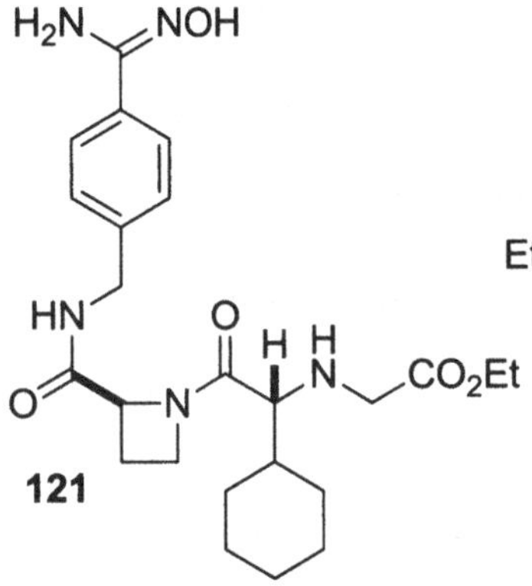

121 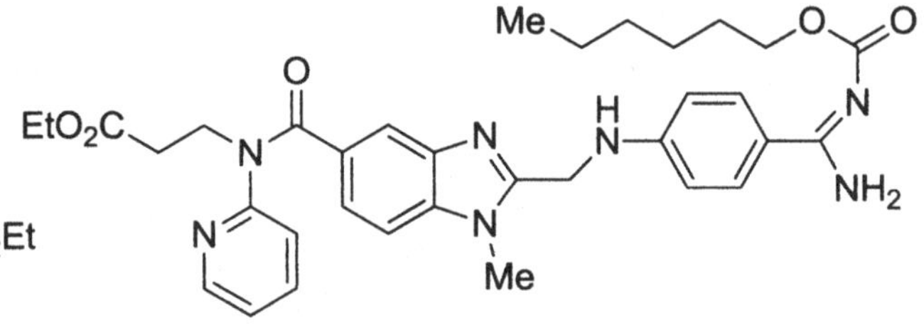   **122**

     Despite these successes, direct thrombin inhibition can lead to undesirable effects. Although direct thrombin inhibition can lead to a reduction in fibrin deposition, it can also reduce platelet activation and promote an unstable fibrin mesh that could result in a higher bleeding risk and a narrow therapeutic window. In addition, the rebound ischemia phenomenon has been seen with both direct (argatroban) and indirect (heparins) thrombin inhibitors. This phenomenon is thought to occur due to the continual thrombin generated within the thrombus by FXa in the prothrombinase complex after treatment has been discontinued, thus resulting in the recurrence of prothrombotic events.[289]

5.2.2   Factor Xa Inhibitors

Factor Xa is a structurally closely related enzyme to thrombin. It cleaves the zymogen prothrombin to thrombin. Although it is capable of cleaving prothrombin on its own, physiologically relevant rates of thrombin production are only achieved following the calcium-dependent assembly of the prothrombinase complex (FXa with its cofactor, factor Va) on an activated platelet membrane, leading to a great increase in FXa catalytic activity. FXa sits at the important juncture between the intrinsic and extrinsic pathways in coagulation, and its inhibition represents a unique opportunity for direct, targeted regulation of thrombin formation. Regulating the activity of FXa also provides a means of limiting thrombin production without interrupting the basal level of thrombin necessary for primary hemostasis and should lead to greater efficacy-to-safety ratios. The deficiency in existing therapies coupled with the advantages outlined above has made the search for FXa inhbitors a top priority in the field.

     Factor Xa inhibitors can block FXa either directly, by binding to the FXa active site, or indirectly, via antithrombin.[289]

Indirect factor Xa inhibitors

Fondaparinux (Arixtra, approved by the FDA in 2001) is a synthetic analog of the unique pentasaccharide sequence that mediates the interaction of heparin or LMWH with antithrombin. Once the pentasaccharide-AT complex binds FXa, the pentasaccharide dissociates from AT and can be reutilized. Thus, the indirect inhibitors are catalytic and result in AT-mediated irreversible inhibition of free FXa. Fondaparinux binds AT with high affinity, has excellent bioavailability after subcutaneous injection, and has a plasma half life of 17 h that permits once-daily administration.[290]

Direct factor Xa inhibitors

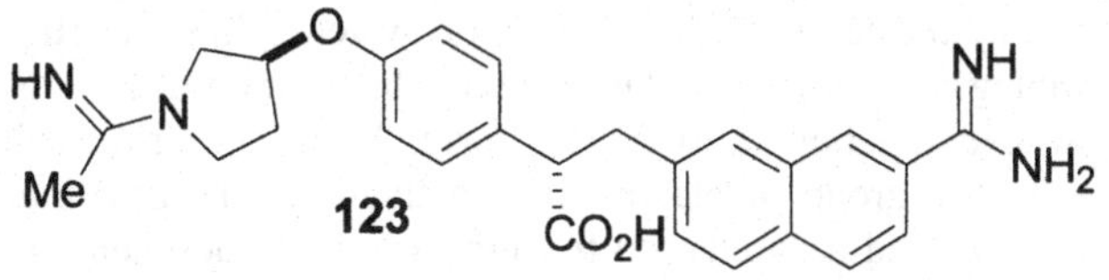

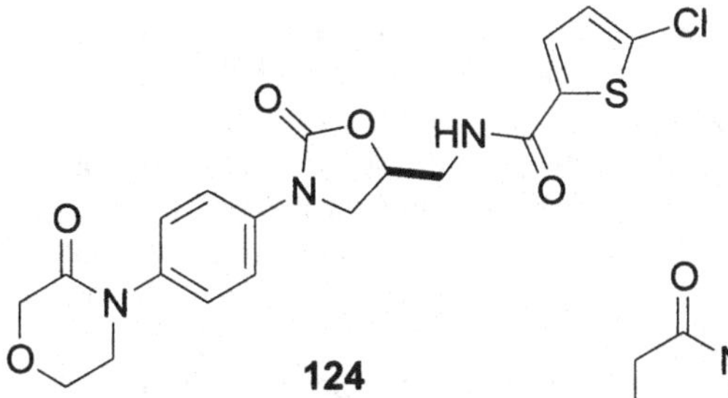

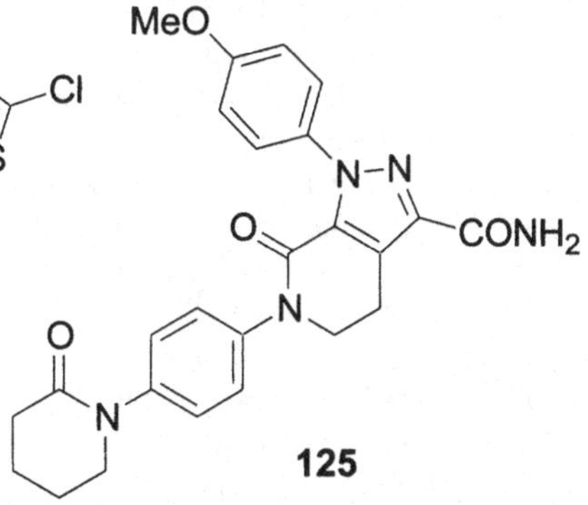

The discovery and development of FXa inhibitors closely mirrors the same approach used for the discovery and development of the thrombin inhibitors. Much of the recent work draws on the initial studies on dibasic bis(amidinoaryl) FXa inhibitors developed by Tidwell.[291] Building on Tidwell's work, in 1994 Daiichi disclosed DX-9065a (123), a potent and selective FXa inhibitor which was inactive against thrombin and was extensively studied in both animals and humans.[292] In clinical trials, it was administered intravenously and shown to be a relatively safe method of anticoagulation during coronary interventions. It quickly became apparent that to develop orally available FXa inhibitors, there should be a search for less polar and less basic replacements of the amidine moiety found in the existing inhibitors. The existing inhibitors suffered from low oral bioavailability, and/or high clearance, and/or short $t_{1/2}$ *in vivo*. In 2005 a group at Bayer Healthcare AG reported on a new series of oxazolidinone inhibitors that were highly potent and selective direct FXa inhibitors with excellent *in vitro* antithrombotic activity. The lead compound in this new series was discovered via an HTS effort.[293] Optimization of this compound led to the identification the neutral 5-chlorothiophene moiety as a suitable replacement for the highly basic amidine moiety in the S_1 subsite of the enzyme. One compound, rivaroxaban (124, Xarelto), was shown to have high bioavailability (60–80%) and is given once or twice daily. Maximum plasma drug levels occur 2 to 3 h after oral administration, and the terminal half-life is 6–9 h. This compound was approved in the EU in 2008 and on July 1, 2011, the FDA approved rivaroxaban for the prophylaxis of deep-vein thrombosis in adults undergoing hip and knee replacement surgery. Apixaban (125), a drug in codevelopment by Bristol-Myers Squibb and Pfizer, is currently in Phase III clinical trials for the prevention of venous thromboembolism.[294]

6   Thrombolytic Agents

Although antiplatelet agents and anticoagulants may prevent the formation of thrombi in the arterial and venous systems, these agents are unable to dissolve clots once they have formed. Blood clots are dissolved by the action of the fibrinolytic system whose function is to remove clots while maintaining the integrity of the vascular system. This system utilizes plasmin, a relatively nonspecific protease whose function is to digest fibrin. Plasmin is released from the inactive proenzyme plasminogen by a group of plasminogen activators. The principal activator is tissue plasminogen activator (t-PA) and it is released from the vascular endothelium. Plasmin activity is regulated by two endogenous and specific inactivators called tissue plasminogen activator inhibitors 1 and 2 (t-PAI-1 and t-PAI-2).

The history of the development of thrombolytic agents reaches as far back as the 1860s; however, the basic knowledge of the natural fibrinolytic system lagged far behind that of the coagulation cascade, leading to a lag in the development of thrombolytics. The history of thrombolytics began in 1861 with the report by von Brucke on the proteolytic activity of human urine. In 1886, Sahli noted urinary proteolytic activity with some specificity for fibrin. The purification and isolation of the fibrinolytic enzyme were achieved in 1947 by MacFarlane and Pilling[295] and in 1951 by Williams.[296] In 1952, Sobel *et al.* coined the name "urokinase" (UK).[297] He and others showed that urokinase does not directly digest fibrin but is an activator of endogenous plasminogen, thereby generating plasmin.

In 1933, William Tillett, a bacteriologist at the New York School of Medicine, found that the β-hemolytic, group C streptococci produced a fibrinolytic agent he termed "fibrinolysin."[298] Milstone, in 1941, showed that the streptococcal fibrinolysin did not directly dissolve fibrin *in vitro*; rather it required a protein co-factor he called "plasma lysing factor" (later proved to be plasminogen).[299] Christensen coined the terms plasminogen and plasmin and renamed fibrinolysin "streptokinase" (SK).[300] Sol Sherry and his team at Bellevue became the first to administer crude SK to humans in early 1947. They treated patients with a variety of lung disorders with intrapleural injections of SK with spectacular results.[301] Purer preparations of SK allowed the first IV infusions in human patients with acute MI to occur in 1958.[302] This same team described the successful use of UK in maintaining a fibrinolytic state in patients in 1965. Subsequent studies demonstrated the effectiveness of both SK and UK in the treatment of pulmonary embolism, acute MI, and venous thromboembolism. Both SK (Streptose) and UK (Abbokinase) were approved for IV use in 1977. SK was approved for deep venous embolism, pulmonary embolism, and arterial thrombosis. Additional indications were approved in the 1980s. Variants of SK began to appear in the 1960s and, in 1979, Smith *et al.* reported the rational design of anistreplase (Eminase), a combination of SK and plasminogen.[303] It was approved for IV use in acute MI by the FDA in 1989.

In 1964, Sherry's group reported the partial purification of t-PA.[304] Fully purified t-PA was prepared by Rijken in 1979.[305] With purified preparations in hand, clinical use evolved rapidly. The first use of natural t-PA in acute MI patients occurred in 1984[306] and led to FDA approval for IV t-PA (Alteplase) for acute MI patients in 1987 and for pulmonary embolism in 1990. Various other second-generation t-PA molecules [e.g., reteplase (Retavase, approved by the

FDA in 1996), tenecteplase (TNKase, approved by the FDA in 2000)] have also reached the market.[307] Hemorrhagic stroke is a rare but serious side effect of thrombolytic therapy.

7   Antianginal Agents

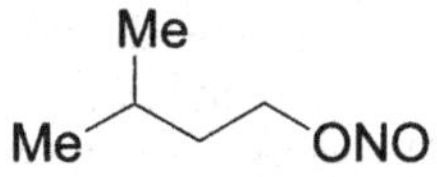

$$\text{Me}$$
$$\text{Me}\text{—}\text{—}\text{—}\text{ONO}$$

126

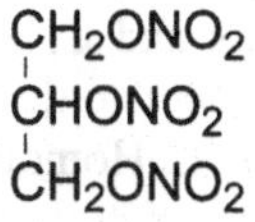

$$CH_2ONO_2$$
$$CHONO_2$$
$$CH_2ONO_2$$

127

The primary symptom of ischemic heart disease is angina. It is characterized by a sudden, severe pain originating in the chest and radiating to the left shoulder and down the left arm. Drug therapy is directed mainly to alleviating and preventing angina attacks by dilating the coronary artery. The mainstay of angina treatment has been the use of organic nitrates as coronary vasodilators. In 1867, Thomas Brunton, a surgeon at Edinburgh Royal Infirmary in Scotland, was the first to show the utility of amyl nitrite (**126**), the first coronary vasodilator, in the treatment of the pain of angina pectoris.[308] Brunton also tried nitroglycerine (**127**), now readily available due to Alfred Nobel's discovery of its value as an explosive. However, when he administered it to himself, he suffered violent headaches and quickly dismissed its future potential. This potential was only realized in 1877, when William Murrell at the Westminster Hospital in London began a series of studies that resulted in the clinical usage of nitroglycerine in the treatment of angina pectoris, a practice that continues to this day. He found that its effects on pain relief were longer lasting than amyl nitrite, the headache was due to overdosage, and sublingual (under-the-tongue) delivery was the best way to deliver the drug.[309] The mechanism of action of all the nitrates is basically the same; they all act as prodrugs for nitric oxide (NO), a powerful vasodilator; however, the precise mechanism that generates NO from the nitrates is still in dispute.

In addition to the organic nitrates, calcium channel blockers and β-blockers are also used as antianginal therapies.

8   Heart Failure Drugs

When all of the treatments we have covered in this chapter fail, the end result is usually congestive heart failure. Once this disease manifests itself, the heart fails to pump blood effectively and is unable to satisfy the metabolic needs of the cardiac tissues. This is as a result of a reduced contractility of the cardiac muscles and a subsequent decrease in CO and increase in blood volume (hence the word congested). Systemic blood pressure and renal flow are both reduced, which often lead to the development of edema in the lower extremities and the lung (pulmonary edema) as well as renal failure. For decades, the dominant drug for the treatment of heart failure has been digitalis. The discovery of digitalis heralded the advent of modern pharmacotherapy. It had long been known that a particular herbal tea, with over twenty ingredients, was an effective treatment for 'dropsy' (edema), but it was unknown what specific ingredient was responsible for the therapeutic effect. In 1775, William Withering, a physician and botanist from Shropshire, England found that the specific ingredient in the herbal tea was the foxglove plant (*Digitalis purpurea*). Over the next nine years, he administered the foxglove plant to a number of patients suffering from "cardiac dropsy," an accumulation of fluid in patients suffering from CHF.[310] At the time, there was little understanding of the origins of the edema and just how digitalis worked. Withering had noted that digitalis was a powerful cardiac stimulant; however, it was not until the early years of the 20th century that a proper understanding of the effects of digitalis on the heart began to emerge. The advent of the electrocardiograph in 1903 allowed the study of the effects of digitalis on the heart. From these studies it was shown that digitalis could be applied to the treatment of certain heart rhythm disorders and CHF. The quest to isolate the active principle from digitalis began in the 1820s when the Société de Pharmacie in Paris offered a cash reward for the isolation of the pure principle. In 1841, E. Homolle and T. Quevenne won the award for the isolation of an impure, but active, crystalline material they called digitaline. In 1875, Scmiedeberg isolated a crystalline compound of high potency called digitoxin which was shown to be the principle cardiotonic glycoside in the leaves of *Digitalis purpurea*. However, it was not until 1928 that the correct structure of digitoxin (**128**) was elucidated by Windaus at Gottingen University; the same group solved the structure of digitalin in 1929.[311] At this time, it was discovered that the powdered leaves of the wooly foxglove (*Digitalis lanata*) had greater pharmacological activity than those of *Digitalis purpurea*. The active principle in this plant was subsequently shown to be a new glycoside that was called digoxin (**129**).[312] Both digitoxin and digoxin are still in use today although they have been supplanted by newer medicines. Their ability to increase cardiac contractility has led to their use in the treatment of CHF; however, nowadays it is only employed after first-line therapy when other agents have failed. Digitalis preparations also have effects on the parasympathetic nervous system and are therefore used for the treatment of various heart rhythm disorders such as atrial fibrillation. The inotropic activity is due to its ability to inhibit the Na^+/K^+ pump and thus increase intracellular calcium levels leading to increased cardiac contractility. This enzyme was discovered in 1957 by Jen Christian Skou of the University of Aarhus in Denmark, who in 1997 shared the Nobel Prize in Chemistry for this discovery.[313]

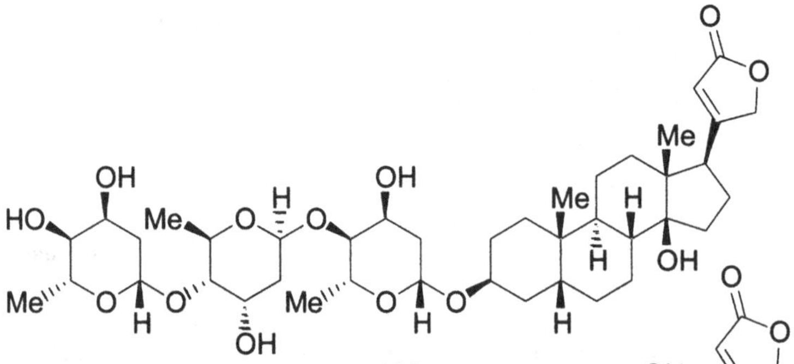

128

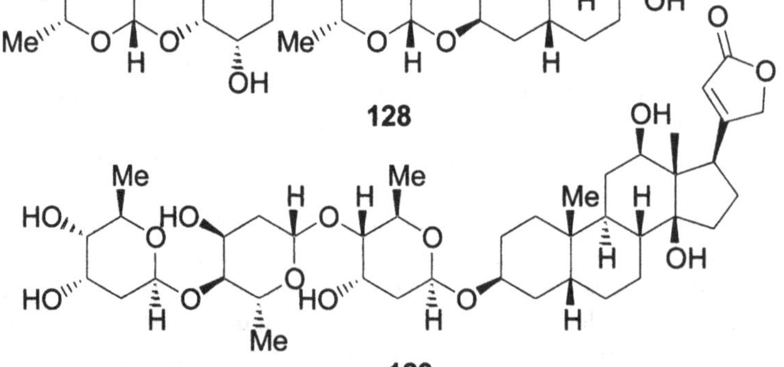

129

Today the standard of care for CHF patients is to delay the progression of heart failure and is currently based on combination therapy utilizing several classes of agents. The main classes of drugs that are currently used include the following: ACE inhibitors, ARBs, aldosterone receptor antagonists, β-blockers, CCBs, diuretics, digitalis, and organic nitrate vasodilators.[314]

There are a myriad of cardiac disorders that require treatment and we have covered the main classes employed. There are numerous drug classes used in the treatment of cardiac arrhythmias. An arrhythmia is an alteration in the normal sequence of electrical impulse activation that leads to contraction of the myocardium. It is manifested as an abnormality in the rate, the site from which the impulses originate, or the conduction through the myocardium. The causes of arrhythmias are numerous and their treatment is complex. The antiarrhythmic drugs used generally occupy four different classes: the local anesthetics (Class I), the β-blockers (Class II), potassium channel blockers (Class III), and calcium channel blockers (Class IV).[315]

9   The Future

This chapter has described the work of many scientists in their quest to tackle the scourge of heart disease. Early treatments sought to support the failing heart by relieving anginal pain and making the failing heart beat harder. The golden era of drug discovery sought to find remedies that controlled the various risk factors (hypercholesterolemia, hypertension) involved in the development of CHD. Today, there is an emphasis that seeks to produce antithrombotic therapies that seek to remove the precipitating factor (clot formation) in the development of the devastating sequelae of MI, stroke, pulmonary embolism, and CHF. One of the more exciting areas of research involves the development of therapies that seek to regenerate muscle tissue in the injured heart. It is hypothesized that CHF could be reversed if new myocardium could be grown in diseased hearts. Early efforts focused on cell-based cardiac repair with the transplantation of autologous skeletal muscle cells (myoblasts) and progenitor cells that normally mediate the regeneration of skeletal muscle. In fact, myoblast grafting has proceeded into clinical trials.[316] In an exciting development, some small molecules have been shown to promote myocardial repair/regeneration by activating cardiac differentiation in human mobilized peripheral blood mononuclear cells (M-PBMCs).[317]

There is no doubt that the search for novel treatments for risk factor modifications such as hypercholesterolemia and hypertension have slowed. Many of the currently available treatments for these risk factors do help some people avoid the development of clinically relevant heart disease but they do not help everybody. In a recent study, the genome was screened for common variants associated with plasma lipids in more than 100,000 individuals of European ancestry with and without CHD. The study identified 95 significantly associated loci (locus: specific location on a gene), with 59 showing genomewide significant associations with lipid traits for the first time. This study shows that drugs targeting a single locus may not be effective in everyone, but more importantly, it has identified new therapeutic opportunities for the prevention of CHD.[318]

References

1.   World Health Organization. *World Health Statistics Quarterly* **1993**, *46*, and *World Health Statistics Quarterly* **1995,** *48.*
2.   World Health Organization. *Fact Sheet No 317*, **2011**.
3.   American Heart Association. *Circulation* **2010**, *121*, e46.
4.   Persidis, A. *Nature Biotech.* **1999,** *17*, 930.
5.   Lichtlen, P. R. *Z. Kardiol.* **2002,** *Suppl.4*, 56.

6.   Virchow, R *Phlogse and Thrombose in Gefassystem Gessamelete Abhandlungen zur Wissenschajtlichen Medicin.* **1856**, p. 458. Frankfurt-am-Main: Meidinger Sohn.
7   Anitschkow, N. N. *Cowdrys Arteriosclerosis,* **1967,** *21,* Blumenthal, H. T. ed., Charles C. Thomas.
8.   Kannel, W. B.; Dawber, T. R.; Kagan, A.; et al. *Ann. Intern. Med.* **1961,** *55,* 33.
9.   Roberts, W. C. *Atherosclerosis* **1992,** *97,* S5-S9.
10.   Muldoon, M. E.; Manuck, S. B.; Matthews, K. A. *N. Eng. J. Med.* **1991,** *324,* 922.
11.   Lipid Research Clinics Program. *JAMA* **1984,** *251,* 365.
12.   Frick, M. B.; Elo, O.; Haapa, K.; et al. *N. Eng. J. Med.* **1987,** *317,* 1237.
13.   Blankenhorn, D. H.; Nessim, S. A.; et al. *JAMA* **1987,** *257,* 3233.
14.   Altschul, R.; Hoffer, A.; Stephen, J. D. *Arch. Biochem.* **1955,** *54,* 558.
15.   Parsons, W. B.; Flinn, J. H. *Arch. Intern. Med.* **1959,** *103,* 783.
16.   Carlson, L. A. *J. Intern. Med.* **2005,** *258,* 94.
17.   Karpe, F.; Frayn, K. N. *Lancet* **2004,** *363,* 1892.
18.   O'Kane, M. J.; Trinick, T. R.; Tynan, M. B.; et al. *Br. J. Clin. Pharmacol.* **1992,** *33,* 451.
19.   Carlson, L. A. *Int. J. Clin. Pract.* **2004,** *58,* 706.
20.   Sood, A.; Arora, R. *J. Clin Hypertens.* **2009,** *11,* 685.
21.   Mandeville, W. H.; Goldberg, D. I. *Curr. Pharm. Des.* **1997,** *3,* 15.
22.   Siperstein, M. D.; Hichols, C. W.; Chaikoff, I. L. *Science* **1953,** *117,* 386.
23.   Tennent, D. M.; Siegel, H.; Zanetti, M. E.; et al. *J. Lipid Res.* **1960,** *1,* 469.
24.   Visintine, R. E.; Michaels, G. D.; Fukayama, G.; et al. *Lancet* **1961,** *2,* 341.
25.   Datta, D. V.; Sherlock, S. *Br. Med. J.* **1963,** 216.
26.   Horan, J. M.; Diluzio, N. R.; Etteldorf, J. N. *J. Pediatr.* **1964,** *64,* 201.
27.   Parkinson, T. M.; Gundersen, K.; Nelson, N. A. *Atherosclerosis* **1970,** *11,* 531.
28.   Bays, H.; Dujovne, C. *Exp. Opin. Pharmacother.* **2003,** *4,* 779.
29.   Takeuchi, A.; Takano, T.; Utsunomiya, H.; *J. Gastroenterol. Hepatol.* **2003,** *18,* 548.
30.   Staels, B. *Postgrad. Med.* **2009,** *121,* 25.
31.   Thorp, J. M.; Waring, W. S. *Nature* **1962,** *194,* 948.
32.   Oliver, M. F.; Heady, J. A.; Morris, J. N.; et al. *Br. Heart J.* **1978,** 40, 1069.
33.   Oliver, M. F.; Heady, J. A.; Morris, J. N.; et al. *Lancet* **1984,** *8403,* 600.
34.   Hodges, R. M. *Proc. Royal Soc. Med.* **1976,** *69(Suppl. 2),* 1.
35.   Sornay, R.; Gurrieri, J.; Tourne, C.; et al. *Arzn.-Forsch.* **1976,** *26,* 885.
36.   Staels, B.; Dallongeville, J.; Auwerx, J.; et al. *Circulation* **1998,** *98,* 2088.
37.   Fruchart, J.-C. *Atherosclerosis* **2009,** *205,* 1.
38.   Duez, H.; Lefebvre, B.; Poulain, P.; et al. *Arterioscler. Thromb. Vasc. Biol.* **2005,** *25,* 585.
39.   Barnhart, J. W.; Sefranka, J. A.; McIntosh, D. D.; Am. *J. Clin. Nutr.* **1970,** *23,* 1229.
40.   Yamamoto, A.; Matsuzawa, Y.; Yokoyama, S.; et al. *Am. J. Cardiol.* **1986,** *57,* 29H.
41.   Yamamoto, A. *J. Atheroscler. Thromb.* **2008,** *15,* 304.
42.   Turley, S. D.; Dietschy, J. M. In *The Liver; Biology and Pathology* Arias, I.; Popper, H.; Schacter, D.; Shafritz, D. A. eds., Raven Press: New York, **1982,** p. 467.
43.   Faust, J. R.; Trzaskos, J. M.; Gaylor, J. L. **1988;** In *The Biology of Cholesterol,* Yeagle, P. L. ed., p. 19–38. Boca Raton, FL.
44.   Brown, M. S.; Goldstein, J. L. *J. Lipid Res.* **1980,** *21,* 505.
45.   Brown, M. S.; Goldstein, J. L. *Science* **1990,** *343,* 425.
46.   Brown, M. S.; Goldstein, J. L. *Cell* **1997,** *89,* 331.
47.   Brown, M. S.; Goldstein, J. L. *J. Lipid Res.* **1984,** *25,* 425.
48.   Curran, G. L.; Azarnoff, D. L. *Arch. Intern. Med.* **1958,** *101,* 685.
49.   Cottet, J.; Mathirat, A.; Redel, J. *Press Med.* **1954,** *62,* 939.
50.   Bargeton, D.; Krumm-Heller, C.; Tricaud, M. E. *Soc. Biol.* **1954,** *148,* 63.
51.   Frederickson, D. S.; Steinberg, D. *Circulation* **1957,** *15,* 391.
52.   Barr, D. P. *J. Chronic Dis.* **1955,** *1,* 63.
53.   Palpoli, F. P.; Allen, R. E.; Schumann, E. L.; et al. In *Proceedings of the 134th National Meeting of the American Chemical Society,* **1958,** p. 12.0 (abs.), Chicago, IL.

54. Blohm, T. R.; Kariya, T.; Laughlin, M.E.; et al. *Fed. Prac.* **1959**, *18*, 369.
55. Avigan, J.; Steinberg, D.; Thompson, M. J. *J. Biol. Chem.* **1960**, *235*, 3123.
56. Laughlin, R. C.; Carey, T. F. *JAMA* **1962**, *181*, 339.
57. Achor, R. W. P.; Winkleman, R. K.; Perry, H. O. *Prac. Staff Meet. Mayo Clin.* **1961**, *36*, 217.
58. Avigan, J.; Steinberg, D.; Thompson, M.J.; et al. *Biochem. Biophys. Res. Commun.* **1960**, *2*, 63.
59. Lupien, P. J.; Moorjani, S.; Brun, D.; et al. *J. Clin. Pharmacol.* **1979**, *19*, 120.
60. Endo, A.; Kuroda, M.; Tsujita, Y. *J. Antibiot.* **1976**, *29*, 1346.
61. Brown, A. G.; Smale, T. C.; King, T. J.; et al. *J. Chem. Soc. Perkin* **1976**, *1*, 1165.
62. Endo, A. *J. Antibiot.* **1979**, *32*, 852.
63. Alberts, A. W.; Chen, J.; Kuron, G.; et al. *Proc. Natl. Acad. Sci. USA* **1980**, *77*, 3957.
64. Serizawa, N.; Serizawa, S.; Nakagawa, K.; et al. *J. Antibiot.* **1983**, *36*, 887.
65. Haruyama, H.; Kuwano, M.; Kinoshita, T.; et al. *Chem. Pharm. Bull.* **1986**, *54*, 1459.
66. MacDonald, J. S.; Gerson, R. J.; Kornbrust, D. J.; et al. *Am. J. Cardiol.* **1988**, *62*, 16J.
67. Chao, Y.; Chen, Y. S.; Hunt, Y. M.; et al. *Eur. J Clin. Pharmacol. Lett.* **1991**, *40*, S311.
68. Pietro, A. A.; Mantell, G. *Cardiovasc. Drugs Rev.* **1990**, *8*, 220.
69. Scandinavian Simvastatin Survival, Study Group. *Lancet* **1994**, *344*, 1383.
70. Sato, A.; Ogiso, A, Noguchi, H.; et al. *Chem. Pharm. Bull.* **1980**, *28*, 1509.
71. Novello, F. C.; Prugh, J. D.; Smith, R. L.; et al. *J. Med. Chem.* **1986**, *28*, 347.
72. Stokker, G. E.; Alberts, A. W.; Anderson, P. S.; et al. *J. Med. Chem.* **1986**, *29*, 170.
73. Roth, B. D.; Ortwine, D. F.; Hoefle, M. L.; et al. *J. Med. Chem.* **1990**, *33*, 21.
74. Kathawala, F. G.; Engstrom, R. G.; Heider, J. G.; et al. *Triangle* **1986**, *25*, 61.
75. Roth, B. D.; Blankley, C. J.; Chucholowski, A. W.; et al. *J. Med. Chem.* **1991**, *34*, 357.
76. Angerbauer, R.; Fey, P.; Huebsch, W.; et al. *US Patent* 5032602, **1989**.
77. Rodriguez, M. L.; Mora, C.; Navarro, J. F. *Ann. Int. Med.* **2000**, *132*, 598.
78. Watanabe, M.; Koike, H.; Ishiba, T.; et al. *Bioorg. Med. Chem.* **1997**, *5*, 437.
79. Pfefferkorn, J. A. *Curr. Opin. Invest. Drugs* **2009**, *10*, 245.
80. Roth, B. D.; Bocan, T. M.; Blankley, C. J.; et al. *J. Med. Chem.* **1991**, *34*, 463.
81. Hsiang, B.; Zhu, Y.; Wang, Z.; et al. *J. Biol. Chem.* **1999**, *274*, 37161.
82. Sacks, F. M.; Pfeffer, M. A.; Moye, L. A.; et al. *N. Engl. J. Med.* **1996**, *335*, 1001.
83. Brown, B. G.; Zhao, X. Q.; Sacco, D. E.; et al. *Circulation* **1993**, *87*, 1781.
84. Brown, B. G.; Hillger, L.; Zhao, X. Q.; et al. *Ann. N.Y. Acad. Sci.* **1995**, *748*, 407.
85. Sliskovic, D. R.; Picard, J. A.; Krause, B. R. *Prog. Med. Chem.* **2002**, *39*, 121.
86. Lee, H. T.; Sliskovic, D. R.; Picard, J. A.; et al. *J. Med. Chem.* **1996**, *39*, 5031.
87. Insull, W.; Koren, M.; Davignon, J.; et al. *Atherosclerosis* **2001**, *157*, 137.
88. Tardif, J-C.; Gregoire, J, L'Allier, P.L.; et al. *Circulation,* **2004**, *110*, 3372.
89. Nissen, S.E.; Tuzcu, E.M.; Brewer, H.B.; et al. *N. Engl. J. Med.* **2006**, *354*, 1253.
90. Meuwese M. C.; de Groot E.; Duivenvoorden R.; et al. *JAMA* **2009**, *30,* 1131.
91. Clader, J. W. *J. Med. Chem.* **2004**, *47*, 1.
92. Altmann, S. W.; Davis, H. R.; Zhu, L.-J.; et al. *Science* **2004**, *303*, 1201.
93. Garcia-Calvo, M.; Lisnock, J, Bull, H. G.; et al. *Proc. Natl. Acad. Sci. USA* **2005**, *102*, 8132.
94. Kastelein, J. J. P.; Akdim, F.; Stroes, E. S. G. *N. Engl. J. Med.* **2008**, *358*, 1431.
95. Rossebø, A. B.; Pedersen, T. R.; Boman, K.; et al. *N. Engl. J. Med.* **2008**, *359*, 1343.
96. Bell, T. A.; Brown, J. M.; Graham, M. J.; et al. *Arterioscler. Thromb. Vasc. Biol.* **2006**, *26*, 1814.
97. Harris, W. S.; Bulchandani, D. *Curr. Opin. Lipidol.* **2006**, *17*, 387.
98. Gordon, T.; Castelli, W. P.; Hjortland, M. C.; et al. *Am. J. Med.* **1977**, *62,* 707.
99. Gordon, D. J.; Probstfield, J. L.; Garrison, R. J.; et al. *Circulation* **1989**, *79*, 8.
100. Adult Treatment Panel III. *JAMA* **2001**, *285*, 2486.
101. Robins, S. J.; Collins, D, Wittes, J. T. et al. *JAMA* **2001**, *285*, 1585.
102. The Bezafibrate Infarction Prevention (BIP) Study. *Circulation* **2000**, *102*, 21.
103. Hulley, S.; Grady, D.; Bush, T.; et al. *JAMA* **2001**, *285*, 1585.
104. Glomset, J. A. *J. Lipid Res.* **1968**, 9, 155.

105.   Ross, R.; Glomset, J. A. *Science* **1973**, *180*, 1332.
106.   Miller, G. J.; Miller, N. E. *Lancet* **1975**, *1*, 16.
107.   Barter, P. J.; Kastelein, J. J. P. *J. Am. Coll. Cardiol.* **2006**, *47*, 492.
108.   Brooks-Wilson, A.; et al. *Nat. Genet.* **1999**, *22*, 336.
109.   Kuivenhoven, J. A.; Pritchard, H.; Hill, J.; et al. *J. Lipid Res.* **1997**, *38*, 191.
110.   Lewis, G. F.; Rader, D. J. *Circ. Res.* **2005**, *95*, 1221.
111.   Huuskonen, J.; Olkkonen, V. M.; Jauhiainen, M.; et al. *Atherosclerosis* **2001**, *155*, 269.
112.   Wang, N.; Lan, D.; Chen, W.; et al. *Proc. Natl. Acad. Sci. USA* **2004**, *101*, 9774.
113.   Silver, D. L.; Wang, N.; Xiao, X.; et al. *J. Biol. Chem.* **2001**, *276*, 25287.
114.   Barter, P. J.; Caulfield, M.; Eriksson, M.; et al. *N. Engl. J. Med.* **2007**, *357*, 2109.
115.   Cannon, C. P.; Shah, S.; Dansky, H. M.; et al. *N. Engl. J. Med.* **2010**, *363*, 2406.
116.   Spillmann, F.; Schultheiss, H.-P.; Tschöpe, C.; et al. *Curr. Pharm. Des.* **2010**, *16*, 1517.
117.   Reddy, S. T.; Van Lenten, B. J. *J. Lipid. Res.* **2009**, *50*, S145.
118.   Carretero, O. A.; Oparil, S. *Circulation* **2000**, *101*, 329.
119.   The Seventh Report of the Joint National Committee on Prevention, Detection, Evaluation and Treatment of High Blood Pressure. **2004**, *NIH Publication No.* 04-5230.
120.   Taylor, D. A.; Abdel-Rahman, A. A. *Adv. Pharmacol.* **2009**, *57*, 291.
121.   Esunge, P. M. *J. Royal Soc. Med.* **1991**, *84*, 621.
122.   Booth, J. *Proc. Royal Soc. Med.* **1977**, *70*, 793.
123.   O'Rourke, M. F. *Hypertension* **1992**, *19*, 212.
124.   Dustan, H. P.; Roccella, E. J.; Garrison, H. H. *Arch. Intern. Med.* **1996**, *156*, 1926.
125.   Oates, J.A.; Brown, N.J. in *Goodman and Gilman's The Pharmacologic Basis of Therpaeutics*, 10th Ed., McGraw Hill: New York, **2001**, p.889.
126.   Schroeder, H. A. *Circulation* **1952**, *5*, 28.
127.   Rubin, A. A.; Roth, F. E.; Weinberg, M. M.; et al. *Science* **1961**, *133*, 2067.
128.   Freis, E. D.; Finnerty, F. A.; Schnaper, H. W.; et al. *Circulation* **1952**, *5*, 20.
129.   Stein, G.A.; Sletzinger, M.; Arnold, H.; et al. *J. Am. Chem. Soc.* **1956**, *78*, 1514.
130.   Spinks, A.; Young, E. W. P.; *Nature* **1958**, *181*, 1397.
131.   Hey, P.; Willey, G. *Br. J. Pharmacol.* **1954**, *9*, 471.
132.   Maxwell, R. A.; Müller, R. P.; Plummer, A. J. *Experientia* **1959**, *15*, 267.
133.   Chopra, R. N.; Gupta, J. C.; Mukerjee, B. *Ind. J. Med. Res.* **1933**, *21*, 261.
134.   Ahlquist, R. P. *Am. J. Physiol.* **1948**, *153*, 586.
135.   Stein, G. A.; Bronner, H. A.; Pfister, K. *J. Amer. Chem. Soc.* **1955**, *77*, 700.
136.   Oates, J. A.; Gillespie, L.; et al. *Science* **1960**, *131*, 1890.
137.   Hartmann, M.; Isler, H. *Naunyn Schmiedebergs Arch. Exp. Pathol. Pharmakol.* **1939**, *192*, 141.
138.   Hoefke, W.; Kobinger, W.; *Arzn.-Forsch.* **1966**, *16*, 1038.
139.   Scholtysik, G.; Lauener, H.; Eichenberger, E.; et al. *Arzn.-Forsch.* **1975**, *25*, 1483.
140.   Baum, T.; Shropshire, A. T.; Rowles, G.; et al. *J. Pharmacol. Exp. Ther.* **1970**, *171*, 276.
141.   Bousquet, P.; Feldman, J.; Schwartz, J. *Br. J. Pharmacol.* **1984**, *230*, 232.
142.   Szabo, B. *Pharm. Ther.* **2002**, *93*, 1.
143.   Hess, H.-J.; Cronin, T. H.; Scriabine, A. *J. Med. Chem.* **1968**, *11*, 130.
144.   Piller, L. B.; Davis, B. R.; Cutler, J. A.; et al. *Curr. Control Trials Cardiovasc. Med.* **2002**, *3*, 10.
145.   Ruffolo, R. R.; Bondinell, W.; Hieble, J. P. *J. Med. Chem.* **1995**, *38*, 3681.
146.   Lands, A. M.; Arnold, A.; McAuliff, J. P.; et al. *Nature* **1967**, *214*, 597.
147.   Emorine, L. J.; Marullo, S.; Briend-Sutren, M. M.; et al. *Science* **1989**, *245*, 1118.
148.   Granneman, J. G. *Am. J. Physiol. Endocrinol. Metab.* **2001**, *43*, E199.
149.   Quirke, V. *Med. Hist.* **2006**, *50*, 69.
150.   Stapleton, M. P. *Tex. Heart Inst. J.* **1997**, *24*, 336.
151.   Powell, C. E.; Slater, I. H. *J. Pharmacol. Exp. Ther.* **1958**, *122*, 480.
152.   Black, J. W.; Crowther, A. F.; Shanks, R. G.; et al. *Lancet* **1964**, *1*, 1080.

153. Prichard, B. N. C. *Postgrad. Med. J.* **1976,** *52,* 35.
154. Lambert, D. M. *Postgrad. Med. J.* **1976,** *52,* 57.
155. Prichard, B. N. C.; Gillam, P. M. S. *Br. Med. J.* **1964,** *2,* 725.
156. Frishman, W. H. *Am. J. Ther.* **2008,** *15,* 565.
157. Reiter, M. J. *Prog. Cardiol. Dis.* **2004,** *47,* 11.
158. Toda, N. *Pharm. Ther.* **2003,** *100,* 215.
159. Pedersen, M. E.; Cockroft, J. R. *Curr. Hypertens. Rep.* **2007,** *9,* 269.
160. Mason, R. P.; Giles, T. D.; Sowers, J. R. *J. Cardiovasc. Pharmacol.* **2009,** *54,* 123.
161. MRC Working Party. *Br. Med. J.* **1992,** *304,* 405.
162. Messerli, F. H.; Grossman, E.; Goldbourt, U. *JAMA* **1998,** *279,* 1903.
163. Elliot, W. J.; Meyer, P. M. *Lancet* **2007,** *369,* 201.
164. Sneader, W. *Drug Discovery: The Evolution of Modern Medicines,* Wiley: Chichester, **1986,** p.153.
165. Vogl, A. *Am. Heart J.* **1950,** *39,* 881.
166. Saxl, P.; Heilig, R. *Wein. Klin. Wochenschr.* **1920,** *33,* 943.
167. Southworth, H. *Proc. Soc. Exp. Biol. Med.* **1937,** *36,* 58.
168. Hober, R. *Proc. Soc, Exp. Biol. Med.* **1942,** *48,* 53.
169. Schwartz, W. B. *N. Engl. J. Med.* **1949,** *240,* 173.
170. Roblin, R. O.; Clapp, J. W. *J. Am. Chem. Soc.* **1950,** *72,* 4890.
171. Beyer, K. H. *Persp. Biol. Med.* **1977,** *20,* 410.
172. Novello, F. C.; Sprague, J. M. *J. Amer. Chem. Soc.* **1957,** *79,* 146.
173. Beaumont, K.; Vaughn, D. A.; Fanestil, D. D. *Proc. Natl. Acad. Sci. USA* **1988,** *85,* 2311.
174. Veterans Cooperative Study Group on Antihypertensive Agents. *JAMA* **1967,** *202,* 1028.
175. Shankar, S. S.; Brater, D. C. *Am. J. Physiol. Renal Physiol.* **2003,** *284,* F11.
176. Sturm, K.; Siedel, W.; Weyer, R. U.S. Patent 3,058,882, **1962.**
177. Rankin, G. O. In *Foye's Principles of Medicinal Chemistry,* 5th ed., Williams, D. A.; Lemke, T. L.; Eds., Lippincott, Williams & Wilkins: Baltimore, **2002,** p. 524.
178. Cella, J. A.; Tweit, R. C. *J. Org. Chem.* **1959,** *24,* 1109.
179. Calhoun, D. A. *Prog. Cardiovasc. Dis.* **2006,** *48,* 387.
180. Sica, D. A. *Heart Failure Rev.* **2005,** *10,* 23.
181. Scholz, H. *Z. Kardiol.* **2002,** *91,* IV34.
182. Lindner, E. *Arzn.-Forsch.* **1960,** *10,* 569.
183. Fleckenstein, A. *Verh. Dtsch. Ges. Inn. Med.* **1964,** *70,* 81.
184. Fleckenstein, A.; Kammermeier, H.; Döring, H. J.; et al. *Z. Kreislauf-Forsch* **1967,** *56,* 716.
185. Bossert, F.; Vater, W. *Naturwissenschaften* **1971,** *58,* 578.
186. Kohlhardt, M.; Bauer, B.; Krause, H.; et al. *Pflugers Arch.* **1972,** *335,* 309.
187. Grossman, E.; Messerli, F. H. *Prog. Cardiovasc. Dis.* **2004,** *47,* 34.
188. Furberg, C. D.; Psaty, B. M.; Meyer, J. V. *Circulation* **1995,** *92,* 1326.
189. Zaman, M. A.; Oparil, S.; Calhoun, D. A. *Nat. Rev. Drug Discov.* **2002,** *1,* 621.
190. Tigerstadt, R.; Bergman, P. G. *Skand. Arch. Physiol.* **1898,** *8,* 223.
191. Goldblatt, H.; Lynch, J.; Hanzal, R. F.; et al. *J. Exp. Med.* **1937,** *59,* 347.
192. Basso, N.; Terragno, N. A. *Hypertension* **2001,** *38,* 1246.
193. Skeggs, L. T.; Kahn, J. R.; Shumway, N. P. *J. Exp. Med.* **1954,** *100,* 363.
194. Schwarz, H.; Bumpus, F. M.; Page, I. H. *J. Am. Chem. Soc.* **1957,** *79,* 5697.
195. Rittel, W.; Iselin, B.; Kappeler, H.; et al. *Helv. Chim. Acta* **1957,** *40,* 614.
196. Erdos, E. G. *FASEB J.* **2006,** *20,* 1034.
197. Ferreira, S. H. In *Hypotensive Peptides. Proceedings of the International Symposium.* Firenze, Italy, **1965.** Erdos, E.G.; Back, N.; Sicuteri, F. eds, p.356, Springer Verlag: New York.
198. Ferreira, S. H.; Bartelt, D. C.; Greene, L. J. *Biochemistry* **1970,** *9,* 2583.
199. Ondetti, M. A.; Williams, N. J.; Sabo, E. F.; et al. *Biochemistry* **1971,** *10,* 4033.

200.   Gavras, H.; Brunner, H. R.; Laragh, J. H.; et al. *N. Engl. J. Med.* **1974,** *291,* 817.

201   Byers, L. D.; Wolfenden, R. *Biochemistry* **1973,** *12,* 2070.

202.   Cushman, D. W.; Cheung, H. S.; Sbo, E. F.; et al. *Biochemistry* **1977,** *16,* 5484.

203.   Ondetti, M. A.; Rubin, B.; Cushman, D. *Science* **1977,** *196,* 441.

204.   Patchett, A. A.; Harris, E.; Tristram, E. W.; et al. *Nature* **1980,** *288,* 280.

205.   Ulm, E. H. *Drug Metab. Rev.* **1983,** *14,* 99.

206.   Krapcho, J.; Turk, C.; Cushman, D. W.; et al. *J. Med. Chem.* **1988,** *31,* 1148.

207.   The Heart Outcomes Prevention Evaluation Study Investigators. *N. Engl. J. Med.* **2000,** *342,* 145.

208.   Visser, L. E.; Stricker, B. H.; van der Velden, J.; et al. *J. Clin. Epidemiol.* **1995,** *48,* 851.

209.   Streeten, D. H.; Anderson, G. H.; Freiburg, J. M.; et al. *N. Engl. J. Med.* **1975,** *292,* 657.

210.   Furukawa, Y.; Kishimoto, S.; Nishikawa, K.; U.S. Patent 4,340,598, **1982.**

211.   Wong, P. C.; Chiu, A. T.; Price, W. A.; et al. *J. Pharmacol. Exp. Ther.* **1988,** *247,* 1.

212.   Chiu, A. T.; Carini, D. J.; Johnson, A. L.; et al. *Eur. J. Pharmacol.* **1988,** *157,* 13.

213.   Duncia, J. V.; Chiu, A. T.; Carini, D. J.; et al. *J. Med. Chem.* **1990,** *33,* 1312.

214.   Weinstock, J.; Keenan, R.M.; Samanen, J.; et al. *J. Med. Chem.* **1991,** *34,* 1514.

215.   Dahlof, B.; Devereux, R. B.; Kjeldsen, S. E.; et al. *Lancet* **2002,** *359,* 995.

216.   Blankley, C. J.; Hodges, J. C.; Klutchko, S. R.; et al. *J. Med. Chem.* **1991,** *34,* 3248.

217.   Webb, R. L.; Schiering, N.; Sedrani, R.; et al. *J. Med. Chem.* **2010,** *53,* 7490.

218.   Szelke, M.; Leckie, B.; Hallett, A.; et al. *Nature,* **1982,** *299,* 555.

219.   Clozel, J. P.; Fischli, W. *Arzn.-Forsch.* **1993,** *43,* 260.

220.   Oefner, C.; Binggeli, A.; Breu, V. et al. *Chem. Biol.* **1999,** *9,* 1397.

221.   Jensen, C.; Herold, P.; Brunner, H. R. *Nat. Rev. Drug Discov.* **2008,** *7,* 399.

222.   Mervaala, E.; Dehmel, B.; Gross, V.; et al. *J. Am. Soc. Nephrol.* **1999,** *10,* 1669.

223.   Pilz, B.; Ashagdarsuren, E.; Wellner, M.; et al. *Hypertension* **2005,** *46,* 569.

224.   Fisher, N. D.; Hollenberg, N. K. *Circulation* **2007,** *116 (Suppl. 2),* 556.

225.   Kirkby, N. S.; Hadoke, P. W.; Bagnall, A. J.; et al. *Br. J. Pharmacol.* **2008,** *153,* 1105.

226.   Saito, Y.; Nakao, K.; Mukoyama, M.; et al. *N. Engl. J. Med.* **1990,** *322,* 205.

227.   Channick, R. N.; Sitbon, O.; Barst, R. J.; et al. *J. Am. Coll. Cardiol.* **2004,** *43,* 62S.

228.   Clozel, M.; Breu, V.; Burri, K.; et al. *Nature* **1993,** *365,* 759.

229.   Clozel, M.; Breu, V.; Gray, G. A.; et al. *J. Pharm. Exp. Ther.* **1994,** *270,* 228.

230.   Furie, B.; Furie, B. C. *N. Engl. J. Med.* **1992,** *326,* 800.

231.   Mackman, N. *Nature* **2008,** *451,* 914.

232.   Brewer, D. B. *Br. J. Haematol.* **2006,** *133,* 251.

233.   MacFarlane, R. G. *Nature* **1964,** *202,* 498.

234.   Davie, E. W.; Ratnoff, O. D. *Science* **1964,** *145,* 1310.

235.   Mueller, R.L.; Scheidt, S. *Circulation* **1994,** *89,* 432.

236.   Weiss, H. J.; Aledort, L. M.; Kochwa, S. *J. Clin. Invest.* **1968,** *47,* 2169.

237.   Vane, J. R. *Nature* **1971,** *231,* 232.

238.   Vane, J. R.; Botting, R. M. *Thromb. Res.* **2003,** *110,* 255.

239.   Sørensen, H. T.; Mellemkjaer, L.; Blot, W. J.; et al. *Am. J. Gastroenterol.* **2000,** *95,* 2218.

240.   Dorsch, M. P.; Lee, J. S.; Lynch, D. R.; et al. *Ann. Pharmacother.* **2007,** *41,* 737.

241.   Michelson, A. D. *Nat. Rev. Drug Disc.* **2010,** *9,* 154.

242.   McTavish, D.; Foulds, D.; Goa, K. L. *Drugs* **1990,** *40,* 238.

243.   Savi, P.; Combalbert, J.; Gaich, C.; et al. *Thromb. Haemostat.* **1994,** *72,* 313.

244.   Bennett, C. L.; Davidson, C. J.; Ratsch, D. W.; et al. *Arch. Intern. Med.* **1999,** *159,* 2524.

245.   Quinn, M. J.; Fitzgerald, D. J. *Circulation* **1999,** *100,* 1667.

246.   Investigators CT. *N. Engl. J. Med.* **2001,** *345,* 494.

247.   Bhatt, D. L.; Topol, E. J. *Nat. Rev. Drug Discov.* **2003,** *2,* 15.

248.   Jakubowski, J. A.; Winters, K. J.; Naganuma, H.; et al. *Cardiol. Drug Rev.* **2007,** *4,* 357.

249.   Seligsohn, U. *Pathophysiol. Haemostat. Thromb.* **2002,** *32,* 216.

250. Zucker, M. B.; Pert, J. H.; Hilgartner, M. W. *Blood* **1966,** *28*, 524.
251. Jackson, D. P.; Morse, E. E.; Zieve, P. D.; et al. *Blood* **1963,** *22*, 827.
252. Nurden, A. T.; Caen, J. P. *Br. J. Haematol.* **1974,** *28*, 253.
253. Phillips, D. R.; Jenkins, C. S. P.; Luscher, E. F.; et al. *Nature* **1975,** *257*, 599.
254. Peerschke, E. L. *Semin. Hematol.* **1985,** *22*, 241.
255. Coller, B. S. *Blood* **1980,** *55*, 169.
256. Coller, B. S.; Peerschke, E. L.; Scudder, L. E.; et al. *J. Clin. Invest.* **1983,** *72*, 325.
257. Coller, B. S *J. Clin. Invest.* **1985,** *76*, 101.
258. Coller, B. S. *Blood* **1985,** *66*, 1456.
259. Tcheng, J. E.; Ellis, S. G.; George, B. S.; et al. *Circulation* **1994,** *90*, 1757.
260. Pytela, R.; Pierschbacher, M. D.; Bennett, J. S.; et al. *Science* **1986,** *231*, 1559.
261. Plow, E. F.; Pierschbacher, M. D.; Ruoslahti, E.; et al. *Blood* **1985,** *66*, 724.
262. Huang, T. F.; Holt, J. C.; Kirby, E. P.; et al. *Biochemistry* **1989,** *28*, 661.
263. Scarborough, R. M.; Rose, J. W.; Hsu, M. A.; et al. *J. Biol. Chem.* **1991,** *266*, 9359.
264. Scarborough, R. M.; Naughton, M. A.; Teng, W.; et al. *J. Biol. Chem.* **1993,** *268*, 1066.
265. Hartman, G. D.; Egbertson, M. S.; Halczenko, W.; et al. *J. Med. Chem.* **1992,** *35*, 4640.
266. Chew, D. P.; Bhatt, D. L.; Sapp, S.; et al. *Circulation* **2001,** *103*, 201.
267. Haycraft, J. B. *Proc. R. Soc. Lond.* **1884,** *36*, 478.
268. Fink, E. *Semin. Thromb. Hemost.* **1989,** *15*, 283.
269. Brinkhous. K. M.; Smith, H. P.; Warner, E. D.; et al. *Am. J. Physiol.* **1939,** *125*, 683.
270. Abildgaard, U. *Scand. J. Clin. Lab. Invest.* **1968,** *21*, 89.
271. Schofield. F. W. *Can. Vet. Rec.* **1922,** *3*, 74.
272. Huebner, C. F.; Link, K. P. *J. Biol. Chem.* **1941,** *138*, 529.
273. Butt, H. R.; Allen, E. V.; Bollman, J. L. *Proc. Staff Meet. Mayo Clin.* **1941,** *16*, 388.
274. Ikawa, M.; Stahman, M. A.; Link, K. P. *J. Am. Chem. Soc.* **1944,** *66*, 902.
275. Link, K. P. *Circulation* **1959,** *19*, 97.
276. Shapiro, S. *Angiology* **1953,** *4*, 380.
277. Clatanoff, D. V.; Triggs, P. O.; Meyer, O. O. *Arch. Int. Med.* **1954,** *94*, 213.
278. Stenflo, J.; Fernlund, P.; Egan, W.; et al. *Proc. Natl. Acad. Sci. USA* **1974,** *71*, 2730.
279. Whitlon, D. S.; Sadowski, J. A.; Suttie, J. W. *Biochemistry* **1978,** *17*, 1371.
280. Salzet, M. *Curr. Pharm. Des.* **2002,** *8*, 493.
281. Sherry, S.; Alkaersig, N.; Fletcher, A. P. *Am. J. Physiol.* **1965,** *209*, 577.
282. Okamoto, S.; Kinjo, K.; Hijikata, A.; et al. *J. Med. Chem.* **1980,** *23*, 827.
283. Okamoto, S.; Hijikata, K.; Kinjo, R.; et al. *Kobe J. Med. Sci.* **1975,** *21*, 43.
284. Okamoto, S.; Hijikata, A.; Kikumoto, R.; et al. *Biochem. Biophys, Res. Commun.* **1981,** *101*, 440.
285. McKeage, K.; Plosker, G. L. *Drugs* **2001,** *61*, 515.
286. Banner, D.W.; Hadvry, P. *J. Biol. Chem.* **1991,** *266*, 20085.
287. Gustafsson, D.; Elg, M.; Lenfors, S.; et al. *Blood Coag. Fibrinol.* **1996,** *7*, 69.
288. Hauel, N. H.; Nar, H.; Priepke, H.; et al. *J. Med. Chem.* **2002,** *45*, 1757.
289. Gould, W. R.; Leadley, R. J. *Curr. Pharm. Des.* **2003,** *9*, 2337.
290. Hirsh, J.; O'Donnell, M.; Eikelboom, J. W. *Circulation* **2007,** *116*, 552.
291. Tidwell, R. R.; Webster, W. P.; Shaver, S. R.; et al. *Thromb. Res.* **1980,** *19*, 339.
292. Nagahara, T.; Yokoyama, Y.; Inamura, K.; et al. *J. Med. Chem.* **1994,** *37*, 1200.
293. Roehrig, S.; Straub, A.; Pohlmann, J.; et al. *J. Med. Chem.* **2005,** *48*, 5900.
294. Wong, P. C.; Crain, E. J.; Xin, B.; et al. *J. Thromb. Haemost.* **2008,** *6*, 820.
295. MacFarlane, R. G.; Pilling, J. *Nature* **1947,** *159*, 779.
296. Williams, J. R. B. *Br. J. Exp. Pathol.* **1951,** *32*, 530.
297. Sobel, G. W.; Mohler, S. R.; Jones, N. W.; et al. *Am. J. Physiol.* **1952,** *171*, 768.
298. Tillett, W. S.; Garner, R. L. *J. Exp. Med.* **1933,** *58*, 485.
299. Milstone, H. *J. Immunol.* **1941,** *42*, 109.
300. Christensen, L. R.; MacLeod, M. *J. Gen. Physiol.* **1945,** *28*, 363.
301. Sherry, S. *J. Am. Coll. Cardiol.* **1989,** *14*, 1085.

302.   Fletcher, A. P.; Alkjaersig, N.; Smyrniotis, F.E.; et al. *Trans. Assoc. Am. Phys.* **1958,** *71*, 287.
303.   Smith, R. A. G.; Dupe, R. J.; English, P. D.; et al. *Nature* **1981,** *290*, 505.
304.   Bachmann, F.; Alkjaersig, N.; Fletcher, A. P.; et al. *Biochemistry* **1964,** *3*, 1578.
305.   Rijken, D. C.; Wijngaards, G.; Zaal-De Jong, M.; et al. *Biochim. Biophys. Acta* **1979,** *580*, 140.
306.   Van de Werf, F.; Ludbrook, P. A.; Bergmann, S. R.; et al. *N. Engl. J. Med.* **1984,** *310*, 609.
307.   Ueshima, S.; Matsuo, O. *Curr. Pharm. Des.* **2006,** *12*, 849.
308.   Brunton, T. L. *Lancet* **1867,** *2*, 97.
309.   Murrell, W. *Lancet* **1879,** *1*, 80.
310.   Lee, M. R. *Proc. R. Coll. Phys. Edinb.* **2001,** *31*,77.
311.   Sneader, W. *Drug Discovery: The Evolution of Modern Medicines*, Wiley: Chichester, **1986,** p. 136.
312.   Smith, S. *J. Chem. Soc.* **1930,** 508.
313.   Skou, J. C. *Biochim. Biophys. Acta* **1957,** *23*, 394.
314.   Mudd, J. O.; Kass, D. A. *Nature* **2008,** *451*, 919.
315.   Pugsley, M. K. *Drug Dev. Res.* **2002,** *55*, 3.
316.   Laflamme, M. A.; Murry, C. E. *Nat. Biotechnol.* **2005,** *23*, 845.
317.   Sadek, H.; Hannack, B.; Choe, E.; et al. *Proc. Natl. Acad. Sci. USA* **2008,** *105*, 6063.
318.   Teslovich, T. M.; Musunuru, K.; Smith, A. V.; et al. *Nature* **2010,** *466*, 707.

6

Diabetes Drugs

Ana B. Bueno, Ana M. Castaño, and Ángel Rodríguez

1 Introduction

1.1 Homeostasis of Glucose

Glucose is the main source of energy for the body. It is distributed via the bloodstream to the different cells. There are three sources of the plasma glucose: diet, glycogenolysis of the liver glycogen stores (hydrolysis of glycogen, a polymer of glucose stored in liver and skeletal muscle), and gluconeogenesis (*de novo* synthesis of glucose from noncarbohydrates, like glycerol, lactate, or amino acids). Despite having periods of fasting and feeding during the day, the levels of plasma glucose in healthy individuals are maintained in a narrow range of 4 to 7 mM. Two hormones are mainly responsible for this tight control: insulin, that reduces the levels of glucose in the bloodstream when they are high, and glucagon, which produces the opposite effect.

Insulin is produced and stored in vesicles in the pancreatic β-cells. When the levels of plasma glucose increase upon consumption of food, glucose is taken up by the β-cell through the glucose transporter protein 2 (GLUT2, Fig. 1). A metabolic process (glycolysis) is then initiated and an increase of ATP concentration is produced. This ATP increase closes the ATP-dependent potassium channels what depolarize the membrane, allowing calcium to enter the cell. The increase in calcium concentration in the cell triggers the migration of the insulin-containing vesicles to the membrane and the release of insulin. Insulin is released from the pancreas in a biphasic manner in response to an increase in blood glucose concentration. The first phase consists of a brief spike lasting ~10 min followed by the second phase, which reaches a plateau at 2–3 h. This process is known as glucose-stimulated insulin secretion (GSIS). The release of insulin produced by the rising of glucose in the blood stream triggers a number of metabolic processes in the body that contribute to the normalization of the levels of glucose (see Section 5.1).[1]

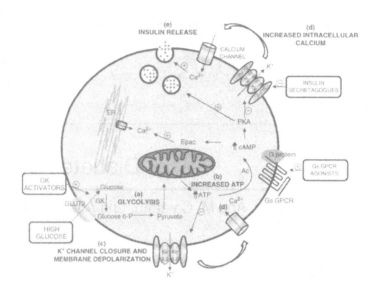

Fig.1. Summary of mechanisms that trigger insulin release in β-cells. 1. GSIS: (a) Glucose enters the β-cell and glycolysis is initiated; (b) an increase of ATP is produced in the mitochondria; (c) ATP closes the potassium channels to the release of potassium and the cell membrane depolarizes; (d) calcium enters the cell through the calcium channels; (e) the increase in calcium concentration in the cytoplasm triggers the release of insulin. 2. Insulin secretagogues. 3. Gs coupled GPCRs activation (GLP-1 agonists, GPR119 agonists). 4. Glucokinase activation.

1.2   *Diabetes*

Diabetes mellitus (DM)[*] can be defined as a group of heterogeneous disorders with the common elements of hyperglycemia and glucose intolerance due to insulin deficiency, impaired effectiveness of insulin action or both.[2] Chronic hyperglycemia is associated with long-term damage of various organs, especially the eyes, kidney, nerves, heart, and blood vessels. The mechanism by which diabetes leads to these complications is complex and involves the direct toxic effects of hyperglycemia, along with the impact of high blood pressure, abnormal lipid levels, and both functional and structural abnormalities of small blood vessels.[3]

     Diabetes mellitus is one of the most common noncommunicable diseases globally and it is certainly becoming an epidemic and a major cause of death not only in developed countries but also in many developing and newly industrialized nations.[3] It has been estimated that the number of people expected to live with diabetes will be around 440 million people in less than 30 years, assuming a global prevalence of 7.8%.[3] However, based on the latest estimations (347 million in 2008),[4] those figures might be even higher.

---

[*] The word "diabetes" comes from Ancient Greek διαβήτης (diabḗtēs) which literally means "a passer-through; a siphon". Aretaeus the Cappadocian (1st century CE) named the disease "diabetes" with the intended meaning "excessive discharge of urine" (see *Online Etymology Dictionary*). The word "mellitus" comes from Latin mellitus, derived from Latin *mel*, meaning "honey; sweetness" (see *MyEtymology*). The term "diabetes mellitus" was coined by Thomas Willis in the XVII century, when he noticed what had already been observed centuries ago, that the urine of diabetics had a sweet taste. World Diabetes Day is the primary global awareness campaign of the diabetes mellitus world and is held on November 14 every year.

The diagnosis of diabetes can be made based on glucose values (fasting or postprandial) and/or HbA1c,[*] with or without classic symptoms.[5] Diabetes mellitus is usually classified on the basis of etiology and clinical presentation of the disorder into four types: type 1, type 2, gestational diabetes, and other specific types.[3]

In order to treat diabetes, several pharmacological approaches have been developed to complement the benefits of diet and appropriate lifestyle.[6] In Sections 2 to 6, the current available drug treatments for DM as well as selected ones that are advancing in clinical trials will be presented.

1.2.1 Type 2 Diabetes

Type 2 diabetes mellitus (T2DM) is the most common type of diabetes, representing more than 80% of the DM cases. It is characterized by a relative insulin deficiency resulting from a reduced sensitivity of tissues to insulin and impairment of insulin secretion from pancreatic β-cells. The muscle, liver, β-cell, fat cell (accelerated lipolysis), gastrointestinal tract (incretin deficiency/ resistance), α-cell (hyperglucagonemia), kidney (increased glucose reabsorption), and brain (insulin resistance), play important roles in the development of glucose intolerance in type 2 diabetic individuals. This highlights the need of multifactorial approaches to the treatment, with an early initiation of therapy, in order to prevent/slow the progressive β-cell failure.[7] Some factors associated with the development of type 2 diabetes are age, family history, diet and inactivity, ethnicity, and less than optimum intrauterine environment during fetal development. The prevalence of T2DM is growing much more than had been estimated, in particular due to the increasing prevalence of obesity, as it can cause insulin resistance. About 90% of T2DM could be attributable to excess weight.[8]

The diagnosis typically occurs after the age of 40 years but could be earlier, especially in populations with high prevalence. Although it depends on the individual patient, T2DM is usually managed on a stepwise basis, which includes lifestyle changes (diet and exercise), oral therapy alone or in combination, as well as injectable therapy as the disease progresses. People with T2DM are not dependent at the earlier stages on exogenous insulin for survival but may require insulin for control of hyperglycemia if not achieved with diet and other medications.

1.2.2   Type 1 Diabetes

Type 1 diabetes mellitus (T1DM) accounts for about 5–10% of all cases of diabetes and affects ~ 4 million people in North America and Europe. This type of diabetes is caused by absolute insulin deficiency resulting from autoimmune destruction of pancreatic β-cells in genetically predisposed individuals.[9,10] The β-cells sense glucose and release insulin to maintain physiological glucose levels within a relatively narrow range. Once those cells are destroyed, patients with type 1 diabetes lose blood glucose control, which can result in both acute conditions (for example, ketoacidosis and severe hyperglycemia) and secondary complications (including heart disease, blindness, and kidney failure).

---

[*] Glycosylated hemoglobin (HbA1c or A1c) is a form of hemoglobin which is measured to estimate the average plasma glucose concentrations during prolonged periods of time. It reflects the degree of glycemic control during the previous 2–3 months. Normal amounts of A1c correspond to normal values of glucose. When glucose increases, the percentage of glycated hemoglobin is also increased. Standard range of values would be 4–6%.

An approximate population incidence of T1DM is 25/100,000 with the highest incidence at the age of 13 to 15.2 years. However, significant ethnic differences have been found, with the lowest incidence in China and Japan (0.4 and 1.6/100,000, respectively) and the highest in Finland (40/100,000).[11,12]

The pathogenesis of T1DM is multifactorial, involving genetic susceptibility, environmental factors, and autoimmune mechanisms. In human linkage studies, genetic susceptibility to T1DM is associated with specific human leukocyte antigen (HLA) class II alleles (at the HLADR and DQ loci).[13] Non-HLA genes have also been associated with T1DM.[14] The geographic variations in the rise of diabetes incidence point to the importance of nongenetic environmental triggers. Possible environmental factors include diet, chemical toxins, and viruses.[15] The third major contributor to T1DM is thought to be a T-cell-mediated immune response to antigens that are components of the pancreatic β-cell, which in turn leads to apoptosis, necrosis, and insulin deficiency. The T-cell-mediated cellular destruction is mediated by the release of cytotoxic molecules[16] and cytokines, including IL-1, IFN-γ and TNF-α, which exacerbate β-cell death by upregulation of Fas and Fas ligand and stimulation of NO and free-radical production.[17]

In addition to the T-cell-mediated process of insulitis, there is a humoral response with B-cell production of antibodies to β-cell antigens. Although it is not believed to be involved in the pathogenesis of the disease, the presence of these antibodies is used as a surrogate marker for the autoimmune process and aids in establishing the clinical diagnosis or to predict future disease onset in high-risk individuals.[18]

Only in the later stages of the disease, when a vast majority of β-cells are destroyed, does one develop glucose intolerance and overt diabetes. Although variable, this progression can be very fast and take only some months. The end result of this autoimmune process is the eventual loss of the vast majority of β-cells and decrease in insulin secretion.

2   Current Therapies for Type 2 Diabetes

2.1   Biguanides

metformin   metformin hydrochloride   phenformin

Guanidine, a natural component of a plant, was found to lower blood glucose in animals in 1918. It was also found to be too toxic for its use as a drug. Metformin and other guanidine derivatives, known as biguanides, were described in the early 1920s and their hypoglycemic effects in animals reported. Priorities of the scientific community then shifted to the insulins, and the interest for biguanides was lost for many years. After World War II, metformin was tested in South Africa for the treatment of influenza.[19] Reviewing these studies, Sterne saw some evidences of hypoglycemia produced by the drug and initiated a research for the treatment of diabetes. He reported his preclinical and clinical findings in 1957.[20] It was approved in some countries in the early 1970s, but its growth faced some challenges as other biguanides, like phenformin, approved at the same time in the U.S., had to be withdrawn from the market because of the risk of lactic acidosis. It is known now that the incidence of lactic acidosis with metformin does not differ from that produced by other hypoglycemic agents. But this concern delayed the approval of metformin in the U.S. until 1995. Nowadays, it is the only commercialized biguanide in most of the world. Current guidelines from the European and American Diabetes Associations recommend metformin as the first line of pharmacological treatment, with addition of other agents when this therapy is no longer sufficient to control glucose levels.[21]

Metformin is a chemically stable small molecule with high water solubility. It is formulated as its HCl salt. The structure of the lowest energy tautomer of metformin as well as the most stable protonated form of its hydrochloride salt have been corrected recently.[22,23] It is metabolically stable, being excreted unchanged by the kidneys.[24]

2.1.1 *Mechanism of Action*

Fifty years after the discovery of metformin, its mechanism of action is not completely clear. It is known that its beneficial effects include decreased glucose output from the liver by decreasing gluconeogenesis, promotion of glucose uptake in skeletal muscle, and lowering of triglycerides and fatty acids.

There is evidence now that it acts indirectly through adenosine monophosphate-activated protein kinase (AMPK) in liver and skeletal muscle.[25] AMPK is an enzyme with energy-sensing capabilities. It is known to detect fluctuations between AMP:ATP and to switch on ATP-producing pathways and to switch off ATP-consuming pathways when the levels of ATP are low and the levels of AMP are high. But the activation of AMPK by metformin is independent of the AMP:ATP ratio.[26] As a result of this activation in the liver, ACC is phosphorylated and inactivated, reducing fatty acid synthesis and increasing fatty acid oxidation. Additionally, AMPK activation reduces the expression of SREBP-1, a transcriptional factor implicated in insulin resistance and related metabolic disorders. The suppression of SREBP-1 results in a reduction of the expression of lipogenetic genes. These combined effects on AMPK result in a reduction of the lipogenetic pathways, reduction of the fatty acid content in liver, and increase in hepatic insulin sensitivity.

AMPK-activation also seems to be responsible for the inhibition of gluconeogenesis produced by metformin. Activation of AMPK increases the expression of the nuclear receptor SHP, which inhibits the expression of PEPCK and Glc-6-Pase, both gluconeogenic enzymes.[27]

The increased uptake of glucose by the muscle also seems to be related to AMPK activation, as metformin has been shown to increase the activity of AMPK in skeletal muscle.[28] AMPK is known to activate the translocation of GLUT4, the glucose transporter expressed in muscle, to the membrane, resulting in an increase in glucose uptake by the muscle. Independently, it has been shown that metformin potentiates the insulin-induced translocation of the transporter to the cell surface of adipocytes.[29]

Metformin lowers A1c by ≈1.5%.[21] The most common adverse effect of metformin is gastrointestinal discomfort. Metformin is weight neutral in contrast with many other antidiabetic drugs that produce body weight gain. It improves the lipid profile in T2DM and has been shown to reduce cardiovascular events.[30]

A number of fixed combinations of metformin with the other drugs discussed in this section are also available to patients when metformin alone is no longer effective.

2.2   *Insulin Secretagogues: Sulfonylureas and Glinides*

Enhancement of insulin secretion from β-cells is of paramount importance in the treatment of T2DM. Glucose is the major physiological trigger of the biosynthesis and secretion of insulin. Ideally, an insulin secretagogue should stimulate insulin secretion in a physiological manner (fast after food intake) and in a glucose-dependent manner (should not induce insulin secretion if plasma glucose concentration is below 60 mg/dL).[31] A number of insulin secretagogues are currently available for the treatment of T2DM.

2.2.1   *Sulfonylureas*

The antihyperglycemic effect of sulfonylureas (SUs) was discovered during the study of sulfonamide antibiotics, when it was found that a SU compound produced hypoglycemia in animals. World War II disrupted the research on this field, and it was not until the mid-1950s when a SU, the first treatment for diabetes after insulin, was approved. Several different drugs (tolbutamide and chlorpropamide, among others) were approved within a few years. These were followed by a second generation of drugs with higher efficacy and lower side effects, such as glipizide, glibenclamide (glyburide), and glimepiride.[32,33] All these drugs have a *p*-substituted phenyl sulfonylurea core. The first-generation drugs are now less commonly used.

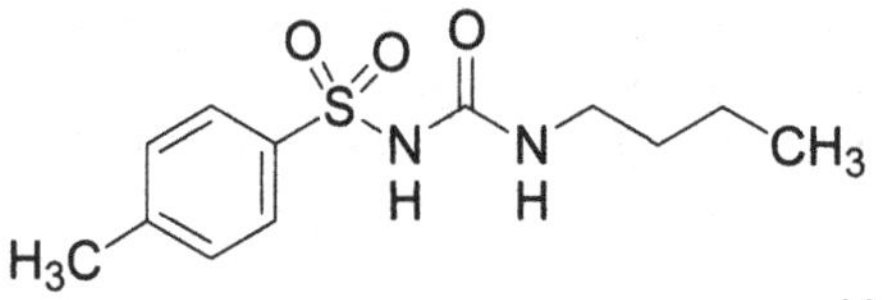

tolbutamide

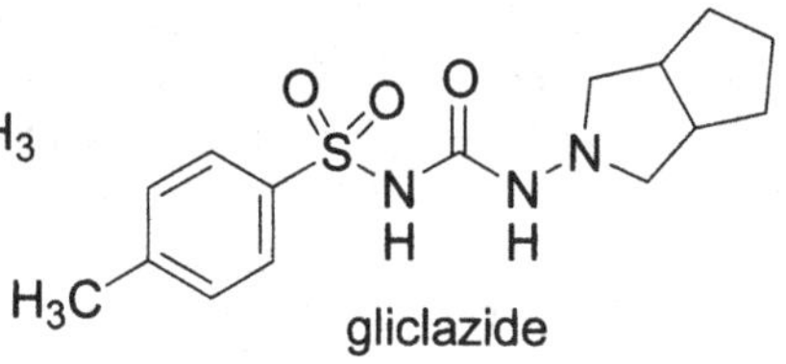

gliclazide

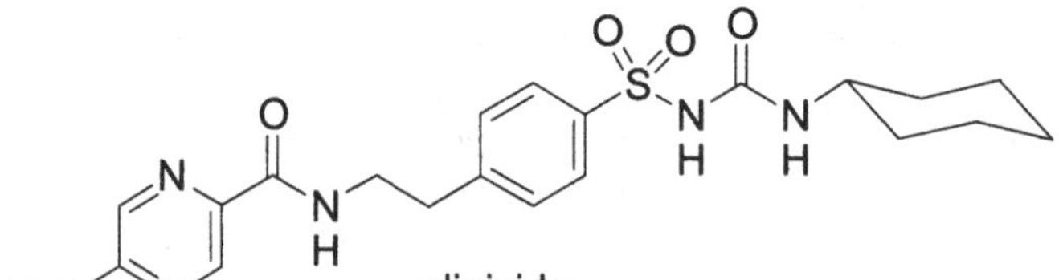

glipizide

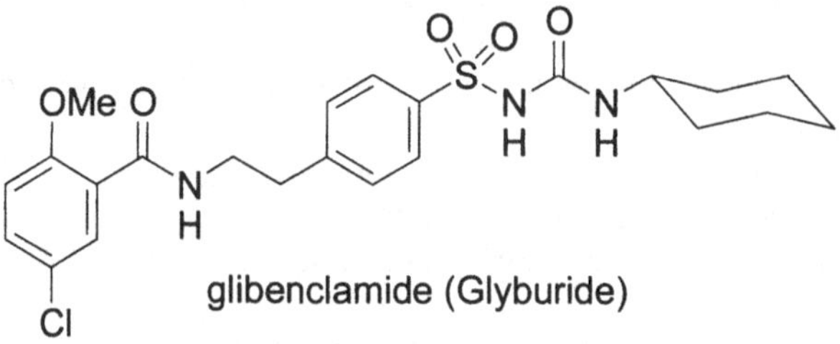

glibenclamide (Glyburide)

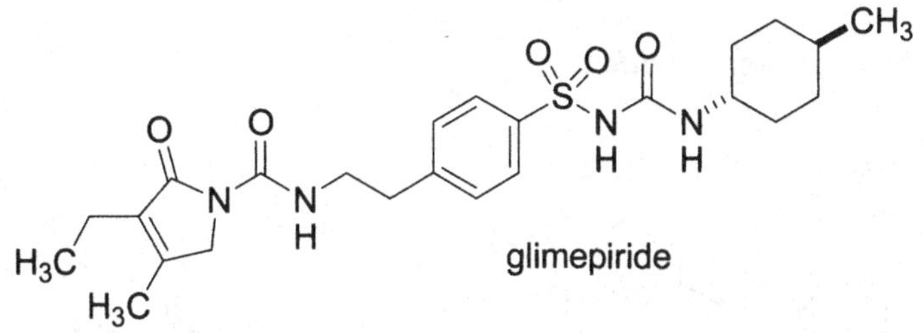

CH3

O O O

H3C

N

O

N
H

H3C
H3C

S
N
H

N
H

glimepiride

Mechanism of action

Sulfonylureas bind to the sulfonylurea receptor SUR1, expressed on the membrane of the β-cells. This receptor is associated with an ATP-dependent potassium channel, Kir6.2 (Fig. 1). The binding to the SUR1-Kir6.2 complex results in closure of the channel and depolarization of the membrane, with subsequent opening of the voltage-dependent calcium channels. The influx of calcium to the cell triggers the exocytosis of insulin from insulin vesicles and the release of insulin.[34] The insulin secretion responses to the administration of sulfonylureas are slow and generally do not match the glucose peak after meals. Sulfonylureas do not restore the loss of the first phase of insulin secretion in T2DM patients, but stimulate in a dose-response manner the second phase of insulin secretion.[35] Sulfonylureas lower the levels of A1c about 1–2%.[21] It is generally believed that the small insulin sensitization effect found after sulfonylurea treatment is due to the improvement in glucotoxicity. They have no effect on plasma lipids.

The mechanism of action of sulfonylureas is somewhat related to the glucose-stimulated insulin secretion (GSIS). The difference is that, during GSIS, the process is regulated by the concentration of the glucose in the cell: Only at high concentrations of glucose is the closure of the channel produced and the secretion of insulin triggered. Thus, no risk of hypoglycemia is associated with the GSIS mechanism.[36] In the case of sulfonylureas, since the release of insulin is independent of the plasma glucose levels, there is a risk of hypoglycemia. Some T2DM patients do not respond well to SU therapy (primary failure) and others become refractory over time (secondary failure). Some factors that predict good response are, among others, being recently diagnosed DM and mild to moderate fasting hyperglycemia.[35]

The main adverse effect of SUs is hypoglycemia. Glyburide is cleared by the kidneys. Elderly and type 2 diabetics with impaired kidney function could suffer severe hypoglycemia. Gliclazide and glimepiride are less associated with hypoglycemia. Another side effect is body weight gain, typically 2–5 kg. Since patients with type 2 diabetes are often obese, this effect can discourage them from taking this medication. Some concerns that SU class may increase cardiovascular mortality were not supported by the UKPDS or ADVANCE study.[30] Whether patients with previous sensitivity to sulfonamide antibiotics are at higher risk of showing an allergic response to sulfonylureas is still controversial, but clinicians should be aware of this.[37]

2.2.2   *Nonsulfonylureas: Glinides*

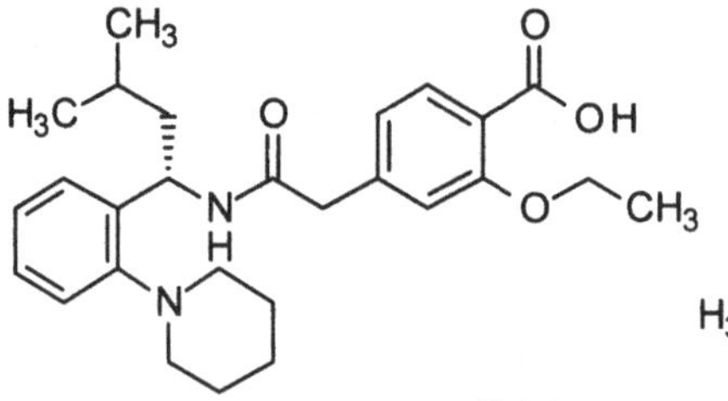

repaglinide 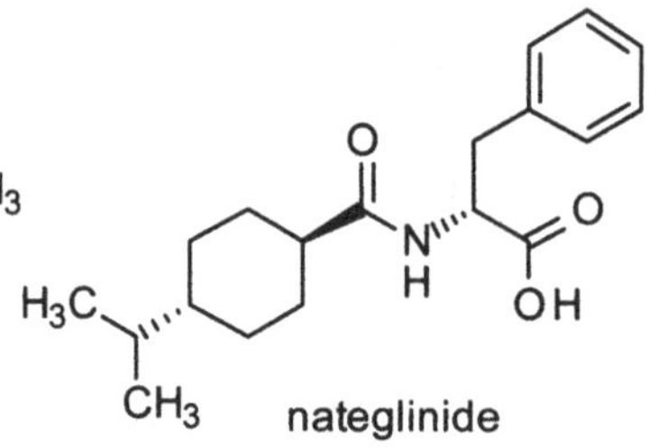   nateglinide

The study of the insulinotropic properties of the nonsulfonyl fragment of glibenclamide (later called meglitinide) led to the discovery of new agents (glinides).[38] Repaglinide, a benzoic acid derivative, was the first drug of this class approved, followed by nateglinide, an amino acid derivative. Mitiglinide, launched in Japan and other countries, is in Phase III clinical studies in the U.S.[39]

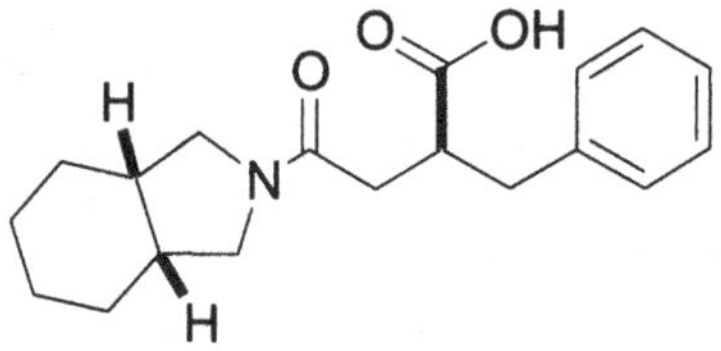

mitiglinide (Phase III US)

Mechanism of action

As sulfonylureas, glinides bind to the SUR1-Kir6.2 complex and close the channel to the efflux of potassium, producing the influx of calcium and release of insulin. The difference with the sulfonylureas is that they bind to a different site of the complex and with different kinetics.[38,40,41]In the case of the glinides, this process is very rapid, particularly with nateglinide, and of very short action, resulting in a fast and brief stimulation of insulin release (nateglinide shows an off-rate at the ATP-dependent potassium channel twice as fast as that of glyburide and glimepiride and five times faster than repaglinide).[42,43] Another difference is that, unlike sulfonylureas, glinides show glucose-dependent insulinotropic effects. Due to its more physiological pattern of insulin secretion, they are indicated for administration at mealtime and are especially suited for the elderly.[32,44] Compared with sulfonylureas and metformin, A1c reduction is similar for repaglinide (1.5%) and slightly lower for nateglinide as 0.5–1.0%.[21] Repaglinide is quickly absorbed and rapidly eliminated through biliary excretion, making it suitable for use in patients with renal impairment.[45] It is still unknown if prolonged use of short-acting insulin secretagogues as glinides can preserve β-cell function and avoid the secondary failure seen with sulfonylureas.[38]

The main side effect is also hypoglycemia, although the incidence is lower than with sulfonylureas, due their kinetics and the fact that they are less effective in stimulating insulin secretion at low glucose concentrations.[32] The risk of weight gain is similar or lower than that of sulfonylureas. Since they are taken at meals, the patients do not have to take the medication if they are going to skip a meal. Thus, this type of medication results in less patients feeling their lifestyle restricted because of need of regular meals, being a treatment especially useful for patients with flexible mealtimes and providing an opportunity to diet to overweight patients.

2.3   α-Glucosidase Inhibitors

Glucose is ingested in the diet as monosaccharide or as part of monosaccharide polymers, like di-, tri- or polysaccharides. These carbohydrates are hydrolyzed to glucose and other monosaccharides in the small intestine by membrane-bound α-glucosidases. They are then absorbed, causing the postprandial glucose elevation. There are two commercialized inhibitors of these enzymes, acarbose and miglitol. Acarbose is a tetrasaccharide analog that is not absorbed and is excreted unchanged in the feces. Miglitol, on the other hand, is a D-glucose analog, derived from the naturally occurring 1-deoxynojirimycin, which also has α-glucosidase inhibitory activity. [46] Miglitol is absorbed in the upper part of the intestine and excreted mainly by kidneys.[47]

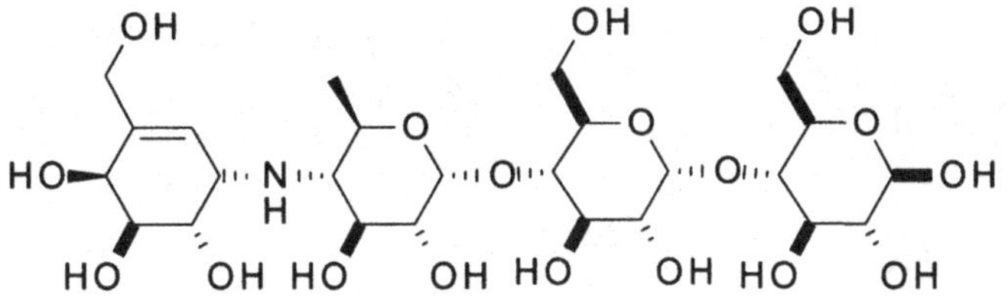

acarbose (Precose)

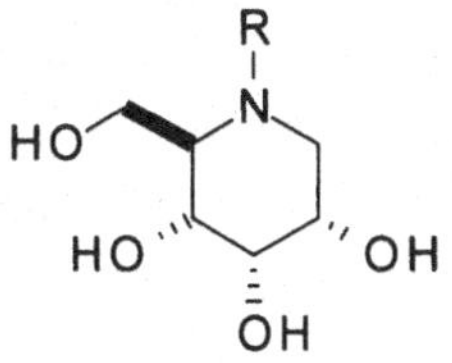

R =CH₂CH₂OH, miglitol
R = H, 2-deoxynojirimycin

2.3.1   Mechanism of Action

α-Glucosidase inhibitors are nonhydrolyzable carbohydrate analogs that compete with the carbohydrates for the active site of the enzyme. The inhibition of the enzymes is reversible with both drugs. Due to the presence of the inhibitors, the saccharides are not hydrolyzed in the small intestine, where most of the absorption takes place, and they are digested in the colon, resulting in a delay and a reduction of the absorption of glucose. Because of their mechanism of action, these antidiabetics have their highest effect if they are administered before meals, and their efficacy depends on the amount of carbohydrates consumed. They do not affect the absorption of the monosaccharides ingested as such in the diet.

α-Glucosidase inhibitors reduce A1c by 0.5–0.8%.[21] They do not produce hypoglycemia or weight gain. Since the digestion of carbohydrates is delayed with these inhibitors, it takes place mainly by bacteria in the colon, producing flatulence, diarrhea, and abdominal discomfort.

2.4   PPARγ Agonists: Thiazolidinediones (TZDs)

The insulin sensitizers thiazolidinediones were discovered from the evolution of an *in vivo* screening for lipid-lowering agents run by Takeda about 30 years ago, when, by serendipity, they discovered the ester AL-294 that effectively reduced plasma glucose and triglyceride levels.[48]

Given that the active form was the carboxylic acid, bioisoster replacement by heterocycles and subsequent structure-activity relationship (SAR) led to the discovery of the antihyperglycemic TZD ciglitazone.[49] The development of ciglitazone was discontinued due to insufficient efficacy. SAR studies revealed that polar modifications on the alkoxy domain improved glucose and lipid-lowering effects.[50] These findings boosted an intensive research from pharmaceutical companies that culminated in the discovery and further development of the first generation of TZDs: troglitazone, rosiglitazone, and pioglitazone. All these thiazolidinediones are racemates as the pure enantiomers undergo rapid racemization under physiological conditions.[51]

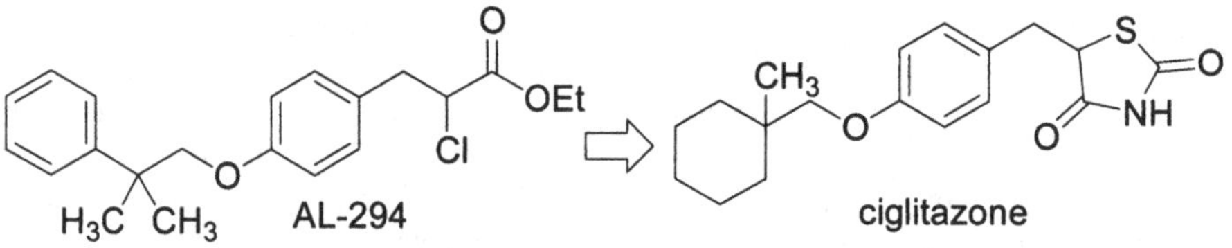

AL-294

ciglitazone

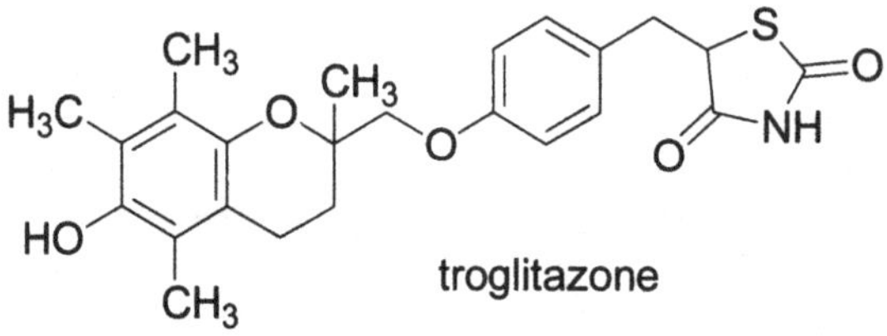

troglitazone

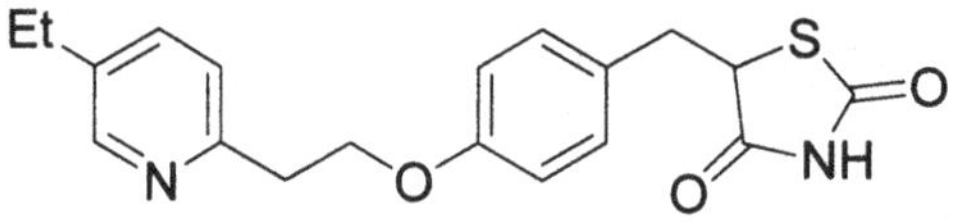

pioglitazone

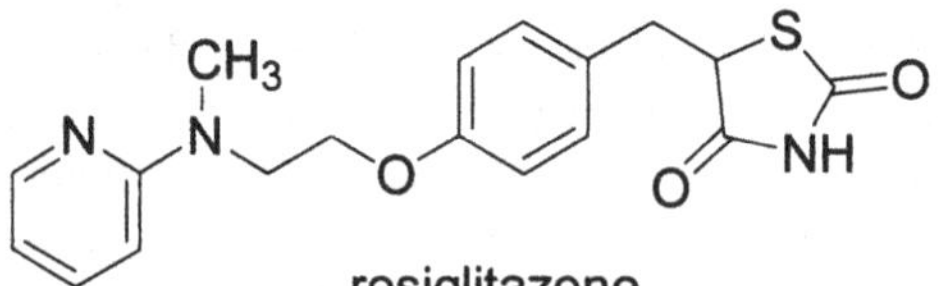

rosiglitazone

2.4.1   Mechanism of Action

Several years after the discovery of the antihyperglycemic action of TZDs, the Spiegelman group discovered the biological target hit by TZDs, PPARγ.[52] It was later found that TZDs exerted their antihyperglycemic action via activation of this receptor.[53] PPARs (peroxisome-proliferator activating receptors, belonging to the family of nuclear receptors) are ligand-activated transcription factors. There are three main subtypes: α, β(δ), and γ. PPARγ, the most investigated subtype, has two isoforms: PPARγ1, widely expressed, and PPARγ2, expressed in adipose tissue. It is involved in adipocyte gene expression and differentiation, playing an important role in insulin sensitization, glucose, and lipid metabolism.[54] The binding of small lipophilic ligands (mainly fatty acids) regulates the activity of this receptor. TZDs bind to the PPARγ-RXR (retinoid X receptor) heterodimer, inducing a change in conformation that leads to stabilization of coactivator recruitment and dissociation of corepressor complex, ultimately triggering the transcription of genes involved in insulin sensitization (and side effects).[55] For more than two decades, there was a huge interest in developing PPARγ agonists for the treatment of T2DM. However, the side effects seen in the clinic, coupled with the difficulty to design PPARγ drugs with a better therapeutic window, led to a decline on the interest for this type of drugs. As will be described later on, renewed interest on this class of targets has arisen from the discovery of PPARγ selective modulators (SPPARγMs) and dual α/γ PPAR ligands.

Two TZDs are currently approved in the U.S. for the treatment of T2DM, pioglitazone and rosiglitazone, both with restrictions due to different side effects. TZDs produce an A1c reduction of about 0.5-1.4%.[21] The positive effects of PPAR ligands are associated with mechanisms that reduce lipotoxicity. PPARγ agonists cause triglyceride accumulation and increase in adipose mass. This adipogenesis is beneficial for the patient, as it produces a redistribution of lipids from visceral fat (insulin resistant) into subcutaneous fat containing small and insulin-responsive adipocytes.[56,57,58] TZDs regulate the production of adipokines that affect insulin action in peripheral tissues: they increase the expression of adiponectin (an insulin-sensitizing factor) and reduce the elevated levels of proinflammatory factors, such as TNFα. As a consequence of these actions, lipolysis is reduced, resulting in lowering of circulating free fatty acids (FFAs). Both FFAs and TNFα are potential mediators of insulin resistance, and the reduction of their levels translates into improved insulin action in muscle and liver (Fig. 2).[59]

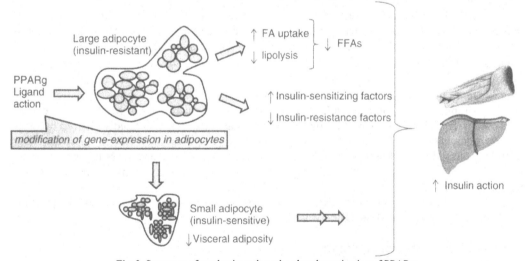

Fig. 2. Summary of mechanisms that take place by activation of PPARγ.

The major side effects associated with PPARγ activators are weight gain and edema. Edema, caused by fluid retention, is the main reason for discontinuation of treatment and can lead to congestive heart failure in predisposed patients. Bone marrow abnormalities that can result in increased fracture rate and potential for increased risk for carcinogenicity are other reported side effects.[55,60]

Despite the number of years that have passed since the discovery of TZDs and its molecular target, there is still ongoing research that can shed light on the understanding of the underlying mechanisms that promote insulin sensitization and/or undesired side effects.[61,62,63] Each TZD activates in a different manner the transcription of genes involved in insulin sensitization and side effects, leading to a variety of efficacy and safety profiles. The diversity of adverse effects suggests that they may be compound specific and not drug class related, making it reasonable to continue research in the field.

2.5   *Incretin-Based Therapies*

It is known that the levels of glucose in the circulation are tightly controlled by insulin and glucagon. But there are other hormones with glucoregulatory action.[64] Among them, gut hormones known as incretins are now considered very important in the control of glucose. The two main incretin hormones are GLP-1 (glucagon-like peptide 1) and GIP (glucose-dependent insulinotropic polypeptide). During meals, these hormones are secreted in the gut by the effect of nutrients and they induce insulin secretion from the pancreas acting through specific GPCRs (G-protein-coupled receptors) present in the surface of the β-cells. This effect only takes place when the levels of glucose are high. The incretins produce an amplification of the insulin secretion that glucose would produce by itself after a meal. This is known as the incretin effect.

Besides the direct effect on insulin secretion, GIP also regulates fat metabolism in healthy people. But it is dubious as a target for diabetes because, although the levels of GIP are normal or slightly elevated in diabetics,[65] it fails to enhance insulin secretion.[66] It is believed to be due to a defective expression of the GIP receptor in diabetics.[67,68]

GLP-1 concentration, on the contrary, is reduced in T2DM,[65,69] while its signaling is intact as has been demonstrated by the normoglycemia obtained in diabetics after infusion of GLP-1.[70] This hormone has other beneficial effects for T2 diabetics: It inhibits glucagon secretion,[71] reduces appetite, and slows gastric emptying.[72] It may have a role in β-cell proliferation and differentiation and in β-cell function. GLP-1 has been shown to inhibit β-cell apoptosis *in vitro*. An increase in β-cell mass has been proven in animal models.[73,74,75]

Despite the great interest in GLP-1 as a potential treatment for T2DM, the half-life of its active form, GLP-1 7–36, released following a meal, is about 2 min. It is rapidly cleaved by DPP-IV (dipeptidylpeptidase IV) to GLP-1 9–36, which does not produce insulin secretion. This peptide is cleared by the kidneys.[76] GLP-1 9–36 was considered for many years the inactive form of GLP-1. There are data now that this peptide also has insulin-like actions independent of the GLP-1 receptor.[77,78]

Two different approaches have been successful in developing GLP-1-based therapies: the GLP-1 receptor agonists and the DPP-IV inhibitors.

2.5.1   *GLP-1 Receptor Agonists*

The GLP-1 receptor is a class B GPCR. Finding small-molecule agonists of class B GPCRs has been elusive for medicinal chemists.[79] The attention of scientists turned to peptides that could have a longer half-life than the natural hormone. Exendin-4 (exenatide) was the first drug developed using this strategy and it was approved in 2005. It is a peptide isolated from the saliva of the Gila monster, which shares a 53% homology with native GLP-1 but has similar affinity for the human GLP-1 receptor. The modification of the first two amino acids of the sequence (where the enzyme cleaves the peptide) from His-Ala (in the native hormone) to His-Gly (in exendin-4) reduces the affinity of the peptide for DPP-IV, showing a half-life of 2–3 h after subcutaneous injection, which makes it suitable for a twice a day (b.i.d.) treatment.[80]

The only other incretin mimetic approved to date is liraglutide. This peptide has the same aminoacid sequence as natural GLP-1, except for the replacement of Arg 28 by a Lys. The free NH_2 of the Lys is then attached to a C16 fatty acid. This long-chain lipid binds noncovalently to albumin, a plasma protein. The formation of this complex protects it from the inactivation by the DPP-IV and from renal clearance. The half-life of liraglutide is 12 h, which makes this drug suitable for a once-a-day (q.d.) subcutaneous injection.

A number of other analogs based on similar strategies are in the clinic, like exenatide long-acting, albiglutide, and lixisenatide.[81]

Mechanism of action

Incretin mimetics act by activating the GLP-1 receptor like the native peptide does. In β-cells, the activation of the G-protein-coupled receptor in the surface of the cell produces the activation of adenylate cyclase by the G-protein in the cytoplasm and consequently the increase of c-AMP production. c-AMP then activates PKA (protein kinase A) and Epac (exchange proteins activated by c-AMP), which are essential for the exocytosis of insulin. This action only takes place at high concentrations of glucose (Fig. 1).[81]

Its action in gastric emptying and/or appetite has been associated to signaling through the GLP-1 receptors present in the brain with subsequent stimulation of efferent tracts of the vagus nerve.[82] This effect is believed to be responsible for the decrease in body weight observed with the use of these drugs.[83,84]

Both peptides exenatide and liraglutide have shown an improvement in β-cell mass and function in preclinical models. In patients in whom the direct measurement effect in β-cell mass is not possible, they improve surrogate markers of β-cell function.[85]

The reduction of A1c with exenatide is 0.5–1% and 1.5–2% with liraglutide. Weight loss is observed with both medications. The main side effect observed with incretin mimetics is nausea, usually observed at the beginning of the treatment. This effect generally disappears after continued treatment.

2.5.2   DPP-IV Inhibitors

GLP-1 is a peptide formed in the gut by cleavage of proglucagon. Its active form, GLP-1 7–36, is then secreted to the blood stream. DPP-IV is an ubiquitous serine protease that rapidly inactivates the hormone by cleaving the first two amino acids, His-Ala. This cleavage produces a 9–36 peptide that does not have insulinotropic effects.

The DPP-IV enzyme also cleaves many other peptides like GIP, PYY, NPY, GLP-2, or RANTES.[86]

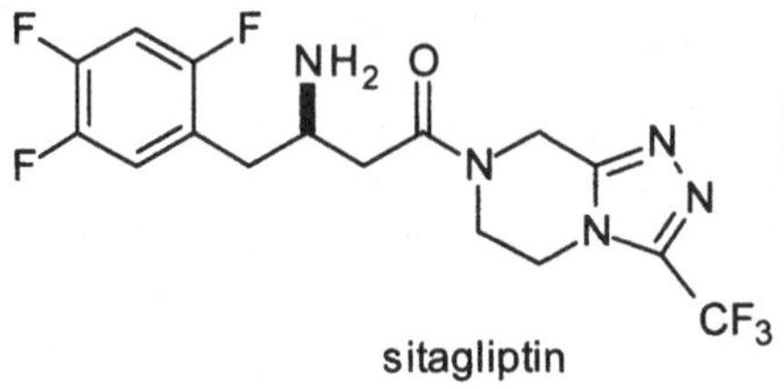

sitagliptin

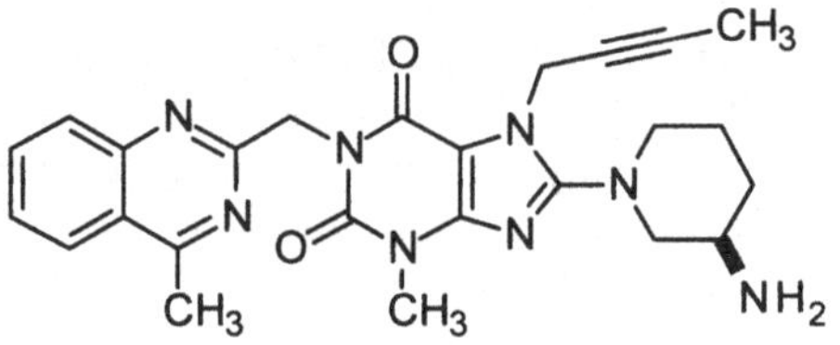

linagliptin

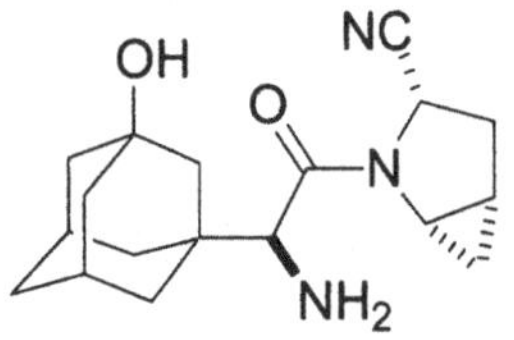

saxagliptin 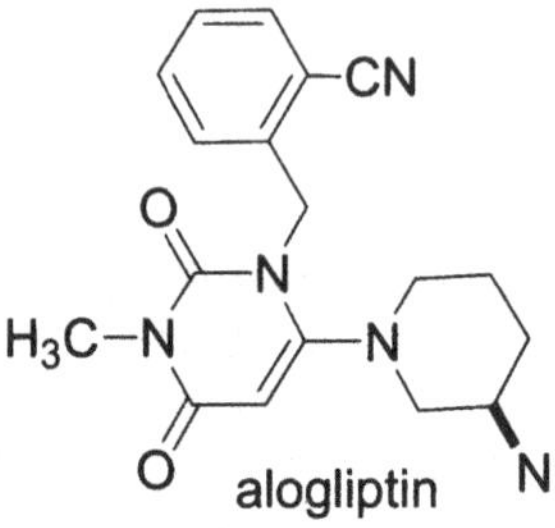   alogliptin 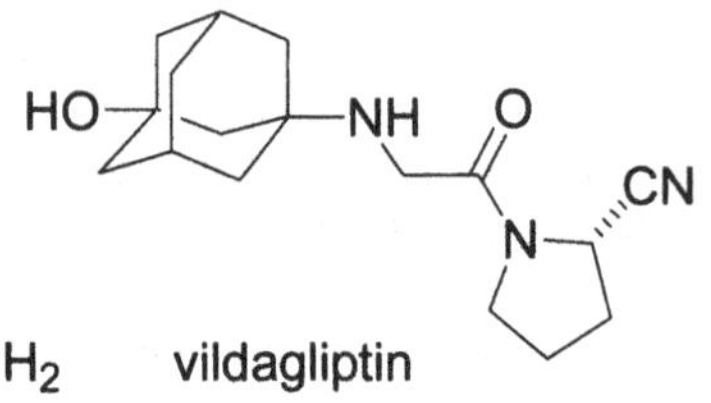   vildagliptin

Sitagliptin was the first oral inhibitor approved in the U.S. in 2006. It was followed by saxagliptin and linagliptin. Vildagliptin (approved in Europe) and alogliptin (approved in Japan) are not approved in the U.S. market yet.

Mechanism of action

DPP-IV is a cell surface serine protease that selectively cleaves the terminal dipeptide from peptides that specifically have proline or alanine as the second amino acid of their sequence. The drugs described in the previous paragraph are competitive inhibitors of DPP-IV that interact with the active site of the enzyme. This inhibition increases the circulating levels of active GLP-1.

They can be divided in peptidomimetics, like sitagliptin, vildagliptin, and saxagliptin, where an α- or β-amino acid core is present, and the non-peptidomimetics, like linagliptin and alogliptin. Among the peptidomimetics, cyanopyrrolidines have a unique mode of action. The nitrile forms a reversible covalent bond with a serine from the catalytic site of the enzyme, generating an amidate of slow dissociation kinetics. The reactivity of this nitrile is also responsible for the instability of this kind of drugs, which generate a 6-atom cycle by intramolecular attack of the nucleophilic amine to the nitrile. The presence of the adamantyl group in the side chain slows down this reaction.[87]

Sitagliptin, linagliptin, and alogliptin have modest metabolism in humans, being excreted mainly as the parent drug. Cyanopyrrolidine derivatives, in contrast, undergo extensive metabolism in humans. All these inhibitors are excreted mainly by the kidneys, except linagliptin, that has an hepatic route of elimination. This difference is believed to be due to the high protein binding of this molecule, which might make difficult the glomerular filtration of the drug.[88]

DPP-IV inhibitors lower A1c levels by 0.6-0.9%.[21] They are in general well tolerated and do not produce hypoglycemia when used as monotherapy. Unlike incretin mimetics, they do not produce body weight loss.

2.6   *Amylin Analog*

Amylin is a 37-amino-acid hormone cosecreted with insulin by β-cells after meals. It is also known as islet amyloid polypeptide (IAPP). It inhibits glucagon secretion, reduces food intake, and slows gastric emptying, contributing to the homeostasis of glucose. Like insulin, it is deficient in type 2 diabetics, and it is absent in type 1 diabetics.[89]

Human amylin is highly amyloidogenic, forming insoluble aggregates that accumulate in the β-cells of T2 diabetics. These deposits are not observed in the pancreas of healthy people. It has been hypothesized that the formation of these amyloid fibers can be associated with the development of the disease.[90,91] Rat amyloid has a very high homology with the human peptide but is nonamyloidogenic.

Human: KCNTATCATQRLANFLVHSSNNFGAILSSTNVGSNTY-NH2
Rat: KCNTATCATQRLANFLVRSSNNLGPVLPPTNVGSNTY-NH2
Pramlintide: KCNTATCATQRLANFLVHSSNNFGPILPPTNVGSNTY-NH2

Pramlintide is an injectable synthetic peptide that does not precipitate and shows similar pharmacokinetic and pharmacodynamic properties to the human amylin.[92] Its structure can be seen as a hybrid between the human and the rat sequences. This drug was approved in 2005 for treatment of type 1 and type 2 diabetes in conjunction with insulin for the control of postprandial glucose. It cannot be coadministered with insulin due to an incompatibility by precipitation of pramlintide at neutral pH, at which insulin is administered.

2.6.1   *Mechanism of Action*

Amylin acts through a family of cell surface receptors. They consist of a common GPCR, the calcitonin receptor (CTR), that is complexed to a family of proteins named RAMPs (receptor activity modifying proteins). The complex of CTR with each of the RAMPs, (RAMP1, RAMP2, and RAMP3) generates a different receptor with a distinct pharmacology (AMY$_1$, AMY$_2$, and AMY$_3$, respectively).[93]

The reduction in food intake seems to depend on a direct action in the area postrema of the brain, where the amylin receptors are highly expressed. The gastric-emptying effect seems to be CNS mediated and depends on the vagus nerve.[92] The inhibitory effect of the glucagon secretion also seems to depend on extrapancreatic pathways.[94]

Pramlintide is administered subcutaneously at mealtimes. It has a half-life of around 48 min. Its primary metabolite, deslys pramlintide, is also active and is cleared by the kidney.

In clinical studies, it reduced A1c by 0.5–0.7%.[21] The weight loss observed over 6 months is 1–1.5 kg. This effect is particularly favorable in obese people using insulin as this treatment is usually accompanied by an increase in body weight.

The main side effect is nausea and tends to decrease after continued treatment.

3   **Other Treatments for Type 2 Diabetes**

Since obesity (particularly abdominal adiposity) and dyslipidemia have a critical role in insulin resistance and the development of T2DM, some therapies targeting appetite, lipid absorption, or metabolism have shown an improvement in the plasma glucose in diabetics. Some of those treatments are discussed in this section.

3.1   *Bile Acid Sequestrants*

Bile acids (BAs) are ambiphilic molecules consisting of a lipophilic sterol core and a side chain with a hydrophilic carboxylic acid. There are a number of BAs that differentiate in the number and position of hydroxyl groups in the sterol core. They are synthesized by the liver from cholesterol and stored in the gallbladder. During digestion, they are secreted to the duodenum where they play an essential role in the digestion of dietary lipids. Around 95% of the BAs are reabsorbed in the ileum and return to the liver *via* the portal vein. This process is known as enterohepatic circulation. BAs are also signaling molecules that activate a nuclear receptor, FXR (farnesoid X receptor), involved in lipid and glucose metabolism, and a G-protein-coupled receptor, TGR5, implicated in energy expenditure and GLP-1 secretion.[95]

BA sequestrants are nonabsorbable cationic polymers that form tight insoluble salt complexes with BAs in the intestine through their carboxylate moiety. The formation of these complexes reduces the reabsorption of the BAs, which are then excreted in the feces.

There are a number of BA sequestrants in the market for the treatment of hypercholesterolemia. During clinical trials with some of these drugs, besides the reduction of plasma cholesterol levels, a reduction of plasma glucose was observed. Although the effect was observed in some degree with other BA sequestrants, only one of them, colesevam, initially approved in the U.S. in 2000 for the reduction of LDL-cholesterol, has also been approved for the treatment of T2DM.[96] It is believed that the modification of the composition of the circulating pool of BAs produced by these sequestrants might have an effect on its interaction with the BA receptors FXR and/or TGR5 and subsequently on glucose homeostasis, but the mechanism by which the glycemia is controlled is still unclear.[97]

3.2   *Orlistat*

The antiobesity agent orlistat is an intestinal lipase inhibitor. It decreases intestinal fat absorption by inhibiting the hydrolysis of dietary fat into free fatty acids and monoglycerides. This effect results in body weight reduction (4–6.5%).[37] Orlistat is limited to use in overweight or obese patients, and since its reduction of A1c is relatively small (0.3–0.9%), it should be used with additional antihyperglycemic agents. Orlistat also produces improvement in total and LDL cholesterol.[98] Its main side effects are gastrointestinal and are directly related to the fat content of the diet.

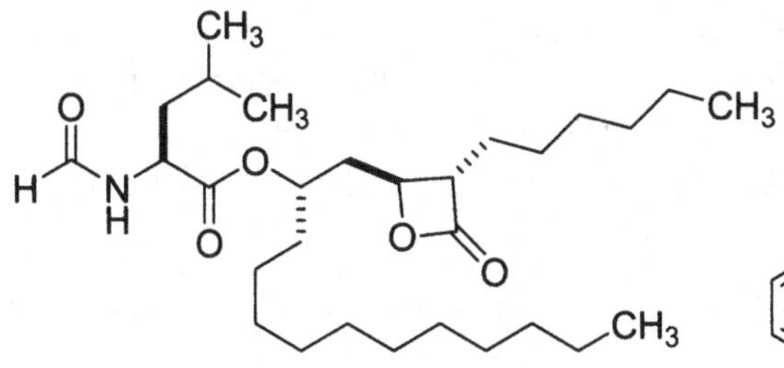

orlistat

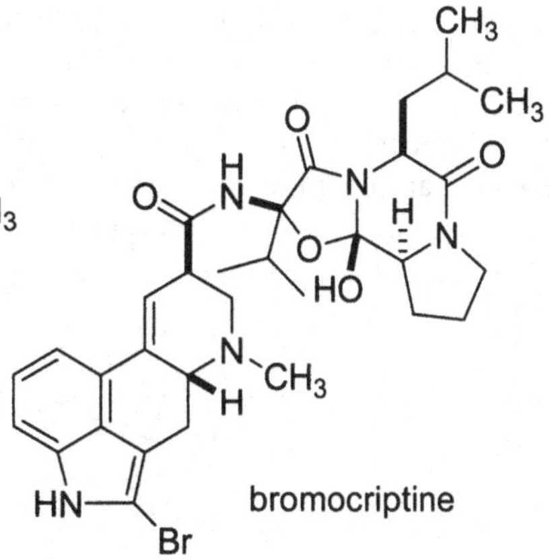

bromocriptine

3.3 Bromocriptine

Bromocriptine, an ergot alkaloid in clinical use for more than 30 years for the treatment of Parkinson's disease, acromegaly and hyperprolactinemia, was approved recently for use in T2DM as a quick-release (QR) formulation.[99,100] Bromocriptine QR functions by targeting central pathways of glucose metabolism, mainly by postsynaptic activation of D2 receptors.[101] The quick-release formulation differs from the standard formulation used for Parkinson's disease and hyperprolactinemia in that QR shows higher bioavailability and faster peak concentration. It is believed that bromocriptine acts by centrally resetting circadian rhythms, thus affecting patterns of food intake and nutrient storage. In T2DM patients, administration of bromocriptine QR early in the morning results in a small improvement of A1c (around 0.6%) with no weight gain or even with a small degree of weight loss and a mild decrease in systolic blood pressure. Benign side effects have been reported (nausea being the most frequent one).

4 Novel Mechanisms of Action: Future Treatments for Type 2 Diabetes

Although a variety of medications and their combinations have been successful in lowering glycemia, they all fail in stopping the progression of the disease, and, at the final stages of the disease, many type 2 diabetics require exogenous insulin to keep their plasma glucose levels under control. For this reason, there is a great need for new and more effective medications with fewer side effects. Most of the current therapies were also developed years before the mechanisms by which they act were known. The knowledge of the biochemical pathways associated with the homeostasis of the glucose and the pathologies associated with it has expanded during the last decade, resulting in a number of new mechanisms of action being tested in the clinic. Some of these emerging approaches are discussed here.[102,103,104]

4.1 Sodium-Glucose Cotransporter 2 (SGLT-2) Inhibitors

The kidney plays a key role in glucose homeostasis. On one side, the kidney is, together with the liver, a contributor to gluconeogenesis, accounting for 20% of the total glucose produced in the fasting state.[105] On the other hand, the kidney is responsible for the filtration and reabsorption of glucose in the proximal convoluted tubule.[106] In a healthy adult, approximately 180 g of glucose is recovered in the glomeruli daily, with less than 1% of it being excreted in the urine. This process contributes to glucose homeostasis and to the retention of calories. Under normal conditions, glucose is not excreted in the urine. When glucose concentration exceeds the so-called "glucose threshold," glucose starts appearing in the urine (glycosuria). Glucose is reabsorbed back to plasma even at high plasma glucose concentration. Thus, inhibiting this process could be a novel mechanism to fight diabetes, independent of insulin secretion and insulin action.[107,108,109]

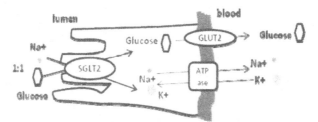

Fig. 3. Mechanism of action of SGLT-2

Sodium-glucose co-transporters (SGLTs) are members of a family of membrane proteins that are responsible for the transport of glucose, vitamins, amino acids, and some ions across the brush-border membrane proximal convoluted tubules in the kidney as well as the intestinal epithelium.[110] There are two major players in the renal glucose reabsorption: SGLT-1 and SGLT-2. SGLT-2 is a low affinity, high capacity transporter, located almost exclusively in the kidney proximal convoluted tubule. It accounts for about 90% of the glucose reabsorption. SGLT-1 is a high-affinity, low-capacity transporter that is located in the gastrointestinal tract and in the late proximal straight tubule; it contributes to about 10% of renal glucose reabsorption. Once plasma glucose has been filtered by the glomeruli in the kidney, it is reabsorbed by the SGLTs across the luminal membranes of the epithelial cells of the proximal tubule. SGLT-2 catalyzes the active transport of glucose against a concentration gradient across the luminal membrane by coupling it with the transport of sodium. The transport of intracellular glucose through the basolateral membrane into the intercellular space (which is in equilibrium with the blood) takes place by glucose transporter protein 2 (GLUT2) (Fig 3). The small amount of glucose that is not reabsorbed by SGLT-2 reaches subsequent segments, where it is reabsorbed by SGLT-1.

The interest in developing inhibitors of SGLTs (discovered in the 1990s) grew with the effects on plasma glucose observed with phlorizin (isolated in 1835 from the bark of apple tree) on animals[111] and man[112] together with the observation that people having familial renal glycosuria (a genetic disorder characterized by glycosuria without hyperglycemia and that involves the gene encoding SGLT-2) had no significant morbidity or changes in life expectancy.[113] Phlorizin development did not advance due to its poor pharmacokinetics (low intestinal absorption and rapid *in vivo* β-glucosidase degradation) and the lack of selectivity *vs.* SGLT-1. Since then, the focus was on the discovery of SGLT-2 selective inhibitors with improved metabolic stability (for example, replacement of the *O*-glycoside by *C*-glycoside).[114] Currently, there are several compounds of this class in Phase II and Phase III clinical studies (see below for a few of the most advanced compounds in development).

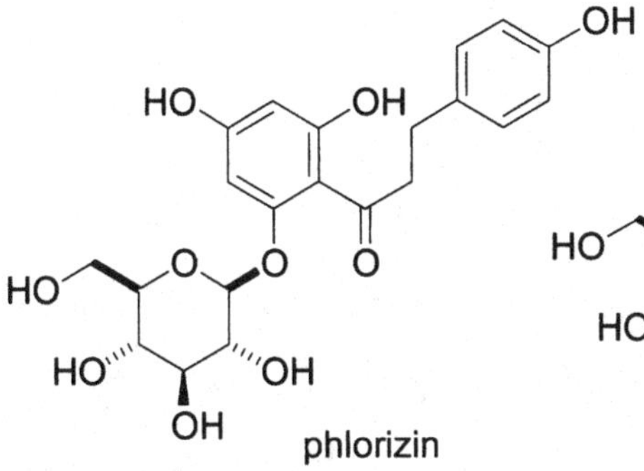

phlorizin

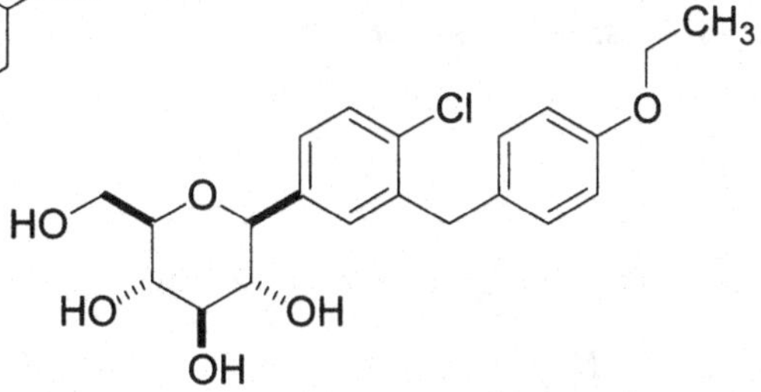

dapagliflozin (Phase III)

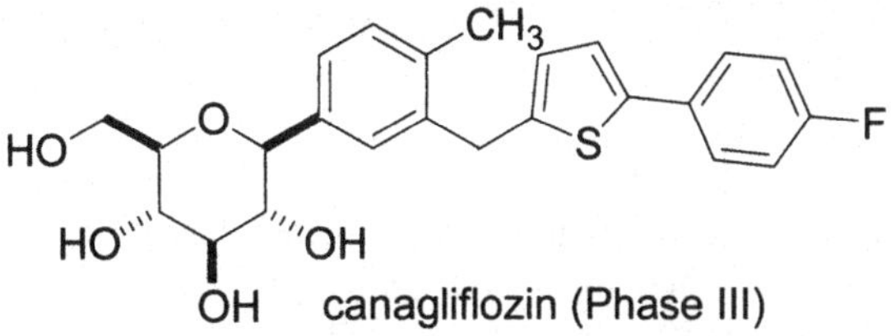

canagliflozin (Phase III)

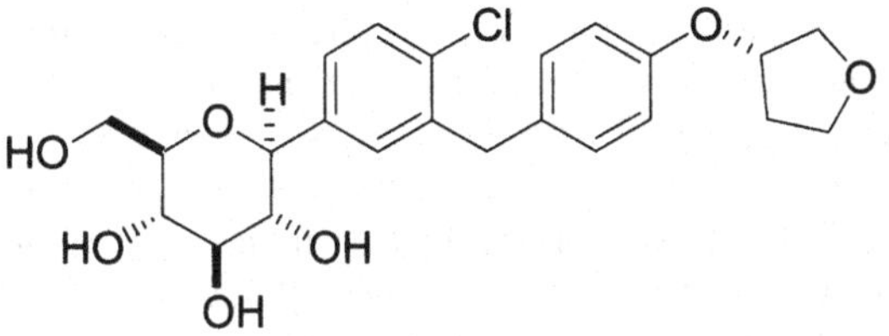

empagliflozin (Phase III)

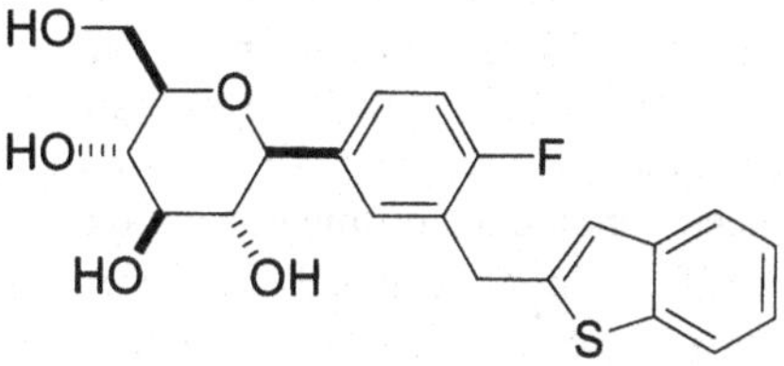

ipragliflozin (Phase II)

Most data published regarding the efficacy and safety of this new class of antidiabetics come from dapagliflozin, which has demonstrated improved control in both fasting and postprandial glucose (alone or coadministered with metformin), with reductions of A1c around 0.5–0.9,[21] and with a significant reduction in body weight.[115,116] Regarding safety considerations for this mechanism, long-term safety studies are lacking. Current studies show this class of compounds to be well tolerated, with no significant events of hypoglycemia. Due to the nature of the mechanism, the concerns have been a potential for increased urinary frequency, urinary tract infections, fungal genitourinary infection, and deterioration of renal function. Long-term studies are required to determine the risk of urinary and genital infections with prolonged therapy.[117]

4.2   *PPARs: Next Generation*

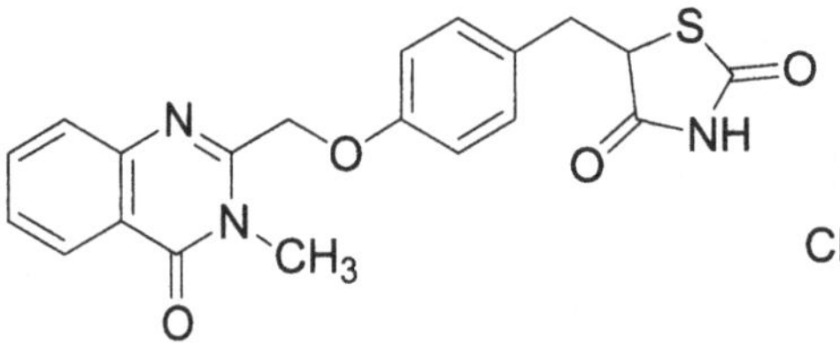

balaglitazone (partial PPARγ agonist)
Phase III

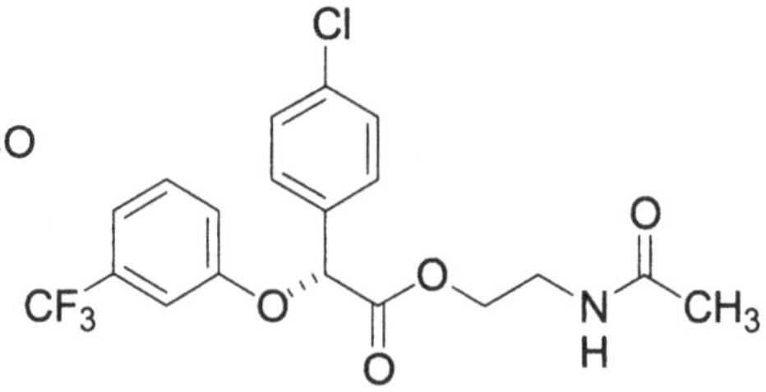

MBX-102 (metaglidasen)
PPARγ modulator (prodrug)
Phase III

During the last decade, there has been intensive research on different ways to activate PPARs such that the antihyperglycemic effect is maintained while the side effects are diminished. Pharmacological studies describe that, while adipogenesis correlates directly with the potency of agonism of the molecules, the insuling-sensitizing effects can be exerted by full and partial agonists as well as by antagonists.[118] A number of structurally diverse PPARγ-selective modulators (SPPARγMs) and partial agonists are currently in advanced phases of development, the expectation being that transcription of genes involved in insulin sensitization are increased whereas those responsible for the side effects are reduced or unaffected.[55,119] One of the most advanced compounds of this generation of PPAR activators is balaglitazone, currently in Phase III studies. Balaglitazone is a PPARγ partial agonist (50% activation) that in animal studies shows similar glucose-lowering effects to pioglitazone but has fewer effects on heart, bone, fat, and fluid retention. Initial clinical data support this hypothesis, showing improved glycemic control with lower body weight gain or edema side effects.[120] However, longer duration studies are needed to confirm a better safety profile than that shown by full agonists.

There has also been significant research on the discovery of PPAR ligands that could simultaneously activate PPARγ and PPARα and/or PPARδ. The discovery that PPARα was the molecular target for members of the fibrate class,[121] which lower triglycerides and increase HDL-cholesterol, made a dual PPARα/γ ligand an opportunity that could be very beneficial for the T2DM population, given the importance of keeping glucose and lipid levels under control.[60] Aside from the potential benefit on lipid control, PPARα is emerging as a therapeutic target in diabetic microvascular damage, which could also help preventing diabetic complications such as retinopathy, neuropathy, and nephropathy.[122] Currently, there are several nonselective PPAR ligands undergoing clinical development.

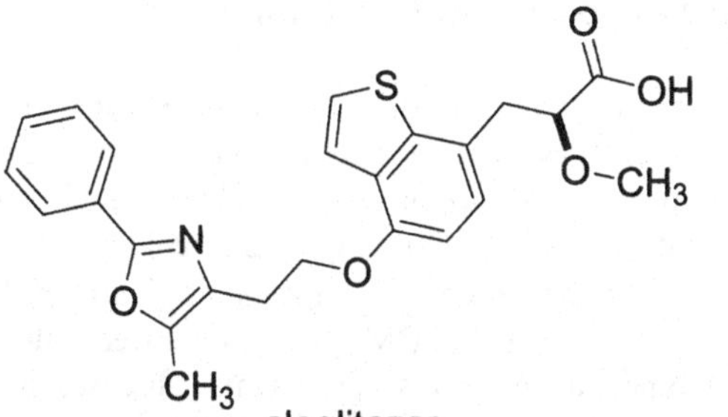

aleglitazar
PPARα and PPARγ agonist
Phase III

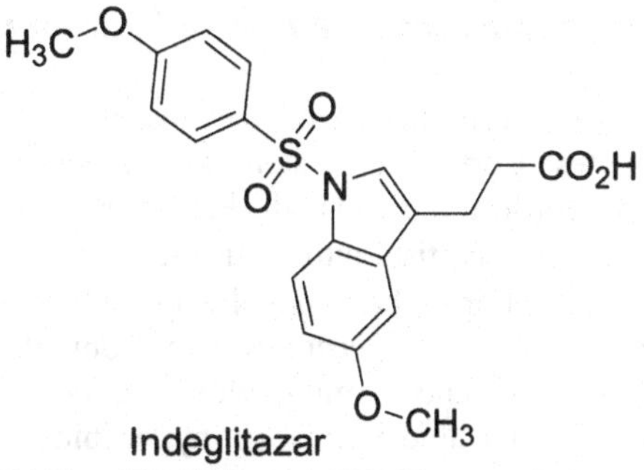

Indeglitazar
PPARα, PPARδ and PPARγ agonist
Phase II

4.3 Glucokinase Activators

Glucokinase (GK) is the sensor of glucose in pancreas and liver. It catalyzes the formation of glucose-6-phosphate from glucose, the first step of glucose metabolism. In the β-cells, it is the rate-limiting step in the GSIS (see Fig. 1). In the liver, it regulates glucose utilization and glycogen synthesis. This enzyme is only activated at high concentration of glucose.[123]

Glucokinase activators are small molecules that bind to an allosteric site of the enzyme increasing its catalytic activity. They produce an increase in glucose uptake by the liver and insulin secretion from the pancreas. Although a few molecules have been terminated in Phase I or II, interest remains in this area, as shown by the number of small molecules in different stages of clinical development.

4.4 GPR119 Agonists

GPR119 is a GPCR expressed in β-cells and enteroendocrine cells (L-cells and K-cells, intestinal cells that secrete GLP-1 and GIP, respectively). It is activated by oleoylethanolamide and other lipids. In the β-cells it stimulates insulin secretion in a glucose-dependent manner, much like GLP-1 agonists do. In the gut, GPR119 agonists increase the secretion of GLP-1 and GIP, a beneficial effect in glycemia that has been discussed in section 2.5. Promising preclinical data in the control of glucose by GPR119 agonists have been shown by different companies.[124] At least 2 compounds are listed as being in phase II of development.

4.5 D-Tagatose

D-Tagatose (an epimer of D-fructose at C-4) is a naturally occurring sugar approved for use as sweetener in food and beverages.[125] It is currently in Phase III studies for diabetes indication. During its development as sweetener, it was found that D-tagatose exhibited antihyperglycemic effect, reducing postprandial glucose excursions, and, in some cases, showing reduced food intake effect and body weight loss. Phase II studies have shown that D-tagatose can reduce A1c 1–2%. The pharmacology of tagatose is still unclear. The major side effects are gastrointestinal (diarrhea, flatulence) and decrease with use.[126] One caution with all these data is the low number of patients treated in each study. Larger Phase III studies will be required to fully assess the potential of D-tagatose as a treatment for diabetes.

4.6   11β-Hydroxysteroid Dehydrogenase Type 1 (11β-HSD1) Inhibitors

The enzyme 11β-HSD1 catalyzes the intracellular conversion of cortisone to cortisol in liver and adipocytes. Cortisol is a steroid hormone that enhances glucose production in the liver and antagonizes insulin-mediated glucose uptake in muscle and adipose tissue. The inverse process (conversion of cortisol into cortisone) is carried out by 11β-HSD2 in kidney and colon. Inhibition of 11β-HSD1 may be a novel approach to treat T2DM. Several inhibitors have advanced to clinical trials.[127, 128] Recent results from clinical studies in T2DM patients carried out with INC13739 are encouraging, with positive outcomes in A1c, fasting plasma glucose, insulin sensitivity, and lipid control (in hyperlipidemic patients), compared to placebo. No drug-related serious adverse events occurred during the study.[129]

4.7   Diacylglycerol Acyltransferase 1 (DGAT-1) Inhibitors

DGAT-1 and DGAT-2 are the enzymes that catalyze the last step of triglyceride synthesis, the coupling of diacyglycerol (DAG) to fatty acyl-CoA. DGAT-1 is expressed ubiquitously, including heart, intestine, and skeletal muscle. Studies on transgenic mice revealed that DGAT-1 is involved in regulating storage of triglycerides, development of obesity, insulin resistance, and fatty acid-induced inflammation.[130] These studies fostered the research on the development of DGAT-1 inhibitors. Clinical trials are ongoing to assess the viability of this approach to treat diabetes and obesity.

5   **Current Therapies for Type 1 Diabetes**

In contrast to T2DM, T1DM is a primary insulinopenic state[131] and the substitution of missing insulin represents the only therapeutic option. Since publication of the Diabetes Control and Complications Trial (DCCT), most of the patients with T1DM are treated using an intensified regimen of basal-bolus therapy.[132] This can be achieved with the use of multiple daily subcutaneous injections or continuous subcutaneous infusion via an insulin delivery system such as an insulin pump.[5, 133]

5.1   Insulin

All patients with type 1 and many patients with advanced type 2 diabetes require treatment with insulin. Insulin is the main anabolic and anticatabolic hormone in mammals, being the key hormone in the regulation of glucose homeostasis.[134] It is secreted by the β-cells of the pancreatic islets of Langerhans in response to an increase in glucose, amino acid and fatty acid levels in the circulation after a meal. Insulin is synthesized as a precursor molecule, preproinsulin, which is subsequently transformed into proinsulin. The β-cell converts proinsulin to insulin and C-peptide,* resulting in the equimolar release of these two peptides.[135]

Human insulin consists of a 21-residue A-chain and a 30-residue B-chain, tethered covalently by two disulfide bridges. An additional disulfide bridge exists in the A-chain (Fig. 4).

---

* C-peptide is used as a marker of insulin secretion, and it is a useful tool to distinguish among different types of diabetes, or to guide treatment decisions. In T1DM, the amount of C-peptide will be decreased, whereas in T2DMwill be normal or high.

At micromolar concentrations, insulin dimerizes; in the presence of zinc, available in the β-cell, it further associates into hexamers. The hexamer is a stable but biologically inactive form of insulin that keeps insulin protected from degradation and, at the same time, readily available for action. Upon extracellular release from the β-cells, the hexamers dissociate into dimers and eventually into monomers, which are the biologically active species.

Insulin exerts its pleiotropic effects on cellular growth and metabolism by binding to its specific receptor in the cell surface, which activates the insulin receptor protein tyrosine kinase and thus causes the phosphorylation of a number of endogenous substrates.[136]

The main actions of insulin are the increase in glucose uptake in skeletal muscle, heart, and adipose tissue and the inhibition of hepatic glucose production, thus serving as the primary regulator of blood glucose concentration. By activation of glycogen synthase and inhibition of glycogen phosphorylase, it blocks glycogenolysis and gluconeogenesis, and stimulates glycogen synthesis, therefore regulating fasting glucose levels. Insulin also stimulates cell growth and differentiation and promotes the storage of substrates in fat, liver, and muscle by stimulating lipogenesis and glycogen and protein synthesis and inhibiting lipolysis, glycogenolysis, and protein breakdown.[1] Insulin resistance or deficiency results in profound dysregulation of these processes and produces elevations in fasting and postprandial glucose and lipid levels. Insulin increases glucose transport in fat and muscle cells by stimulating the translocation of the transporter GLUT4 from intracellular sites to the plasma membrane and the rate of GLUT4-vesicle exocytosis and by slightly decreasing the rate of internalization.[136] Up to 75% of insulin-dependent glucose disposal occurs in skeletal muscle, whereas adipose tissue accounts for only a small fraction.[9]

5.1.1   History

In 1922, Banting, Best, Collip, and MacLeod succeeded in purifying insulin from beef pancreas to treat children dying from T1DM.[137] For many years, treatment was with different types of animal insulin, which had amino acid structures not identical to that of human insulin. Initially, treatment was with regular, crystalline insulin given by subcutaneous injection shortly before meals. A number of attempts were made to reduce the frequency of injections by modifying insulin in order to prolong its duration of action by slowing its rate of absorption from injection sites.[138]

H. C. Hagedorn found a solution in protamine (a fish protein) complexed with insulin and zinc to yield a repository delivery form which, when injected subcutaneously, slowly released insulin for up to several days.[139] When injected subcutaneously, protamine is degraded by proteolytic enzymes, and the insulin is released and absorbed. The duration of action of the preparation is roughly twice as long as that of regular unmodified insulin. He continued his work and, in 1946, developed neutral protamine Hagedorn (NPH) insulin.

Another method of extending the action of insulin was by the addition of an excess of zinc ions in variable amounts to form amorphous crystalline suspensions, the so-called lente insulins.[140] By varying the zinc concentration, it was possible to create suspensions of zinc crystallized insulin with different rates of release of insulin from the subcutaneous repository.[141] However, NPH insulin has variable absorption leading to unpredictable insulin peaking, which can result in hypoglycemia. It also has a limited duration of action so it often does not last through the night. There was also variability in insulin levels with the lente and ultra lente insulins, resulting in unpredictable hypoglycemia.

The discovery of the amino acid sequence of insulin by Frederick Sanger[142] and the elucidation of the three-dimensional structure of the insulin molecule by Hodgkin[143] together with the discovery of the insulin receptor and detailed maps of insulin receptor interaction enabled scientists to modify the insulin molecule to achieve therapeutic benefit.[144] The insulin analogs have been specifically designed using a systematic approach based on a detailed knowledge of biochemistry and physiological actions of insulin (Fig. 4).[138]

5.2   *Insulin Analogs*

The pharmacodynamic profiles of conventional insulins, including short-acting regular insulin, intermediate, and longer acting insulins, are suboptimal. Human insulin is released directly from the pancreas into the hepatic portal circulation, and thus, the structure and physical properties of unmodified insulin are not best suited to achieving physiological replacement by subcutaneous injection. Following subcutaneous injection of soluble insulin, the hexameric zinc-containing insulin complex undergoes a series of dissociations and dissolutions. The resulting monomeric molecules are then absorbed through the capillary bed and lymphatic vessels in the surrounding subcutaneous tissue. There are several other factors that can delay or accelerate insulin absorption; even under controlled conditions, absorption is variable. The quest for more physiological insulin replacement was therefore directed at the development of short-acting analogs to mimic postprandial insulin secretion and long-acting analogs to mimic basal or background insulin secretion. Recombinant DNA technology has made it possible to produce human insulin biosynthetically and, more recently, to modify the insulin molecule for potential therapeutic and physiological advantage.[138]

5.2.1   *Rapid-Acting Insulin Analogs*

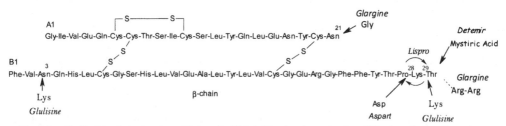

Fig. 4. Insulin (double-chain structure with two disulfide bridges) and structural modifications for insulin analogs.

Insulin mapping has shown that the positions B26–30 in the region of the insulin molecule are not critical for binding to the insulin receptor but are important in mediating the formation of insulin dimers and hence hexamers. This sequence has been a focus for amino acid substitutions in order to encourage the formation of insulin monomers. Several approaches have been tried: charge repulsion, decrease interface hydrophobicity, removal of metal ion binding site, interference with hydrophobic contacts, and β-sheet formation or steric hindrance (Fig. 4).[138,145]

Although the rapid-acting analogs available differ in terms of molecular structure, they exhibit similar pharmacokinetic and pharmacodynamic characteristics, and there is no obvious theoretical advantage of one over another. All three are used in the treatment of both type 1 and type 2 diabetes. The pharmacokinetic and pharmacodynamic profiles of all three and the activity profiles following preprandial subcutaneous injection more closely mimic the normal physiological insulin response than soluble human insulin.[138]

Insulin lispro

Insulin lispro differs from human insulin by the substitution of proline with lysine at position 28 and lysine with proline at position 29 of the insulin B-chain.[146] These substitutions result in a diminished tendency of the insulin molecule to self-associate. The change of the two amino acids in positions 28 and 29 was inspired by the structure of insulin-like growth factor (IGF)-1* which unlike insulin does not tightly aggregate as hexamers. This may explain the higher affinity of insulin lispro for the IGF-1 receptor, relative to human insulin, reported in some studies.[147]

This insulin analog is injected 5–15 min before or immediately after a meal, resulting in better postprandial glucose control and less frequent late hypoglycemia than human insulin.[148] Pharmacokinetic studies have shown that insulin lispro has peak activity at approximately 1 h and duration of action between 3 and 4 h.[149] A number of studies have shown that treatment with lispro (administered immediately before meals) results in a significant improvement in postprandial glucose levels with a lower rate of hypoglycemic events compared with regular human insulin (injected 30–40 min before meals).[150,151]

Insulin aspart

In insulin aspart, the proline at position B28 is substituted with the charged aspartic acid.[152] This reduces self-association of the molecule, allowing only weak dimeric and hexameric formation and thereby rapid dissociation after subcutaneous injection.

Receptor interaction kinetic studies have shown that insulin aspart behaves essentially like human insulin with regard to both the insulin and IGF-1 receptor with a similar potency to that of human insulin. Insulin aspart is absorbed twice as fast as regular insulin and reaches a maximum concentration in plasma of approximately twice that of human insulin. Its activity profile is very similar to that of human lispro[152] with pharmacokinetic and pharmacodynamic profiles identical in adult patients with type 1 diabetes.[153]

A similar improvement to that of lispro in postprandial glucose control has been shown for insulin aspart.[154,155] Lower rates of hypoglycemia are also seen with insulin aspart when compared with regular human insulin.[156]

Insulin glulisine

Insulin glulisine is the last rapid-acting human insulin analog produced by recombinant DNA. In insulin glulisine, asparagine at position 3 of the insulin B-chain is replaced by lysine, and lysine at position 29 is replaced by glutamate. It is thought that this latter substitution is predominantly responsible for its pharmacokinetic properties. Studies indicate that glulisine has a very comparable pharmacokinetic and pharmacodynamic profile to insulin lispro.[157, 158]

---

* IGF-1 or insulin-like growth factor-1 is a hormone with a molecular structure similar to insulin. It plays a key role in growth during childhood and has some anabolic effects in adults. It exerts its actions by binding to a specific receptor (IGF-1R), but it can also bind to the insulin receptor. The same happens with insulin, which can also bind to IGF-1R. Certain insulin analogs may lead to greater activation of this receptor when compared with human insulin, thus potentially increasing adverse mitogenic effects.

It has been claimed to have a faster onset of action than other short-acting analogs, due to its zinc-free formulation[159] although there is no evidence that this might pose a clinical advantage. The reduction in hypoglycemic episodes with glulisine is similar to that obtained with lispro.[156]

5.2.2   Basal (Long-Acting) Insulin Analogs

Until the recent introduction of basal insulin analogs, NPH insulin was the most frequently used basal insulin, usually administered in the evening. It is characterized by peaks in plasma insulin concentrations 5 to 10 h after administration, increased risk of hypoglycemia during the night, and duration of action of approximately 12 to 18 h that may contribute to hyperglycemia in the morning. Common differences in crystal size and inadequate resuspension make absorption kinetics and dosing precision with NPH insulin variable and result in unpredictable glucose levels.[160]

In order to create long-acting insulin analogs, the following sites can be targeted: the subcutaneous depot, the circulation and the interstitial compartment. The modifications that have been tried and successfully adopted include the modification of the isoelectric point of insulin (insulin glargine) and acylation with hydrophobic residues, which results in the binding to albumin (insulin detemir).[138]

Insulin glargine

Insulin glargine (A21-Gly-B31-Arg-B32-Arg-human insulin) was the first long-acting insulin analog available. It is produced by recombinant DNA technology using *Escherichia coli* plasmid DNA.[161]

The structure of insulin glargine differs from that of human insulin by glycine substitution for an asparagine at position A21 and addition of arginine to the carboxyl terminal of the B chain at positions B31 and B32.[162] These amino acid modifications change the isoelectric point of the molecule closer to neutral (from pH 5.4 to 6.8), making the molecule more soluble within the acidic environment of the vial but insoluble at the neutral pH of the injection site. When injected subcutaneously as an acidic solution (pH 4.0), insulin glargine forms microprecipitates at the injection site, greatly slowing its absorption. These microprecipitates slowly dissociate to hexamers, dimers and finally monomers capable of being absorbed across the capillaries. This process is slow but predictable.[163] Zinc added to the insulin glargine formulation further stabilizes the molecule and helps delay absorption time. As a result of delayed absorption, the duration of effect of insulin glargine is prolonged.

The onset of action of insulin glargine is about 1.5 h. The insulin serum concentration/time profile is relatively constant over 18 h, with a slow decrement in the time period of 20–24 h.[163] In contrast to NPH human insulin and lente insulin, there is no pronounced peaking of insulin concentration with insulin glargine. [163,164] Insulin glargine has a 60% lower binding affinity to the insulin receptor and 1.5 times faster dissociation than does human insulin. However, it has a 6.5 times increased binding affinity for the IGF-1 receptor.[149]

Glargine has been compared with NPH in a number of studies, both in type 1 and type 2 diabetes. Overall, no superiority in terms of HbA1c lowering was shown for insulin glargine *vs.* NPH, although a significant reduction in the relative risk for symptomatic and overall hypoglycemia with glargine was observed, both in type 1 and type 2 diabetes.[165,166,167,168] This reduction in hypoglycemia episodes occurs primarily at night due to the flatter pharmacokinetic profile of insulin glargine.[169] Insulin glargine has also less weight gain compared with NPH insulin.[170]

Insulin detemir

Insulin detemir is a soluble derivative of human insulin [LysB29(N-tetradecanoyl) des (B30) human insulin], in which the threonine residue at position B30 of the human insulin molecule has been removed and a 14-carbon fatty acid side chain has been attached to position B29. Due to its faster dissociation from the insulin receptor and, thus, a lower molar potency than human insulin, insulin detemir is formulated at a 4-fold higher molar concentration than human insulin. The protracted action of insulin detemir is mediated by the strong self-association of insulin detemir hexameric molecules at the injection site and albumin binding via the myristic fatty acid side chain. More than 98% of insulin detemir in the bloodstream is albumin bound, which makes detemir more slowly distributed to peripheral target tissues compared to NPH insulin.[171] The addition of the fatty acid also allows insulin detemir to be formulated as a solute in a neutral liquid solution, which does not precipitate during administration or absorption. Because precipitation and dissolution of a precipitate are unpredictable processes, the mechanism of protracted absorption of insulin detemir may contribute to reduced variability in insulin detemir action.[172]

Pharmacodynamic clamp studies have shown that insulin detemir has an average duration of action of up to 20 h, which can be significantly longer than that of NPH insulin.[173]

In a similar fashion to what was previously described for glargine, detemir has been shown to improve fasting blood glucose levels, but with no clinically significant difference in HbA1c compared with NPH. However, use of detemir is associated with a consistently lower risk of hypoglycemia, particularly nocturnal hypoglycemia in both types 1 and 2 DM.[165,174,175] A consistent finding from clinical trials with detemir is that it is associated with less weight gain in patients with both type 1 and type 2 diabetes compared with both NPH and insulin glargine. Several mechanisms behind it have been proposed: reduced risk of hypoglycemia with insulin detemir and a subsequent decrease in defensive snacking by patients; more influence on hepatocytes than peripheral tissues, thus effectively suppressing hepatic glucose output without promoting lipogenesis in the periphery;[176] and also an increase in satiety signals within the central nervous system.[177]

Insulin lispro protamine suspension

Insulin lispro protamine suspension (ILPS) or neutral protamine lispro (NPL) insulin is a protamine-based, intermediate-acting insulin formulation of the analog insulin lispro. It was developed for use within insulin lispro mixtures because an exchange between soluble insulin lispro and protamine-bound human insulin within human NPH precludes prolonged storage of mixtures of these insulins.[178]

The pharmacokinetic characteristics of ILPS are similar to other basal insulins such as those of NPH insulin, with onset of action within 1 to 4 h after injection, a peak at hour 6, and duration of action of about 15 h.[178] In a longer clamp study, the duration of action was similar to other basal analogs.[179]

In a clinical trial comparing 2 basal bolus treatment schedules in patients with type 1 and type 2 diabetes, ILPS has demonstrated better glycemic control and similar incidence of hypoglycemic events than did NPH.[180]

Several clinical trials have assessed glycemic efficacy and safety of ILPS versus insulin glargine and insulin detemir in T2DM. Overall, there were no differences in efficacy, and there was a benefit with regards to hypoglycemias for patients with glargine or detemir, although the number of episodes was low. This benefit was significant when the number of daily injections of ILPS was two instead of one.[181,182,183] Similar results were observed in T1DM when it was compared with insulin detemir in basal bolus therapy.[184]

Premixed insulins and insulin analogs

Both rapid-acting insulin analogs and conventional human insulins are also available in premixed preparations with rapid and intermediate insulin activity, with different proportions of rapid and intermediate-acting components. Fixed-premixed combinations may simplify the insulin regimen and reduce the number of daily injections and errors in accuracy when free-mixing. In comparison with conventional insulin mixtures, analog mixtures exhibit a more rapid onset and a shorter duration of action. Biphasic insulin analogs reduce postprandial glucose more effectively than their human insulin counterparts.[185]

6   Future Treatments for Type 1 Diabetes

6.1   Developments in Insulin Area

6.1.1   Alternative Ways of Delivering Insulin

Several companies are currently developing alternative and noninvasive routes for insulin delivery. Although some of them are already in the market (buccal), the majority are still under development (Phases II–III). These routes include aerosol formulations for buccal absorption based on different delivery technologies, oral gel capsule formulation of unmodified recombinant human insulin, rectal suppository formulations, transdermal patches, or inhaled formulations.[186]

The major limitation in the formulation of oral insulin is gastric enzymatic degradation resulting in poor gastrointestinal absorption of the insulin molecule. Modified insulin molecules use steric interference to resist enzymatic attacks. The use of conjugate technology results in increased chemical stability, enhanced absorption, and improved resistance to enzymatic degradation, thus prolonging activity.[187]

The intrapulmonary route offers a large surface area of drug delivery. However, certain factors are known to influence the efficiency of aerosolized insulin. Asthma, interstitial lung disease, smoking, and exercise can all interfere with deposition of aerosolized insulin particles into the alveoli and circulation, thereby reducing the bioavailability of the drug.[188]

Although an inhaled insulin was commercialized some years ago, it remained available for a limited period of time, and no other inhaled insulin has been approved for use to date. There is one inhaled insulin product in development: technosphere insulin. It consists of an insulin inhalation powder which is dosed with a specific delivery system. The particles also contain diketopiperazine, which together with insulin is rapidly cleared from the lungs. About half of the load from the cartridge is delivered to the lungs and uniformly distributed.[189] The clinical trials have shown a pharmacokinetic profile which meets prandial insulin needs in diabetic patients. It has been shown to be efficacious in postprandial and overall glycemic control, with up to 4 years of follow-up. The risk of hypoglycemia is relatively low as well as the weight gain when compared with subcutaneous mealtime insulins.[190]

6.1.2 Enhancement of Insulin Delivery (Hyaluronidase Delivery, Phase II)

Recombinant human hyaluronidase (rHuPH20) is a genetically engineered soluble hyaluronidase approved by the U.S. Food and Drug Administration as an adjuvant to increase the absorption and dispersion of other injected drugs. The use of rHuPH20 technology for delivery of recombinant insulin and insulin analogs (as Insulin-PH20 and Analog-PH20, respectively) is being tested for the potential treatment of diabetes mellitus. Clinical findings have demonstrated that co-administration of rHuPH20 with insulin or an insulin analog achieved faster systemic absorption, reduced inter- and intrapatient variability of insulin absorption, and faster metabolic effects compared with injection of either insulin formulation alone.[191] When added to human insulin, the results for postprandial blood glucose and overall glycemic control were similar to insulin lispro.[192]

6.1.3 Insulin Degludec (Phase III)

Insulin degludec (NN-1250, LysB29Nε-hexadecandioyl-γ-Glu deB30 human insulin, Degludec) is a next-generation, long-acting, basal insulin analog for the potential treatment of type 1 and type 2 diabetes. It forms soluble multihexamers upon subcutaneous injection, resulting in a depot from which the insulin is continuously and slowly absorbed. In Phase II trials, it has been shown to provide a comparable glycemic control to insulin glargine, with reduced rates of hypoglycemia.[193] Its long duration has been tested also at more prolonged intervals between injections, with similar results.[194]

DegludecPlus is a soluble coformulation of insulin degludec (70%) and insulin aspart (30%). This mixture provides both basal and rapid-acting insulin analogs in one injection, thus targeting fasting and postmeal plasma glucose. A proof-of-concept trial showed that it can provide a similar overall glycemic control to glargine, with a better postdinner glucose control when administered before the evening meal.[195]

6.2 Approaches to Fight β-Cell Destruction

As described above, T1DM is an autoimmune disease and for that reason for a number of years there has been a huge interest in therapies which interfere with the process of β-cell destruction. The therapies that are being tested can be divided into primarily immuno-modulatory or primarily immunosuppressive, with some therapies combining both features. The focus today is on three major pathways of the immune attack directed at the β-cell: autoantigen vaccination, T-cell modulatory approaches, and innate immune system modulation.[196] These potential treatments are being tested in clinical trials for efficacy and safety. The results of these studies will provide the evidence for the role of any of these drugs in the prevention and/or treatment of T1DM.

6.2.1 Abatacept (Phase II)

At the time of clinical onset of T1DM, a significant amount of insulin-producing β-cells are destroyed, but as many as 10–20% are still capable of insulin production. CTLA4-Ig (abatacept) inhibits a crucial stimulatory pathway in the activation of T-cells. CTLA4-Ig regulates T-cell function but does not deplete T-cells. Therefore, its safety profile might be better than other immunosuppressive agents. Costimulation modulation with abatacept has been shown to slow reduction in β-cell function over 2 years. The beneficial effect suggests that T-cell activation still occurs around the time of clinical diagnosis of type 1 diabetes. Further research will define the potential role of this drug in prevention or early treatment of type 1 diabetes.[197]

6.2.2   GAD (Phases II–III)

Glutamic acid decarboxylase assists in the conversion of L-glutamic acid to GABA. It exists in two distinct forms: GAD-65, common in human and rat islets, and GAD-67, which predominates in mouse islets.[198]

GAD-65 is a protein of 585 amino acids and is related to diabetes in humans.[199] The specific function of GAD in pancreatic islets is unknown, as it is the reason why it is a major autoantigen in autoimmune diabetes. Detection of autoantibodies to GAD has been regarded as a sign of a rather early phase of the autoimmune process preceding type 1 diabetes and has become one of the most important markers to predict type 1 diabetes.[200,201]

GAD-alum (rhGAD-65; recombinant human glutamic acid decarboxylase 65-kDa isoform) is used as an antigen-specific immune modulator. Previous studies have shown that it may slow or prevent autoimmune destruction of pancreatic islet cells by introducing immune tolerance. If the immune attack can be halted in a patient with recent-onset T1DM, residual insulin secretion may be maintained.

The efficacy of GAD treatment with respect to preserving residual insulin secretion in recent-onset type 1 diabetes patients has shown preservation of residual insulin secretion in type 1 patients (10-18 years of age), being more pronounced in patients with short duration of the disease.[202,203] Further analyses suggest that GAD-alum treatment can induce GAD-65 specific T-cells with regulatory features.[204] In children, it induced an early T-helper 2 immune-enhanced response to GAD-65 followed by a wider spectrum of cytokines at 3 and 9 months.[205]

6.2.3   Rituximab (Phases II–III)

Rituximab is approved by the FDA for the treatment of B-lymphocyte lymphoma. Research has shown that rituximab might be helpful in treating other conditions caused by T-cells and B-cells, including type 1 diabetes. The B-cells may be what trigger the T-cells to attack. Rituximab has been investigated to see if it can help lower the number of immune B-cells, thus preventing the destruction of any remaining insulin-producing β-cells that remain at diagnosis. It has been shown that at 1 year the mean AUC for the level of C-peptide was significantly higher in the rituximab than in the placebo group. The rituximab group also had significantly better glycemic control and required less insulin. Between 3 and 12 months, the rate of decline in C-peptide levels in the rituximab group was significantly less than that in the placebo group.[206]

6.2.4   Diapep277 (Phases II–III)

Heat-shock proteins (hsp) are potent activators of the innate immune system.[207] DiaPep277 is a major T-cell epitope of human heat-shock protein 60 (HSP60). It consists of 24 amino acids derived from the C-terminal region (positions 437–460) of hsp60, manufactured as a synthetic injectable peptide. Two original cysteines at positions 6 and 11 of the sequence were replaced by valines for stability without changing the immunological properties of the native p277 sequence.[207] The amino acid sequence of DiaPep277 is VLGGGVALLRVIPALDSLTPANED.

In contrast to hsp60, the peptide DiaPep277 has no effect on TLR4 but only activates anti-inflammatory effectors through TLR2. TLR2 promotes cell adhesion, inhibits migration and modulates cytokine secretion to a TH2 cytokine profile causing a shift from an inflammatory to a regulatory response.[207] It is expected to prevent further destruction of the β-cells by stimulating regulatory responses without causing immunological suppression. Combined analysis of the Phase II trials conducted so far show a significant preservation of β-cell function in adults without adverse effects, but HbA1c was not changed. [207]

6.2.5 Monoclonal Antibodies Targeting CD3: Otelixizumab and Teplizumab (Phases II–III)

Monoclonal antibodies (mAbs) that target CD3 work by blocking the function of effector T-cells, which mistakenly attack and destroy insulin-producing β-cells while stimulating regulatory T-cells, which protect against effector T-cell damage, thus preserving the normal ability of β-cells to make insulin.

The limitations identified with the antihuman CD3 mAb OKT3 when used for treatment of transplant rejection were addressed with the creation of humanized non-FcR binding antibodies. Two specific humanized non-FcR binding anti-CD3 antibodies have been studied in patients with new-onset T1DM: teplizumab and otelixizumab. These two antibodies appear to have decreased cytokine release potential because of the changes in the Fc regions but maintain their immunomodulatory effects.

Some studies have suggested that the modified anti-CD3 mAbs affect lymphocyte trafficking rather than cause depletion of T-cells.[208] The mechanism appears to involve induction of immune tolerance, possibly mediated by regulatory T-cells, rather than chronic immune suppression. Studies to prolong the clinical effects and to establish the optimal timing for intervention are now in progress.

ChAglyCD3 or otelixizumab is a nonmitogenic, aglycosylated monoclonal antibody that contains a humanized γ1 heavy chain and a rat/human chimeric λ light chain. It is directed against CD3ε, which is a part of the CD3/T-cell receptor (TCR) complex present on T-lymphocytes. Keymeulen et al. showed that the mAb treatment group had a greater C-peptide response.[209]

hOKT3γ1 (Ala-Ala), teplizumab, is a humanized version of the mouse monoclonal OKT3 antibody, retaining the same binding region of OKT3 but with amino acids at positions 234 and 235 of the human IgG1 Fc changed to alanine, resulting in decreased Fc binding.[210] In individuals with recent-onset type 1 diabetes, the C-peptide response was sustained at 12 months in the group that received anti-CD3 mAb.[211, 212] Also, the glycosylated hemoglobin levels and the insulin requirements were lower in the treatment group. Further results showed a reduced rate of decline in insulin production and a decrease in insulin requirements for more than 2 years after diagnosis.[208]

In summary, randomized clinical trials have shown an ability to reduce the loss of insulin production over the first 2 years of the disease. In addition, the need for exogenous insulin to maintain glucose control has been reduced. However, these agents alone do not restore normal glucose control, and future approaches will likely require combinations of agents with complementary immune or metabolic activity.[213]

6.2.6   Lisofylline (LSF, Phase II)

Lisofylline [LSF, 1-(5-*R*-hydroxyhexyl)-3,7-dimethylxanthine] is an anti-inflammatory agent that protects β-cells from the cytokine-induced dysfunction and reduces the onset of T1DM in preclinical animal models.[214]

The molecule has a low potency, poor oral bioavailability, and short half-life. For that reason, LSF analogs that maintain the side chain (5-*R*-hydroxyhexyl) constant while substituting nitrogen-containing heterocyclic substructures for the xanthine moiety have also been tested. In addition, an injectable formulation of LSF is under development. The span of potential conditions to be treated include T1DM and intravenous treatment of related complications, including post islet cell transplant therapy, as well as the potential treatment of T2DM.

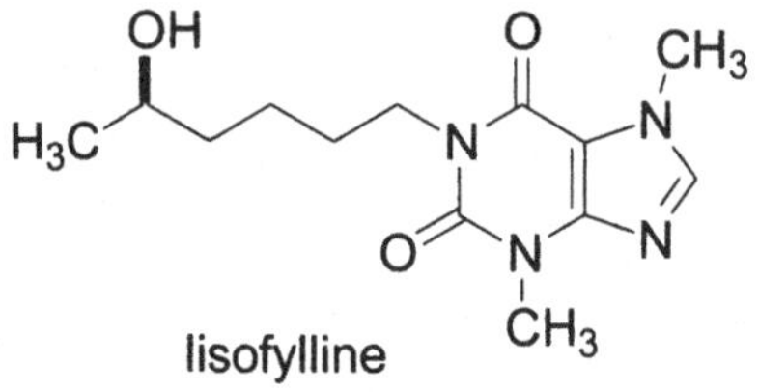

lisofylline

7   Future Prospects for New Diabetes Drugs

Diabetes mellitus and its associated complications have reached an epidemic stage in developed and emerging countries. As shown throughout this chapter, despite the number of pharmacological approaches available, there is no cure for diabetes, nor are there treatments that clearly slow down the disease progression.[215] An incredible amount of research is in progress, both in the discovery and development arenas, with the goals of (1) understanding the mechanisms causing the disease and (2) developing drugs that could effectively cure diabetes. Technical advances are helping in this race,[216, 217] which also need to tackle the control of macro- and microvascular consequences of diabetes. The large number of molecules under development offers an opportunity that will hopefully mature in the next few years.

Acknowledgment:

The authors thank Dr. H. Broughton, Dr. M. Coghlan, Dr. V. Koivisto, and Dr. M. Michael for their comments.

References

(1)   Saltiel, A. R.; Kahn, C. R. *Nature* **2001**, *414*, 799.

(2)   Zimmet, P.; Cowie, C.; Ekoe, J. M.; Shaw, J. E. In *Int. Textbook of Diabetes Mellitus,* DeFronzo, R. A.; Ferrannini, E.; Keen, H.; Zimmet, P. eds., Wiley: Hoboken, NJ, **2004**, p 3.

(3)   IDF Diabetes Atlas. www.diabetesatlas.org, Accessed, 2011-07-11.

(4)   Danaei, G.; Finucane, M. M.; Lu, Y.; Singh, G. M.; Cowan, M. J.; Paciorek, C. J.; Lin, J. K.; Farzadfar, F.; Khana, Y.-H.; Stevens, G. A.; Rao, M.; Ali, M. K.; Riley, L. M.; Robinson, C. A.; Ezzati, M. *Lancet* **2011**, *378*, 31.

(5)   American Diabetes Association. Standards of medical care in diabetes 2011, *Diabetes Care* **2011**, *34 (Suppl 1)*, S11.

(6)   Intensive Lifestyle Can Have A Profound Effect On Diabetes: Robertson, R. P.; Kendall, D. M.; Seaquist, E. R. *Nature Rev. Endocrinol.* **2010**, *6*, 128.

(7)   DeFronzo R. A. *Diabetes* **2009**, *58*, 773.

(8)   Hossain, P.; Kawar, B.; El Nahas, M. *N. Engl. J. Med.* **2007**, *356*, 213.

(9)   Klip, A.; Paquet, M. R. *Diabetes Care* **1990**, *13*, 228.

(10)   Bluestone, J. A.; Herold, K.; Eisenbarth, G. *Nature* **2010**, *464*, 1293.

(11)   Balkau, B.; Eschwege, E. In: Pickup, J. C.; Williams G, Eds. *Textbook of Diabetes*. Blackwell Science: Oxford, **2003**, 2.1–2.13.

(12)   EURODIAB ACE Study Group. *Lancet* **2000**, *355*, 873.

(13)   Eisenbarth, G. S. *J. Clin. Endocrinol. Metab.* **2007**, *92*, 2403.

(14)   Steck, A. K.; Zhang, W.; Bugawan, T. L.; Barriga, K. J.; Blair, A.; Erlich, H. A.; Eisebarth, G. S.; Norris, J. M.; Rewers, M. J. *Diabetes* **2009**, *58*, 1028.

(15)   Akerblom, H. K.; Knip, M. *Diabetes Metab. Rev.* **1998**, *14*, 31.

(16)   Garcia-Ocana, A.; Vasavada, R. C.; Cebrian, A.; Reddy, V.; Takane, K. K.; Lopez-Talavera, J.-C.; Steward, A. F. *Diabetes* **2001**, *50*, 2752.

(17)   Kawasaki, E.; Gill, R. G.; Eisenbarth, G. S. In: *Endocrine and Organ Specific Autoimmunity*, Eisenbarth, G. S., Ed., R.G. Landes: Austin, TX, **1999**, 149.

(18)   Barker, J. M.; Barriga, K. J.; Yu, L.; Miao, D.; Erlich, H. A.; Norris, J. M.; Eisenbarth, G. S.; Rewers, M. *J. Clin. Endocrinol. Metab.* **2004**, *89*, 3896.

(19)   Garcia, E. Y. *J. Phillipine Med. Assoc.* **1950**, *26*, 287.

(20)   Sterne, J. *Maroc Médical* **1957**, *36*, 1295.

(21)   Nathan, D. M.; Buse, J. B.; Davidson, M. B.; Ferrannini, E.; Holman, R. R.; Sherwin, R.; Zinman, B. *Diabetes Care* **2009**, *32*, 193.

(22)   Bharatam, P. V.; Patel, D. S.; Iqbal, P. *J. Med. Chem.* **2005**, *48*, 7615.

(23)   Patel, D. S.; Bharatam, P. V. *Chem. Commun.* **2009**, *9*, 1064.

(24)   Robert, F.; Fendri, S.; Hary, L.; Lacroix, C.; Andréjak, M.; Lalau, J. D. *Diabetes Metab.* **2003**, *29*, 279.

(25)   Zhou, G.; Myers, R.; Li, Y.; Chen, Y.; Shen, X.; Fenyk-Melody, J.; Wu, M.; Ventre, J.; Doebber, T.; Fujii, N.; Musi, N.; Hirshman, M. F.; Goodyear, L. J.; Moller, D. E. *J. Clin. Invest.* **2001**, *108*, 1167.

(26)   Hawley, S. A.; Gadalla, A. E.; Olsen, G. S.; Hardie, D. G. *Diabetes* **2002**, *51*, 2420.

(27)   Kim, Y. D.; Park, K.; Lee, Y.; Park, Y.; Kim, D.; Nedumaran, B.; Jang, W. G.; Cho, W.; Ha, J.; Lee, I.; Lee, C. C. H. *Diabetes* **2008**, *57*, 306.

(28)   Musi, N.; Hirsham, M. F.; Nygren, J.; Svanfeldt, M.; Bavenholm, P.; Rooyackers, O.; Zhou, G.; Williamson, J. M.; Liungvist, O.; Efendic, S.; Moller, D. E.; Thorell, A.; Goodyear, L. *Diabetes* **2002**, *51*, 2074.

(29)   Matthaei, S.; Reinbold, J. P.; Hamann, A; Benecke, H.; Häring, H. U.; Greten, H.; Klein, H. H. *Endocrinol.* **1993**, *133*, 304.

(30)   UK Prospective Diabetes Study (UKPDS). *Group Lancet* **1998**, *352*, 854.

(31)   Mauvais-Jarvis, F.; Andreelli, F.; Hanaire-Broutin, H.; Charbonnel, B.; Girard, J. *Diabetes Metab.* **2001**, *27*, 415.

(32)   Perfetti, R.; Ahmad, A. *Trends Endocrinol. Metab.* **2000**, *11*, 218.

(33)   There is ambiguity in the assignation of several of these drugs to different generations: Raptis, S. A.; Dimitriadis, G. D. *Exp. Clin. Endocrinol. Diabetes* **2001**, *109*, S265.

(34)   Rorsman, P. *Br. J. Diabetes Vasc. Dis.* **2005**, *5*, 187.

(35)   Moneva, M. H.; Dagogo-Jack, S. *Curr. Drug Targets* **2002**, *3*, 203.

(36)   Kokil, G. R.; Rewatkar, P. V.; Verma, A.; Thareja, S.; Naik, S. R. *Curr. Med. Chem.* **2010**, *17*, 4405.

(37)   Cheng, A. J. J.; Fantus, I. G. *CMAJ* **2005**, *172*, 213.

(38)   Dornhorst, A. *Lancet* **2001**, *358*, 1709.

(39)   Malaisse, W. J. *Expert Opin. Pharmacother.* **2008**, *9*, 2691.

(40)   Winkler, M.; Stephan, D.; Bieger, S.; Kühner, P.; Wolff, F.; Quast, U. *J. Pharmacol. Exp. Ther.* **2007**, *322*, 701.

(41)   Hansen, A. M. K.; Hansen, J. B.; Carr, R. D.; Ashcroft, F. M.; Wahl, P. *Br. J. Pharmacol.* **2005**, *144*, 551.

(42)   Hu, S.; Boettcher, B. R.; Dunning, B. E. *Diabetologia* **2003**, *46 (Suppl. 1)*, M37.

(43)   Recent studies show that nateglinide promotes GLP-1 release from the intestinal L cells. This may be another mechanism by which nateglinide restores early-phase insulin. Kitahara, Y.; Miura, K.; Yasuda, R.; Kawanabe, H.; Ogawa, S.; Eto, Y. *Biol. Pharm. Bull.* **2011**, *34*, 671.

(44)   Keilson, L.; Mather, S.; Walter, Y. H.; Subramanian, S.; McLeod, J. F. *J. Clin. Endocrinol. Metab.* **2000**, *85*, 1081.

(45)   Massi-Benedetti, M.; Damsbo, P. *Expert Opin. Investig. Drugs* **2000**, *9*, 885.

(46)   Legler, G. *Adv. Carbohydr. Chem. Biochem.* **1990**, *48*, 319.

(47)   Scott, L. J.; Spencer, C. M. *Drugs* **2000**, *59*, 521.

(48)   Kawamatsu, Y.; Asakawa, H.; Saraie, T.; Mizuno, K.; Imamiya, E.; Nishikawa, K.; Hamuro, Y. *Arzneim-Forsch* **1980**, *30*, 751.

(49)   Sohda, T.; Mizuno, K.; Imamiya, E.; Sugiyama, Y.; Fujita, T.; Kawamatsu, Y. *Chem. Pharm. Bull.* **1982**, *30*, 3580.

(50)   Cho, N.; Momose, Y. *Curr. Topics Med. Chem.* **2008**, *8*, 1483.

(51)   Bharatam, P. V.; Khanna, S. *J. Phys. Chem. A* **2004**, *108*, 3784.

(52)   Distel, R. J.; Ro, H. S.; Rosen, B. S.; Groves, D. L.; Spiegelman, B. M. *Cell* **1987**, *49*, 835.

(53)   Lehmann, J. M.; Moore, L. B.; Smith-Oliver, T. A.; Wilkison, W. O.; Willson, T. M.; Kliewer, S. A. *J. Biol. Chem.* **1995**, *270*, 12953.

(54)   Auwerx, J. *Diabetologia* **1999**, *42*, 1033.

(55)   Doshi, L. S.; Brahma, M. K.; Bahirat, U. A.; Dixit, A. V.; Nemmani, K. V. S. *Expert Opin. Investig. Drugs* **2010**, *19*, 489.

(56)   Okuno, A.; Tamemoto, H.; Tobe, K.; Ueki, K.; Mori, Y.; Iwamoto, K.; Umesono, K.; Akanuma, Y.; Fujiwara, T.; Horikoshi, H.; Yazaki, Y.; Kadowaki, T. *J. Clin. Invest.* **1998**, *101*, 1354.

(57)   Boden, G. Cheung, P.; Mozzoli, M.; Fried, S. K. *Metabolism* **2003**, *52*, 753.

(58)   Picard, F.; Auwerx, J. *Ann. Rev. Nutr.* **2002**, *22*, 167.

(59)   Moller, D. *Nature* **2001**, *414*, 821.

(60)   Feldman, P. L.; Lambert, M. H.; Henke, B. R. *Curr. Topics. Med. Chem.* **2008**, *8*, 728.

(61)   Choi, J. H.; Banks, A. S.; Estall, J. L.; Kajimura, S.; Bostroem, P.; Laznik, D.; Ruas, J. L.; Chalmers, M. J.; Kamenecka, T. M.; Blueher, M.; Griffin, P. R.; Spiegelman, B. M. *Nature* **2010**, *466*, 451.

(62)   Lu, M.; Sarruf, D. A.; Talukdar, S.; Sharma, S.; Li, P.; Bandyopadhyay, G.; Nalbandian, S.; Fan, W. Q.; Gayen, J. R.; Mahata, S. K.; Webster, N. J.; Schwartz, M. W.; Olefsky, J. M. *Nat. Med.* **2011**, *17*, 618.

(63)   Ryan, K. K.; Li, B.; Grayson, B. E.; Matter, E. K.; Woods, S. C.; Seeley, R. J. *Nat. Med.* **2011**, *17*, 623.

(64)   Aronoff, S. L.; Berkowitz, K.; Shreiner, B.; Want, L. *Diabetes Spectrum* **2004**, *17*, 183.

(65)   Vilsboll, T.; Krarup, T.; Deacon, C. F.; Madsbad, S.; Holst, J. J. *Diabetes* **2001**, *50*, 609.

(66)   Vilsboll, T.; Krarup, T.; Madsbad, S.; Holst, J. J. *Diabetologia* **2002**, *45*, 1111.

(67)   Holst, J. J.; Gromada, J.; Nauck, M. A. *Diabetologia* **1997**, *40*, 984.

(68)   Younan, S. M.; Rashed, L. A. *Gen. Physiol. Biophys.* **2007**, *26*, 181.

(69)   Lugari R.; Dei Cas A.; Ugolotti, D.; Finardi, L.; Barilli, A. L.; Ognibene, C.; Luciani, A.; Zandomeneghi, R.; Gnudi, A. *Horm. Metab. Res.* **2002**, *34*, 150.

(70) Rachman, J.; Gribble, F. M.; Barrow, B. A.; Levy, J. C.; Buchanan, K. D.; Turner, R. C. *Diabetes* **1996**, *45*, 1524.

(71) Dunning, B. E.; Foley, J.; Ahren, B. *Diabetologia* **2005**, *48*, 1700.

(72) Drucker, D. J. *Cell Metab.* **2006**, *3*, 153.

(73) Drucker D. J. *Mol. Endocrinol.* **2003**, *17*, 161.

(74) Holst, J. J. *Physiol. Rev.* **2007**, *87*, 1409.

(75) Tomas, E.; Wood, J. A.; Stanojevich, V.; Habener, J. F. *Diabetes Obes. Metab.* **2011**, *13*, 26.

(76) Holst, J. J.; Deacon, C. F. *Diabetes* **1998**, *47*, 1663.

(77) Abu-Hamdah, R.; Rabiee, A.; Meneilly, G. S.; Shannon, R. P.; Andersen, D. K.; Elahi, D. *J. Clin. Endocrinol. Metab.* **2009**, *94*, 1843.

(78) Tomas, E.; Habener, J. F. *Trends Endocrinol. Metab.* **2010**, *21*, 59.

(79) Hoare, S. R. J. *Drug Discov. Today* **2005**, *10*, 417.

(80) Barnett, A. *Expert Opin. Pharmacother.* **2007**, *8*, 2593.

(81) Ahren, B. *Nature Rev. Drug Discov.* **2009**, *8*, 369.

(82) Rachman, J.; Gribble, F. M.; Barrow, B. A.; Levy, J. C.; Buchanan, K. D.; Turner, R. C. *Diabetes* **1996**, *45*, 1524.

(83) Buse, J. B.; Henry, R. R.; Han, J.; Kim, D. D.; Fineman, M. S.; Baron A. D. *Diabetes Care* **2004**, *27*, 2628.

(84) DeFronzo, R. A.; Ratner, R. E.; Han, J.; Kim, D. D.; Fineman, M. S.; Baron, A. D. *Diabetes Care* **2005**, *28*, 1092.

(85) Piya, M. K.; Tahrani, A. A.; Barnett, A. H. *Br. J. Clin. Pharmacol.* **2010**, *70*, 631.

(86) Mentlein, R. *Regul. Pept.* **1999**, *85*, 9.

(87) Villhauer, E. B.; Brinkman, J. A.; Naderi, G. B.; Burkey, B. F.; Dunning B. E.; Prasad, K.; Mangold, B. L.; Russell, M. E.; Hughes, T. H. *J. Med. Chem.* **2003**, *46*, 2774.

(88) Deacon, C. F. *Diabetes Obes. Metab.* **2011**, *13*, 7.

(89) Edelman, S.; Maier, H.; Wilhelm, K. *BioDrugs* **2008**, *22*, 375.

(90) Hoppener, J. W. M.; Lips, C. J. M. *Int. J. Biochem. Cell Biol.* **2006**, *38*, 726.

(91) Lorenzo, A.;Yankner, B. A. *Neurobiol. Alzh. Dis.* **1996**, 89.

(92) Schmitz, O.; Brock, B.; Rungby, J. *Diabetes* **2004**, *53 (Suppl. 3)*, S233.

(93) Morfis, M.; Tilakaratne, N.; Furness, G. B.; Chirstopoulos, G.; Werry, T. D.; Chirstopoulos, A.; Sexton, P. M. *Endocrinol.* **2008**, *149*, 5423.

(94) Silvestre, R. A.; Rodríguez-Gallardo, J.; Jodka, C.; Parkes, D. G.; Pittner, R. A.; Young, A. A.; Marco, J. *Am J. Physiol. Endocrinol. Metab.* **2001**, *280*, E443.

(95) Houten, S. M.; Watanabe, M.; Auwerx, J. *EMBO J.* **2006**, *25*, 1419.

(96) Bell, D. S. H.; O'Keefe, J. H. *Diabetes Obes. Metab.* **2009**, *11*, 1114.

(97) Prawitt, J.; Caron, S.; Staels, B. *Curr. Diab. Rep.* **2011**, *11*, 160.

(98) Kelley, D. E.; Bray, G. A.; Pi-Sunyer, F. X.; Klein, S.; Hill, J.; Miles, J.; Hollander, P. *Diabetes Care*, **2002**, *25*, 1033.

(99) Holt, R. I. G.; Barnett, A. H.; Bailey, C. J. *Diabetes Obes. Metab.* **2010**, *12*, 1048.

(100) Via, M. A.; Chandra, H.; Arak, T.; Potenza, M. V.; Skamagas, M. *Diabetes, Metabolic Syndrome and Obesity: Targets and Therapy* **2010**, *3*, 43.

(101) For a recent review on targeting CNS for the treatment of T2DM, see: Sandoval, D. A.; Obici, S.; Seeley, R. J. *Nature Rev. Drug Disc.* **2009**, *8*, 386.

(102) Given the large number of approaches currently under development for the treatment of T2DM (235 medicines to treat diabetes and related conditions were at different stages of development in 2010, according to the 2010 Report by the America's Pharmaceutical Research Companies), the author's inclusion criteria for this section have been new approaches that have reached Phase II or III as of Q1 2011 and: (1) Positive Phase II clinical data are available or (2) Phase II with no clinical data but with significant activity in the field according to: Carpino, P.A.; Goodwin, B. *Expert Opin. Ther. Patents* **2010**, *20*, 1627.

(103)   Tahrani, A. A.; Bailey, C. J.; Del Prato, S.; Barnett, A. H. *Lancet* **2011**, *378*, 182.
(104)   Aicher, T. D.; Boyd, S. A.; McVean, M.; Celeste, A. *Expert Rev. Clin. Pharmacol.* **2010**, *3*, 209.
(105)   Gerich, J. E.; Woerle, H. J.; Meyer, C.; Stumvoll, M. *Diabetes Care* **2001**, *24*, 382.
(106)   Abdul-Ghani, M. A.; DeFronzo, R. A. *Endocr. Pract.* **2008**, *14*, 782.
(107)   For recent reviews on the field see: Bailey, C. J. *Trends Pharmacol. Sci.* **2011**, *32*, 63. Bailey, C. J.; Day, C. *Br. J. Diabetes Vasc. Dis.* **2010**, *10*, 193.
(108)   Neumiller, J. J.; White, J. R. Jr.; Campbell, R. K. *Drugs* **2010**, *70*, 377.
(109)   Chao, E. C.; Henry, R. R. *Nature Rev. Drug Discov.* **2010**, *9*, 551.
(110)   Lee, Y. J.; Han, H. J. *Kidney Int. Suppl.* **2007**, *72*, S27.
(111)   Ehrenkranz, J. R. L.; Lewis, N. G.; Kahn, G. R.; Roth, J. *Diabetes Metab. Res. Rev.* **2005**, *21*, 31.
(112)   Chasis, H.; Jolliffe, N.; Smith, H. *J. Clin. Invest.* **1933**, *12*, 1083.
(113)   Calado, J; Soto, K.; Clemente, C.; Correia, P.; Rueff, J. *Hum. Genet.* **2004**, *114*, 314.
(114)   Kinne, R. K. H; Castaneda, F. *Handbook Exp. Pharmacol* **2011**, *203*, 105.
(115)   Katsiki, N.; Papanas, N.; Mikhailidis, D. P. *Expert Opin. Investig. Drugs* **2010**, *19*, 1581.
(116)   Jadoon, K.; Idris, I. *Clin. Med. Insights: Therapeutics* **2011**, *3*, 185.
(117)   Kipnes, M. S. *Clin. Invest.* **2011**, *1*, 145.
(118)   Cock, T. A.; Houten, S. M.; Auwerx, J. *EMBO Rep.* **2004**, *5*, 142.
(119)   See, for example for MBX-102: Gregoire, F. M.; Zhang, F.; Clarke, H. J.; Gustafson,T. A.; Sears, D. D.; Favelyukis, S.; Lenhard, J.; Rentzeperis, D.; Clemens, L. E.; Mu, Y.; Lavan, B. E. *Mol. Endocrinol.* **2009**, *23*, 975.
(120)   Henriksen, K.; Byrjalsen, I.; Qvist, P.; Beck-Nielsen, H.; Hansen, G.; Riis, B. J.; Perrild, H.; Svendsen, O. L.; Gram, J.; Karsdal, M. A.; Christiansen, C. *Diabetes Metab. Res. Rev.* **2011**, *27*, 392.
(121)   Issemann, I.; Green, S. *Nature* **1990**, *347*, 645.
(122)   Hiukka, A.; Maranghi, M.; Matikainen, N.; Taskinen, M.-R. *Nat. Rev. Endocrinol.* **2010**, *6*, 454.
(123)   Pal, L. *DDT* **2009**, *14*, 784.
(124)   Jones, R. M.; Leonard, J. N.; Buzard, D. J.; Lehmann, J. *Expert Opin. Ther. Patents* **2009**, *19*, 1339.
(125)   Levin, G. V.; Zehener, L. R.; Saunders, J. P. *Am. J. Clin. Nutr.* **1995**, *62 (Suppl)*, 1161S.
(126)   Espinosa, I.; Fogelfeld, L. *Expert Opin. Investig. Drugs* **2010**, *19*, 285.
(127)   Boyle, C. D. *Curr. Opin. Drug Discov. Develop.* **2008**, *11*, 495.
(128)   Hollis, G.; Huber, R. *Diabetes Obesity Metab.* **2011**, *13*, 1.
(129)   Rosenstock, J.; Banarer, S.; Fonseca, V. A.; Inzucchi, S. E.; Sun, W.; Yao, W. W.; Hollis, G.; Flores, R.; Levy, R.; Williams, W. V.; Seckl, J. R.; Huber, R. *Diabetes Care* **2010**, *33*, 1516.
(130)   Birch, A. M.; Buckett, L. K.; Turnbull, A. V. *Curr. Opin. Drug Disc. Develop.* **2010**, *13*, 489.
(131)   Expert Committee on the Diagnosis and Classification of Diabetes Mellitus. Report of the Expert Committee on the Diagnosis and Classification of Diabetes Mellitus. *Diabetes Care* **2002**, *25*, S5.
(132)   Steffes, M. W.; Chavers, B. M.; Molitch, M. E.; Cleary, P. A.; Lachin, J. M.; Genuth, S.; Nathan, D. M.; Engel, S.; Martinez, H.; Shamoon, H.; et al. The Writing Team for the Diabetes Control and Complications Trial/Epidemiology of Diabetes Interventions and Complications Research Group. *J. Am. Med. Assoc.* **2003**, *290*, 2159.
(133)   Canadian Diabetes Association. Clinical Practice Guidelines for The Prevention and Management of Diabetes in Canada, *Can. J. Diabetes* **2008**, *32*, S1.
(134)   González- Sánchez, J. L.; Serrano-Ríos, M. *Drug News Perspect.* **2007**, *20*, 527.
(135)   Poitout, V.; Stein, R.; Rhodes, C. J. In *Int. Textbook of Diabetes Mellitus*, DeFronzo R. A.; Ferrannini E.; Keen H.; Zimmet P. eds., Wiley: Hoboken, NJ, 2004, p. 97.
(136)   Pessin, J. E.; Thurmond, D. C.; Elmendorf, J. S.; Coker, K. J.; Okada, S. *J. Biol. Chem.* **1999**, *274*, 2593.
(137)   Banting, F. G.; Best, C. H.; Collip, J. B.; Campbell, W. R.; Fletcher, A. A. *CMAJ* **1922**, *12*, 141.
(138)   Sheldon, B.; Russell-Jones, D.; Wright, J. *Diabetes Obes. Metab.* **2009**, *11*, 5.
(139)   Hagedorn, H. C.; Jensen, B. N.; Krarup, N. B.; Wodstrug, I. *J. Am. Med. Assoc.* **1936**, *106*, 177.

(140)   Hallas-Moller, K. *Diabetes* **1956,** *5,* 7.

(141)   Murray, I.; Wilson, R. B. *Br. Med. J.* **1953,** *2,* 1023.

(142)   Ryle, A. P.; Sanger, F.; Smith, L. F.; Kitai, R. *Biochem. J.* **1955,** *60,* 541.

(143)   Hodgkin, D. C. *Diabetes* **1972,** 1131.

(144)   De Meyts, P.; Van Obberghen, E.; Roth, J.; Brandenburg, D.; Wollmer, A. *Nature* **1978,** *273,* 504.

(145)   Vajo, Z.; Fawcett, J.; Duckworth, W. *Endocrinol. Rev.* **1994,** *22,* 706.

(146)   Howey, D. C.; Bowsher, R. R.; Brunelle, R. L.; Woodworth, J. R. *Diabetes* **1994,** *43,* 396.

(147)   Kurtzhals, P.; Schaffer, L.; Sorensen, A.; Kristensen, C.; Jonassen, I.; Schmid, C.; Trub, T. *Diabetes* **2000,** *49,* 999.

(148)   Chase, H. P.; Lockspeiser, T.; Peery, B.; Shepherd, M.; MacKenzie, T.; Anderson. J.; Garg, S. K. *Diabetes Care* **2001,** *24,* 430.

(149)   Torlone, E.; Fanelli, C.; Rambotti, A. M.; Kassi, G.; Modarelli, E.; Di Vincenzo, A.; Epifano, L.; Ciofetta, M.; Pampanelli, S.; Brunetti, P.; Bolli, G. B. *Diabetologia* **1994,** *37,* 713.

(150)   Pfutzner, A.; Kustner, E.; Forst, T.; Schulze-Schleppinghoff, B.; Trautmann, M. E.; Haslbeck, M.; Schatz, H.; Beyer, J. *Exp. Clin. Endocrinol. Diabetes* **1996,** *104,* 25.

(151)   Anderson, J. H. Jr.; Brunelle, R. L.; Koivisto, V. A.; Pfutzner, A.; Trautmann, M. E.; Vignati, L.; DiMarchi, R. *Diabetes* **1997,** *46,* 265.

(152)   Mudaliar, S.-R.; Lindberg, F. A.; Joyce, M.; Beerdsen, P.; Strange, P.; Lin, A.; Henry, R. R. *Diabetes Care* **1999,** *22,* 1501.

(153)   Hedman, C. A.; Lindston, T.; Arnqvist, H. J. *Diabetes Care* **2001,** *24,* 1120.

(154)   Raskin, P.; Guthrie, R. A.; Leiter, L.; Riis, A.; Jovanovic, L. *Diabetes Care* **2000,** *23,* 583.

(155)   Brunner, G. A.; Hirschberger, S.; Sendlhofer, G.; Wutte, A.; Ellmerer, M.; Balent, B.; Schaupp, L.; Krejs, G. J.; Pieber, T. R. *Diabet. Med.* **2000,** *17,* 371.

(156)   Siebenhofer A.; Plank J.; Berghold A.; Jeitler K.; Horvath, K.; Narath, M.; Gfrerer, R.; Pieber, T. R. *Cochrane Database Syst. Rev.* **2006.**

(157)   Garg, S. K.; Ellis, S. L.; Ulrich, H. *Expert Opin. Pharmacother.* **2005,** *6,* 643.

(158)   Becker, R. H. A.; Frick, A. D.; Burger, F.; Scholtz, H.; Potgieter, J. H.: *Exp. Clin. Endocrinol. Diabetes* **2005,** *113,* 292.

(159)   Arnolds, S.; Rave, K.; Hövelmann, U.; Fischer, A.; Sert-Langeron, C.; Heise, T. *Exp. Clin. Endocrinol. Diabetes* **2010,** *118,* 662.

(160)   Jehle, P. M.; Micheler, C.; Jehle, D. R.; Breitig, D.; Boehm, B. O. *Lancet* **1999,** *354,* 1604.

(161)   Bahr, M.; Kolter, T.; Seipke, G.; Eckel, J. *Eur. J. Pharmacol.* **1997,** *320,* 259.

(162)   Bolli, G. B.; Owens, D. R. *Lancet* **2000,** *356,* 443.

(163)   Lepore, M.; Pampanelli, S.; Fanelli, C.; Porcellati, F.; Bartocci, L.; Di Vincenzo, A.; Cordoni, C.; Costa, E.; Brunetti, P.; Bolli, G. B. *Diabetes* **2000,** *49,* 2142.

(164)   Owens, D. R.; Coates, P. A.; Luzio, S. D.; Tinbergen, J. P.; Kurzhals, R. *Diabetes Care* **2000,** *23,* 813.

(165)   Horvath, K.; Jeitler, K.; Berghold, A.; Ebrahim, S. H.; Gratzer, T. W.; Plank, J.; Kaiser, T.; Pieber, T. R.; Siebenhofer, A. *Cochrane Database Syst. Rev.* **2007.**

(166)   Ratner, R. E.; Hirsch, I. B.; Neifing, J. L.; Garg, S. K.; Mecca, T. E.; Wilson, C. A. *Diabetes Care* **2000,** *23,* 639.

(167)   Yki-Jarvinen, H.; Dressler, A.; Ziemen, M. *Diabetes Care* **2000,** *23,* 1130.

(168)   Riddle, M. C.; Rosenstock, J.; Gerich, J. *Diabetes Care* **2003,** *26,* 3080.

(169)   Fulcher, G. R.; Gilbert, R. E.; Yue, D. K. *Intern. Med.* **2005,** *35,* 536.

(170)   Rosenstock, J.; Schwartz, S. L.; Clark, C. M. Jr.; Park, G. D.; Donley, D. W.; Edwards, M. B. *Diabetes Care* **2001,** *24,* 631.

(171)   Markussen, J.; Havelund, S.; Kurtzhals, P.; Andersen, A. S. *Diabetologia* **1996,** *39,* 281.

(172)   Havelund, S.; Plum, A.; Ribel, U.; Jonassen, I.; Volund, A.; Markussen, J.; Kurtzhals, P. *Pharm. Res.* **2004,** *21,* 1498.

(173)   Plank, J.; Bodenlenz, M.; Sinner, F.; Magnes, C.; Gorzer, E.; Regittnig, W.; Endahl, L. A.;
        Draeger, E.; Zdravkovic, M.; Pieber, T. R. *Diabetes Care* **2005**, *28*, 1107.
(174)   Hermansen, K.; Davies, M.; Derezinski, T.; Martinez, R. G.; Clauson, P.; Home, P. *Diabetes Care*
        2006, *29*, 1269.
(175)   Philis-Tsimikas, A.; Charpentier, G.; Clauson, P.; Martinez Ravn, G.; Roberts, W. L.;
        Thorsteinsson, B. *Clin. Ther.* **2006**, *28*, 1569.
(176)   Hermansen, K.; Davies, M. *Diabetes Obes. Metab.* **2007**, *9*, 209.
(177)   Tschritter, O.; Hennige, A. M.; Preissl, H.; Porubska K.; Schafer S. A.; Lutzenberger W.; Machicao
        F.; Birbaumer N.; Fritsche A.; Haring H-U. *PLoS One* **2007**, *2*, e1196.
(178)   Roach, P.; Woodwoth, J. R. *Clin. Pharmacokinet.* **2002**, *41*, 1043.
(179)   Hompesch, M.; Ocheltree, S. M.; Wondmagegnehu, E.; Morrow, L.; Kollmeier, A.; Campaigne,
        B.; Jacober, S. J. *Curr. Med. Res. Opin.* **2009**, *25*, 2679.
(180)   Roach, P.; Strack, T.; Arora, V.; Zhao, Z. *Int. J. Clin. Pract.* **2001**, *55*, 177.
(181)   Fogelfeld, L.; Dharmalingam, M.; Robling, K.; Jones, C.; Swanson, D.; Jacober, S. *Diabetic Med.*
        2010, *27*, 181.
(182)   Strojek, K.; Shi, C.; Carey, M. A.; Jacober, S. J. *Diabetes Obes. Metab.* **2010**, *12*, 916.
(183)   Esposito, K.; Ciotola, M.; Maiorino, M.; Gualdiero, R.; Schisano, B.; Ceriello, A.; Beneduce, F.;
        Feola, G.; Giugliano, D. *Ann. Intern. Med.* **2008**, *149*, 531.
(184)   Chacra, A.; Kipnes, M.; Ilag, L.; Sarwat, S.; Giaconia, J.; Chan, J. *Diabetic Med.* **2010**, *27*, 563.
(185)   Garber, A. J.; Ligthelm, R.; Christiansen, J. S.; Liebl, A. *Diabetes, Obes. Metab.* **2007**, *9*, 630.
(186)   Owens, D. R. *Nat. Rev. Drug Discov.* **2002**, *1*, 529.
(187)   Still, J. G. *Diabetes Metab. Res. Rev.* **2002**, *18* , S29.
(188)   Laube, B. L. *Chest* **2001**, *120*, 99.
(189)   Cassidy, J. P.; Amin, N.; Marino, M.; Gottfried, M.; Meyer, T.; Sommerer, K.; Baughman, R. A.
        Pharm. Res. **2011**, *28*, 2157.
(190)   Neumiller, J. J.; Campbell, R. K.; Wood, L. D. *Ann. Pharmacother.* **2010**, *44*, 1231.
(191)   Muchmore, D. B.; Vaughn, D. B.; *J. Diabetes Sci. Technol.* **2010**, *4*, 419.
(192)   Hompesch, M.; Muchmore, D. B.; Morrow, L.; Vaughn, D. E. *Diabetes Care* **2011**, *34*, 666.
(193)   Birkeland, K. I.; Home, P. D.; Wendisch, U.; Ratner, R. E.; Johansen, T.; Endahl, L. A.; Lyby, K.;
        Jendle, J. H.; Roberts, A. P.; DeVries, J. H.; Meneghini, L. F. *Diabetes Care* **2011**, *34*, 661.
(194)   Zinman, B.; Fulcher, G.; Rao, P. V.; Thomas, N.; Endahl, L. A.; Johansen, T.; Lindh, R.; Lewin,
        A.; Rosenstock, J.; Pinget, M.; Mathieu, C. *Lancet* **2011**, *377*, 924.
(195)   Heise, T.; Tack, C. J.; Cuddihy, R.; Davidson, J.; Gouet, D.; Liebl, A.; Romero, E.; Mersebach, H.;
        Dykiel, P.; Jorde, R. *Diabetes Care* **2011**, *34*, 669.
(196)   Eldor, R.; Kassem, S.; Raz, I. *Diabetes Metab. Res. Rev.* **2009**, *25*, 316.
(197)   Orban, T.; Bundy, B.; Becker, D. J.; DiMeglio, L. A.; Gitelman, S. E.; Goland, R.; Gottlieb, P. A.;
        Greenbaum, C. J.; Marks, J. B.; Monzavi, R.; Moran, A.; Raskin, P.; Rodriguez, H.; Russell, W. E.;
        Schatz, D.; Wherrett, D.; Wilson, D. M.; Krischer, J. P.; Skyler, J. S., and the Type 1 Diabetes
        TrialNet Abatacept Study Group. *Lancet* **2011**, *378,* 412.
(198)   Jun, H. S.; Khil, L.Y.; Yoon, J. W. *Cell Mol. Life Sci.* **2002**, *59*, 1892.
(199)   Ludvigsson, J. *Expert Opin. Biol. Ther.* **2010**, *10*, 787.
(200)   Atkinson, M. A.; Kaufman, D. L.; Campbell, L.; Gibbs, K. A.; Shah, S. C.; Bu, D. F.; Erlander, M.
        G.; Tobin, A. J.; Maclaren, N. K. *Lancet* **1992**, *339*, 458.
(201)   Bingley, P. J.; Bonifacio, E.; Williams, A. J.; Genovese, S.; Bottazzo, G. F.; Gale, E. A. *Diabetes*
        1997, *46*, 1701.
(202)   Ludvigsson, J.; Faresjo, M.; Hjorth, M.; Axelsson, S.; Cheramy, M.; Pihl, M.; Vaarala, O.;
        Forsander, G.; Ivarsson, S.; Johansson, C.; Lindh, A.; Nilsson, N-O.; Aman, J.; Ortgvist, E.;
        Zerhouni, P.; Casas, R. *N. Engl. J. Med.* **2008**, *359*, 1909.
(203)   Ludvigsson, J.; Hjorth, M.; Cheramy, M.; Axelsson, S.; Pihl, M.; Forsander, G.; Nilsson, N.-O.;
        Samuelsson, B.-O.; Wood, T.; Aman, J.; Ortgvist, E.; Casas, R. *Diabetologia* **2011**, *54*, 634.

(204)   Hjorth, M.; Axelsson, S.; Rydén, A.; Faresjö, M.; Ludvigsson, J.; Casas, R. *Clin. Immunol.* **2011**, *138*, 117.

(205)   Axelsson, S.; Hjorth, M.; Akerman, L.; Ludvigsson, J.; Casas, R. *Diabetes Metab. Res. Rev.* **2010**, *26*, 559.

(206)   Pescovitz, M. D.; Greenbaum, C. J.; Krause-Steinrauf, H.; Becker, D.; Gitelman, S. E.; Goland, R.; Gottlieb, P. A.; Marks, J. B.; McGee, P. F.; Moran, A. M.; Raskin, P.; Rodriguez, H.; Schatz, D. A.; Wherrett, D.; Wilson, D. M.; Lachin, J. M.; Skyler, J. S. *N. Engl. J. Med.* **2009**, *361*, 2143.

(207)   Fischer, B.; Elias, D.; Bretzel, R. G.; Linn, T. *Expert Opin. Biol. Ther.* **2010**, *10*, 265.

(208)   Herold, K. C.; Gitelman, S.; Greenbaum, C.; Puck, J.; Hagopian, W.; Gottlieb, P.; Sayre, P.; Bianchine, P.; Wong, E.; Seyfert-Margolis, V.; Bourcier, K.; Bluestone, J. A. *Clin. Immunol.* **2009**, *132*, 166.

(209)   Keymeulen, B.; Vandemeulebroucke, E.; Ziegler, A. G.; Mathieu, C.; Kaufman, L.; Hale, G.; Gorus, F.; Goldman, M.; Walter, M.; Candon, S.; Schandene, L.; Crenier, L.; De Block, C.; Seigneurin, J-M.; De Pauw, P.; Pierard, D.; Weets, I.; Rebello, P.; Bird, P.; Berrie, E.; Frewin, M.; Waldmann, H.; Bach, J-F.; Pipeleers, D.; Chatenoud, L. *N. Engl. J. Med.* **2005**, *352*, 2598.

(210)   Herold, K. C.; Bluestone, J. A.; Montag, A. G.; Parihar, A.; Wiegner, A.; Gress, R. E.; Hirsch, R. *Diabetes* **1992**, *41*, 385.

(211)   Herold, K. C.; Hagopian, W.; Auger, J. A.; Poumian-Ruiz, E.; Taylor, L.; Donaldson, D.; Gitelman, S. E.; Harlan, D. M.; Xu, D.; Zivin, R. A.; Bluestone, J. A. *N. Engl. J. Med.* **2002**, *1*, 1692.

(212)   Herold, K. C.; Gitelman, S. E.; Masharani, U.; Hagopian, W.; Bisikirska, B.; Donaldson, D.; Rother, K.; Diamond, B.; Harlan, D. M.; Bluestone, J. A. *Diabetes* **2005**, *54*, 1763.

(213)   Kaufman, A.; Herold, K. C. *Diabetes Metab. Res. Rev.* **2009**, *25*, 302.

(214)   Cui, P.; Macdonald, T. L.; Chen, M.; Nadler, J. L. *Bioorg. Med. Chem. Lett.* **2006**, *16* , 3401.

(215)   Metabolic surgery is emerging as a potential treatment for T2 diabetics. However, not being a drug treatment, it was out of the scope of this review. See reference 103.

(216)   Muller, G. *Pharmacology* **2010**, *85*, 168.

(217)   Wang, T. J.; Larson, M. G.; Vasan, R. S.; Cheng, S.; Rhee, E. P.; McCabe, E.; Lewis, G. D.; Fox, C. S.; Jacques, P. F.; Fernandez, C.; O'Donnell, C. F.; Carr, S. A.; Mootha, V. K.; Florez, J. C.; Souza, A.; Melander, O.; Clish, C. B.; Gerszten, R. E. *Nature Med.* **2011**, *17*, 448.

7

CNS Drugs

John A. Lowe, III

1   Introduction

The unfathomable complexity of the human brain presents a formidable challenge to the discovery of new drugs for central nervous system (CNS) disease. Our survey of these drugs will describe the prominent role serendipity has played in their genesis and will make it clear that testing in human patients often initiated their discovery. Indeed, studies of their mechanism of action form the basis for much of what we know about mental disease. Animal models based on the prototype drugs allowed the discovery of numerous improved versions. But clinical observations in human patients have remained the most reliable method for CNS drug discovery to this day.

The DSM-IV-TR manual (*Diagnostic and Statistical Manual of Mental Disorders*, 4th Edition, Text Revision, 2000) provides detailed descriptions of the diseases of the CNS, to which the reader is referred for a thorough account of this subject.[1] It has served as the basis for diagnosis in psychiatry for the past two decades and is now undergoing a revision in preparation for the fifth edition in 2013. As it represents the current view of CNS diseases, even those that are controversial, it will serve as the basis for classifications of diseases in this chapter. DSM-IV-TR describes many categories of CNS diseases, but we will concentrate on the following ones: schizophrenia and other psychotic disorders, mood disorders (including depression and bipolar disorder), anxiety disorders, substance-related disorders (substance abuse), analgesia, epilepsy, attention-deficit and hyperactivity disorder (ADHD), and neurodegenerative disorders.

Drugs that act in the CNS have played an important cultural and therapeutic role throughout human history. Alcohol, nicotine, marijuana, and opium, as well as many of the common CNS drugs, are recognizable elements of our culture. The following survey is not an exhaustive list or catalog of CNS drugs, but rather an overall perspective. We will focus on the background and history of the discovery of the major drugs in each disease category and review their uses and the light they shed on the etiology of the disease they treat.

2   Antipsychotic Drugs

For countless generations, treatment for schizophrenia consisted of isolation, restraint, lobotomy, and/or denial. Although one can get a glimpse today of the burden of this disease on society in the street people of major American cities, it is hard now to imagine the time when 1–2% of the entire population suffered from this disease without the benefit of psychotherapeutic medication.

Schizophrenia is characterized by thought disorders, generally delusions of persecution and hallucinations, as well as deficits in cognition and memory, classified as positive and cognitive symptoms, respectively.[2] The disease can progress to negative symptoms, such as a complete lack of social engagement, and often involves suicidal ideation and other affective symptoms. Given its complexity and the enormous panorama of these four symptom domains, present in unique combinations in each patient, the difficulty of finding suitable treatment is understandable.

The revolution in psychiatry brought about by the first antipsychotic drugs generated a vigorous effort to explain their efficacy, which was eventually traced to the neurotransmitter dopamine.[3] Controlling the overactivity or underactivity of dopaminergic transmission seems to be the key for controlling the positive symptoms of the disease.[4] Though this comes at a price, as described below, it makes patients easier to control and even allows them some function, though only rarely returning them to a normal life in society. More recent efforts to understand the basis for the disease have identified some of its apparent genetic basis, but, while neurodevelopmental defects are thought to be involved, the basis for the disease remains a largely unsolved problem.

2.1   *Typical Antipsychotics—The First Generation*

Like many classes of currently available drugs, the first antipsychotic agents were discovered through observations in human subjects. The original chemical structures go back to the studies of Daniel Bovet with the phenothiazine template and their activity as antihistamines, work for which he was awarded the Nobel Prize in Medicine in 1957.[5] Bovet began work on phenothiazine-based dye structures in the 1930s, following the discovery of the antibiotic sulfanilamide, which was based on another class of dye compounds. His first success came with a compound labeled F929, which protected mice from lethal histamine-induced anaphylaxis. While further research went on to develop the antihistamine drugs used for treating allergic rhinitis, the observation of sedative effects in some of the early compounds led to continued work in this latter direction.

Chlorpromazine (Thorazine) was first prepared at Rhone-Poulenc in 1950 and provided it to physicians to assess its clinical effects in 1951.[6] Its remarkable ability to calm psychotic patients was observed serendipitously and reported by Delay and Deniker in 1952, and it was subsequently introduced into medical practice as the drug Thorazine.[7] This development opened a new era with the realization that chemotherapy could play a crucial role in the treatment of CNS disease. It allowed many patients to leave the hospital and resume a life in society. It was quickly followed by the development of numerous similar drugs, described below.

Chlorpromazine also helped bring about a revolution in our understanding of the central nervous system through research into the drug's mechanism of action. This work enabled the discovery of the role of the neurotransmitter dopamine in regulating reward, cognitive behavior, and mood through its receptors in the brain. It also specifically identified the D_2 receptor, one of five dopamine receptor subtypes, as the relevant pharmacological target for chlorpromazine, and all the subsequent marketed antipsychotic drugs and helped build a tradition of animal testing based on behaviors induced by pharmacological treatment to mimic human disease.

The drugs that followed chlorpromazine made structural changes to the pendant basic amine group as well as the template, as shown below[8]:

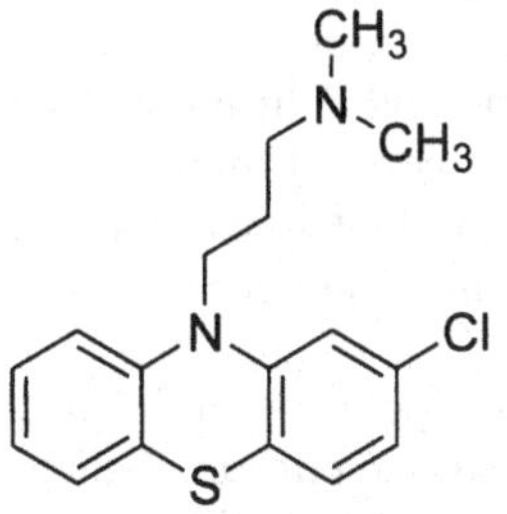

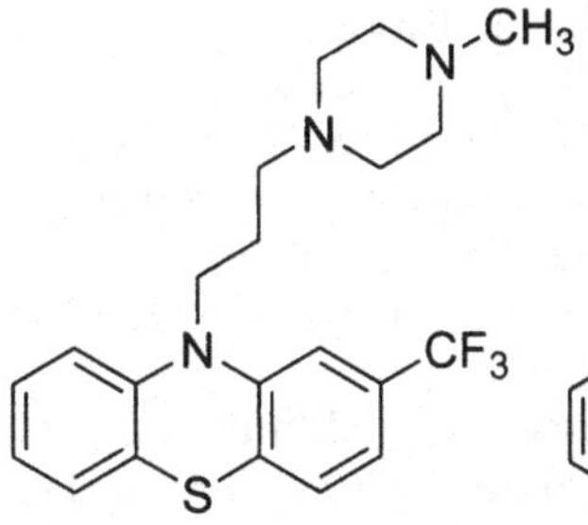

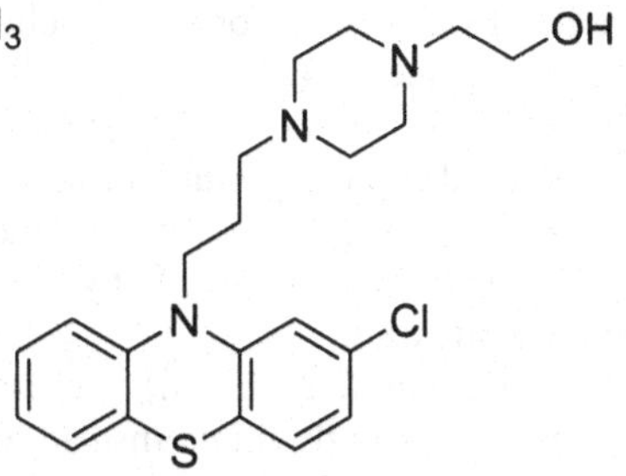

chlorpromazine (Thorazine)   trifluperazine (Stelazine) perphenazine (Trilafon)

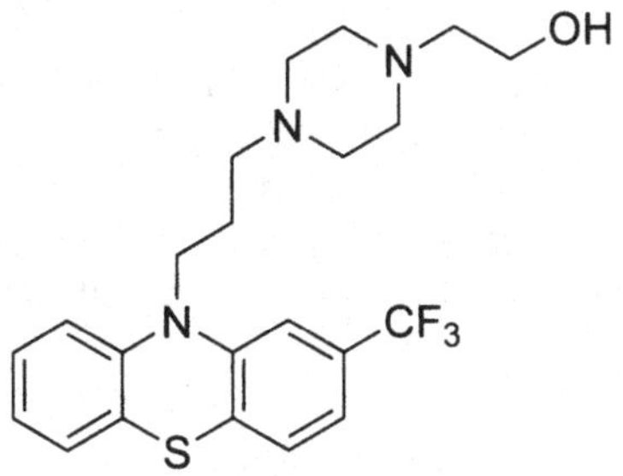

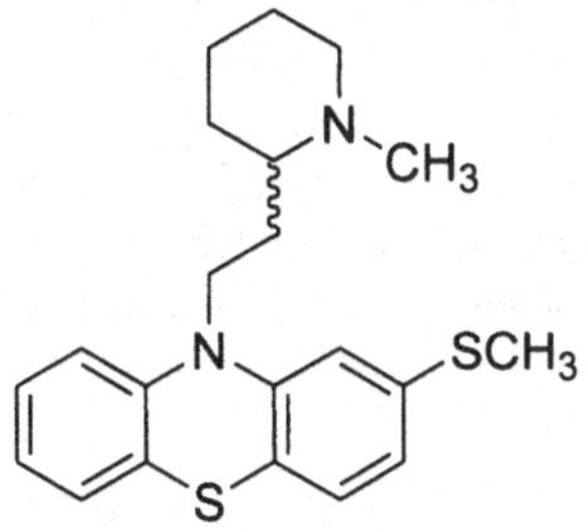

fluphenazine (Sinqualone)   thioridazine (Melleril)

The discovery of the antipsychotic drug haloperidol illustrates the methodology used to find a new structural class of antipsychotic agents, which has been applied many times since. Following the discovery of the antipsychotic efficacy of chlorpromazine, Paul Janssen designed animal models to reproduce the symptoms of schizophrenia and replicate the therapeutic effects of chlorpromazine and then screened a series of compounds that would imitate and improve on the prototype. The efficiency of this process was such that, in one year, 1958, haloperidol was synthesized, tested in humans, and shown to be as effective as chlorpromazine but far more potent.[9] Approval for marketing was granted in Europe and subsequently in the U.S. in 1967 and haloperidol became a mainstay of antipsychotic therapy in the years that followed.

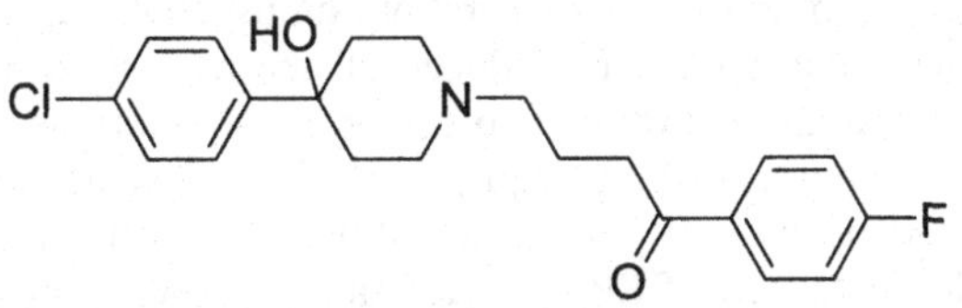

haloperidol (Haldol)

2.2   Atypical Antipsychotics—The Second Generation

Despite the profound change in psychiatric medicine that resulted from the introduction of the first-generation antipsychotic agents, side effects plague their long-term use. These include extrapyramidal side effects, or EPS (including Parkinsonism, tremor, and lack of motor coordination), tardive dyskinesia (a persistent, disfiguring motor effect with loss of muscle control), and neuroleptic malignant syndrome (which includes fever, muscle rigidity, and delirium and which may be fatal). These issues prompted a concerted effort in the pharmaceutical industry to find antipsychotic drugs with a better side-effect profile as a follow-up to the first-generation agents. The effort was stimulated by the discovery of clozapine, the first so-called "atypical" antipsychotic drug.

Clozapine (Clozaril) was first prepared in the early 1960s as a potential antidepressant but was subsequently found instead to be an antipsychotic and introduced into clinical medicine in 1971.[10] It was the first antipsychotic virtually devoid of EPS and the resulting tardive dyskinesia and has thus been classified as the first atypical, or second-generation, antipsychotic drug. In addition, it shows unique therapeutic benefits, with efficacy in disease resistant to treatment with other antipsychotic agents, and an ability to decrease the incidence of suicide, a serious issue with this disease. Despite this very promising profile, it had to be withdrawn from the market due to rare but fatal incidents of agranulocytosis, a depletion of white blood cells. Because of its favorable therapeutic profile, however, it was reinstated in the clinic in 1989, provided blood monitoring was employed to address potentially fatal consequences of agranulocytosis. Another downside to clozapine is the tendency to cause weight gain, eventually leading to obesity and symptoms of diabetes. It remains nonetheless an agent with a unique therapeutic benefit.

clozapine (Clozaril)   olanzapine (Zyprexa)

The discovery of clozapine stimulated efforts to replicate its therapeutic benefits and lack of EPS in a new drug, especially following its initial withdrawal from the market.[11] Its pharmacological profile, however, is complex enough to have stymied all subsequent efforts to reproduce it exactly, and consequently, the other second-generation antipsychotic agents, while they improve on the first generation, have never quite reproduced all the benefits of clozapine. One obvious approach to reproducing clozapine's profile is to mimic its structure as closely as possible, an effort that resulted in the development of olanzapine, with its thiophene-for-benzene ring bioisosteric replacement.[12] Olanzapine has demonstrated antipsychotic efficacy at much lower doses than clozapine and reproduces its lack of EPS. In addition, its lower dose may be a factor in its lack of the agranulocytosis side effect, although it retains the weight gain and metabolic issues of the prototype. Even though olanzapine does not seem to reproduce all the benefits of clozapine in clinical practice, it does provide valuable therapeutic benefit and has been the leading selling antipsychotic drug for many years.[13]

One of the main rationales for the lack of motor side effects with clozapine is its ratio of binding activity at the 5-HT$_2$ receptor for serotonin compared with that at the dopamine D$_2$ receptor.[11] In theory, then, a favorable ratio of 5-HT$_2$/D$_2$ blockade will ameliorate D$_2$-receptor-mediated EPS. The first drug to use this rationale was risperidone, another discovery from Paul Janssen's laboratory, using many of the same techniques used to discover haloperidol. Risperidone (Risperdal) retains the 4-substituted piperidine of the prototype but cyclizes the butyrophenone, shortening the side chain. Its profile is similar to the prototype, but its favorable 5-HT$_2$/D$_2$ ratio presumably plays a role in its reduced motor side-effect profile.[14]

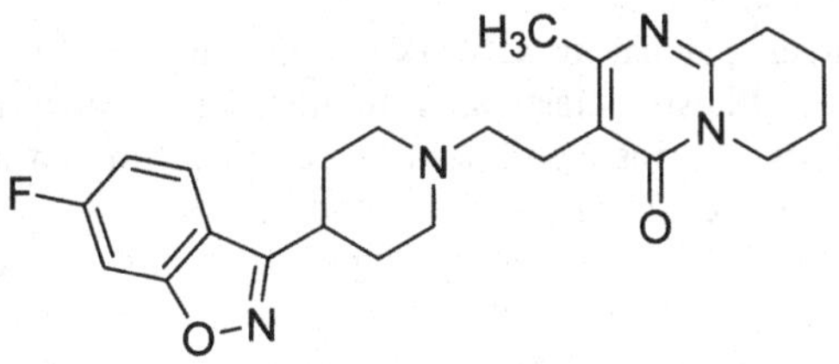

risperidone (Risperdal) 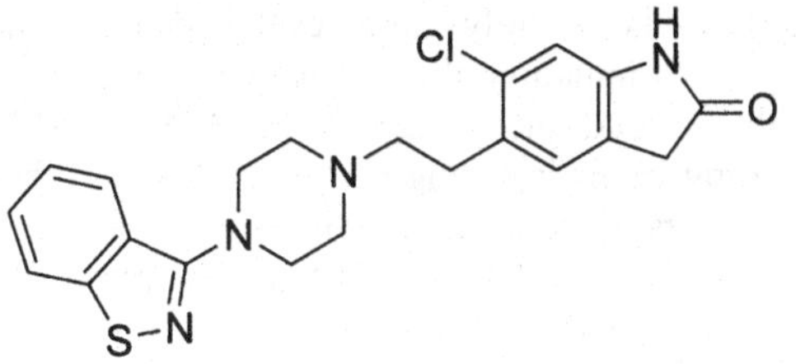   ziprasidone (Geodon)

Ziprasidone (Geodon) starts with a benzisothiazolyl piperazine, originally designed as a mimic of the core structure of clozapine,[15] and adds a novel oxindole side chain.[16] Its profile is dominated by its potent 5-HT$_2$ receptor blockade, and it has the expected low side-effect liability.[17] In addition, it does not induce weight gain and other metabolic abnormalities seen in many of the other atypical agents, possibly due to its weak H$_1$-receptor blockade,[18] and this benefit may help many patients.

Quetiapine (Seroquel) harkens back to the original phenothiazine series but adds an element to its template from clozapine and then adds an extended version of the side chain from fluphenazine.[19] Its activity in animal models of schizophrenia and motor side effects predicts a favorable profile,[20] confirmed by its efficacy as an antipsychotic and its position as a leading selling drug. In addition, it has branched out as a treatment for bipolar disorder.

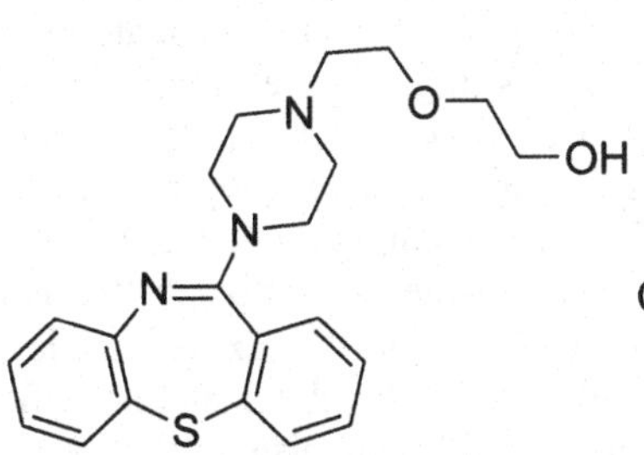

quetiapine (Seroquel) 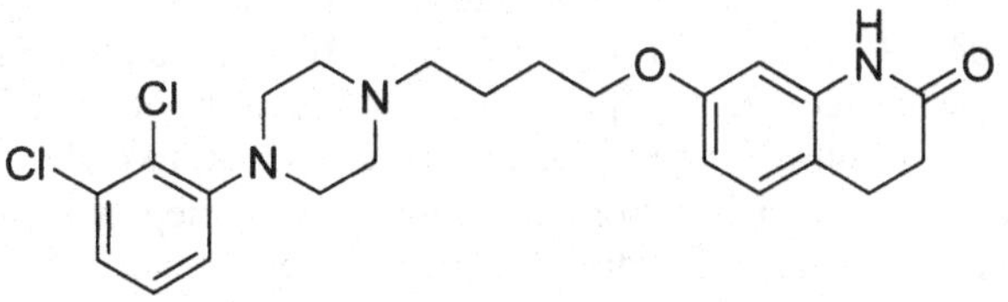   aripiprazole (Abilify)

Aripiprazole (Abilify) defies the profile of all the preceding antipsychotics with partial agonism, rather than antagonism, at the D$_2$-receptor. Combined with potent 5-HT$_2$ receptor antagonism, this activity is a different approach to reducing the first generation's motor side effects.[21] Its efficacy and lack of weight gain and EPS make it another of the leading selling drugs in this category.

2.3   *Recent Development of Newer Agents*

With so many marketed antipsychotic drugs that work through D_2-receptor antagonism, research has begun to investigate new mechanisms that could offer improvements over existing drugs. Potential benefits of these approaches include reduced side effects and/or treatment of additional symptom domains, such as negative and cognitive symptoms. Several potential mechanistically novel agents currently being developed are summarized below.

Stimulation of the dopamine system involves a signaling network that includes elevation of the second messenger cyclic AMP (cAMP). Phosphodiesterases regulate dopaminergic signaling by hydrolyzing cAMP to AMP. Inhibitors of phosphodiesterases can therefore act as modulators of the dopamine signaling network. There are 22 types of phosphodiesterase enzymes, and one of these, PDE10A, is a primary player in regulating dopaminergic signaling in areas of the brain such as the striatum that are involved in schizophrenia. PDE10A inhibitors have shown activity in animal models of schizophrenia, and compound MP-10 shown below has been advanced to clinical trials.[22]

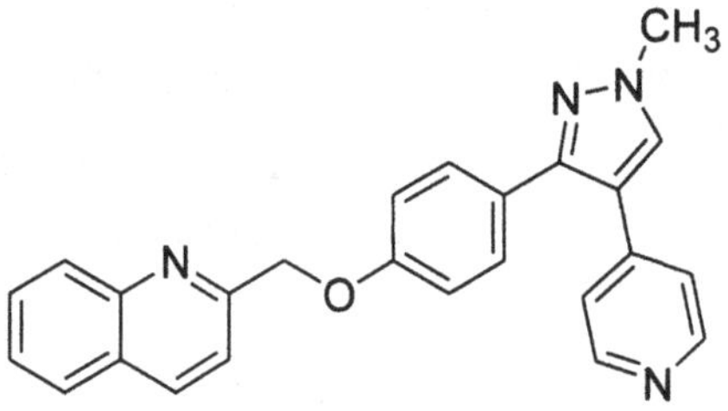

MP-10

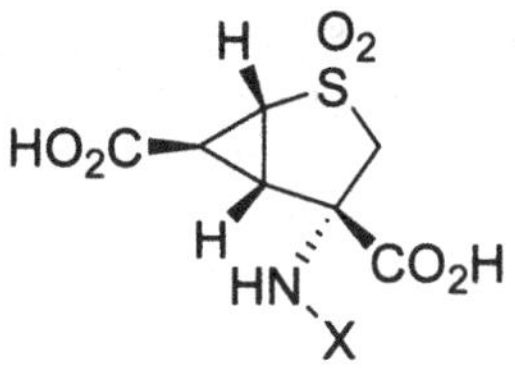

LY404039, X = H
LY2140023, X = $COCH(NH_2)CH_2CH_2SCH_3$

LY404039 is a selective agonist for metabotropic glutamate receptors belonging to subgroup II, which includes mGluR2 and mGluR3.[23] These receptors regulate glutamate release presynaptically, and may play a role in schizophrenia since deficits in glutamatergic transmission are thought to underlie the etiology of the disease. A Phase II clinical trial with a prodrug of LY404039 coded LY2140023, showed efficacy against positive and negative symptoms with no significant side effects, representing considerable potential for a novel antipsychotic treatment.[24,25]

Another of the glutamate receptors that may offer a novel target for schizophrenia is the AMPA (α-amino-3-hydroxy-5-methyl-4-isoxazolepropanoic acid) receptor.[26] Expression levels of these receptors are reduced in schizophrenic brain and may underlie the glutamatergic deficits of the disease.[27] Rather than risk side effects that could result from direct agonism of the receptor, AMPA positive allosteric modulators (PAMs) offer a better alternative by binding to an allosteric site on the receptor to slow agonist dissociation and thereby increase AMPA signaling.[28] An example of this approach is the so-called AMPAkine CX-516, which has shown positive results in a clinical trial in schizophrenia in combination with olanzapine.[29]

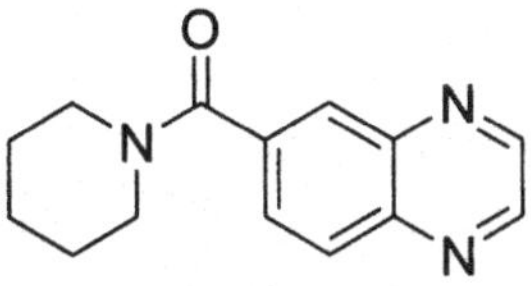

CX-516

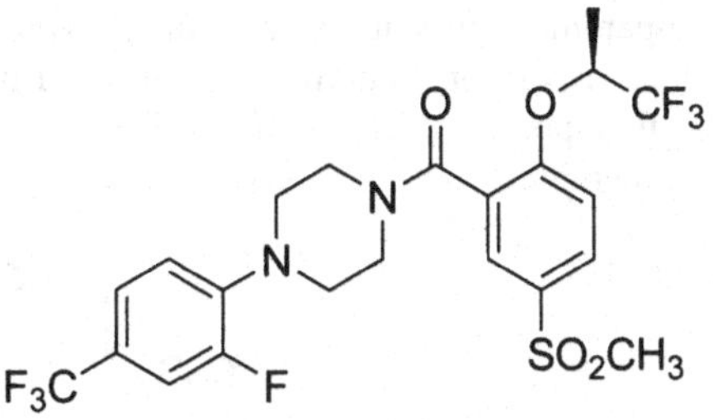

RG1678

A third type of receptor for glutamate is the NMDA (*N*-methyl-D-aspartate) receptor, which controls mechanisms of learning and memory and may be important for deficits in reasoning and cognition in schizophrenia. Glycine is an obligatory cotransmitter at the NMDA receptor, controlling its sensitivity to glutamate, and therefore, elevating glycine levels should improve cognitive function.[30] Inhibiting glycine uptake by the type 1 glycine transporter (GlyT1) has been shown to increase NMDA transmission to improve mechanisms of learning and memory.[31] Among several reported GlyT1 inhibitors, RG1678 has demonstrated clinical efficacy by relieving negative symptoms in a Phase II clinical trial.[32]

Animal models[33] and human genetic studies[34] indicate the alpha-7 nicotinic receptor (α7NR) may offer a novel approach to improving cognitive function in schizophrenia. Another intriguing feature of this receptor is its potent affinity for the β-amyloid protein involved in Alzheimer's disease, suggesting a role in neurodegenerative disease as well.[35] A clinical trial with the α7NR agonist GTS-21 (DMBX-A) showed improvements in certain aspects of learning and memory in schizophrenic patients.[36] The selective α7NR agonist A-582941 activates areas in rat brain involved in attention and working memory.[37] α7NR positive allosteric modulators, such as PNU-120596, have the advantage of increasing receptor activation with less tendency for receptor desensitization and loss of activity.[38]

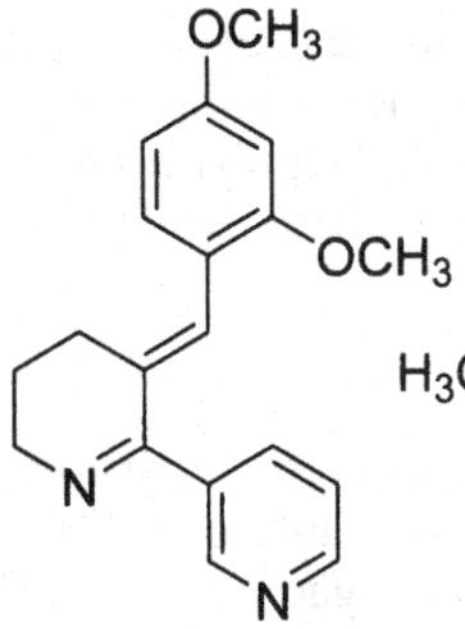

GTS-21 (DMBX-A)

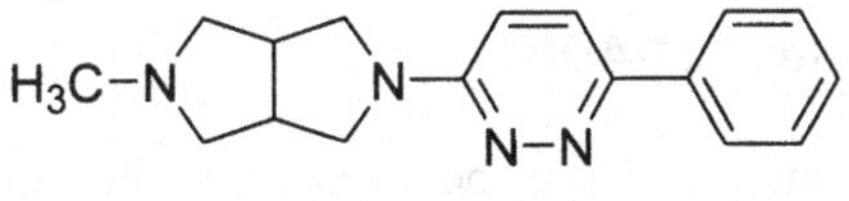

A-582941

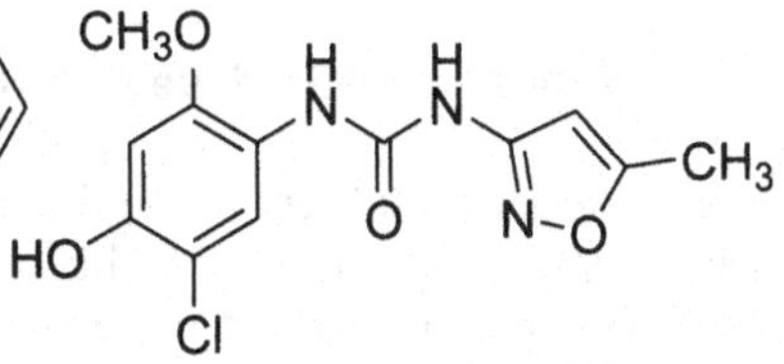

PNU-120596

Numerous 5-HT$_6$ receptor antagonists are in development as cognitive enhancers which may be useful in treating the cognitive deficits in schizophrenia.[39] In particular, the potent 5-HT$_6$ receptor antagonist SB-399885 increases acetylcholine levels in rat prefrontal cortex, an index of memory potentiation, and improves performance in learning models.[40] In addition, this compound increases dopamine efflux in rat prefrontal cortex in combination with marketed antipsychotic agents such as risperidone, another indication of its cognitive-enhancing properties.[41] These data suggest the therapeutic potential of a 5-HT$_6$ receptor antagonist either alone or in combination with current therapy.

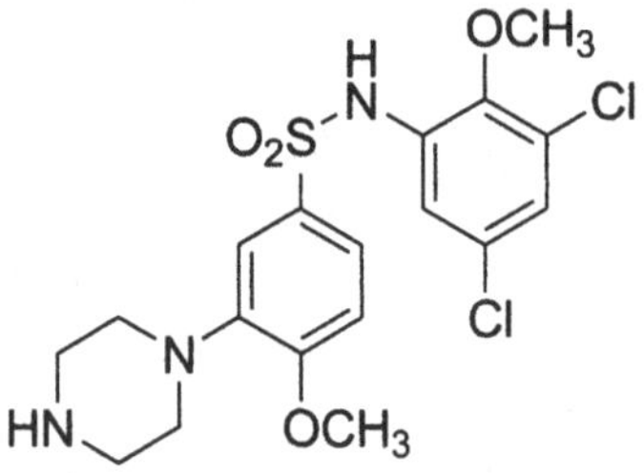

SB-399885

3   Antidepressant Drugs

Depression, like psychosis, was traditionally treated with isolation or psychotherapy and was surrounded by a social stigma that made seeking help very difficult. More properly known as major depressive disorder, or unipolar depression to distinguish it from bipolar depression, it is characterized by a despondent mood, involving low self-esteem and loss of appetite for pleasurable or interesting aspects of life.[42] Cognition, memory, and sleep difficulties are common, and more severe cases can lead to suicide, making treatment an imperative to save the patient's life in such cases. While psychotherapy and electroconvulsive shock therapy (ECT) have been used in the past, the advent of antidepressant drugs changed the way society viewed this disease, offering hope to countless patients.

The discovery and development of the antidepressant drugs have many parallels with the story of the antipsychotic drugs. The same tricyclic nucleus that led to the discovery of the first antipsychotic drugs also figures in the history of the antidepressants, and the next generation of antidepressants, like the atypical antipsychotics, addressed the side effects in the first generation. Finally, both areas have reached the same point in their development, with the second-generation drugs dominating medical practice in their respective areas, while various new approaches are under investigation. In contrast to the antipsychotic drugs, however, the first antidepressants came from a completely different chemical area, the hydrazide class of monoamine oxidase inhibitors.

3.1   Monoamine Oxidase Inhibitors (MAOIs)

The serendipitous discovery of the first antidepressants began in 1951 with observations regarding patients treated with the new hydrazide class of antituberculosis agents, such as isoniazid.[43,44] These patients were reported to be far too happy, and their elevated mood suggested a trial in patients suffering from depression, ironically in the same hospital and by the same physician, Jean Delay, who played a role in discovering the antipsychotic drugs in the following year.[45] While the success of these early trials was not as revolutionary as the antipsychotic trials that occurred at about the same time, these agents nonetheless established a new modality for therapy. Unfortunately, their side effects, especially hepatotoxicity and drug-drug interactions, led to market withdrawals and the tricyclic class eventually superseded them.   Phenelzine and tranylcypromine are nonetheless still occasionally used in clinical practice. While research in this area has continued, aimed primarily at reducing side effects by, e.g., more specific inhibition of either MAO-A or MAO-B, much of this work has led to drugs in other areas (such as selegiline for Parkinson's disease as described below).

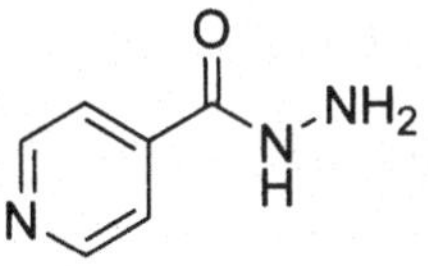

isoniazid (Laniazid) 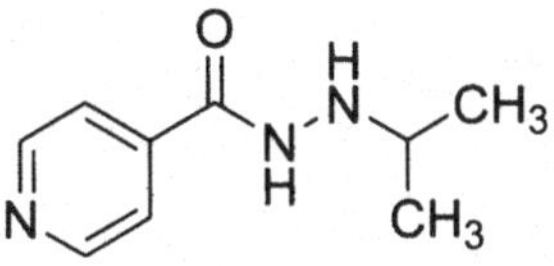   iproniazid (Ipronin) 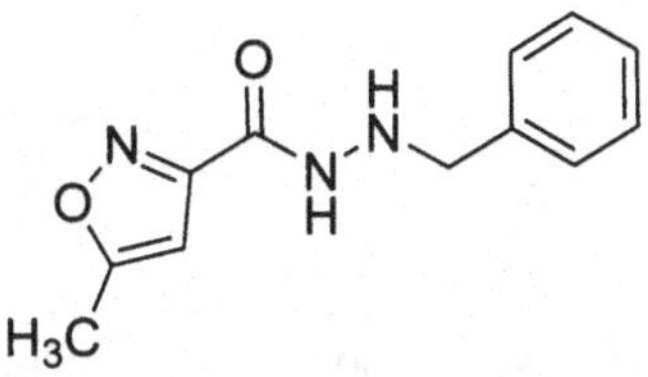   isocarboxazid (Marplon)

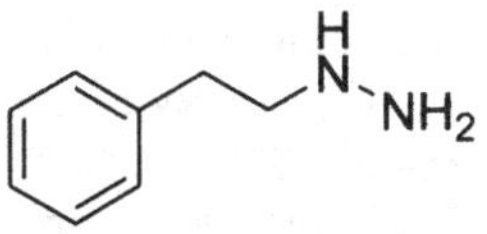

phenelzine (Nardil) 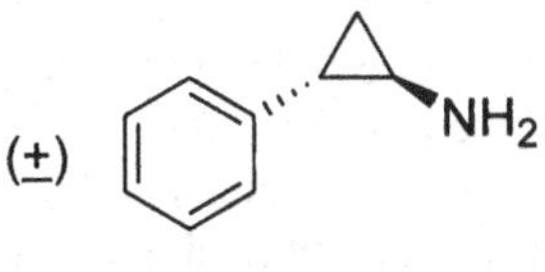   tranylcypromine (Parnate)

3.2   Tricyclic Antidepressants (TCAs)

Following the discovery of the tricyclic antipsychotic agents, research into further structural variations on the phenothiazine nucleus and subsequent careful clinical study identified the dibenzazepine imipramine (Tofranil), which had been prepared in 1951, as an effective antidepressant.[46]   This disclosure prompted a search for structural variations with improved properties, leading to several tricyclic antidepressants (TCAs), many of which are still marketed.

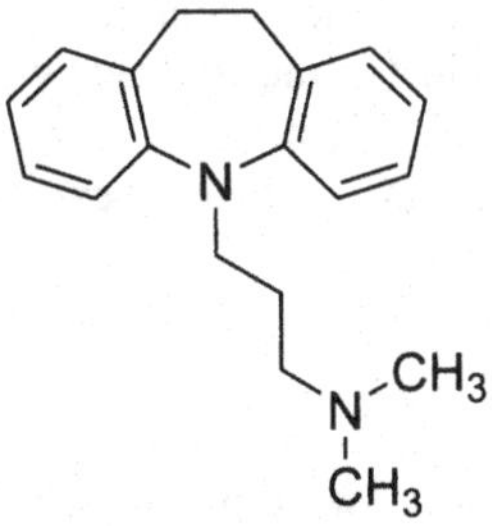

imipramine (Tofranil) 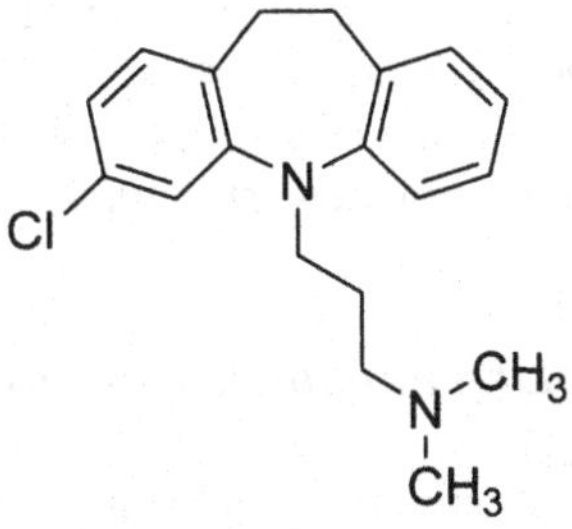   clomipramine (Anafranil) 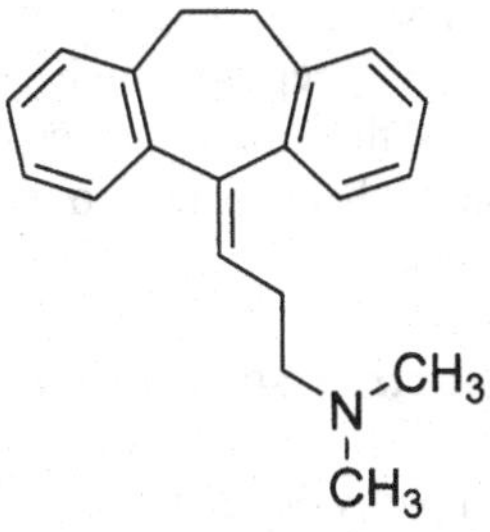   amitriptylline (Elavil)

The primary metabolites of the tricyclic antidepressants are the dealkylated secondary amines, such as nortriptylline (Pamelor), the primary metabolite of amitriptylline (Elavil). These are also marketed and have a similar pharmacological profile, being responsible for most of the efficacy of the tertiary amine parent drug.

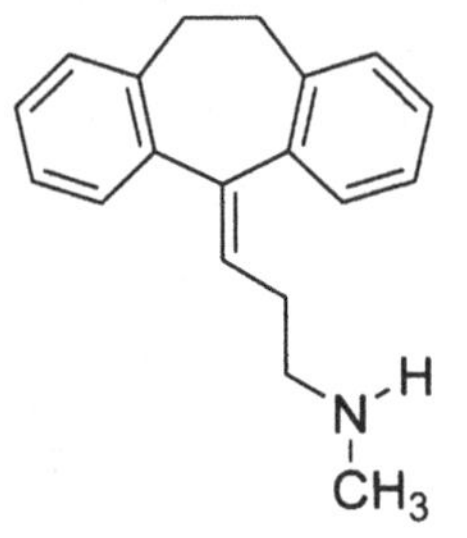

nortriptylline (Pamelor) 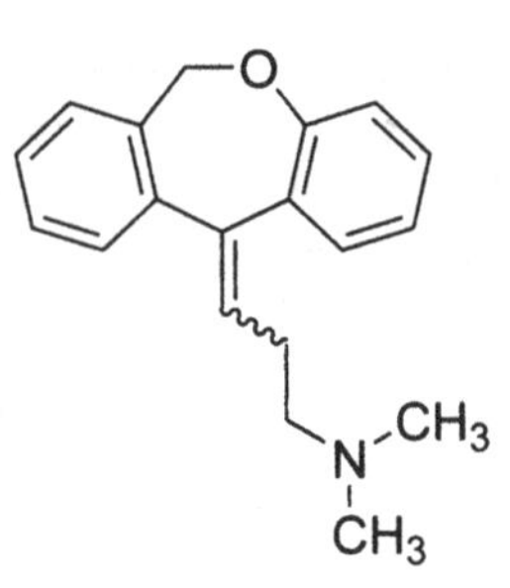   doxepin (Sinequan)

The TCAs have also found additional clinical applications. For example, in addition to its use as an antidepressant, doxepin (Sinequan) is also used to treat insomnia due to its potent H_1-receptor antagonism and resulting sedating activity, under the brand name Silenor.

As in the case of the antipsychotics, the serendipitous discovery of the MAOIs and TCAs opened the way to a mechanistic understanding of depression that would have been otherwise unlikely. The storage, release, and breakdown of the neurotransmitters, in particular noradrenaline and later serotonin, had already been established in the work of von Euler,[47] Axelrod,[48] and others. Subsequently, Axelrod[49] and Carlsson[50] showed that imipramine and structurally related drugs block neuronal uptake of adrenaline and serotonin. This work led to the so-called monoamine hypothesis of depression, which states that depression results from a deficiency of serotonergic, noradrenergic, and/or dopaminergic transmission.[51] Hence the efficacy of antidepressant drugs results from their blockade of uptake (TCAs) or breakdown (MAOIs) of adrenaline and serotonin. Although this hypothesis has been challenged and other hypotheses are currently being pursued, it was nonetheless useful in developing the next generation of antidepressant drugs.

While the side effects of the first-generation drugs, such as the sedating effects noted above for doxepin or cholinergic effects such as dry mouth, are clinically manageable, they nonetheless have led to significant drop-out rates from clinical trials.[52] The potent affinity of the TCAs for CNS receptors, such as the H_1- and M_1- receptors for histamine and acetylcholine, respectively, is thought to be responsible for many of their observed side effects. This need for better-tolerated therapy led to research into new structural variations and the next class of antidepressants.

3.3   Selective Serotonin Reuptake Inhibitors (SSRIs)

The research paths that led to the next generation of antidepressant drugs were based on screening in animal models that mimicked the activity of imipramine, in analogy to the process Paul Janssen used to discover haloperidol. For example, imipramine blocks the hypothermia in mice induced by p-chloroamphetamine (PCA), which causes monoamine depletion.[53] Using this model, Pfizer screened a series of tetralin compounds that led to their discovery of the structurally novel antidepressant sertraline (Zoloft).[54] In contrast, the group at Lilly used hypothermia in mice induced by apomorphine, an effect also blocked by imipramine, to screen compounds based on the structure of lead compound diphenhydramine, leading to the novel antidepressant fluoxetine (Prozac).[55] Both groups also measured the activity of their drugs in blocking serotonin uptake, and less potently noradrenaline uptake, into rat synaptosomes, establishing their mechanism of action.   The term SSRI (selective serotonin reuptake inhibitor) was subsequently coined to designate these drugs. Each group's efforts removed the off-target activity of the TCAs, reducing side effects and providing very successful drugs.

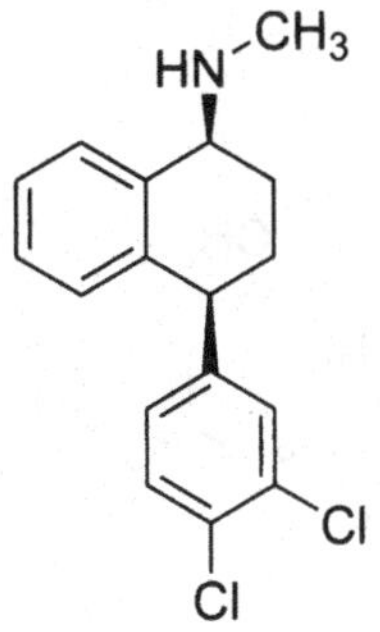

sertraline (Zoloft) 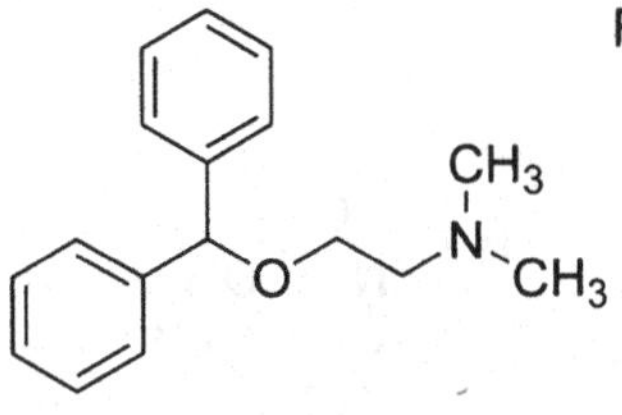   diphenhydramine 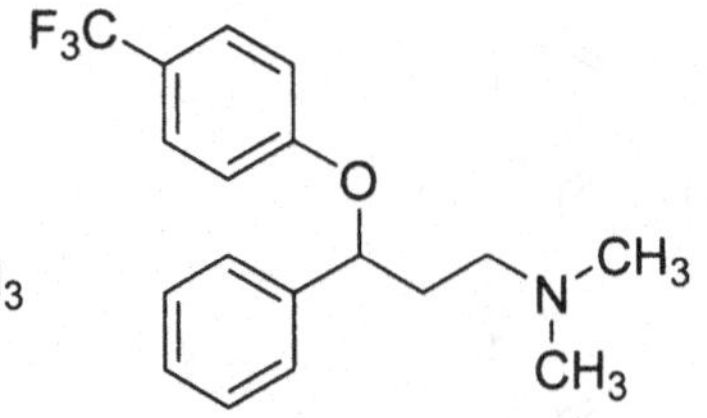   fluoxetine (Prozac)

Other compounds discovered by similar research include paroxetine (Paxil), discovered at Ferrosan, again using the PCA assay,[56] and eventually marketed by GSK.[57] Citalopram, and its *S*-isomer escitalopram, began as a program at H. Lundbeck A/S in the 1960s to find bicyclic versions of the first-generation antidepressants,[58] with eventual SAR studies that optimized the aromatic portions of the template.[59] Both of these drugs maintain the SSRI profile and resulting low incidence of side effects while affording antidepressant efficacy.

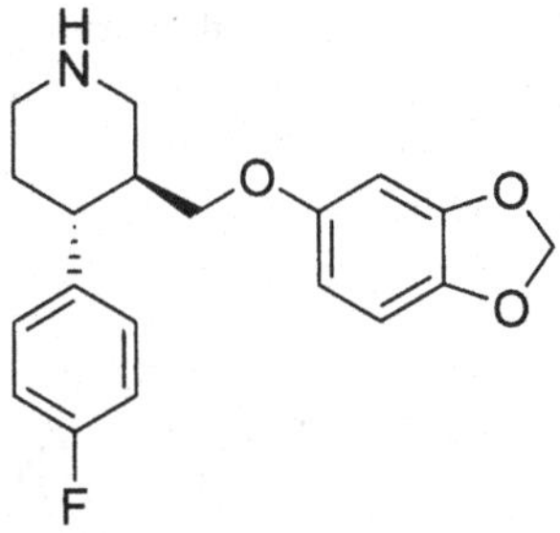

paroxetine (Paxil) 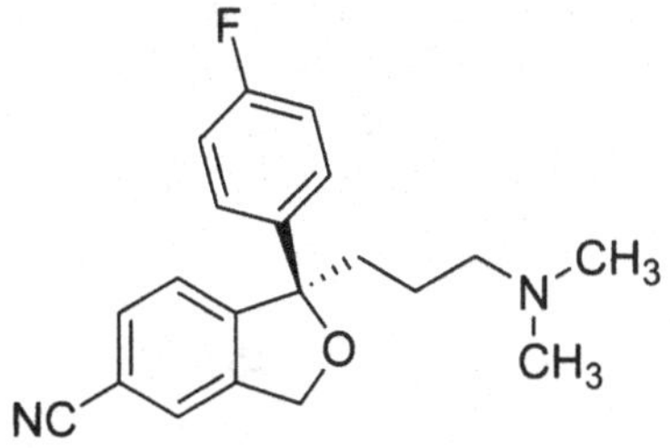   escitalopram (Lexapro)

3.4   *Combined Noradrenaline/Serotonin Reuptake Inhibitors (SNRIs)*

Although the SSRIs have proven to be effective antidepressants, 30% of patients do not respond to SSRI treatment, suggesting the need for additional approaches. Adding back noradrenaline uptake blockade to afford a profile more similar to the first-generation agents affords effective drugs referred to as SNRIs. The SNRI duloxetine (Cymbalta) was discovered as a follow-up to the program at Lilly that had originally found fluoxetine. Both the medicinal chemistry[60] and pharmacology[61] of the racemate have been described, and the (+)-enantiomer was developed as duloxetine. Duloxetine is marketed as an analgesic as well as an antidepressant. Discovered and marketed somewhat earlier, the SNRI venlafaxine (Effexor) arose from a program seeking a novel analgesic based on the structure of the mixed opiate agonist/antagonist ciramadol. Venlafaxine was designed to reduce the number of stereocenters but failed to afford the desired analgesic activity. However, venlafaxine was later submitted for antidepressant testing based on its resemblance to structures of antidepressants that were known at the time, and its activity led to a successfully marketed drug.[62]

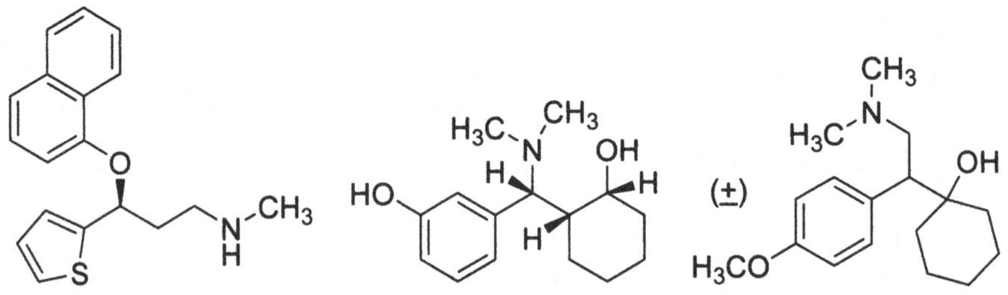

duloxetine (Cymbalta)   ciramadol   venlafaxine (Effexor)

3.5   *Novel Approaches*

The indoleamine-based hormone melatonin, secreted from the pineal gland, controls circadian rhythm in mammals by binding to two G-protein-coupled receptors MT_1 and MT_2. The French company Servier explored the potential for antidepressant activity of MT_1/MT_2 agonists based on the finding that circadian rhythm is typically dysfunctional in depression. Agomelatine (Valdoxan), a structural mimic of melatonin, is also a 5-HT_{2c} antagonist in addition to its potent melatonin agonist activity.[63] Despite a lengthy approval process, the drug was eventually approved in 2009 for marketing as an antidepressant. This is the only additional mechanism of action beyond monoaminergic potentiation that has yet been shown effective in treating depression, but its ultimate efficacy will only be defined in time.

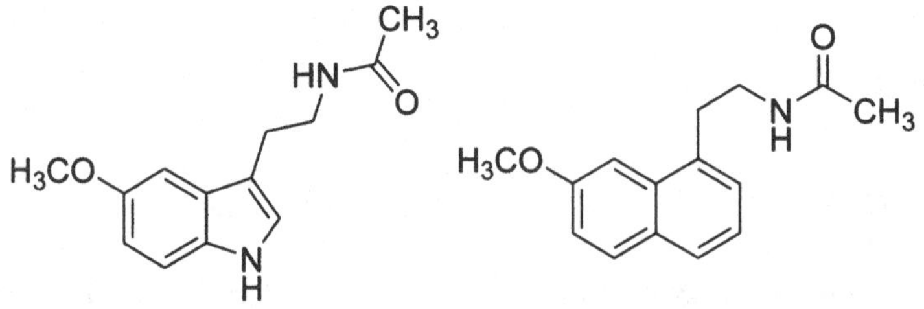

melatonin   agomelatine (Valdoxan)

The lactam compound rolipram was originally discovered in another rodent hypothermia model, in this case induced by reserpine (vide infra), but it works by a completely different mechanism than the TCAs, inhibition of PDE4.[64] This enzyme is another member of the phosphodiesterase class of enzymes that was reviewed above in the case of PDE10A. PDE4 is a calcium-independent, cAMP-selective member of this family and is responsible for homeostatic control of cAMP formed in response to monoaminergic transmission. PDE4 inhibitors thus potentiate monoaminergic signaling and hence provide the same effect as the TCAs, although with very different side effects. In this case, nausea and vomiting are the principal issues with PDE4 inhibitors. Thus, although it showed evidence of efficacy in small-scale clinical trials, rolipram's side effects precluded demonstration of efficacy in larger trials,[65] and it never reached the market. The same fate befell follow-up compounds developed as potent PDE4 inhibitors, such as cilomilast[66] and piclamilast.[67]

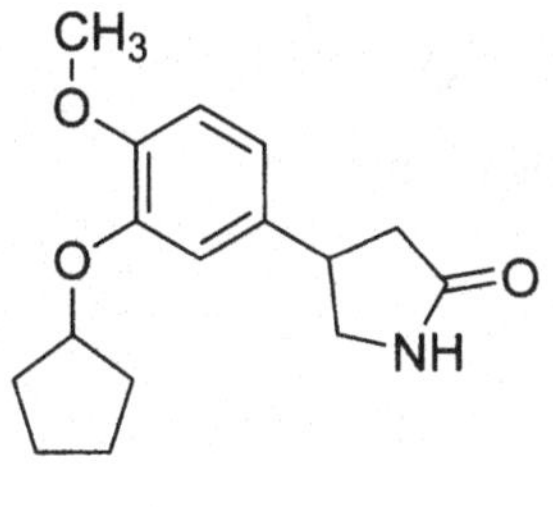

rolipram 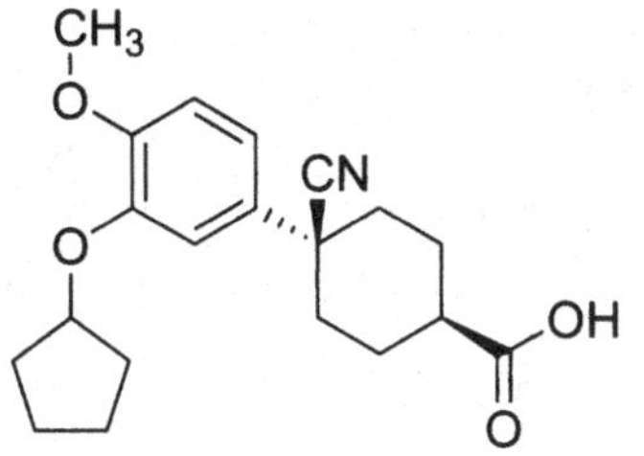   cilomilast 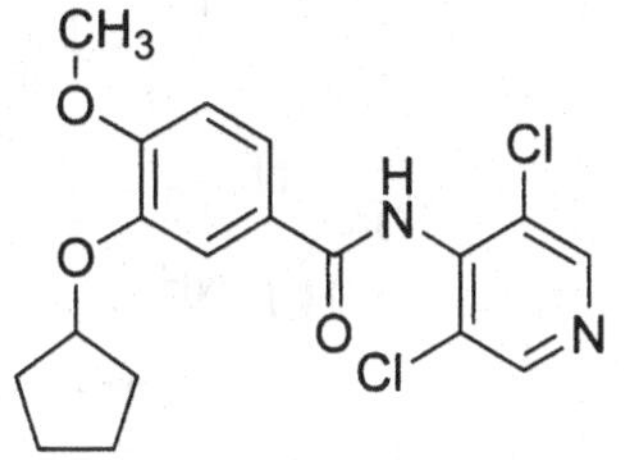   piclamilast

The NK-1 receptor antagonists were originally discovered as drugs in search of a disease, based on their hypothesized antiinflammatory activity. The first NK-1 receptor antagonist, CP-96,345,[68] was discovered at Pfizer in 1988 by screening in a receptor-binding assay, and the structure was subsequently simplified to a viable clinical candidate CP-122,721.[69] In subsequent work at Merck and GSK, the piperidine ring was modified to a morpholine and piperazine, respectively, and the benzyl side modified to accommodate the new SAR that resulted from these changes, providing aprepitant[70] and vestipitant.[71] The potential for these drugs in depression was discovered serendipitously in observations of animal behavior.[72] Although the first two drugs initially showed promising antidepressant efficacy in the clinic, subsequent large-scale trials failed to support these findings.[73] Aprepitant was found, however, to be an effective antiemetic in patients undergoing cancer chemotherapy[74] and is currently marketed for that indication following approval in 2003.[75]

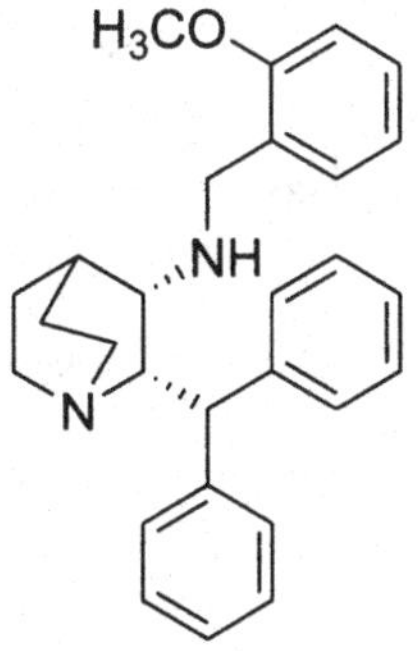

CP-96,345 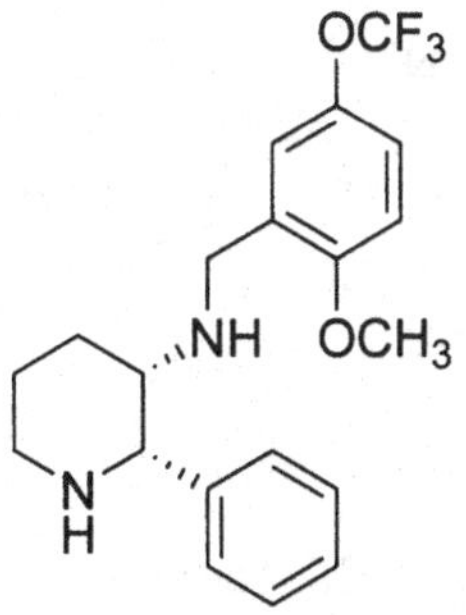   CP-122,721

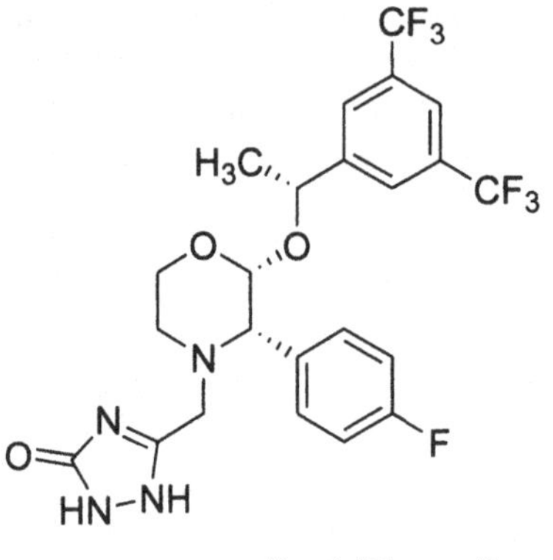

aprepitant (Emend) 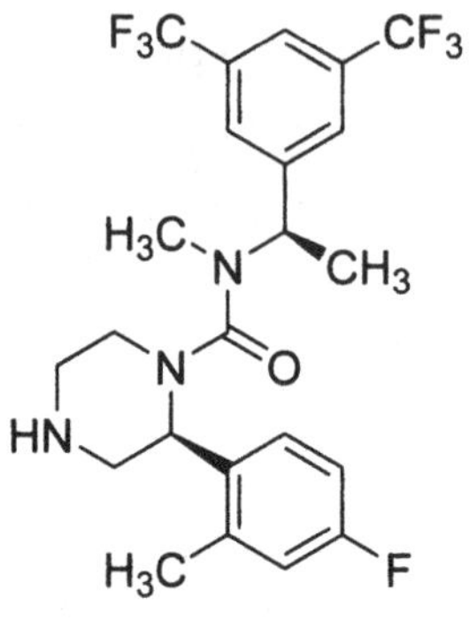   vestipitant

Corticotropin-releasing hormone (CRH) is a 41-amino-acid neuropeptide secreted by the hypothalamus that binds to its receptor, CRH1, in the pituitary, causing release of adrenocorticotropin (ACTH), which then regulates the activity of the adrenal gland in effecting the flight-or-fight response. This HPA (hypothalamic-pituitary-adrenal) axis is hypothesized to play a role in anxiety and depression.[76] The first CRH1 receptor antagonist, CP-154526,[77] discovered at Pfizer in 1991, again by screening in an *in vitro* receptor-binding assay, was developed to test this hypothesis.[78] Several structurally related compounds were subsequently reported following this discovery, including CP-316311,[79] R121919,[80] and pexacerfont.[81] Clinical trials with R121919 initially appeared to indicate success with this approach,[82] but again, large-scale clinical trials of CP-316,311 and pexacerfont failed to demonstrate efficacy.[83,84] Exploration of other indications, including irritable bowel syndrome (IBS), continue with this class of compounds.[85]

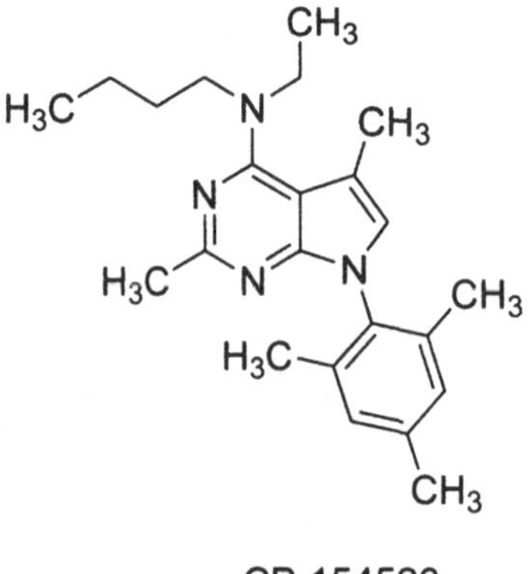

CP-154526 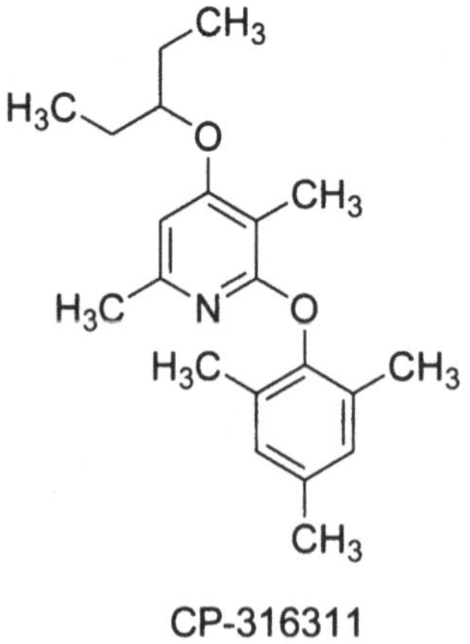   CP-316311

R121919   pexacerfont

Although it has been difficult to find new classes of antidepressants that work by a novel mechanism, research in this area has uncovered important aspects of depression that may lead to new therapies.

4   Drugs for Epilepsy and Bipolar Disorder

Although it seems surprising to cover these two diseases in the same section, both epilepsy and bipolar disorder involve dysregulation of neuronal excitability, at least as far as the drugs used to treat them seem to indicate. Epilepsy has a long history in human culture, having long been associated with demonic possession and other supernatural phenomena. It has a manifold of variations but is fundamentally characterized by seizures that lead to loss of consciousness.[86] The role of neuronal overexcitability is underscored by the finding that mutations in ion channels controlling neuronal excitation cause some of its variations.[87]

The causes of bipolar disorder are far less clear. While it has historically attracted less attention than the other CNS diseases, its impact has been felt in the many famous people it has purportedly afflicted.[88] Bipolar disorder is characterized by mood swings, from mania, or elevated mood resulting in abnormal behavior, to depression. Bipolar depression, in which the depressive phase of the disease predominates, is poorly understood and difficult to diagnose. In fact, correct diagnosis of bipolar disorder is typically delayed by many years as other diagnoses are made and the patient receives inappropriate treatments.[89] Aside from dysregulation of neuronal excitability, little is understood of the disease's etiology, and genetic studies are beginning in an effort to unravel its genetic basis.[90] The following survey of drugs for these diseases centers on the antiepileptics, but these drugs provide a spectrum of efficacy, from those effective against epilepsy only to those effective in bipolar disease only, and many in between.

4.1   Older Antiepileptics

The prototype antiepileptic barbiturate phenobarbital (Luminal) was synthesized by Emil Fischer at Bayer early in the 20th century after the discovery that the diethyl analog, barbital (Veronal) showed potent sedative activity.[91] Phenobarbital was originally prescribed for epilepsy patients in hopes that its sedative activity would be helpful, and its antiepileptic activity was only discovered serendipitously. Given its superiority over the only available therapy at the time, potassium bromide, it quickly became the leading drug in the field and was only displaced once its side effects were better appreciated and newer drugs, such as phenytoin and carbamazepine, became available. Barbituates act by opening the chloride ion channel that is gated by the inhibitory neurotransmitter γ-aminobutyric acid (GABA), which is also the target through which benzodiazepines (BZDs) exert their antiepileptic and anxiolytic effects, although BZDs act to potentiate GABA activity rather than directly open the channel and also bind to a different site on the channel, as explained in more detail below.

R = Et, barbital (Veronal)
R = Ph, phenobarbital (Luminal)   phenytoin (Dilantin) valproic acid (Depakote)

The discovery of phenytoin was based on the development of an animal model of epilepsy, in which convulsions are generated using an electric current.[92] This model is able to distinguish between phenobarbital and other barbituates such as barbital that are ineffective as anticonvulsants. Beginning in the 1930s, an academic lab screened compounds that generally resembled phenobarbital in having a pendant phenyl ring, discovered the activity of phenytoin (Dilantin), originally prepared in 1908, and went on to prove its efficacy in epilepsy patients.[93] Its activity-dependent block of sodium channel ion currents[94] leads to a reduction in neuronal excitation, accounting for its therapeutic efficacy.

Eymard et al. have described their serendipitous discovery of the anticonvulsant properties of valproic acid (Depakote). Because of its ability to solubilize lipophilic compounds, they used it as a vehicle for testing the efficacy of compounds in preventing pentylenetetrazole-induced seizures and discovered the activity remained when they used the vehicle alone.[95] It became widely prescribed as a treatment for epilepsy[96] and was subsequently discovered, again serendipitously due to observations in bipolar disease patients being treated for seizures, to be effective in treating mania.[97] Its mechanism of action is not fully understood, but it appears to act as an anticonvulsant by increasing GABA-ergic transmission.[98] The basis for its antimanic activity, however, has been harder to determine, since it is also an inhibitor of histone deacetylases (HDACs), enzymes that regulate chromatin and hence gene transcription.[99] It also shares potential mechanisms of action with the other mood stabilizers lithium and carbamazepine, including effects on inositol levels[100] and inhibition of prolyl oligopeptidase.[101]

Carbamazepine (Tegretol) is another drug originally discovered to be an antiepileptic and subsequently found to be effective in treating mania. It was originally prepared as part of the first wave of tricyclic drugs, and its antiepileptic properties were discovered in the 1960s.[102] Like phenytoin, it is a sodium channel blocker,[103] but in addition, it also possesses antimanic activity as a mood stabilizer, which was again discovered serendipitously, in Japan in the 1970s.[104]

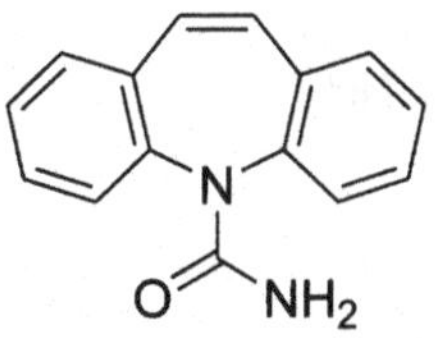

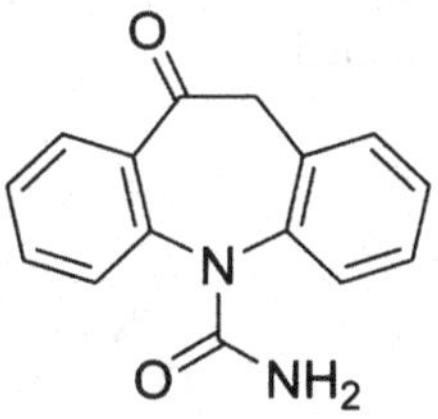

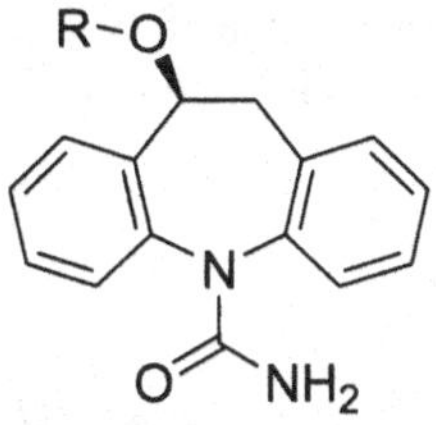

carbamazepine (Tegretol) oxcarbazepine (Trileptal) R = H, eslicarbazepine
R = Ac, eslicarbazepine acetate

Oxcarbazepine (Trileptal) is structurally related to carbamazepine but is metabolically more stable, exchanging the olefin, a metabolic hotspot, for a ketone, which affords fewer drug-drug interactions. It is otherwise similar to carbamazepine in its mechanism of action and efficacy in epilepsy and bipolar disorder.[105] It is metabolized to eslicarbazepine, which, administered as the acetate prodrug, has a similar therapeutic profile.[106]

4.2   Newer Antiepileptics

Lamotrigine (Lamictal) was originally developed as a dihydrofolate reductase inhibitor, hypothetically for treating convulsions,[107] and then serendipitously discovered to work at calming mania in two patients.[108] Subsequent study proved its therapeutic efficacy in epilepsy and bipolar disorder.[109] Lamotrigine is also a sodium channel blocker[110] and may have additional mechanisms of action, such as effects on glutamate release.

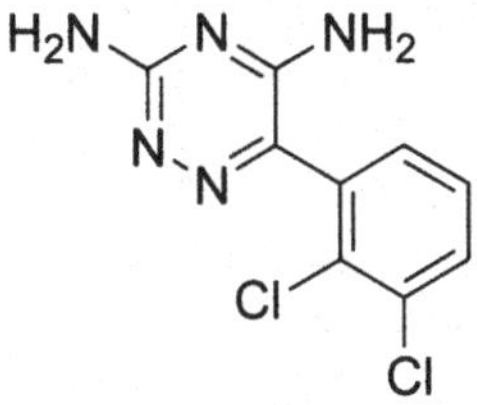

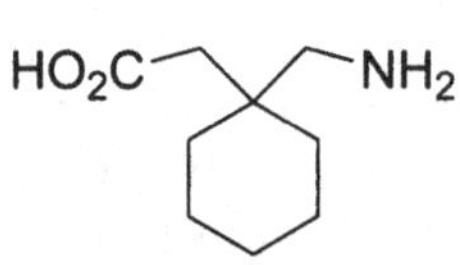

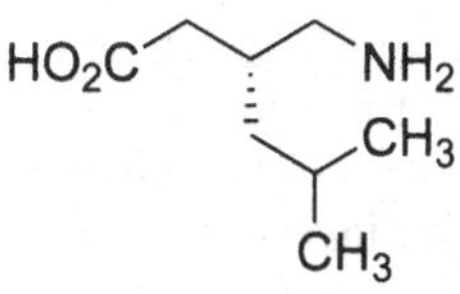

lamotrigine (Lamictal)   gabapentin (Neurontin)   pregabalin (Lyrica)

It may seem surprising that gabapentin (Neurontin) is a CNS-active drug, since it is an amino acid. It is a substrate for the L amino acid transporter, which ensures its uptake into the brain, where is binds to the $\alpha2\delta$ subunit of the calcium channel and thus regulates neuronal excitation.[111] In addition to its efficacy in epilepsy, it is also used to treat neuropathic pain. The structurally related compound pregabalin (Lyrica) was discovered by the Silverman group at Northwestern University in a quest to find activators of glutamic acid decarboxylase, the biosynthetic enzyme that produces GABA.[112] Pregabalin, however, turns out to work in analogy with gabapentin by binding to $\alpha2\delta$.[113] It is also more potent than gabapentin and is currently used to treat neuropathic pain, although an application for treating generalized anxiety disorder was recently turned down.

Zonisamide (Zonegran) was discovered in Japan in the early 1970s as an anticonvulsant and it subsequently demonstrated clinical efficacy in epilepsy.[114] Its mechanism of action is still not completely clear, although it does block T-type calcium channels,[115] and it affects other channels, including the GABA channel.[116]

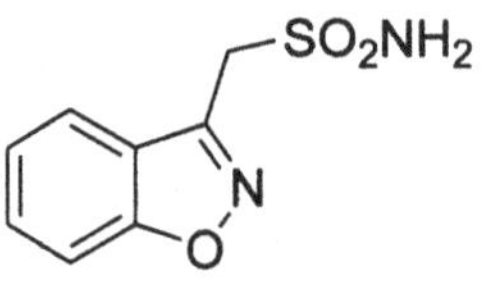

zonisamide (Zonegran) 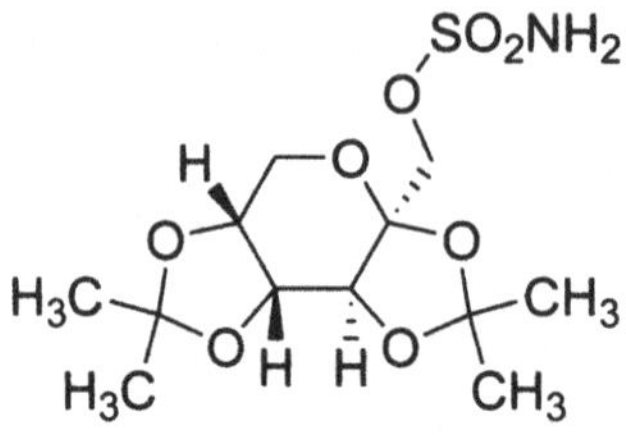   topiramate (Topamax)

Topiramate (Topamax), while also a sulfonamide, is structurally distinct from zomisamide and indeed is quite structurally unique in this category, being a modified diketal of fructose. It was discovered by Bruce Maryanoff at McNeil Pharmaceutical in 1979[117] and reached the market as a treatment for epilepsy.[118] Again, its mechanism of action is not fully determined, but it does enhance chloride ion flux at the GABA channel.[119] Topiramate is also prescribed for migraine[120] and has been studied in bipolar disorder[121] and obesity.[122]

The discovery of tiagabine (Gabatril) originates with Krogsgaard-Larsens's observation that nipecotic acid, a structural analog of GABA, inhibits GABA uptake,[123] with the R-(–)-isomer being the more potent enantiomer. It is interesting that he also found that some of the constituents of betel nut, a plant preparation used traditionally as a mild stimulant with psychotropic properties, resemble the structure of nipecotic acid.[124] Several companies worked on improving the potency of nipecotic acid by adding lipophilic groups to the nitrogen, culminating in tiagabine[125] at Novo and CI-966 at Warner-Lambert.[126] Although CI-966 was discontinued due to adverse effects in the clinic,[127] tiagabine successfully reached the market as a novel antiepileptic drug. Its pharmacological target is the type 1 GABA transporter, GAT-1,[128] with the resulting increase in levels of the inhibitory transmitter GABA responsible for its anticonvulsant effects.[129] It is now being investigated for other indications, including panic disorder.[130]

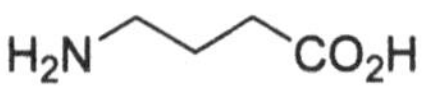

γ-aminobutyric acid (GABA) 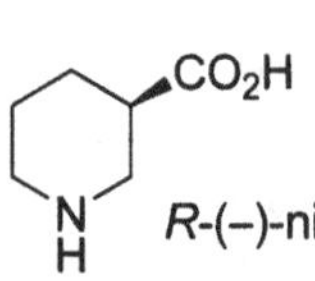   R-(–)-nipecotic acid 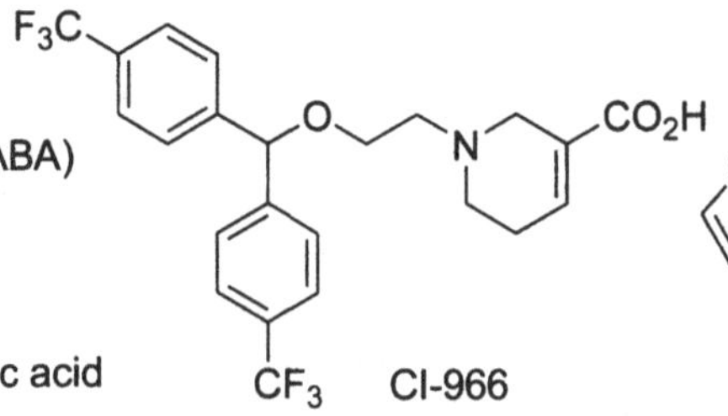   CI-966 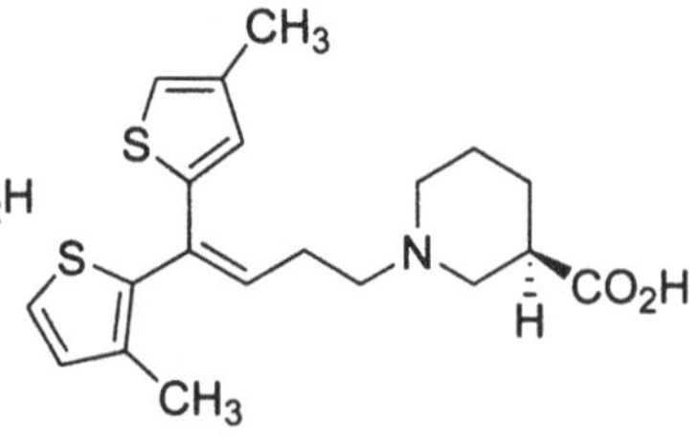   tiagabine (Gabatril)

The final entry in this section is a drug without useful anticonvulsant activity but which is one of the mainstays of treatment for bipolar disorder. Lithium, usually administered as lithium carbonate, was discovered serendipitously by John Cade.[131] He was initially working with uric acid as a potentially interesting component of urine from mentally ill patients but found that the calming effects of uric acid that he observed in guinea pigs were due to his use of its lithium salt.

He subsequently found that lithium citrate or carbonate effectively calmed his patients with mania. The mechanism of action of lithium is still being investigated, with one hypothesis being that it dampens excessive neurotransmitter signaling by interfering with the regulatory activity of the inositol second messenger system.[132] Although it is an inhibitor of myo-inositol-1-phosphatase,[133] which recycles inositol from its various phosphorylated forms that serve as second messengers, it is not clear that this activity is responsible for its therapeutic effects.

5   Anxiolytic Drugs

Although anxiety strikes a familiar chord with many people due to the stresses of everyday life, it is classified as a disease when it becomes exaggerated or irrational. While situational anxiety may result from extreme stresses of catastrophic personal tragedy, generalized anxiety disorder (GAD) is a life-long chronic state of excessive worry.[134] Its etiological basis is still unknown, although many causative factors have been established,[135] and the drugs that treat it address only its symptoms.

5.1   Benzodiazepines

The discovery of the benzodiazepine class of anxiolytics begins with a major role for serendipity. As described by Leo Sternbach, who first prepared chlordiazepoxide (Librium), the initial member of this class of compounds, while working at Roche, the compound was screened merely to provide data for publication and found to promote sedation and muscle relaxation, in analogy with the barbituates.[136] Subsequent testing in the clinic established its anxiolytic activity, and both chlordiazepoxide and a follow-on, diazepam (Valium), were successfully marketed. As mentioned above, benzodiazepines facilitate GABA activity at the chloride-ion channel, a major inhibitory neurotransmitter mechanism in the CNS.

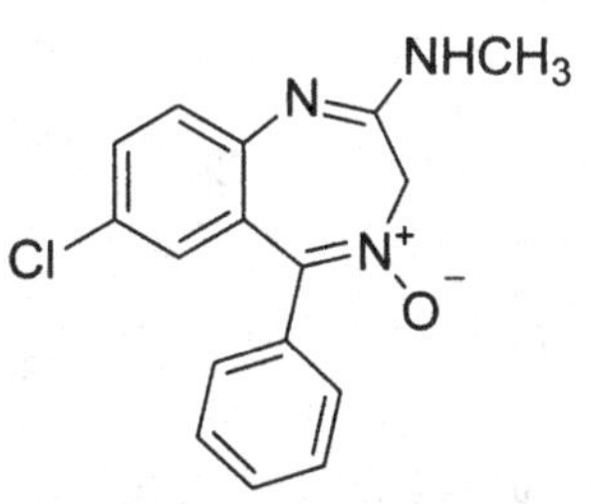

chlordiazepoxide (Librium) 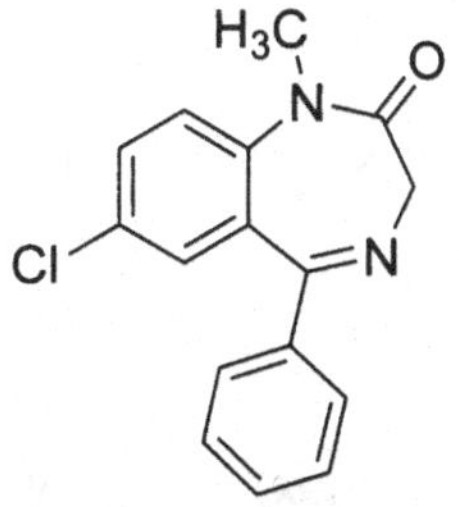   diazepam (Valium)

Numerous follow-on versions were also developed,[137] including clonazepam (Klonopin),[138] which is used for treatment of acute seizures in epilepsy. Its long-term use, however, is circumscribed because of the development of tolerance and side effects that can occur following discontinuation of therapy.[139] It shares this property with other long-acting benzodiazepines. Flunitrazepam (Rohypnol), another potent member of this class, has a shorter half-life and is used to treat insomnia.[140] It is not approved for use in the U.S., but is one of the street drugs supposedly involved in "date rape."[141,142] Lorazepam (Ativan) is a potent, fast-acting drug, used originally for treating anxiety, as well as insomnia.[143] Its side effects, however, include tolerance and dependence,[144] and thus the recommended use is for no more than two to four weeks. These effects are associated with the decreased expression of the pharmacological target of the benzodiazepines, the GABA-A receptor, upon chronic administration.[145] These three drugs illustrate some of the benefits and challenges of the benzodiazepine drugs as used in clinical practice.

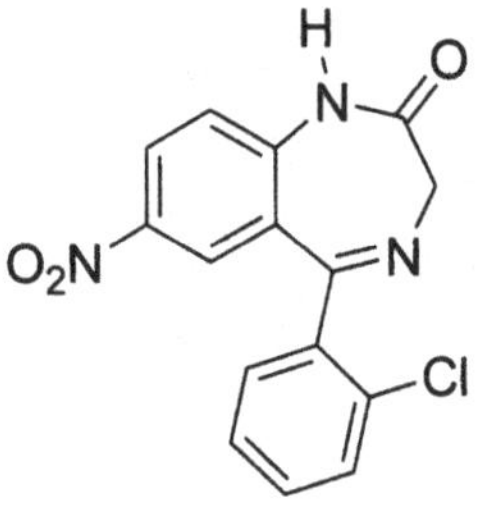

clonazepam (Klonopin) 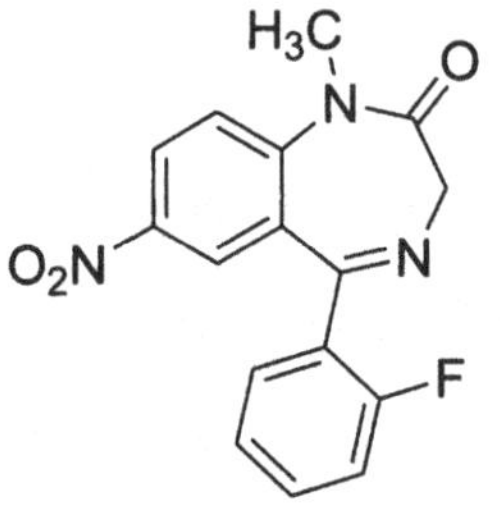   flunitrazepam (Rohypnol) 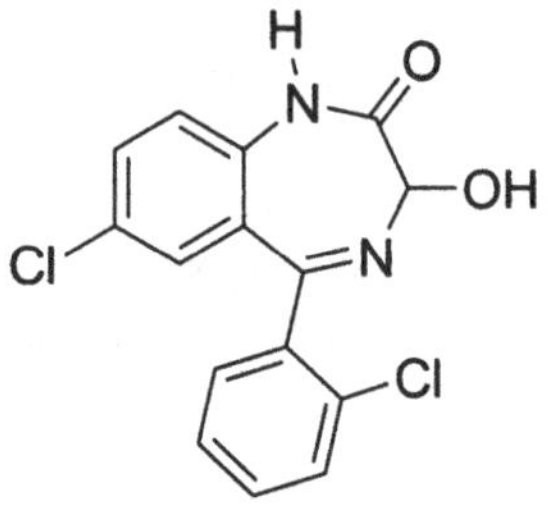   lorazepam (Ativan)

The discovery that the addition of an imidazo or triazolo ring also affords potent activity in this series led to several new drugs. For example, alprazolam (Xanax)[146] is a potent, short-acting drug used primarily to treat situational anxiety and panic attacks.[147] The short half-life of midazolam (Versed) makes it suitable as an anesthetic for brief surgical procedures such as colonoscopy.[148] Triazolam (Halcion), on the other hand, is used to treat insomnia based on its hypnotic properties, although its short half-life can limit its utility in this indication.[149]

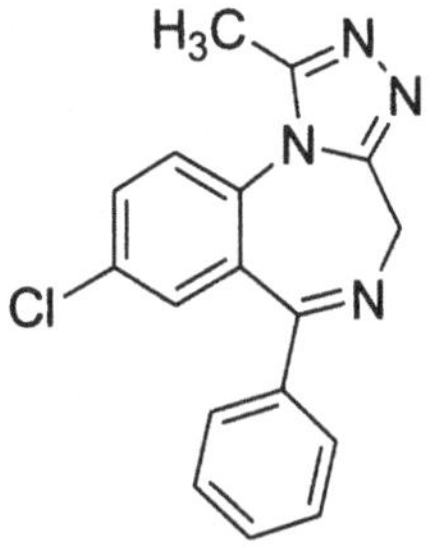

alprazalam (Xanax) 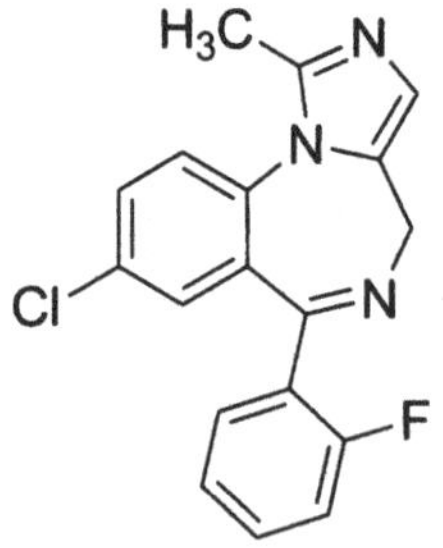   midazolam (Versed) 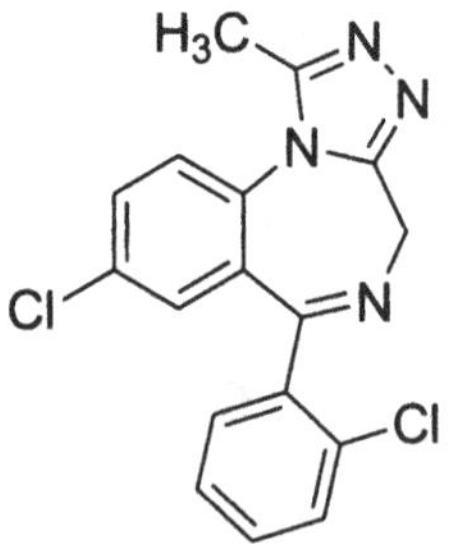   triazolam (Halcion)

The benzodiazepines, along with the earlier barbiturate drugs, were instrumental in discovering their pharmacological target: the GABA-A receptor for the inhibitory neurotransmitter γ-aminobutyric acid.[150] The structural complexity of the GABA-A receptor, which consists of five subunits, primarily from among various α (1–6), β (1-4), γ (1-4), δ, and ε subtypes, helps explain the pharmacological complexity and the variety of potential therapeutic outcomes of the benzodiazepines.[151] By facilitating the action of GABA at its receptor, benzodiazepines augment inhibitory neurotransmission and dampen excitation, resulting in efficacy in the treatment of disorders such as epilepsy and anxiety. While both BZDs and barbiturates bind to allosteric sites separate from the GABA binding site, barbiturates can open the channel and increase chloride ion flow regardless of the presence of GABA.[152] This action accounts for their greater efficacy in epilepsy but also for their side effect profile.

Flumazenil (Anexate) is a useful tool discovered during the SAR development of the BZDs that acts as an antagonist of the GABA-potentiating action of the BZDs.[153] While it has proconvulsant effects on its own, it is useful in reversing the effects of BZD overdose. With a better understanding of the target for the BZDs, it has been possible to develop agents for more specific conditions. For example, although they are structurally unrelated to the BZDs, the so-called non-benzodiazepine hypnotics bind selectively to GABA-A receptors containing an α_1 subunit, such as the widely expressed $(\alpha_1)_2(\beta_2)_2\gamma_2$ GABA-A receptor,[154] which is responsible for the hypnotic effect of the BZDs.[155] In addition to their efficacy in treating insomnia, nonBZDs such as zolpidem (Ambien) and zaleplon (Sonata) also produce less disruption of GABA-A receptor expression and thus may have a lower propensity to cause side effects such as dependence.[156]

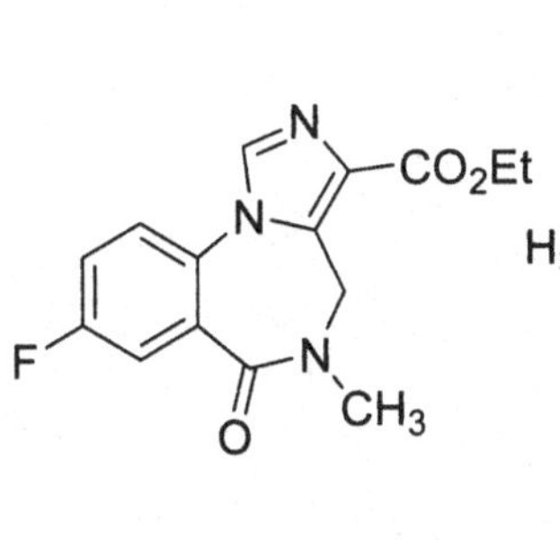

flumazenil (Anexate) 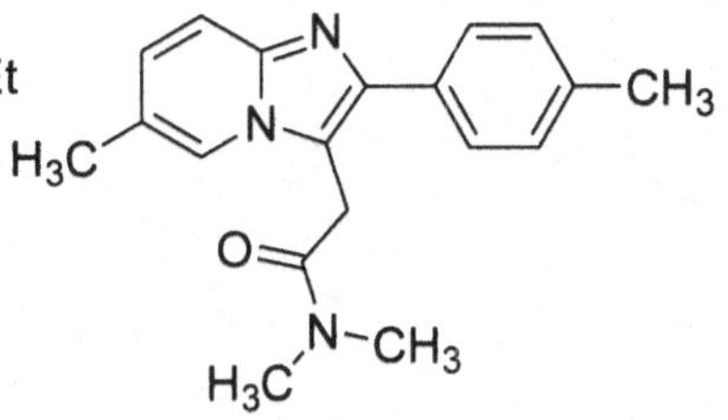   zolpidem (Ambien) 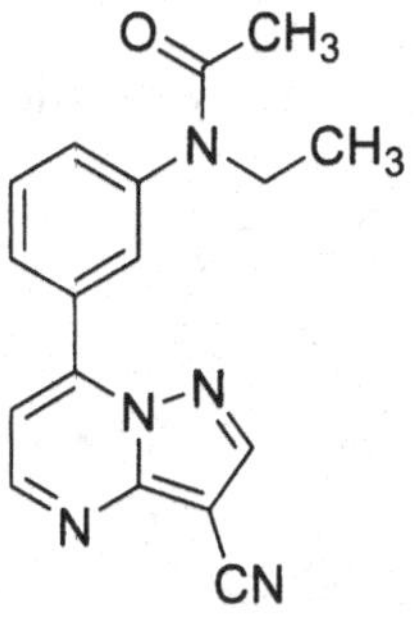   zaleplon (Sonata)

5.2   Novel Anxiolytics

Buspirone (Buspar) was originally designed as an antipsychotic agent, showing activity comparable to chlorpromazine in animal models of psychosis and sedation.[157] It failed in the clinic for this indication[158] but was later found to be an anxiolytic and approved for marketing in 1986.[159] Its pharmacological target appears to be the 5-HT$_{1A}$ receptor, where it acts as a partial agonist.[160] Analogous structures, including gepirone[161] and ipsapirone,[162] have also shown evidence of anxiolytic activity, although their onset is delayed compared with that of diazepam, and they have yet to gain FDA approval.

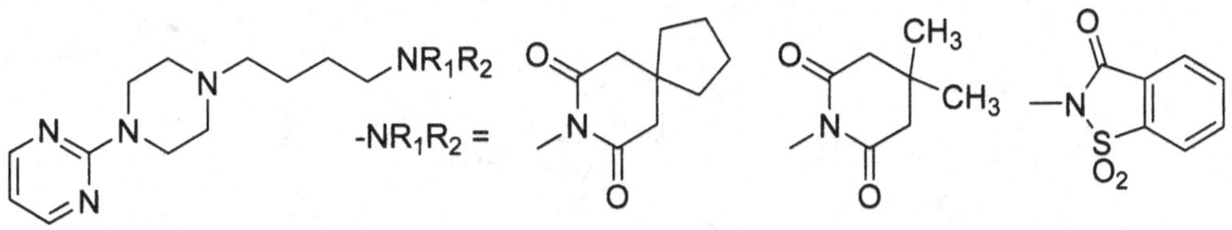

buspirone (Buspar)   gepirone (Arixa)   ipsapirone

6   Centrally Acting Analgesic Drugs

6.1   Opiates

Opium's medicinal properties were evidently discovered before the beginning of recorded history. Morphine, the principal component of opium responsible for its analgesic and euphoriant properties, was isolated by Serturner in 1817.[163] The structure was formally established in 1952 by Marshall Gates' total synthesis.[164] Parenthetically to this story, the experiment of William Henry Perkin to prepare it by distilling a mixture that mimicked its molecular formula, leading instead to the discovery of the dye mauveine, is well known.[165] Snyder and Pert reported the characterization of the pharmacological target for morphine, the μ-opiate receptor, in 1973.[166] Although morphine binds to all the opiate receptor subtypes, including the δ and κ receptors, its analgesic activity is mediated by the μ-opiate receptor.[167] Heroin, the diacetyl derivative of morphine, originally intended as an improved and safer analgesic,[168] turned out to be a prodrug for morphine with an even greater addiction liability.[169] Other alkaloids found in opium with analgesic activity mediated by this receptor include codeine and thebaine.

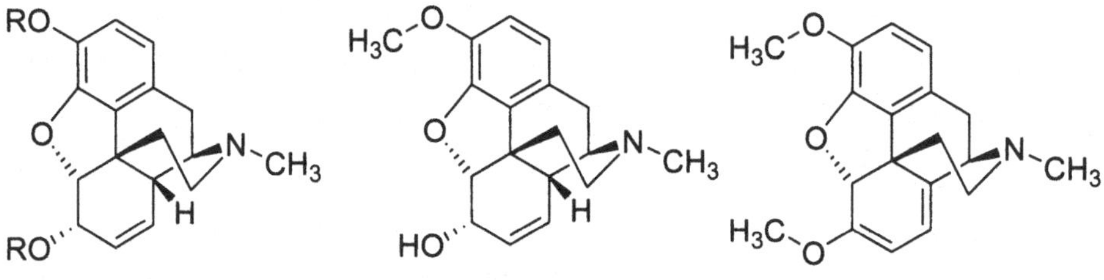

R = H, morphine (Roxanol)
R = COCH₃, heroin   codeine   thebaine

Thebaine is also the source of a number of semisynthetic opiate drugs, also used as analgesics, such as oxycodone (OxyContin), oxymorphone (Opana), and etorphine (Immobilon). The latter is at least 100-times more potent than morphine and is fatal to humans, so it is only used in tranquilizing large animals such as elephants.[170]

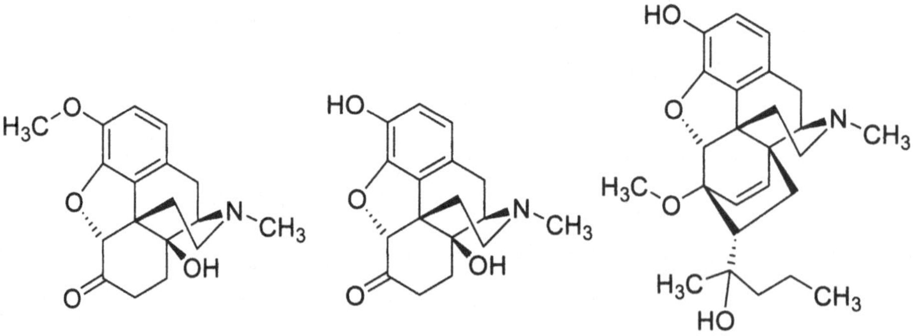

oxycodone (OxyContin)   oxymorphone (Opana)   etorphine (Immobilon)

In addition to the naturally occurring opioid drugs, there are several synthetic analgesics that bind to the same target. Meperidine (Demerol) was originally synthesized at I. G. Farben in 1932 and expected to be an antispasmodic, in analogy with the muscarinic agents after which it was modeled. But designs sometimes go awry, and in this case, meperidine's potent analgesic activity was discovered later in 1940,[171] and its mechanism of action established much later as the same as the opioids, agonist activity at the μ-opiate receptor. Although its efficacy supposedly came without the side effects of the other opiate drugs, its potent affinity for the dopamine and norepinephrine transporters gives it a pharmacological activity similar to cocaine and actually carries a greater potential for addiction.[172] In addition, its drug-drug interactions can be serious, and hence it is no longer widely used clinically.

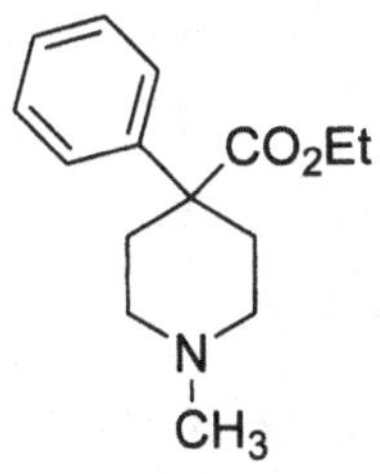

meperidine (Demerol) 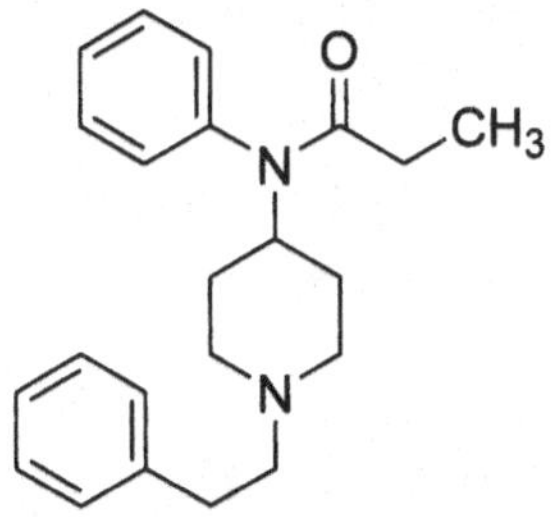   fentanyl (Duragesic) 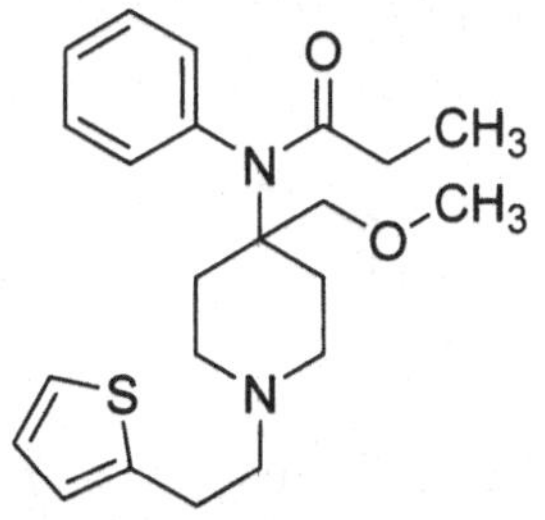   sufentanil (Sufenta)

Meperidine did, however, inspire the synthesis of fentanyl (Duragesic) by Paul Janssen, who used his favorite 4-substituted piperidine pharmacophore to model its features and discovered a far more potent analgesic.[173] Janssen then followed up with the discovery of an even more potent analgesic sufentanil (Sufenta).[174]

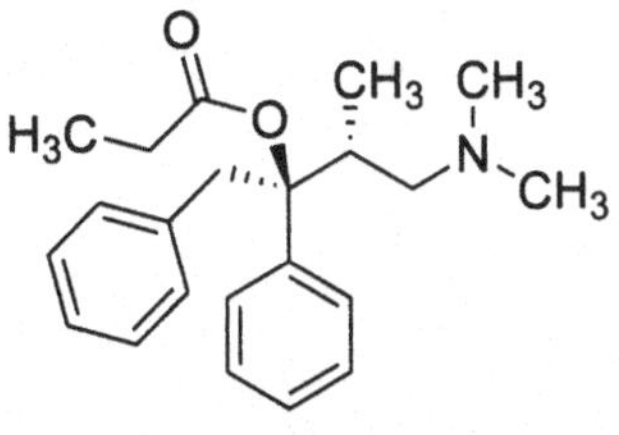

propoxyphene (Darvon) 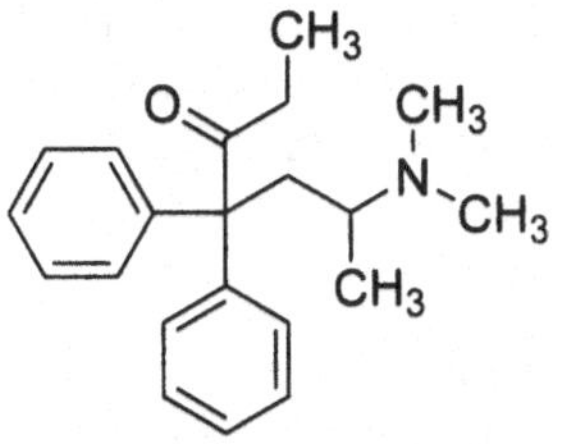   methadone (Dolophine) 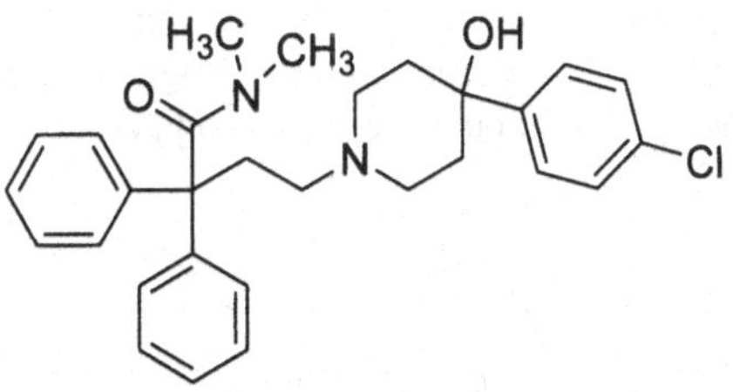   loperamide (Imodium)

The opioid pharmacophore has been simplified still further with acyclic varieties, such as propoxyphene (Darvon), discovered at Lilly in the 1940s and developed as an analgesic.[175] Although widely used, propoxyphene has been recently removed from clinical practice due to limited efficacy and a history of troublesome side effects.[176] Methadone (Dolophine) was discovered by scientists at I. G. Farben[177] as a remedy for Germany's opium shortage during World War II and marketed in the U.S. by Lilly. It eventually became a treatment for heroin detoxification, although it has not been a satisfactory solution to the problem of opiate addiction.[178] Finally, Janssen Labs developed an acyclic compound in this series with high affinity for the μ-opiate receptor based on the 4-substituted piperidine pharmacophore loperamide (Imodium). Loperamide is not an analgesic, however, since it does not cross the blood-brain barrier but instead is useful in treating diarrhea based on its antimotility effects, present as a side effect, constipation, with the other opiates.

In addition to μ-opiate agonists, the μ-opiate antagonists naloxone (Narcan), naltrexone (Revia), and nalmefene (Revex) have found medical utility. While naloxone is used to reverse opiate-induced side effects from overdose,[179] such as occurs in heroin addicts, the latter two drugs are used in controlling alcohol abuse,[180,181] as they are postulated to lessen the reward associated with alcohol consumption, encouraging abstinence.

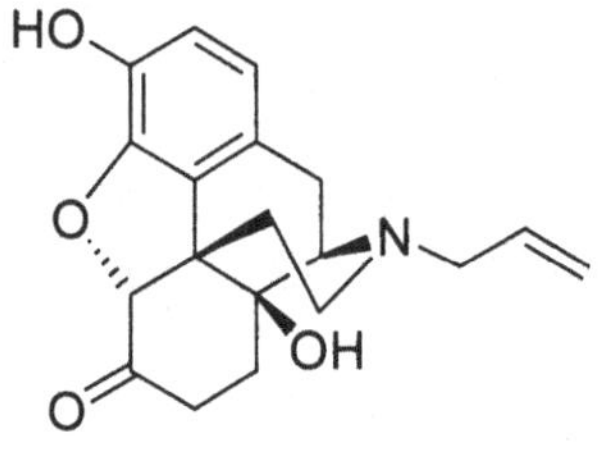

naloxone (Narcan) 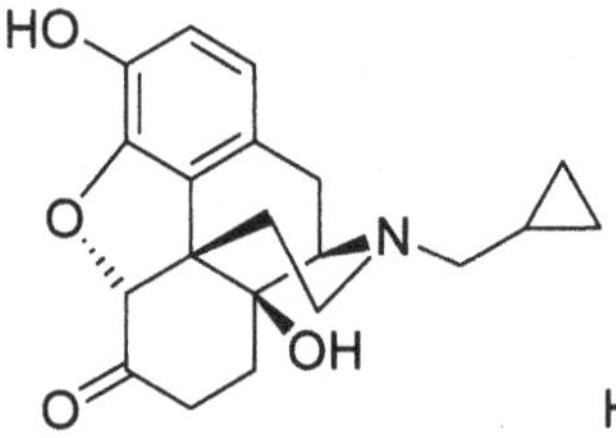   naltrexone (Revia) 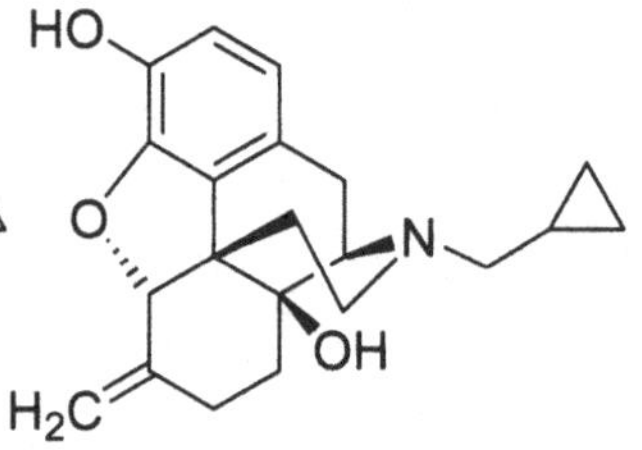   nalmefene (Revex)

6.2   Miscellaneous CNS Analgesics

While many other drugs are reputed to be centrally acting analgesics, including marijuana, three drugs are worth considering here. The first is ketamine (Ketanest, the more active S-isomer), developed at Parke–Davis in 1962.[182] It is used primarily as a general anesthetic[183] but is also an analgesic,[184] although its side effects limit its overall utility. Its pharmacological target is thought to be the NMDA receptor for the excitatory neurotransmitter glutamate,[185] or possibly an opiate receptor, but recent evidence implicates the neuronal pacemaker HCN1 channel in its anesthetic effects.[186] Recent findings suggest it may have a novel and quite robust antidepressant effect, which could be quite useful in patients who do not respond to the SSRIs.[187]

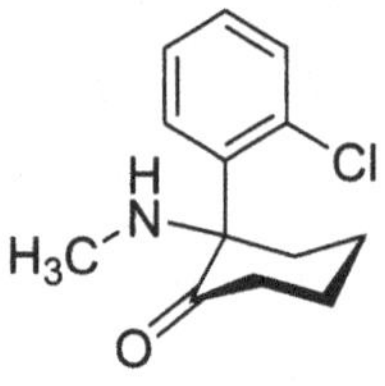

(S)-ketamine (Ketanest) 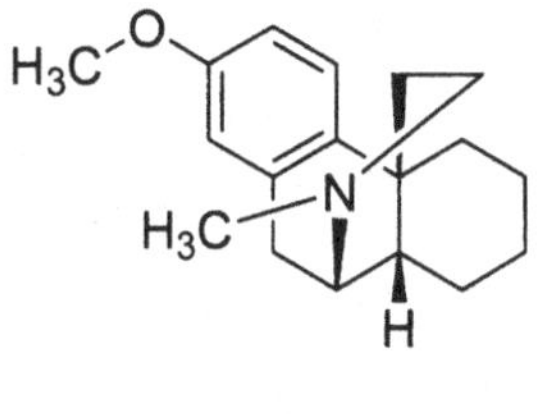 dextromethorphan (DXM) 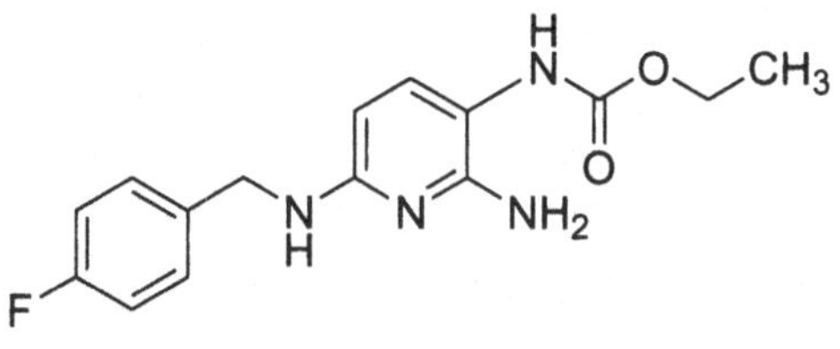   flupirtine (Katadolon)

Dextromethorphan (DXM) was originally marketed as an antitussive (cough suppressant, in popular cough syrups such as Robitussin),[188] but its efficacy has been questioned in more recent studies.[189] In addition, its abuse liability is similar to that of ketamine, possibly due to its activity at the same pharmacological target, the NMDA receptor, further curbing its utility.[190] The target for the chronic back pain drug flupirtine (Katadolon) is the Kv7.2/5 potassium ion channel, which controls neuronal excitability.[191] Flupirtine has been marketed in Europe since 1984 and is being investigated for the treatment of fibromyalgia in the U.S.,[192] and similar analogs are being developed for epilepsy.[193]

7   Drugs for Treating Substance Abuse and ADHD

Ironically, these two areas are related because drugs used to treat attention deficit hyperactivity disorder (ADHD) are often abused as stimulants, and hence both areas will be covered in this section.

7.1   Substance Abuse

The previous section covered drugs used for treating heroin and alcohol addiction, based on μ-opiate receptor antagonism. Treatments for smoking cessation typically focus on the receptor for the active ingredient in cigarette smoke–nicotine, or its downstream targets. One obvious drug used to wean smokers off their habit is nicotine itself, in various forms such as a gum (Nicorette) or patch (Nicoderm). These "nicotine replacement" therapies show some limited efficacy in treating smoking cessation.[194]

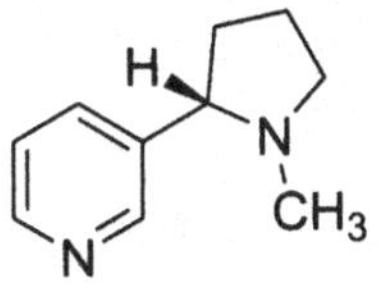

nicotine (Nicorette) 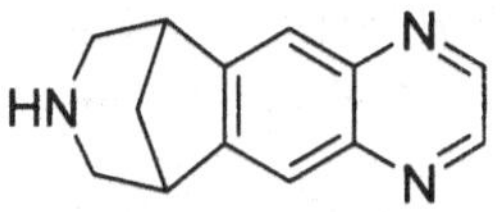   varenicline (Chantix) 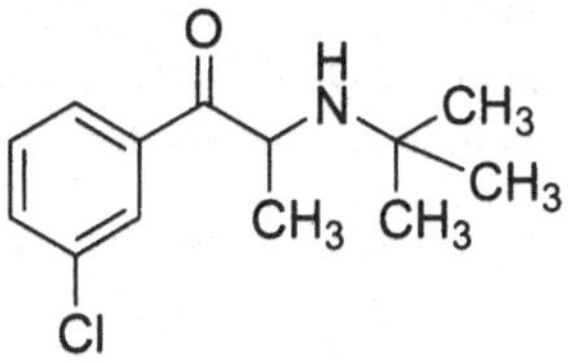   bupropion (Zyban)

The first nonnicotine drug for smoking cessation acting through nicotine receptors is varenicline (Chantix). The compound was synthesized by Jotham Coe at Pfizer[195] and developed for marketing in 2006.[196] Its efficacy is thought to result from its partial agonism at the α4β2 nicotine receptor subtype, through which nicotine exerts its reinforcing, addictive effects, thus acting as a replacement that, however, is not itself addictive while at the same time blocking the need to smoke.[197] Its success was tempered by reports of suicidal ideation while using the drug, and a black box warning was added in 2009.[198]

Bupropion (Zyban), a combined dopamine and norepinephrine reuptake inhibitor originally marketed for treating depression as Wellbutrin,[199] has also found use in treating smoking cessation. Bupropion satisfies the craving for nicotine by acting downstream from nicotine receptors and substituting for it, analogous to replacement therapy with nicotine.[200] One of its drawbacks is the ability to lower seizure threshold.[201] Like other antidepressants, it carries a black box warning regarding suicidal ideation.[202]

7.2   Attention Deficit Hyperactivity Disorder (ADHD)

ADHD is often difficult to diagnose, as it begins in childhood, with its multitude of behavioral issues, and often persists into adulthood; in addition, it has an uncertain etiology.[203] It is characterized by the co-occurrence of attention dysfunction and hyperactivity, it is more prevalent in boys than in girls, and its diagnosis is often controversial.[204]   Drugs used to treat ADHD originate from the stimulant class of drugs, including some classified as potential substances of abuse. It seems paradoxical that stimulants would treat a hyperactivity disorder, but these drugs actually help focus attention, thus addressing a major unmet need in ADHD patients, in addition to reducing hyperactivity.   Stimulant drugs have a long history outside of mainstream clinical medicine, and only three of them have found reliable use in treating ADHD.

Amphetamine was originally synthesized in 1887 and marketed in the 1930s as a bronchodilator and later for depression, narcolepsy, and obesity in the U.S. as Dexedrine and Benzedrine.[205] Its more widespread use as a stimulant by the military and others led to a better appreciation of its addictive liability, and it was finally regulated in 1970.[206] Its utility in ADHD was discovered serendipitously in the 1930s during a failed trial as a headache remedy for children when it nevertheless improved their behavior and performance.[207] It is currently approved as a treatment for ADHD and narcolepsy in a mixed-salt form marketed as Adderall. A salt form with lysine, which provides a once-a-day dosage form due to its slow absorption in the GI tract, is marketed for ADHD as Vyvanse.[208]   Its mechanism of action combines action on both the vesicular monoamine transporter VMAT-2 and the dopamine transporter (DAT). By serving as a VMAT-2 substrate, it displaces and thus releases dopamine, and by serving as a DAT substrate, it effects reverse transport of dopamine.[209] In addition to this two-fold action of dopamine elevation, it also elevates levels of norepinephrine and serotonin, and the latter has been implicated as a possible contributor to its therapeutic benefit in ADHD.[210] The methyl analog methamphetamine (Desoxyn) is also FDA approved for the treatment of ADHD, although it is more widely abused than amphetamine, in part due to improved brain penetration.[211]

R = H, dextroamphetamine (Dexedrine)
R = CH_3, methamphetamine (Desoxyn) methylphenidate (Ritalin) atomoxetine (Strattera)

Methylphenidate (Ritalin) was originally synthesized in 1944[212] and its stimulant properties disclosed in 1954.[213] Its development as a stimulant began to accelerate in the 1990s as the ADHD diagnosis became more prevalent.[214] Although it is sold as a mixture of stereoisomers, the D-*threo* isomer is the active compound, being about 10 times more active than the L-*threo* isomer.[215] Like amphetamine, methylphenidate's efficacy in ADHD results from elevating levels of dopamine and norepinephrine, although in this case it blocks the DAT rather than serving as a substrate.[216] Atomoxetine (Strattera) was developed in the same program that developed the antidepressant fluoxetine (vide supra), with the difference that atomoxetine selectively inhibits the norepinephrine transporter (NET) rather than the serotonin transporter (SERT).[217] Elevating norepinephrine and dopamine, which is also a substrate for NET, in the frontal cortex is thought to improve attention in ADHD patients.[218]

8   Drugs for Neurodegenerative Diseases

Neurodegenerative diseases have only recently received much attention from the medical profession, as they were long felt to be just an unavoidable consequence of old age. Alzheimer's disease (AD) is the most prevalent of these diseases, involving progressive memory loss and behavioral difficulties leading to death.[219] Its prevalence is almost 50% over the age of 85,[220] and its etiology is poorly understood, although deposition of amyloid plaques and neurofibrillary tangles in AD brain are hallmarks of the disease. It is known to involve neurodegeneration in cholinergic regions of the brain important for memory and cognition, such as the basal forebrain and cerebral cortex.[221]

Parkinson's disease (PD) sometimes involves progressive cognitive dysfunction, but its most characteristic symptoms are motor effects such as resting tremor, rigidity, and shuffling gait.[222] One theory attributes its etiology to aggregation of alpha-synuclein, the main component of Lewy bodies, a characteristic deposit in PD brain.[223] Several other factors, including the protein kinase LRRK2[224] and the ubiquitin-protein ligase Parkin,[225,226] have also been implicated in PD. External factors, such as environmental toxins and brain trauma from injury or disease, are also thought to play a role in the disease.[227] These factors lead to the loss of dopaminergic neurons in the substantia nigra region of the brain and progressive disability leading to death.

Huntington's disease is one of the so-called triplet-repeat diseases, being an autosomal dominant genetic disorder caused by an increase in the length of CAG repeats that code for polyglutamine in the amino-terminal region of the Huntington protein Htt.[228] While there is a more severe, juvenile, form of the disease, it typically afflicts patients in middle age with progressive loss of motor control, a characteristic chorea, or dance like movement disorder, and cognitive dysfunction leading to death.[229] Despite the known association of Htt with HD and alpha-synuclein with PD suggesting avenues to halt disease progression, treatments for all three diseases are limited to symptomatic relief.

8.1   Alzheimer's Disease (AD)

physostigmine (Antilirium)   neostigmine (Prostigmin)

Inhibitors of acetylcholinesterase (AChE), the enzyme responsible for the breakdown of the neurotransmitter acetylcholine (ACh), were long known from traditional medicine. Physostigmine (Antilirium) is one of the principal components of the Calabar bean, used for generations in traditional African culture in the "ordeal trial" to judge the guilt of an accused criminal.[230,231] If the accused vomits the concoction of Calabar beans quickly enough, he will survive, thus proving his innocence. The cholinergic potentiating effect of physostigmine has proven valuable in Western medicine, both as an antagonist for atropine in ophthalmology and, along with the analog neostigmine (Prostigmin), for recovery from anesthesia. The basis for its inhibition of AChE is the carbamate group, which acylates the active site serine in the enzyme that hydrolyzes the acetyl group from ACh.

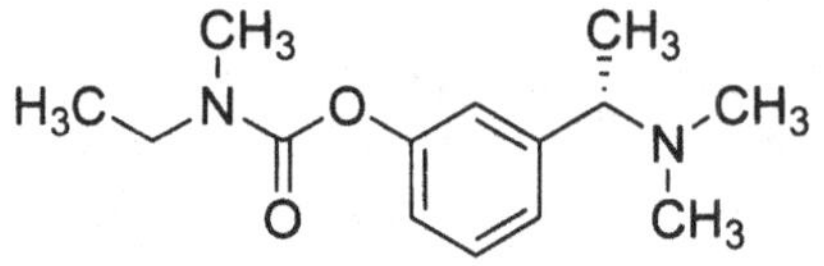

rivastigmine (Exelon)

Given that the deficit in cholinergic transmission was one of the first aspects of Alzheimer's disease (AD) to be identified, it is surprising that the first AChE inhibitor to be proven in clinical trials for AD did not come from this structural class. Instead, this hypothesis was tested with a synthetic AChE inhibitor, tacrine (Cognex), originally synthesized in the 1940s in Australia in an attempt to find new antibacterials for the war effort.[232]   The drug was subsequently nicknamed THA and shown to be an AChE inhibitor. W. K. Summers reported the first clinical trial with THA in 17 AD patients in 1986.[233] The alleviation of cognitive deficits in the patients in this trial led to many subsequent clinical studies, some of which were also successful. Tempering the enthusiasm for the first successful treatment of AD, however, was the finding of THA's hepatic toxicity.[234] Thus Summers' studies were more important for confirming the cholinergic hypothesis of AD and setting the stage for the subsequently marketed AD drugs.[235]

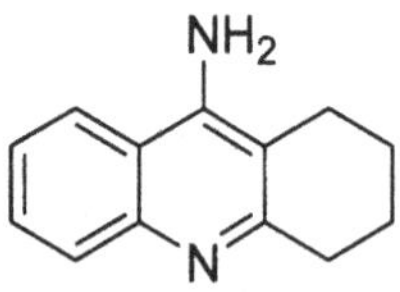

tacrine (Cognex) 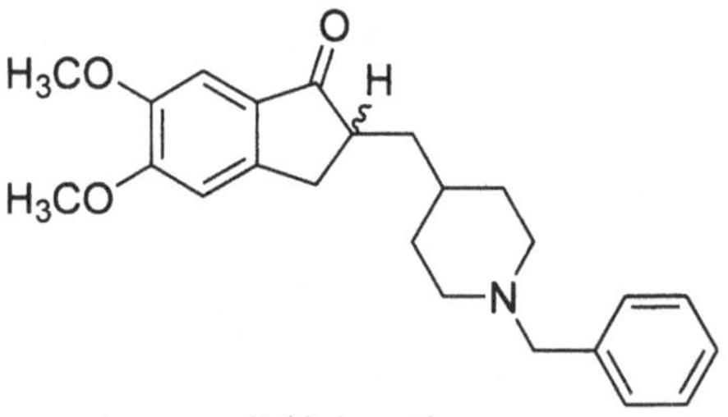   donepezil (Aricept) 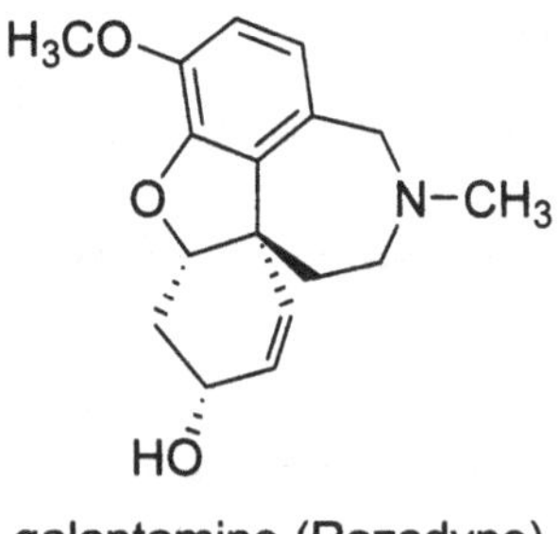   galantamine (Razadyne)

Following on the discovery of tacrine's beneficial effects on cognition in AD patients, donepezil (Aricept) was the next drug to reach the market.[236] Its better safety profile enabled it to become the market leader and supplant tacrine. The next drug returned to the carbamate series with rivastigmine (Exelon), which has demonstrated efficacy in AD,[237] and also is approved for treating symptoms of dementia in Parkinson's disease (PD).[238] The final entry is galantamine (Razadyne), a natural product from the Caucasian snowdrop originally studied in the Soviet Union and found to be an AChE inhibitor.[239] It subsequently demonstrated efficacy in treating cognitive symptoms of AD[240] and was approved in 2006.[241]

While they improve cognition for a limited period of time, the AChE inhibitors have minimal effects on slowing disease progression. Another mechanism appears to have an advantage in this area, exemplified by the NMDA receptor antagonist memantine (Namenda).

R = H, amantadine (Symmetrel)
R = CH$_3$, memantine (Namenda)

The original discovery in this structural class was actually the parent, amantadine (Symmetrel), which was found to block replication of influenza virus by binding to the virus's M2 protein, required for viral entry into host cells. After its approval in 1966,[242] it was found serendipitously to improve cognition in a PD patient treated for the flu and was subsequently shown in larger clinical trials to be an effective treatment for PD.[243] Memantine (Namenda) was patented by Lilly in 1968, and with the discovery of the efficacy of amantadine, it was tried in PD patients and found to be similarly effective.[244] Considerable effort to characterize the pharmacological target for this effect identified the NMDA receptor, a prime mediator of cognition and memory, as a likely candidate, although several others remain in contention. Memantine's uncompetitive block of the NMDA receptor ion channel seems to be important both for avoiding the side effects of other NMDA blockers and for affording its efficacy.[244] AD was a logical target for a cognitive-enhancing agent, and memantine has proven effective there as well, especially when used in conjunction with AChE inhibitors such as donepezil; it may also slow disease progression.[245]

8.2   Parkinson's Disease (PD)

Cotzias describes in detail the early events in the discovery of the efficacy of L-DOPA (Levodopa) in the treatment of PD.[246] The study of L-DOPA for PD began with the observations of Arvid Carlsson in the 1950s that whereas administering dopamine itself to animals after creating a CNS deficiency in dopamine did not restore its levels, this could be accomplished by administration of the biosynthetic precursor DOPA, which is actively transported into the brain and converted to dopamine by DOPA-decarboxylase.[247] In this work, Carlsson was able to reverse the symptoms of reserpine treatment in animals, which resemble Parkinsonism and result from monoamine depletion in the brain. The subsequent discovery by Hornykiewicz that dopamine levels are greatly reduced in the PD brain connected Carlsson's work to the actual disease,[248] stimulating clinical studies that initially met with some success.[249] The utility of L-DOPA, however, proved difficult to replicate, and ultimate confirmation of efficacy required careful administration of high doses of L-DOPA by Cotzias, who successfully treated 8 out of 16 PD patients.[250] Side effects remained a major obstacle in L-DOPA therapy, however, and these were finally alleviated by coadministration of carbidopa (Lodosyn, the combination is tradenamed Sinemet), which blocks DOPA-decarboxylase's conversion of L-DOPA to dopamine in the periphery.[251] As a further enhancement of efficacy, entacapone (COMTan), an inhibitor of catechol-O-methyl transferase, an enzyme that converts L-DOPA to its 3-O-methyl ether, which is not transported into the brain, is coadministered with Sinemet.[252] The combination of entacapone with L-DOPA/carbidopa is tradenamed Stalevo. Although this therapy can dramatically reverse many of the symptoms of

PD, dyskinesia and other side effects remain, and the therapy does not slow disease progression, which inexorably leads to loss of efficacy in the later stages of the disease.

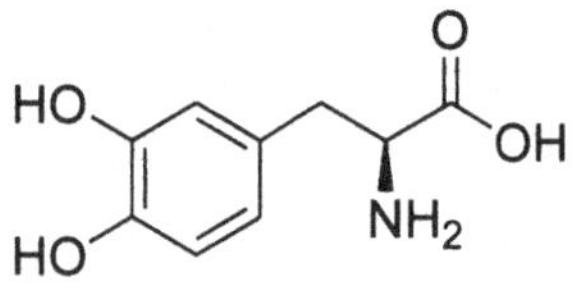

L-DOPA (Levodopa) 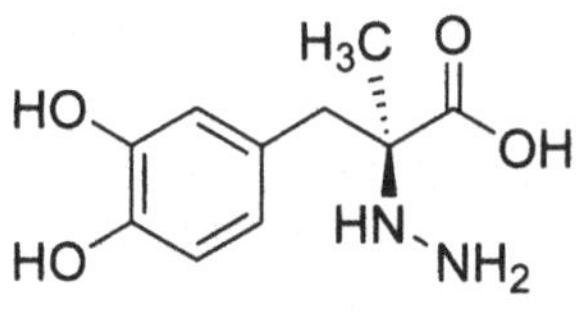   carbidopa (Lodosyn) 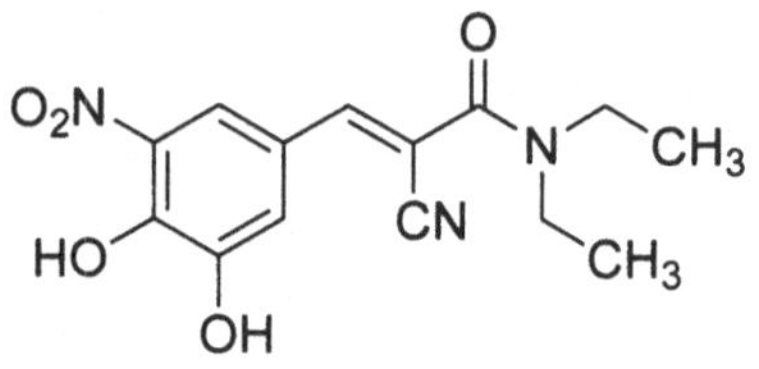   entacapone (COMTan)

A similar therapeutic strategy involves the use of dopamine agonists as a replacement for the lack of available dopamine. Apomorphine was originally discovered as a breakdown product from heating morphine in strong acid, but it shares none of morphine's pharmacology and is instead a dopamine agonist, primarily at the D_2 through D_5 receptors.[253] Cotzias was also responsible for defining the therapeutic benefit for apomorphine, especially in relation to L-DOPA.[254] Side effects with apomorphine are similar to those seen with L-DOPA, including nausea and vomiting, and are countered with a dopamine receptor antagonist, domperidone.[255] In addition, many of the motor side effects of apomorphine and other dopamine agonists, as well as those induced by chronic L-DOPA treatment, may result from receptor down-regulation typically seen in animals with such chronic treatments, although these receptor interactions are very complex,[256] and this situation greatly complicates PD therapy.

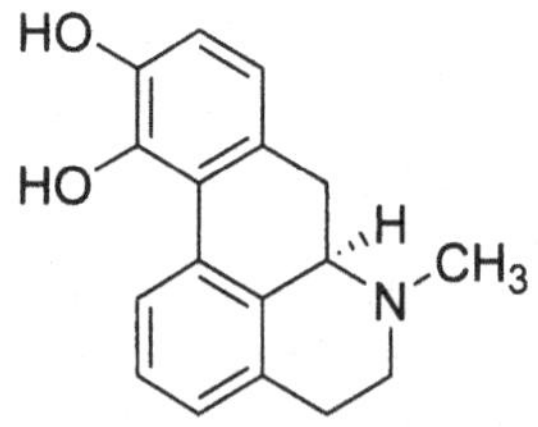

apomorphine (Apokyn) 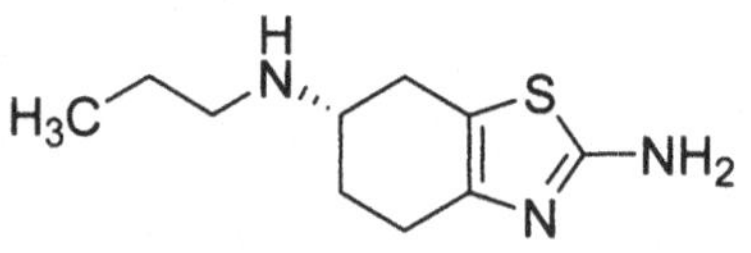   pramipexole (Mirapex) 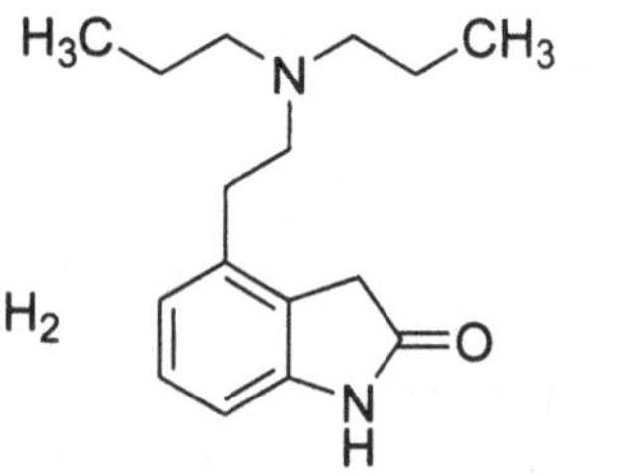   ropinirole (Requip)

Several additional dopamine agonists have been developed to improve efficacy and better manage side effects in treating PD, including pramipexole (Mirapex), originally synthesized as a structural analog of apomorphine, where the aminothiazole serves to mimic the catechol.[257] Its efficacy in PD is similar to L-DOPA and may be better in early PD patients, while its side-effect profile is also similar.[258] Another is ropinirole (Requip), originally discovered as a selective D_2 receptor dopamine agonist in which the oxindole ring is the catechol mimetic.[259] While its efficacy is comparable to L-DOPA, it shows a reduction in side effects, especially in early PD patients.[260] Both of these latter drugs are indicated for the treatment of restless leg syndrome as well as PD.

MAO-B inhibitors are also used to manage the symptoms in PD, although these drugs differ from the MAO inhibitors reviewed earlier that are used to treat depression. Again, their mechanism of action is inhibition of the breakdown of dopamine by MAO-B, and they include selegiline and rasagiline. Joseph Knoll discovered selegiline in collaboration with the Hungarian company Chinoin as an alternative to methamphetamine and the MAO inhibitors known at the time.[261] Its structure combines elements of both, and it acts as a selective MAO-B inhibitor at low doses, although its metabolites include methamphetamine and its metabolites. Birkmayer established its efficacy in PD, in which it reduces the required dose of L-DOPA and treats some of the motor dyskinesia,[262] although claims that it slows disease progression have not been substantiated.[263] Discovered and developed by Teva, rasagiline shows similar benefits in the treatment of PD[264] and is being investigated for its potential in slowing disease progression.[265]

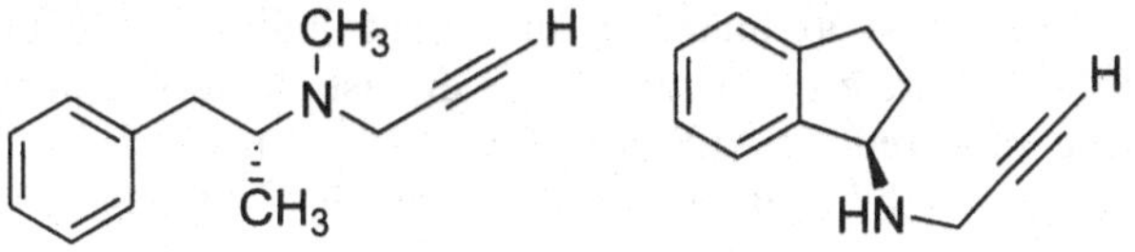

selegiline (Eldepryl)   rasagiline (Azilect)

8.3   Huntington's Disease (HD)

Reserpine (Serpasol) is an Indian herbal medicine from the snakeroot plant that was isolated and characterized by Schlittler in the early 1950s.[266] It was known from its traditional history as an antihypertensive, and this activity was confirmed in Western medicine. Like many other drugs being investigated at the time, it was also tested as a potential antipsychotic drug but failed to establish itself as a reliable therapy. In further testing during this time, however, it did demonstrate a benefit in controlling chorea in HD patients.[267] Its side effects, especially reserpine-induced depression, as well the advent of safer drugs such as chlorpromazine for psychosis and propranolol for treating hypertension, curtailed its clinical usage.[268] So despite its efficacy in HD, it was eventually phased out in favor of tetrabenazine (vide infra).[269] The basis for its activity is depletion of monoamines by blocking the vesicular monoamine transporters VMAT-1 and VMAT-2, which load neurotransmitter amines such as histamine, dopamine, and norepinephrine into synaptic vesicles in endocrine cells and neurons, respectively.[270]

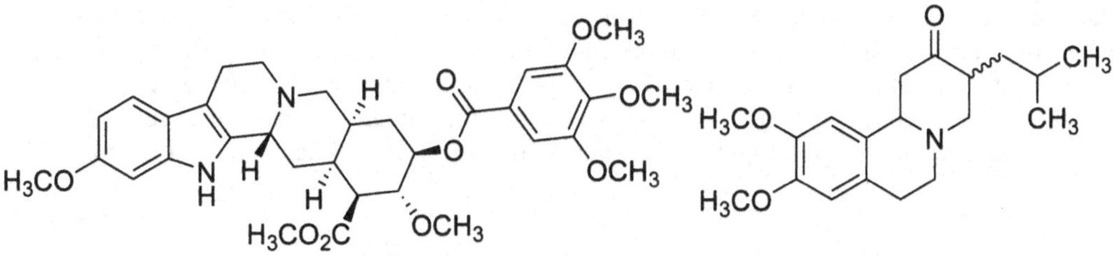

reserpine (Serpasol)   tetrabenazine (Xenazine)

Tetrabenazine (Xenazine) was synthesized in the 1950s and designed as a potential antipsychotic drug during the era of the discovery of the tricyclic antipsychotics and antidepressants.[271] While not a commercial success as an antipsychotic, it nevertheless proved to be a suitable substitute for reserpine in treating the choretic movement disorders that occur in HD.[272] In 2008, it became the first drug approved by the FDA for the treatment of chorea associated with HD.[273] Unlike reserpine, it selectively targets the neuronal transporter VMAT-2,[270] which may explain its reduced side-effect profile. Thus its monoamine-depleting activity in the CNS enables control of the hyperactivity that leads to involuntary hyperkinetic movement disorders.

8.4   Amyotrophic Lateral Sclerosis (ALS)

ALS, also known as Lou Gehrig's disease after the famous New York Yankees baseball player who died from the disease in 1941, is a progressive neurodegenerative disease of motor neurons.[274] Although the cause(s) are unknown, the recent finding that mutations in the gene for superoxide dismutase (SOD1) are associated with the disease suggests that oxidative damage is a factor.[275] The only drug approved for ALS is riluzole (Rilutek), which is postulated to work by reducing glutamate levels through increased glutamate uptake,[276] although other mechanisms have been invoked.[277] Riluzole was originally discovered as an anticonvulsant in the 1980s[278] and was subsequently tested in ALS patients, demonstrating a slowing of progression of some ALS symptoms and improved survival.[279]

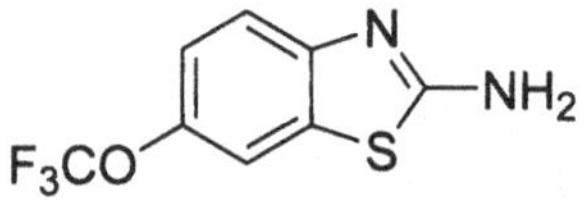

riluzole (Rilutek)

8.5   New Directions in Neurodegenerative Disease

Much of the current work in the field of CNS drug discovery is focused on cognitive enhancement and slowing neurodegenerative disease progression, because of the large unmet medical need and the consequent commercial potential in these areas. Covering all these efforts is beyond the scope of the present chapter, but a few new directions will be mentioned. The family of compounds exemplified by piracetam and levetiracetam has been studied for many years, and their mechanism of action is still undetermined.[280] Piracetam (Nootropil) was discovered in 1964, and it is marketed for treating mild seizures termed myoclonus.[281] It has been reported to have cognition enhancing effects,[282] although there are a number of controversial aspects of this application.[283] Levetiracetam (Keppra) was discovered in 1992 using an audiogenic seizure model to find improved versions of piracetam.[284] It has been approved for treating partial onset seizures,[285] but while it too is reputed to have cognitive enhancing properties, this has been difficult to prove.[286]

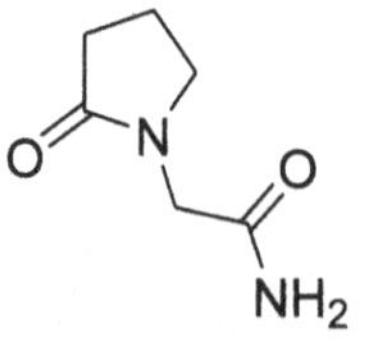

piracetam (Nootropil) 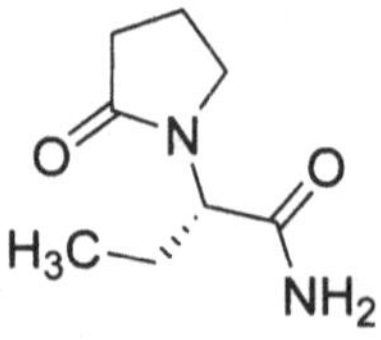   levetiracetam (Keppra)

Probably the most extensive effort in the CNS drug discovery area is currently focused on AD, with several approaches under investigation.[287] One strategy is to lower the β-amyloid deposits characteristic of AD that seem to play a critical role in neuronal toxicity by the inhibition of γ-secretase, an enzyme that produces the Aβ42 form of the protein in these deposits which is reputed to be the toxic moiety. Aside from antibody and vaccine approaches, two small molecules have advanced to Phase III clinical trials. The first is semagacestat, which has demonstrated lowering of Aβ levels in plasma and cerebrospinal fluid in Phase II studies. A second compound, R-flurbiprofen (Tarenflurbil), failed to demonstrate efficacy in its Phase III trials despite promising results from earlier studies. These results illustrate the slow progress made to date with this approach.

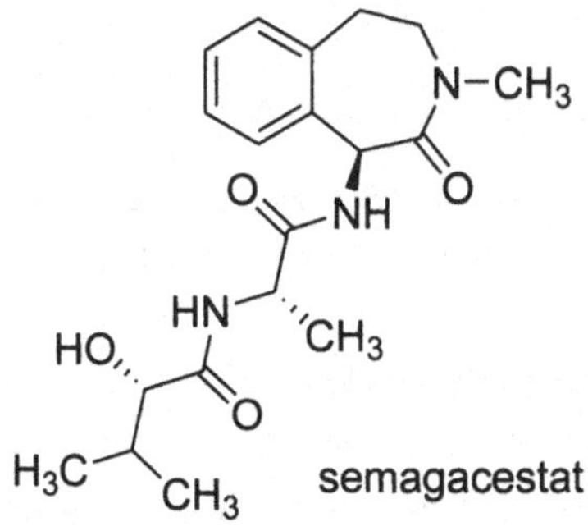

semagacestat 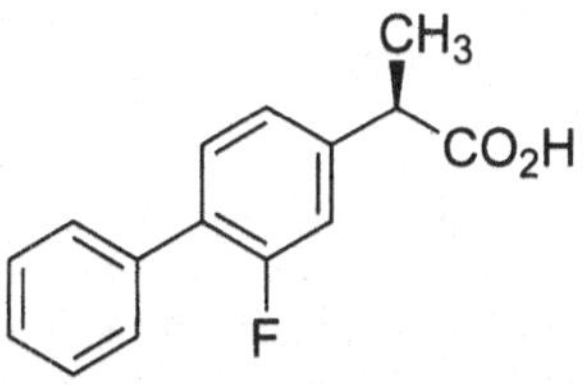   fluribiprofen (Tarenflurbil)

A number of other approaches have failed to advance compounds very far in the clinic. For example, developing inhibitors of the other enzyme responsible for β-amyloid production, β-secretase, has been hampered by the difficulty of identifying brain penetrant compounds, given that the enzyme is an aspartyl protease.[288] In addition, the other hallmark in AD pathology is neurofibrillary tangles containing hyperphosphorylated tau protein, normally a component of microtubules.[289] Protein kinase inhibitors are being studied preclinically to address this, but this approach is also challenged in finding brain penetrant compounds.[290]

9   Future Prospects for New CNS Drugs

The sequencing of the human genome has offered the potential to facilitate identification of genes involved in human disease,[291] including CNS disorders. Given the complexity of CNS diseases, however, it may seem surprising that one gene will be identified that will lead to a significant advance or cure in this area. But enthusiasm for this approach is strong, and an excellent review of the history and future directions in schizophrenia drug discovery recounts its complexity and the promise of new targets from molecular genetics studies.[292] As an example of the potential of genetics studies to uncover new targets, the authors describe the discovery of the DISC-1 gene, discovered in a large Scottish family with a history of psychotic disorders.[293] Along with its binding partners, NDEL1 and NDE1,[294] DISC1 appears to play a role both in neurodevelopmental abnormalities in schizophrenia that begin in utero, and also later in life in the behavioral dysfunction that more obviously characterizes the disease.[295] A great deal of work remains to elucidate the functions of DISC1 in an effort to identify new targets for potential schizophrenia drugs, but in the meantime, it is also becoming increasingly clear that there is considerably more genetic complexity involved in the etiology of the disease. For example, a recent study found multiple deletions and duplications, often in pathways involving neurodevelopment or glutamate transmission, in a survey of schizophrenia patients, emphasizing this complexity.[296]

The situation for other CNS diseases, such as anxiety and depression is even less clear. Although the numerous novel approaches currently under investigation suggest the potential for therapeutic advances in this area, the discovery of valid targets for these diseases is daunting. The crucial unmet medical need in the neurodegenerative disease area demands a serious commitment of resources to find better treatment for victims of these diseases. But until and unless our understanding of the molecular pathology of these diseases improves greatly, we will be dependent on serendipitous clinical observation for progress in this field, as we have for most of the CNS drugs on the market.

References

1   *DSM-IV: Diagnostic and Statistical Manual of Mental Disorders*, American Psychiatric Association: Washington DC, 1994.
2   Hirsch, S. R.; Weinberger, D. R. *Schizophrenia*, Blackwell Science Ltd: Malden, MA, 1995.
3   Davis, K.; Kahn, R.; Ko, G.; Davidson, M. *Am. J. Psychiatry* **1991**, *148*, 1474.
4   Haracz, J. *Schizophrenia Bull.* **1982**, *8*, 438.
5   Timmerman, H. *Pharm. World Sci.* **1989**, *11*, 146.
6   Healy, D. In *The Creation of Psychopharmacology*, Harvard University Press: Cambridge, 2004.
7   Ban, T. A. *Neuropsychiatr. Dis. Treat.* **2007**, *3*, 495.
8   Rees, L. *Br. Med. J.* **1960**, *2*, 522.
9   Granger, B.; Albu, S. *Ann. Clin. Psychiatry: Official J. Am. Acad. Clin. Psychiatrists* **2005**, *17*, 137.
10   Fitton, A.; Heel, R. *Drugs* **1990**, *40*, 722.
11   Grunder, G.; Hippius, H.; Carlsson, A. *Nat. Rev. Drug Discov.* **2009**, *8*, 197.
12   Chakrabarti, J. K.; Horsman, L.; Hotten, T. M.; Pullar, I. A.; Tupper, D. E.; Wright, F. C. *J. Med. Chem.* **1980**, *23*, 878.
13   Bhana, N.; Foster, R.; Olney, R.; Plosker, G. *Drugs* **2001**, *61*, 111.
14   Marder, S.; Meibach, R. *Am. J. Psychiatry* **1994**, *151*, 825.
15   Yevich, J.; New, J.; Smith, D.; Lobeck, W.; Cart, J.; Minielli, J.; Eison, M.; Taylor, D.; Riblet, L.; Temple, D. *J. Med. Chem.* **1986**, *29*, 359.
16   Howard, H.; Lowe III, J.; Seeger, T.; Seymour, P.; Zorn, S.; Maloney, P.; Ewing, F.; Newman, M.; Schmidt, A.; Furman, J. *J. Med. Chem.* **1996**, *39*, 143.
17   Seeger, T.; Seymour, P.; Schmidt, A.; Zorn, S.; Schulz, D.; Lebel, L.; McLean, S.; Guanowsky, V.; Howard, H.; Lowe, J. *J. Pharm. Exp. Ther.* **1995**, *275*, 101.
18   Kim, S.; Huang, A.; Snowman, A.; Teuscher, C.; Snyder, S. *Proc. Nat. Acad. Sci.* **2007**, *104*, 3456.
19   Warawa, E. J.; Migler, B. M.; Ohnmacht, C. J.; Needles, A. L.; Gatos, G. C.; McLaren, F. M.; Nelson, C. L.; Kirkland, K. M. *J. Med. Chem.* **2001**, *44*, 372.
20   Migler, B.; Warawa, E.; Malick, J. *Psychopharmacology* **1993**, *112*, 299.
21   Davies, M. A.; Sheffler, D. J.; Roth, B. L. *CNS Drug Rev.* **2004**, *10*, 317.
22   Verhoest, P. R.; Chapin, D. S.; Corman, M.; Fonseca, K.; Harms, J. F.; Hou, X.; Marr, E. S.; Menniti, F. S.; Nelson, F.; O'Connor, R.; Pandit, J.; Proulx-LaFrance, C.; Schmidt, A. W.; Schmidt, C. J.; Suiciak, J. A.; Liras, S. *J. Med. Chem.* **2009**, *52*, 5188.
23   Rorick-Kehn, L. M.; Johnson, B. G.; Burkey, J. L.; Wright, R. A.; Calligaro, D. O.; Marek, G. J.; Nisenbaum, E. S.; Catlow, J. T.; Kingston, A. E.; Giera, D. D.; Herin, M. F.; Monn, J. A.; McKinzie, D. L.; Schoepp, D. D. *J. Pharmacol. Exp. Ther.* **2007**, *321*, 308.
24   Patil, S. T.; Zhang, L.; Martenyi, F.; Lowe, S. L.; Jackson, K. A.; Andreev, B. V.; Avedisova, A. S.; Bardenstein, L. M.; Gurovich, I. Y.; Morozova, M. A.; Mosolov, S. N.; Neznanov, N. G.; Reznik, A. M.; Smulevich, A. B.; Tochilov, V. A.; Johnson, B. G.; Monn, J. A.; Schoepp, D. D. *Nat. Med.* **2007**, *13*, 1102.

25   Snyder, E. M.; Murphy, M. R. *Nat. Rev. Drug Discov.* **2008**, *7*, 471.
26   Lynch, G. *Curr. Opin. Pharmacol.* **2006**, *6*, 82.
27   Meador-Woodruff, J. H.; Healy, D. J. *Br. Res. Rev.* **2000**, *31*, 288.
28   Nagarajan, N.; Quast, C.; Boxall, A. R.; Shahid, M.; Rosenmund, C. *Neuropharmacology* **2001**, *41*, 650.
29   Goff, D. C.; Leahy, L.; Berman, I.; Posever, T.; Herz, L.; Leon, A. C.; Johnson, S. A.; Lynch, G. *J. Clin. Psychopharmacol.* **2001**, *21*, 484.
30   Coyle, J. T.; Tsai, G. *Psychopharmacology* **2004**, *174*, 32.
31   Martina, M.; Gorfinkel, Y.; Halman, S.; Lowe, J. A.; Periyalwar, P.; Schmidt, C. J.; Bergeron, R. *J. Physiol.* **2004**, *557*, 489.
32   Pinard, E. A., D.; Borroni, E.; Fischer, H.; Hainzl, D.; Jolidon, S.; Moreau, J.-L.; Narquizian, R.; Nettekoven, M.; Norcross, R.; Stalder, H.; Thomas, A.; Wettstein, J. G. In *239th American Chemical Society National Meeting* San Francisco, California, 2010.
33   Young, J. W.; Crawford, N.; Kelly, J. S.; Kerr, L. E.; Marston, H. M.; Spratt, C.; Finlayson, K.; Sharkey, J. *Eur. Neuropsychopharmacol.* **2007**, *17*, 145.
34   Leonard, S.; Gault, J.; Hopkins, J.; Logel, J.; Vianzon, R.; Short, M.; Drebing, C.; Berger, R.; Venn, D.; Sirota, P.; Zerbe, G.; Olincy, A.; Ross, R. G.; Adler, L. E.; Freedman, R. *Arch. Gen. Psychiatry* **2002**, *59*, 1085.
35   Wang, H.-Y.; Lee, D. H. S.; Davis, C. B.; Shank, R. P. *J. Neurochem.* **2000**, *75*, 1155.
36   Olincy, A.; Stevens, K. E. *Biochem. Pharmacology* **2007**, *74*, 1192.
37   Thomsen, M. S.; Mikkelsen, J. D.; Timmermann, D. B.; Peters, D.; Hay-Schmidt, A.; Martens, H.; Hansen, H. H. *Neurosci.* **2008**, *154*, 741.
38   Hurst, R. S.; Hajos, M.; Raggenbass, M.; Wall, T. M.; Higdon, N. R.; Lawson, J. A.; Rutherford-Root, K. L.; Berkenpas, M. B.; Hoffmann, W. E.; Piotrowski, D. W.; Groppi, V. E.; Allaman, G.; Ogier, R.; Bertrand, S.; Bertrand, D.; Arneric, S. P. *J. Neurosci.* **2005**, *25*, 4396.
39   Liu, K. G.; Robichaud, A. J. *Drug Dev. Res.* **2009**, *70*, 145.
40   Hirst, W. D.; Stean, T. O.; Rogers, D. C.; Sunter, D.; Pugh, P.; Moss, S. F.; Bromidge, S. M.; Riley, G.; Smith, D. R.; Bartlett, S.; Heidbreder, C. A.; Atkins, A. R.; Lacroix, L. P.; Dawson, L. A.; Foley, A. G.; Regan, C. M.; Upton, N. *Eur. J. Pharmacol.* **2006**, *553*, 109.
41   Li, Z.; Huang, M.; Prus, A. J.; Dai, J.; Meltzer, H. Y. *Br. Res.* **2007**, *1134*, 70.
42   Fava, M.; Kendler, K. *Neuron* **2000**, *28*, 335.
43   Bloch, R.; Dooneief, A.; Buchberg, A.; Spellman, S. *Ann. Int. Med.* **1954**, *40*, 881.
44   Loomer, H.; Saunders, J.; Kline, N. *Psychiatric Res. Rep.* **1957**, *8*, 129.
45   Delay, J.; Buisson, J. *J. Clin. Exp. Psychopathol.* **1958**, *19*, 51.
46   Kuhn, R. *Am. J. Psychiatry* **1958**, *115*, 459.
47   Von Euler, U. *Science* **1971**, *173*, 202.
48   Axelrod, J. *Science* **1971**, *173*, 598.
49   Hertting, G.; Axelrod, J.; Whitby, L. *J. Pharmacol. Exp. Ther.* **1961**, *134*, 146.
50   Carlsson, A. *J. Pharm. Pharmacol.* **1970**, *22*, 729.
51   Wong, M.-L.; Licinio, J. *Nat. Rev. Drug Discov.* **2004**, *3*, 136.
52   Wilson, K.; Mottram, P. *Int. J. Geriatr. Psych.* **2004**, *19*, 754.
53   Meek, J.; Fuxe, K.; Carlsson, A. *Biochem. Pharmacol.* **1971**, *20*, 707.
54   Koe, B. K.; Weissman, A.; Welch, W. M.; Browne, R. G. *J. Pharmacol. Exp. Ther.* **1983**, *226*, 686.
55   Wong, D. T.; Perry, K. W.; Bymaster, F. P. *Nat. Rev. Drug Discov.* **2005**, *4*, 764.
56   Christensen, J. US Patent 4,007,196 (1977).
57   Tang, S. W., Helmeste, D. *Expert Opin. Pharmacother.* **2008**, *9*, 787.
58   Petersen, P. V. L., N.; Hansen, V.; Huld, T.; Hjortkjær, J.; Holmblad, J.; Nielsen, I. M.; Nymark, M.; Pedersen, V.; Jørgensen, A.; Hougs, W. *Acta Pharmacol. Tox.* **1966**, *24*, 121.
59   Bigler, A. J., Boegesoe, K. P.; Toft, A.; Hansen, V. *Eur. J. Med. Chem.* **1977**, *12*, 289.
60   Bymaster, F. P.; Beedle, E. E.; Findlay, J.; Gallagher, P. T.; Krushinski, J. H.; Mitchell, S.; Robertson, D. W.; Thompson, D. C.; Wallace, L.; Wong, D. T. *Bioorg. Med. Chem. Lett.* **2003**, *13*, 4477.

61   Wong, D. T.; Robertson, D. W.; Bymaster, F. P.; Krushinski, J. K.; Reid, L. R. *Life Sci.* **1988**, *43*, 2049.
62   Yardley, J. P.; Husbands, G. E. M.; Stack, G.; Butch, J.; Bicksler, J.; Moyer, J. A.; Muth, E. A.; Andree, T.; Fletcher, H. *J. Med. Chem.* **1990**, *33*, 2899.
63   Millan, M. J.; Gobert, A.; Lejeune, F.; Dekeyne, A.; Newman-Tancredi, A.; Pasteau, V.; Rivet, J. M.; Cussac, D. *J. Pharmacol. Exp. Ther.* **2003**, *306*, 954.
64   Wachtel, H. *Neuropharmacol.* **1983**, *22*, 267.
65   Scott, A. I. F.; Perini, A. F.; Shering, P. A.; Whalley, L. J. *Eur. J. Clin. Pharmacol.* **1991**, *40*, 127.
66   Norman, P. *Expert Opin. Ther. Pat.* **1999**, *9*, 1101.
67   Ashton, M.; Cook, D.; Fenton, G.; Karlsson, J.; Palfreyman, M.; Raeburn, D.; Ratcliffe, A.; Souness, J.; Thurairatnam, S.; Vicker, N. *J. Med. Chem.* **1994**, *37*, 1696.
68   Snider, R.; Constantine, J.; Lowe 3rd, J.; Longo, K.; Lebel, W.; Woody, H.; Drozda, S.; Desai, M.; Vinick, F.; Spencer, R. *Science* **1991**, *251*, 435.
69   McLean, S.; Ganong, A.; Seymour, P.; Bryce, D.; Crawford, R.; Morrone, J.; Reynolds, L.; Schmidt, A.; Zorn, S.; Watson, J. *J. Pharmacol. Exp. Ther.* **1996**, *277*, 900.
70   Hale, J. J.; Mills, S. G.; MacCoss, M.; Finke, P. E.; Cascieri, M. A.; Sadowski, S.; Ber, E.; Chicchi, G. G.; Kurtz, M.; Metzger, J.; Eiermann, G.; Tsou, N. N.; Tattersall, F. D.; Rupniak, N. M. J.; Williams, A. R.; Rycroft, W.; Hargreaves, R.; MacIntyre, D. E. *J. Med. Chem.* **1998**, *41*, 4607.
71   Di Fabio, R.; Griffante, C.; Alvaro, G.; Pentassuglia, G.; Pizzi, D. A.; Donati, D.; Rossi, T.; Guercio, G.; Mattioli, M.; Cimarosti, Z.; Marchioro, C.; Provera, S.; Zonzini, L.; Montanari, D.; Melotto, S.; Gerrard, P. A.; Trist, D. G.; Ratti, E.; Corsi, M. *J. Med. Chem.* **2009**, *52*, 3238.
72   Kramer, M. S.; Cutler, N.; Feighner, J.; Shrivastava, R.; Carman, J.; Sramek, J. J.; Reines, S. A.; Liu, G.; Snavely, D.; Wyatt-Knowles, E.; Hale, J. J.; Mills, S. G.; MacCoss, M.; Swain, C. J.; Harrison, T.; Hill, R. G.; Hefti, F.; Scolnick, E. M.; Cascieri, M. A.; Chicchi, G. G.; Sadowski, S.; Williams, A. R.; Hewson, L.; Smith, D.; Carlson, E. J.; Hargreaves, R. J.; Rupniak, N. M. *Science* **1998**, *281*, 1640.
73   Keller, M.; Montgomery, S.; Ball, W.; Morrison, M.; Snavely, D.; Liu, G.; Hargreaves, R.; Hietala, J.; Lines, C.; Beebe, K.; Reines, S. *Biol. Psychiatr.* **2006**, *59*, 216.
74   Warr, D. G.; Hesketh, P. J.; Gralla, R. J.; Muss, H. B.; Herrstedt, J.; Eisenberg, P. D.; Raftopoulos, H.; Grunberg, S. M.; Gabriel, M.; Rodgers, A.; Bohidar, N.; Klinger, G.; Hustad, C. M.; Horgan, K. J.; Skobieranda, F. *J. Clin. Oncol.* **2005**, *23*, 2822.
75   Patel, L.; Lindley, C. *Expert Opin. Pharmacother.* **2003**, *4*, 2279.
76   Gold, P. W.; Chrousos, G. P. *Proc. Assoc. Am. Phys.* **1999**, *111*, 22.
77   Chen, Y. L.; Mansbach, R. S.; Winter, S. M.; Brooks, E.; Collins, J.; Corman, M. L.; Dunaiskis, A. R.; Faraci, W. S.; Gallaschun, R. J.; Schmidt, A.; Schulz, D. W. *J. Med. Chem.* **1997**, *40*, 1749.
78   Schulz, D. W.; Mansbach, R. S.; Sprouse, J.; Braselton, J. P.; Collins, J.; Corman, M.; Dunaiskis, A.; Faraci, S.; Schmidt, A. W.; Seeger, T.; Seymour, P.; Tingley, F. D.; Winston, E. N.; Chen, Y. L.; Heym, J. *Proc. Natl. Acad. Sci. USA* **1996**, *93*, 10477.
79   Chen, Y. L.; Braselton, J.; Forman, J.; Gallaschun, R. J.; Mansbach, R.; Schmidt, A. W.; Seeger, T. F.; Sprouse, J. S.; Tingley, F. D.; Winston, E.; Schulz, D. W. *J. Med. Chem.* **2008**, *51*, 1377.
80   Chen, C.; Grigoriadis, D. E. *Drug Develop. Res.* **2005**, *65*, 216.
81   Gilligan, P. J.; Clarke, T.; He, L.; Lelas, S.; Li, Y.-W.; Heman, K.; Fitzgerald, L.; Miller, K.; Zhang, G.; Marshall, A.; Krause, C.; McElroy, J. F.; Ward, K.; Zeller, K.; Wong, H.; Bai, S.; Saye, J.; Grossman, S.; Zaczek, R.; Arneric, S. P.; Hartig, P.; Robertson, D.; Trainor, G. *J. Med. Chem.* **2009**, *52*, 3084.
82   Zobel, A. W.; Nickel, T.; Kunzel, H. E.; Ackl, N.; Sonntag, A.; Ising, M.; Holsboer, F. *J. Psychiatr. Res.* **2000**, *34*, 171.
83   Binneman, B.; Feltner, D.; Kolluri, S.; Shi, Y.; Qiu, R.; Stiger, T. *Am. J. Psychiatr.* **2008**, *165*, 617.
84   Coric, V.; Feldman, H. H.; Oren, D. A.; Shekhar, A.; Pultz, J.; Dockens, R. C.; Wu, X.; Gentile, K. A.; Huang, S.-P.; Emison, E.; Delmonte, T.; D'Souza, B. B.; Zimbroff, D. L.; Grebb, J. A.; Goddard, A. W.; Stock, E. G. *Depress. Anxiety* **2010**, *9999*, n/a.

85   Sweetser, S.; Camilleri, M.; Linker Nord, S. J.; Burton, D. D.; Castenada, L.; Croop, R.; Tong, G.;
        Dockens, R.; Zinsmeister, A. R. *Am. J. Physiol. Gastr. L.* **2009**, *296*, G1299.
86   Lennox, W.; Lennox, M. *Epilepsy and Related Disorders*; Little, Brown, 1960.
87   Lerche, H.; Jurkat-Rott, K.; Lehmann-Horn, F. *Am. J. Med. Genet.* **2001**, *106*, 146.
88   Belmaker, R. H. *New Engl. J. Med.* **2004**, *351*, 476.
89   Hirschfeld, R.; Lewis, L.; Vornik, L. *J. Clin. Psychiatr.* **2003**, *64*, 161.
90   Dick, D.; Foroud, T.; Flury, L.; Bowman, E.; Miller, M.; Rau, N.; Moe, P.; Samavedy, N.; El-
        Mallakh, R.; Manji, H. *Am. J. Hum. Gene.* **2003**, *73*, 107.
91   Scott, D. *The History of Epileptic Therapy: An Account of How Medication Was Developed*;
        Informa HealthCare, 1993.
92   Putnam, T.; Merritt, H. *Science* **1937**, *85*, 525.
93   Merritt, H.; Putnam, T. *Arch. Neurol. Psychiatr.* **1938**, *39*, 1003.
94   Matsuki, N.; Quandt, F.; Ten Eick, R.; Yeh, J. *J. Pharmacol. Exp. Ther.* **1984**, *228*, 523.
95   Meunier, H.; Carraz, G.; Meunier, Y.; Eymard, P.; Aimard, M. *Therapie* **1963**, *18*, 435.
96   Johannessen, C. U.; Johannessen, S. I. *CNS Drug Rev.* **2003**, *9*, 199.
97   Henry, T. *Psychopharmacol. Bull.* **2003**, *37*, 5.
98   Macdonald, R.; McLean, M. *Adv. Neurol.* **1986**, *44*, 713.
99   Phiel, C.; Zhang, F.; Huang, E.; Guenther, M.; Lazar, M.; Klein, P. *J. Biol. Chem.* **2001**, *276*,
        36734.
100   Lubrich, B.; Van Calker, D. *Neuropsychopharmacol.* **1999**, *21*, 519.
101   Williams, R. S. B.; Cheng, L.; Mudge, A. W.; Harwood, A. J. *Nature* **2002**, *417*, 292.
102   Ketter, T.; Wang, P.; Post, R. *Essentials Clin. Psychopharmacol.* **2006**, 367.
103   Willow, M.; Gonoi, T.; Catterall, W. *Mol. Pharmacol.* **1985**, *27*, 549.
104   Okuma, T.; Kishimoto, A. *Psychiatr. Clin. Neurosci.* **1998**, *52*, 3.
105   Grant, S.; Faulds, D. *Drugs* **1992**, *43*, 873.
106   McCormack, P.; Robinson, D. *CNS Drugs* **2009**, *23*, 71.
107   Messenheimer, J. *Clin. Neuropharmacol.* **1994**, *17*, 548.
108   Weisler, R. H.; Calabrese, J. R.; Bowden, C. L.; Ascher, J. A.; DeVeaugh-Geiss, J.; Evoniuk, G. *J.
        Aff. Disorders* **2008**, *108*, 1.
109   Goa, K.; Ross, S.; Chrisp, P. *Drugs* **1993**, *46*, 152.
110   Helen, C.; Kamp, D.; Harris, E. *Epilepsy Res.* **1992**, *13*, 107.
111   Brown, J.; Dissanayake, V.; Briggs, A.; Milic, M.; Gee, N. *Anal. Biochem.* **1998**, *255*, 236.
112   Silverman, R. B. *Angew. Chem. Intl. Ed.* **2008**, *47*, 3500.
113   Taylor, C.; Angelotti, T.; Fauman, E. *Epilepsy Res.* **2007**, *73*, 137.
114   Peters, D.; Sorkin, E. *Drugs* **1993**, *45*, 760.
115   Suzuki, S.; Kawakami, K.; Nishimura, S.; Watanabe, Y.; Yagi, K.; Scino, M.; Miyamoto, K.
        Epilepsy Res. **1992**, *12*, 21.
116   Leppik, I. *Seizure* **2004**, *13*, S5.
117   Maryanoff, B. E.; Nortey, S. O.; Gardocki, J. F.; Shank, R. P.; Dodgson, S. P. *J. Med. Chem.***1987**,
        30, 880.
118   Shank, R. P.; Gardocki, J. F.; Vaught, J. L.; Davis, C. B.; Schupsky, J. J.; Raffa, R. B.; Dodgson, S.
        J.; Nortey, S. O.; Maryanoff, B. E. *Epilepsia* **1994**, *35*, 450.
119   White, H. S.; Brown, S. D.; Woodhead, J. H.; Skeen, G. A.; Wolf, H. H. *Epilepsy Res.* **1997**, *28*,
        167.
120   Mackey, C. *Nat. Rev. Drug Discov.* **2010**, *9*, 265.
121   Chengappa, K.; Gershon, S.; Levine, J. *Bipolar Disorders* **2008**, *3*, 215.
122   McElroy, S.; Hudson, J.; Capece, J.; Beyers, K.; Fisher, A.; Rosenthal, N. *Biol. Psychiatr.* **2007**,
        61, 1039.
123   Johnston, G.; Krogsgaard-Larsen, P.; Stephanson, A.; Twitchin, B. *J. Neurochem.* **2006**, *26*, 1029.
124   Johnston, G.; Krogsgaard-Larsen, P.; Stephanson, A. *Nature* **1975**, *258*, 627.
125   Andersen, K. E.; Braestrup, C.; Groenwald, F. C.; Joergensen, A. S.; Nielsen, E. B.; Sonnewald,
        U.; Soerensen, P. O.; Suzdak, P. D.; Knutsen, L. J. S. *J. Med. Chem.* **1993**, *36*, 1716.

126   Pavia, M.; Lobbestael, S.; Nugiel, D.; Mayhugh, D.; Gregor, V.; Taylor, C.; Schwarz, R.; Brahce, L.; Vartanian, M. *J. Med. Chem.* **1992**, *35*, 4238.

127   Taylor, C. P. *Drug Develop. Res.* **1990**, *21*, 151.

128   Borden, L. A.; Dhar, T. G. M.; Smith, K. E.; Weinshank, R. L.; Branchek, T. A.; Gluchowski, C. *Eur. J. Pharm-Molec. Ph.* **1994**, *269*, 219.

129   Suzdak, P.; Jansen, J. *Epilepsia* **2005**, *36*, 612.

130   Rupprecht, R.; Zwanzger, P. *Der Nervenarzt* **2003**, *74*, 543.

131   Cade, J. *Med. J. Aust. II* **1949**, *2*, 349.

132   Berridge, M.; Downes, C.; Hanley, M. *Cell*, **1989**, *59*, 411.

133   Hallcher, L.; Sherman, W. *J. Biol. Chem.* **1980**, *255*, 10896.

134   Beck, A.; Emery, G.; Greenberg, R. *Anxiety Disorders and Phobias: A Cognitive Perspective*; Basic Books: Cambridge, MA, 2005.

135   Mineka, S.; Oehlberg, K. *Acta Psychol.* **2008**, *127*, 567.

136   Sternbach, L. H. *J. Med. Chem.* **1979**, *22*, 1.

137   Jacobsen, E. *Psychopharmacol.* **1986**, *89*, 138.

138   Browne, T. *Arch. Neurol.* **1976**, *33*, 326.

139   Rosenberg, H.; Tietz, E.; Chiu, T. *Epilepsia* **2007**, *30*, 276.

140   Mattila, M.; Larni, H. *Drugs* **1980**, *20*, 353.

141   Schwartz, R.; Weaver, A. *Clin. Pediatr.* **1998**, *37*, 321.

142   Schwartz, R.; Milteer, R.; LeBeau, M. *South. Med. J.* **2000**, *93*, 558.

143   Cohn, J.; Wilcox, C. *Pharmacother.*, *4*, 93.

144   Nutt, D.; Costello, M. *Life Sci.* **1988**, *43*, 1045.

145   Kang, I.; Miller, L. *Br. J. Pharmacol.* **1991**, *103*, 1285.

146   Hester, J. B.; Rudzik, A. D.; Kamdar, B. V. *J. Med. Chem.* **1971**, *14*, 1078.

147   Chouinard, G.; Annable, L.; Fontaine, R.; Solyom, L. *Psychopharmacol.* **1982**, *77*, 229.

148   Dundee, J.; Halliday, N.; Harper, K.; Brogden, R. *Drugs* **1984**, *28*, 519.

149   Pakes, G.; Brogden, R.; Heel, R.; Speight, T.; Avery, G. *Drugs* **1981**, *22*, 81.

150   Macdonald, R.; Olsen, R. *Ann. Rev. Neurosci.* **1994**, *17*, 569.

151   Whiting, P.; Bonnert, T.; McKernan, R.; Farrar, S.; Le Bourdelles, B.; Heavens, R.; Smith, D.; Hewson, L.; Rigby, M.; Sirinathsinghji, D. *Ann. NY Acad. Sci.* **2006**, *868*, 645.

152   Rho, J.; Donevan, S.; Rogawski, M. *J. Physiol.* **1996**, *497*, 509.

153   Brogden, R.; Goa, K. *Drugs* **1988**, *35*, 448.

154   Verkman, A. S.; Galietta, L. J. V. *Nat. Rev. Drug Discov.* **2009**, *8*, 153.

155   Crestani, F.; Martin, J.; Mehler, H.; Rudolph, U. *Br. J. Pharmacol.* **2000**, *131*, 1251.

156   Doble, A. *Journal of Psychopharmacol. (Oxford, England)* **1999**, *13*, S11.

157   Wu, Y.-H.; Rayburn, J. W.; Allen, L. E.; Ferguson, H. C.; Kissel, J. W. *J. Med. Chem.* **1972**, *15*, 477.

158   Yevich, J. P.; New, J. S.; Smith, D. W.; Lobeck, W. G.; Catt, J. D.; Minielli, J. L.; Eison, M. S.; Taylor, D. P.; Riblet, L. A.; Temple, D. L. *J. Med. Chem.* **1986**, *29*, 359.

159   Goa, K.; Ward, A. *Drugs* **1986**, *32*, 114.

160   Taylor, D. *FASEB J.* **1988**, *2*, 2445.

161   Rickels, K.; Schweizer, E.; DeMartinis, N.; Mandos, L.; Mercer, C. *J. Clin. Psychopharmacol.* **1997**, *17*, 272.

162   Cutler, N.; Sramek, J.; Keppel Hesselink, J.; Krol, A.; Roeschen, J.; Rickels, K.; Schweizer, E. *J. Clin. Psychopharmacol.* **1993**, *13*, 429.

163   Huxtable, R. J.; Schwarz, S. K. W. *Mol. Interv.* **2001**, *1*, 189.

164   Gates, M.; Tschudi, G. *J. Am. Chem. Soc.* **1952**, *74*, 1109.

165   Ball, P. *Nature* **2006**, *440*, 429.

166   Pert, C. B.; Snyder, S. H. *Proc. Natl. Acad. Sci. USA* **1973**, *70*, 2243.

167   Sora, I.; Takahashi, N.; Funada, M.; Ujike, H.; Revay, R. Ä.; Donovan, D. Ä.; Miner, L. Ä.; Uhl, G. Ä. *Proc. Natl. Acad. Sci. USA* **1997**, *94*, 1544.

168   Wright, C. R. A. *J. Chem. Soc.* **1874**, *27*, 1031.

169   Inturrisi, C.; Max, M.; Foley, K.; Schultz, M.; Shin, S.; Houde, R. *New Engl. J. Med.* **1984**, *310*, 1213.

170   Bentley, K. W.; Hardy, D. G. *J. Am. Chem. Soc.* **1967**, *89*, 3281.

171   Michaelis, M.; Schölkens, B.; Rudolphi, K. *N.-S. Arch. Pharmacol.* **2007**, *375*, 81.

172   Izenwasser, S.; Newman, A. H.; Cox, B. M.; Katz, J. L. *Eur. J. Pharmacol.* **1996**, *297*, 9.

173   Stanley, T. *J. Pain and Symptom Management* **1992**, *7*, S3.

174   Niemegeers, C.; Schellekens, K.; Van Bever, W.; Janssen, P. *Arznei.-Forschung* **1976**, *26*, 1551.

175   Gruber, C. M., Jr. *J. Am. Med. Assoc.* **1957**, *164*, 966.

176   Lipman, A. *J. Pain. Palliat. Care Pharmacother.* **2009**, *23*, 104.

177   Bockmuhl, M.; Ehrhart, G. *Liebigs Ann. Chem.* **1949**, *561*, 52.

178   Stimmel, B.; Goldberg, J.; Rotkopf, E.; Cohen, M. *J. Am. Med. Assoc.* **1977**, *237*, 1216.

179   Sporer, K. *Ann Intern. Med.* **1999**, *130*, 584.

180   O'Malley, S.; Jaffe, A.; Chang, G.; Schottenfeld, R.; Meyer, R.; Rounsaville, B. *Arch. Gen. Psychiatr.* **1992**, *49*, 881.

181   Mason, B.; Salvato, F.; Williams, L.; Ritvo, E.; Cutler, R. *Arch. Gen. Psychiat.* **1999**, *56*, 719.

182   Sinner, B.; Graf, B. in *Modern Anesthetics*, 2008, p. 313.

183   Lanning, C.; Harmel, M. *Annu. Rev. Med.* **1975**, *26*, 137.

184   Schmid, R.; Sandler, A.; Katz, J. *Pain* **1999**, *82*, 111.

185   Hirota, K.; Lambert, D. *Br. J. Anaesth.* **1996**, *77*, 441.

186   Chen, X.; Shu, S.; Bayliss, D. A. *J. Neurosci.* **2009**, *29*, 600.

187   Berton, O.; Nestler, E. *Nat. Rev. Neurosci.* **2006**, *7*, 137.

188   Matthys, H.; Bleicher, B.; Bleicher, U. *J. Int. Med. Res.* **1983**, *11*, 92.

189   Lee, P.; Jawad, M.; Eccles, R. *J. Pharm. Pharmacol.* **2000**, *52*, 1137.

190   Desai, S.; Aldea, D.; Daneels, E.; Soliman, M.; Braksmajer, A.; Kopes-Kerr, C. *J. Am. Board Fam. Med.* **2006**, *19*, 320.

191   Miceli, F.; Soldovieri, M. V.; Martire, M.; Taglialatela, M. *Curr. Opin. Pharmacol.* **2008**, *8*, 65.

192   Wulff, H.; Castle, N.; Pardo, L. *Nat. Rev. Drug Disc.* **2009**, *8*, 982.

193   Xiong, Q.; Gao, Z.; Wang, W.; Li, M. *Trends Pharmacol. Sci.* **2008**.

194   Stead, L. F.; Perera, R.; Bullen, C.; Mant, D.; Lancaster, T. In *Cochrane Db. Syst. Rev.* **2008**, Issue 1.

195   Coe, J.; Brooks, P.; Vetelino, M.; Wirtz, M.; Arnold, E.; Huang, J.; Sands, S.; Davis, T.; Lebel, L.; Fox, C. *J. Med. Chem* **2005**, *48*, 3474.

196   Jimenez-Ruiz, C.; Berlin, I.; Hering, T. *Drugs* **2009**, *69*, 1319.

197   Schroeder, S.; Sox, H. *Ann. Intern. Med.* **2006**, *145*, 784.

198   FDA. http://www.fda.gov/Drugs/DrugSafety/PublicHealthAdvisories/ucm169988.htm, Ed. 2009.

199   Ascher, J.; Cole, J.; Colin, J.; Feighner, J. *J. Clin. Psychiatry* **1995**, *56*, 395.

200   Holm, K.; Spencer, C. *Drugs* **2000**, *59*, 1007.

201   Johnston, J.; Lineberry, C.; Ascher, J.; Davidson, J.; Khayrallah, M.; Feighner, J.; Stark, P. *J. Clin. Psychiatr.* **1991**, *52*, 450.

202   Kondro, W. *Can. Med. Assoc. J.* **2004**, *171*, 837.

203   Faraone, S.; Biederman, J. *Biol. Psychiatr.* **1998**, *44*, 951.

204   Zwi, M.; Ramchandani, P.; Joughin, C. *Br. Med. J.* **2000**, *321*, 975.

205   Rasmussen, N. *J. Hist. Med. All. Sci.* **2006**, *61*, 288.

206   Yoshida, T. *Use and Misuse of Amphetamines: An International Overview*; Harwood Academic Publishers: Amsterdam, 1997.

207   Wood, J. G. a. Z., Robert, H. *Beyond Behav.* **2001**, *Fall 2001*, 39.

208   Rosack, J. *Psychiatric News* **2007**, *42*, 1.

209   Moore, K. *Biol. Psychiatr.* **1977**, *12*, 451.

210   Gainetdinov, R. R.; Wetsel, W. C.; Jones, S. R.; Levin, E. D.; Jaber, M.; Caron, M. G. *Science* **1999**, *283*, 397.

211   Cho, A.; Melega, W. *J. Addic. Dis.* **2001**, *21*, 21.

212   Panizzon, L. *Helv. Chim. Acta* **1944**, *27*, 1748.

213   Meier, R.; Gross, F.; Tripod, J. *J. Mol. Med.* **1954**, *32*, 445.
214   Safer, D.; Zito, J.; Fine, E. *Pediatrics* **1996**, *98*, 1084.
215   Heal, D.; Pierce, D. *CNS Drugs* **2006**, *20*, 713.
216   Volkow, N.; Wang, G. *Am. J. Psychiatr.* **1998**, *155*, 1325.
217   Seneca, N.; Guly·s, B.; Varrone, A.; Schou, M.; Airaksinen, A.; Tauscher, J.; Vandenhende, F.; Kielbasa, W.; Farde, L.; Innis, R. *Psychopharmacol.* **2006**, *188*, 119.
218   Bymaster, F.; Katner, J.; Nelson, D.; Hemrick-Luecke, S.; Threlkeld, P.; Heiligenstein, J.; Morin, S.; Gehlert, D.; Perry, K. *Neuropsychopharmacol.* **2002**, *27*, 699.
219   McKhann, G.; Drachman, D.; Folstein, M.; Katzman, R.; Price, D.; Stadlan, E. *Neurol.* **1984**, *34*, 939.
220   Evans, D.; Funkenstein, H.; Albert, M.; Scherr, P.; Cook, N.; Chown, M.; Hebert, L.; Hennekens, C.; Taylor, J. *J. Am. Med. Assoc.* **1989**, *262*, 2551.
221   Coyle, J.; Price, D.; DeLong, M. *Science* **1983**, *219*, 1184.
222   Jankovic, J. *J. Neurol. Neurosur. Ps.* **2008**, *79*, 368.
223   Goedert, M. *Nat. Rev. Neurosci.* **2001**, *2*, 492.
224   Zimprich, A.; Biskup, S.; Leitner, P.; Lichtner, P.; Farrer, M.; Lincoln, S.; Kachergus, J.; Hulihan, M.; Uitti, R.; Calne, D. *Neuron* **2004**, *44*, 601.
225   Lucking, C.; Durr, A.; Bonifati, V.; Vaughan, J.; De Michele, G.; Gasser, T.; Harhangi, B.; Meco, G.; Denefle, P.; Wood, N. *New Engl. J. Med.* **2000**, *342*, 1560.
226   Shimura, H.; Hattori, N.; Kubo, S.; Mizuno, Y.; Asakawa, S.; Minoshima, S.; Shimizu, N.; Iwai, K.; Chiba, T.; Tanaka, K. *Nat. Genet.* **2000**, *25*, 302.
227   Warner, T.; Schapira, A. *Ann. Neurol.* **2003**, *53*, S16.
228   Andrew, S.; Goldberg, Y.; Kremer, B.; Telenius, H.; Theilmann, J.; Adam, S.; Starr, E.; Squitieri, F.; Lin, B.; Kalchman, M. *Nat. Genet.* **1993**, *4*, 398.
229   Walker, F. *Lancet* **2007**, *369*, 218.
230   Dworacek, B.; Rupreht, J. *Int. Congr. Ser.* **2002**, *1242*, 87.
231   Proudfoot, A. *Toxicol. Rev.* **2006**, *25*, 99.
232   Albert, A. *The Acridines: Their Preparation, Physical, Chemical, And Biological Properties And Uses*; Edward Arnold & Co.: London, 1951.
233   Summers, W.; Majovski, L.; Marsh, G.; Tachiki, K.; Kling, A. *New Engl. J. Med.* **1986**, *315*, 1241.
234   Hammel, P.; Larrey, D.; Bernuau, J.; Kalafat, M.; FrÈneaux, E.; Babany, G.; Degott, C.; Feldmann, G.; Pessayre, D.; Benhamou, J. *J. Clin. Gastroenterol.* **1990**, *12*, 329.
235   Summers, W.; Tachiki, K.; Kling, A. *Eur. Neurol.* **1989**, *29*, 28.
236   Birks, J.; Harvey Richard, J. In *Cochrane Db. Syst. Rev.* **2006**, 3.
237   Jann, M. *Pharmacother.* **2000**, *20*, 1.
238   FDA. *www.fda.gov/NewsEvents/Newsroom/PressAnnouncements/2006/ucm108680.htm* **2006**, *P06-88*.
239   Heinrich, M.; Teoh, L. *J. Ethnopharmacol.* **2004**, *92*, 147.
240   Scott, L.; Goa, K. *Drugs* **2000**, *60*, 1095.
241   Birks, J. *Cochrane Db. Syst. Rev.* **2006**, Issue 1.
242   Maugh, T. *Science* **1976**, *192*, 130.
243   Schwab, R.; England Jr, A.; Poskanzer, D.; Young, R. *J. Am. Med. Assoc.* **1969**, *208*, 1168.
244   Lipton, S. A. *Nat. Rev. Drug Discov.* **2006**, *5*, 160.
245   Robinson, D. M.; Keating, G. M. *Drugs* **2006**, *66*, 1515.
246   Cotzias, G. *J. Am. Med. Assoc.* **1969**, *210*, 1255.
247   Carlsson, A.; Lindqvist, M.; Magnusson, T. *Nature* **1957**, *180*, 1200.
248   Hornykiewicz, O. *Wien. Klin. Wochenschr.* **1963**, *75*, 309.
249   Birkmayer, W.; Hornykiewicz, O. *Eur. Arch. Psychiatr. Clin. Neurosci.* **1962**, *203*, 560.
250   Cotzias, G.; Papavasiliou, P.; Gellene, R. *New Engl. J. Med* **1969**, *280*, 337.
251   Calne, D. *New Engl. J. Med.* **1993**, *329*, 1021.
252   Holm, K.; Spencer, C. *Drugs* **1999**, *58*, 159.

253   Millan, M. J.; Maiofiss, L.; Cussac, D.; Audinot, V. r.; Boutin, J.-A.; Newman-Tancredi, A. *J. Pharmacol. Exp. Ther.* **2002**, *303*, 791.
254   Duby, S.; Cotzias, G.; Papavasiliou, P.; Lawrence, W. *Arch. Neurol.* **1972**, *27*, 474.
255   Corsini, G.; Zompo, M.; Gessa, G.; Mangoni, A. *Lancet* **1979**, *313*, 954.
256   Creese, I.; Sibley, D. *Annu. Rev. Pharmacol. Toxicol.* **1981**, *21*, 357.
257   Schneider, C. S.; Mierau, J. *J. Med. Chem.* **1987**, *30*, 494.
258   Biglan, K.; Holloway, R. *Expert Opin. Pharmacother* **2002**, *3*, 197.
259   Eden, R. J.; Costall, B.; Domeney, A. M.; Gerrard, P. A.; Harvey, C. A.; Kelly, M. E.; Naylor, R. J.; Owen, D. A. A.; Wright, A. *Pharmacol. Biochem. Behav.* **1991**, *38*, 147.
260   Rascol, O.; Brooks, D.; Korczyn, A.; De Deyn, P.; Clarke, C.; Lang, A. *New Engl. J. Med.* **2000**, *342*, 1484.
261   Knoll, J. *J. Neural Transm. Supp.* **1986**, *22*, 75.
262   Birkmayer, W.; Knoll, J.; Riederer, P.; Youdim, M.; Hars, V.; Marton, J. *J. Neural Transm.* **1985**, *64*, 113.
263   Heinonen, E.; Rinne, U. *Acta Neurol. Scand.* **2009**, *80*, 103.
264   Stern, M.; Marek, K.; Friedman, J.; Hauser, R.; LeWitt, P.; Tarsy, D.; Olanow, C. *Movement Disord.* **2004**, *19*, 916.
265   Olanow, C.; Rascol, O.; Hauser, R.; Feigin, P.; Jankovic, J.; Lang, A.; Langston, W.; Melamed, E.; Poewe, W.; Stocchi, F. *New Engl. J. Med.* **2009**, *361*, 1268.
266   Dorfman, L.; Huebner, C.; MacPhillamy, H.; Schlittler, E.; St. AndrÈ, A. *Cell. Mol. Life Sci.* **1953**, *9*, 368.
267   Lazarte, J.; Petersen, M.; Baars, C.; Pearson, J. *Proc. Staff Meet. Mayo Clinic* **1955**, *30*, 358.
268   Lopez-Munoz, F.; Bhatara, V.; Alamo, C.; Cuenca, E. *Actas Esp. Psiquiatr.* **2004**, *32*, 387.
269   Rosenblatt, A.; Ranen, N.; Nance, M.; Paulsen, J. *New York: Huntington's Disease Society of America* **1999**, *41*, 1.
270   Erickson, J.; Eiden, L.; Schafer, M.; Weihe, E. *J. Mol. Neurosci.* **1995**, *6*, 277.
271   Kenney, C.; Jankovic, J. *Expert Rev. Neurother.* **2006**, *6*, 7.
272   Ondo, W.; Tintner, R.; Thomas, M.; Jankovic, J. *Clin. Neuropharmacol.* **2002**, *25*, 300.
273   Hayden, M.; Leavitt, B.; Yasothan, U.; Kirkpatrick, P. *Nat. Rev. Drug Discov.* **2009**, *8*, 17.
274   Cheah, B.; Vucic, S.; Krishnan, A.; Kiernan, M. *Curr. Med. Chem.* **2010**, *17*, 1942.
275   Shaw, C.; Enayat, Z.; Chioza, B.; Al-Chalabi, A.; Radunovic, A.; Powell, J.; Leigh, P. *Ann. Neurol.* **2004**, *43*, 390.
276   Azbill, R.; Mu, X.; Springer, J. *Brain Res.* **2000**, *871*, 175.
277   Kretschmer, B. D.; Kratzer, U.; Schmidt, W. J. *N.-S. Arch. Pharmacol.* **1998**, *358*, 181.
278   Mizoule, J.; Meldrum, B.; Mazadier, M.; Croucher, M.; Ollat, C.; Uzan, A.; Legrand, J. J.; Gueremy, C.; Le Fur, G. *Neuropharmacol.* **1985**, *24*, 767.
279   Bensimon, G.; Lacomblez, L.; Meininger, V. *New Engl. J. Med.* **1994**, *330*, 585.
280   Malykh, A.; Sadaie, M. *Drugs* **2010**, *70*, 287.
281   Genton, P.; Guerrini, R.; Remy, C. *Pharmacopsychiatry* **1999**, *32*, 49.
282   Tariska, P.; Paksy, A. *Orvosi hetilap* **2000**, *141*, 1189.
283   Rose, S. P. R. *Nat. Rev. Neurosci.* **2002**, *3*, 975.
284   Rogawski, M. *Br. J. Pharmacol.* **2008**, *154*, 1555.
285   Dooley, M.; Plosker, G. *Drugs* **2000**, *60*, 871.
286   Genton, P., Van Vleymen, B. *Epileptic Disord.* **2000**, *2*, 99.
287   Citron, M. *Nat. Rev. Drug Discov.* **2010**, *9*, 387.
288   Fan, Y.; Unwalla, R.; Denny, R. A.; Di, L.; Kerns, E. H.; Diller, D. J.; Humblet, C. *J. Chem. Inf. Model.* **2010**.
289   Crowther, R. *Curr. Opin. Struct. Biol.* **1993**, *3*, 202.
290   Brunden, K. R.; Trojanowski, J. Q.; Lee, V. M. Y. *Nat. Rev. Drug Discov.* **2009**, *8*, 783.
291   Collins, F.; McKusick, V. *J. Am. Med. Assoc.* **2001**, *285*, 540.
292   Sawa, A.; Snyder, S. H. *Science* **2002**, *296*, 692.

293   Millar, J. K.; Wilson-Annan, J. C.; Anderson, S.; Christie, S.; Taylor, M. S.; Semple, C. A. M.; Devon, R. S.; Clair, D. M. S.; Muir, W. J.; Blackwood, D. H. R.; Porteous, D. J. *Hum. Mol. Genet.* **2000,** *9,* 1415.
294   Burdick, K. E.; Kamiya, A.; Hodgkinson, C. A.; Lencz, T.; DeRosse, P.; Ishizuka, K.; Elashvili, S.; Arai, H.; Goldman, D.; Sawa, A.; Malhotra, A. K. *Hum. Mol. Genet.* **2008,** *17,* 2462.
295   Chubb, J. E.; Bradshaw, N. J.; Soares, D. C.; Porteous, D. J.; Millar, J. K. *Mol. Psychiatr.* **2007,** *13,* 36.
296   Walsh, T.; McClellan, J. M.; McCarthy, S. E.; Addington, A. M.; Pierce, S. B.; Cooper, G. M.; Nord, A. S.; Kusenda, M.; Malhotra, D.; Bhandari, A.; Stray, S. M.; Rippey, C. F.; Roccanova, P.; Makarov, V.; Lakshmi, B.; Findling, R. L.; Sikich, L.; Stromberg, T.; Merriman, B.; Gogtay, N.; Butler, P.; Eckstrand, K.; Noory, L.; Gochman, P.; Long, R.; Chen, Z.; Davis, S.; Baker, C.; Eichler, E. E.; Meltzer, P. S.; Nelson, S. F.; Singleton, A. B.; Lee, M. K.; Rapoport, J. L.; King, M.-C.; Sebat, J. *Science* **2008,** *320,* 539.

8

Cancer Drugs

Narendra B. Ambhaikar

1   Introduction

The term cancer is used to describe diseases that are characterized by unrestrained or unregulated growth of cells leading to the formation of malignant tumors and spread of abnormal cells eventually causing death. Each year, 7.6 million people die from cancer worldwide.[1] In the year 2007, 13% of the total deaths worldwide were due to cancer.[2] As one of the leading causes of death in both the developed and the developing countries, cancer is expected to bring about an annual worldwide death toll to over 11 million by the year 2030.[3,4] In the United States, it is the second most common cause of death, the first one being the cardiovascular diseases. Although the past century has witnessed some of the greatest discoveries toward the treatment of this destructive infirmity, its cure still remains a challenge and continues to require extensive research. Internal factors such as inherited mutations, hormones, or immune conditions as well as external or environmental factors like tobacco, diet, radiation, and infections can both lead to cancer.[5] Cancer is typically treated surgically, through chemotherapy, hormone therapy, biological as well as targeted therapies. While some cancers are a strongly hereditary consequence, most of them are not and a large number result from genetic damage that has taken place during one's lifetime. The damage could be due to internal factors such as hormones or metabolism of nutrients within cells or external factors mentioned above. Many types of cancer can be completely prevented by incorporating lifestyle-related changes. Most cancers are not of hereditary origin and lifestyle factors such as dietary habits, smoking, alcohol consumption, and infections greatly influence their development.[6] While prevention is certainly the best option, it is not yet completely understood. This chapter describes drugs that have found wide utility among cancer patients.

Almost anyone can develop and suffer from cancer. The risk of a positive diagnosis of cancer increases with age with most cases generally observed in middle-aged or older adults. About 78% of all cancers are diagnosed in adults above 55 years of age. According to the National Institutes of Health (NIH), the overall cost estimate of cancer in 2010 was $263.8 billion that included direct medical expenditures, indirect morbidity costs such as lost productivity due to illness and indirect mortality cost including lost productivity due to premature death.[7] The economic load of these diseases is astounding with health care costs estimated at $50–150 billion per year worldwide. With over 110 types of cancer, they can all be classified into four categories depending on the tissue involved: *carcinoma, lymphoma, leukemia,* and *sarcoma. Carcinomas* are the most common, with 85–90% of all cancers falling into this category. They are tumors that originate in epithelial tissue such as skin, breast, lung, prostate, stomach, colon, ovary, and many other organs. *Leukemia* is the cancer of the blood, bone marrow, and liver. *Sarcomas*, the rarest of all four types, are tumors arising from cells in connective tissue, bone, or muscle. The four most common cancers are lung, breast (almost exclusively in women), colon/colorectal, and prostate (in men). With the exceptionally vast information on the developments in the field of cancer therapy that already exists, the current review attempts to delineate a general idea of cancer drugs that have entered the market and focuses on some of the recent developments in the field. Here, only a portion of the cancer drug market is covered with emphasis on landmark and recent discoveries. It is important to point out that despite the introduction of so many drugs, cancer still remains a daunting challenge to overcome, simply because of various difficulties such as the complexity of the disease, the adverse side effects of the drugs, and the development of clinical resistance to these drugs. With these challenges it is clear that research in the field will continue to offer exciting opportunities as new discoveries are made.

The development and growth of cancer disease is a highly complex process that has been the subject of intense investigation for over a century, with several research journals dedicated to this subject alone. At a molecular level, cellular dysregulation in cancer takes place *via* (i) dysregulation of the cell cycle by abnormal signaling or (ii) evasion of tumor suppression mechanism and apoptotic cell death.[8] The cell has been understood to go through a number of stages during its lifetime. G_0 is the quiescent, or resting, state which progresses to the mitotic M stage that ultimately brings about cell division. During this progression there are multiple states involved: G_0, G_1, S, G_2, and M. A restriction point R during the G_1 state can determine absolutely whether or not the cell stays in G_0 or early G_1 or proceeds to the S phase. The S phase involving DNA replication inevitably results in the M stage followed by cell division. Retinoblastoma (pRb) proteins acting as R-point gatekeepers allow passage when extensively phosphorylated. Phosphorylation levels are determined by signals outside the cell *via* attachment of signaling molecules to cell surface receptors and they decide cell cycle progression. The signals activate a cascade of tyrosine kinase signaling that results in hyperphosphorylation of the R-point proteins, which are tumor suppressor proteins. These proteins perform many functions, of which a key role is to prevent excessive cell growth in normal cells. However, when cancer occurs, these proteins are found to be dysfunctional in several cases. Normal cells reproduce themselves exactly, at the right time, stick together at the appropriate location, self-destroy if necessary, and become specialized when needed. On the other hand, cancer cells keep proliferating, do not obey signals from other cells, do not stick together and remain immature.

During the 1970s two classes of genes were discovered in connection with cancer: *the oncogenes* and *the tumor suppressor genes*. Genes possessing a potential to cause cancer are called oncogenes. These are mutated forms of genes that cause normal cells to grow out of control and become cancer cells and they are expressed at high levels in tumor cells. Identification of oncogenes involved in the initiation and progression of tumors has generated targets for the development of new anticancer drugs. Normal genes (for example, pRb and p53) that slow down cell division, repair DNA errors, and tell cells when to die are tumor-suppressor genes. These, when mutated to cause a loss in function can cause cancer. Alterations in oncogenes, tumor-suppressor genes, and microRNA genes causing cancer are generally somatic (relating to the body) events, although germ-line mutations (transmittable to offspring) can incline a person to inborn or familial cancer.[9] Most evidence indicates a multistep process of sequential alterations in several (rather than just one) oncogenes, tumor suppressor genes, or microRNA genes in cancer cells. Cancer cells multiply rapidly. Cytotoxic chemotherapy differentially affects cells that are rapidly growing, preferentially killing them. However, adverse effects of these drugs including diarrhea, mucosal ulcers, hair loss, and others occur during therapy. Side effects occur because while exerting the desired action of killing cancer cells, cytotoxic chemotherapeutic agents also have an effect on rapidly growing normal cells present in various organs. The cause of cancer as a disease was not known for a long time. Despite several investigations in the twentieth century from various scientists across the world suggesting that carcinogens and/or viruses can cause cancer, there is enough proof today to support the statement that cancer is ultimately a disease of genes.[10] At present the understanding is that cancer is a multistep process, characterized by mutations in several oncogenes and the loss of function of tumor-suppressing genes.

2   Historical Perspective of Cancer Drugs

Although the concept of treating cancer goes back at least 500 years when preparations of silver, zinc, and mercury were used, the usefulness was not yet documented. When it comes to cancer therapy, it is the latter half of the past 100 years of monumental discoveries that seems to overshadow all of the years before. In fact, some decades ago cancer of any kind was considered almost untreatable. Ironically the first effective anticancer drug nitrogen mustard resulted originally from a war gas rather than a medicine.[11] Sulfur mustard gas, a scourge of World War II and used as chemical warfare, began to open new avenues for treating cancer. Since this discovery, there has been an immense growth in the number of drugs functioning through a variety of different mechanisms prescribed to treat cancers.

Systematic treatment of cancer *via* chemotherapy began only after World War II, when thousands of mustard compounds were synthesized and tested. Several of these compounds have become drugs over the years, thereby benefiting mankind. During this time, another class of chemotherapeutic agents called the antimetabolites was discovered. Some examples of this class today include methotrexate and aminopterin that interfere with folate synthesis. These compounds operate *via* interfering with the synthesis of DNA and halt replication or cause mistakes during DNA replication, which ultimately leads to cell death. Success of this concept has led to the discovery of several compounds in this class. With the advent of the idea of combination therapy in the 1960s, there were remarkable improvements in patients. This was especially the case in patients with leukemia in which combination chemotherapy provided the first cures. The logic behind this concept was: whereas becoming resistant to a single agent requires just one or a few mutations, becoming resistant to multiple agents that attack different targets would require more mutations. The 1960s also witnessed the development of the vinca alkaloids with Eli Lilly introducing the spindle poisons or tubulin inhibitors vincristine (Oncovin) and vinblastine that could be used to treat a several cancers. Other natural products like paclitaxel and camptothecin exhibiting powerful anticancer properties with novel mechanisms of action were also discovered in the 1960s. However, given the complex nature of cancer and its numerous subtypes, it took several years of extensive research investigation to understand their effects on the patient populations until they found approved clinical use among the patient populations. Paclitaxel and camptothecin derivatives, discussed later in this chapter were approved in the 1990s and have been prescribed to treat cancer of various types. In addition to the above-mentioned therapies, a very effective treatment for some cancers such as prostate cancer and breast cancer has been hormone therapy. Drug classes like the estrogen receptor antagonists, aromatase inhibitors, LHRH (luteinizing hormone-releasing hormone) analogs as well as inhibitors since the 1970s have greatly aided treatment of prostate and breast cancers. The impact of hormones on cancer growth has been studied, leading to the development of new drugs enabling researchers to look at new ways to use drugs and reduce the risk of developing breast and prostate cancers.

In the past three decades a great deal of knowledge of the oncogenic processes occurring at the molecular level has been garnered by the scientific community. This understanding has enabled the rational design of targeted agents for cancer therapy. Targeted therapy is a term employed to the new generation of anticancer drugs that interfere with specific molecular targets expressed by the tumor cell or in its microenvironment, that play a key role in tumor growth, progression, or survival.[13] Targeted therapies comprising monoclonal antibodies and small molecule inhibitors have significantly changed the treatment of cancer over the past 10 years. These drugs are routinely employed in therapy for many common malignancies, including breast, colorectal, lung, and pancreatic cancers as well as lymphoma, leukemia, and multiple myeloma.[14] Several new drugs encompassing small molecules and monoclonal antibodies directly impacting oncogene products have been developed. There will be more such drugs to come, given the recent progress in producing small molecules capable of inhibiting the enzymatic activity of ABL, KIT, EGFR, and ERBB2. The mechanisms of action and the toxicities of these targeted therapies differ greatly from the conventional cytotoxic chemotherapeutic approaches. In 2008, of the new anticancer drugs approved by the U.S. FDA since 2000, 15 were targeted therapies while only 5 were traditionally chemotherapeutic agents.

As innovative cancer research progresses further in pursuit of improving the quality of lives of patients, new drugs discovered are expected to be more specific, with minimal side-effects and devoid of clinical resistance. The developments that have taken place over the past century certainly indicate this due to the ground-breaking advancements made in a variety of diverse areas such as molecular biology, physics, analytical technologies, natural products isolation and synthetic chemistry. All of them have contributed to the new drugs seen in pharmacies in the present day. These developments have enabled physicians to prescribe the appropriate cancer drug for type of cancer. Physicians have thus gone from being able to do very little to treat cancer patients some decades ago to accomplishing unprecedented survival rates. Today, the area of oncology and related drug discovery remains an exciting field posing many challenges ahead. But it is also one that will lead to landmark developments, thus greatly benefiting humankind with life-saving drugs. A discussion of many of the important types of cancer drugs that have been widely used follows, with some emphasis on the newer classes.

3   Antimetabolites

Antimetabolites are compounds that interfere with the normal metabolism of an organism and cause its death. In the context of cancer treatment, compounds that interfere with the synthesis of nucleic acids are antimetabolites. Some of them inhibit the production of deoxyribonucleoside triphosphates that are the immediate precursors for DNA synthesis and therefore stop the replication process. They are of various types such as folate antagonists, pyrimidine antagonists, purine antagonists, sugar-modified analogs, and ribonucleotide reductase inhibitors. Folate antagonists as well as the pyrimidine and purine compounds are of great importance and will be discussed in this section.

3.1   Folate Antagonists

Cancer cells divide rapidly. Compounds interfering with folate metabolism can be used to treat cancer, which is the underlying principle of antifolates. Folates play a key role in cellular metabolism; however, the core component folic acid in its fully oxidized state is not useful to mammalian cells. To act as a coenzyme requires its reduction over two successive steps *via* dihydrofolate reductase (DHFR) to form 7,8-dihydrofolate and then 5,6,7,8-tetrahydrofolate (THF). The fully reduced product is capable of picking up and transferring single carbon units in many metabolic processes by attaching at the N-5 or N-10 position. Thus cofactors of THF are involved in the transfer reactions of the one-carbon unit used in the biosynthesis of nucleic and amino acids, including methylation of dUMP to dTMP.[15]   DHFR is thus a well-studied and essential enzyme in the folate pathway. Inhibition of DHFR occurs by stopping the synthesis of DNA, RNA, and proteins, thus stopping cell growth and making it an important target in the world of drug discovery to find antibacterial, antifungal, antimalarial, and anticancer folates. While the number of antifolates synthesized has increased over the past ten years, anticancer antifolates have been synthesized to a less extent than antimicrobial agents in drug discovery. Antifolates are the oldest of the antimetabolite class of anticancer agents and were one of the first modern anticancer drugs.

The first clinically useful antifolate aminopterin (AMT) or 4-amino-folic acid was disclosed in the year 1947. Sidney Farber at the Harvard Medical School demonstrated its activity through remissions in childhood leukemia. It was found to work by competing for the folate binding site of DHFR. However, it showed substantial toxicity and soon it was replaced with methotrexate (MTX). MTX remains, with one limited exception, the only antifolate anticancer agent in clinical use to this date.[16]

aminopterin (AMT)   methotrexate (MTX)

An antifolate structurally similar to folic acid is pemetrexed (Alimta), discovered by Edward Taylor at Princeton University and afterward developed by Eli Lilly and Company. It was approved in the United States for the treatment of advanced non-small-cell lung cancer (NSCLC), as a multitargeted antifolate cytotoxic agent.[17]   Pemetrexed exerts anticancer activity *via* the inhibition of three enzymes: thymidylate synthase (TS), dihydrofolate reductase (DHFR) and glycinamide ribonucleotide formyl transferase (GARFT). It has been used as initial therapy in combination with cisplatin as monotherapy after prior chemotherapy and also as maintenance therapy with metastatic non-squamous non-small cell lung cancer (NSCLC).

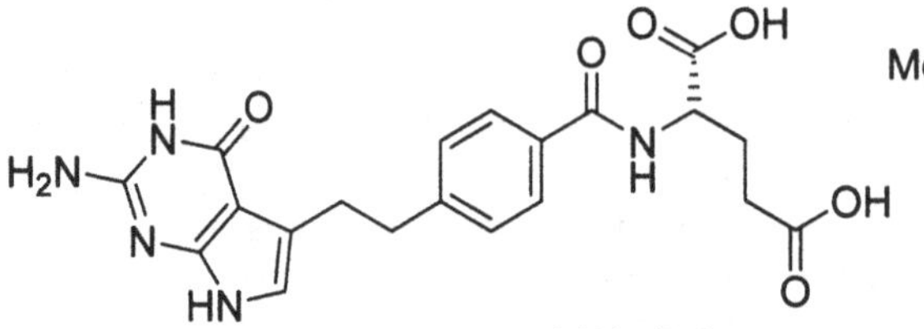

pemetrexed (Almita) 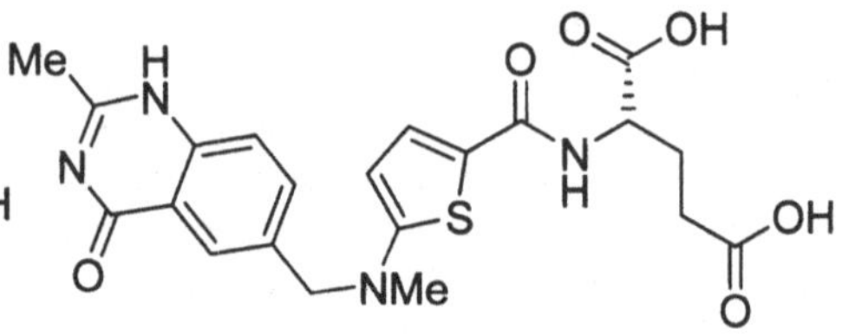   raltitrexed

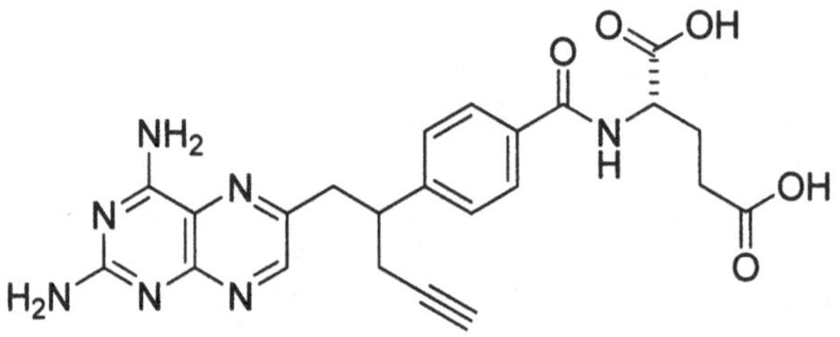

pralatrexate (Fotolyn)

Raltitrexed (Tomudex, ZD1694) is a novel quinazoline folate analog that selectively inhibits thymidylate synthase. Such inhibition causes damage to single- and double-stranded DNA leading to the observed cytotoxicity observed. This antitumor activity of raltitrexed has been used in the treatment of colorectal cancer. Similarly pralatrexate (Fotolyn) is a novel targeted antifolate that has been approved by the FDA in 2009 for the treatment of peripheral T-cell lymphoma (PTCL).[18] Discovered through a medicinal chemistry collaboration between scientists at the Memorial Sloan-Kettering Cancer Center, Stanford Research Institute (SRI), and the Southern Research Institute in pursuit of a targeted therapy, this compound has been found to target the reduced folate carrier in cancer cells. Tumor cells, like normal cells take in natural folate as well as methotrexate. However, unlike noncancerous cells, tumor cells do so *via* a plasma membrane transporter RFC-1. Thus they looked for an agent that tumor cells could use to bring in methotrexate without losing the ability to inhibit folic acid metabolism and eventually found pralatrexate. Pralatrexate is a selective antifolate with better efficacy and less toxicity than other therapeutic regimens. However, it can still cause toxicity in the form of mucositis and stomatitis as well as bone marrow suppression.

3.2   *Purine and Pyrimidine Antimetabolites*

Analogs of nucleotide precursors have been proven to be a useful class of anticancer agents. Purines and pyrimidines as antimetabolites are an important class of compounds used for treating a variety of cancer. They may share several structural and biochemical characteristics, although each compound can display unique activities that render it a useful drug.[19] It is indeed striking that purine and pyrimidine nucleoside analogs with similar structural features and sharing metabolic pathways as well as elements of mechanism of action show such diversity in their clinical activities.[20] Since molecular targets exist in both tumor cells and normal host tissues, selectivity of these agents is not very distinct, although it may be attributed to differences in metabolism and proliferative states between them. Despite the toxicity of these molecules limiting their utility, they play an important role in the treatment of cancer and many still may get approved in the coming years. Antimetabolite molecules that have been approved are various thiopurines, fluoropyrimidines, and the deoxynucleoside analogs. Structures of some of these molecules are shown below.

6-Mercaptopurine, also called purinethol, was first synthesized and developed in 1952 by Gertrude Elion and George Hitchings at the Burroughs-Wellcome pharmaceutical company as an intermediate toward 6-aminopurione. This intermediate was found to exhibit better biological profile than the desired target molecules. It was first approved in 1953 for the treatment of childhood lymphocytic leukemia, where it is curative and still a standard treatment.[21] An analog of hypoxanthine, it is a good substrate for hypoxanthine/guanine phosphoribosyl transferase. It works by interfering with nucleic acid synthesis and inhibition of purine metabolism. It is used, usually in combination with other drugs, in the treatment of or in remission maintenance programs for leukemia. It competes with hypoxanthine and guanine for the inhibition of enzyme hypoxanthine-guanine phosphoribosyltransferase forming thioinosinic acid. 6-Mercaptopurine inhibits several reactions involving inosinic acid, such as the conversion of inosinic acid to xanthylic acid and the conversion to adenylic acid *via* adenylosuccinate. In addition, 6-methylthioinosinate is formed by the methylation of thioinosinate.

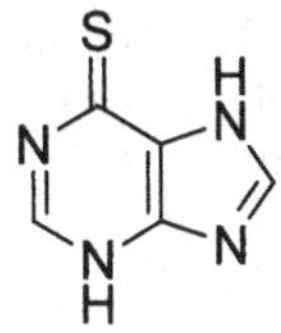

mercaptopurine 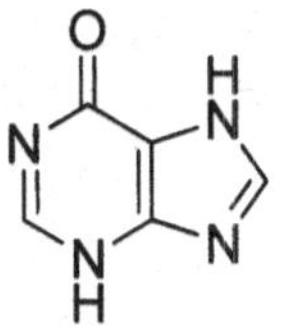   hypoxanthine 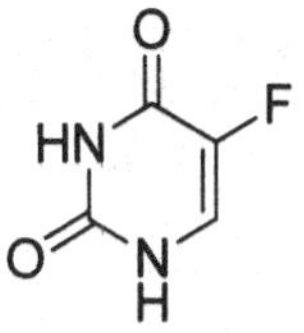   5-fluorouracil

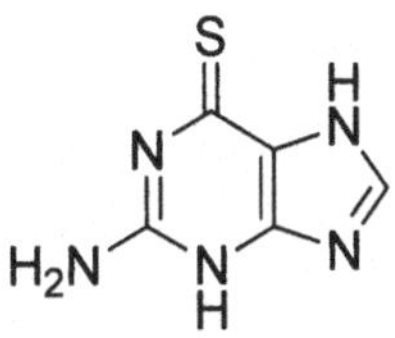

thioguanine 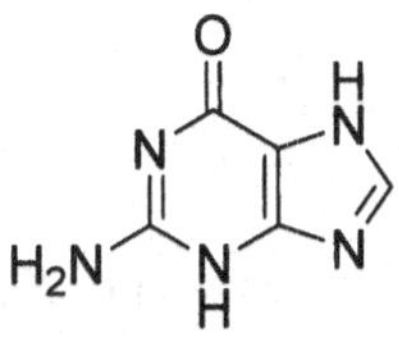   guanine 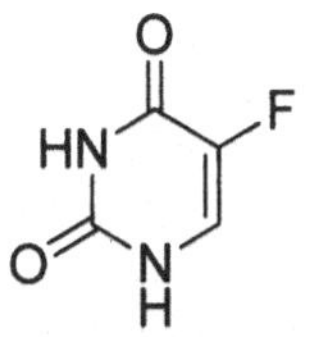   uracil

Thioguanine is another antimetabolite, inhibiting DNA/RNA synthesis but exhibiting a less complex metabolism mechanism than mercaptopurine. Like mercaptopurine, thioguanine is also a substrate for hypoxanthine/guanine phosphoribosyl transferase and is also methylated by *S*-methyl transferase however the mechanism of cytotoxicity is believed to be mainly due to its incorporation into DNA and subsequent DNA damage.[22] It is mainly used in the treatment of acute lymphoblastic leukemia (ALL). While mercaptopurine has been a standard therapy for long-term continuing treatment for childhood lymphoblastic leukaemia,[23] both mercaptopurine and thioguanine are prodrugs with no intrinsic anticancer activity. In regards to metabolic pathways, thioguanine forms thioguanine nucleotides directly. On the other hand, mercaptopurine forms intermediate metabolites, which are major substrates for thiopurine methyltransferase, thus resulting in the formation of methyl metabolites at the expense of thioguanine nucleotides.[24] Thus, intracellular thioguanine nucleotides seem to be formed more reliably with thioguanine administration than with 6-mercaptopurine.

5-Fluorouracil (Adrucil, Carac, Efudix, Efudex, and Fluoroplex) is a pyrimidine antimetabolite that was originally synthesized more than 50 years ago as an analog of the pyrimidine uracil.[25]   It continues to be widely used in the treatment of common human malignancies, including carcinoma of the colon, breast, and skin.[26] It is one of the first anticancer drugs designed on the basis of biochemical information. (i) That fluorine atom is almost similar to a hydrogen atom in size, (ii) that C–F bond strength is much higher than C–H bond, (iii) the thymidylate synthase inhibition and (iv) the utilization of uracil by rat hepatoma cells rather than normal liver cells all put together allowed Heidelberger et al.[27] to hypothesize that 5-fluorouracil could potentially act as an anticancer agent.   In 1957, Charles Heidelberger developed and patented 5-fluorouracil, an antimetabolite that became widely used in cancer chemotherapy.[28] Charles Heidelberger, who had earlier found that fluorine in fluoroacetic acid inhibited a vital enzyme, sought a collaboration with Robert Duschinsky and Robert Schnitzer at Hoffman-La Roche to synthesize fluorouracil. During nucleic acid replication, 5-fluorouracil acts as uracil. While 5-fluorouracil is akin to the uracil form without carrying out the same chemistry, the drug inhibits RNA replication. Thus it stops RNA synthesis, preventing the proliferation of cancerous cells.   Although a well-tolerated drug, 5-fluorouracil shows bone marrow and gastrointestinal mucosa toxicities much like other antineoplastic drugs. It has a relatively narrow therapeutic window, with the effective and toxic doses not being very different. The drug has been most commonly administered intravenously, typically as a bolus or as an infusion over hours to days.

capecitabine (Xeloda)

Oral fluorinated pyrimidines have been prescribed in Japan for over two decades. They have been used as primary and adjuvant therapies for several types of cancers such as advanced and localized colorectal, gastric, breast, lung, head, and neck cancers. Capecitabine signifies the first in this class to gain approval by regulatory authorities outside Japan.[29] This prodrug of 5-fluorouracil that was launched as Xeloda by Roche in 1999 is administered orally, exhibiting almost 100% bioavailability.[30] It is prescribed for the treatment of metastatic colon cancer and metastatic breast cancer.[31] A sequence of events involving (i) carboxylesterases in the lever to generate 5′-deoxy-5-fluorocytidine, (ii) cytidine deaminase, and (iii) 5′-deoxy-5-fluorouridine which is a good substrate for nucleoside kinases, ultimately leads to 5-fluorouracil, thus demonstrating the use of capicatibine as a prodrug. Capicatibine has two advantages over 5-fluorouracil, namely ease of administration (orally vs. IV) and an enhanced therapeutic effect. An earlier molecule also approved by the FDA in the 1970s for the treatment of colorectal cancer is fluorodeoxyuridine (floxuridine).[32]

Deoxynucleoside analogs are useful as antimetabolites in the treatment of cancer, one of the earliest examples being cytarabine, also called AraC, a deoxycytidine analog that was approved in 1969 for the treatment of acute leukemias and marketed by Upjohn.[32] Deoxycytidine analogs undergo a relatively less complex metabolism than the thiopurines and fluoropyrimidines. Gemcitabine (Gemzar) is a very well known deoxycytidine that was approved by the FDA and European Agency for the Evaluation of Medicinal Products (EMEA) for non-small cell lung cancer (NSCLC), breast cancer, ovarian cancer, and adenocarcinoma of the pancreas.

cytarabine (AraC)   gemcitabine (Gemzar)

Gemcitabine was originally investigated for its antiviral effects but has since been developed for cancer treatment.[32] It was first approved in 1996 as the hydrochloride as first-line treatment for patients with locally advanced or metastatic adenocarcinoma of the pancreas for patients previously treated with fluorouracil. It has been proven to be more effective than 5-fluorouracil in patients with advanced pancreatic cancer in terms of survival duration and clinical status. Typical adverse effects of this compound are myelosuppression, transient elevation of hepatic enzymes, and nausea and vomiting.

4   Alkylating Agents

The alkylating agents are a well-established and much understood class of compounds and much used for the treatment of cancer since World War I. These are essentially compounds that react with electron-rich atoms in biological molecules to form covalent bonds and are typically the classical alkylating compounds such as nitrogen mustards, nitrosoureas, and alkyl sulfonate.[33] The other categories include platinum-based compounds and nonclassical alkylators such as the carbazines. Many of them continue to be used to treat a variety of cancers. Much is known about this class of compounds and therefore only few, fairly recent examples have been included. Alkylators act directly on the DNA during all phases of the cell cycle. While doing so they cross-link the N-7-guanine residues and cause DNA strand breaks, consequences of which are abnormal base pairing, inhibition of cell division, and eventually apoptosis.[34] Major problems associated with these compounds that restrict their clinical use are wide-ranging adverse side effects and drug resistance, despite the development of several strategies to overcome them.

Although thousands of nitrogen mustards were the subject of research to discover them as anticancer agents, only about five continue to remain in use. Cyclophosphamide (Cytoxan, Endoxan) is one of the earliest members of the nitrogen mustards and a compound that undergoes activation in the liver mediated by cytochrome P-450-mediated microsomal oxidation. Developed by Asta Werke AG it was introduced in the late 1950s and has since been used to slow the progression of several cancers such as lymphomas and leukemia's, multiple myeloma, neuroblastoma, and retinoblastoma as well as ovarian and breast cancers. Another compound, ifosamide, also a prodrug developed by Asta Werke AG in the 1960s, has been used in testicular cancer as well as some sarcomas.[35]   Many more compounds exist and some, like mechlorethanmine, chlorambucil,[36] and mephalan,[37] are shown below.

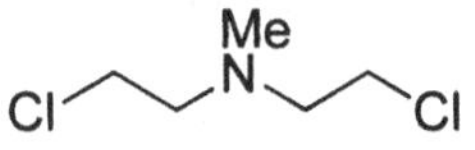

mechlorethamine 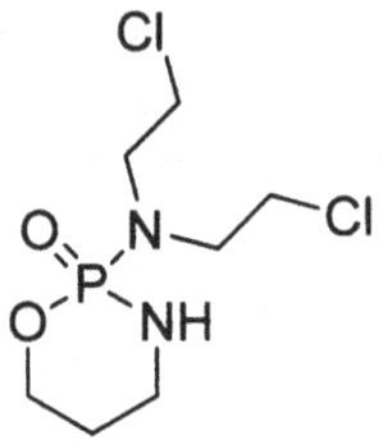   cyclophosphamide 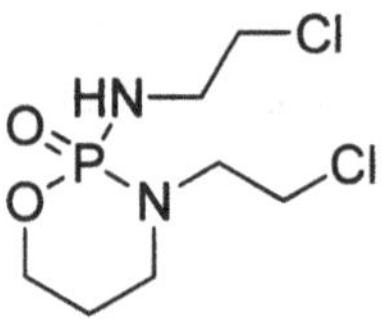   ifosamide

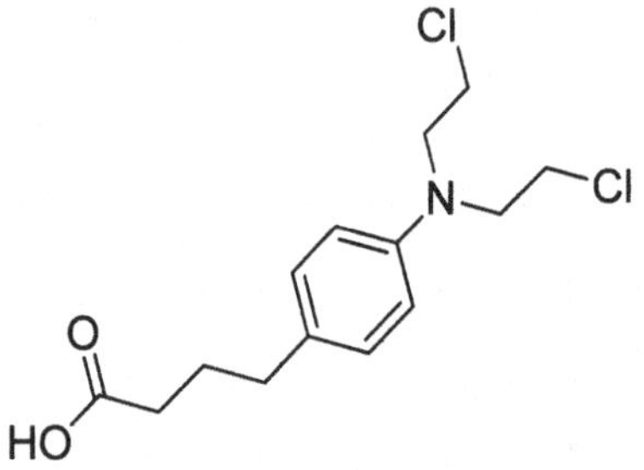

chlorambucil 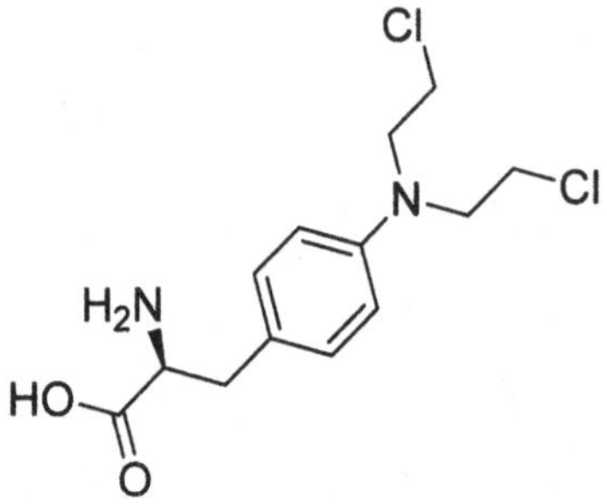   mephalan

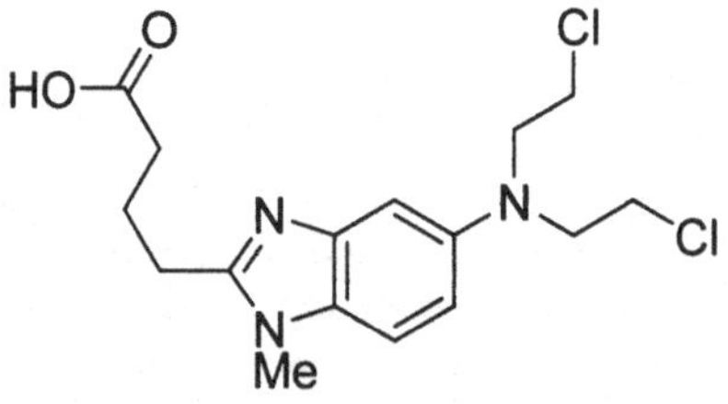

bendamustine (Ribomustine)

Bendamustine is an old nitrogen mustard compound that has recently found additional utility. An interesting and unique cytotoxic agent with structural similarities to alkylating agents and antimetabolites, bendamustine is non-cross resistant with alkylating agents and other drugs *in vitro* and in the clinic. Initial clinical research was conducted in East Germany about 30 years ago revealed its potential in indolent non-Hodgkin's lymphoma (NHL).[38] Bendamustine is approved in Germany for the treatment of patients with indolent NHL, chronic lymphocytic leukemia (CLL) and multiple myeloma. It has been found to be superior to chlorambucil in untreated patients suffering from CLL. Bendamustine hydrochloride (Treanda) from Cephalon was approved in 2008 by the FDA for marketing in the U.S. Treanda can be used for the treatment of indolent B-cell NHL that has progressed during or within six months of treatment with rituximab or a rituximab containing regimen.

A remarkable class of alkylating agents are the aziridines that have been used in cancer treatment. Two very well known anticancer antibiotics are mitomycin C and thiotepa. In 1956, shortly after mitomycins A and B were isolated from *Streptomyces caespitosus,* mitomycin C was found to be present in the same strain. Mitomycin C ultimately was found to be a valuable anticancer agent because of its uniquely superior activity against solid tumors and reduced toxicity as compared to the natural derivatives. It comprises the aziridine, quinone, and carbamate substructures organized in a compact pyrrolo-[1,2-*a*]-indole structure, presenting the unusually favorable ability to cross-link DNA with high efficiency and specificity.[39]

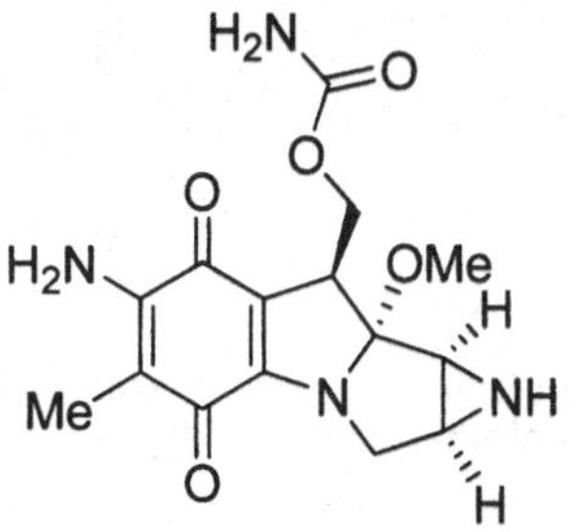

mitomycin C (Mutamycin)    thiotepa

Mitomycin is an alkylating agent used in combination with other agents for the treatement of cervical, stomach, breast, bladder, head, neck, and lung cancer. Since its approval by the FDA in 1974 for use in gastric and pancreatic carcinomas when combined with other chemotherapeutic agents, mitomycin found much use in order to treat a variety of tumors, especially in the 1980s. It is a natural product used to treat breast cancer and cancers of the gastrointestinal tract.[40] The aziridine ring exerts its cytotoxicity *via* cross-linking of DNA.[41] Mitomycin C gets reduced in cells due to the reactivity of the carbon-1 atom in the three-membered ring for external nitrogen nucleophiles on guanylic acid in DNA.[42] Subsequently there is displacement of the activated carbamate group on the 10-carbon atom by the amino nitrogen of a guanylic acid on the complementary DNA strand to produce an interstrand DNA cross-link.[43] Mitomycin is highly efficacious but suffers from adverse side effects. Common toxicities include anorexia, vomiting, and myelosuppression.[44] Another problem associated with this compound is clinical resistance. Despite the high efficacy in the treatment of solid tumors, acquired or intrinsic drug resistance of tumor cell populations leads to refractory malignant target tissue and a rather narrow utility of this antineoplastic drug. Mitomycin C has also been recommended to be used topically rather than intravenously for a variety of indications such as bladder and intraperitoneal cancers and during eye surgery.

The nitrosoureas such as carmustine, lomustine, and streptozocin are another class of alkylators that have been used for the past several decades to treat cancers. The *N*-alkyl- and *N*-(2-haloethyl)-*N*-nitrosoureas represent one of the generally most useful classes of anticancer agents, with a wide range of activities against various leukemias and solid tumors. Many nitrosoureas have been reviewed elsewhere in great detail.[45] Under physiological conditions, these compounds undergo a process that involves their decomposition to produce alkylating agents, which can happen *via* different mechanisms, one of them possibly being a base-catalyzed decomposition to a chloroethyl diazonium moiety that reacts with DNA, forming interstrand DNA cross-link.[46]

Triazines and hydrazines, another group of compounds, also belong to the class of alkylating agents. These compounds decompose spontaneously or metabolize to produce alkyl diazonium intermediates that can alkylate DNA. For example procarbazine and dacarbazine are both used in the treatment of Hodgkin's disease. Procarbazine is a component of combination regimens used for the treatment of primary brain tumors[47] while dacarbazine has been used in the treatment of melanoma.[48]

5   Platinum Complexes

cisplatin

carboplatin
(Paraplatin)

oxaliplatin
(Eloxatin)

Platinum complexes have been the subject of much discussion in several reviews due their high value in cancer chemotherapy.[49]   With the discovery of cisplatin or *cis*-diamminedichloro-platinum(II) (*cis*-[PtCl$_2$(NH$_3$)$_2$] by Barnett Rosenberg at Michigan State University in the 1960s back when platinum complexes were found to inhibit cell division, this field of chemotherapy witnessed landmark improvements and positively impacted the lives of cancer patients. It was approved for the treatment of testicular and ovarian cancer in 1978. Since then, cisplatin has been well utilized in the management of metastatic testicular cancer, ovarian tumors, bladder cancer, oropharyngeal carcinoma, bronchogenic carcinoma, cervical carcinoma, lymphoma, osteosarcoma, melanoma, bladder carcinoma, and neuroblastoma.[50] The anticancer property of cisplatin stems from its ability to form cross-links between and within the DNA strands in the nuclei of dividing cells, akin to the nitrogen mustards. This happens after intravenous administration when it undergoes hydrolysis forming platinum cation, which being positively charged coordinates to the nitrogen atoms of the purine base of DNA. Other platins also work similarly. Cisplatin has been one of the top-selling anticancer agents since its launch. Unfortunately, there are serious adverse effects associated with the use of cisplatin, and these include conditions such as nausea, vomiting, myelotoxicity, nephrotoxicity, neurotoxicity, and emetogensis. Due to limited aqueous solubility it is administered as an intravenous infusion. Depending on the tumor being treated, a typical intravenous dose of cisplatin is several milligrams to hundreds of milligrams per day for a few weeks. Like many other chemotherapeutic agents, another problem that cisplatin poses is that some tumors have natural resistance to it while others develop resistance after initial treatment. These drawbacks have led to the search for improved forms of platinum-based antitumor drugs.

Oxaliplatin is the first clinically approved platinum compound that was developed due to a proven absence of cross-resistance in cisplatin-resistant mouse lymphocytic leukemia cells,[51] presumably due to the 1,2-diaminocyclohexane ligand (DACH). It was approved for the treatment of advanced colorectal cancer in combination with fluorouracil.[52]   Another platinum-based anticancer agent is carboplatin (Paraplatin) in the treatment of advanced-stage ovarian cancer introduced in the year 1989 and also for the palliative treatment of ovarian cancer recurrent after prior chemotherapy. Carboplatin shows reduced toxicity and good potency against cancers that are not treated with cisplatin such as leukemia and lung cancer.

6   Plant- and Marine-Based Natural Products

A variety of natural products or their derivatives have been approved for the treatment of many cancers.[53] Nature has provided great inspiration for the discovery and development of these compounds. However typical impediments to the development are the limited availability of the material as well as structural complexity. Typically the odds of a natural product directly being approved for the treatment of cancer are very low. Yet a great deal of information can be gathered toward the development of analogs. Within the area of cancer, quite a few new commercialized drugs have been derived from natural sources, by structural modification of natural compounds, or simply by the development of new compounds, intended to follow a natural compound as the model. The search for improved cytotoxic agents continues to be an important line in the discovery of modern anticancer drugs. Extensive structural variety of natural compounds and their bioactivity potential have paved the way for "lead" compounds for improvement of their therapeutic potential by molecular modification. Additionally, semisynthetic processes of new compounds, obtained by molecular modification of the functional groups of lead compounds, are able to generate structural analogs with greater pharmacological activity and with fewer side effects.[54]

vinorelbine (Navelbine)

Much is known about the earliest and well-studied alkaloids vincristine and vinblastin, isolated from *Catharanthus roseus* (Madagascar periwinkle) to treat cancers. They act by inhibiting microtubule formation, by binding to tubulin a structural protein which is the basic building block of microtubules. Both agents were approved in the United States for clinical use in the 1960s. They have been used for Hodgkin's lymphoma, testicular cancer, Kaposi's sarcoma, and many more cancers.

Microtubules are protein polymers that are responsible for various aspects of cellular morphology and movement. Their major component is tubulin. Microtubules are long hollow cylinders made of α- and β-tubulin dimers and they contribute to many key cellular processes, serving as structural components of cells and also participating in intracellular transport processes. A single microtubule comprises thirteen parallel protofilaments forming a hollow structure with a minus (–) end which is usually stabilized by attachment to an organizing center and a plus (+) end where growth or shrinkage of microtubule takes place. Formation of mitotic spindle by microtubules is responsible for the segregation of chromosomes during cell division, meaning that their inhibition will end cell division,[55] at the metaphase stage of the cell cycle. Tubulin continues to be a major therapeutic target for the development of clinically useful anticancer drugs.[56]

Vinca alkaloids distinctively exert their influence as anticancer agents. After binding to tubulin, they particularly form paracrystalline aggregates of equimolar drug–tubulin dimers.[57] This kinetically complex binding is governed by such factors as ionic strength, magnesium ion concentration as well as interaction with non-tubulin proteins, and nucleotide concentrations. Vinblastine and vincristine are well-known alkaloids to treat cancer and will not be discussed here.[58] However, it is important to mention vinorelbine (Navelbine, 5-norhydrovinblastine), a unique semisynthetic analog of vinblastine which differs structurally from other semisynthetic vinca alkaloids with the modifications being on the catharanthine ring instead of the vindoline ring. Vinorelbine was invented as a semisynthetic vinca alkaloid in the 1980s and subsequently gained approval from the FDA in 1995 for the treatment of metastatic breast cancer and non-small cell lung cancer (NSCLC). It is a relatively new semisynthetic vinca alkaloid with demonstrated and predictable efficacy accompanied by a favorable toxicity profile in first-line use as well as previously treated patients. Vinorelbine is active in both small-cell lung cancer (SCLC) as well as NSCLC and it also a well-tolerated therapy for metastatic breast cancer.[59] In terms of dosage, its administration depends on the type of cancer. Administered as a 10-mg-base/ml injection, it is well tolerated in patients with NSCLC. Here a dose-limiting concern is leucopenia, although there appears to be a minimal peripheral neurotoxicity.[60] In case of metastatic breast carcinoma after failure with anthracyclines and taxanes as well as in patients with taxane-refractory metastatic breast carcinoma. These data suggest that vinorelbine and taxanes are not cross-resistant, and weekly vinorelbine is an active salvage therapy.[61] Vinorelbine induces cytotoxicity by inhibiting microtubule assembly and has demonstrated significant activity in patients with pretreated breast, non-small-cell lung, and ovarian carcinomas as well as lymphoma.

The plants of the genus *Taxus* create diterpenes called taxanes which have become an integral part of chemotherapy since the mid-1990s. The taxanes also operate as anticancer agents at the microtubule. However, contrary to other microtubule antagonists, they do so by disrupting the equilibriums between free tubulin and microtubule by driving it toward the assembly rather than disassembly, thus preventing the microtubule polymer from falling apart. The net result is the inability of chromosomes to attain a metaphase spindle configuration causing blockage of mitosis and eventually apoptosis. Later on it was found that although these effects might have a role in chemotherapeutic actions of taxanes as well as vinca alkaloids, at lower concentrations, microtubule-targeted drugs could suppress microtubule dynamics with no change in the microtubule mass. Such action could also lead to mitotic blocking and apoptosis.[62] Taxanes are potent inhibitors of microtubular depolymerization,[63] causing mitotic arrest in the G2M phase of the cell cycle.[64] The parent molecule paclitaxel (Taxol) is a powerful cytoxic natural product isolated from the Pacific Yew tree *Taxus brevifolia* that was discovered in the late 1960s as a consequence of a National Cancer Institute screening program conducted at the Research Triangle Institute) by Monroe Wall and Mansukh Wani. Years of work and extensive research resulted in its clinical development (by Bristol-Myers Squibb) and ultimately its approval in 1992. During development inadequate supply of material precluded rigorous investigation in the initial stages. Semisynthetic methods employing 10-deacetyl baccatin III also isolated from needles of yew trees in reasonable quantities have enabled its production.

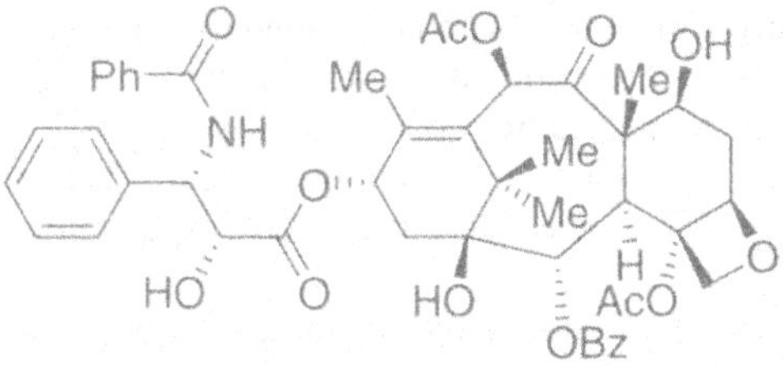

paclitaxel (Taxol)

Since its first approval by the FDA in 1992, paclitaxel has subsequently been approved worldwide for a variety of cancers such as ovarian, breast, and lung cancers and AIDS-related Kaposi's sarcoma. Insoluble in aqueous medium, paclitaxel is administered typically in the form of an injection in purified Cremophor EL (polyoxyethylated castor oil) and dehydrated alcohol. This conventional mode of administration over the years has been found to compromise the high activity that is expected. It has also led to adverse events including hypersensitivity, increased myelosuppression, and neuropathy.[65] 130-Nanometre albumin-bound paclitaxel (*nab*-paclitaxel) or Abraxane is a water-soluble formulation of paclitaxel[66] Developed by Abraxis BioScience (now Celgene), it was approved by the FDA in 2005 for the treatment of metastatic breast cancer. This formulation achieves the purpose of elimination of hypersensitivity reactions, the need for special tubing, reducing the time to administer the drug, improving tolerance, and enhancing the clinical efficacy. *nab*-Paclitaxel is thus an improved form of paclitaxel on account of its enhanced antitumor activity.

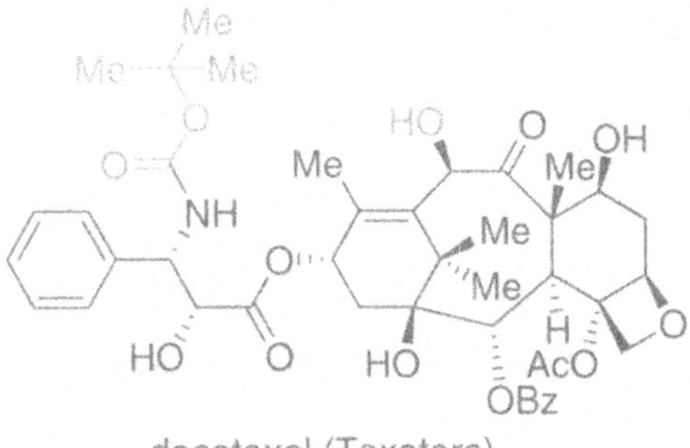

docetaxel (Taxotere)

Docetaxel (Taxotere) is a semisynthetic version of paclitaxel and was approved by the FDA in the U.S. in 1996. It was launched by Rhone Poulenc Rorer (now Sanofi Aventis) for the treatment of locally advanced or metastatic breast cancer. Structurally, it possesses the same backbone as paclitaxel, the only two differences being (i) a hydroxyl group instead of an acetate at the C10 position and (ii) *tert*-butyl carbamate ester replacing the benzyl amide on the phenylpropionate side chain. Like paclitaxel, it is administered intravenously and has been recommended for the treatment of breast, hormone refractory prostate and other non-small-cell cancers such as metastatic non-small cell lung cancer (NSCLC) after failure of prior platinum-based chemotherapy. Efforts continue so as to render it useful in the treatment of other cancers such as gastric cancer.[67] This taxoid analog is a potent inhibitor of microtubular depolymerization, thereby leading to mitotic arrest in cells. It can also counter the effects of the antiapoptotic protein Bcl-2 in prostate cancer cells. Docetaxel in combination with prednisone is approved for the treatment of hormone-refractory metastatic prostate cancer in the U.S. and E.U. The suggested dosage of docetaxel in men suffering from hormone-refractory metastatic prostate cancer is 75 mg/m^2 as a 1-h intravenous infusion once every three weeks as a combination with oral prednisone 5 mg twice daily every day.[68]

As can be expected with a majority of chemotherapeutic agents, docetacel is not without its adverse side effects due to its cytotoxicity. These include alopecia, fatigue, diarrhoea, grade 3/4 neutropenia, sensory neuropathy, nail changes, and stomatitis.

A problem associated with taxane therapy, whether monotherapy or in combination with other agents for treatment of a variety of tumors in the breast, prostate, ovary, lung, head and neck, has been the fact that paclitaxel and docetaxel both act as a substrate for multidrug resistant proteins.[69] These proteins relate to constitutive and acquired resistance to the antitumor activity of taxanes. Drug resistance is a much recognized impediment in cancer treatment and any drug that can possibly overcome common mechanisms of resistance would be an appreciable development. Cabazitaxel is a promising novel antineoplastic taxoid compound developed recently by Sanofi-Aventis.[70] It has been specifically approved for the treatment of a cancer called metastatic castration-resistant prostate cancer (mCRPC) in patients who received prior treatment with docetaxel. It is recommended as a second-line chemotherapy option for advanced prostate cancer based on the result of Phase III trial that showed that cabazitaxel is the first chemotherapeutic agent to demonstrate a survival benefit in mCRPC since the approval of docetaxel. The U.S. FDA approved this candidate in June 2010 while the European Medicines Agency approved it in January 2011 based on the encouraging results in the clinical trials mentioned above.

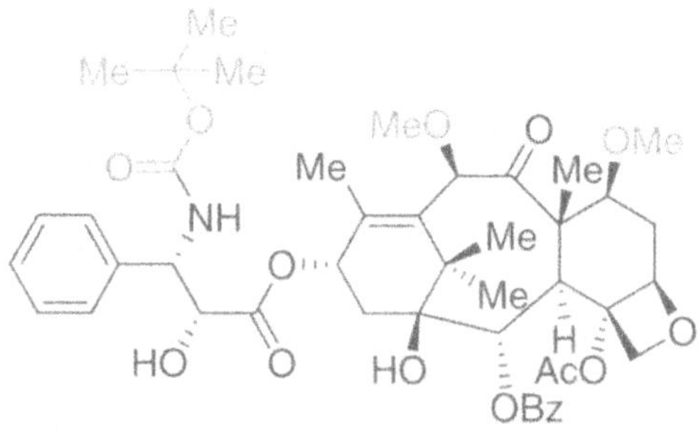

cabazitaxel (Jevtana)

Natural products have time and again inspired the discovery and development of landmark molecules in a variety of therapeutic areas. Once identified for their exceptional biological activity, sometimes their structural complexity can present enormous challenges, for example in paclitaxel or epothilone. Difficulties posed by complex natural product include structure elucidation and developing synthetic strategies to study analogs as well as to ensure adequate preclinical and clinical supplies. This very complexity is often seen as an opportunity rather than a challenge and can lead to scientific discoveries. Molecules such as ixabepilone and eribulin epitomize this statement.

Anthracyclines and taxanes are recommended and are widely administered to patients with hormone-sensitive metastatic breast cancer. Resistance to these compounds can often build up with prolonged use in adjuvant therapy and they do not remain a viable option when the cancer progresses. Ixabepilone (Ixempra, BMS-247550) is a recent example of a natural product analog. It is an analog of the natural product epothilone B. Epothilones are known to stabilize microtubules resulting in cell cycle arrest and apoptosis. It is indicated for the treatment of locally advanced or metastatic breast cancer in the U.S.

ixabepilone (Ixempra)

Ixabepilone has been recommended for the treatment of metastatic or locally advanced breast cancer in combination with capecitabine after failure of an anthracycline and a taxane, or as monotherapy after failure of an anthracycline, a taxane, and capecitabine.[71] It induces microtubule polymerization and stabilization by binding to β-tubulin subunits resulting in cell cycle arrest and apoptosis. There are limited options for treatment of advanced breast cancer after failure of or resistance to anthracyclines or taxanes. Ixabepilone is a semisynthetic lactam analog of the natural product epothilone B, which, like the taxanes, stabilizes microtubules resulting in cell cycle arrest and cell. It is administered as an intravenous infusion.[72]

The development of eribulin is a heroic effort from recent times. It is a synthetic compound found to mimic the anticancer behaviour of halichondrin B (halB), a natural product isolated from the marine sponge *Halichondria okadai*.[73] Halichondrin B was discovered in 1986 and found to exhibit potent tumor-fighting activity. Unfortunately its isolation was difficult, thereby limiting its availability. Additionally, halichondrin B possesses an extremely complicated structure comprising 32 chiral centers, making it nearly impossible for any chemist to produce it from scratch at the time. Yoshito Kishi at Harvard University and his team succeeded in completing the synthesis; however, the quantities synthesized were limited.[74] Collaborative investigations carried out by Eisai and Harvard University indicated that halichondrin B's anticancer activity resides in its right-half macrocyclic lactone moiety. Eisai performed medicinal chemistry and pharmaceutical optimization of halichondrin B right-half analogs resulting in the discovery of Eribulin. Synthesis of hundreds of analogs continued and ultimately eribulin, possessing a much less complex structure than halichondrin B, was discovered. Yet eribulin was a challenging molecule to construct on commercial scale due to the 19 asymmetric centers it contains. With 62 steps for its production it is a remarkably long synthesis for a molecule that has gone a commercial and is prescribed for cancer treatment. The discovery and development of eribulin are indeed an extraordinary feat.[75]

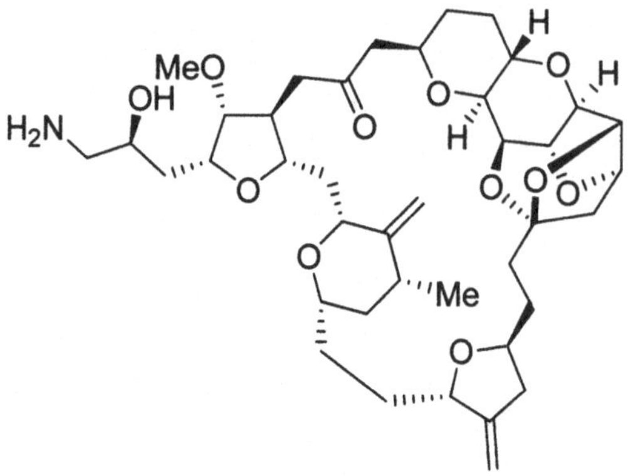

eribulin (Halaven)

In terms of its mechanism of action, eribulin displays uniqueness. As mentioned earlier in this section, vinca alkaloids bind to the plus ends of microtubules as well as along the sides and the taxanes bind to β-tubulin subunits inside the microtubule. Eribulin exerts its effect by binding to the plus ends of microtubules and suppressing microtubule growth, without corresponding effects on microtubule shortening, and by inducing the formation of non-productive tubulin aggregates.[76,77] Such action on microtubule dynamics is distinct from other classes. Thus it is a non-taxane microtubule dynamics inhibitor with tubulin-based antimitotic activity and antitumor cytotoxicity.[78] Eribulin mesylate has been approved in the US and EU for the treatment of patients with locally advanced or metastatic breast cancer who have previously received at least two chemotherapeutic regimens for the treatment of metastatic disease. Prior therapy should have included an anthracycline and a taxane in either the adjuvant or metastatic setting. 1.23 mg/m^2 eribulin (or 1.4 mg/m^2 eribulin mesylate) is administered intravenously by infusion over 2–5 minutes on days 1 and 8 of a 21-day cycle during treatment. Some common unfavorable side effects include asthenia, neutropenia, alopecia, peripheral neuropathy and nausea. It is the first single-agent therapy to demonstrate overall survival benefit in patients with advanced breast cancer.[79]

7   Topoisomerase Inhibitors

Topoisomerases are a family of nuclear enzymes that manipulate DNA topology, such as knots, tangles, and catenanes, remaining on DNA after replication or transcription. The topoisomerase inhibitors represent a distinct class of chemotherapy drugs with a distinct and highly specific mechanism of action. Through an inhibition of the nuclear enzyme topoisomerase I or II they inhibit mechanisms which allow normal DNA replication to occur, leading ultimately to cell death.[80]

The enzymes topoisomerases I and II are critical for DNA function and cell survival. These enzymes have been identified as cellular targets for several clinically active anticancer drugs. Discovered in 1971 but understood well, later in the 1980s, these enzymes have gained greater appreciation over the past two decades.[81] There are two main types of topoisomerases: topoisomerase I and topoisomerase II. Topoisomerase I is a very important enzyme from a normal DNA replication viewpoint. In its physiological state in the chromosome, with the supercoiled and tightly packed DNA helix in chromatin, replication requires transient relaxation and unwinding of the parent DNA. Relaxation and unwinding are required so as to allow the replication fork to proceed down the DNA strand and serve as a template for synthesis of new strands of DNA. Transient cleavage of the DNA can achieve this without the need to create extreme torsional stress on the parent DNA. It also allows passage of newly synthesized DNA without any permanent entanglement in the parent strands. Topoisomerase I promotes this process through a reversible trans-esterification reaction leading to a covalent intermediate form with the tyrosine of the enzyme bound to the 3' end of the DNA strand.[82] This complex is temporary and normally lasts just long enough to enable passage of the newly formed strand through the cut, after which the cleaved complex gets re-sealed by topoisomerase I.

Topoisomerase I operates on a single DNA strand. Another enzyme, called topoisomerase II, catalyzes the opening and closing of two strands of DNA rather than a single strand, either simultaneously or sequentially. It functions by creating transient single-stranded nicks in DNA supercoils relieving torsional strain that has accumulated during DNA replication and transcription.[83] Topoisomerase II is one of the most abundant chromatin protein after histones.[84] Topoisomerase II inhibitors (anthracyclines, epipodophyllotoxins, etc.) have been known to be active against several types of tumors.[85] Unfortunately, treatment with these inhibitors often leads to the development of the multidrug resistance. Because topoisomerase II-active drugs have several different modes of action, different mechanisms of resistance such as decreased activation as well as enhanced detoxification have also been implicated. In contrast to topoisomerase II, topoisomerase I does not depend on a cell cycle and, therefore, it is a favored cellular target in the development of drugs in oncology.[86]

7.1   Topoisomerase I Inhibitors

Perhaps the best known member of this class of compounds is camptothecin, with its discovery in the early 1960s as an anticancer drug with a unique mechanism of action. Inhibition of DNA topoisomerase I has provided a novel aspect to chemotherapeutic research. Camptothecin is a natural alkaloid that was first isolated from the stem wood of the Chinese tree *Camptotheca acuminata* during the search for steroids through screening of thousands of plants. During this investigation the extract from *C. acuminate* was found to exhibit antitumor activity, which later on culminated in the identification of camptothecin as a possible antitumor agent. Further studies revealed that it is selectively cytotoxic to S-phase cells, arresting cells in the G$_2$ phase and inducing fragmentation in the chromosomal DNA.[87,88] Despite the strong antitumor activity, unpredictable and severe adverse effects such as myelosuppression, vomiting, diarrhea, and hemorrhagic cystitis led to an abandonment of its clinical development in 1972.[89] The poor solubility exhibited by camptothecin is complicated by a pH-dependant equilibrium hydrolysis and lactonization reaction.[90] However, with the late 1980s witnessing the identification of topoisomerase I as the distinctive site of action of camptothecin, there was a renewed appreciation[91] of camptothecin as well as a search for its analogs.[92] Since then, extensive development over the years has led to an understanding of camptothecin related issues such as poor solubility and toxicity. This has resulted in the discovery of compounds such as topotecan and irinotecan.

Quite a few topoisomerase I inhibitors have been on the market and have shown significant activity against a variety of tumors.[93] They are typically not substrates for either the multidrug-resistant P-170 glycoprotein or the multidrug-resistance-associated protein. The encouraging toxicity against solid tumors combined with manageable adverse toxicity effects of topoisomerase I-active drugs seems to suggest their great potential in cancer therapy.[94] Topotecan (Hycamptin, SmithKline) was first approved by the FDA in May 1996 for the treatment of metastatic carcinoma of the ovary in previously treated patients.[95]   Recently it has also been approved for the treatment of cervical cancer. In such patients, topotecan serves as an additional therapeutic alternative when the disease progresses after initial platinum-based treatment. This water-soluble semisynthetic derivative of camptothecin, with established antineoplastic activity in a wide range of cell culture and xenograft systems, was also approved for second-line therapy in small-cell lung cancer (SCLC). A more recent development has been the approval of Hycamtin (topotecan) capsules for the treatment of relapsed small-cell lung cancer (SCLC) by the FDA. Topotecan is the first camptothecin analog to have been approved for clinical use by the US Food and Drug Administration (FDA). Its aqueous solubility is due to its side-chain at carbon 9 of the A ring. It inhibits replication of rapidly dividing cells by disrupting the normal function of the nuclear enzyme topoisomerase I. Topoisomerase I induces single-strand breaks in supercoiled DNA, and once the torsional strain is released the enzyme reseals the break and disassociates from the DNA. It is administered in various forms as an injection or orally.

topotecan (Hycamtin)   irinotecan (Camptosar)

In terms of side effects neutropenia is the principal dose-limiting toxicity of topotecan.[96] The gastrointestinal adverse effects of topotecan, including diarrhea, nausea, and vomiting, can be successfully controlled with standard supportive care measures. A shorter drug exposure time and lower cellular accumulation of topotecan as a result of its strong affinity for the efflux transporter P-glycoprotein is possibly the reason for the relatively less gastrointestinal adverse effects with this drug. In ovarian cancer, it is typically administered as 1.5 mg/m^2 by intravenous infusion over 30 minutes daily for 5 consecutive days, starting on day 1 of a 21-day course.

While topotecan was the first camptothecin analog to be approved by the FDA in 1996, irinotecan was the first water-soluble semisynthetic derivative of camptothecin to enter clinical trials. Irinotecan was first approved in Japan in 1994 for therapeutic indications such as small cell and non-small-cell lung cancers, cervix, and ovaries. In 1995, it was approved in Europe as a second-line agent for colon cancer, about a year before European approval of topotecan. It was approved by the FDA in the US in 1996 as Camptosar to treat advanced colorectal cancer that was refractory to fluorouracil.[97] The structure of irinotecan is characterized by a bulky dipiperidino side-chain linked to the camptothecin molecule *via* a carboxyl-ester bond that causes a reduction in the anticancer activity. Carboxylesterases typically present in the liver and gastrointestinal tract can cleave this side chain to yield metabolite SN-38 or 7-ethyl-10-hydroxycamptothecin, which is about 1000-fold more potent in inhibition of topoisomerase 1 than irinotecan and is thus the predominantly active form of the drug.[98] While neutropenia is often seen with irinotecan administration, diarrhea associated with this drug is severe. Irinotecan-induced gastrointestinal toxicities do not respond well to conventional treatment and this dose-limiting toxicity still remains a clinical problem for patients receiving this drug. Irinotecan is administered in higher doses than topotecan. For example in patients with advance colorectal cancer, the recommended initial dose of irinotecan in the U.S. is 125 mg/m^2 administered as a 90-min intravenous infusion The recommended treatment regimen is 1 weekly dose for 4 weeks followed by a 2-week drug-free interval, while in Europe it is 350 mg/m^2 once every 3 weeks administered as a $\geq$ 30-minute intravenous infusion.

7.2   *Topoisomerase II Inhibitors*

Topoisomerase II enzymes are a broad class of antineoplastic drugs with a wide spectrum of activity against cancers. Research in this area began several decades ago. They are further classified into various types such as podophyllotoxins, anthracyclines, and anthraquinones as well as acridines. They are called topoisomerase II poisons because they cause accumulation of DNA double-strand breaks, dislocating the normal catalytic enzyme cycle by stabilizing the intermediate "cleavable" enzyme-DNA complex which eventually leads to drug-induced apoptosis or cell cycle arrest in the G_1, S, and G_2 phases. In addition to the poisons, there are small molecules or catalytic inhibitors of topoisomerase II that regulate topoisomerase II activity by preventing binding of topoisomerase II to DNA, blocking the ATP-binding site or inhibiting the cleavage reaction rather than interfering with the cleavable complex.[99]

Although several potent topoisomerase II inhibitors are available for the treatment of cancers, their clinical use appears to diminish with the introduction of newer and specific anticancer therapies. Yet there is continued research in this therapeutic area because: many forms of cancer are devastating enough to require a cytotoxic agent to curb rapid tumor growth; combination therapies require cytotoxics; the emergence of clinical drug resistance in targeted therapeutics; and approved topoisomerase II inhibitors produce adverse side reactions. Some existing topoisomerase II inhibitors will be briefly discussed below.

Podophyllotoxins are a well-known class of topoisomerase II inhibitors that do not intercalate DNA since their discovery in the 1960s, with one such compound etoposide (VP-16, Eposin, Etopophos) being approved by the FDA in 1983. Etoposide is a semisynthetic derivative of podophyllotoxin in the treatment of refractory testicular cancer, small-cell lung cancer, non-Hodgkin's lymphoma, and acute myelogenous leukemia (AML). Administered intravenously or orally, it has been widely used over the past forty years; however, it poses toxicity concerns such as myelosuppression, nausea, vomiting, and alopecia.[100] An analog of the same compound is teniposide that is used in pediatric patients with acute leukemia.[101]

Doxorubicin (Adriamycin) and daunorubicin (Cerubidine) are some of the earlier anthracycline topoisomerase II inhibitors that act as anticancer drugs. With the only difference between the two being a single hydroxyl group, they differ greatly in their activities. Doxorubicin, introduced in the U.S. in 1974, is effective against a range of cancers such as several lymphomas and leukemias as well as breast, ovarian, bladder, stomach, and bladder cancers.[102] Daunorubicin discovered a little earlier but introduced later than doxorubicin has been mostly used to treat AML.

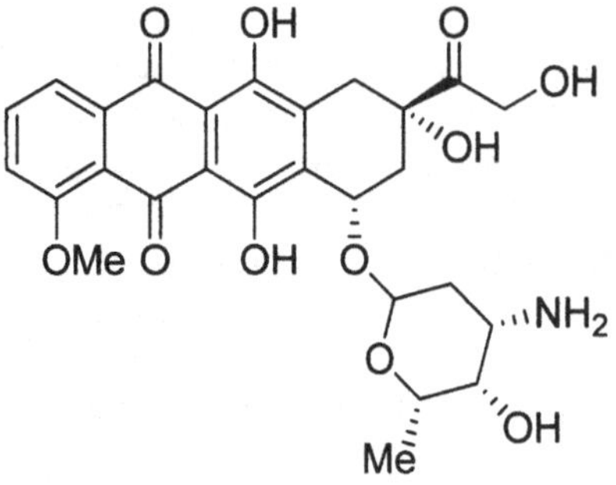

doxorubicin (Adriamycin) 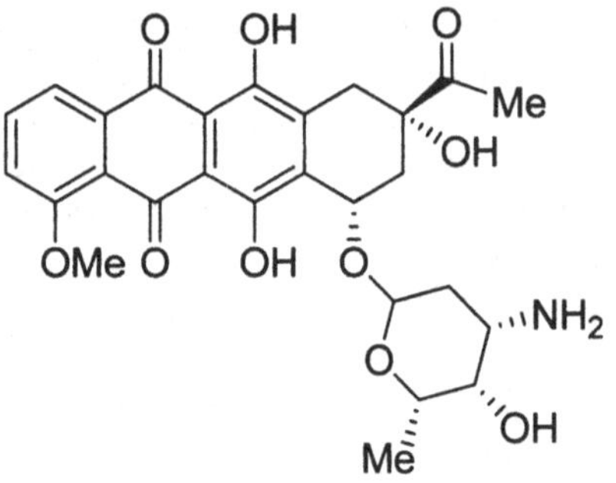   daunorubicin (Cerubidine)

Short-term toxicity includes myelosuppression and gastrointestinal toxicity. Long-term survivors are at risk of cardiac toxicity and secondary leukemia. Since the introduction of these two anthracycline anticancer antibiotics, several related anthracyclines have been developed over the years, including idarubicin (Idamycin), epirubicin (Ellence), and valrubicin (Valstar). These compounds act by intercalating with DNA so as to form a temporary complex with DNA and topoisomerase II. This complexation hinders re-ligation and leads to tumor cell apoptosis.[103] Unfortunately, tumors treated with anthracyclines can become multidrug resistant and even cross-resistant to a many structurally different drugs.[104]

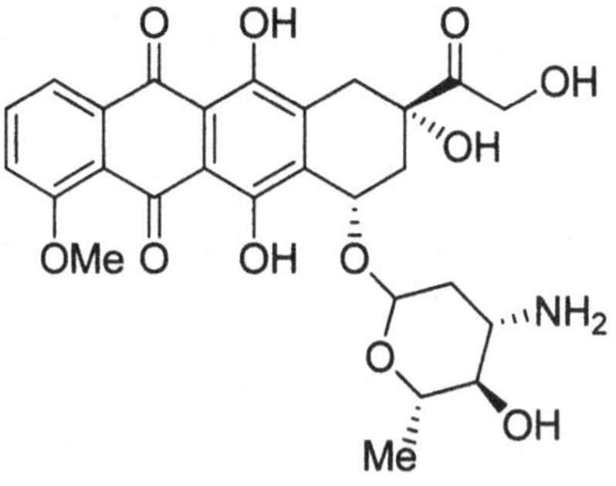

epirubicin (Ellence)

Epirubicin (Ellence, Pharmacia & Upjohn, now Pfizer) is another anthracycline that is recommended for use as a component of adjuvant therapy in patients with evidence of auxiliary node tumor involvement following resection of primary breast cancer. Although it was discovered in 1980, it was finally approved by the FDA in 1999. It is meant to be administered as an intravenous infusion, the dose being 25 or 50 mg/m^2. Epirubicin possesses an epimeric carbon on the sugar at 4″-position of the anthracycline antibiotic doxorubicin and has been used alone or in combination with other cytotoxic agents in the treatment of a variety of malignancies. It is a cell cycle phase nonspecific anthracycline compound that exhibits most of its cytotoxic effects in the S and G$_2$ phases. It has a lower propensity to produce cardiotoxic effects than doxorubicin. Its recommended maximum cumulative dose is almost double that of doxorubicin, which allows more treatment cycles in possibly higher doses.[105] There are many more topoisomerase II inhibitors belonging to the anthracycline or anthracenedione class, such as idarubicin[106] and mitoxantrone,[107] to name a few that operate similarly and have added great value in the treatment of cancer but also pose some toxicity concerns.

8   Antitumor Antibiotics

Cytotoxic antibiotic compounds based on natural products have always had a special place in cancer chemotherapy since the discovery of the actinomycins in 1940 by Waksman.[108] Actinomycins D(C1) and C3 are highly effective chemotherapeutics in the treatment of Wilms' tumor, trophoblastic tumors, and rhabdonyosarcoma. Some well-known compounds include doxorubicin, mitomycin, actinomycin, and bleomycin in this category. They operate through different mechanisms, and therefore mitomycin and doxorubicin have been discussed in the earlier sections on alkylating ageints (mitomycin) and topoisomerase II inhibitors (doxorubicin) in this chapter. It is important to point out that all these compounds have found various applications either alone or as part of a combination therapy.

8.1   Bleomycin

First isolated in 1962 by Hamao Umezawa and co-workers and later introduced as anticancer drugs in Japan by Nippon Kayaku in 1969 and in the U.S. by Bristol Laboratories (now Bristol Myers Squibb) in 1973, the bleomycins are a class of glycopeptide antibiotics that are water-soluble and found in the cultures of *Streptomyces verticillus*. The various types differ only in their terminal amine functional groups. The term bleomycin indicates a commercial preparation: a mixture of bleomycin A_2 and bleomycin B_2. A unique feature of this compound is its effectiveness in causing minimal bone-marrow depression.[109] An effective antineoplastic drug especially when used in combination with other cytostatic drugs including cisplatin and vinblastine, bleomycin binds to and damages the DNA of tumor cells. It causes fewer side effects than most other antitumor drugs.[110] It is a frequently administered antitumor agent that causes single- and double-strand breaks in cellular DNA *in vivo* and *in vitro,* resulting in the genomic instability of damaged cells. Bleomycin has been found to bring about an increase in the reactive oxygen species resulting in oxidative stress and pulmonary fibrosis. Bleomycin forms a complex with an iron ion (Fe^{2+}), an oxygen molecule (O_2), and a single-electron reductant to form a reactive oxygen radical which reacts with DNA. Thus radical chemistry almost certainly participates in bleomycin mechanisms that presumably involve an oxygen-centered DNA derivative.[111] Further, bleomycin induces apoptosis and senescence in epithelial and nonepithelial cells of the lung. Bleomycin is a chemotherapy drug usually given to treat testicular cancer, lymphoma, and cancers of the head and neck. It has been widely utilized in the treatment of metastatic testicular cancer, where it is a part of chemotherapy that involves bleomycin, etoposide, and cisplatin (BEP). It is structurally complex, and this complexity has precluded the possibility of studying well modified analogs and limited most of the modifications at either the *C*-terminal amine or the *N*-terminal aminoalaninamide moiety by either directed biosynthesis or semisynthesis. Synthetic production of such compounds contributes to the cost and limits the practicality in pharmaceutical applications. Although a number of analogs have been synthesized in the past two decades, none has improved properties. Bleomycin exhibits a biological effect via a sequence-selective, metal-dependent oxidative cleavage of DNA and RNA in the presence of oxygen.[112,113] It is typically administered intravenously or intramuscularly. Bleomycin is also used to treat Hodgkin's disease and non-Hodgkin's lymphoma, testicular cancer, and malignant pleural effusion. Bleomycin does not cause myelosuppression, promoting its wide application in combination chemotherapy. Some limitations of this compound are the early development of drug resistance and cumulative pulmonary toxicity.

bleomycin (Blenoxane)

9   Tyrosine Kinase Inhibitors (TKIs)

Kinases are a large family of enzymes responsible for phosphorylation, a process that involves transferring phosphate groups from high-energy donor molecules such as ATP to certain substrates.[114] A protein kinase inhibitor is a compound that specifically blocks the action of one or more protein kinases, which typically happens on serine, threonine, or tyrosine. Inhibition of a kinase can elicit a real physiological response. Several scientists have contributed to understanding the nature and functions of these enzymes. Oncology drug discovery has benefited immensely from the developments in the understanding of kinases as targets for small organic molecules and more so than other therapeutic areas over the past decade. With over 500 kinases encoded in the human genome and with approximately eighty inhibitors advancing to some stage of clinical evaluation as of 2009, this is an area of drug discovery that has shown tremendous potential to offer sophisticated cancer therapy despite its complexity.[115] A majority of these kinases have been investigated for the treatment of cancer that has led to the development of highly selective small molecules with favorable pharmaceutical properties.[116]

    Conventionally chemotherapeutic agents have a narrow therapeutic index with the response being mostly in the form of brief and unpredictable relief rather than being completely curative, albeit with multiple side effects. In other words, it does not discriminate between normal cells and tumor cells well. Lack of proper discrimination by traditional agents is manifested in the form of toxic side effects. Targeted therapy such as kinase inhibition on the other hand is oriented toward cancer specific molecules as well as signaling pathways. A variety of targeted therapies with antitumor activity in human cancer cell lines and xenograft models are used to produce specific responses, slow down development of the disease, and often improve survival of patients with advanced malignancies. Kinase inhibitors therefore show a broader therapeutic window, more specificity, and less toxicity or more limited nonspecific toxicities than conventional therapy. Tyrosine kinase inhibitors compete with the ATP binding site of the catalyst domain of several oncogenic tyrosine kinases. As small molecules, their oral activity and favorable safety profile can be easily combined with other forms of chemotherapy or radiation therapy.[117]

The past fifteen years have witnessed revolutionary advances in both recognizing the role of kinases in malignancies as well as characterizing inhibitors that block their activity as therapeutic agents. Difficulties need to be overcome in order to utilize this field to its fullest potential so that newer and more specific drugs reach the market for improved therapy.[118] For example, a thorough molecular characterization of the role of kinase in the pathophysiology of cancers remains a challenge. Another problem to overcome is the emergence of resistance which seems almost unavoidable today. Kinase inhibitor resistance resulting from selection for mutant alleles or upregulation of alternative signaling pathways is commonly encountered and it calls for a need to develop multiple inhibitors of different kinases with their synergistic combinations. Most kinase inhibitors discovered to date are ATP competitive and present hydrogen bonds (one to three) to the amino acids in the hinge region of the kinase. Thus they mimic the hydrogen bonds that are normally formed by the adenine ring of ATP. A majority of these inhibitors do not use the ribose binding site or the triphosphate binding site of the ATP. There are a few types of kinase inhibitors: Type 1, Type 2, allosteric and covalent. Type 1 kinase inhibitors include most of the known ATP-competitive molecules and recognize the active conformation of the kinase which otherwise favors phosphotransfer.[119] Type 1 compounds possess heterocyclic substructures. Type 2 kinase inhibitors recognize the inactive conformation of the kinases and some examples in this class are imitinib, nilotinib, and sorafenib. Allosteric inhibitors bind at an allosteric site outside the ATP site. Covalent inhibitors form irreversible, covalent bonds to the kinase active site through a reaction with the nucleophilic cysteine residue.[120,121]

Most of the clinically approved kinase inhibiting drugs for cancer therapy have been discovered as a consequence of rational drug design. Since the discovery of imatinib to treat chronic myeloid leukemia (CML), a number of developments have taken place that has enabled understanding of the disease. Prior to the discovery and launch of imatinib (Gleevec), the treatment options for chronic myeloid leukemia (CML) were hematopoietic stem cell transplantation (HSCT), hydroxycarbamide, and interferon-α. Despite the great value that they demonstrate in the treatment, these options do pose certain limitations. A less complicated approach in the form of an orally administered drug could potentially offer great value toward treating CML. One such drug, a tyrosine kinase inhibitor, imatinib is the product of a ground-breaking drug discovery program spearheaded by researcher Nicholas Lydon at Novartis, oncologist Brian Druker at the Oregon Health Sciences University, and Charles L. Sawyers at Memorial Sloan-Kettering Cancer Center in the 1990s. Their contributions led to the discovery of imatinib, the first tyrosine kinase inhibitor that was approved by the FDA for the treatment of CML. The peripheral blood and bone marrow show abnormalities including an increased granular leukocyte count in CML. Initially named STI-571, imatinib is specifically designed to inhibit the breakpoint cluster region (BCR)-Abelson (ABL) fusion protein that results from the chromosomal abnormality known as the Philadelphia chromosome, a factor responsible for CML.[122,123] Imatinib is approved in numerous countries worldwide for the treatment of newly diagnosed Philadelphia chromosome-positive (Ph+) chronic-phase CML, Ph+ accelerated-phase or blast-crisis CML, and in patients with Ph+ chronic-phase CML who have failed to respond to interferon therapy.[124]

Table : Kinase Inhibitor Drugs Marketed for The Treatment of Various Types Of Cancer

Drug	Target	Cancer Type Treated	Launch Year
Small molecules			
Imatinib (Gleevec)	Bcr-Abl, KIT-positive	Ph+ chronic myeloid leukemia, gastrointestinal stromal tumor	2001
Gefitinib (Iressa)	EGFR	Breast cancer, non-small-cell lung cancer	2003
Erlotinib (Terceva)	EGFR	Non-small-cell lung cancer	2004
Sorafenib (Nexavar)	C-Raf, B-Raf, V600E BRAF, FLT-3, c-KIT, VEGFRs 2 and 3, PDGFR- β	Advanced renal cell carcinoma, hepatocellular carcinoma	2005
Sunitinib (Sutent)	Receptor tyrosine kinase	Renal cell carcinoma (RCC), gastrointestinal stromal tumor (GIST)	2006
Dasatinib (Sprycel)	Multi- BCR/ABL, Src family TKI	Chronic myeloid leukemia or Philadelphia +ve acute lymphoblastic leukemia	2006
Lapatinib (Tykerb/Tyverb)	EGFR, HER2/neu	Metastatic breast cancer	2007
Nilotinib (Tasigna)	Bcr-Abl, KIT, LCK, EPHA3	Imatinib-resistant Philadelphia +ve CML	2007
Pazopanib (Votrient)	VEGFR-1, 2 and 3; PDGFRα/β; c-kit	Renal cell carcinoma soft-tissue carcinoma	2009
Crizotinib (Xalkori)	ALK, ROS1	Non-small-cell lung cancer	2011
Ruloxitinib (Jakafi)	JAK1, JAK2	Myelofibrosis	2011
Vemurafenib (Zelboraf)	B-Raf	Melanoma with B-Raf mutation	2011
Axitinib (Inlyta)	VEGFRs 1-3, PDGFR, cKIT	Renal cell carcinoma	2012
Monoclonal antibodies			
Rituximab (Rituxan)	CD20	CD20+, low-grade B-cell non-Hodgkin's lymphoma, follicular lymphoma, other B-cell lymphomas	1997
Trastuzumab (Herceptin)	HER2/neu	Metastatic breast cancer	1998
Bevacizumab (Avastin)	VEGF	Colorectal cancer, nonsquamous non-small-cell lung cancer, metastatic renal cell cancer, glioblastoma multiforme	2004
Cetuximab (Erbitux)	EGFR	head and neck cancer, metastatic colorectal cancer	2004
Panitumumab (Vectibix)	EGFR	Metastatic colorectal cancer	2006

Imatinib is effective and generally well tolerated in patients with Ph+ CML, especially in patients with early-stage chronic-phase CML where it has been found to be more effective than interferon therapy with cytarabine. The inadvertent activation of the Abelson tyrosine kinase (Abl) causes chronic myelogenous leukemia (CML). Mechanistically, as a kinase inhibitor, imatinib is a competitive inhibitor of ATP, binding at the ATP site, blocking ATP binding and thereby inhibiting kinase activity.[125] It binds to an inactive conformation of Abl. The orally administered imatinib is absorbed rapidly, and with bioavailability being 98%, attaining plasma concentrations are reached after 2–4 h. Since its launch by Novartis in 2001, imatinib has been approved for many other types of cancer. In patients with late-stage CML, a problem observed with this compound has been primary or acquired resistance.

The proto-oncogene c-Kit or tyrosine-protein kinase Kit, or CD117, is a protein that in humans is encoded by the KIT gene. Activating mutations in this gene are associated with rare gastrointestinal stromal tumor (GIST) and some other cancers.[126] Platelet-derived growth factor receptors (PDGFR) are cell surface tyrosine kinase receptors. An interesting feature of imatinib is that it also inhibits the c-KIT and PDGFR tyrosine kinases. Given the fact that dysregulation of c-KIT or PDGFR- α plays a role in the rare gastrointestinal stromal tumor (GIST), this attribute of imatinib has been studied extensively. Imatinib has shown activity in patients with unresectable metastatic GISTs, including patients unresponsive to standard chemotherapy. In 2002 imatinib mesylate was approved for the treatment of GIST. Over the years it has been approved for various other indications. For example, in the U.S. and E.U., it has been approved for the treatment of adults with relapsed or refractory Philadelphia chromosome positive acute lymphoblastic leukemia.[127]

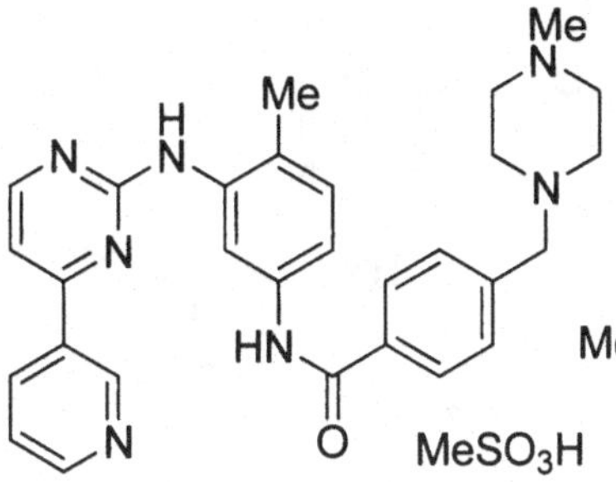

imatinib (Gleevec)

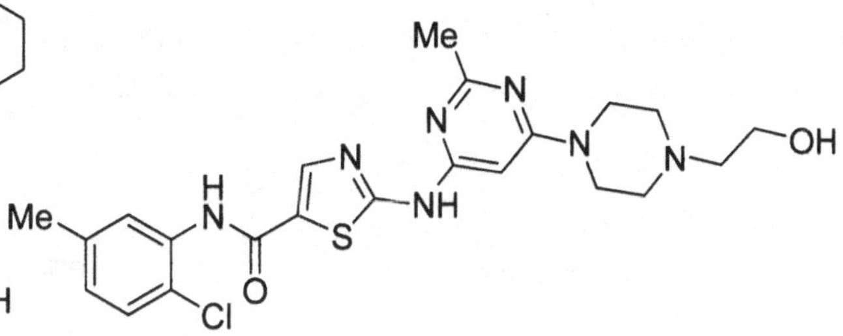

dasatinib (Sprycel)

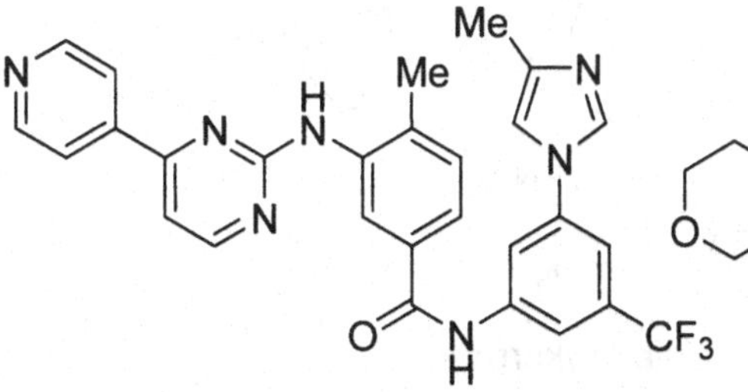

nilotinib (Tasigna)

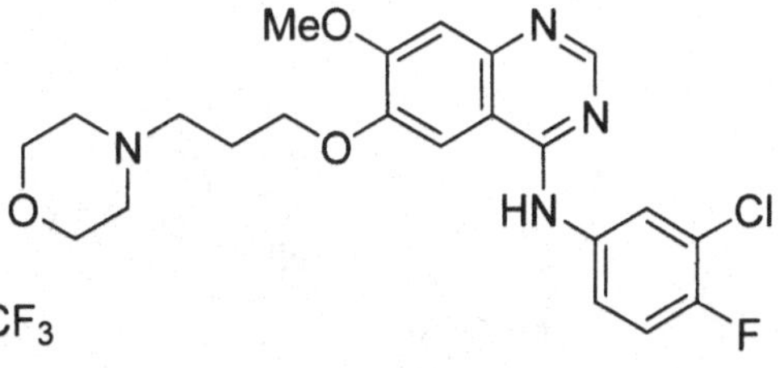

geftinib (Iressa)

The discovery of imatinib resistance mutations has led to development of several new ATP-competitive inhibitors (Table 1) such as dasatinib. Dasatinib, a thiazole-containing molecule, has a lower cellular EC_{50} than imatinib and brings about inhibition of all known BCR-ABL1 mutants. Another molecule is nilotinib, which has 20-fold higher cellular activities and inhibits most of BCR-ABL1 mutants. Some patients fail to respond to therapeutic regimens of imatinib (primary resistance), while others may lose a previously established response (secondary resistance).[128] Dasatinib (Sprycel), initially called BMS-354825, is an orally administered small molecule. It is an inhibitor of multiple tyrosine kinases, including BCR-ABL and SRC family kinases, which is indicated for the treatment of adults with newly diagnosed chronic-phase chronic myeloid leukaemia (CML), CML (chronic-, accelerated- or blast-phase) with resistance or intolerance to prior therapy, including imatinib, or Philadelphia chromosome-positive (Ph+) acute lymphoblastic leukaemia (ALL) with resistance or intolerance to prior therapy.[129] Nilotinib is another molecule that was recently approved for imatinib-resistant or imatinib-intolerant, Ph+ CML in the chronic or accelerated phase. This aminopyrimidine derivative works by inhibiting BCR-ABL tyrosine kinase, including most imatinib resistant BCR-ABL mutants.[130] Adverse events upon treatment with this drug are typically rash, nausea, pruritus, headache, and fatigue.[131] Some studies suggest that nilotinib may be useful to overcome dasatinib resistance in some patients.

Epidermal growth factor receptor (EGFR), the cell surface receptor family, comprises four transmembrane tyrosine kinase growth factor receptors: EGFR (ErbB1) (EGFR/HER1), ErbB2 (HER2/neu), ErbB3 (HER3), and ErbB4 (HER4).[132] The EGFR signal transduction pathways are responsible for the regulation of several neoplastic processes such as cell cycle progression, apoptosis inhibition, tumor cell movement, invasion, and metastasis.

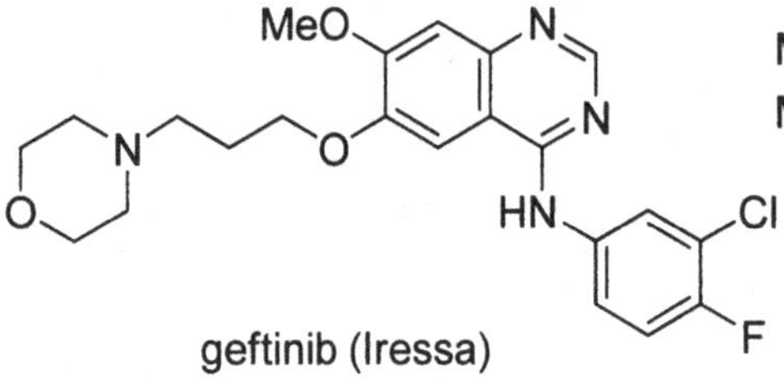

geftinib (Iressa) 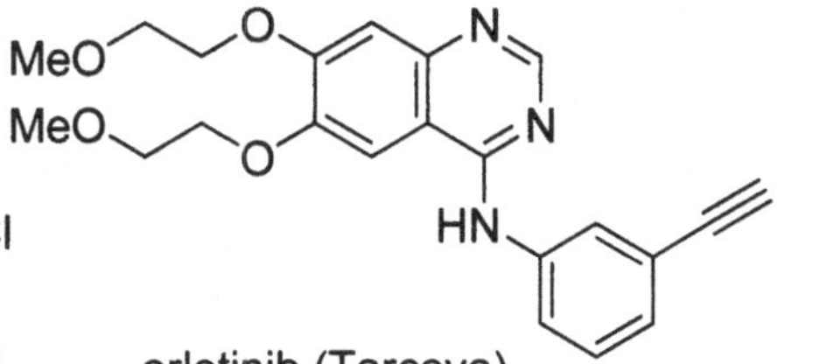   erlotinib (Tarceva)

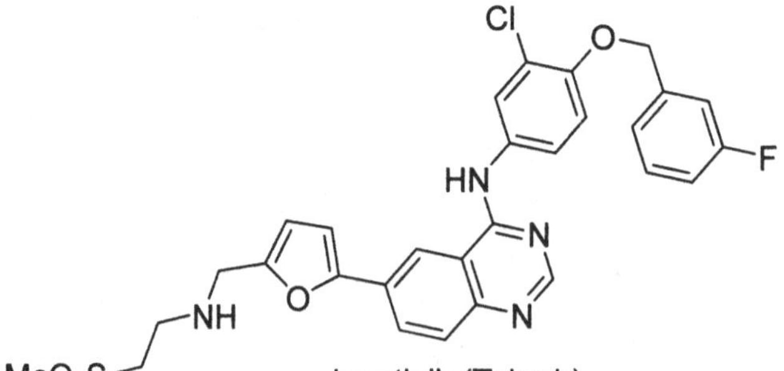

lapatinib (Tykerb)

Gefitinib, the first selective EGFR tyrosine kinase inhibitor, prevents autophosphorylation of EGFR in tumor cell lines and xenografts.[133] It has been approved for the treatment of patients with non-small-cell lung cancer after failure of both platinum-based or docetaxel chemotherapies. It is very selective with respect to inhition of EGFR and adverse effects albeit mild, are seen after its administration. They include nausea, vomiting, pruritus, anorexia, asthenia, and dry skin. A concerning side effect is interstitial lung disease the highest rate of which was observed in Japan. Erlonitinib (Tarceva or OSI774) is another orally administered potent, reversible, and selective inhibitor of EGFR (ErbB1) tyrosine kinase. It was approved by the FDA for the treatment of advanced refractory metastatic non-small-cell lung cancer in 2004. Erlotinib has also been under investigation for a variety of other tumor types such as pancreatic and colon cancer in combination with chemotherapy.[134] Typical side effects include acneiform skin rash and diarrhea as well as headache, mucositis, and anemia.[135] Similarly, lapatinib (GW-572016, Tykerb) from GlaxoSmithKline is also a reversible and specific receptor tyrosine kinase inhibitor of both ErbB1 and ErbB2[136] that was approved by the FDA in 2007 for the treatment of breast cancer and solid tumors.

Angiogenesis is the process of developing new blood vessels from existing ones and is responsible for tumor cell growth, survival, invasion, and metastasis. Of the various growth factors fostering angiogenesis, vascular endothelial growth factor (VEGF) is perhaps the chief one that has been implicated on account of its specificity as an endothelial cell mitogen and also because of the fact that many tumor cells produce it in physiologically significant amounts.[137] With four isoforms of VEGF (A, B, C, and D) and three types of VEGF receptors (VEGFRs) having emerged as anticancer targets acting in the microenvironment, VEGF-targeted therapy may inhibit tumor growth by blocking new vessel growth.[138] VEGF-targeted drug discovery over the years, has resulted in the development of promising therapy. For example, sunitinib (SU11248, Sutent) is a selective inhibitor of several tyrosine kinases associated with VEGFR-1 and VEGFR-2 as well as fms-like tyrosine kinases 3, KIT, and platelet-derived growth factor receptors.[139] A randomized Phase III trial revealed sunitinib to be more effective than IFNα in first-line metastatic renal cell carcinoma. Clinical activity was demonstrated in neuroendocrine, colon, and breast cancers in Phase II studies, whereas definitive efficacy has been demonstrated in advanced renal cell carcinoma and in imatinib-refractory GISTs, leading to U.S. Food and Drug Administration approval of sunitinib in 2006 for treatment of these two diseases. Sorafenib (Nexavar) is a novel dual-action Raf kinase and VEGFR inhibitor that inhibits cell proliferation and angiogenesis.[140] It is an inhibitor of VEGF and PDGF receptor kinases, available for the treatment of renal cell carcinoma.

sunitinib (Sutent)   sorafenib (Nexavar)

Research on kinases inhibitors continues to enable treatment of more cancers, since the discovery and launch of imatinib. More recently, crizotinib was introduced as a small-molecule kinase inhibitor and approved by the FDA for the treatment of locally advanced or metastatic non-small-cell lung cancer (NSCLC) that is anaplastic lymphoma kinase (ALK) positive. Determination of this reponse is carried out by a companion FDA-approved test.[141] The test helps determine if a patient has the abnormally expressed ALK gene, which causes cancer development and growth. Originally developed as an inhibitor of mesenchymal epithelial growth factor (c-Met)/ hepatocyte growth factor receptor (HGFR), it was later shown to be a potent inhibitor of ALK.[142]

Crizotinib works *via* blockade of kinases, including the protein produced by ALK. It was granted accelerated approval by the FDA in 2011 for NSCLC that is ALK positive, patients with this form of lung cancers being typically nonsmokers.[143] The aforementioned anticancer drugs represent some of the well-known small-molecule approved kinase inhibitors. More examples are listed in the Table 1, all of which happen to be validated targets for cancer therapy.

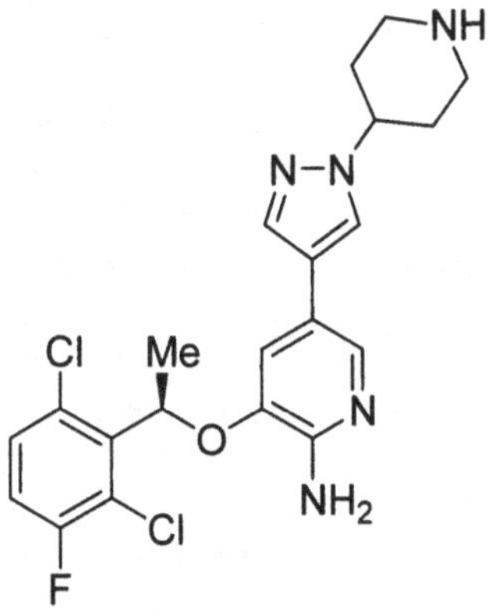

crizotinib (Xalkori)

9.1   *Monoclonal Antibodies*

First reported in 1975, the monoclonal antibodies are generally produced by a single clone of the B-cell, ensuring homogeneity. The earlier antibodies were created by fusing B cells that were obtained from mice immunized with human lymphoma cells. However, they were found to exhibit adverse effects such as allergic reaction and a diminishing efficacy.[144] Technological developments in the field of recombinant DNA have enabled the invention of chimeric antibodies (65–90% human), and partially humanized (95%) and recently fully humanized antibodies.[145] In addition to small molecules mentioned in the preceding section, significant and landmark work in the field of monoclonal antibodies has led to their approval by regulatory agencies for the treatment of various cancers. These are typically administered intravenously. Monoclonal antibodies are very specific agents with moderate toxic effects. However, a restriction that they pose is their size that limits their tumor penetration into the brain. Some examples that are currently on the market are presented below.

Rituximab (Rituxan and MabThera)

Rituximab is the first monoclonal antibody to be approved by the FDA. Its approval in 1997 heralded a new era of cancer treatment that is of great significance. It is a chimeric monoclonal antibody that acts against the CD20 protein on normal and malignant B-lymphocytes. It produces antibody dependent cell and complement-mediated cytotoxicity in these cells. It binds to the CD20 antigen, which is expressed on normal B-lymphocytes and on 90% of B-lymphocyte-derived non-Hodgkin's lymphomas with high selectivity and affinity. It is particularly effective in patients with relapsed or refractory indolent lymphoma, particularly those with follicular histology.[146] It has been indicated for relapsed or refractory indolent non-Hodgkin's lymphoma. It has been found that prolonged treatment with rituximab in patients with follicular lymphoma results in a continuous reduction in circulating B-lymphocytes. It is administered as an intravenous infusion with a dose of 375 mg/m^2 body surface area, with the frequency of administration being once every 2 months for a maximum period of 2 years.[147] Frequent side effects include transient flu-like symptoms during first infusion while a serious side effect is severe cytokine release syndrome during first infusion. Rituximab shows a reasonable toxicity profile as a single-agent maintenance therapy causing minimal unexpected adverse events compared with induction therapy. When administered intravenously, rituximab in combination with chemotherapy has been reported to improve overall survival in patients with newly diagnosed and relapsed indolent lymphoma compared with chemotherapy alone.[148] Rituximab was approved recently in 2011 in the U.S. and the EU as single-agent maintenance therapy for patients with previously untreated follicular lymphoma achieving a response to induction therapy with rituximab in combination with chemotherapy.

Trastuzumab (Herceptin)

Herceptin is a bioengineered human monoclonal antibody for breast cancer. In the mid-1980s following the identification of the genes namely the human epidermal growth-factor-2 (HER-2) by Ullrich of Genentech and the identical *neu* gene by Weinberg at MIT, their link with breast cancer was recognized by oncologist Slamon at UCLA. It was then established that blocking the HER-2/*neu* gene *via* a small molecule or an antibody would mean arresting cancer growth. An agent found among a panel of 100 murine monoclonal antibodies at Genentech to attain this task was the most potent clone muMAb 4D5. Subsequently in 1990, this antibody was humanized successfully to eventually obtain a chimeric antibody chMAb 4D5 which over time went to become trastuzumab (Herceptin), which was 95% human and 5% murine–a remarkable feat accomplished in a short span of time. Herceptin gained approval from the FDA for the treatment of patients with metastatic breast cancer whose tumors overexpress the HER-2 protein and who have received one or more chemotherapy regimens for their metastatic disease. This was the first approved monoclonal antibody for erbB2-overexpressing metastatic breast cancer, serving as a proof-of-principle to support target-specific growth factor receptors and bestowing clinical benefit to patients. As it binds to the surfaces of erbB2-overexpressing cancer cells, it appears that its mechanism of action may involve antibody-dependent cellular cytotoxicity (ADCC).[149] Cells treated with trastuzumab get arrested during the G1 phase of the cell cycle. Trastuzumab does not cause the typical toxic effects that are generally produced by chemotherapy, such as nausea, vomiting, hair loss and bone marrow toxicity.[150] It has been effective when used alone as well as in combination with other chemotherapeutic agents. As a combination it can enhance the median time toward progression of the disease and thus improve survival chances. Some common adverse effect, with >35% incidence are fever, chills, pain, asthenia, and nausea. However, an important concern with this therapy, especially in preexisting cardiac conditions and when used in combination with anthracycline chemotherapeutic treatment is cardiotoxicity.[151]

Cetuximab (Erbitux)

Cetuximab is a human-mouse chimeric anti-EGFR monoclonal antibody indicated in patients with EGFR-expressing mCRC. It binds competitively to the extracellular domain of the EFGR sterically inhibiting dimerization and thus the tumor growth as well as metastasis. It is administered intravenously as a monotherapy or in combination with irinotecan for the treatment of head-neck and colorectal cancer.[152]   It has been approved for the first- or subsequent-line treatment of squamous cell carcinoma of the head and neck (SCCHN) either as monotherapy or in combination with radiotherapy.[153]   It has also been approved for the second- or further-line treatment of metastatic colorectal cancer (mCRC) either as monotherapy in patients who cannot tolerate irinotecan.   Panitumumab and matuzumab are also anti-EGFR monoclonal antibodies showing preclinical activity against several tumors. Bevacizumab (Avastin) is another very well known monoclonal antibody introduced in 2004 as the first selective angiogenesis inhibitor. It brings about blockage of VEGF action and is effective for treating metastatic colorectal cancer in combination with chemotherapeutic agents.[154]   It has been approved for a variety of indications such as colorectal cancer, nonsquamous non-small-cell lung cancer, metastatic renal cell cancer, and glioblastoma multiforme (brain tumor).

10   Hormones

The endocrine system is vital for the overall growth and functioning of the human body. Unfortunately, sometimes malignancies can also arise in organs that are influenced by endocrine-system-secreted hormones.[155] Endocrine-related cancers in humans are typically gender specific in terms of hormonal responsiveness.   They typically include cancers of the breast, prostate, pituitary, testes, ovary, and neuroendocrine system. Other hormone-dependent cancers elsewhere in the body also fall in this category. Breast and prostate cancers are well-understood cancers of this type and are a leading cause of cancer death in women and men, respectively, in the United States.   These hormone-dependent tumors are generally treated effectively by antihormone therapies.   Sadly, emergence of resistance is not uncommon and especially so for locally advanced and metastatic tumors, where resistance is expected.   The following discussion on hormone therapy includes some recent and more effective drugs that have proven to be dependable.

10.1   Antiestrogens

Estrogen the female hormone, is capable of promoting the growth of breast cancer cells. Tamoxifen is the classic estrogen receptor antagonist for over 30 years.   It is a compound that was initially developed by Imperial Chemical Industries (ICI) in search of a birth control pill but was found to effective as an anticancer drug, receiving approval from the FDA in 1977. Eventually, it became one of the most prescribed anticancer drugs in the world.   Being a partial agonist as well as an antagonist of the estrogen receptor, it is classified as a selective estrogen receptor modulator (SERM).[156]   It has been used for the systemic treatment of patients with breast cancer for nearly three decades. In a majority of the cases, estrogen typically sets off breast cancer upon surgery, radiation, and chemotherapy. Estrogen-dependent breast cancer is often hormonally treated with tamoxifen for five years. Tamoxifen was also approved by the FDA in 1998 as a palliative medicine in women at high risk of developing breast cancer. However, success of the treatment depends essentially on the presence of the estrogen receptor (ER) in the breast carcinoma. About half of patients with advanced ER-positive disease immediately fail to respond to tamoxifen. In patients that respond, the disease ultimately progresses to a resistant phenotype.

tamoxifen (Nolvadex)   raloxifene (Evista)

A newer SERM is raloxifene (Evista), a second-generation compound in this class that functions as an estrogen antagonist on breast and uterine tissues and an estrogen agonist on bone. It reduces the risk of invasive breast cancer in postmenopausal women at a high risk of suffering from breast cancer as well as osteoporosis.[157] It is orally administered, upon which it is absorbed rapidly, undergoing extensive first-pass glucuronidation and enterohepatic cyclization with absolute bioavailability being only 2%. It does not seem to be metabolized by the cytochrome P450 system, however it is extensively distributed in the body. Compared to tamoxifen, raloxifene shows lower incidences of thromboembolic events, cataracts, leg cramps, and vasomotor, gynaecological, and bladder symptom, but more severe incidences of musculoskeletal symptoms, dyspareunia, and weight gain. Until the approval of raloxifene in 2007, tamoxifen was the only drug available for the prevention of breast cancer. Raloxifene is the second agent in addition to tamoxifen that has been approved in the U.S. for the prevention of breast cancer, for reducing the risk of invasive breast cancer in postmenopausal women with osteoporosis, and in postmenopausal women at high risk of invasive breast cancer.

10.2   Androgens

Partial estrogen agonists such as tamoxifen exhibit beneficial effects such as delaying the progression of osteoporosis. However, they may also cause endometrial hyperplasia or carcinoma as well as thrombosis and possibly produce withdrawal symptoms in such patients. An agent that was designed to overcome the problems associated with estrogen agonists is fulvestrant (ICI 182,780 Faslodex). It is an estrogen antagonist or a pure antiestrogen compound possessing no agonist activity.[158] Fulvestrant is a selective estrogen receptor downregulator (SERD) that was approved by the FDA in 2002 to treat hormone receptor-positive metastatic breast cancer in postmenopausal patients with disease progression following antiestrogen therapy. The approved dose is 250-mg monthly injections. Hormone-sensitive breast cancer patients that have previously responded to tamoxifen may possibly receive added benefit from a second endocrine anticancer in case of progression or relapse upon tamoxifen treatment.

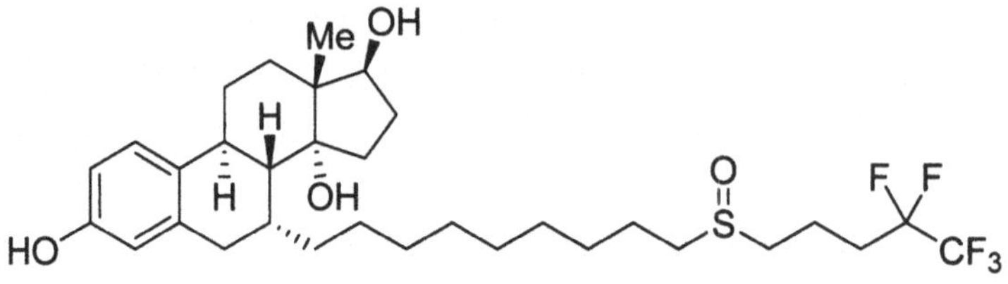

fulvestrant (Faslodex)

The mechanism of action of fulvestrant involves its binding to estrogen receptor (ER). This stops the binding of endogenous estrogens, preventing the ER-mediated gene transcription events that effect tumor cell proliferation. Unlike the ER antagonists such as tamoxifen, fulvestrant downregulates ER and PgR expression. Such a unique mechanism not only enhances the estrogen ablative activity of fulvestrant but also reduces the likelihood of cross-resistance with other endocrine agents to some extent.[159] This distinct mechanism of action is the reason for fulvestrant to offer some tolerability advantages and also provide a treatment alternative for patients with a history of relapse upon prior endocrine therapy.

10.3   Aromatase Inhibitors

Aromatase is an enzyme complex catalyzing the conversion of C_{19} steroids or androgens to C_{18} steroids or phenolic estrogens, thus bringing about aromatization. It comprises two proteins: nicotinamide adenine dinucleotide phosphate (NADPH-cytochrome) P450 reductase and P450$_{arom}$, which is a hemoprotein and is responsible for the biosynthesis of estrogens. One-third of all breast carcinomas are estrogen dependent and will revert upon the deprivation of estrogen. Reduction of estrogen levels is a promising target for breast carcinoma treatment in both premenopausal and postmenopausal women. In postmenopausal women, the ovary constitutes the main supply of estrogens and circulating estrogens. The estradiol (E2 estrogen) levels in tumors in postmenopausal women is 10- to 40-fold higher than those in their serum.[160] As mentioned earlier, tamoxifen is one such agent that can do so at the estrogen receptor level. An alternative approach is to inhibit the biosynthesis of estrogen *via* an agent that would inhibit the cytochrome P450 enzyme aromatase and thereby prevent the conversion of androgens to estrogens. Such an agent is called aromatase inhibitor (AI). The first of this class of compounds was aminoglutethiamide introduced in the late 1970s.[161] Despite its putative efficacy as a second option to tamoxifen, the use of this first-generation nonsteroidal compound was limited by its toxicity and lack of selectivity for the aromatase enzyme. Subsequently formestane, a steroidal compound that showed antitumor effects in patients who failed to respond to aminoglutethimide, was introduced in 1993.[162] Formestane is a second-generation, steroidal and more selective AI based on the androgenic compound androstenedione, a steroid. However, it suffers from extensive first-pass metabolism and therefore is administered twice a month *via* an intramuscular injection to attain optimal inhibition.

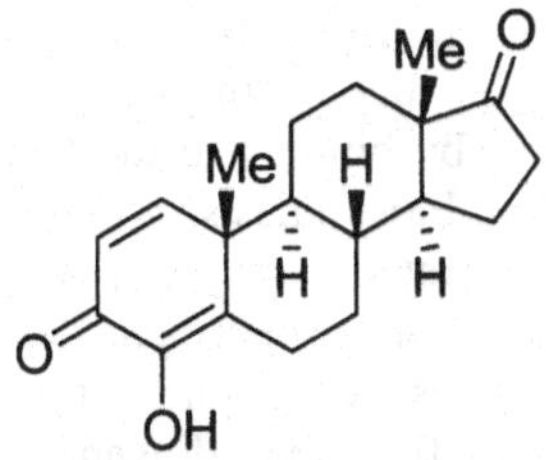

formestane (Lentaron) 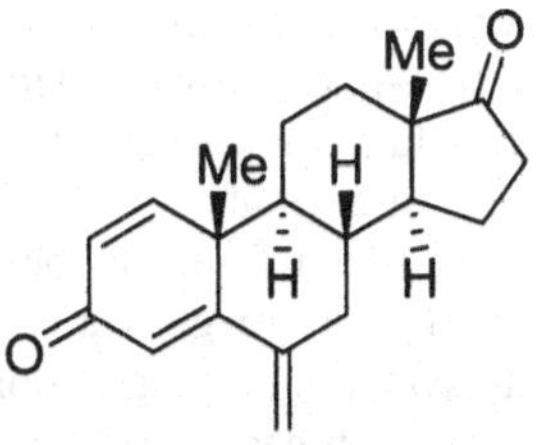   exemestane (Aromasin)

Exemestane is a third-generation aromatase inhibitor that was introduced for use in postmenopausal women with advanced, hormone-responsive breast carcinoma. It is the only third-generation compound with a steroidal backbone.[163]   Both formestane and exemestane are classified as Type 1 aromatase inhibitors, based on their steroidal nature as well as their irreversible binding to the aromatase enzyme complex. They both cause permanent inactivation even after the drug is cleared from the circulation. Exemestane was first approved by the FDA in 1999. In 2005 it was approved by the FDA for adjuvant treatment of postmenopausal women with estrogen-receptor-positive early breast cancer who have received two to three years of tamoxifen and have been switched to exemestane for completion of a total of five consecutive years of adjuvant hormonal therapy.[164]   It is typically well tolerated in early-stage or advanced breast cancer patients.   Adverse events are generally mild to moderate and of a comparable nature irrespective of the stage of disease. Overall, hot flashes are possibly the most frequent adverse event, much like in the case of tamoxifen therapy.[165]

Nonsteroidal compounds anastrozole and letrozole are classified as Type II aromatase inhibitor because they inhibit the conversion of androgens to estrogens competitively and reversibly. Like exemestane, both are administered orally at once-daily doses.[166]

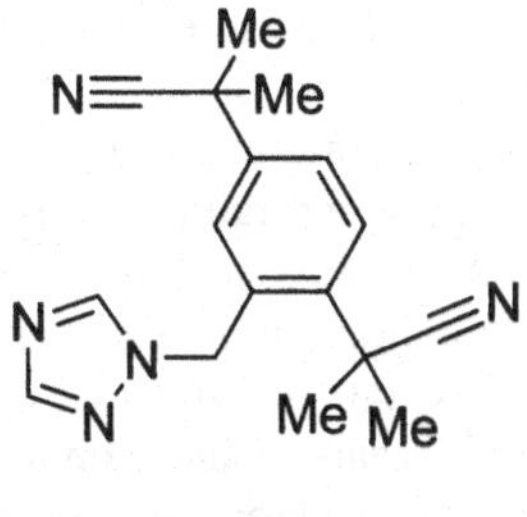

anastrozole (Arimidex) 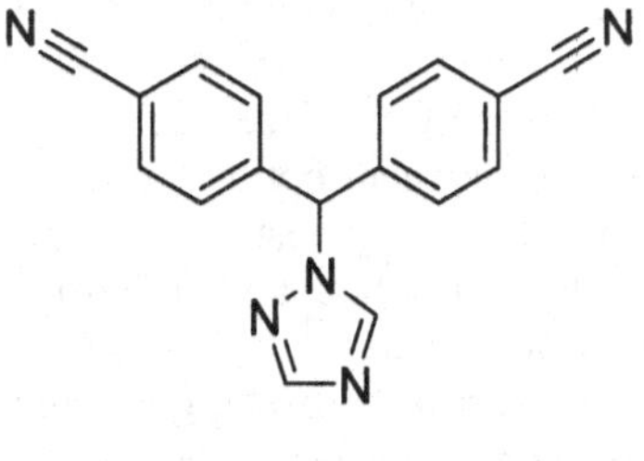   letrozole (Femara)

Anastrozole (Arimidex) is a Type II aromatase inhibitor that was approved in 1996 for the adjuvant treatment of postmenopausal women with hormone receptor-positive early breast cancer. It works through inhibition of aromatase-mediated conversion of adrenal androgens to estrogen. It is considerably more effectual than tamoxifen for time to tumor recurrence and the odds of a primary contralateral tumor as a first event. Its typical side effects are hot flushes, musculoskeletal disorders, fatigue, mood disturbances, and nausea/vomiting.[167]   It is worth noting that anastrozole is a specific inhibitor of aromatase and does not act on other enzyme systems. Anastrozole is administered orally with the suggested dose in postmenopausal women with early breast cancer being 1-mg once daily (taken orally).[168]

Letrozole (Femara) is a highly selective, nonsteroidal, third-generation aromatase inhibitor approved for first-line and extended adjuvant therapy in postmenopausal women with hormone response, early-stage breast cancer. It works by binding to the heme component of the cytochrome P450 subunit of aromatase, thus inhibiting biosynthesis in the body.[169] Treatment of patients with letrozole has been found to be well tolerated for up 5 years, with fewer adverse events than tamoxifen. These are generally characteristic of estrogen deprivation such as hot flushes, arthralgia, night sweating, myalgia, bone fractures. A dose of 2.5 mg letrozole is generally well tolerated in early-stage breast cancer patients in first-line adjuvant therapy as well as in extended adjuvant therapy following 5 years of tamoxifen adjuvant therapy. Letrozole was approved by the FDA in 2001 for the first-line treatment of postmenopausal women with locally advanced or metastatic breast cancer. Letrozole is more potent than aminoglutethimide, anastrozole, and formestane against human aromatase.

Aromatase inactivators/inhibitors are now an established alternative treatment option to tamoxifen for use in postmenopausal women. They reduce estrogen production by inhibiting the activity of aromatase, a key enzyme involved in the synthesis of the majority of nonovarian estrogen. The newer generation non-steroidal compounds anastrozole and letrozole have shown better efficacy than tamoxifen as first-line treatments and megestrol as second-line therapy. According to Buzdar et al., with the advent of aromatase inhibitors the place of tamoxifen as the gold standard for the first-line treatment of postmenopausal women with advanced breast carcinoma appears to have been challenged by the recent new-generation aromatase inhibitors.[170] Non-steroidal aromatase inhibitors compete with endogenous ligands for the active site forming a strong yet reversible, coordinate bond to the heme iron to exclude both ligands and oxygen from the enzyme. Despite the different methods of Type I and Type II inhibition of aromatase, the ultimate result of both is a potent suppression of the enzyme.

10.4   Gonadotropin-Releasing Hormone Agonists

Prostate cancer is one of the most common cancers in men. Prostate cancer is the most common malignancy in men; it accounts for between one-fifth and one-quarter of all newly diagnosed cancers in the U.K., E.U. and U.S. Over the years, the death rate due to this cancer has reduced, partly due to early-stage treatment options including surgery and radiation, along with late-stage options that rely on suppressing testosterone production, such as orchiectomy or hormonal ablation therapy.[171] Patients often favour hormonal ablation therapies over orchiectomy as a means of suppressing testosterone.[172,173] Growth of the prostate gland is unique since it continues throughout life with the size of a prostate gland in men 65 years or older on average two to three times larger than that of a 20-year-old man. Androgens can favor carcinogenesis development simply by promoting repeated rounds of cell division. There is a complex balance between the rate of cell multiplication and apoptosis.

The pituitary gonadotropin-releasing hormone (GnRH), also known as lutenizing hormone-releasing hormone (LHRH), is produced in the hypothalamic area of the brain under the influence of norepinephrine, dopamine, histamine, and other neurotransmitters. GnRH was discovered in 1971 and subsequently several GnRH agonists and antagonists were developed to treat a variety of therapeutic indications. The production of GnRH happens in a pulsatile fashion, with the pulses occurring at 60 to 90 min,[174] with GnRH having a short half-life of 3–4 min. GnRh acts on pituitary gonadotropes expressing GnRH receptors (GnRHRs) to signal synthesis and secretion of the gonadotropin hormones, namely the lutenizing hormone (LH) and the follicle-stimulating hormone (FSH). Thus maintenance of LH and FSH release is controlled by GnRH pulse frequency and amplitude, suggesting that reproductive function and precise hormonal control are regulated by GnRH.[175] This process leads to production of testosterone and fosters excessive growth in case of prostate cancer. Androgens seem to contribute to the development of prostate cancer.[176] Hence androgen deprivation treatment has become a standard palliative remedy for advanced or metastatic disease.[177]

GnRH agonist therapy that includes long-acting synthetic GnRH agonists has been used widely for about 30 years to reduce serum testosterone levels. For example, leuprorelin, gosorelin, buserelin, histerelin, and triptorelin all have been used for hormone-responsive cancers such as breast cancer and prostate cancer, including the palliative treatment of advanced prostate cancer as well as breast cancer. They have been used in androgen deprivation therapy (ADT), which has been demonstrated to slow down growth of prostate cancer or even shrink prostate tumors by containing testosterone levels.

leuprolide acetate (Lupron, Eligard)   AcOH

Leuprorelin, also called leuprolide acetate, is a synthetic nonapeptide and a potent GnRHR agonist administered as an injection for a variety of applications, including the palliative treatment of early-stage and advanced prostate cancer. It was first approved by the FDA in 1985 for the treatment of advanced prostate cancer and has continued to receive approvals for its different dosages from various other regulatory bodies since then. This superagonist is more potent than natural GnRH peptide due to its enhanced affinity for GnRH receptors and a longer half-life of 3 h than that of endogenous GnRH (3–4 min).[178]

These GnRH receptor agonists, such as leuprolide when used to treat prostate cancer, result in the inhibition of leutenizing hormone production, leading to control of testosterone and dihydrotestosterone (DHT). On the whole leuprolide acetate is considered to be safe and tolerable. There is however a serious concern with such compounds as leuprolide, and it has to do with the initial stimulatory effect it produces.[179] Initially, GnRH agonists like leuprolide stimulate production of LH, leading to a surge of testosterone and DHT for 5 to 12 days before inhibition of LH. Initially, however, GnRH agonists stimulate LH production, which in turn causes a surge of testosterone and DHT for 5 to 12 days before the inhibition of LH. Such androgen surge of male hormones can cause a flare reaction (clinical flare or flare-up). Clinical flare is often painful and always dangerous, often precipitating such clinical symptoms as bone pain, compression of a nerve root, spinal cord compression, or blockage of one or both ureters. GnRH agonists are similar in structure and function to the natural GnRH, but they are as much as 60 times more potent than the natural hormone. However, the adverse clinical effects observed warrant significant caution in the use of agonists in many patients.

10.5   Gonadotropin-Releasing Hormone Antagonists

GnRH antagonists can provide an additional but important therapy in prostate cancer. These compounds are expected to be devoid of the initial androgen stimulation characteristics of GnRH agonists that have been mentioned in the preceding section. One such comound is abarelix. It was the first GnRH receptor antagonist in a sustained-duration formulation to progress through clinical studies.[180] This antagonist blocks GnRH and inhibits LH production, which in turn causes a rapid suppression of testosterone and DHT. Unlike the GnRH agonists, however, GnRH antagonists do not cause an initial stimulation of LH production, testosterone, or DHT. This lack of testosterone surge prevents a temporary worsening of the cancer.[181] However, abarelix does exhibit undesirable properties such as induction of its immediate-onset systemic allergic reactions in a low proportion of patients.

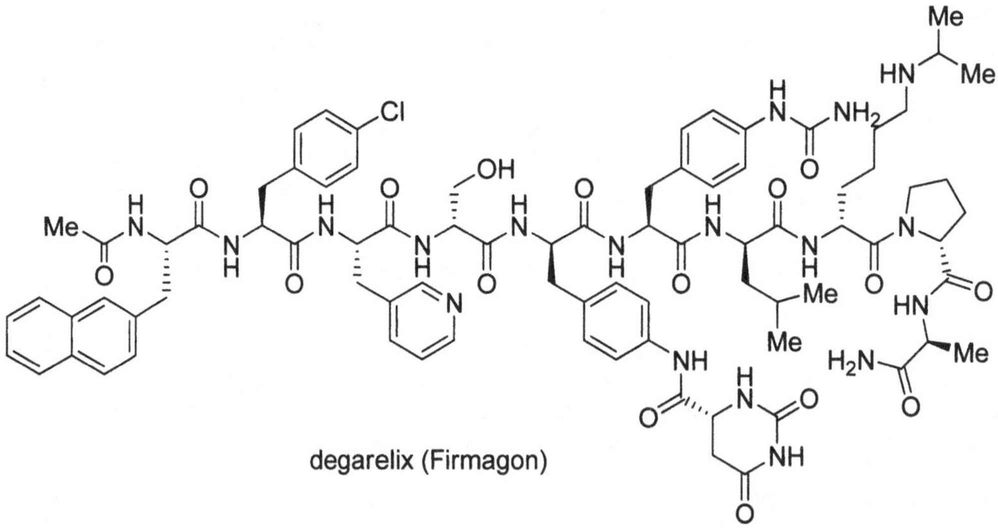

degarelix (Firmagon)

Degarelix is also a GnRH antagonist from Ferring Pharmaceuticals that was approved by the FDA in the year 2008 for the treatment of prostate cancer. Degarelix is a gonadotropin-releasing hormone antagonist (GnRH receptor blocker) with immediate onset of action, suppressing gonadotropins, testosterone, and prostate-specific antigen (PSA) in prostate cancer.[182] A novel GnRH receptor blocker with weak histamine-releasing properties, degarelix also shows rapid and profound testosterone suppression compared to existing GnRH antagonists. When administered subcutaneously, it immediately and very effectively blocks GnRH receptors in the pituitary gland. The net result is a quick and sustained suppression of gonadotropin secretion minus the preliminary stimulation of the gonadotropic axis.[183] Degarelix has been proven to induced testosterone suppression more rapidly than leuprolide, median serum testosterone levels of ≤ 0.5 ng/mL being brought about in 3 days in degarelix recipients but not until day 28 in leuprolide recipients.[184] In terms of dosage, the initial dose is 240 mg followed by a maintenance dose of 80 mg every 28 days. Additionally subcutaneously administered degarelix has been found to be associated with rapid, profound, and sustained suppression of serum testosterone and prostate-specific antigen (PSA) without evidence of testosterone surges or microsurges. One year of treatment with subcutaneous degarelix has been documented to be well tolerated.[185] Adverse effects of this drug include injection site reactions, hot flushes (flashes), weight gain, and increase in serum levels of hepatic transaminases and γ-glutamyl transferase. These adverse effects are mostly related to the drug administration route and the problems associated with antiandrogen therapy and not the drug *per se*.

11 Histone Deacetylase (HDAC) Inhibitors

The histone deacetylase inhibitors are a recent class of anticancer compounds that have been found to maintain regulation of gene expression, induction of cell death, apoptosis, and cell cycle arrest. The modification of tumor suppressor genes, oncogens, or tumorigenesis causing cancer is due to inhibition of transcription rather than mutations. Epigenetics–the study of changes in gene expression that are not caused by DNA sequence changes–has been extensively studied of late.[186] Chromatin is a dynamic complex composed of DNA, histones, and nonhistone proteins. Transcriptional regulation involves multiple enzymatic modifications of histones, such as acetylation and methylation, which affect the accessibility of regulatory proteins to the DNA. The fine equilibrium between histone acetylation and deacetylation is maintained by histone acetyltransferases (HATs) and histone deacetylases (HDACs) maintain a fine equilibrium between histone acetylation and deacylation. In tumor cells this equilibrium is disturbed. HDACs have been known to catalyze the removal of the acetyl group from *N*-Ac-lysine side chains of histone tails and other proteins. These enzymes play an important role in the complex epigenetic regulation of cellular processes and therefore are important targets in cystic fibrosis and neurodegenerative disorders. HDAC inhibitors also reactivate gene expression to induce cell cycle arrest and apoptosis, thus making themselves potentially useful as anticancer agents.[187] During tumorigenesis, gene silencing can be attained by two identified molecular mechanisms: either *via* aberrant methylation or histone deacetylation. Histone deacetylation can silence genes through chromatin modification and deacetylation of histone lysine residues by different types of HDACs.[188] The consequence of this activity is the compaction of the chromatin structure and tight folding of the nucleosome causing gene silencing. Histone acetylation first observed in the 1960s is associated with an open chromatin structure causing activated transcription while histone deacetylation works opposite, thus affecting both the structure of chromatin and gene expression. Alteration of HDACs has been observed in solid tumors as well as hematological malignancies.[189] Such epigenetic alteration is reversible as a result of which HDACs now happen to be attractive targets for cancer therapy. Thus HDACs have a key role to play in the epigenetic regulation of gene expression. Recently, quite a few such compounds offering a new concept in cancer therapy have entered development with two compounds entering the market for treating refractory cutaneous T-cell lymphoma.[190] With over 18 HDACs identified in humans so far, they have been categorized into three main classes based on similarity to yeast HDACs: Zn^{2+} containing Class I (includes HDAC1, 2, 3 and 8) and Class II (includes HDAC 4, 5, 6, 7, 9 and 10); Class III is characterized by sir2-related proteins containing SIRT1 to 7 and Class IV (includes HDAC11).[191] Likewise several types of HDAC inhibitors can be classified according to their structures, such as hydroxamate, cyclic peptide, aliphatic acids, and benzamide.

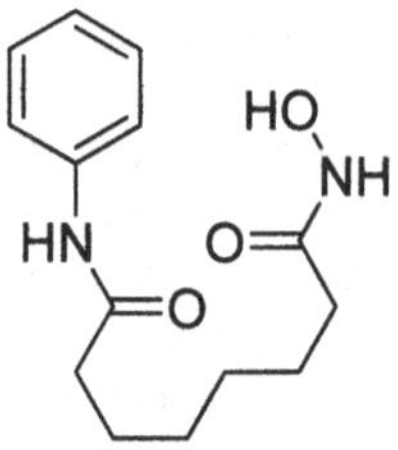

vorinostat or
suberoylanilide hydroxamic
acid (SAHA, Zolinza)

The discovery of the first approved HDAC inhibitor was a result of efforts to develop the first generation hybrid polar compounds that could induce the differentiation of transformed cells, thus showing potential as anticancer agents.[192] Vorinostat or suberoylanilide hydroxamic acid (SAHA, Zolinza) is an HDAC (Classes I and II) inhibitor inducing histone and protein acetylation and altering gene expression. It was approved for the treatment of advanced and refractory primary cutaneous T-cell lymphoma.[193] The molecule was granted orphan drug status by the FDA and was approved in the United States in October 2006 after priority review. Vorinostat is the first-in-class HDAC inhibitor to be introduced in the market and was launched by Merck.[194] It induces growth arrest, differentiation, or apoptosis in a variety of transformed cells. The anticancer effects of vorinostat are believed to be due to drug-induced accumulation of acetylated proteins, including the core nucleosomal histones and other proteins. The marketed product Zolinza in the form of 100-mg capsules has been approved for the treatment of cutaneous manifestations in patients with cutaneous T-cell lymphoma (CTCL), a class of non-Hodgkin's lymphoma with progressive, persistent, or recurrent disease on or following two systemic therapies. The most common adverse reactions are diarrhea, fatigue, nausea, thrombocytopenia, anorexia, and dysgeusia.[195] The tumor selectivity of HDAC inhibitors like vorinostat may be attributed to its induction of oxidative damage. Vorinostat selectively induces the generation of reactive oxygen species in transformed cells in association with diminished induction of the antioxidant enzyme thioredoxin.[196]

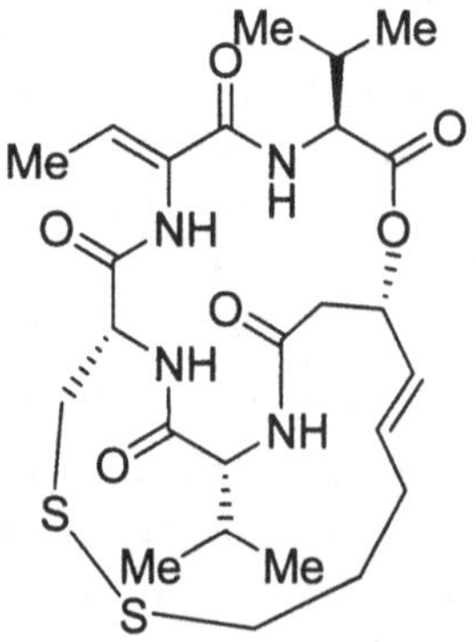

romidepsin
(FK228, Istodax)

Romidepsin is another approved histone deacetylase inhibitor with high inhibitory activity for Class I histone deacetylases. It is a macro-bicyclic depsipeptide, initially isolated from *Chromobacterium violaceum*.[197] Upon administration of this drug, reductive cleavage of the disulfide (S-S) bond leads to the formation of the active but less stable compound that is responsible for the anticancer properties. Thus romidepsin is a natural prodrug, wherein reduction of an intramolecular disulfide bond greatly enhances its inhibitory activity. The sulfide bond is rapidly reduced in cells by cellular reducing activity.[198] In 2009 the FDA granted approval of romidepsin (Istodax) for injection by Gloucester Pharmaceuticals Inc. for the treatment of cutaneous T-cell lymphoma (CTCL) or peripheral T-cell lymphoma (PTCL) in patients who have received at least one prior systemic therapy.

The recommended dose and schedule of romidepsin are 14 mg/m^2 intravenously over 4 h on days 1, 8, and 15 of a 28-day cycle. The mechanism of action of romidepsin also involves inhibition of histone deacetylation, leading to various antineoplastic effects, including cell cycle arrest and apoptosis. Adverse side effects upon administration of this compound are leukopenia, lymphopenia, granulocytopenia, thrombocytopenia, fatigue, and anemia.[199] Fatigue, nausea, anorexia, and vomiting are common adverse events, but serious cardiac adverse events have occurred in some patients with metastatic neuroendocrine tumors.

12   Miscellaneous Cancer Drugs

Multiple myeloma is a type of cancer caused by the accumulation of malignant plasma cells in the bone marrow. It remains asymptomatic for several years and attains its symptomatic stage indicated by bone pain in patients. The consequences of this accumulation are complications that result in bone marrow failure, skeletal destruction, increased plasma volume and viscosity, suppression of normal immunoglobulin production, anemia, and vulnerability to infections as well as kidney failure.[200] This disorder affects a significant portion of the world population. For quite some time it has been considered incurable with earlier existing therapies not a creating durable response. Recently, with the introduction of novel therapies such as the proteasome inhibitor bortezomib (Velcade) and the immunomodulatory agents thalidomide and analog lenalidomide, the attitude toward multiple myeloma has been positive.

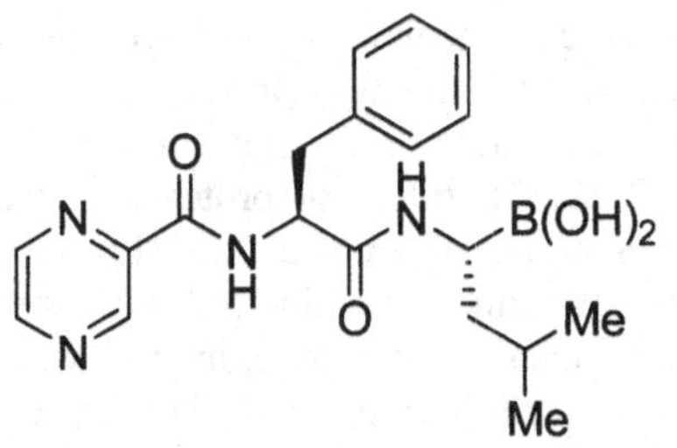

bortezomib (Velcade)

Bortezomib (Velcade, PS-341) is the first FDA-approved proteasome inhibitor for the treatment of multiple myeloma and mantle cell lymphoma.[201] It is a powerful therapeutic agent and has found utility in other types of cancers due to its ability to preferentially induce toxicity and apoptosis in tumor cells without affecting healthy cells. It was developed and introduced in 2003 by Millennium Pharmaceuticals for the treatment of refractory multiple myeloma, following its accelerated approval as a single agent by the FDA. Later, in 2005, it received approval from the FDA for the treatment of multiple myeloma patients who received at least one prior therapy and in 2008 for the initial treatment of multiple myeloma. In 2006 it also received approval for the treatment of patients with mantle cell lymphoma who have received at least one prior therapy. It has also received approval from the European Medicines Agency (EMEA) for use as monotherapy for the treatment of multiple myeloma who have received at least one prior therapy.[202]

Regulation of key cellular functions like cell cycle progression, proliferation, differentiation, and apoptosis requires protein degradation by an assembly of protein subunits called proteasomes.[203] Any aberration in this proteosomal degradation alters normal dynamics of cellular processes. Such alteration causes uncontrolled cell cycle progression with reduced apoptosis thus constituting the typical cancer cell phenotype. In terms of its mechanism of action, bortezomib acts by inhibiting proteasome. This modified dipeptidiyl boronic acid analog binds reversibly to the N-terminus threonine residue with a high affinity to the β-subunit of the multicatalytic enzyme 26S proteosome. The 26S proteasome enzyme is found in the nucleus and cytoplasm of eukaryotic cells and known to be associated with the degradation of ubiquinated proteins.[204] Inhibitory protein kappa B (IκB), among several other important proteins is a key protein that can accumulate or get dysregulated upon the inhibition of 26S proteasome. Normally, the nuclear factor (NF)-κB gets activated during cellular stress, promoting growth, survival and drug resistance of multiple myeloma cells. Hyperactivity of the NF-κB pathway is a feature of many cancers, such as melanoma and multiple myeloma (MM). Bortezomib inhibits the degradation of IκB, thus preventing the activation of nuclear factor (NF)-κB. The disruption of these cellular processes causes arrest of the G2M phase cell cycle, leading to apoptosis.[205] Bortezomib acts in the bone marrow of the microenvironment by preventing the binding of the myeloma cells to the bone marrow stromal cells, thus inhibiting growth, inducing apoptosis, and overcoming drug resistance in human multiple myeloma cells.[206]

Bortezomib, when administered as a single agent, shows limited efficacy. When combined with other therapeutic agents, an enhancement in the efficacy is observed. Adverse side effects, including diarrhea, fatigue and thrombocytopenia, have been observed in patients with lymphoma. In regards to dosage, bortezomib is potent. For example, as a single-agent treatment for the treatment of multiple myeloma, the recommended dosage is 1.3 mg/m^2 and was approved for usage in patients who had previously received at least two prior treatments and have continued disease progression during their last treatment. Bortezomib resistance in tumors has been a challenge, as in the case of many anticancer drugs, and it has made understanding the mechanism of action somewhat difficult. Better inhibitors of proteasome that can overcome some of the existing challenges are therefore essential to attain successful treatment of cancer.

Another recent drug is thalidomide (Thalidomid, Celgene), an anticancer agent that has recently reemerged as a valuable molecule. Originally, in 1956 when it was launched as a sedative agent for treating morning sickness, it resulted in a worldwide tragedy from its teratogenic effects in newly born babies caused by malformation of the fetus.[207] Since then it has been banned all over the world. However, decades later it has re-emerged for a variety of different treatments including FDA approval for treating erythema leprosum nodosum (ENL) in 1998.[208] Thalidomide and its derivatives have showed great promise as anticancer drugs to control several malignancies. They have been used as therapy for metastatic prostate cancer, multiple myeloma, HIV-related ulcers, Kaposi's sarcoma, weight loss, and body wasting associated with HIV. In 2006, thalidomide was approved by the FDA for treatment of newly diagnosed multiple myeloma with dexamethasone.[209]   As a single agent it has shown clinical efficacy in multiple myeloma.[210] Adverse effects due to thalidomide treatment include sedation, constipation, neuropathy, and thromboembolism along with fetal abnormalities.

thalidomide (Thalomide)   lenalidomide (Revlimid)

Thalidomide and its analogs operate *via* similar mechanisms of action. They inhibit angiogenesis, can induce apoptosis, and arrest growth.[211]   Irregular generation of angiogenesis factors such as cytokines and tumor necrosis factor (TNF)-α has been understood as a potential mediator of myelodysplastic syndromes (MDSs) and multiple myeloma. Pathological features of early stage MDS include accelerated apoptotic death of hematopoietic progenitors, including erythroid progenitor, caused by an overproduction of cytokines, including TNF, interleukin (IL)-6, and IL-1 in bone marrow. An impact of angiogenesis in the pathogenesis and progression of MDS is evident from the fact that the bone marrow microenvironment has a direct correlation between measured microvascular density and myeloblast percentage. Thalidomide analogs block increased production of the vascular endothelial growth factor (VEGF), IL-6, and TNF. As far as multiple myeloma is concerned, they prevent the union of the myeloma cells with bone marrow stromal cells and protect against apoptosis.[212,213]

Modification of the thalidomide structure by replacing the phthaloyl group with 4-aminoisoindolin-1-one in search of more potent and less toxic immunomodulators led to the discovery of lenalidomide. Lenalidomide (Revlimid, Celgene) was approved by the FDA in 2005 for the treatment of patients with transfusion-dependent anemia due to low or intermediate-1 risk myelodysplastic syndrome with deletion 5q31 with or without karyotype abnormalities. A little later, in 2006, it also received approval as a combination therapy with dexamethasone to treat patients with multiple myeloma who had received at least one prior therapy. Lenalidomide does not cause significant sedation, constipation, or neuropathy but does lead to significant myelosuppression, unlike thalidomide.[214]   Lenalidomide shows substantial improvement in the overall survival in myeloma without the toxicity of thalidomide. It thus lacks common adverse effects associated with thalidomide and also shows activity in patients with thalidomide-resistant disease. It is administered orally as a capsule formulation of 5, 10, or 25 mg.

13   Conclusion

An increased understanding of the molecular, structural, and biological features of tumors is necessary for a productive development of drugs in targeted therapies. The various drugs mentioned and discussed in this chapter represent some of the important therapies that are being used in the treatment of cancer. Since the discovery of the nitrogen mustard compounds as anticancer agents in the 1940s, cancer therapy has come a long way. The field of targeted therapies in cancer has witnessed great developments, primarily due to the safety and efficacy of the treatment. Given the complexity of cancer, it is evident that research in pursuit of developing better cancer drugs with minimal side effects and clinical resistance will continue in the future.

References

1.   http://www.cdc.gov/Features/WorldCancerDay, accessed April 2012.
2.   Shewach, D. *Chem. Rev.* **2009**, *109*, 2859.
3.   Jemal, A.; Bray, F.; Center, M. M.; Ferlay, J.; Ward, E.; Forman, D. *CA Cancer J. Clin.* **2011**, *61*, 69.
4.   Mattmann, M. E.; Stoops, S. L.; Lindsley, C. W. *Expert Opin. Ther. Pat.* **2011**, *21*, 1309.
5.   Anand, P.; Kunnumakara, A. B.; Sundaram, C.; Harikumar, K. B.; Tharakan, S. T.; Lai, O. S.; Sung, B.; Aggarwal, B. B. *Pharm. Res.* **2008**, *25*, 2097.
6.   Irigaray, P.; Newby, J. A.; Clapp, R.; Hardell, L.; Howard, V.; Montagnier, L.; Epstein, S.; Belpomme, D. *Biomed. Pharmacother.* **2007**, *61*, 640.
7.   Cancer Facts & Figures 2011. http://www.cancer.org/acs/groups/content/@epidemiologysurveilance/documents/document/acspc-029771.pdf, accessed April 2012.
8.   Corey, E. J.; Czakó, B.; Kürti, L. *Molecules and Medicine,* 1st ed.; Wiley Interscience: Hoboken, NJ, 2007; pp. 184–200.
9.   Croce, C. M. *N. Engl. J. Med.* **2008**, *358*, 502.
10.   Li, J. J. *Laughing Gas, Viagra, and Lipitor: The Human Stories Behind The Drugs We Use*, 1st ed.; Oxford University Press: New York, NY, 2006; Chapter 1.
11.   Pratt, W. B.; Ruddon, R. W.; Ensminger, W. D.; Maybaum, J. *The Anticancer Drugs, 2nd ed.;* Oxford University Press: New York, NY, 1994.
12.   Sawyers, C. *Nature* **2004**, *432*, 294.
13.   Gerber, D. E. *Am. Fam. Phys.* **2008**, *77*, 311.
14.   Kompis, I. M.; Islam, K.; Then, R. L. *Chem. Rev.* **2005**, *105*, 593.
15.   McGuire, J. J. *Curr. Pharm. Des.* **2003**, *9*, 2593.
16.   Baldwin, C. M.; Perry, C. M. *Drugs* **2009**, *69*, 2279.
17.   Lansigan, F.; Foss, F. M. *Drugs* **2010**, *70*, 273.
18.   Parker, W. B. *Chem. Rev.* **2009**, *109*, 2880.
19.   Plunkett, W.; Gandhi, V. *Cancer Chemother. Biol. Response Modif.* **2001**, *19*, 21.
20.   Elion, G. B. *Science* **1989**, *244*, 41.
21.   Karran, P. **2007**, *79*, 153.
22.   Nelson, J. A. *Cancer Bull.* **1992**, *44*, 470.
23.   Evans, W. E.; Relling, M. V. *Leuk. Res.* **1994**, *18*, 811.
24.   Diasio, R. B. *Drugs* **1999**, *58*, 119.
25.   Diasio, R. B.; Harris, B. E. *Clin. Pharmacokinet.* **1989**, *16*, 215.
26.   Heidelberger, C.; Chaudhuri, N. K.; Danenberg, P.; Mooren, D.; Griesbach, L.; Duschinsky, R.; Schnitzer, R. J.; Pleven, E.; Scheiner, J. *Nature* **1957**, *179*, 663.
27.   '150 Years of Advances Against Cancer - 1940s-1950s' on http://www.cancer.gov/aboutnci/overview/150-years-advances/page4, accessed April 2012.
28.   Pazdur, R. *Drugs* **1999**, *58*, 77.
29.   Walko, C. M.; Lindley, C. *Clin. Ther.* **2005**, *27*, 23.
30.   Dooley, M.; Goa, K. L. *Drugs* **1998**, *58*, 69.
31.   Longley, D. B, Harkin, D. P., Johnston, P. G. *Nat Rev. Cancer* **2003**, *3*, 330.
32.   (a) Haddow, A.; Timmis, G. M. *Lancet* **1953**, *1*, 207. (b) Noble, S.; Goa, K. L. *Drugs* **1997**, *54*, 447.
33.   Ralhan, R.; Kaur, J. *Expert Opin. Ther. Patents* **2007**, *17*, 1061.
34.   (a) Antman, K. H.; Elias, A.; Ryan, L. *Semin. Oncol.* **1990**, *17*, 68. (b) Loehrer, P. J. Sr.; Lauer, R.; Roth, B. J.; Williams, S. D.; Kalasinski, L. A.; Einhorn, L. H. *Ann. Intern. Med.* **1988**, *109*, 540.

35.   (a) The French Cooperative Group on Chronic Lymphocytic Leukemia. *Blood* **1990**, *75*, 1422. (b) Rundles R. W.; Grizzle, J.; Bell, W. N.; Corley, C. C.; Frommeyer, Jr., W. B.; Greenberg, B. G.; Huguley, Jr., C. M.; James, III, G. W.; Jones, Jr., R.; Larsen, W. E.; Loeb, V.; Leone, L. A.; Palmar, J. G.; Riser, Jr., W. H.; Wilson , S. J. *Am. J Med.* **1959**, *27*, 424. (c) Harding, M.; Kennedy, R.; Mill, L.; MacLean, A.; Duncan, I.; Kennedy, J.; Soukop, M.; Kaye, S. B. *Br. J. Cancer* **1988**, *58*, 640.

36.   (a) Costa, G.; Engle, R. L. Jr.; Schilling, A. et al. *Am. J. Med.* **1973**, *54*, 589. (b) Barlogie, B.; Jagannath, S.; Dixon, D. O.; Cheson, B.; Smallwood, L.; Hendrickson, A.; Purvis, J.; Bonnem, E.; Alexanian, R. *Blood* **1990**,*76*, 677. (c) Young, R. C.; Walton, L. A.; Ellenberg, S. S.; Homesley, H. D.; Wilbanks, G. D.; Decker, D. G.; Miller, A.; Park, R.; Major, M. D. *N. Engl. J. Med.* **1990**, *322*, 1021.

37.   Cheson, B. D.; Rummel, M. J. *J. Clin. Oncol.* **2009**, *27*, 1492.

38.   (a) Planting, A. S.; Schellens, J. H.; van der Burg, M. E.; Boer-Dennert, M.; Winograd, B.; Stoter, G.; Verweij, J. *Anticancer Drugs* **1999**, *10*, 821. (b) Verweij, J.; Schellens, J. H. M.; Loo, T. L.; Pinedo, H. M. *Cancer Chemother. Biother.: Principles Practice* **1996**, 409.

39.   (a) Lyss, A. P.; Luedke, S. L.; Einhorn, L.; Luedke, D. W.; Raney, M. *Oncology* **1989**, *46*, 357. (b) Menichetti, E. T.; Silva, R. R.; Tummarello, D.; Miseria, S.; Torre, U.; Celerino, R. *Tumori* **1989**, *75*, 473. (c) Wils, J.; Bleiberg, H. *Eur. J. Cancer Clin. Oncol.* **1989**, *25*, 3.

40.   (a) Dorr, R. T.; Bowden, G. T.; Alberts, D. S.; Liddil, J. D. *Cancer Res.* **1985**, *45*, 3510. (b) Kennedy, K. A.; McGuirl, J. D.; Leondaridis, L.; Alabaster, O. *Cancer Res.* **1985**, *45*, 3541.

41.   Borowy-Borowski, H.; Lipman, R.; Chowdary, D.; Tomasz, M. *Biochemistry* **1990**, *29*, 2992.

42.   (a) Lyss, A. P.; Luedke, S. L.; Einhorn, L.; Luedke, D. W.; Raney, M. *Oncology* **1989**, *46*, 357. (b) Menichetti, E. T.; Silva, R. R.; Tummarello, D.; Miseria S.; Torre, U.; Celerino, R. *Tumori* **1989**, *75*, 473. (c) Wils J.; Bleiberg, H. Eur. J. Cancer Clin. Oncol. **1989**, *25*, 3.

43.   Doll, D. C.; Weiss, R. B.; Issell, B. F. *J. Clinical Oncol.* **1985**, *3*, 276.

44.   (a) Gnewuch, C. T.; Sosnovsky, G. *Chem. Rev.* **1997**, *97*, 829. This is a comprehensive review describing nitrosoureas in detail citing key reviews therein. (b) DeVita, V. T.; Carbone, P. P.; Owens, A. H. Jr.; Gold, G. L.; Krant, M. J.; Edmonson, J. *Cancer Res.* **1965**, *25*, 1876.

45.   (a) Colvin, M.; Brundrett, R. B.; Cowens, W.; Jardine, I.; Ludlum, D. B. *Biochem. Pharmacol.* **1976**, *25*, 695. (b) Kohn, K. W. *Cancer Res.* **1977**, *37*, 1450.

46.   Levin, V. A.; Silver P.; Hannigan, J.; Wara, W. M.; Gutin, P. H.; Davis, R. L.; Wilson, C. B. *Int. J. Radiat. Oncol. Biol. Phys.* **1990**, *18*, 321.

47.   (a) Flaherty, L. E.; Redman, B. G.; Chabot, G. G.; Martino, S.; Gualdoni, S. M.; Heilbrun, L. K.; Valdivieso, M.; Bradley, E. C. *Cancer* **1990**, *65*, 2471. (b) Kirkwood, J. M.; Ernstoff, M. S.; Giuliano, A.; Gams, R.; Robinson, W. A.; Costanzi, J.; Pouillart, P.; Speyer, J.; Grimm, M.; Spiegel, R. *J. Natl. Cancer Inst.* **1990**, *82*, 1062.

48.   (a) Wong, E.; Giandomenico, C. M. *Chem. Rev.* **1999**, *99*, 2451. (b) Reedijk, J. *Chem. Rev.* **1999**, *99*, 2499. (c) Reedijk, J. *Pure Appl. Chem.* **1987**, *59*, 181.

49.   Weiss, R. B.; Christian, M. C. *Drugs* **1993**, *46*, 360.

50.   Burcheranal, J. H.; Kalaher, K.; Dew, K.; Lokys, L.; Gale, G. *Biochimie* **1978**, *60*, 961.

51.   Lokich, J. *Cancer Invest.* **2001**, *19(7)*, 756.

52.   Cragg, G. M.; Grothaus, P. G.; Newman, D. J. *Chem. Rev.* **2009**, *109*, 3012.

53.   Gordaliza, M. *Clin. Trans. Oncol.* **2007**, *9*, 767.

54.   Alberts, B.; Bray, D.; Lewis, J.; Raff, M.; Roberts, K.; Watson, J. D. *Molecular Biology of the Cell*, 2nd ed., Garland Publishing: New York, 1989, pp. 652–661.

55.   Jordan, M. A.; Wilson, L. *Nat. Rev. Cancer* **2004**, *4*, 253. For a recent example see: Pons, V.; Beaumont, S.; Tran, M. E.; Dau, H.; Iorga, B. I.; Dodd, R. H. *ACS Med. Chem. Lett.* **2011**, *2*, 565.

56.   (a) Bensch, K. G.; Malawista, S. E. *J. Cell. Biol.* **1969**, *40*, 95. (b) Bryan, J. *J. Mol. Biol.* **1972**, *66*, 157.

57.   Guéritte, F.; Fahy, J. In *The Vinca Alkaloids in Anticancer Agents from Natural Products*, Cragg, G. M.; Kingston, D. G. I.; Newman, D. J., eds., CRC Press: Boca Raton, FL, 2005, pp 123–135.

58.   Sørensen, J. B. *Drugs* **1992**, *44*, 60.

59.	Urso, R.; Nencini, C.; Giorgi, G.; Fiaschi, A.I *Eur. Rev. Med. Pharmacol. Sci.* **2007**, *11*, 413.
60.	Zelek, L. Barthier S.; Riofrio, M.; Riofrio, F. K.; Rixe, O.; Delord, J. P.; LeCesne, A.; Spielmann, M. *Cancer* **2001**, *92*, 2267.
61.	Jordan, M. A.; Wilson, L. *Nature Rev. Cancer* **2004**, *4*, 253.
62.	(a) Ringel, I.; Horwitz, S. B. *J. Natl. Cancer Inst.* **1991**, *83*, 288. (b) Diaz, J. F.; Andreu, J. M. *Biochemistry* **1993**, *32*, 2747.
63.	Cortes, J. E.; Pazdur, R. *J. Clin. Oncol.* **1995**, *13*, 2643.
64.	Winer, E. P.; Berry, D. A.; Woolf, S.; Duggan, D.; Kornblith, A.; Harris, L. N.; Michaelson, R. A.; Kirshner, J. A.; Fleming, G. F.; Perry, M. C.; Graham, M. L.; Sharp, S. A.; Keresztes, R.; Henderson, I. C.; Hudis, C.; Muss, H.; Norton, L. *J. Clin. Oncol.* **2004**, *22*, 2061.
65.	Ibrahim, N. *Drugs* **2006**, *66*, 949.
66.	Deeks, E. D.; Scott, L. J. *Drugs* **2007**, *67*, 1893.
67.	McKeage, K.; Keam, S. J. *Drugs* **2005**, *65(16)*, 2287.
68.	Attard, G.; Greystoke, A.; Kaye, S.; De Bono, J. *Pathol. Biol.* **2006**, *33*, 421.
69.	 Villanueva, C.; Bazan, F.; Kim, S.; Demarchi, M.; Chaigneau, L.; Thiery-Vuillemin, A.; Nguyen, T.; Cals, L.; Dobi, E.; Pivot, X. *Drugs* **2011**, *71*, 1251.
70.	Beslija, S.; Bonneterre, J.; Burstein, H.; Cocquyt, V.; Gnant, M.; Goodwin, P.; Heinemann, V.; Jassem, J.; Köstler, W. J.; Krainer, M.; Menard, S.; Petit, T.; Petruzelka, L.; Possinger, K.; Schmid, P.; Stadtmauer, E.; Stockler, M.; Van Belle, S.; Vogel, C.; Wilcken, N.; Wiltschke, C.; Zielinski, C. C.; Zwierzina, H. *Ann. Oncol.* **2007**, *18*, 215.
71.	Moen, M. D. *Drugs* **2009**, *69*, 1471.
72.	Towle, M. J.; Salvato, K. A.; Budrow, J.; Wels, B. F.; Kuznetsov, G.; Aalfs, K. K.; Welsh, S.; Zheng, W.; Seletsky, B. M.; Palme, M. H.; Habgood, G. J.; Singer, L. A.; Dipietro, L. V.; Wang Y.; Chen, J. J.; Quincy, D. A.; Davis, A.; Yoshimatsu, K.; Kishi, Y.; Yu, M. J.; Littlefield, B. A. *Cancer Res.* **2001**, *61*, 1013.
73.	For a review describing successful approaches to the halichondrins: Jackson, K. L.; Henderson, J. A.; Phillips, A. J. *Chem. Rev.* **2009**, *109*, 3044.
74.	Ledford, H. *Nature* **2010**, *468*, 608.
75.	Bai, R.; Paull, K. D.; Herald, C. L.; Malspeis, L.; Pettit, G. R.; Hamel, E. *J Biol Chem.* **1991**, *266*, 15882.
76.	Jordan, M. A.; Wilson, L. *Nat. Rev. Cancer* **2004**, *4*, 253.
77.	Perry, C. M. *Drugs* **2011**, *71*, 1321.
78.	Cortes, J.; O'Shaughnessy, J.; Loesch, D.; Blum, J. L.; Vahdat, L. T.; Petrakova, K.; Chollet, P.; Manikas, A.; Diéras, V.; Delozier, T.; Vladimirov, V.; Cardoso, F.; Koh, H.; Bougnoux, P.; Dutcus, C. E.; Seegobin, S.; Mir, D.; Meneses, N.; Wanders, J.; Twelves, C. *Lancet* **2011**, *377 (9769)*, 914.
79.	Hall, G. D.; Perren, T. J. *Rev. Gynaecolog. Practice* **2002**, *2*, 29.
80.	Wang, J. C. *Annuitev Biochem.* **1985**, *54*, 665.
81.	Rothenberg, M. L. *Ann. Oncol.* **1997**, *8*, 837.
82.	Cased, M.; Amadei, A.; Camilloni, G.; Di Mauro, E. *Biochemistry* **1990**, *29*, 8152.
83.	Roca, J. *Nucleic Acids Res.* **2009**, *37*, 721.
84.	Bailly, C. *Chem. Rev.* **2012**, *17*, 364.
85.	Bradbury, B. J.; Pucci, M. J. *Curr. Opin. Pharmacol.* **2008**, *8, 574.*
86.	Howwitz, S. B.; Chang, C. K.; Grollman, A. P. *Mol. Pharmacol.* **1971**, *7*, 632.
87.	Kessel, D.; Bosmann, H. B.; Lohr, K. *Biochem. Biophys. Acta* **1972**, *269*, 210.
88.	Ulukan, H.; Swaan, P. W. *Drugs* **2002**, *62*, 2039.
89.	Luzzio, M. J.; Besterman, J. M.; Emerson, S. D. L.; Evans, M. G.; Lackey, K.; Leitner, P. L.; McIntyre, G.; Morton, S. B.; Myers, P. L.; Peel, M.; Sisco, J. M.; Sternbach, D. D.; Tong, W.-Q.; Truesdale, A.; Uehling, D. E.; Vuong, A.; Yates, J. *J. Med. Chem.* **1995**, *38, 395.*
90.	Hsiang, Y. H.; Liu, L. F. *Cancer Res.* **1988**, *48*, 1722.
91.	Wani, M. C.; Nicholas, A. W.; Wall, M. E. *J. Med. Chem.* **1986**, *29*, 2358.
92.	Pommier, P. *Nature Rev. Cancer* **2006**, *6*, 789.
93.	Sinha, B. K. *Drugs* **1995**, *49*, 11.

94.   Takimoto, C. H.; Arbuck, S. G. *Oncology* **1997**, *11*, 1635.
95.   Costin, D.; Potmesil, M. *Adv. Pharmacol.* **1994**, *29B*, 51.
96.   Pizzolato, J. F.; Saltz, L. B. *Lancet* **2003**, *361*, 2235.
97.   Kawato, Y.; Aonuma, M.; Hirota, Y.; Kuga, H.; Sato, K. *Cancer Res.* **1991**, *51*, 4187.
98.   Lee, M. T.; Bachant, J. *DNA Repair* **2009**, *8*, 557.
99.   Hainsworth, J. D.; Greco, F. A. *Ann. Oncol.* **1995**, *6*, 325.
100.   Rozencweig, M.; Von Hoff, D. D.; Henney, J. E.; Muggia, F. M. *Cancer* **1977**, *40*, 334.
101.   Nadas, J.; Sun, D. *Expert Opin. Drug Discov.* **2006**, *1*, 549.
102.   Dautant, A.; Langlois, d'Estaintot, B.; Gallois, B.; Brown, T.; Hunter, W. N. *Nucleic Acids Res.* **1995**, *23*, 1710.
103.   Gottesman, M. M.; Fojo, T.; Bates, S. E. *Nat. Rev. Cancer* **2002**, *2*, 48.
104.   Plosker, G. L.; Faulds, D. *Drugs* **1993**, *45*, 788.
105.   Hollingshead, L., M.; Faulds, D. *Drugs* **1991**, *42*, 690.
106.   Faulds, D.; Balfour, J. A.; Chrisp, P.; Langtry, H. D. *Drugs* **1991**, *41*, 400.
107.   Hollstein, H. *Chem. Rev.* **1974**, *74*, 625.
108.   Petering, D. H.; Byrnes, R. W.; Antholine, W. E. *Chem.-Biol. Interactions* **1990**, *73*, 133.
109.   Galm, U.; Hager, M. H.; Van Lanen, S. G.; Ju, J.; Thorson, J. S.; Shen, B. *Chem. Rev.* **2005**, *105*, 739.
110.   Burger, R. M. *Chem. Rev.* **1998**, *98*, 1153.
111.   (a) Boger, D. L.; Cai, H. *Angew. Chem. Int. Ed. Engl.* **1999**, *38*, 448. (b) Hecht, S. M. *J. Nat. Prod.* **2000**, *63*, 158.
112.   Burger, R. M. *Chem. Rev.* **1998**, *98*, 1153.
113.   Blume-Jensen, P.; Hunter, T. *Nature* **2001**, *411*, 355.
114.   Zhang, J.; Yang, P. L.; Gray, N. *Nature Rev.* **2009**, *9*, 28.
115.   Davies, S. P.; Reddy, H.; Carvano, M.; Cohen, P. *Biochem. J.* **2000**, *351*, 95.
116.   Arora, A.; Scholar, E. M. *JPET* **2005**, *315*, 971.
117.   Pearson, M.; Fabbro, D. *Expert Rev. Anticancer Ther.* **2004**, *4*, 1113.
118.   Liu, Y.; Gray, N. S. *Nature Chem. Biol.* **2006**, *2*, 358.
119.   Cohen, M. S.; Zhang, C.; Shokat, K. M.; Taunton, J. *Science* **2005**, *308*, 1318.
120.   Kwak, E. L.; Sordella, R.; Bell, D. W.; Godin-Heymann, N.; Okimoto, R. A.; Brannigna, B. W.; Harris, P. L.; Driscoll, D. R.; Fidas, P.; Lynch, T. J.; Rabindran, S. K.; McGinnis, J. P.; Wissner, A.; Sharma, S. V.; Isselbacher, K. J.; Settleman, J.; Haber, D. A. *Proc. Natl. Acad. Sci. USA* **2005**, *102*, 7765.
121.   Lyseng-Williamson, K.; Jarvis, B. *Drugs* **2001**, *61*, 1765.
122.   Deininger, M.; Buchdunger, E.; Druker, J. *Blood* **2005**, *105*, 2640.
123.   Moen, M. D.; McKeage, K.; Plosker, G. L.; Siddiqui, M. A. A. *Drugs* **2007**, *67*, 299.
124.   Schindler, T.; Bornmann, W.; Pellicena, P.; Miller, W. T.; Clarkson, B.; Kuriyan, J. *Science* **2000**, *289 (5486)*, 1938.
125.   Hirota, S.; Isozaki, K.; Moriyama, Y.; Hasihimoto, K.; Nishida, T.; Ishiguro, S.; Kawano, K.; Hanada, M.; Kurata, A.; Takeda, M.; Muhammad Tunio, G.; Matsuzawa, Y.; Kanakura, Y.; Shinomura, Y.; Kitamura, Y. *Science* **1998**, *279(5350)*, 577.
126.   Cross, S. A.; Lyseng-Williamson, K. A. *Drugs* **2007**, *67*, 2645.
127.   Moen, M. D.; McKeage, K.; Plosker, G. L.; Siddiqui, M. A. A. *Drugs* **2007**, *67*, 299.
128.   McCormack, P. L.; Keam, S. J. *Drugs* **2011**, *71*, 1771.
129.   (a) Weisberg, E.; Manley, P. W.; Breitenstein, W. *Cancer Cell* **2005**, *7*, 129. (b) Weisberg, E.; Manley, P.; Mestan, J.; Cowan-Jacob, S.; Ray, A.; Griffin, J. D. *Br. J. Cancer* **2006**, *94(12)*, 1765.
130.   Plosker, G. L.; Robinson, D. M. *Drugs*, **2008**, *68*, 449.
131.   Ranson, M.; Hammond, L. A.; Ferry, D.; Kris, M.; Tullo, A.; Murray, P. I.; Miller, V.; Averbuch, S.; Ochs, J.; Morris, C.; Feyereislova, A.; Swaisland, H.; Rowinsky, E. K. *J. Clin. Oncol.* **2002**, *20*, 2240.
132.   Arteaga, C. L.; Johnson, D. H. *Curr. Opin. Oncol.* **2001**, *6*, 491.
133.   Schiller, J. H. *Semin. Oncol.* **2003**, *30*, 49.

134.   Ranson, M. *Br. J. Cancer* **2004**, *90*, 2250.
135.   Rusnak, D. W.; Lackey, K.; Affleck, K.; Wood, E. R.; Alligood, K. J.; Rhosdes, N.; Keith, B. R.; Murray, D. M.; Knight, W. B.; Mullin, R. J.; Gilmer, T. *Mol. Cancer Ther.* **2001**, *1*, 85.
136.   Cao, Y. *Semin. Cancer Biol.* **2004**, *14*, 139.
137.   Press, M. F.; Lenz, H.-J. *Drugs,* **2007**, *67*, 2045.
138.   Chow, L. Q. M.; Eckhardt, S. G. *J. Clin. Oncol.* **2007**, *25,* 7 884.
139.   Wilhelm, S. M.; Carter, C.; Tang, L.; Wilike, D.; McNabola, A.; Rong, H.; Chen, C.; Zhang, X.; Vincent, P.; McHugh, M. Cao, Y.; Shujath, J.; Gawlak, S.; Eveleigh, D.; Rowley, B.; Liu, L.; Adnane, L.; Lynch, M.; Auclair, D.; Taylor, I.; Gedrich, R.; Voznesensky, A.; Riedl, B.; Post, L. E.; Bollag, G.; Trail P. A. *Cancer Res.* **2004**, *64*, 7099.
140.   Shaw, A. T.; Yasothan, U.; Kirkpatrick, P. *Nat. Rev. Drug Discov.* **2011**, *10*, 897.
141.   Shaw, A.T.; Solomon B. *Clin Cancer Res.* **2011**, *17*, 2081.
142.   FDA Press Release,
        http://www.fda.gov/NewsEvents/Newsroom/PressAnnouncements/ucm269856.htm, accessed April 2012.
143.   Winter, G.; Harris, W. J. *Immunol. Today* **1993**, *14*, 243.
144.   Levene, A. P.; Singh G.; Palmieri, C. *J. R. Soc. Med.* **2005**, *98(4)*, 146.
145.   Onrust, S. V.; Lamb, H. M.; Barman Balfour, J. A. *Drugs* **1999**, *58*, 79.
146.   Croxtall, J. D. *Drugs* **2011**, *71*, 885.
147.   Vidal, L.; Gafter-Gvili, A.; Leibovici, L.; Dreyling, M.; Ghielmini, M.; Hsu Schmitz, S.-F.; Cohen, A.; Shpilberg, O. *J. Natl. Cancer Inst.* **2009**, *101*, 248.
148.   (a) Clynes, R. A.; Towers, T. L.; Presta, L. G.; Ravetch, J. V. *Nat. Med.* **2000**, *6*, 443. (b) Repka, T.; Chiorean, E. G.; Gay, J.; Herwig, K. E.; Kohl, V. K.; Yee, D.; Miller, J. S. *Clin. Cancer Res.* **2003**, *9*, 2440. (c) Gennari, R.; Menard, S.; Fagnoni, F.; Ponchio, L.; Scelsi, M.; Tagliabue, E.; Castiglioni, F.; Villani, L.; Magalotti, C.; Gibelli, N.; Olivero, B.; Ballardini, B.; Da Prada, G.; Zambelli, A.; Costa, A. *Clin. Cancer Res.* **2004**, *10*, 5650.
149.   Vogel, C. L.; Cobleigh, M. A.; Tripathy, D.; Gutheil, J. C.; Harris, L. N.; Fehrenbacher, L.; Slamon, D. J.; Murphy, M.; Novotny, W. F.; Burchmore, M.; Shak, S.; Stewart, S. J.; Press M. *J. Clin. Oncol.* **2002**, *20*, 719.
150.   McKeage, K.; Perry, C. M. *Drugs* **2002**, *62*, 209.
151.   Reynolds, N. A.; Wagstaff, A. J. *Drugs* **2004**, *64*, 109.
152.   Prewett, M. C.; Hooper, A. T.; Bassi, R.; Ellis, L. M.; Waksa, H. W.; Hicklin, D. J. *Clin. Cancer Res.* **2002**, *8*, 994.
153.   McCormack, P. L.; Keam, S. J. *Drugs* **2008**, *68,* 487.
154.   Rau, K.-M.; Kang, H.-Y.; Cha, T.-L.; Miller, S. A.; Hung, M.-C. *Endocrine-Related Cancer* **2005**, *12*, 511.
155.   Dorssers, L. C.; Van der Flier, S.; Brinkman, A.; van Agthoven, T.; Veldscholte, T.; Berns, E. M.; Klijn, J. G.; Beex, L. V.; Foekens, J. A. *Drugs* **2001**, *61*, 1721.
156.   Moen, M. D.; Keating, G. M. *Drugs* **2008**, *68*, 2059.
157.   Howell, A.; Osborne, C. K.; Morris, C.; Wakeling, A. E. *Cancer* **2000**, *89*, 817.
158.   Dauvois, S.; White, R.; Parker, M. G. *J. Cell Sci.* **1993**, *106*, 1377.
159.   Samojlik, E.; Santen, R. J.; Worgul, T. J. *Steroids* **1982**, *39*, 497.
160.   Wells, S. A.; Santen, R. T.; Lipton, A.; Haagensen, D. E., Jr.; Ruby, E. J.; Harve, H.; Dilley, W. G. *Ann. Surg.* **1978**, *187*, 475.
161.   (a) Murray, R.; Pitt, P. *Breast Cancer Res. Treat.* **1995**, *35*, 249. (b) Geisler, J.; Johannessen, D. C.; Anker, G.; Lonning, P. E. *Eur. J. Cancer* **1996**, *32A*, 789.
162.   Saji, S.; Toi, M. *Expert Opin. Emerging Drugs* **2002**, *7*, 303.
163.   http://www.accessdata.fda.gov/drugsatfda_docs/appletter/2005/020753s006ltr.pdf, accessed April 2012.
164.   Deeks, E. D.; Scott, L. J. *Drugs* **2009**, *69*, 889.
165.   Buzdar, A. U.; Robertson, J. F. R.; Eirmann, W.; Nabholtz, J.-M. *Cancer* **2002**, *95(9)*, 2006.

166. Wellington, K.; Faulds, D. M. *Drugs* **2002**, *62*, 2483.
167. Baum, M.; Budzar, A. U.; Cuzick, J.; Forbes, J.; Houghton, J. H.; Klijn, J. G.; Sahmoud, T.; ATAC Trialists' Group. *Lancet* **2002**, *359 (9324)*, 2131.
168. Scott, L. J.; Keam, S. J. *Drugs* **2006**, *66*, 353.
169. Buzdar, A. U.; Robertson, J. F. R.; Eirmann, W.; Nabholtz, J.-M. *Cancer* **2002**, *95*, 2006.
170. Wilson, A. C.; Meethal, S. V.; Bowen, R. L.; Atwood, C. S. *Expert Opin. Invest. Drugs* **2007**, *16*, 1.
171. McLeod, D. G. *Urology* **2003**, *61*, 3.
172. Schally, A. V.; Reddinng, T. W.; Comaru-Schally, A. M. *Cancer Treat. Rep.* **1984**, *68*, 281.
173. Lincoln, D. W. *Gonadotropin-Releasing Hormone (GnRH): Basic Physiology.* In: *Endrocrinology*, DeGroot, L. J. *et al.*, eds., W. B. Saunders, Co.: Philadelphia, PA, 1997, 142–151.
174. Schally, A. V.; Coy, D. H.; Arimura, A. *Int. J. Gynaecol. Obstet.* **1980**, *18*, 318.
175. Cook, T.; Sheridan, W. P. *Oncologist* **2000**, *5*, 162.
176. Perlmutter, M. A.; Lepor, H. *Rev. Urol.* **2007**, *9(Suppl. 1)*, S3.
177. (a) Sennello, L. T.; Finley, R. A.; Chu, S. Y.; Jagst, C.; Max, D.; Rollikns, D. E.; Tolman, K. G. *J. Pharm. Sci.* **1986**, *75*, 158. (b) Periti, P.; Mazzei, T.; Mini, E. *Clin. Pharmacokinet.* **2002**, *41*, 485.
178. Kastrup, E., Ed. *Drug Facts and Comparisons,* Wolters Kluwer: St. Louis, MO, 2003.
179. Tomera, K.; Gleason, D.; Gittelman, M.; Moseley, W.; Zinner, N.; Murdoch, M.; Menon, M.; Campion, M.; Garnick, M. B.; Brito, C. G.; Brosman, S.; Centeno, A.; Childs, S.; Feldman, R.; Friedel, W.; Glode, M.; Kahn, R.; Klimberg, I.; McMurray, J.; Sharifi, R.; Steidle, C. *J. Urol.* **2001**, *165*, 1585.
180. Cook, T.; Sheridan, W. P. *Oncologist* **2000**, *5*, 162.
181. Van Poppel, H.; Tombal, B.; de la Rosette, J. J.; Persson, B.-E.; Jensen, J.-K.; Olesen, T. K. *Eur. Urol.* **2008**, *54*, 805.
182. (a) Broqua, P.; Riviere, P. J.-M.; Conn, P. M.; Rivier, J. E.; Aubert, M. L.; Junien, J.-L. *J. Pharmacol. Exp. Ther.* **2002**, *301*, 95. (b) Jiang, G.; Stalewski, J.; Galyean, R.; Dykert, J.; Schteingart, C.; Broqua, P.; Aebi, A.; Aubert, M. L.; Semple, G.; Robson, P.; Akinsanya, K.; Haigh, R.; Riviere, P.; Trojnar, J.; Junien, J. L.; Rivier, J. E. *J. Med. Chem.* **2001**, *44*, 453. (c) Schwach, G.; Oudry, N.; Delhomme, S.; Lück, M.; Lindner, H.; Gurny, R. *Eur. J. Pharm. Biopharm.* **2003**, *56*, 327. (d) Schwach, G.; Oudry, N.; Giliberto, J. P.; Broqua, P.; Lück, M.; Lindner, H.; Gurny, R. *Eur. J. Pharm. Biopharm.* **2004**, *57*, 441.
183. Frampton, J. E.; Lyseng-Williamson, K. A. *Drugs* **2009**, *69*, 1967.
184. Van Poppel, H.; Tombal, B.; de la Rosette, J. J.; Persson, B.-E.; Jensen, J.-K.; Kold Olesen, T. *Eur. Urol.* **2008**, *54*, 805.
185. Bird, A. *Nature* **2007**, *447*, 396.
186. Bolden, J. E.; Peart, M. J.; Johnstone, R. W. *Nat. Rev. Drug Discov.* **2006**, *5*, 769.
187. (a) Mariadason, J. M. *Epigenetics* **2008**, *3(1)*, 28. (b) Glozak, M. A.; Seto, E. *Oncogene* **2007**, *26*, 5420.
188. Dokmanovic, M.; Clarke, C.; Marks, P. A. *Mol. Cancer Res.* **2007**, *5*, 981.
189. (a) Marks, P. A.; Breslow, R. *Nat. Biotechnol.* **2007**, *25*, 84. (b) Piekarz, R. L.; Frye, R.; Prince, H. M.; Kirschbaum, M. H.; Zain, J.; Allen, S. L.; Jaffe, E. S.; Ling, A.; Turner, M.; Peer, C. J.; Figg, W. D.; Steinberg, S. M.; Smith, S.; Joske, D.; Lewis, I.; Hutchins, L.; Craig, M.; Fojo, A. T.; Wright, J. J.; Bates, S. E. *Blood* **2011**, *117*, 5827.
190. (a) Xu, W. S.; Parmigiani, R. B.; Marks, P. A. *Oncogene* **2007**, *26*, 5541. (b) Carew, J. S.; Giles, F. J.; Nawrocki, S. T. *Cancer Lett.* **2008**, *269*, 7.
191. Richon, V. M.; Webb, Y.; Merger, R.; Sheppard, T.; Jursic, B.; Ngo, L.; Civoli, F.; Breslow, R.; Rifkind, R. A.; Marks, P. A. *Proc. Natl. Acad. Sci. USA* **1996**, *93*, 5705.
192. Yang, X. J.; Seto, E. *Oncogene* **2007**, *26*, 5310.
193. Grant, S.; Easley, C.; Kirkpatrick, P. *Nature Rev. Drug Discov.* **2007**, *6*, 21.
194. Ma, X.; Ezzeldin, H. H.; Diasio, R. B. *Drugs* **2009**, *69*, 1911.
195. Trachootham, D.; Zhou, Y.; Zhang, H.; Demizu, Y.; Chen, Z.; Pelicano, H.; Chiao, P. J.; Achanta, G.; Arlinghaus, R. B.; Liu, J.; Huang, P. *Cancer Cell* **2006**, *10*, 241.

196.   Ueda, H.; Nakajima, H.; Hori, Y.; Fujita, T.; Nishimura, M.; Goto, T.; Okuhara, M. *J Antibiot.* **1994**, *47*, 301.
197.   Furumai, R.; Matsuyama, A.; Kobashi, N.; Lee, K. H.; Nishiyama, M.; Nakajima, H.; Tanaka, A.; Komatsu, Y.; Nishino, N.; Yoshida, M.; Hironouchi, S. *Cancer Res.* **2002**, *62*, 4916.
198.   Yang, L. P. H. *Drugs* **2011**, *71*, 1469.
199.   Kyle, R. A.; Rajkumar, S. V. *N. Engl. J. Med.* **2004**, *351*, 1860.
200.   Richardson, P. G.; Hideshima, T.; Anderson, K. C. *Cancer Control* **2003**, *10*, 361.
201.   Curran, M. P.; McKeage, K. *Drugs* **2009**, *69*, 859.
202.   Shahshahan, M. A.; Beckley, M. N.; Jazirehi, A. R. *Am. J. Cancer Res.* **2011**, *1*, 913.
203.   Adams, J.; Palombella, V. J.; Sausville, E. A.; Johnson, J.; Destree, A.; Lazarus, D. D.; Maas, J.; Pien, C. S.; Prakash, S.; Elliott, P. J. *Cancer Res.* **1999**, *59*, 2615.
204.   Chen, C.; Edelstein, L. C.; Gelinas, C. *Mol. Cell. Biol.* **2000**, *20*, 2687.
205.   Hideshima, T.; Richardson, P.; Chauhan, D. Palombella, V. J.; Eliott, P. J.; Adams, J.; Anderson, K. C. *Cancer Res.* **2001**, *61*, 3071.
206.   Ali, I.; Wani, W. A.; Saleem, K.; Haque, A. *Curr. Drug Ther.* **2012**, *7*, 13.
207.   Raje, N.; Anderson, K. *N. Engl. J. Med.* **1999**, *341*, 1606.
208.   Shah, S. R.; Thu, M. T. *Drugs* **2007**, *67*, 1869.
209.   Singhal, S.; Mehta, J.; Desikan, R.; Ayers, D.; Robserson, P.; Eddlemon, P.; Munshi, N.; Anaissie, E.; Wilson, C.; Dhodapkar, M.; Zeddis, J.; Barlogie, B. *N. Engl. J. Med.* **1999**, *341*, 1565.
210.   D'Amato R.; Loughnan, M.; Flynn, E.; Folkman, J. *Proc. Natl. Acad. Sci. USA* **1994**, *91*, 4082.
211.   Bartlett, J.; Dredge, K.; Dalgleish, A. *Nat. Rev. Cancer* **2004**, *4*, 314.
212.   Richardson, P.; Anderson, K. *J. Clin. Oncol.* **2004**, *22*, 3212.
213.   List, A. F.; Dewald, G.; Bennett, J.; Giagounidis, A.; Raza, A.; Feldman, E.; Powell, B.; Greenberg, P.; Thomas, D.; Stone, R.; Reeder, C.; Wride, K.; Patin, J.; Schmidt, M.; Zeldis, J.; Knight, R. *N. Engl. J. Med.* **2006**, *355*, 1456.
214.   Shah, S. R.; Tran, T. M. *Drugs* **2007**, *67*, 1869.

9

Anti-Inflammatory and Immunomodulatory Drugs

Jeremy J. Edmunds

1   Introduction

Inflammation is a localized response to tissue damage or infection often associated with edema, redness, and pain. These observable effects are in fact a protective response to external stimuli that ultimately leads to a healing process. In an inflammatory situation such as that induced by a microbial agent the body initiates an immune response that is typically classified as either innate (natural) or adaptive (acquired). The innate immune response is characterized by cells and biochemical pathways that respond in a nonspecific manner to infectious agents. These responses involve many cell types that express recognition molecules such as Toll-like receptors (TLRs) that are poised to react quickly to an invading insult. The adaptive immune system by contrast involves a limited number of principally T- and B-lymphocytes that express antigen-specific receptors on their cell surfaces. This response is one that is dependent upon the context in which activating signals are received (i.e., strength of signal, availability of costimulatory signals, cytokine milieu) that results in the differentiation of effector T-cells and the production of antibodies that are specific to the bacterium or virus. This process develops over several days because of the need for clonal expansion before an effective response can be mounted. Fortunately the adaptive response produces long-lived memory cells that can rapidly express their effector function when reexposed to the particular antigen.[1-3] Since these processes are fundamental to the inflammatory response, it is important to understand the mechanistic biology associated with each response. With that background the anti-inflammatory and immunomodulatory drugs can be characterized by their associated mechanism of action (MOA).

1.1   Innate Immune Response

In the early phase of the inflammatory response, dendritic cells and resident macrophages phagocytose the noxious antigens and present appropriate peptide epitopes in the context of major histocompatibility complex (MHC) molecules on the surface of the cell. The activated antigen-presenting cells increase production of reactive oxygen molecules such as hydrogen peroxide and secrete cytokines such as IFN-γ,[4,5] IL-1 and TNFα which serve to signal other cells.[6] The endothelial cells that line nearby blood vessels respond to these inflammatory cytokines by secreting chemokines and expressing selectins and integrin ligands on their cell surface.[7]   These proteins allow the circulating neutrophils expressing selectin ligands to adhere and roll along the surface of the blood vessel. With further stimulus, the neutrophil releases integrin from its intracellular stores allowing the cell to rapidly associate with its binding partner intercellular adhesion molecule (ICAM) on endothelial cells.[8] This interaction slows cell rolling on the endothelial surface and initiates the process of extravasation across the basement membrane into the surrounding tissue. These neutrophils are the principal effectors that phagocytose bacteria, viruses, and cellular debris and can release enzymatic granules to damage the pathogenic species.[9]

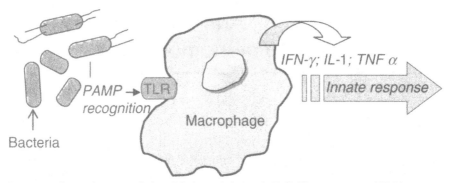

Figure 1. Macrophages ingest cellular debris and through Toll-like receptors (TLR) can recognize specific pathogen-associated molecular patterns (PAMP) to initiate an innate immune response. This innate response provides an immediate defense to an infection and is present in all forms of plants and animals.

In addition to phagocytic macrophages and neutrophils there is another cell type, the natural killer cell (NK), which rapidly proliferates in inflamed tissue.[10] Aside from producing cytokines, these cells destroy other infected cells that express unusual carbohydrates or lack class I MHC molecules. This cell destruction occurs by way of delivery of granzyme B into a targeted cell or through interaction of the Fas ligand on the NK cell with Fas protein which causes the cell to die. Thus the cooperative initiation and propagation of innate inflammatory responses is coordinated by macrophages, neutrophils, and natural killer cells such that the inflammatory insult is eliminated.[11]

1.2   *Adaptive Immune Response*

Dendritic cells (DCs) serve to bridge the innate and adaptive immune response. DCs carry out a critical function in lymphocyte activation through their role in presenting antigen to naïve cells.[12] Immature, tissue-resident DC phagocytose foreign antigen which begins a process of maturation and eventually leads to DC migration to regional lymph nodes. In the lymph nodes T-cells filter through the node probing thousands of DCs for specific antigen/MHC combinations which, in the context of appropriate costimulatory signals, can lead to T-cell activation and proliferation.[13] CD4+ helper T-cells (Th) can then support B-cell activation and antibody isotype switching in the lymph nodes and spleen.[14]   CD8+ cytolytic T-cells (CTLs) have the ability to destroy virally-infected host cells[15] and may also contribute to protective immunity against bacteria. DCs are one of the only antigen-presenting cells (APC) that can provide the requisite costimulatory signal for activating naïve lymphocytes. Over 95% of T-cells in circulation have the α/β T-cell receptor (TCR) and express either the CD8 or CD4 coreceptor. CTL that express CD8 and Th (T helper) cells that express CD4 assess the MHC Class I and MHC Class II, respectively, for cognate antigen. Once the TCR engages with its cognate antigen presented by the MHC, the CD8 or CD4 coreceptors stabilize the TCR-MHC interaction and strengthen the TCR signal.[16] This engagement upregulates the expression of costimulatory receptors such as CD40L proteins on the T-helper cell, which interacts with CD40 on dendritic cells.[17] As a result of this CD40 interaction the dendritic cell produces soluble cytokines such as IL-12 and IL-23, which promotes T-cell survival and differentiation and upregulates expression of costimulatory molecules.[18] The final connection required for activation of T-cells is provided by B7 proteins (CD80 and CD86) expressed on the surface of APC and the CD28 receptor on the T-cell surface.[19] Once this costimulation is complete, the T-helper cell is fully activated, undergoes clonal expansion in response to IL-2,[20] and eventually leaves the lymph node and migrates to nonlymphoid tissue where it can respond to infectious agents.

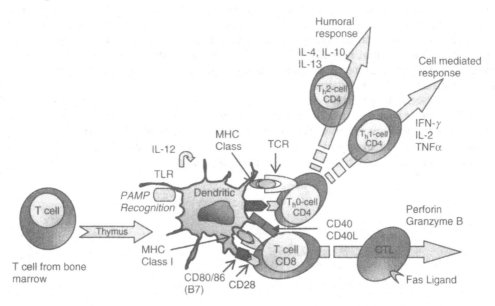

Figure 2. T-cells within the thymus are activated by antigen presenting cells that have traveled from the site of infection with the pathogen displayed on the appropriate MHC. Upon further cross-linking of CD28, T-cells are activated and initiate the adaptive immune response.

At sites of infection or foreign antigen, Th cells may be restimulated by APCs and begin to secrete cytokines as mature effector helper T-cells.[21] The Th1 subset secretes IL-2 (growth factor that stimulates proliferation of NK and cytotoxic T lymphocytes), IFN-γ (primes macrophages and class switches B-cells to IgG3 for opsonizing viruses and bacteria), and TNF (activates primed macrophages and NK cells) whereas the Th2 subset secretes IL-4 (growth factor for B-cells and class switching to IgE), IL-5 (encourages IgA antibodies), and IL-10 (enhances B-cell survival, proliferation and antibody production). These cytokines act locally at the site of injury to orchestrate the appropriate response for destroying virus, bacteria, or parasites.

CTL activation proceeds on a similar path to Th cell activation following activation by DCs presenting appropriate antigen and a costimulatory signal.[22] At the site of infection Th1 cells provide sufficient IL-2 growth factor for proliferation of CTLs and the engagement with the target cell membrane leads to apoptosis. Apoptosis may occur by two mechanisms:[23,24]

1) Delivery of perforin and granzyme B,
2) Fas ligand (FasL) binding to Fas protein on the target cells.

The Th cell also plays a critical role in the activation of naïve B-cells, which need to be activated before they can produce antibodies that participate in the anti-inflammatory or immune response. Upon B-cell receptor (BCR) engagement and cross-linking with appropriate antigen, the Th cell provides an appropriate costimulatory signal by interaction of the CD40 ligand with CD40 on the surface of the B-cell.[25] In a T-helper-cell-independent mechanism, a B-cell that experiences a high degree of BCR cross-linking, with an antigen containing multiple repeating epitopes will proliferate.[26] Further signaling from cytokines such as IFN-γ is required to complete B-cell activation and production of antibodies. This latter mechanism provides an opportunity for B-cells to be activated by substances such as lipids and carbohydrates, a process distinct from the protein–antigen-mediated T-cell stimulation.

For B-cells to produce secreted antibodies, or soluble forms of the B-cell receptor, plasma cells need to be formed.[27] These antibodies can then bind to a microbe which makes them easier targets for phagocytes and activation of the complement system.[28] Alternatively the B-cell becomes a long-lived memory cell, which is critical to rapidly defend against a subsequent reinfection with a pathogen that has previously been eliminated.[29]

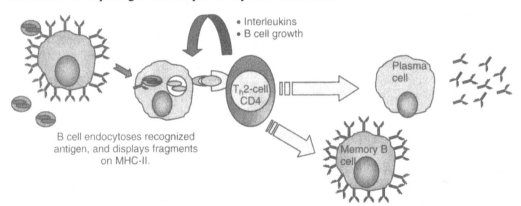

Figure 3. Once a B-cell experiences a matched antigen to the B-cell receptor, it endocytoses the antigen and processes into smaller fragments. These are displayed on the MHC-II and recognized by CD4+ Th2 cells. With secretion of interleukins and growth factors the B-cell becomes activated and matures into antibody-producing plasma cells, or alternatively memory B-cells that would be able to respond rapidly to the same infectious agent.

Antibodies are classically thought of as Y-shaped proteins where the tip of the Y contains variable amino acids that are specific and complementary to the target or antigen to which it binds.[30] The antibody is a structurally complex polypeptide composed of two dimers of heavy chain (50-kD) and light-chain (25-kD) peptides resulting in a 150-kD molecule. The arms of the 'Y' are referred to as the fragment antigen binding (Fab) region comprise the variable heavy (V_H), constant heavy 1 (C_H1) and the variable light (V_L) and constant light (C_L) chains. Within the variable light and variable heavy chains there are 3 loops of β-strands and these loops are referred to as complementarity-determining regions (CDRs) which define the antigen to which they bind. A pair of heavy chains completes the base of the Y or fragment-crystallizable (Fc) region of the antibody. For IgG1, the Fc region comprises of the constant heavy 2 (C_H2) and constant heavy 3 (C_H3) chains, although whether there are 2 or 3 constant heavy chains is dependent on the class of the antibody.[31,32]

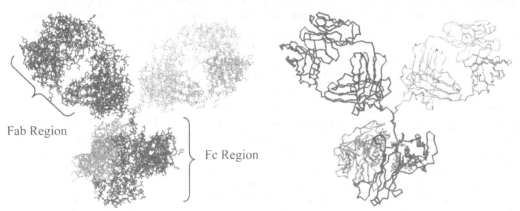

Figure 4. Crystallographic structure of a prototypical intact IgG1 monoclonal antibody (1IGY).

Figure 5. IgG1 showing backbone of amino acids by chain.

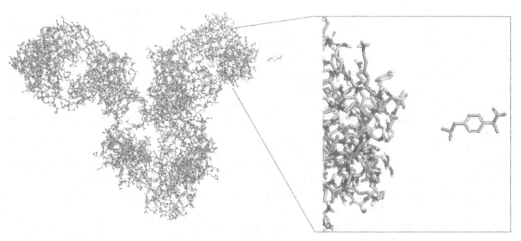

Figure 6. Comparison of typical IgG1 therapeutic antibody with the NSAID ibuprofen (top right)

Figure 7. Expansion of IgG1 antibody and ibuprofen

The antibodies produced from plasma cells are able to bind to appropriate antigen through their directed variable antigen binding region (Fab region), which leaves the constant region (Fc region) of the antibody capable of binding to Fc receptors on the cell surface of, for example, macrophages. They are comprised of two light chains and two heavy chains that are bound by sulfhydryl linkages.[33] Each chain has a constant region that is similar amongst all antibodies and is responsible for interacting with the immune system to facilitate a cell-mediated response. The Fc region allows classification of antibodies as IgG, IgA, IgM, IgE, and IgD[34] that control what immune cells the antibodies bind to and the effector function.[35,36] When an antibody targets a macrophage to a bacterium, the macrophage is now able to engage a previously unrecognizable infectious agent. Antibody binding enhances the phagocytotic activity of the macrophage and thus the target can be rapidly engulfed and eliminated by the macrophage.

IgM – Antibodies: These pentameric IgG-like antibodies initiate the complement cascade production of C3b by bringing several C1 complexes into close proximity through Fc region binding.[37] This process allows the antibody to bind some bacteria-expressed epitope through the Fab region and target the complement cascade by Fc binding. Class switch from IgM to IgG isotype is one of the first steps in initiating a high-affinity B-cell memory response.

IgG – Antibodies: These highly abundant circulating gamma globulins account for 75% of those found in serum and exhibit a number of different subclasses.[38,39] The IgG1 bind macrophages and neutrophils through interaction with the Fc region which serves to deliver these phagocytic cells to the inflammatory mitogen. The IgG2 is the predominant antibody isotype produced in response to some polysaccharide antigens and thus responds to infections with bacteria. With IgG3 antibodies, NK cells are stimulated upon Fc binding and are targeted by the antibody Fab binding to the appropriate infectious agent.

IgA – Antibodies: These dimeric-like IgG antibodies serve to protect the gastrointestinal, reproductive and respiratory mucosal surfaces from an inflammatory or infectious insult.[40] The Fc region of these antibodies is linked, which prevents complement activation but provides resistance to the acids and enzymes of the digestive tract. These antibodies coat the pathogen, which are then eliminated through the mucus or feces.

IgE – Antibodies: Mast cells bind to the Fc region of these antibodies and with sufficient density of Fab binding to an allergen are triggered to degranulate. Histamine is released, which increases capillary permeability so that fluid can escape into the tissues and is recognized as anaphylactic shock and even death if the fluid release affects multiple major organs.[41]

Once B-cells are activated and have proliferated they enter the maturation phase that involves class switching, somatic hypermutation, and plasma cell or memory cell definition. The ability of antibodies to class switch under the influence of specific cytokines[42] provides the adaptive immune system with a broad repertoire of responses tailored for different pathogens in specific tissues.

Following class selection, the antibodies can undergo somatic hypermutation, a process wherein the Fab region is modified or mutated under the influence of cytokines. This process helps B-cells avidity with tighter binding antibodies receiving more profound re-stimulating signals. Enhanced engagement with the BCR leads to enhanced proliferation and thus selects for antibodies with higher affinity for the antigen.[43]

While the immune system is eliminating pathogenic microbes, it must avoid a response that results in excessive damage to self-tissues. Critical to this ability is the recognition of particular features of pathogens that mark the toxin as different from self and is termed self-tolerance. The failure of self-tolerance is thought to underlie many autoimmune diseases and therefore over the past several decades the adaptive immune system has been the focus of research efforts.[44,45] More recently it has been proposed that rather than viewing autoimmunity as an over-zealous response to host antigens by the adaptive system, autoimmunity could result from an over-zealous response to exogenous and endogenous ligands by the innate system. While the innate immune response is the first line of defense to infections, this system interacts with the adaptive immune system through antigen-presenting macrophages or dendritic cells and thus also participates in autoimmune diseases.[46,47]

With chronic inflammation the progressive destruction of tissue by cells involved in the immune response results in diseases such as rheumatoid arthritis, asthma, multiple sclerosis, and atherosclerosis. These diseases involve the infiltration and activation of many inflammatory immune cells that can further stimulate other cells at the site of the inflammation. Thus anti-inflammatory and immunomodulatory drugs have evolved to influence the inflammatory genes and cells that are common, and on occasion those that are specific, to these chronic diseases.

The mechanism by which anti-inflammatory agents act allows categorization of the drugs into three principle classes: non-steroidal anti-inflammatory drugs (NSAIDs), steroidal anti-inflammatory drugs, and disease-modifying immunomodulatory drugs.

2   Arachidonic Acid Cascade

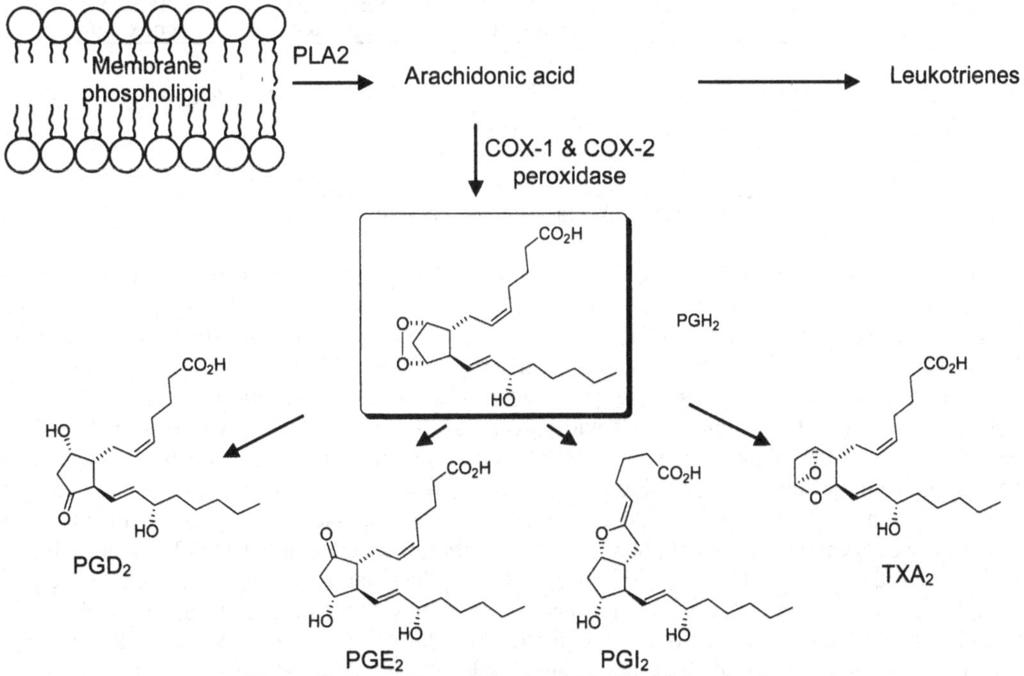

Figure 8. Overview of the arachidonic acid cascade.

A popular pharmacological approach to modulating pain and inflammation is enabled through modulation of the prostaglandin and leukotriene biosynthetic pathways.[48] With prostaglandins the biosynthesis is initiated by hydrolysis of arachidonic acid esterified with membrane glycerophospholipids by the esterase phospholipase A_2 (PLA$_2$). Cyclooxygenases (COX-1 and COX-2) oxidize the liberated arachidonic acid to form prostaglandin G_2 (PGG$_2$), which is reduced by a peroxidase to afford prostaglandin H_2 (PGH$_2$). A variety of synthase enzymes then produce the five principal bioactive prostaglandins: prostaglandin D_2 (PGD$_2$),[49,50] prostaglandin E_2 (PGE$_2$),[51,52] prostacyclin (PGI$_2$), thromboxane A_2 (TXA$_2$), and prostaglandin $F_2\alpha$ alpha (PGF$_{2\alpha}$). These prostaglandins are lipid-derived compounds that act like hormones and thus modulate physiological systems such as the cardiovascular system and the immune system. The actions are mediated by the active prostaglandins agonizing their respective G-protein-coupled receptors (D prostanoid receptor, E prostanoid receptor 1–4, prostacyclin receptor, thromboxane receptor, and F prostanoid receptor).[53]

2.1   Cyclooxygenase (COX) Inhibitors

Salicylates	Fenamic acids	Acetic acids	Propionic acids	Oxicams
Aspirin	Flufenamic acid	Diclofenac	Fenoprofen	Droxicam
Diflunisal	Meclofenamic acid	Etodolac	Flurbiprofen	Isoxicam
Salsalate	Mefenamic acid	Indomethacin	Ibuprofen	Lornoxicam
	Tolfenamic acid	Ketorolac	Ketoprofen	Meloxicam
		Nabumetone	Loxoprofen	Piroxicam
		Sulindac	Naproxen	Tenoxicam

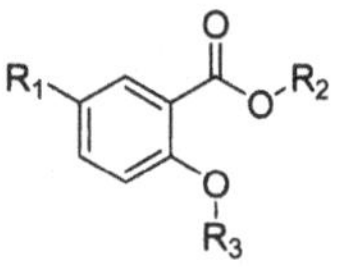

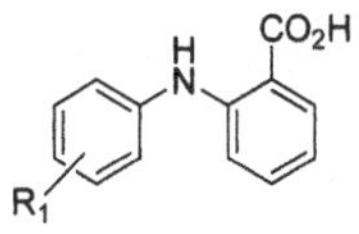

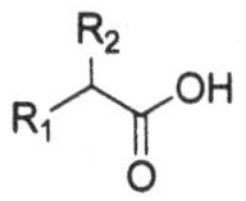

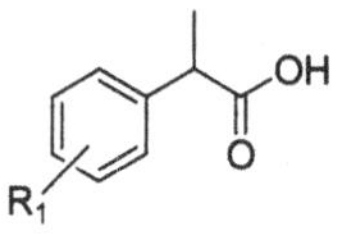

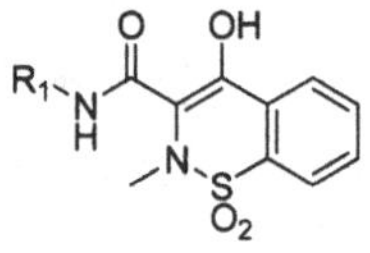

Figure 9. NSAIDs classified by chemical similarity.

The common NSAIDs acetylsalicylic acid (Aspirin),[54–56] ibuprofen (Advil), and naproxen (Aleve) inhibit the two forms of cyclooxygenase (COX-1 and COX-2) and as such represent one of the most widely used classes of medication directed toward reducing pain, swelling, and fever. The best-selling analgesic is acetaminophen (Tylenol), which, although exhibiting weak anti-inflammatory activity, is a potent antipyretic and lacks the gastrotoxicity associated with other NSAIDs.[57] In spite of acetaminophen's widespread use, the mechanism of action is unclear. It is believed that acetaminophen activates TRPA1 via the electrophilic metabolites N-acetyl-p-benzoquinoneimine and p-benzoquinone and thereby reduces calcium and sodium currents in sensory neurons.[58] With respect to more conventional targets connected with pain pathways such as the cyclooxygenases, acetaminophen is a weak inhibitor of COX-1 and COX-2 but an excellent inhibitor of the poorly characterized COX-3. Others have argued that acetaminophen could directly reduce cyclooxygenase-2 to the inactive or resting form in cells through an ability to modulate the oxidative potential of the low intracellular levels of hydrogen peroxide. By contrast, inflammatory sites with high hydrogen peroxide levels are relatively inert to the actions of acetaminophen, thus explaining its poor anti-inflammatory activity.[59]

　　The efforts to discover NSAIDs beyond aspirin can be traced to the 1950s with significant research being conducted by Geigy, Merck Sharpe and Dohme, Parke-Davis, and Boots Pharmaceutical companies. Studies at these companies led to the development of such legendary NSAIDs as indomethacin and ibuprofen. Quite remarkably there are some 50 NSAID preparations which include fenamates (1950s), acetic acids (1960s), propionates (1970s), and the oxicams (1980s). Many of these NSAIDs were only marketed for a brief period, prior to their withdrawal as agents with improved therapeutic indices replaced them. Despite such structural diversity these compounds all share the common characteristic of inhibiting arachidonic acid conversion to the pharmacologically active prostaglandins.

2.2   *Cyclooxygenase-2 (COX-2) Inhibitors*

Acetaminophen (Tylenol), which has analgesic properties comparable to aspirin but weaker anti-inflammatory properties, displays notably different inhibitory activity toward COX preparations from different tissues. This unusual inhibition of COX activity was also apparent with indomethacin and was particularly apparent in enzyme preparations prepared from rabbit brain tissues.[60] With further work the inflammation-inducible form of COX-2 was subsequently identified and found to be expressed in the brain. The COX-2 promoter contained elements reminiscent of genes that are induced during cellular stress and inhibited by glucocorticoids. Thus, with COX-2 being the predominant inflammatory-inducible gene, the hypothesis was formed that a selective COX-2 inhibitor would be an ideal drug, possessing the anti-inflammatory action but lacking the gastrointestinal ulceration and platelet aggregation side effects associated with COX-1 production of prostaglandins and thromboxanes. Contrary to this hypothesis; however, were COX-1 deficient mice that exhibited less ulceration when treated with indomethacin than wild-type controls.[61] In addition, homozygous mice lacking COX-2 showed a normal inflammatory response when treated with several agonists, apparently contradicting the hypothesis that COX-2 was the predominant inflammatory enzyme.[62] Despite these limitations the discovery of the second form of cyclooxygenase (namely COX-2) revitalized the field of anti-inflammatory drug discovery and stimulated the hunt for COX-2-selective agents.[63,64]

The first COX-2-selective compounds to be identified were DuP697 and NS-398 which surprisingly were already in clinical development when COX-2 was discovered. These compounds displayed limited ulcerogenic activity in animal models and with the availability of recombinant COX-2 were shown to be 80- and 1000-fold selective relative to COX-1, respectively. Although the development of both of these compounds was ultimately discontinued, they did provide the impetus to continue structure-activity relationship (SAR) studies directed toward selective COX-2 inhibitors with large teams of medicinal chemists at Merck or Searle.

The first of the COX-2 selective compounds to come to market was celecoxib. The discovery of this drug followed classical SAR studies which revealed four key attributes:

1) The sulfonamide provided superior *in vitro* and *in vivo* activity and efficacy relative to the methyl sulfone
2) The aryl ring at the pyrazole C-5 provided the greatest flexibility to modify solubility, lipophilicity, pKa, ionizability, and COX-1-to-COX-2 selectivity
3) The 3-position was optimally substituted by either trifluoromethyl or difluoromethyl, and
4) The 1,5-substitution of the pyrazole with aryl rings was optimal[65]

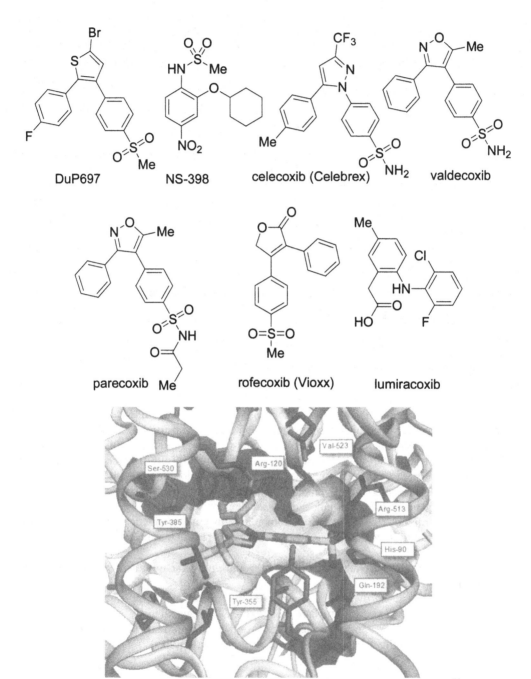

Figure 10. Crystal structure of celecoxib bound to the COX-2 active site.[66]

These SAR trends are rationalized by X-ray crystallography that revealed the larger active site of COX-2 relative to COX-1 is appropriately accommodated with the phenyl sulfonamide of celecoxib. This *para*-sulfonamide interacts with Val-523 of the so-called side pocket and is positioned for hydrogen bonds with Arg-513, Gln-192, and His-90. The 4-methyl-substituted phenyl optimally interacts with Tyr-385 and Ser-530, which serve to restrict the binding site. With the binding site of COX-2 being mostly hydrophobic, van der Waals contacts dominate the remaining binding interactions. The central ring serves only to dictate the orientation of the aromatic rings and as such can be replaced by a number of heterocycles. Not surprisingly are the subsequent COX-2 inhibitors from Searle that incorporated such core switches as an isoxazole to afford parecoxib (Dynastat),[67] which is an injectable water-soluble prodrug of valdecoxib (Bextra).[68]

As the analogs were evaluated *in vivo* it was observed that some of the earlier leads had half-lives in the range of several days to a week. To address this issue a number of metabolic sites were introduced in a systematic fashion to enhance metabolic turnover to inactive compounds. With celecoxib the 4-hydroxymethyl and 4-carboxylic acid metabolites, which exhibit no COX-1 and limited COX-2 inhibitory activity, are produced by oxidation of the corresponding 4-methyl substituent. Thus the half-life in rats was reduced from 117 h to 3.5 h by replacing a 4-chloro-substituent with a 4-methyl group (celecoxib).

When evaluated *in vivo* in rats these pyrazoles exhibited comparable oral anti-inflammatory activity to standard NSAIDs, good analgesic activity, and most notably no GI side effects at 200 mg/kg. In clinical studies celecoxib revealed a half-life of 12 h and analgesic activity at a dose of 100 mg, which was equivalent to aspirin.

At Merck a similar impetus to create a COX-2 inhibitor resulted in the discovery of rofecoxib (MK-966, Vioxx)[69] via a process that relied upon:

1)   A methylsulfone substituent that was preferred over methylsulfonamide to maximize COX-2 selectivity
2)   The removal of the C-2 substituent (thiophene and DuP697 numbering) to enhance COX-2 selectivity
3)   The replacement of the thiophene with other heterocycles to enhance absorption[70]

When rofecoxib was evaluated *in vivo* it also demonstrated equal efficacy compared to indomethacin in pain and inflammation models but without the ulcerogenic properties of indomethacin. When compared to celecoxib in human whole-blood assays, rofecoxib exhibited a 38-fold selectivity favoring COX-2 over COX-1, whereas the ratio for celecoxib was 6.3-fold.

The comparison of *in vitro* selectivity for compounds like celecoxib and rofecoxib relative to traditional NSAIDs is complicated by enzyme kinetics.[71] With celecoxib, initial binding to COX-1 and COX-2 is competitive and reversible, which is revealed by similar Ki's in assays that evaluate these early inhibitor time points. Likely of more relevance is the subsequent slow dissociation of celecoxib from COX-2 that is observed as the COX-2 enzyme presumably rearranges to achieve a more enduring enzyme/inhibitor complex. This behavior is inferred from analysis of kinetic parameters that measured the rate of substrate turnover with preincubation of the inhibitor. As the preincubation time of inhibitor and enzyme was increased from 2 s to 120 s, substrate turnover significantly decreased. This slow time-dependent process, which occurs after the initial binary complex of COX-2 or COX-1 with inhibitor, is specific for COX-2. It is this kinetic behavior that is the largest contributor to the potency and selectivity of celecoxib. To explain this time-dependent inhibition a number of mutations within the catalytic site of celecoxib reveal contributions from Arg-120 and the side pocket. As a result of these interactions some NSAIDs exhibit similar kinetic behaviors to each isoform of COX (e.g., ibuprofen and indomethacin), with time-dependent (indomethacin) and non-time-dependent inhibition (ibuprofen). Rofecoxib exhibits similar kinetic behavior to celecoxib, although inhibition of COX-1 was only evident in the kinetic *in vitro* assays that employed very low concentrations of arachidonic acid. In the case of rofecoxib both intact inhibitor and active COX-2 could be recovered after time-dependent inhibition, demonstrating that inhibition was not due to irreversible covalent inactivation of the enzyme.

In 2004 Merck voluntarily withdrew rofecoxib from the market worldwide due to increased risk of cardiovascular events associated with the drug.[72] Given the absence of COX-1 inhibition rofecoxib is thought to lack the beneficial cardioprotective qualities ascribed to low-dose aspirin that are mediated by COX-1 inhibition. In platelets aspirin inhibits the production of thromboxane, a pro-aggregatory prostaglandin, by inhibition of COX-1.[73] Furthermore within endothelial cells production of prostacyclin, an anti-aggregatory prostaglandin, is unaffected, thus aspirin in net provides an antithrombotic vascular prostaglandin environment. COX-2-selective inhibitors are known to reduce prostacyclin synthesis by as much as 50%. Combined with limited effect on thromboxane production the COX-2 inhibitors are biased toward creating a prothrombotic environment, and thus increased risk of cardiovascular events is expected.

An additional entry into the field of COX-2 inhibitors is lumiracoxib (Prexige), which exhibits exceptional selectivity. In fact, by aggregating some studies and acknowledging the inherent difficulty of comparing data across assays, the lumiracoxib selectivity ratio of 700-fold compares favorably with rofecoxib (272-fold) and celecoxib (30-fold). The clinical significance of this is likely limited because celecoxib at a supra-therapeutic (3- to 6-times the therapeutic) dose of 600 mg (b.i.d.) did not influence the normal mechanism of platelet aggregation.[74] Since platelets lack COX-2, the synthesis of thromboxane A_2 must be mediated by COX-1 and thus a lack of effect on this prostaglandin production supports the premise that celecoxib is COX-1 sparing even at these supratherapeutic doses.

Lumiracoxib is related to diclofenac, making it a member of the arylalkanoic acid class of NSAIDs. This COX-2 inhibitor was approved in the E.U. in 2006 but was withdrawn from the market in several countries the following year in the wake of a handful of serious liver adverse events. In December 2009, Novartis submitted to European regulatory authorities a genetic biomarker that can identify patients who are genetically susceptible to the liver-related side effects of lumiracoxib and thus exclude them from treatment.[75] In 2011, the European regulatory application for lumiracoxib, with the brand name Joicela, was withdrawn. In other countries where the product is available, recommendations for pretreatment genetic testing is being implemented.

When the cardiovascular safety profile of NSAIDs (naproxen, ibuprofen, diclofenac, celecoxib, etoricoxib, rofecoxib, lumiracoxib[76]) was evaluated by combining data from 31 trials in 11,6429 patients, there was little evidence to suggest any of the drugs are safe in terms of cardiovascular events.[77] While safety profiles varied with respect to the outcome, rofecoxib and lumiracoxib had the highest rate ratios for myocardial infarction, whereas naproxen and diclofenac had the lowest. Beyond thromboxane A_2 and prostacyclin synthesis, endothelial function, nitric oxide production, and renal effects such as volume retention could all contribute to adverse cardiovascular outcomes. Thus it is likely that one not only needs to consider the COX selectivity profile but also the half-life,[78] kinetics of inhibition, tissue distribution, frequency of administration, and co-administered drugs when assessing the potential risk of cardiovascular events. Furthermore, despite the fact that naproxen seems to be the safest analgesic for patients with osteoarthritis with respect to cardiovascular outcomes, this advantage has to be weighed against the risk of gastrointestinal toxicity, the very reason these COX-2-selective compounds were created.[79]

2.3   *Prostaglandin I₂ (Prostacyclin)*

In addition to the NSAIDs there are other anti-inflammatory drugs that modulate the pathophysiological role of prostaglandins. Chief among those are the prostacyclin and PGE_2 receptor agonists. PGI_2 drugs such as epoprostenol, which is actually PGI_2 or prostacyclin, have seen utility in the treatment of pulmonary hypertension. While not classically thought of as an inflammatory disease there is accumulating evidence to suggest that inflammation plays a significant role in the pathogenesis of the disease.[80,81] Endothelial dysfunction is a classical hallmark of pulmonary hypertension such that the reduced availability of vasodilators and increased vascular tone result in increased pressure. Increased amounts of inflammatory cytokines and inflammatory cell infiltrate have been detected in the lungs of patients with idiopathic pulmonary hypertension. Thus the action of prostacyclin on endothelial cells and platelets, resulting in smooth-muscle relaxation and vasodilation, coupled with antiaggregatory effects on platelets, serves to reduce pulmonary pressure and improve exercise capacity.

Given the short plasma half-life of prostacyclin (3 to 5 min), analogs of prostacyclin (iloprost and treprostinil) were prepared. They are administered by inhalation or by continuous infusion. With treprostinil, reduction of cytokine production (IL-6, IL-1-β, GMCSF) in a concentration-dependent manner has been observed in macrophages from humans that are stimulated with LPS, implying a beneficial anti-inflammatory property of the drug. Beraprost is an oral prostacyclin mimetic that was first approved in 1992 (Japan) for treatment of peripheral vascular disease. With antiplatelet, antithrombotic, and anti-inflammatory effects, beraprost sodium is used in the prevention and treatment of pulmonary hypertension and in the prevention of thrombosis.[82]

iloprost   treprostinil   beraprost

2.4   *Prostaglandin E₂*

PGE_2 is generated in settings of inflammation and exerts potent immunomodulatory effects that are dependent on cell population and the context of activation. The diverse roles of PGE_2 may be accounted for in part by the existence of four receptors designated EP_1, EP_2, EP_3, EP_4 and the heterogeneity of coupling of these receptors to intracellular signal transduction pathways. In rheumatoid arthritis and osteoarthritis, prostaglandin E_2 (PGE_2) is elevated, causing inflammation, pain, and joint swelling.[83]

Other than the use of dinoprostone for induction of labor, there is limited clinical experience with modulators of PGE_2. Studies are ongoing with EP_1 antagonists as analgesics, EP_2 agonists to enhance bone formation, EP_3 antagonists as antiplatelet agents for protection from myocardial infarction, EP_4 agonists for bone repair, and EP_4 antagonists for migraine. Selective suppression of microsomal prostaglandin E_2 synthase (mPGES)-1 has been considered as an alternative mechanism to relieve pain and inflammation; however, research and development are still at an early stage.[84]

dinoprostone

2.5   *Thromboxane A_2*

Allergic diseases such as bronchial asthma, allergic rhinitis, and atopic dermatitis require inflammation as the basis of their pathophysiology.[85] As a result of improved understanding of the mechanism of allergic inflammation at a molecular level, new therapeutics to combat allergic diseases have been sought.

These diseases are characterized by elevated serum immunoglobulin (Ig) E levels and hypersensitivity to normally innocuous antigens or allergens. These allergens can be presented by allergen-processing cells (such as macrophages or dendritic cells) to $CD4^+$ T-cells. These cells are polarized to the major cellular subsets (Th1, Th2, Th17, Th9, and T_{FH}) and importantly the Th2 cells produce IL-4 (promotes differentiation and proliferation of Th_2 cells), IL-5 (eosinophil activation and recruitment), and IL-13 (promotes airway hyper-responsiveness and mucus hyperproduction). IL-4 and IL-13 stimulate immunoglobulin class-switching and thus IgE production. Binding of IgE to Fc-ε receptor 1 (FcεR1) and cross-linking on the surface of mast cells and basophils lead to the rapid release of histamine, leukotrienes (LTs), and prostaglandin D_2. The chemokines that are also produced attract the eosinophils to the allergic lesion that then release proteases and leukotrienes, causing edema and epithelial cell damage.[86] Current therapeutic approaches seek to control symptoms and thus include use of glucorticoids and antihistamines, leukotriene inhibitors, thromboxane A_2 inhibitors, Th_2 cytokine inhibitors, mast cell stabilizers, inhaled β₂-agonists, anticholinergics or methylxanthines.

Thromboxane A_2 is produced mostly by mast cells in allergic diseases and can stimulate proliferation of airway smooth muscle. The airway disease asthma is characterized by chronic inflammation of the airways induced by the various cellular responses of inflammatory cells such as eosinophils, basophils, mast cells, macrophages, and T-helper cells. Within Japan a TXA_2 synthetase inhibitor (ozagrel) and two TXA_2 receptor antagonists (ramatroban and seratrodast) are employed for the treatment of bronchial asthma. The efficacy of these agents for asthma is still under discussion, but the effectiveness of ramatroban in allergic rhinitis is noteworthy.

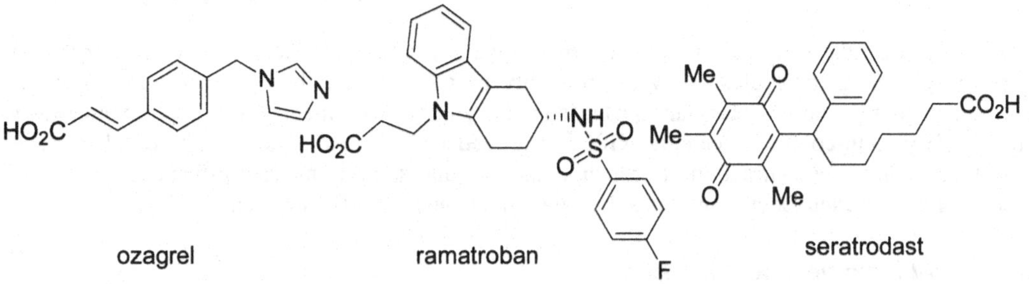

ozagrel   ramatroban   seratrodast

3   Leukotriene Pathway Inhibitors

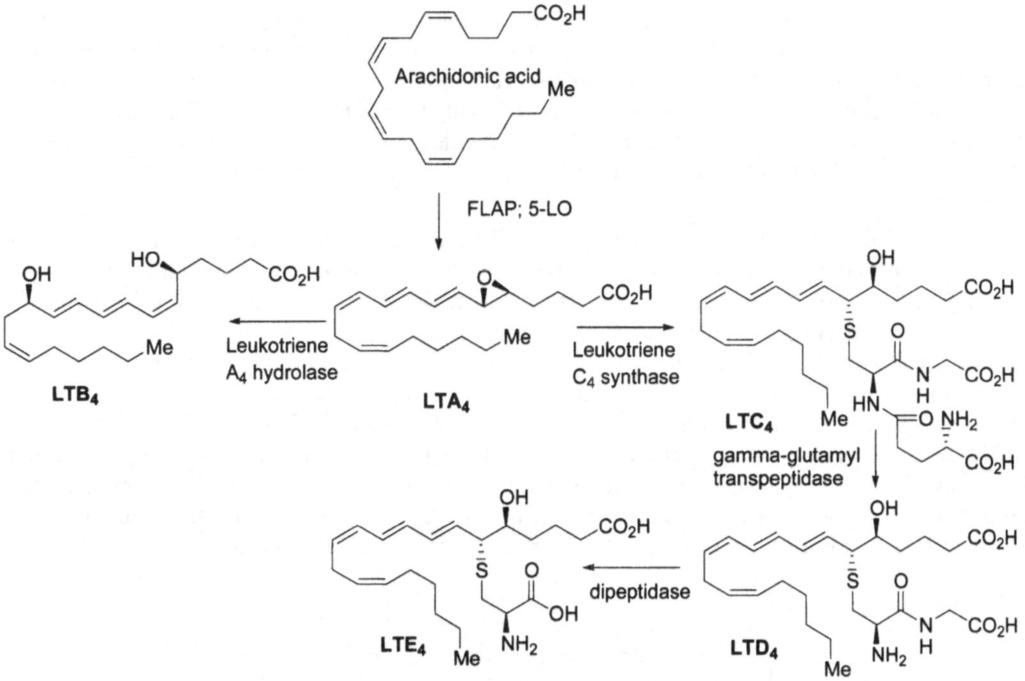

Figure 11. Leukotriene biosynthesis

Arachidonic acid metabolism via the 5-lipoxygenase pathway leads to the production of leukotrienes such as LTA$_4$, LTB$_4$, LTC$_4$, LTD$_4$, LTE$_4$, and LTF$_4$. The sequence of metabolic events is initiated by myeloid cell activation wherein 5-lipoxygenase translocates to the nuclear membrane where it associates with the arachidonic acid presenting 5-lipoxygenase activating protein (FLAP). The epoxide LTA$_4$ is the pivotal intermediate that can undergo hydrolysis to produce LTB$_4$ or conjugation with glutathione to produce LTC$_4$. The name *leukotriene* derives from the source of the lipids, i.e., *leuko*cytes and the presence of three conjugated double bonds or *triene*. The lipids LTC_4, LTD_4, and LTE_4 are termed cysteinyl leukotrienes because of the presence of a cysteine residue in their structures.[87]

In fact, these lipids are released from myeloid cells and likely act as autocrine- and paracrine-signaling molecules. They are probably best known for their role in atopic diseases wherein their immunoregulatory and proinflammatory effects are mediated through extracellular specific G-protein-coupled receptors (GPCRs) termed $CysLT_1$ and $CysLT_2$. These cell surface receptors mediate the contraction of bronchial and vascular smooth muscle, influence secretion of mucus, and recruit leukocytes and indeed eosinophils to sites of inflammation.

3.1   *5-Lipoxygenase Inhibitors*

Arachidonic acid metabolism by 5-lipoxygenase was linked to the formation of the so called slow reaction substance of anaphylaxis, or now more commonly known as leukotrienes, as early as 1977.[88,89] These lipids elicited a long-lasting contraction of the guinea pig ileum and were known to contract airway smooth muscle, stimulate mucus secretion, and enhance vascular permeability. With the realization that the leukotrienes mimicked many of the features of asthma, a number of companies began investigation of inhibitors of leukotriene biosynthesis with the hope of developing drugs for asthma. These efforts led to the development of a 5-lipoxygenase inhibitor (Abbott's zileuton) and the LTD_4 antagonists montelukast, zafirlukast, and pranlukast.[90]

Zileuton (Zyflo) is an orally active 5-lipoxygenase inhibitor that resulted from elaboration of compounds that were described in the literature to be inhibitors of the 5-lipoxygenase cascade.[91] Of particular interest to Abbott were agents that inhibited the enzyme by binding to the active site ferric ion. Early compounds included A-61442, which contained a classical hydroxamate chelator pharmacophore. Upon the realization that the hydroxamate was rapidly metabolized, a number of *N*-hydroxy ureas were prepared. The superior bioavailability of these analogs was reflected by effective inhibition of LTE_4 synthesis. The preferred compound A-64077 or zileuton selectively inhibited LTE_4 production with an ED_{50} of 3 mg/kg while not inhibiting TXA_2/TXB_2 formation at doses up to 70 mg/kg. As expected, zileuton inhibited bronchospasm in the guinea pig induced by arachidonic acid or antigen challenge. Zileuton had no effect on LTA_4 hydrolase or glutathione transferase, further confirming its good selectivity.

A-61442   zileuton

Zileuton is currently used to prevent symptoms associated with asthma such as chest tightness, wheezing, and coughing as a consequence of its ability to inhibit swelling and mucus production in the airways. The extended-release version allows twice-a-day preventive treatment, providing 2400 mg of drug. Zileuton is not indicated for the reversal of bronchospasm and hepatic function must be monitored initially because of elevated liver enzymes that are seen in 2% of patients.

3.2   Leukotriene Receptor Antagonists

The first LTD_4 inhibitor approved for marketing was pranlukast (Azlaire).[92] This compound originated from a combination of pharmacophores expressed by the chromone carboxylic acid FPL 55712 and the lead acrylamide. Replacement of the carboxylic acid with an isosteric tetrazole and optimization of the lipid backbone afforded the competitive antagonist of the cysteinyl leukotrienes LTC_4, LTD_4, and LTE_4. In clinical studies, pranlukast dosed at 225 mg twice daily appeared to be as effective as montelukast 10 mg once daily and zafirlukast given at 40 mg twice daily to adults with mild to moderate asthma.

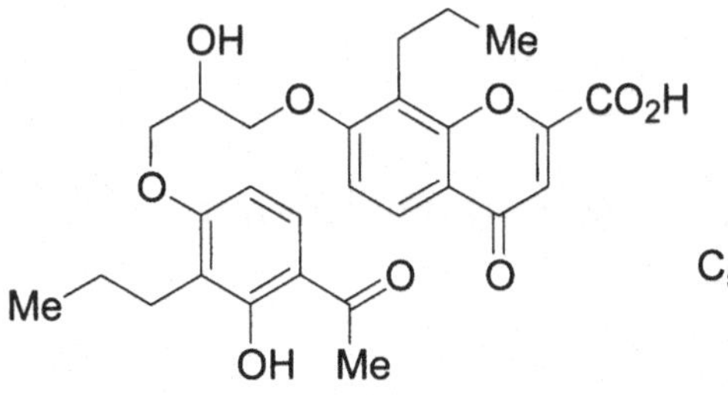

FPL 55712

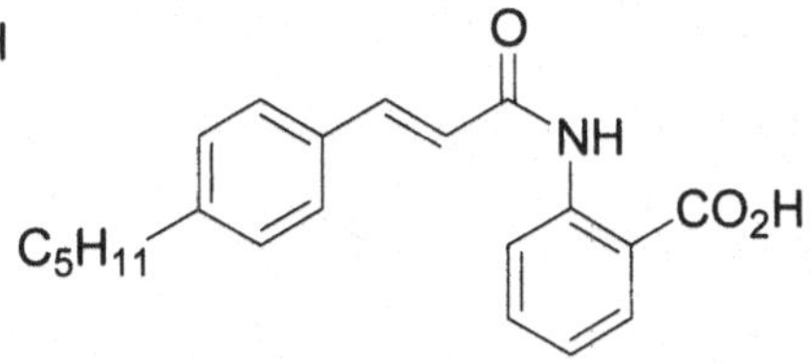

(E)-2-(3-(4-pentylphenyl)acrylamido)benzoic acid

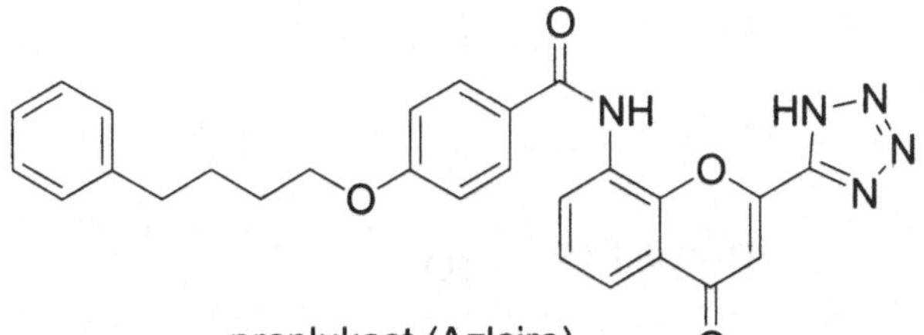

pranlukast (Azlaire)

During studies on LTD_4 receptor antagonists at Merck-Frosst it was noted that peroxisomal enzyme induction (PEI) in rats or mice was significant and liver weight increased.[93] Since peroxisomal proliferators in general were associated with liver tumor formation, efforts were directed toward eliminating this undesirable activity. Throughout these studies the profound ability of the β-substituent to influence PPAR activity was noted and is best illustrated by comparing analogs (X) LTD_4 IC_{50} 0.4 nM; PEI 109% @ 400 mg/kg p.o. 4 days) with (Y) LTD_4 IC_{50} 0.3 nM; PEI 24% @ 400 mg/kg p.o. 4 days and the compound (Z) LTD_4 IC_{50} 0.5 nM; PEI 16% @ 400 mg/kg p.o. 4 days which ultimately became montelukast (Singulair).[94]

Zafirlukast (Accolate) is a leukotriene CysLT1 (LTD_4) antagonist that was launched in 1996 by AstraZeneca for asthma. This compound originated from SAR studies that suggested the indole template served to present an acidic arylsulfonyl amide and lipophilic cyclopentylurethane for appropriate recognition by the LTD_4 receptor. ICI-204,219 (zafirlukast)[95] emerged as the clinical development compound of choice because of its excellent *in vitro* profile coupled with improved bioavailability relative to earlier LTD_4 antagonists such as ICI-198,615.[96,97] It is a potent competitive antagonist of LTD_4 and LTE_4, with equal affinity, and is able to reverse the bronchospasm induced by LTD_4 or LTE_4s in guinea pigs. Thus the preclinical data supported development of zafirlukast for asthma when used in a therapeutic or a prophylactic manner.

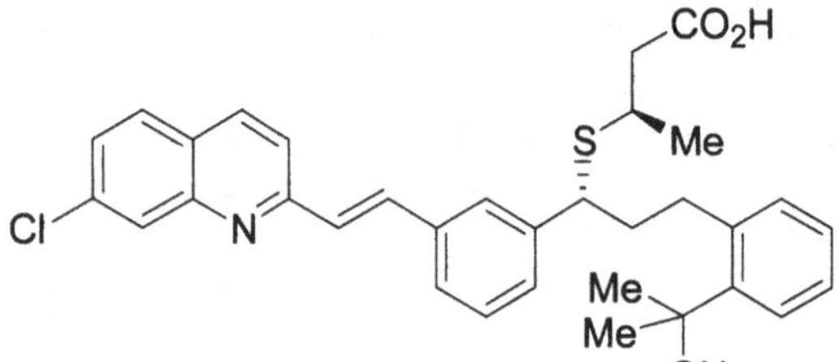

(X)

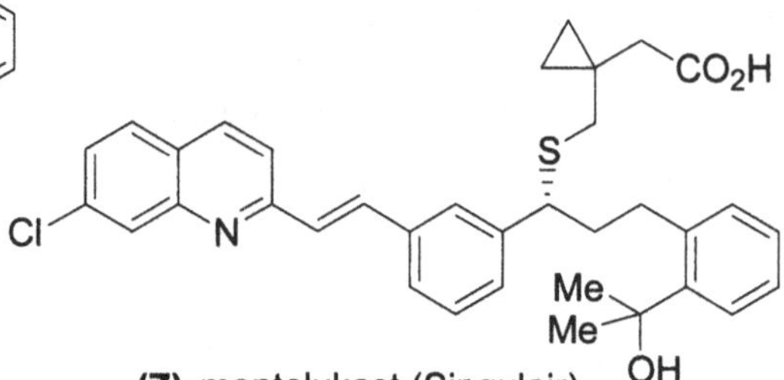

(Z) montelukast (Singulair)

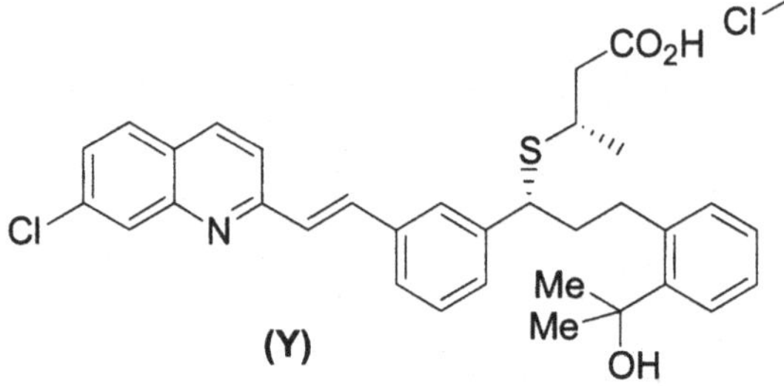

(Y)

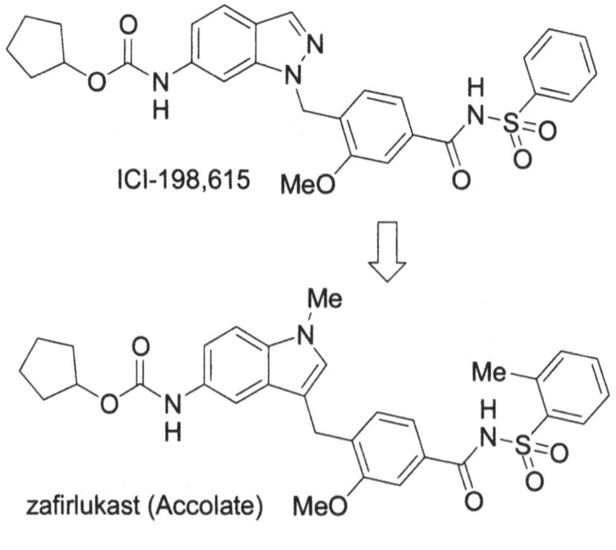

ICI-198,615   MeO

zafirlukast (Accolate)   MeO

The mainstay of asthma treatment is low-dose inhaled corticosteroids which reduce inflammation and improve overall control. Given the reluctance of some patients to use steroids, the leukotriene inhibitors might be preferred because of their additional long-lasting protection and lack of tolerance development. As such leukotriene antagonists have been evaluated as a first-line or add-on asthma therapy over a 2 year period in the United Kingdom.[98] These studies revealed that leukotriene receptor antagonists and inhaled glucocorticoids are equivalent as first-line controller therapy. As add-on therapy, leukotriene receptor antagonists were similarly effective as long-acting β-agonists when added to inhaled glucocorticoids.

4   Antihistamines

The chemical compound 2-(1*H*-imidazol-5-yl)ethanamine was first synthesized as early as 1907 and subsequently named histamine, meaning an amine present in all tissues. This short-acting biogenic amine is produced by the enzyme histidine-decarboxylase where the highest activity is detected in basophils and mast cells, two cell types that are well known for their role in allergic disorders and inflammation. It was the first mediator to be associated with allergic reaction and where therapeutic inhibition proved useful in the management of allergic inflammatory diseases.

Histamine receptor characterization revealed four receptors termed H_1 through H_4, which are all expressed on leukocytes. In addition to predominantly H_1 receptor-mediated actions on smooth muscle, vascular permeability and allergic responses histamine is involved in H_2-mediated gastric acid secretion, neurotransmission in the CNS largely by the H_3 receptor and modulation of the immune system through the H_1 and H_4 receptor.[99] In general, anti-inflammatory effects are mediated by the H_1 receptors, whereas such effects as lymphocyte proliferation, induction of suppressor cells, and neutrophil chemotaxis can be mediated by the H_2 receptor. Not surprisingly histamine is involved in the regulation of cytokines and can inhibit the release of IL-1, IL-2, IL-4, IL-6R, IL-12, interferon-γ, and TNFα (mediated by H_2 receptor) and even increase release of IL-6 principally by the H_1 and H_2 receptors.[100,101]

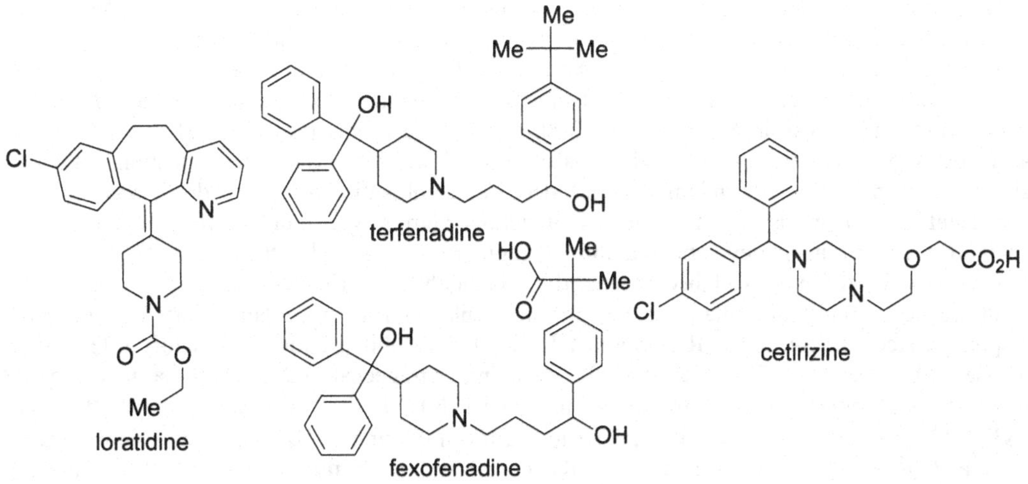

Well known are the H_1 antagonists such as diphenhydramine (Benadryl) that still enjoy widespread use despite side effects of dry mouth and sedation. Second-generation inhibitors that were actively investigated in the 1980s sought to limit the CNS side effect of sedation by designing peripherally restricted compounds and include such marketed compounds as loratidine, terfenadine, fexofenadine, and cetirizine. Some studies have suggested that these second-generation antihistamines have additional anti-inflammatory activity.[102] Cetirizine, for example, inhibits the expression of intercellular adhesion molecule-1 (ICAM-1) on the surface of epithelial cells, which limits leukocyte recruitment and activation. Decarboethoxy-loratidine, the active metabolite of loratidine, inhibits the production of the neutrophil chemoattractant IL-8 from stimulated human mast cells and thus inhibits one of the initiators of allergic inflammation.[103] In comparison, studies of H_1 antagonists (loratidine, cetirizine) with H_2 antagonists, such as ranitidine, TNFα secretion was inhibited in stimulated human leukemic mast cells (HMC-1) by all compounds but with the greatest effect for loratidine. Given that these effects have been observed *in vitro*, their importance to the clinical efficacy of these H_1 antagonists at pharmacological doses is unclear.

Current attention within the pharmaceutical industry is directed toward the newest members of the histamine receptor family, namely H_3 and H_4. The anti-inflammatory and antinociceptive properties of the histamine H_3 receptor[104] was demonstrated preclinically with BP 2-94, a potent and selective azomethine prodrug of (R)-α-methylhistamine, a histamine H_3 agonist.[105] In rodent preclinical studies the three distinct phases of an inflammatory response can be evaluated. In particular the first phase that is caused by increased vascular permeability results in the exudation of fluids from the blood into the interstitial space and can be reproduced by injection of carrageenan into the rat paw. In this model the edema preventing effect of BP 2-94 was observed during the prostaglandin phase, about 1 h after carrageenan injection, and was similar to that achieved with indomethacin. Indeed, the combination of the two drugs was additive, consistent with their distinct mechanisms of action. Currently a number of H_3 agonists and antagonists are in clinical development, although none have advanced to the market.[106,107]

The H_4 receptor is highly expressed in eosinophils, leukocytes, mast cells, T-cells, and dendritic cells, suggesting that this receptor may play a role in inflammatory and immune responses. Accordingly, the H_4R antagonist JNJ777120 exhibited anti-inflammatory effects in a mouse asthma model. However, studies involving carrageenan-induced edema in rat paws revealed only limited efficacy with the H_4R antagonist JNJ777120.[108] More recent data report that JNJ777120 is a partial agonist with respect to β-arrestin binding to H_4R and thus complicates *in vitro* and *in vivo* correlations and hence the understanding of the relevance of histamine H_4 receptor to inflammatory diseases.[109]

BP 2-94   JNJ777120

The plethora of different effects exerted by histamine through its four receptors and splice variants suggests a need for immunologists to reevaluate the effects of histamine beyond the traditional pharmacological properties. It is commonly acknowledged that the H_1 receptor activation results in proinflammatory activity of immune cells, whereas H_2 receptors appear to be suppressors of inflammatory and effector functions. The discovery of the H_4 receptor provides an opportunity for therapeutic approaches designed to block H_1 and H_4 signaling or the development of selective H_1 and H_4 antagonists for such diseases as asthma and RA.[110]

5   Corticosteroids

The use of steroids as a treatment for inflammatory diseases can be traced to the intriguing observation that arthritic patients who became pregnant and had enlarged adrenal glands exhibited a rapid amelioration of arthritic symptoms. These observations led Hench to first hypothesize that a substance elaborated by the adrenal glands was responsible for remission of arthritis. Some two decades later the first published report of the efficacy of cortisone in the treatment of rheumatoid arthritis, which appeared in 1949, was so well-received that the following year Hench and colleagues earned a Nobel Prize in Medicine and Physiology. The glucocorticoid receptor to which cortisone binds is a member of the intracellular hormone receptor superfamily that, upon binding cortisone or other glucocorticoids, acts on targets in the nucleus to modulate gene expression.[111]

The ligand-bound glucocorticoid receptor (GR) operates as a transcription factor which is able to recruit various cofactor proteins (coactivators and corepressors) that can activate or repress the transcription of respective target genes. This ability to recruit co-activators and corepressors is sensitive to the conformation of GR that is conferred by ligand binding. A GR agonist like cortisol will induce a GR conformation that shifts and positions helix 12 so that it provides a particular complementary binding surface for coactivators. This can be contrasted with a GR antagonist such as mifepristone (or RU 486) that destabilizes helix 12 packing and thus provides a distinct and alternative surface for corepressor binding.

For asthma, a disease that has seen widespread treatment with glucocorticoid agonists, the airway epithelial cells exposed to IL-1β, TNFα, or endotoxin activate the proinflammatory transcription factor NF-kappaB which binds to specific regions of DNA. Upon binding with cAMP-response-element-binding protein the complex is able to acetylate core histones as a consequence of their histone acetyltransferase (HAT) activity.[112] Upon acetylation of the N-terminal lysine residue of the histone, the DNA is able to unwind and participate in gene transcription dictated by RNA polymerase II.

With glucocorticoid treatment, the glucocorticoid receptors in the cytoplasm, which are normally bound to heat shock protein-90 (HSP-90), shed their inhibitory molecular chaperones and transport across the nuclear membrane into the cell nucleus in a process known as translocation. Within the nucleus the glucocorticoid–ligand complex can function in 3 ways:[113,114]

(i) Transcriptional activation: The liganded glucocorticoid receptor undergoes homodimerization, binds to glucocorticoid response elements (GREs), and leads to the transcription of anti-inflammatory genes such as mitogen-activated protein kinase phosphatase (MKP-1), IL-1 receptor, IL-10, and inhibitor of NFkappaB (IkB-alpha). This occurs through a mechanism involving enhanced HAT activity.

(ii) cis-Repression: Here the liganded glucocorticoid receptor undergoes homodimerization and binds to GREs and represses transcription of genes. Typically this leads to the negative side effects of steroid treatment such as osteoporosis, reflected by suppression of osteocalcin by osteoblasts.

(iii) Transcriptional repression: Within the nucleus the activated transcription complex comprising of the monomeric glucocorticoid receptor bound to its ligand and coactivator recruits histone deacetylase 2. Through reversal of histone acetylation, the chromatin complex reverts to its resting closed form which precludes RNA polymerase II binding to DNA and thus limits expression of inflammatory cytokines (IL-1, IL-2, IL-6, IL-13, TNF), chemokines, adhesion proteins, and other inflammatory proteins (COX-2, cPLA2).

From an adaptive immune perspective, glucocorticoid treatment leads to a rapid depletion of circulating T-cells due to a combination of effects including enhanced flux to the lymphatic system, induction of apoptosis, inhibition of T-cell growth factors, and impaired release of cells from lymphoid tissues. With the key antigen-presenting dendritic cells glucocorticoid treatment is known to suppress dendritic cell maturation and function. The cells are effectively converted to a state of hyporesponsiveness to T-cells and can even induce formation of T-reg cells. Glucocorticoids also enhance macrophage phagocytosis of apoptotic cells through induction of the S/Mer tyrosine kinase-dependent apoptotic pathway. In addition, during the differentiation of monocytes to macrophages glucocorticoids create a highly anti-inflammatory hyperphagocytotic macrophage. The suppression of dendritic cell derived antigen presentation and downregulation of costimulatory molecules is also likely to contribute to the glucocorticoid agonist's efficacy in inflammatory diseases.[115] An alternative mechanism, by which corticosteroids are thought to act, includes the induction of phospholipase A_2 inhibitory proteins, which are also known as lipocortins. These proteins control the biosynthesis of prostaglandins and leukotrienes, as discussed previously, by inhibiting the release of their common precursor, arachidonic acid.[116]

cortisone   cortisol (hydrocortisone)   betamethasone

The remarkable efficacy of cortisone spurred many medicinal chemistry groups to pursue the design of cortisone analogs devoid of the metabolic side effects such as weight gain or increased blood glucose levels. Given that the synthetic sequences were lengthy and there were no rational design guidelines to follow, progress was dictated by stepwise improvements enabled by synthetic chemistry.[117] Microbial preparations, isolation of alpha and beta stereoisomers, halogen incorporation, and introduction of a C-1,2 alkene are a few of the sequences which resulted in surprisingly active steroids. For example, prednisone (C-11 ketone) was first isolated in March 1954 and the active metabolite prednisolone (C-11 beta hydroxy) shortly afterwards. This analog exhibited enhanced glucocorticoid activity, with reduced mineralocorticoid activity. The first dose of prednisolone was administered to an arthritic woman in August 1954. Prednisone was launched in March 1955 at a 5-mg dose for use in arthritis, with prednisolone being introduced shortly thereafter! Dexamethasone is approximately 4-times more potent than prednisone and is available as IV, oral, nasal, ophthalmic, and topical creams for treatment of diseases ranging from multiple myeloma to psoriasis. Although glucocorticoids are associated with osteoporosis, fluid retention, and hyperglycemia with chronic therapy, more acute therapy is remarkably efficacious and explains widespread use of steroidal therapy.

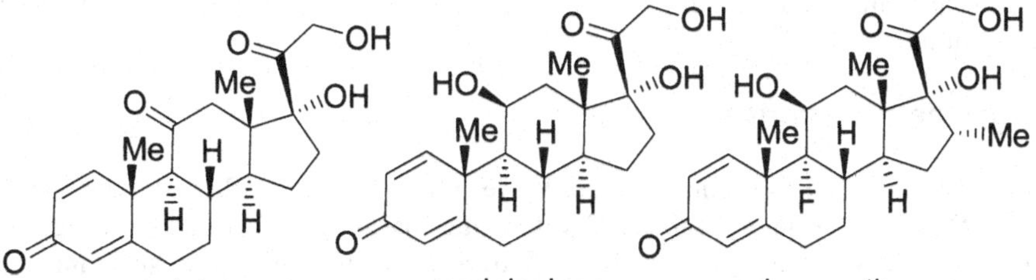

prednisone   prednisolone   dexamethasone

Other glucocorticoids are available wherein their chemical structures are modified to affect overall lipophilicity to favor uptake in tissues and lower systemic bioavailability for treatment of asthma or psoriasis. Typically assays were employed that characterized differentiation features such as the morphological response of fibroblasts to irritants, thymus size reduction, diminution of granuloma formation in response to cotton implants, and a rat ear assay that assessed ear weight gain. Such examples include budesonide, beclomethasone, triamcinolone, fluticasone, flunisolide, and mometasone furoate, which have seen widespread use to treat asthma or rhinitis.

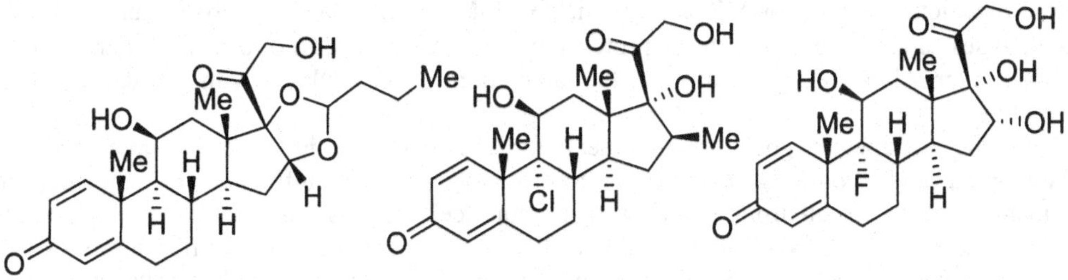

budesonide   beclomethasone   triamcinolone

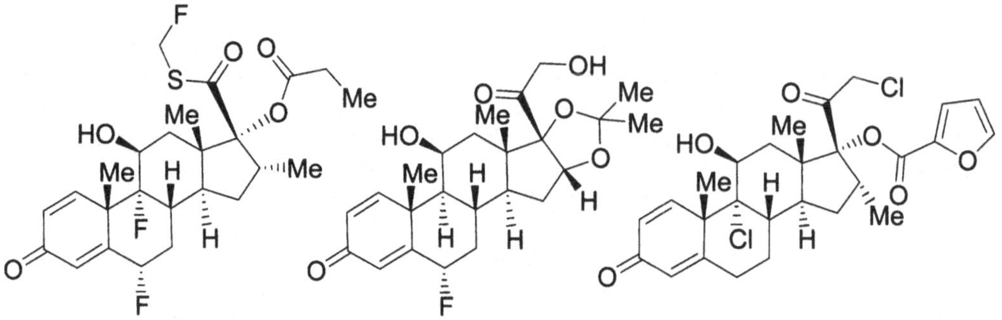

fluticasone propionate   flunisolide   mometasone furoate

Efforts were also directed towards discovering topical agents for treatment of dermatological diseases. Since cortisone was ineffective when administered topically, hydrocortisone became the basis of subsequent analogs. Selection of appropriate functional groups to improve topical potency with increased penetrability, solubility, and percutaneous absorption was critical, since simply formulating potent compounds in standard creams or ointment vehicles were disappointing.[118]

6   Rheumatoid Arthritis

Rheumatoid arthritis (RA) is a chronic disorder resulting from an inflamed synovium wherein 80% of all patients are partially disabled 12 years after disease onset. It ranks among the top ten major chronic diseases in western countries with over 1% of the population suffering from pain, stiffness, and swelling of the small joints of the hands, wrists, and feet that characterize the disease.[119] This debilitating systemic inflammatory disorder leads to local bone and cartilage degradation with panus formation. The hypertrophic synovial membrane consists of hyperplastic synoviocytes and inflammatory T, B, macrophage, mast, and endothelial cells which release cytokines and enzymes that invade and damage bone and cartilage.[120] Typically small joints are affected and lose their shape and alignment manifesting as pain and loss of movement. The primary cause of rheumatoid arthritis is unknown and likely involves genetic, environmental, and other epigenetic factors.[121]

While there is clearly an imbalance between pro- and anti-inflammatory cytokine activities that favors joint damage, it is unclear which cytokine may be the best target for clinical intervention. The proinflammatory cytokines TNFα, IL-1, and IL-17 act synergistically to release matrix metalloproteinases (MMPs), in particular MMP-1 and MMP-3, from fibroblasts and macrophages, and degrade structural proteins of the cartilage. TNFα and IL-1 can promote RANKL expression, which results in the activation of osteoclasts with subsequent bone destruction.

The therapeutic goals of treatment for rheumatoid arthritis hinge upon reducing inflammation and protecting the joints from damage.[122] The inflammatory signatures of rheumatoid arthritis result from an influx of immune cells that is regulated by chemokine and adhesion molecule expression in the synovial tissue. The resident synovial fibroblasts express genes rendering themselves resistant to apoptosis and angiogenesis, which results in synovial hyperplasia. The cartilage and bone surrounded by this inflamed synovium suffer from an imbalance between catabolic and anabolic pathways induced by cytokines but mediated by aggrecanases and collagenases. Since inflammation is the driving force for this structural damage, the longer and more intense the inflammatory insult, the more structural damage results.

This inflammation of the synovial membrane or synovitis may reside in joints that appear normal upon simple inspection but are evident by high-resolution ultrasound and magnetic resonance imaging (MRI). TNF-blocking therapies halt progression of structural damage and are more effective in this regard than agents such as methotrexate.

Within the synovium monocytes that have infiltrated because of on-going inflammation are able to differentiate into osteoclasts under the influence of the two cytokines macrophage colony-stimulating factor (M-CSF) and RANKL.[123] The osteoclasts resorb bone by attaching to its surface and secreting MMPs 1, 9, 13, 14 and cathepsin K that contribute to further bone matrix degradation.

Not satisfied with destruction of bone, the inflamed synovium also provides aggrecanases such as aggrecanase-5, which target the proteoglycan of the cartilage. Upon the loss of proteoglycan the MMPs such as MMP-3 then degrade the collagen backbone.

The management of rheumatoid arthritis includes pharmacological and surgical approaches. There is no cure for rheumatoid arthritis, but rather goals of treatment seek to induce and maintain remission through control of inflammation. Halting joint destruction and reversing disability are critical to improving quality of life.

For the vast majority of patients with rheumatoid arthritis the first line of treatment is the synthetic oral disease modifying anti-rheumatic drugs (DMARD) such as methotrexate, sulfasalazine, and leflunomide, where patients typically attain a very low index of disease activity or even remission within 3 to 6 months. The anchor DMARD is methotrexate principally because of familiarity with the efficacy and side effects of the drug but also because of additive effects that are seen when codosed with antirheumatic biologics. Sulfasalazine, leflunomide and intramuscular gold (e.g., auranofin) are typically used when there is intolerance of methotrexate, and notably clinical studies have not shown inferiority to methotrexate. Because the antimalarial drugs chloroquine and hydroxychloroquine do not inhibit structural damage sufficiently, they are likely only useful for patients who have very mild disease or show contraindications to the other drugs.

The American College of Rheumatology (ACR) proposed a system for the classification of the extent of arthritis. The criteria, which include morning stiffness, number of rheumatoid nodules under the skin, joint erosions apparent by X-ray, and blood tests for rheumatoid factor, help judge disease severity. To measure effectiveness of therapy physicians evaluate the percentage of patients that achieve ACR20 (i.e., a reduction of 20% or more in number of swollen and tender joints plus improvement in three or more of pain, physician or self-assessment of disease activity, and self-assessment of disability, acute phase reactant levels: C-reactive protein (CRP), or erythrocyte sedimentation rate (ESR). Alternatively, or in addition, the disease activity score of 28 joints (DAS 28) is reported as a continuous scale (0–7+) based on 4 components:

1) Number of tender joints upon touching (TEN)
2) Number of swollen (SW) joints
3) Erythrocyte sedimentation rate (or CRP)
4) General health self assessment (SA: 0–100)

With this assessment and the equation below, high disease activity is typically scored at values >5.1, low disease activity at values < .2, and patients in remission with scores of <2.6.

$$DAS28 = 0.56 \times \sqrt{TEN28} + 0.28 \times \sqrt{SW28} + 0.7 \times \ln{(ESR)} + 0.014 \times SA$$

Patients with rheumatoid arthritis who are treated with immunosuppressive drugs are susceptible to infections. Pulmonary infections are particularly prevalent in those patients that also smoke. Some drugs such as corticosteroids and COX inhibitors contribute to increased cardiovascular mortality and in the case of NSAIDs can lead to additional gastrointestinal complications. Chronic systemic inflammation leads to atherosclerosis in rheumatoid arthritis patients, which may result in an increased risk of myocardial infarction. In fact, cardiovascular and respiratory diseases are the most common cause of mortality in RA patients, and these outcomes significantly correlate to levels of rheumatoid factor, an autoantibody to the Fc region of IgG, which reflects autoimmune burden.

6.1   Methotrexate

The importance of methotrexate as an oral DMARD is evident by the fact that since the approval of methotrexate for the treatment of rheumatoid arthritis in 1980, subsequent clinical studies of new rheumatoid arthritis therapies evaluate effectiveness in combination with, or relative to, this drug. Given that methotrexate was originally conceived as an agent for oncology, it is no surprise that inhibition of cellular proliferation is fundamental to its therapeutic usefulness. By reducing purine and pyrimidine biosynthesis, methotrexate acts to limit proliferation of cells pertinent to rheumatic disease.[124] In rheumatoid arthritis the effective dose is substantially lower than that used in oncology and thus it is hypothesized that the anti-inflammatory properties of methotrexate may differ from those for treating malignancies. In a classical sense, methotrexate is known to be an inhibitor of dihydrofolate reductase, an enzyme involved in the folate synthesis metabolic pathway.[125] Dihydrofolate reductase catalyzes the conversion of dihydrofolic acid to the active tetrahydrofolic acid, which is the central single-carbon donor in several reactions involved in the *de novo* synthesis of purine and pyrimidine precursors of DNA and RNA. Therefore methotrexate is cytotoxic to cells that rapidly replicate their DNA such as malignant and myeloid cells. The inhibition of dihydrofolate reductase also limits spermine and spermidine synthesis because they require methionine and *S*-adenosylmethionine which are generated by the active methyl donor methyltetrahydrofolate. These amines are typically elevated in rheumatoid arthritis patients' synovial fluid and act as a source of deleterious ammonia and hydrogen peroxide.

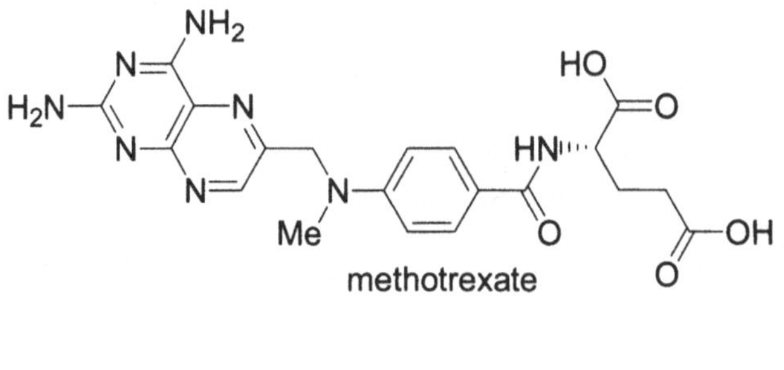

methotrexate

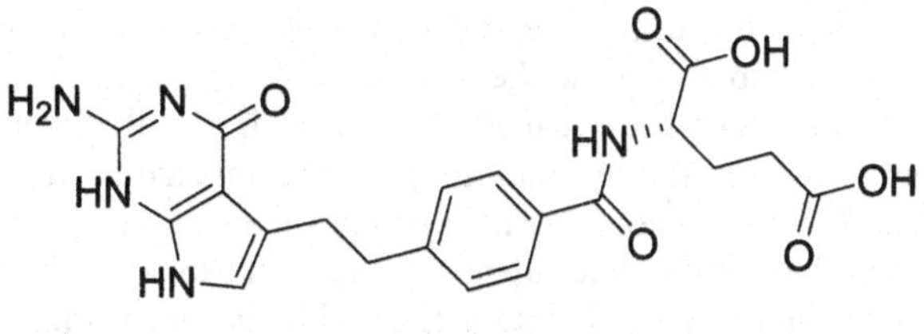

pemetrexed

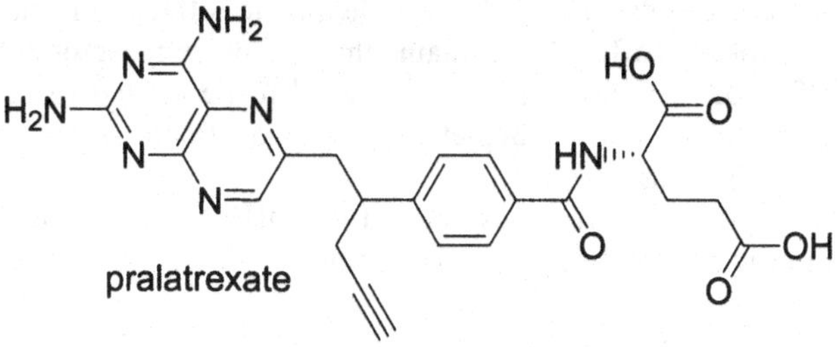

pralatrexate

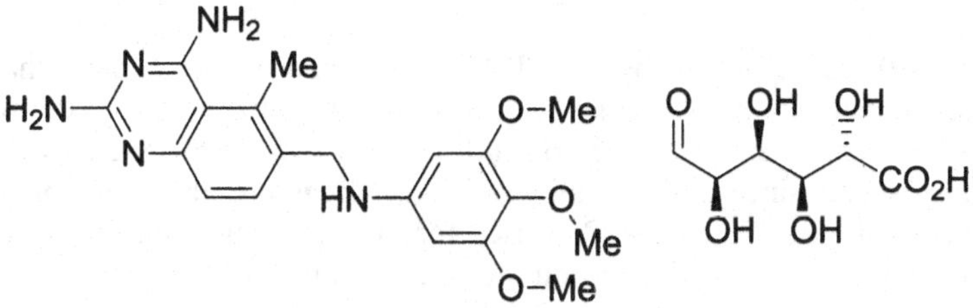

trimetrexate glucuronate

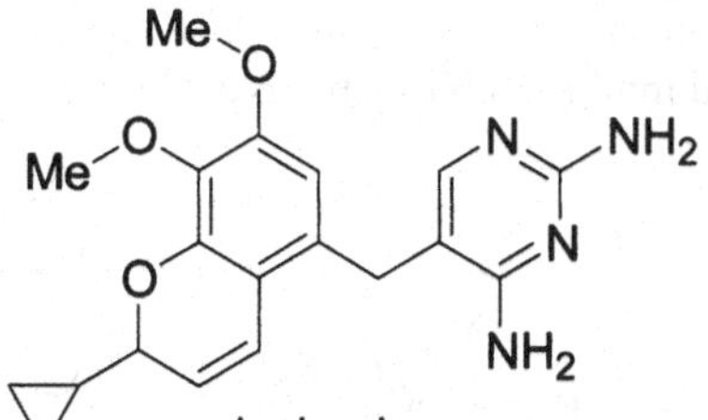

icalaprim

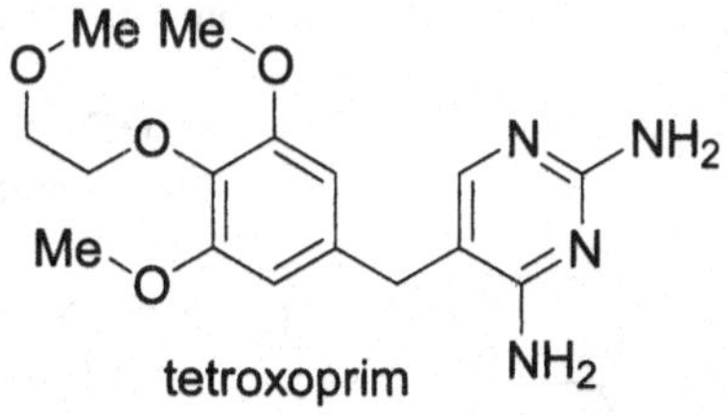

tetroxoprim

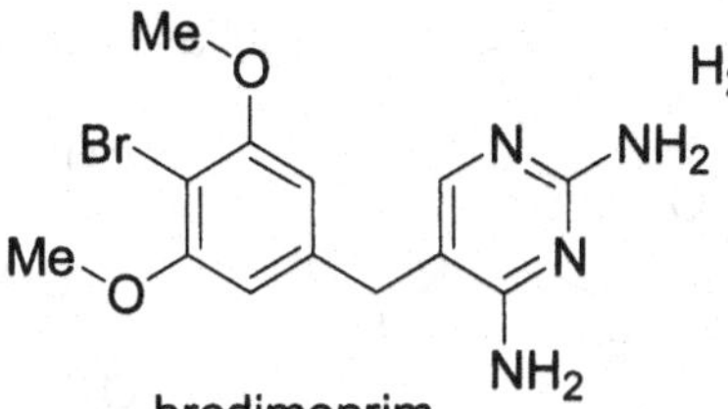

brodimoprim

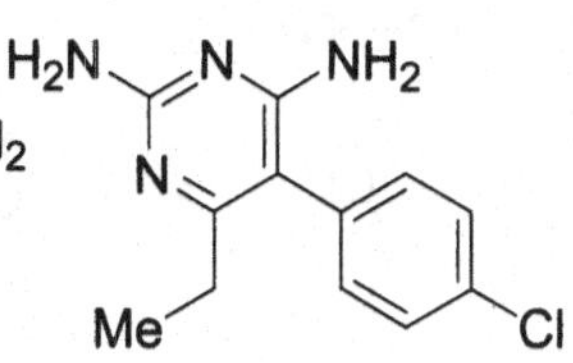

pyrimethamine

Of particular interest however is that the pharmacodynamic activity of methotrexate treatment (once a week and a low dose in rheumatoid arthritis patients) far exceeds the drug's plasma half-life. The *in vivo* metabolites of methotrexate such as polyglutamate persist in tissues where they inhibit purine synthesis by inhibiting aminoimidazolecarboxamidoribonucleotide (AICAR) transformylase. The rise in concentration of the substrate AICAR results in inhibition of two deaminases (adenosine deaminase and adenosine monophosphate deaminase) and thus leads to elevated levels of adenosine. Through interaction with the adenosine A_{2a}, and possibly A_3 receptors, adenosine not only inhibits T-cell function but also promotes T-cell tolerance.[126] These anergic T-cells are no longer able to proliferate and no longer produce IL-2, IFN-γ, and TNFα upon challenge. Not surprisingly, then, the antirheumatic effects of methotrexate may be diminished in patients drinking coffee that contains the adenosine receptor antagonist caffeine.[127] The major adverse effect of methotrexate can be argued to be related to its mechanism as the hepatotoxicity is thought to result from depletion of hepatic folate stores and accumulation of methotrexate pyroglutamates in the liver.[128]

A number of dihydrofolate reductase inhibitors (DHFR) have been launched since 1953 or are in the preregistration stage (2011); however, they are principally anti-infective therapy and oncology drugs.

6.2   *Sulfasalazine*

Sulfasalazine (SASP) was approved by the FDA in 1950 and is a compound that releases two active ingredients, namely 5-aminosalicylic acid (5-ASA) and sulfapyridine (SP) upon diazo bond cleavage, by action of bacteria in the large bowel. It is likely that the 5-ASA provides the majority of the therapeutic benefit since other 5-aminosalicylate drugs are also commonly used to treat inflammatory bowel disease. When SASP is used for the treatment of ulcerative colitis (UC) 5-aminosalicylic acid is present in high concentrations in the lumen of the colon where it can inhibit cycloxygenase and lipoxygenase, thus reducing the production of the mucosal-injuring and pro-inflammatory prostaglandins and leukotrienes.[129] Furthermore, transcriptional profiling studies in patients revealed that the therapeutic benefit of sulfasalazine for ulcerative colitis might, at least in part, be attributed to downregulation of proinflammatory cytokine mRNA (IL-2) expression as a consequence of 5-ASA's PPAR-γ agonist-mediated inhibition NF-kappaB activation.

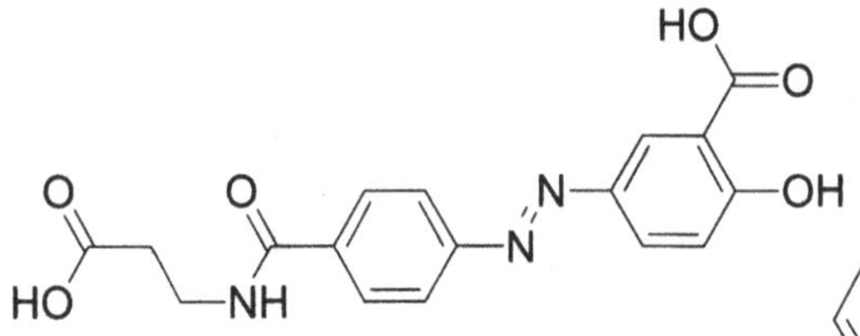

basalazide 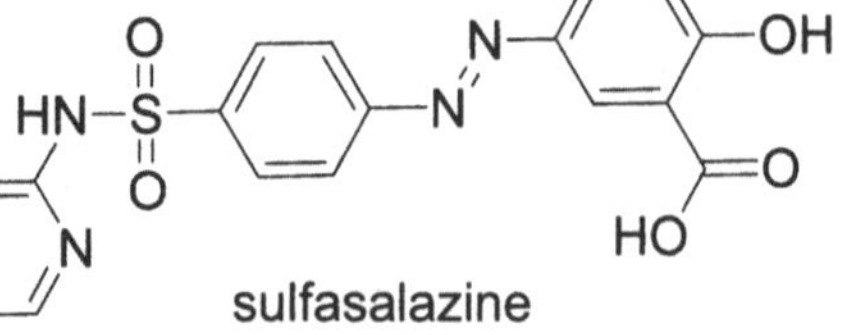   sulfasalazine

H₂N mesalazine   olsalazine

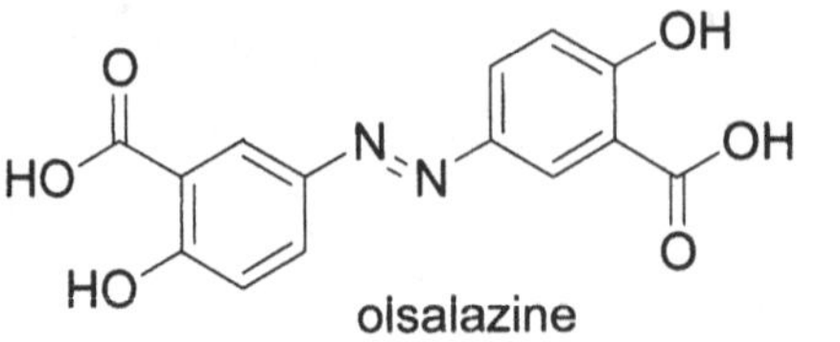

6.3   *Minocycline*

The modest efficacy of minocycline, a broad-spectrum tetracycline antibiotic, in rheumatoid arthritis (The MIRA trial) has been considered a result of its concentration in inflammatory cells and inhibition of protein arginine deiminase (PAD4). This enzyme catalyzes the conversion of arginine residues to citrulline and thus the disregulation of PAD4 activity in rheumatoid arthritis patients leads to auto-antibodies to these citrullinated proteins. Other tetracycline derivatives (doxycycline, tetracycline, and chlortetracycline) were evaluated for PDA4 inhibitory activity with chlortetracycline being identified as the most potent (IC_{50} 100 µM). More classically, their ability to act as oxygen radical scavengers and inhibit activity and synthesis of MMPs have equally been argued to contribute to their anti-inflammatory activity. While these antibiotics have demonstrated benefit related to symptom relief and joint swelling relief, and even laboratory measures of inflammation, there was no measurable effect on joint damage. Thus minocycline is preferably used in combination with other drugs that are likely to have a more beneficial effect at suppressing bone catabolism.

chlortetracycline   minocycline

6.4   *Hydroxychloroquine*

Antimalarial drugs have been used for many years to treat rheumatoid arthritis and systemic lupus erythematosus (SLE).[130] Although the efficacy of the antimalarial drugs chloroquine (launched 1934) and hydroxychloroquine is lower than methotrexate, they remain popular choices because of their good safety profile. From a mechanism-of-action perspective, they are thought to have an inhibitory effect on Toll-like receptor (TLR)-signaling which would account for their efficacy in rheumatic diseases.[131]

Toll-like receptors are a family of receptors on the plasma membrane or intracellular compartments of monocytes, macrophages, mast, dendritic, and B-cells, which initiate an inflammatory response by recognition of structurally conserved molecules derived from pathogens.[132] These receptors recognize so-called pathogen-associated molecules which are critical to pathogen function and hence are highly conserved. TLRs can signal via exogenous ligands such as those produced by microbes or endogenous ligands by engagement with either Toll homodimers or heterodimers that may use coreceptor proteins for full ligand engagement. When activated, the TLRs recruit one of several adapter proteins (MyD88, Tirap, Trif, Tram, and SARM) in the cytoplasm to propagate the signal.[133] Kinases such as IRAK1, IRAK4, and TBK1 can amplify the signal leading to the induction or suppression of a number of inflammatory genes.

TLR agonists can induce the expression of TNFα, IL-6, IL-1β, and IFN-β and anti-inflammatory cytokines (IL-10, TGF-β). The proinflammatory cytokines activate surrounding cells to produce chemokines which recruit inflammatory cells. Within the skin, as the primary line of defense against pathogens, TLR activation leads to enhanced phagocytic ability of macrophages. Dendritic cells respond to TLR activation with upregulation of MHCI and MHCII molecules and costimulatory CD80 and CD86 molecules which, after migration to lymph nodes, allow efficient antigen presentation to T-cells and an adaptive response. Thus TLRs not only produce an early inflammatory or antimicrobial response to pathogen but also initiate a more durable adaptive immune response.

Toll-like receptor signaling and in particular TLR9 deliver a second signal to B-cells, resulting in full activation and production of rheumatoid factor. Chloroquine can block rheumatoid factor production through endosomal acidification and thus inhibit TLR signaling. Furthermore, proinflammatory cytokine and chemokine production from synovial fibroblasts can be blocked by hydroxychloroquine action on TLR3. This drug acts only on endosomally located TLRs as the cell surface TLRs such as 2 and 4 are not inhibited by chloroquine or hydroxychloroquine. Chloroquine interferes with the phosphorylation of extracellular signal-regulated kinases (ERKs) and prevents activation of mitogen-activated protein kinases. This ERK inhibition is sufficient to inhibit LPS-induced TNF production in human peripheral blood mononuclear cells. In composite this effect on TLR signaling underscores the importance of the innate immune system to the pathogenesis of rheumatoid arthritis and suggests that targeting endosomally located TLRs may yield additional immune therapeutics.

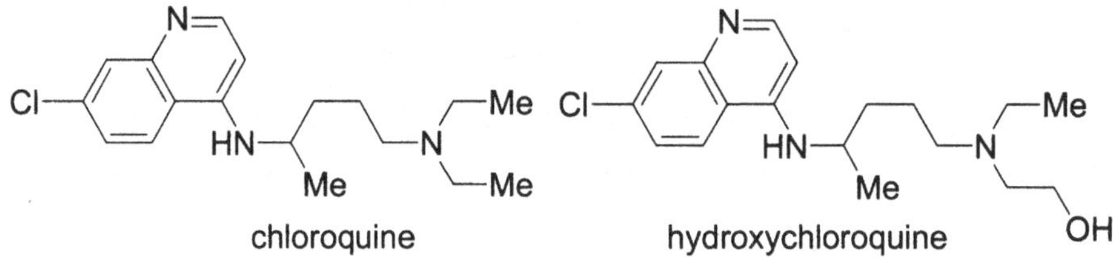

chloroquine   hydroxychloroquine

6.5   Leflunomide

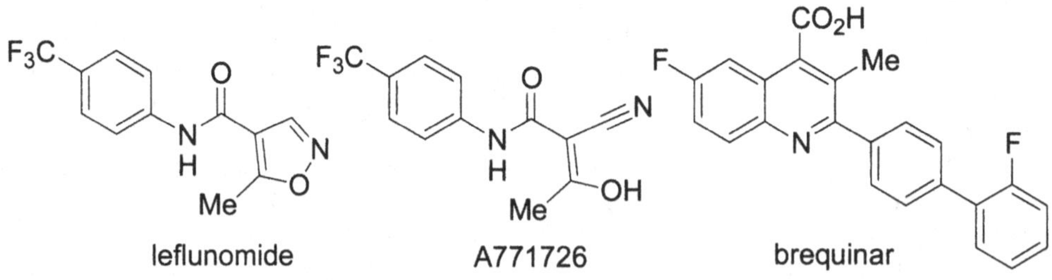

leflunomide   A771726   brequinar

Leflunomide was introduced in 1998 for patients with active rheumatoid arthritis and in 2004 was approved in Europe for the treatment of patients with active psoriatic arthritis. After oral administration leflunomide is converted to its active metabolite A771726, which is an antiproliferative agent for mononuclear and T-cells.[134] It is a competitive inhibitor of the ubiquinone binding site of dihydroorotate dehydrogenase (DHODH), an enzyme that catalyses the rate-limiting step of *de novo* pyrimidine biosynthesis. DHODH catalyzes the reduction of dihydroorotate (DHO) to orotate (ORO). Electrons are transferred by the oxidation of flavin mononucleotide (FMN) to yield dihydroflavin mononucleotide (FMNH2). The reduction of ubiquinone then results in the regeneration of flavin mononucleotide. The net depletion of cellular stores of pyrimidine results in the observed antiproliferative effect and hence explains its utility in arthritis in a manner similar to methotrexate.

Brequinar sodium is a noncompetitive (with respect to the substrate dihydroorotic acid) inhibitor of both the DHO and ubiquinone binding sites of the DHODH enzyme.[135] This immunosuppressive agent inhibits antigen-induced lymphocyte proliferation through inhibition of DNA and RNA synthesis as described above. Adverse effects; however, included thrombocytopenia, leucopenia, and stomatitis that appeared to be dose related.[136] Since there was intersubject variability in the clearance of brequinar, close monitoring for toxicity would have been required and hence development was discontinued. Despite the similarity of inhibiting the same enzymes, leflunomide and brequinar exhibit widely different therapeutic indices, implying additional off-target activity or tissue distribution and accumulation of brequinar.

7 Osteoarthritis

Osteoarthritis (OA) is primarily a degenerative joint disease characterized by progressive loss of cartilage with subchondral bone sclerosis. Changes in synovial fluid volume and reduced lubricant viscosity are often accompanied by joint pain, joint stiffness, and limitations of movement. OA is the most common type of arthritis with a prevalence that includes 60% of men and 70% of women over the age of 65 years. Unlike other joint diseases, OA is usually restricted to particular joints (knee, hand, shoulder, hips), which makes nonsystemic drug treatment a useful option.[137] Because cartilage has no blood supply of its own, the synovial fluid is fed by an intricate system of capillaries and lymphatics that provide cell-mediated and humoral immunity. Furthermore, the synovium has an extensive neural network which provides nociceptive feedback and controls synovial blood volume.

The cartilage breakdown characteristic of OA is driven by a multitude of cytokines, growth factors, and proteases. The matrix metalloproteinases (MMPs) play a crucial role in cleaving cartilage macromolecules such as type II collagen. Bone remodeling and new bone formation are observed in the form of osteophytes, which protrude into the joint space. These bony structures affect joint function and even the health of associated cartilage.

Delivery of steroidal drugs directly in to the joint allows sufficiently high concentrations of the drug to be achieved at the expense of the discomfort and pain associated with injection into an already sensitive joint. Drugs are cleared from the synovium by lymph drainage which can cause residence time to be particularly short (1–5 h). Thus emphasis is currently directed toward sustained-release formulations that can maintain therapeutic levels of drug for several weeks or months. Glucocorticoids that are applied as suspensions (betamethasone dihydrogen phosphate, dexamethasone acetate, triamcinolone acetonide) persist in the joint longer than the soluble glucocorticoids due to their poor solubility and long dissolution time. Alternatively, hyaluronic acid (HA) supplementation provides additional symptom relief through enhanced lubrication of the joint. This high-molecular-weight polysaccharide is normally present in joints at a high concentration (0.35 g/100 mL) and provides viscoelasticity and lubrication. HA supplementation administered intra-articularly also provides inhibition of PGE_2 synthesis and influences leukocyte proliferation, migration, and phagocytosis. HA can also stimulate production of tissue inhibitors of metalloproteinases (TIMP-1) and inhibit neutrophil-mediated cartilage degradation.[138]

8   Chronic Inflammatory Arthritis and Gout

Gout is a chronic type of inflammatory arthritis caused by the buildup of uric acid in the joints, leading to swelling, inflammation, and intense pain of the joints and lower extremities.[139,140] It is characterized by a sudden onset of pain and inflammation that has a tendency to recur. The hyperuricemia (>6.8 mg/dL) results from the liver producing more uric acid than the body can excrete, or from a diet rich in meat, seafood, fructose-sweetened beverages, and beer.[141] Dietary purines such as guanosine (from beer) promote uric acid overproduction as do mutations in enzymes involved in purine metabolism. The majority of gout patients experience renal uric acid hypoexcretion which is aggravated by genetics, high alcohol consumption, decreased glomerular filtration rate (GFR), diuretics, low-dose aspirin, and niacin. Monosodium urate crystals and calcium pyrophosphate dihydrate crystals liberated from soft-tissue deposits mediate activation of the NLRP3 inflammasome and overproduction of IL-1β.[142] Innate immune engagement of the crystals by Toll-like receptors (TLR-2 and TLR-4) promotes phagocytosis and protease release, generation of reactive oxygen species, and lowering of intracellular potassium levels which lead to inflammasome activation.[143]

Uric acid is the end product of purine nucleotide catabolism in which xanthine oxidase plays a key role. This enzyme catalyzes the oxidation of hypoxanthine to xanthine and can further catalyze the oxidation of xanthine to uric acid. Not surprisingly treatment of gout seeks to limit uric acid accumulation or improve its excretion.

8.1   *Colchicine*

Colchicine's medicinal value was identified in the first century A.D. and physicians have been prescribing colchicine for treating acute gout since 1810.[144] In August 2009, colchicine was approved by the FDA for the treatment of acute flares of gout! While trial data support its use in acute gout, overdosage is associated with nausea, vomiting, and diarrhea and significant toxicity including neurotoxicity, bone marrow damage and even death.

During gouty inflammation, monosodium urate phagocytosis by macrophages and mast cells leads to the activation of the NLRP3 inflammasome and release of IL-1β. Upon binding of IL-1β to the endothelial IL-1 β receptor type 1 there is activation and release of potent chemokines such as IL-8 for neutrophil recruitment. Once neutrophils reach the site of inflammation they release reactive oxygen species, hydrolytic enzymes, and IL-1β that perpetuate the inflammatory response. Colchicine exhibits multiple mechanisms as a result of its microtubule binding that in concert explain its therapeutic effect. By inhibiting the synthesis of TNFα by macrophages, the priming effect of TNFα on neutrophils, before their activation by monosodium urate, is limited. Multiple chemotaxis mechanisms including LTB4 formation, IL-8 production and adhesion molecule expression are all inhibited such that neutrophils cannot hone to the site of inflammation. Finally, neutrophil function is inhibited through the inhibition of cell division and proliferation that occurs as a result of colchicine binding to microtubules. Because colchicine has no effect on formation or dissolution of monosodium urate crystals, it is often coprescribed with agents that lower uric acid levels or block production of uric acid.

8.2 Uricosuric Agents

Uricosuric agents such as probenecid, sulfinpyrazone, and benzbromarone enhance renal uric acid excretion primarily by inhibiting urate anion reabsorption by proximal renal tubule epithelial cells. Probenecid blocks the kidney's organic anion transporter (OATP), which would normally recover uric acid from the urine and return it to the circulation. Since probenecid can also competitively inhibit renal excretion of drugs, its use must be carefully considered with patients on other background therapy. Benzbromarone is structurally related to the antiarrhythmic amiodarone, with clinical trials suggesting it is superior to allopurinol and probenecid. The uricosuric activity of benzbromarone is mediated by the active metabolite 6-hydroxybenzbromarone, which has a half-life of 30 h. While reports of adverse events are infrequent, some cases of serious hepatotoxicity have likely contributed to benzbromarone's limited availability.[145]

8.3 Xanthine Oxidase

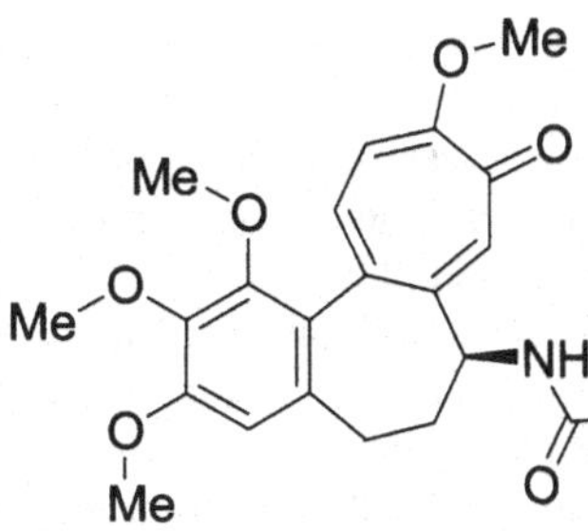

colchicine

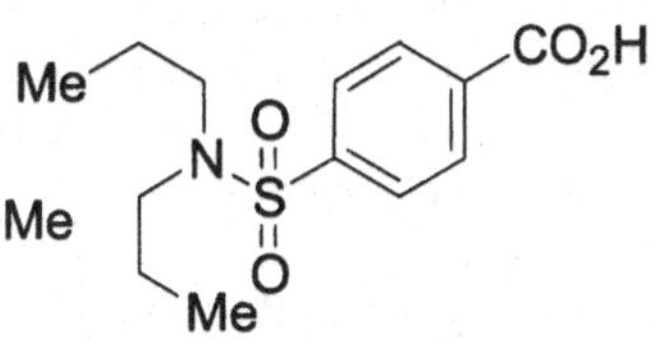

probenecid

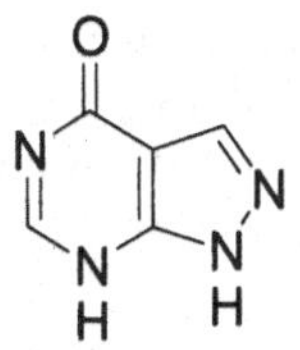

allopurinol

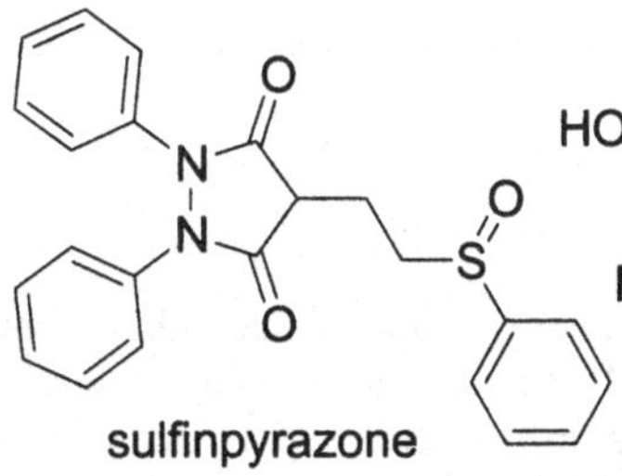

sulfinpyrazone

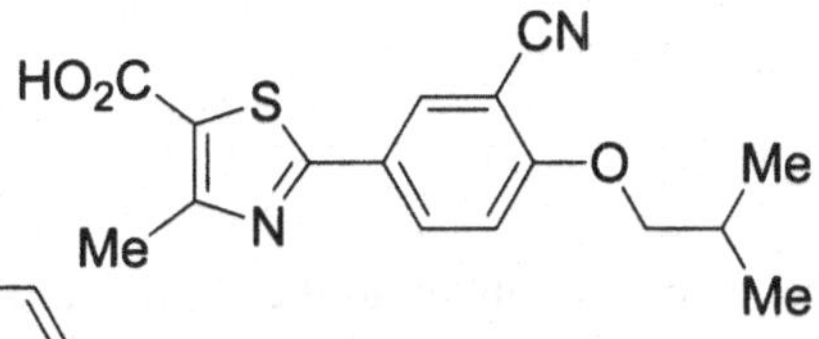

febuxostat

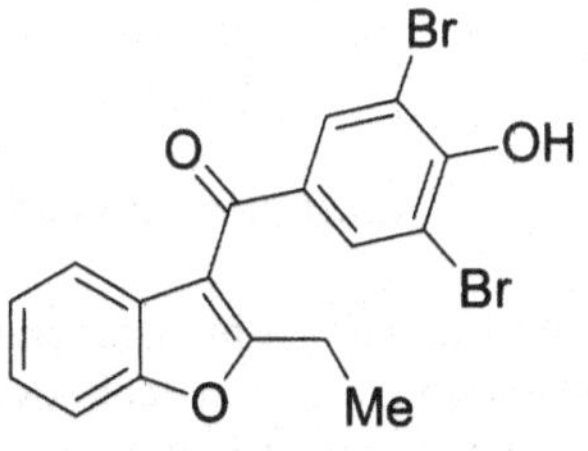

benzbromarone

Inhibitors of uric acid production include allopurinol, which is used to normalize serum urate levels to less than 6.0 mg/dL through progressive dose titration from a starting dose of 100 mg daily to a maximum of 800 mg daily. As a xanthine oxidase inhibitor allopurinol prevents the oxidation of hypoxanthine and xanthine which would result in the production of uric acid. Since hypoxanthine and xanthine levels are increased, they are converted to the closely related purine ribotides such as adenosine and guanosine monophosphates. Through increased levels of these ribonucleotides, a feedback inhibition of amidophosphoribosyltransferase results in limiting purine biosynthesis. Thus xanthine oxidase inhibition ultimately decreases both uric acid formation and purine synthesis. Febuxostat can be used in patients with allopurinol hypersensitivity or those that have intolerance or treatment failure. This selective xanthine oxidase inhibitor is administered at doses between 40 and 80 mg daily and can achieve target serum urate levels in patients with preserved renal function.[146]

8.4   Urate Oxidase

An alternative mechanism to reduce uric acid levels involves oxidation of uric acid, through the intermediacy of 5-hydroxyisourate, to the more soluble and renally cleared allantoin. The enzyme responsible for this oxidation, urate oxidase, is absent in humans and other primates and thus can be administered as a protein drug called rasburicase. A PEGylated form of rasburicase is also available as pegloticase, although the immunogenicity of these uricases limits their tolerability and efficacy.[147]

9   Multiple Sclerosis

Multiple sclerosis (MS) is a chronic inflammatory disease associated with central nervous system demyelination of neurons.[148] The disease follows a relapsing and remitting (RRMS) course in which patients experience symptoms for a period of time and then these symptoms resolve. In the secondary or later stages of the disease patients may enter a progressive phase of the disease, after about 10 years, which is typically associated with greater neurodegeneration. Generally an environmental trigger is thought to cause, in genetically susceptible individuals, an immune response directed toward self that results in inflammation, demyelination, and dysfunctional neuronal repair. Within the periphery, T-cells are activated by antigen-presenting cells and are then able to transmigrate the blood-brain barrier of the CNS with the help of adhesion molecule expression. Thus therapy is typically directed toward inhibition of T-cell activation, suppression of T-cell, B-cell and macrophage proliferation, inhibition of antigen presentation, reduction of cell trafficking, and inhibition of pro-inflammatory cytokine production.

Traditional therapy for MS includes short courses of IV or oral corticosteroids which serve to decrease inflammation in nerve tissues. More chronic treatment involves the use of IFN-β (1a and 1b),[149] glatiramer acetate,[150] and natalizumab.[151]

9.1   Natalizumab (Tysabri)

Natalizumab is a humanized monoclonal antibody to the alpha-4 integrin subunit that inhibits the interaction between leukocyte-expressed alpha-4 beta-1 (VLA-4) cell surface receptor and its respective ligand found on brain endothelium (vascular cell adhesion molecule 1, VCAM1).[152,153] This inhibition prevents T-cell transmigration across the blood-brain barrier and has shown substantial efficacy at suppressing relapse rates in MS patients. By reducing the numbers of activated disease-causing myelin-reactive T-cells in the central nervous system, natalizumab prevents the cascade of immune reactions that result in myelin destruction.

Additional alpha-4 beta-7 inhibiting properties of Tysabri resulted in approval for Crohn's disease as a result of inhibition of T-leukocyte trafficking to the gut. The alpha-4 beta-7 integrin on leukocytes (except neutrophils) binds to the mucosal vascular addressing cell adhesion molecule 1 (MAdCAM-1) in the gastrointestinal endothelium, preventing passage of leukocytes into the gut.

Because several cases of progressive multifocal leukoencephalopathy (PML), a rare demyelinating neurological disorder caused by reactivation of the very common John Cunningham virus in immune-compromised individuals, were reported, natalizumab was temporarily withdrawn from the market. The drug is currently marketed with safety warnings about PML, and is considered appropriate for patients who could benefit from its ability to slow the progression of MS and improve neurological function when they are not helped by other treatments for MS.[154]

9.2   IFN-Beta

Avonex (IFN-β 1a) is a recombinant version of human interferon-beta that consists of a glycosylated polypeptide consisting of 166 amino acid residues of approximate molecular weight 225,000 daltons, produced from genetically engineered cultured mammalian cells, Chinese hamster ovary (CHO) cells. Interferon-beta (IFN-β) is normally produced by fibroblasts and exhibits both antiviral and antiproliferative activity. By binding to its receptor, IFN-β induces a complex transcriptional response that reduces antigen presentation and proliferation of T-cells. It is also thought that interferon-beta improves the integrity of the blood-brain barrier and thus limits immune cell transmigration into the CNS while increasing levels of the anti-inflammatory IL-10 in the cerebrospinal fluid. Side effects such as flulike symptoms and injection site reactions from the required once-weekly dosing regimen coupled with no perceived improvement often lead to patient discontinuations.[155]

Rebif (IFN-β 1a) is administered by subcutaneous injection three times per week, as opposeD to Avonex, which is administered once per week by intramuscular injection. In a comparative study of Rebif and Avonex over 64 weeks Rebif was demonstrated to be superior to Avonex as measured by reduced MRI lesion activity and prevention of relapses.[156]

Betaseron (IFN-β 1b) is a form of IFN-β that is produced in a bacterial expression system which generates a different sequence of amino acids and posttranslational modifications and thus is designated IFN-β 1b. The transcriptional response to IFN-β 1b and IFN-β 1a IS similar, and the observed difference in biological response or clinical effect likely results from the differences in dosing frequency. Betaseron is started at 0.0625 mg and increased over 6 weeks to 0.25 mg dosed subcutaneously every other day.[157]

9.3   Mitoxantrone

In 2000, Serono launched mitoxantrone hydrochloride as an intravenous infusion for the reduction of neurological disability in patients with chronic progressive multiple sclerosis.[158] It intercalates into the growing DNA double helix, causing cross-links and strand breaks. Mitoxantrone also interferes with RNA and is a potent inhibitor of topoisomerase II, an enzyme responsible for uncoiling and repairing damaged DNA. Mitoxantrone has been shown in vitro to inhibit immune cell proliferation (B-cell, T-cell, and macrophage) and secretion of cytokines such as IFN-γ,[159] TNFα, and IL-2. However, the drug is considered a second-line therapy due to cardiomyopathy concerns that serve to limit the lifetime dosage of the drug.[160]

9.4   Glatiramer Acetate (Copaxone)

Glatiramer acetate is a tetrameric oligopeptide randomly composed of L-alanine, L-tyrosine, L-glutamic acid, and L-lysine which was launched in 1996 as an injection for the treatment of relapsing-remitting multiple sclerosis.[161] It is thought that glatiramer acetate is an immunochemical mimic of myelin basic protein, a putative autoantigen in MS. By binding to

MHC class II molecules on antigen-presenting cells the drug competes with various myelin antigens in their presentation to T-cells. Glatiramer acetate also promotes T-helper 2 (Th2) cell development and these cells can migrate to the brain and secrete anti-inflammatory cytokines such as IL-4, IL-10, and TGF-β. In general, glatiramer acetate has clinical benefits similar to IFN-β and is best prescribed early in the course of the disease and may safely enhance outcomes when used with other immunomodulators.[162]

9.5   *Fingolimod*

Fingolimod (FTY720, Gilenya) was discovered through structurE-activity studies on the fungal metabolite myriocin, whereby analogs were shown to reduce circulating levels of CD4[+], CD8[+] T-cells, and B-cells.[163] Through binding to S1P$_1$ receptors, fingolimod-phosphate (the active phosphate metabolite of fingolimod generated *in vivo* by sphingosine kinase 2) alters lymphocyte trafficking and inhibits egress of T-cells from lymph nodes. Upon continued fingolimod dosing the G-protein-coupled S1P$_1$ receptors internalize and thus the lymphocytes are unable to respond to the sphingosine concentration gradient that influences exit of these cells from the lymphoid organs. Impaired or leaky blood-brain barrier function is also characteristic of MS and beneficial effects of fingolimod on maintaining the integrity or permeability of the barrier have been demonstrated in rodent models. In net, inhibition of egress of reactive T-cells and maintenance of the blood-brain barrier may be invoked to explain the beneficial effects of fingolimod in MS. Given the efficacy and safety profile coupled with once-daily oral administration, fingolimod offers substantial benefits to patients.

mitoxantrone   fingolimod

10   **Transplantation**

The immune system is able to recognize tissues or grafts that are non-self and enact effector mechanisms that lead to the rejection of the graft. With the use of cyclosporine, which was introduced in 1978, the immunosuppressive action of the drug largely overcame the issue of graft rejection. This immunosuppressive therapy may also lead to an increased risk for infection, a significant and expected limitation of therapy. For kidney transplant patients, as an example, 1-year graft survival rates have improved due to a decline in incidence of acute rejections. Long-term graft survival and morbidity are however adversely affected by bacterial infection such as septicemia. Viral infections are also high and, in particular, BK virus (polyomavirus from the urine of a renal transplant patient with the initials B. K.) is increasingly leading to kidney allograft dysfunction. In response to this, early diagnosis of BK virus infection is essential, and prompt reduction in immunosuppressant therapy is associated with better outcomes.[164] Unfortunately rapid reduction in immunosuppression may result in insufficient control of immunity and may lead to acute graft rejection.[165,166]

10.1   Azathioprine

Azathioprine is among the oldest pharmacological immunosuppressive agents that are in use today for the prevention of organ rejection following organ transplant.[167] Its legacy can be traced back to the early 1960s when 6-mercaptopurine was reported to be effective in treating rheumatoid arthritis. Azathioprine was developed as a prodrug that is activated *in vivo* by glutathione or other thiol compounds to generate 6-mercaptopurine, which is then enzymatically converted to thioinosinic acid intracellularly. Mercaptopurine (brand name Purinethol) can become incorporated into replicating DNA and also block the *de novo* pathway of purine synthesis. Because lymphocytes lack a salvage pathway for purine synthesis, T-cell and B-cell proliferation is inhibited, which reduces the number of cytotoxic T-cells and plasma cells in circulation and thus suppresses immune responses.

10.2   Cyclosporine

In 1958 Sandoz established a screening program to identify antifungal compounds that could be isolated from soil samples that may contain microorganisms. A sample from *Hardanger Vidda* in Norway yielded the fungus *Tolypocladium inflatum* that synthesized metabolites that could be extracted from the mycelium. The isolate possessed a narrow spectrum of antifungal activity against *Aspergillus niger* and *Neurospora crassa* but more importantly displayed immunosuppressive activity. The most active component of the isolate was termed cyclosporine A (CsA) and was characterized as a cyclic undecapeptide.[168] When cyclosporine was evaluated in rats that had experimental allergic encephalomyelitis, an autoimmune disease model of multiple sclerosis, marked improvement was seen. In an adjuvant induced arthritis model in rats a similar improvement in signs and symptoms were observed when tested therapeutically or prophylactically.

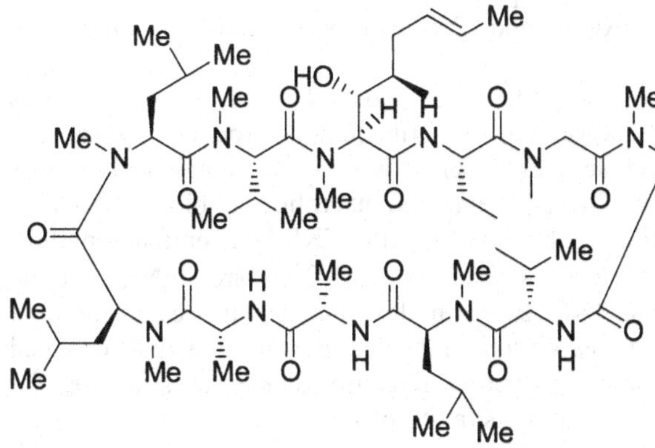

cyclosporine

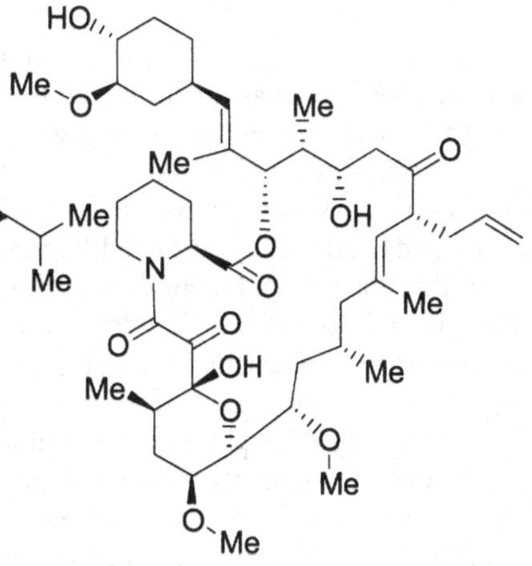

tacrolimus

Before the discovery of cyclosporine, immunosuppressive therapy involved the use of azathioprine, which acts by blocking all cells in mitosis. Cyclosporine demonstrated selective activity on lymphoid spleen cells and thus lacked the myelotoxicity evident in patients treated with azathioprine. Since its introduction in 1978 cyclosporine went on to revolutionize organ transplantation following successful demonstration of good protection from rejection. This naturally occurring cyclic peptide binds to the cytosolic protein cyclophilin or immunophilins. The cyclosporine–cyclophilin complex interacts with the calcium-binding protein calmodulin and with the serine/threonine phosphatase calcineurin. Uninhibited calcineurin dephosphorylates the cytosolic subunit of nuclear factor of activated T-cell (NFAT) in activated T-cells whereupon NFAT activates transcription of the IL-2 gene. Thus inhibition of IL-2 release prevents lymphocyte proliferation and potentially reactive T-cells from maturing into cytotoxic T-cells. This mechanism of action makes cyclosporine an attractive drug for autoimmune diseases such as psoriasis, rheumatoid arthritis, uveitis, lupus, and atopic dermatitis. Indeed, in a study of rheumatoid arthritis patients who started cyclosporine at doses of less than 4 mg/kg/day, none of the patients developed nephropathy or any sign of renal function deterioration. This is particularly important for cyclosporine because doses used in initial clinical studies were high and thus lowering the dose and monitoring cyclosporine concentration limited the deleterious nephrotoxicity side effect.[169] Cyclosporine can cause strong vasoconstriction by activation of the renin–angiotensin system (RAS), reduced synthesis of nitric oxide, increased sympathetic nervous system activity, production of thromboxane A_2, and synthesis of endothelin. The renal vasoconstriction leads to decreased glomerular filtration, elevated uric acid, and volume expansion, which can lead to irreversible renal damage.

In the early 1990s a new microemulsion formulation of cyclosporine was introduced to address the large inter- and intra-individual variability in bioavailability. This new formulation, Neoral, showed better and more predictable pharmacokinetics and dose linearity in cyclosporine exposure. The most common side effects are high blood pressure (25% of patients) and kidney problems (50% of patients), which explains its relatively limited use in a chronic disorder such as rheumatoid arthritis.

10.3   Tacrolimus

While cyclosporine has been successfully used to suppress the rejection of transplanted tissue, adverse events such as nephrotoxicity, hepatotoxicity, and CNS disturbances have encouraged the quest for alternative immunosuppressants.

Tacrolimus (FK506, Fujimycin), for example, was isolated from the soil fungus *Streptomyces tsukubaensis* using extensive screening of fermentation products to identify compounds, inhibiting the mixed lymphocyte reaction (MLR).[170] *In vitro*, tacrolimus activity was some 100-times more potent than cyclosporine. This 23-membered macrolide antibiotic lactone,[171] when bound to the intracellular receptor FK-506-binding protein (FKBP-12 or macrophilin), suppresses the production of cytokines IL-1, 2, 3, 4, 6, 7, 8, GM-CSF, and IFN-γ and the expression of the IL-2 receptor. The FK506/FKBP complex binds with calcineurin, a serine/threonine phosphatase, to inhibit the translocation of nuclear factor of activated T-cell (NFAT) into the nucleus and thus inhibit cytokine production. Because calcineurin is required for multiple processes throughout the body, systemic administration of tacrolimus is associated with significant side effects. Prolonged therapeutic blood levels of tacrolimus are thus associated with hypertension, nephrotoxicity, psychiatric disturbances, and hyperlipidemia.

The impressive immunosuppressive efficacy of both tacrolimus and cyclosporine and some adverse systemic toxicity have encouraged exploration of topical administration of these agents. Cyclosporine, although very lipophilic, is unable to penetrate the skin and has been ineffective as a topical agent despite excellent efficacy orally in psoriasis and atopic dermatitis.[172] Tacrolimus, however, can be applied as an ointment (Protopic) and is used to treat eczema and in particular atopic dermatitis without the undesirable skin thinning seen with steroid use.

10.4   Sirolimus (Rapamycin)

Rapamycin is a powerful immunosuppressant macrolide that was isolated from the bacterium *Streptomyces hygroscopicus* in a soil sample from Easter Island, an island also known as Rapa Nui (hence the name rapamycin) in the 1970s.

Sirolimus is not a calcineurin inhibitor, but it has a similar suppressive effect on the immune system. The rapamycin-FKBP12 complex inhibits the mammalian target of rapamycin (mTOR) pathway by directly binding the mTOR Complex1 (mTORC1).[173-175] mTOR is also called FRAP (FKBP-rapamycin-associated protein), RAFT (rapamycin and FKBP target), RAPT1, or SEP and is a serine/threonine protein kinase that controls cell growth, proliferation, and cell survival.[176] Rapamycin and its analogs bind to a domain separate from the catalytic site to block a subset of mTOR functions. Raptor (regulatory-associated protein of mTOR) works in conjunction with the mTOR protein to regulate cell size in response to nutrient levels. Rapamycin interferes with this system causing T-cells to stop growing and dividing, thus repressing rejection.

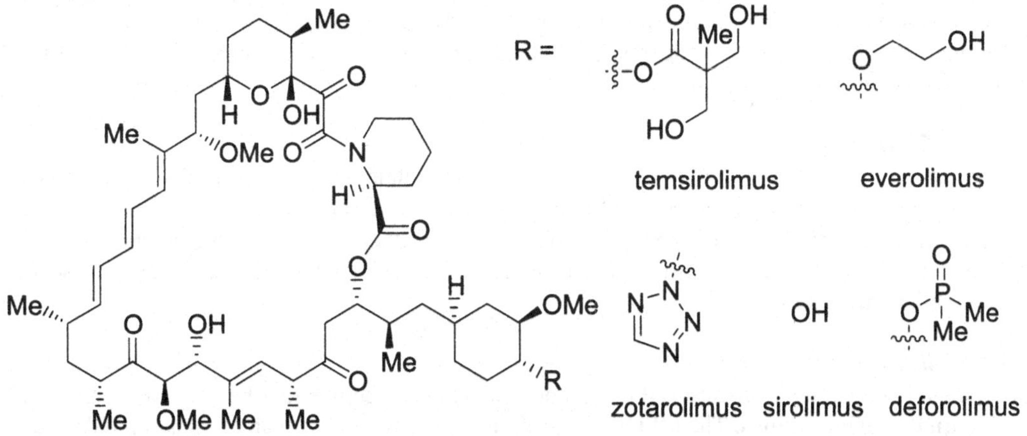

The chief advantage rapamycin has over calcineurin inhibitors is its low kidney toxicity. Transplant patients maintained on calcineurin inhibitors tend to develop impaired kidney function or even chronic renal failure. Rapamycin is therefore particularly advantageous in patients with kidney transplants for hemolytic-uremic syndrome, as this disease is likely to recur in the transplanted kidney if a calcineurin inhibitor is used.

However, on October 7, 2008, the FDA approved safety labeling revisions for rapamycin to warn of the risk for decreased renal function associated with its use. Furthermore, lung toxicity is a serious complication associated with rapamycin therapy, especially in the case of lung transplants or in patients that have underlying lung disease.[177]

In order to facilitate intravenous infusion of rapamycin, a dihydroxymethyl propionic acid ester prodrug was prepared (temsirolimus) which is rapidly converted to rapamycin upon injection.

Similarly, everolimus demonstrates improved bioavailability by incorporation of a hydroxyethyl chain substitution at C-40, whereas deforolimus has a phosphine oxide at the same position of the lactone ring, and zotarolimus incorporates a tetrazole.

11   Biological Agents That Suppress Cytokine Production or Signaling

The first biologic agents to treat rheumatoid arthritis were the inhibitors of tumor necrosis factor (TNF).[178] This cytokine is important in the pathogenesis of rheumatoid arthritis principally because of its proinflammatory effects.[179,180] These effects include lymphocyte activation and the proliferation of synovial fibroblasts, a cell type that is thought to be important in both the initiation and perpetuation of the disease. TNFα also acts to increase cytokines and chemokine and metalloproteinase activity and helps supply the developing panus through enhanced angiogenesis. Within the diseased joint, TNFα stimulates the resorption of bone and cartilage and prevents bone formation. These effects are mediated by binding the type 1 TFR receptor (p55) and the type 2 receptor (p75). They are found on immune, endothelial, and inflammatory cells. The level of TNFα in rheumatoid arthritis patients has been correlated to the severity of the disease and thus neutralizing the effects of this cytokine through inhibition of production or through blocking the receptor effectively treats rheumatoid arthritis. These agents, including etanercept, infliximab, adalimumab, golimumab, and certolizumab pegol, are very effective at improving the signs and symptoms and at slowing or preventing the structural damage associated with the disease.[181] The synergistic combination of these agents with methotrexate leads to an almost complete reversal of bone destruction.

11.1   Etanercept

Etanercept (Enbrel) is a dimeric fusion protein that contains two extracellular TNF binding domains of p75 linked to the Fc region of human IgG1 that effectively sequesters TNFα, or can be thought of as a decoy receptor for TNFα. It was launched in 1999 for reducing the signs and symptoms of rheumatoid arthritis and delaying structural damage in moderate to severe rheumatoid arthritis patients.[182]

11.2   Infliximab

Infliximab (Remicade) is an anti-TNFα monoclonal antibody comprising human constant regions and murine variable regions. The off rate for infliximab is slow and with an elimination half-life of 210 h, the antibody has a long duration of action. Infliximab can bind to different species of TNF, including free TNF, transmembrane TNF (tmTNF), and TNF that may be bound to its receptor. Infliximab has been licensed by the FDA since 1998 for the treatment of rheumatoid arthritis and Crohn's disease.[183]

11.3   Adalimumab (Humira)

One of the problems with chimeric and humanized antibodies in treating chronic disease is the propensity for which neutralizing antibodies are generated to these nonnative proteins.[184] A fully human antibody should be less immunogenic and thus therapeutic levels of the drug would be maintained upon prolonged dosing. The fully human anti-TNF antibody adalimumab (Humira) is a recombinant human IgG1 protein that was launched in 2003 in the US for the treatment of patients with rheumatoid arthritis. The antibody can lyse TNF-expressing cells in the presence of complement as a function of its IgG1 kappa subtype. The half-life of adalimumab is 10 to 20 days, and it is administered subcutaneously in doses ranging from 12.5 to 80 mg every other week. Subsequent to 2003, the list of approved indications expanded, as with other TNFα inhibitors, to include psoriatic arthritis, ankylosing spondylitis, Crohn's disease, and psoriasis.

11.4   *Golimumab*

Golimumab (Simponi) is a human monoclonal antibody IgG1 kappa that binds to the soluble and membrane-bound forms of TNFα.[185] The discovery of golimumab relied upon the use of genetically engineered mice that, upon exposure to the antigen human TNF, allowed human-derived variable and constant region antibodies to be isolated. Similar to other TNFα inhibitors, golimumab binds to soluble and membrane-bound TNF, thereby inhibiting TNF signaling. It was approved for the treatment of rheumatoid arthritis, psoriatic arthritis, and ankylosing spondylitis by the FDA in 2009 with subcutaneous dosing every 4 weeks. There lacks any clear differentiation for golimumab relative to competing TNF inhibitors with respect to efficacy or safety and thus selection of appropriate agents is best guided by patient, physician, and payer preferences.

11.5   *Certolizumab Pegol*

Certolizumab Pegol (Cimzia) is the first PEGylated humanized Fab fragment that antagonizes TNFα.[186] The Fab fragment is conjugated to polyethylene glycol (~40 kDa) which prolongs the plasma half-life. Due to the absence of the Fc receptor the product does not activate the complement cascade or direct cytotoxicity. Cimzia was approved in the U.S. in 2008 for the treatment of adult patients with Crohn's disease and in 2009 for rheumatoid arthritis.

Crohn's is one of the two major forms of inflammatory bowel disease which manifests with the symptoms of pain and diarrhea because of chronic inflammation of the gastrointestinal tract. In principle, T-cell-mediated response to mucosal microflora leads to the production of proinflammatory cytokines such as TNFα. As such, anti-TNFα antibodies were investigated in Crohn's disease with infliximab being approved for this indication in 1998. Since the safety and efficacy are similar to other anti-TNFα inhibitors approved for Crohn's, the use of certolizumab pegol is likely dictated by cost and the convenience of a once-monthly home injection maintenance dosing schedule. Alternatively nonresponders to anti-TNFα inhibitors may benefit from alternative agents such as natalizumab, a biologic that inhibits T-cell trafficking to the gut.

Physicians may switch between anti-TNF agents to address therapeutic limitations resulting from variation in bioavailability and antidrug antibodies experienced by patients on particular anti-TNF drugs. The formation of antibodies to TNF biologic agents is a significant problem because it reduces the efficacy by lowering the plasma concentration of the drug. Fortunately the concomitant use of methotrexate reduces the incidence of antibody formation.

In recent years there has been a steady increase in biologic drugs that are antirheumatic and act by a mechanism distinct from the classical anti-tumor necrosis factor (TNF) agents.[187] Thus rather than switching patients from one anti-TNF agent to another, physicians now have the ability to approach the disease from a different mechanistic vantage point.[188-190]

12   B-Cell Therapy

Over the last ten years, there has been renewed interest in targeting therapies that limit B-cell activity, particularly in rheumatoid arthritis.[191] B-cell differentiation to rheumatoid factor producing plasma cells and the production of antibodies to citrullinated peptides are prevalent in rheumatoid arthritis patients.

12.1   Rituximab

Rituximab (Rituxan) is a chimeric human–mouse monoclonal antibody to CD20, which is a glycosylated phosphoprotein found on the surface of B-cells. The mechanism of action of rituximab relies upon both antibody-dependent cellular cytotoxicity (ADCC) and complement-dependent cytotoxicity (CDC) that is able to deplete B-cells.[192] The binding of rituximab to CD20$^+$ cells activates the complement cascade which results in C3b being deposited on cells in close proximity to bound rituximab. Once C5b combines with C6, C7, C8, and C9 the membrane attack complex (MAC) forms transmembrane channels leading to cell lysis and death. This chimeric anti-CD20 monoclonal antibody was the first B-cell-depleting agent approved for rheumatoid arthritis. The antibody was approved in combination with methotrexate in the U.S. and Europe in 2006 for use after failure of at least one anti-TNF. It is an IV treatment and inhibits the progression of joint damage in rheumatoid arthritis over 2 years. Rituximab depletes all B-cell subsets except plasma cells (which lack CD20 expression). Infection rates are similar to those observed with anti-TNF therapy.

12.2   Ocrelizumab

Ocrelizumab is a humanized monoclonal anti-CD20 antibody which binds to an overlapping but different epitope of the extracellular domain of CD20 than that bound by rituximab. *In vitro* ocrelizumab produces less complement-dependent cytotoxicity and more ADCC than rituximab and thus in sum is expected to produce less severe infusion reactions and be less immunogenic.[193]

12.3   Ofatumumab

Ofatumumab is a human monoclonal antibody administered subcutaneously that binds to the small loop on the CD20 molecule as opposed to the large loop epitopes that bind rituximab and ocrelizumab, resulting in rapid and sustained peripheral B-cell depletion.[194]

12.4   Belimumab (Benlysta)

The cytokine B-cell-activating factor (BAFF), also known as B lymphocyte stimulator (BLyS), which is a 285-amino acid glycoprotein, is the natural ligand for three TNF receptors: BAFF-R, transmembrane activator and calcium-modulating cyclophilin ligand receptor (TACI), and B-cell maturation antigen (BCMA).[195] Signaling through these receptors stimulates B-cells to undergo proliferation and to counter apoptosis. Belimumab (Benlysta) is a fully human IgG1 monoclonal antibody to BAFF/BLyS that is not being developed for rheumatoid arthritis due to modest clinical response. It has however been approved in the U.S., Canada, and Europe for treatment of systemic lupus erythmatosis (SLE) despite being only marginally effective.[196]

13   Cytotoxic T-lymphocyte Antigen 4 (CTLA4)

13.1   Abatacept (Orencia):

T-cell activation results from interaction of the T-cell receptor with antigen presented by the major histocompatibility complex of the antigen-presenting cell (APC) and a second costimulatory signal mediated by CD28.[197] Abatacept is a fusion protein that combines the extracellular domain of cytotoxic T-lymphocyte antigen 4 (CTLA4) and a portion of the Fc domain of human IgG1. CTLA-4 (CD152) is expressed only transiently by activated T-cells and has higher affinity to B7-1 (CD80) and B7-2 (CD86) than CD28. With abatacept this extracellular domain of CTLA-4 binds to CD80/86 of the APC and thus scavenges the potential costimulatory signal from CD86 to CD28. Through this mechanism, T-cell activation is inhibited and thus proliferation and cytokine production is suppressed. To ablate the antibody-dependent cell-mediated cytotoxicity, the Fc receptor of the IgG1 was modified by replacing cysteines at positions 130, 136, and 139 to serine and proline at position 148 to serine. Abatacept was the first agent targeting the costimulatory signal required for full T-cell activation and was approved in the U.S. and Europe in 2005 for treatment of rheumatoid arthritis in patients with an inadequate response to TNF inhibitors.

13.2   Belatacept

Belatacept (Nulojix) is a second-generation CTLA4 fusion protein that enjoys two amino acid substitutions (L140E and A29Y) that give rise to slower dissociation rate constants from both CD80 and CD86.[198] So while CTLA4 binds with much higher avidity to CD80 (2500-fold) and CD86 (500-fold) relative to CD28, the CTLA4-Ig abatacept is surprisingly 100-fold less potent at inhibiting CD86-dependent costimulation than CD80. Thus in an effort to enhance the B7-2 (CD86) avidity and thus inhibition of T-cell activation belatacept was generated with 4-fold greater avidity for CD86 and 2-fold enhanced avidity for CD80 relative to abatacept. This enhanced in vitro activity translated to more effective immunosuppression in vivo, with recent approval for use in kidney transplant patients.

14   Interleukins

The success of TNFα inhibitors as biologic DMARDs and the incidence of TNFα non-responders have spurred interest in discovering inhibitors of other cytokines implicated in the pathogenesis of rheumatoid arthritis. IL-6 levels are raised in the synovium of rheumatoid arthritis patients, and these elevations have been correlated to measurements of disease. In preclinical species, it has been shown that antibodies to the IL-6 receptor inhibit collagen-induced arthritis in mice, suggesting an important role for this cytokine in autoimmune diseases.

14.1   Tocilizumab

Tocilizumab (Actemra) is an antibody directed toward the IL-6 receptor.[199] As such it competitively inhibits the binding of IL-6 to its membrane-bound receptor and soluble form of the IL-6 receptor. The IL-6 receptor is associated with the signal-transducing membrane protein gp130 which triggers downstream signaling events. Tocilizumab is administered by IV infusion every 4 weeks and was approved in Europe (2009) and the U.S. (2010) for moderate to severe rheumatoid arthritis patients that have failed one or more DMARDs or TNF antagonists. Physicians monitor for decreased neutrophil counts and increased lipid or liver enzyme levels because of the induction of a hepatic acute-phase response. The increase in liver enzyme levels is particularly evident in combination therapy with methotrexate but does not necessarily lead to drug-induced hepatitis. Since it is not possible to predict whether a particular patient would best respond to a TNF-inhibitor or IL-6 receptor antagonist, the perception about tocilizumab's safety profile will likely affect prescribing decisions.[200]

14.2   Ustekinumab

Ustekinumab (Stelara) is a human monoclonal antibody directed to the p40 subunit of IL-12 and IL-23 cytokines that was approved in the US and Europe in 2009 for treatment of patients with moderate to severe plaque psoriasis.[201] This fully human IgG1 kappa monoclonal antibody prevents IL-12 and IL-23 binding to the IL-12 receptor β1 on the surface of immune cells. The major pathogenic features of psoriasis involve abnormal keratinocyte differentiation, keratinocyte hyperproliferation, and tissue infiltration by inflammatory cells. Psoriasis is characterized by scaly plaques on the skin and affects the physical and emotional well-being and indeed quality of life when particularly severe.[202] The excessive proliferation of skin cells, keratinocytes, can be triggered by T-cells migrating to the dermis of the skin and releasing TNFα and IL-17. Dendritic cells can also migrate to the epidermis and release IL-12 as a consequence of TLR4 activation, which causes keratinocytes to proliferate. TNFα inhibitors, as discussed earlier, have considerably improved the treatment of moderate to severe psoriasis but are generally prescribed only after drugs traditionally used to treat this disease such as vitamin D analogs, glucocorticoids, retinoids, and immune-suppressive macrocycles have failed.[203] Ustekinumab is as effective as anti-TNF antibodies for the treatment of psoriasis and long-term data indicate a stable clinical response. From a practical perspective the drug is only administered 4- to 5-times a year, which should lead to good compliance, assuming long-term safety data are reasonable.

Serendipity has played a large role in identifying agents that are effective at treating psoriasis. With methotrexate it was observed that patients who suffered from rheumatoid arthritis and psoriasis often experienced resolution of their plaques. Cyclosporine treatment for psoriasis was similarly introduced when a renal transplant patient witnessed the additional benefit of their plaques being eliminated. Finally, for vitamin D analogs, a patient who was being treated for osteoporosis saw resolution of their skin disorder.

The interleukin-1 (IL-1) family of proinflammatory cytokines, of which there are at least ten members, includes the two cytokines IL-1α and IL-1β.[204,205] These cytokines are synthesized as 31-kDa pro-proteins that in the case of ProIL-1β is cleaved by the IL-1β-converting enzyme ICE or caspase-1 to generate mature IL-1β of 17 kDa. On binding to IL-1 receptors both IL-1α and IL-1β induce similar effects through activation of c-Jun N-terminal kinase (JNK) and p38 kinase. With activation of the transcription factors AP-1 and NF-kB, a whole host of chemokines, cytokines, cell-adhesion molecules, and proteases are liberated which contribute to the inflammatory process. Since increased production of IL-1β or the naturally occurring IL-1Ra has been demonstrated in the circulation of RA patients, a number of different approaches have been investigated to limit the proinflammatory and tissue destructive roles of IL-1.

14.3   Anakinra

Anakinra (Kineret) is a 153-amino-acid protein which differs from the naturally occurring IL-1 receptor antagonist (IL-1RA) by the addition of an N-terminal methionine residue.[206,207] By competitively binding to the IL-1 receptor type 1 on T-cells and the IL-1 receptor type II on polymorphonuclear leukocytes and B-cells with avidity equal to IL-1, anakinra is able to inhibit any intracellular response induced by IL-1. Since most cells express IL-1 receptors, it is anticipated that a 100- to 1000-fold excess of anakinra is necessary to completely inhibit IL-1-induced signaling. Anakinra is administered subcutaneously, which results in a maximal concentration occurring within three to seven hours, and a half life of only 4 to 6 h. To date, anakinra is mainly administered to RA patients that have failed TNFα inhibitors and is generally considered to exert poorer clinical benefit presumably as a result of low systemic exposure.

14.4 Canakinumab

Canakinumab (Ilaris) is an IgG kappa monoclonal antibody that binds to human IL-1β with high affinity at an epitope that is actually distinct from that involved in the IL-1β and IL-1 receptor type 1 interface.[207,208] Despite this, the canakinumab/IL-1β complex is unable to bind to cell surface receptors and thus IL-1β signaling is interrupted. Canakinumab has no reactivity for IL-1RA and was approved by the FDA in 2009 for the treatment of cryopyrin-associated periodic syndromes (CAPS), which is a rare inherited autoimmune disease associated with oversecretion of IL-1.[209,210] Patients with this syndrome typically experience inflammation of the skin, bones, and joints with accompanied fatigue and fever. Daily injections of anakinra can ameliorate the symptoms, as can subcutaneous administration of canakinumab every eight weeks.

14.5 Rilonocept

Rilonacept, also known as IL-1 Trap and Arcalyst, is a fusion protein comprised of the extracellular domains of the IL-1 receptor type 1 and accessory protein (IL-1-RAcP) combined with the Fc region of human IgG1.[211,212] Rilonacept can bind to IL-1β but unlike canakinumab can also bind to IL-1α. Rilonacept was approved for the treatment of CAPS disorders in 2008 by the FDA administered subcutaneously with a 320-mg loading dose and then followed weekly by a half loading dose. Development for treating RA was discontinued when clinical studies revealed IL-1 blockade had limited benefit in RA.[213]

15 Safety

As a result of suppressing the immune system it is common to see bacterial and fungal infections in patients on any of the anti-TNF biologics.[214] Reactivation of latent tuberculosis (TB) has led to the initiation of screening procedures prior to dosing with these biologics, with labeling for these biologics containing boxed warnings about the potential for infection.[215] Since most cases develop shortly after initiation of therapy, it is generally thought that this is due to reactivation of latent TB.[216] TNFα is considered important for maintaining the integrity of the granuloma which houses the TB within its confines. When the granuloma is compromised there is a breach in the integrity of the cellular compartment which normally confines the viable and latent bacilli. Classically, $CD4^+$ T-cells are important mediators of immunity to *M. tuberculosis* because they secrete cytokines that activate macrophages. It is also apparent that the human CD8 T-cells subset $CD45RA^+$ effector memory (T_{EMRA}, $CCR7^-$, $CD45RA^+$) are capable of killing *M. tuberculosis*-infected cells through release of their granule-containing perforin and granulysin, and also indirectly through cytokine secretion and activation of macrophages. However, the membrane-bound TNF on these cells results in TNF-antibody drugs such as infliximab and adalimumab engaging in T_{EMRA} cell depletion. This depletion and the inactivation of immune cells result in inefficient control of mycobacterial growth and ultimately the spread of the *M. tuberculosis* infection.

16   Summary

It is remarkable that, from the humble beginnings where inflammation was observed as redness, swelling and pain and treated with extract from the willow tree in 5th century BC, we have progressed to treating inflammation with antibodies derived from live cells. The intervening years witnessed a greater understanding of the molecular pathways involved in acute and chronic inflammation and the intricate role of the immune system in abrogating inflammatory responses. Synthetic compounds display a unique ability to unravel the intricacies of signal transduction pathways mediated by cellular activation that leads to the production of proinflammatory cytokines and thus provide useful therapies. The requirement for small-molecular-weight synthetic compounds to be systemically available upon oral dosing and then freely permeable to cell membranes limits the complexity and ultimately the specificity of the drug. This is of course revealed by the therapeutic index of the drug and the convenience of oral dosing. Alternatively biologic agents typically display exquisite selectivity as a result of the molecular interaction not being limited by the constraints of oral absorption. These high-molecular-weight compounds that ideally are administered once a month subcutaneously are however more expensive to manufacture in a reproducible manner and consequently are available at a premium price. Fortunately the patient has a number of therapeutic options which are expanding as a result of our greater understanding of the disease-associated biologic processes. The future likely holds the option of combining small-molecule and biologic drugs to provide synergistic benefit to the patient suffering from diseases that result from multiple independent biologic mechanisms.

References

(1)   Sompayrac, L. *How The Immune System Works*, Wiley-Blackwell: Hoboken, NJ,, 2008.

(2)   Mahmoudi, M. *Immunology Made Ridiculously Simple*; Medmaster: Miami, FL, 2009.

(3)   Janeway, C. T. P.; Walport, M.; Shlomchik, M. *Immunobiology: The Immune System In Health and Disease*, Garland Science Publishing: Oxford, UK, 2005.

(4)   Billiau, A.; Matthys, P. *Cytokine Growth Factor Rev.* **2009**, *20*, 97–113.

(5)   Hall, J. C.; Rosen, A. *Nat. Rev. Rheumatol.* **2010**, *6*, 40–49.

(6)   Akira, S.; Takeda, K. *Nat. Rev. Rheumatol.* **2004**, *4*, 499–511.

(7)   Springer, T. A. *Nature* **1990**, *346*, 425–434.

(8)   Yang, L.; Froio, R. M.; Sciuto, T. E.; Dvorak, A. M.; Alon, R.; Luscinskas, F. W. *Blood* **2005**, *106*, 584–592.

(9)   Segal, A. W. *Ann. Rev. Immunol.* **2005**, *23*, 197.

(10)   Lanier, L. L. *Nat. Immunol.* **2001**, *2*, 23–27.

(11)   Moretta, A. *Nat. Rev. Rheumatol.* **2002**, *2*, 957–965.

(12)   Banchereau, J.; Briere, F.; Caux, C.; Davoust, J.; Lebecque, S.; Liu, Y. J.; Pulendran, B.; Palucka, K. *Nat. Rev. Rheumatol.* **2000**, *18*, 767–811.

(13)   Banchereau, J.; Steinman, R. M. *Nature* **1998**, *392*, 245–252.

(14)   Mosmann, T.; Coffman, R. *Nat. Rev. Rheumatol.* **1989**, *7*, 145–173.

(15)   Lieberman, J.; Shankar, P.; Manjunath, N.; Andersson, J. *Blood* **2001**, *98*, 1667.

(16)   Werlen, G.; Hausmann, B.; Naeher, D.; Palmer, E. *Science* **2003**, *299*, 1859.

(17)   O'Sullivan, B.; Thomas, R. *Curr. Opin. Hematol.* **2003**, *10*, 272–278.

(18)   Clatza, A.; Bonifaz, L. C.; Vignali, D. A. A.; Moreno, J. *J. Immunol.* **2003**, *171*, 6478–6487.

(19)   Slavik, J. M.; Hutchcroft, J. E.; Bierer, B. E. *Immunol. Res.* **1999**, *19*, 1–24.

(20)   Appleman, L. J.; Berezovskaya, A.; Grass, I.; Boussiotis, V. A. *J. Immunol.* **2000**, *164*, 144.

(21)   O'Garra, A. *Immunity* **1998**, *8*, 275–283.

(22)   Zhang, S.; Zhang, H.; Zhao, J. *Biochem. Biophys. Res. Commun.* **2009**, *384*, 405–408.

(23)   Andrade, F.; Roy, S.; Nicholson, D.; Thornberry, N.; Rosen, A.; Casciola-Rosen, L. *Immunity* **1998**, *8*, 451–460.

(24)   Cohen, J. J.; Duke, R. C.; Fadok, V. A.; Sellins, K. S. *Nat. Rev. Rheumatol.* **1992**, *10*, 267–293.

(25)   Noelle, R. J.; Roy, M.; Shepherd, D. M.; Stamenkovic, I.; Ledbetter, J. A.; Aruffo, A. *Proc. Natl. Acad. Sci. U. S. A.* **1992**, *89*, 6550.

(26)   Vos, Q.; Lees, A.; Wu, Z. Q.; Snapper, C.; Mond, J. *Immunol. Rev.* **2000**, *176*, 154–170.

(27)   Calame, K. L. *Nat. Immunol.* **2001**, *2*, 1103–1108.

(28)   Carroll, M. C. *Nat. Rev. Rheumatol.* **1998**, *16*, 545–568.

(29)   McHeyzer-Williams, L. J.; McHeyzer-Williams, M. G. *Ann. Rev. Immunol.* **2005**, *23*, 487–513.

(30)   Nelson, A. L.; Dhimolea, E.; Reichert, J. M. *Nat. Rev. Drug Discov.* **2010**, *9*, 767–774.

(31)   Fan, Z.-C.; Shan, L.; Goldsteen, B. Z.; Guddat, L. W.; Thakur, A.; Landolfi, N. F.; Co, M. S.; Vasquez, M.; Queen, C.; Ramsland, P. A.; Edmundson, A. B. *J. Mol. Recognit.* **1999**, *12*, 19–32.

(32)   Harris, L. J.; Skaletsky, E.; McPherson, A. *J. Mol. Biol.* **1998**, *275*, 861–872.

(33)   Winter, G.; Griffiths, A. D.; Hawkins, R. E.; Hoogenboom, H. R. *Nat. Rev. Rheumatol.* **1994**, *12*, 433–455.

(34)   Chen, K.; Cerutti, A. *Curr. Opin. Immunol.* **2011**, *23*, 345–352.

(35)   Kracker, S.; Durandy, A. *Immunol. Lett.* **2011**, *138*, 97–103.

(36)   Bell, C. G. H. *Biochem. Soc. Trans.* **1998**, *26*, S200.

(37)   Mershon, K. L.; Morrison, S. L. Chapter 16. Antibody-Complement Interaction in *Therapeutic Monoclonal Antibodies: From Bench to Clinic,* An, Z. ed., Wiley: Hoboken, NJ, 2009, pp. 373–383.

(38)   Sandlie, I.; Michaelsen, T. E. Choosing and Manipulating Effector Functions In *Antibody Engineering: A Practical Approach*, McCafferty, J.; Hoogenboom, H. R.; Chiswell, D. J., IRL Press: Oxford, UK, 1996, pp 187–202.

(39)   Edmundson, A. B.; Guddat, L. W.; Rosauer, R. A.; Andersen, K. N.; Shan, L.; Fan, Z.-C. Three-Dimensional Aspects of IgG Structure and Function, In *The Antibodies*, Maurizio Zanetti, M.; Capra, D. J., eds, Harwood: 1995; Vol. 1, pp. 41–100.

(40)   Mason, K. L.; Huffnagle, G. B.; Noverr, M. C.; Kao, J. Y. *Adv. Exp. Med. Biol.* **2008**, *635*, 1–14.

(41)   Valenta, R.; Mittermann, I.; Werfel, T.; Garn, H.; Renz, H. *Trends Immunol.* **2009**, *30*, 109–116.

(42)   Konforte, D.; Simard, N.; Paige, C. J. *J. Immunol.* **2009**, *182*, 1781–1787.

(43)   Cerutti, A.; Puga, I.; Cols, M. *Trends Immunol.* **2011**, *32*, 202–211.

(44)   Hoyne, G. F. *Clin. Dev. Immunol.* **2011**, 294968, 294969 pp.

(45)   Nurieva, R. I.; Liu, X.; Dong, C. *Immunol. Rev.* **2011**, *241*, 133–144.

(46)   Meda, F.; Folci, M.; Baccarelli, A.; Selmi, C. *Cell. Mol. Immunol.* **2011**, *8*, 226–236.

(47)   Dai, R.; Ahmed, S. A. *Transl. Res.* **2011**, *157*, 163–179.

(48)   Folco, G.; Murphy, R. C. *Pharmacol. Rev.* **2006**, *58*, 375–388.

(49)   Arima, M.; Fukuda, T. *Korean J. Intern. Med.* **2011**, *26*, 8–18.

(50)   Pettipher, R.; Hansel, T. T.; Armer, R. *Nat. Rev. Drug Discov.* **2007**, *6*, 313–325.

(51)   Korotkova, M.; Jakobsson, P.-J. *Front. Inflammation Pharmacol.* **2011**, *1*, 1–8.

(52)   Vassiliou, E.; Jing, H.; Ganea, D. *Cell. Immunol.* **2003**, *223*, 120–132.

(53)   Hata, A. N.; Breyer, R. M. *Pharmacol. Ther.* **2004**, *103*, 147–166.

(54)   Moriarty, L. M.; Lally, M. N.; Carolan, C. G.; Jones, M.; Clancy, J. M.; Gilmer, J. F. *J. Med. Chem.* **2008**, *51*, 7991–7999.

(55)   Vane, J. R.; Botting, R. M. *Thromb. Res.* **2003**, *110*, 255–258.

(56)   Botting, R. M. *Pharmacol. Rep.* **2010**, *62*, 518–525.

(57)   Lanas, A. *Rheumatology* **2010**, *49*, ii3–ii10.

(58)   Andersson, D. A.; Gentry, C.; Alenmyr, L.; Killander, D.; Lewis, S. E.; Andersson, A.; Bucher, B.; Galzi, J.-L.; Sterner, O.; Bevan, S.; Högestätt, E. D.; Zygmunt, P. M. *Nat. Commun.* **2011**, *2*, 551.

(59)   Toussaint, K.; Yang, X. C.; Zielinski, M. A.; Reigle, K. L.; Sacavage, S. D.; Nagar, S.; Raffa, R. B. *J. Clin. Pharm. Ther.* **2010**, *35*, 617–638.

(60)   Lysz, T. W.; Needleman, P. *J. Neurochem.* **1982**, *38*, 1111–1117.

(61)   Langenbach, R.; Morham, S. G.; Tiano, H. F.; Loftin, C. D.; Ghanayem, B. I.; Chulada, P. C.; Mahler, J. F.; Lee, C. A.; Goulding, E. H.; Kluckman, K. D.; Kim, H. S.; Smithies, O. *Cell* **1995,** *83,* 483–492.

(62)   Morham, S. G.; Langenbach, R.; Loftin, C. D.; Tiano, H. F.; Vouloumanos, N.; Jennette, J. C.; Mahler, J. F.; Kluckman, K. D.; Ledford, A.; Lee, C. A.; Smithies, O. *Cell* **1995,** *83,* 473–482.

(63)   Flower, R. J. *Nat. Rev. Drug Discov.* **2003,** *2,* 179–191.

(64)   DeWitt, D. L. *Mol. Pharmacol.* **1999,** *55,* 625–631.

(65)   Penning, T. D.; Talley, J. J.; Bertenshaw, S. R.; Carter, J. S.; Collins, P. W.; Docter, S.; Graneto, M. J.; Lee, L. F.; Malecha, J. W.; Miyashiro, J. M.; Rogers, R. S.; Rogier, D. J.; Yu, S. S.; Anderson, G. D.; Burton, E. G.; Cogburn, J. N.; Gregory, S. A.; Koboldt, C. M.; Perkins, W. E.; Seibert, K.; Veenhuizen, A. W.; Zhang, Y. Y.; Isakson, P. C. *J. Med. Chem.* **1997,** *40,* 1347–1365.

(66)   Wang, J. L.; Carter, J.; Kiefer, J. R.; Kurumbail, R. G.; Pawlitz, J. L.; Brown, D.; Hartmann, S. J.; Graneto, M. J.; Seibert, K.; Talley, J. J. *Bioorg. Med. Chem. Lett.* **2010,** *20,* 7155–7158.

(67)   Talley, J. J.; Bertenshaw, S. R.; Brown, D. L.; Carter, J. S.; Graneto, M. J.; Kellogg, M. S.; Koboldt, C. M.; Yuan, J.; Zhang, Y. Y.; Seibert, K. *J. Med. Chem.* **2000,** *43,* 1661–1663.

(68)   Talley, J. J.; Brown, D. L.; Carter, J. S.; Graneto, M. J.; Koboldt, C. M.; Masferrer, J. L.; Perkins, W. E.; Rogers, R. S.; Shaffer, A. F.; Zhang, Y. Y.; Zweifel, B. S.; Seibert, K. *J. Med. Chem.* **2000,** *43,* 775–777.

(69)   Chan, C. C.; Boyce, S.; Brideau, C.; Charleson, S.; Cromlish, W.; Ethier, D.; Evans, J.; Ford-Hutchinson, A. W.; Forrest, M. J.; Gauthier, J. Y.; Gordon, R.; Gresser, M.; Guay, J.; Kargman, S.; Kennedy, B.; Leblanc, Y.; Leger, S.; Mancini, J.; O'Neill, G. P.; Ouellet, M.; Patrick, D.; Percival, M. D.; Perrier, H.; Prasit, P.; Rodger, I.; Tagari, P.; Therien, M.; Vickers, P.; Visco, D.; Wang, Z.; Webb, J.; Wong, E.; Xu, L. J.; Young, R. N.; Zamboni, R.; Riendeau, D. *J. Pharmacol. Exp. Ther.* **1999,** *290,* 551–560.

(70)   Prasit, P.; Wang, Z.; Brideau, C.; Chan, C. C.; Charleson, S.; Cromlish, W.; Ethier, D.; Evans, J. F.; Ford-Hutchinson, A. W.; Gauthier, J. Y.; Gordon, R.; Guay, J.; Gresser, M.; Kargman, S.; Kennedy, B.; Leblanc, Y.; Leger, S.; Mancini, J.; O'Neill, G. P.; Ouellet, M.; Percival, M. D.; Perrier, H.; Riendeau, D.; Rodger, I.; Tagari, P.; Therien, M.; Vickers, P.; Wong, E.; Xu, L. J.; Young, R. N.; Zamboni, R.; Boyce, S.; Rupniak, N.; Forrest, M.; Visco, D.; Patrick, D. *Bioorg. Med. Chem. Lett.* **1999,** *9,* 1773–1778.

(71)   Gierse, J. K.; Koboldt, C. M.; Walker, M. C.; Seibert, K.; Isakson, P. C. *Biochem. J.* **1999,** *339,* 607–614.

(72)   Davies, N. M.; Jamali, F. *J. Pharm. Pharm. Sci.* **2004,** *7,* 332–336.

(73)   Goldenberg, N. A.; Jacobson, L.; Manco-Johnson, M. J. *Ann. Intern. Med.* **2005,** *142,* 506–509.

(74)   Leese, P. T.; Hubbard, R. C.; Karim, A.; Isakson, P. C.; Yu, S. S.; Geis, G. S. *J. Clin. Pharmacol.* **2000,** *40,* 124–132.

(75)   Singer, J. B.; Lewitzky, S.; Leroy, E.; Yang, F.; Zhao, X.; Klickstein, L.; Wright, T. M.; Meyer, J.; Paulding, C. A. *Nat. Genet.* **2010,** *42,* 711–714.

(76)   Schnitzer, T. J.; Burmester, G. R.; Mysler, E.; Hochberg, M. C.; Doherty, M.; Ehrsam, E.; Gitton, X.; Krammer, G.; Mellein, B.; Matchaba, P.; Gimona, A.; Hawkey, C. J. *Lancet* **2004,** *364,* 665–674.

(77)   Trelle, S.; Reichenbach, S.; Wandel, S.; Hildebrand, P.; Tschannen, B.; Villiger, P. M.; Egger, M.; Juni, P. *BMJ* **2011,** *342,* c7086.

(78)   Brune, K.; Hinz, B. *Scand. J. Rheumatol.* **2004,** *33,* 1–6.

(79)   Chen, Y. F.; Jobanputra, P.; Barton, P.; Bryan, S.; Fry-Smith, A.; Harris, G.; Taylor, R. S. *Health Technol. Assess.* **2008,** *12,* 1–278, iii.

(80)   Tuder, R. M.; Voelkel, N. F. *J. Lab. Clin. Med.* **1998,** *132,* 16–24.

(81)   Raychaudhuri, B.; Bonfield, T. L.; Malur, A.; Hague, K.; Kavuru, M. S.; Arroliga, A. C.; Thomassen, M. J. *Clin. Immunol.* **2002,** *104,* 191–198.

(82)   Melian, E. B.; Goa, K. L. *Drugs* **2002,** *62,* 107–133.

(83)   Kojima, F.; Matnani, R. G.; Kawai, S.; Ushikubi, F.; Crofford, L. J. *Inflamm. Regener.* **2011,** *31,* 157–166.

(84) Jakobsson, P.-J.; Engblom, D.; Ericsson-Dahlstrand, A.; Blomqvist, A. *Curr. Med. Chem.: Anti-Inflamm. Anti-Allergy Agents* **2002**, *1*, 167–175.

(85) Nagai, H.; Teramachi, H.; Tuchiya, T. *Allergol. Int.* **2006**, *55*, 35–42.

(86) Godessart, N. *Ann. N. Y. Acad. Sci.* **2005**, *1051*, 647–657.

(87) Esser, J.; Gehrmann, U.; Salvado, M. D.; Wetterholm, A.; Haeggstroem, J. Z.; Samuelsson, B.; Gabrielsson, S.; Scheynius, A.; Raadmark, O. *FASEB J.* **2011**, *25*, 1417–1427.

(88) Hedi, H.; Norbert, G. *J. Biomed. Biotechnol.* **2004**, 99–105.

(89) Lewis, R. A.; Austen, K. F.; Drazen, J. M.; Clark, D. A.; Marfat, A.; Corey, E. J. *Proc. Natl. Acad. Sci. U. S. A.* **1980**, *77*, 3710–3714.

(90) Bernstein, P. R. *Am. J. Respir. Crit. Care Med.* **1998**, *157*, S220–225; discussion S225–226, S247–228.

(91) Bell, R. L.; Young, P. R.; Albert, D.; Lanni, C.; Summers, J. B.; Brooks, D. W.; Rubin, P.; Carter, G. W. *Int. J. Immunopharmacol.* **1992**, *14*, 505–510.

(92) Keam, S. J.; Lyseng-Williamson, K. A.; Goa, K. L. *Drugs* **2003**, *63*, 991–1019.

(93) Labelle, M.; Belley, M.; Gareau, Y.; Gauthier, J. Y.; Guay, D.; Gordon, R.; Grossman, S. G.; Jones, T. R.; Leblanc, Y.; et al. *Bioorg. Med. Chem. Lett.* **1995**, *5*, 283–288.

(94) Markham, A.; Faulds, D. *Drugs* **1998**, *56*, 251–256.

(95) Krell, R. D.; Aharony, D.; Buckner, C. K.; Keith, R. A.; Kusner, E. J.; Snyder, D. W.; Bernstein, P. R.; Matassa, V. G.; Yee, Y. K.; Brown, F. J.; et al. *Am. Rev. Respir. Dis.* **1990**, *141*, 978–987.

(96) Matassa, V. G.; Maduskuie, T. P., Jr.; Shapiro, H. S.; Hesp, B.; Snyder, D. W.; Aharony, D.; Krell, R. D.; Keith, R. A. *J. Med. Chem.* **1990**, *33*, 1781–1790.

(97) Snyder, D. W.; Giles, R. E.; Keith, R. A.; Yee, Y. K.; Krell, R. D. *J. Pharmacol. Exp. Ther.* **1987**, *243*, 548–556.

(98) Price, D.; Musgrave, S. D.; Shepstone, L.; Hillyer, E. V.; Sims, E. J.; Gilbert, R. F. T.; Juniper, E. F.; Ayres, J. G.; Kemp, L.; Blyth, A.; Wilson, E. C. F.; Wolfe, S.; Freeman, D.; Mugford, H. M.; Murdoch, J.; Harvey, I. *N. Engl. J. Med.* **2011**, *364*, 1695–1707.

(99) Zampeli, E.; Tiligada, E. *Br. J. Pharmacol.* **2009**, *157*, 24–33.

(100) Lippert, U.; Moller, A.; Welker, P.; Artuc, M.; Henz, B. M. *Exp. Dermatol.* **2000**, *9*, 118–124.

(101) Igaz, P.; Novak, I.; Lazar, E.; Horvath, B.; Heninger, E.; Falus, A. *Inflamm. Res.* **2001**, *50*, 123–128.

(102) Mullol, J.; Roca-Ferrer, J.; Alobid, I.; Pujols, L.; Valero, A.; Xaubet, A.; Bernal-Sprekelsen, M.; Picado, C. *Clin. Exp. Allergy* **2006**, *36*, 52–58.

(103) Mullol, J.; de, B. C. F.; Martinez-Anton, M. A.; Mendez-Arancibia, E.; Alobid, I.; Pujols, L.; Valero, A.; Picado, C.; Roca-Ferrer, J. *Respir. Res.* **2011**, *12*, 23.

(104) Sirois, J.; Menard, G.; Moses, A. S.; Bissonnette, E. Y. *J. Immunol.* **2000**, *164*, 2964–2970.

(105) Rouleau, A.; Stark, H.; Schunack, W.; Schwartz, J.-C. *J. Pharmacol. Exp. Ther.* **2000**, *295*, 219–225.

(106) Berlin, M.; Boyce, C. W.; de, L. R. M. *J. Med. Chem.* **2011**, *54*, 26–53.

(107) Rzodkiewicz, P.; Wojtecka-Lukasik, E.; Szukiewicz, D.; Schunack, W.; Maslinski, S. *Inflammation Res.* **2010**, *59*, S187–S188.

(108) Coruzzi, G.; Adami, M.; Guaita, E.; de, E. I. J. P.; Leurs, R. *Eur. J. Pharmacol.* **2007**, *563*, 240–244.

(109) Seifert, R.; Schneider, E. H.; Dove, S.; Brunskole, I.; Neumann, D.; Strasser, A.; Buschauer, A. *Mol. Pharmacol.* **2011**, *79*, 631–638.

(110) Jutel, M.; Akdis, M.; Akdis, C. A. *Clin. Exp. Allergy* **2009**, *39*, 1786–1800.

(111) Coutinho, A. E.; Chapman, K. E. *Mol. Cell. Endocrinol.* **2011**, *335*, 2–13.

(112) Barnes, P. J. *Eur. Respir. J.* **2006**, *27*, 413–426.

(113) Hu, X.; Du, S.; Tunca, C.; Braden, T.; Long, K. R.; Lee, J.; Webb, E. G.; Dietz, J. D.; Hummert, S.; Rouw, S.; Hegde, S. G.; Webber, R. K.; Obukowicz, M. G. *Endocrinology* **2011**, *152*, 3123–3134.

(114) Cosio, B. G.; Torrego, A.; Adcock, I. M. *Arch. Bronconeumol.* **2005**, *41*, 34–41.

(115) Moser, M.; De Smedt, T.; Sornasse, T.; Tielemans, F.; Chentoufi, A. A.; Muraille, E.; Van Mechelen, M.; Urbain, J.; Leo, O. *Eur. J. Immunol.* **1995**, *25*, 2818–2824.

(116)   Peers, S. H.; Smillie, F.; Elderfield, A. J.; Flower, R. J. *Br. J. Pharmacol.* **1993**, *108*, 66–72.

(117)   Herzog, H.; Oliveto, E. P. *Steroids* **1992**, *57*, 617–623.

(118)   Katz, M.; Gans, E. H. *J. Pharm. Sci.* **2008**, *97*, 2936–2947.

(119)   van Vollenhoven, R. F. *Discov. Med.* **2010**, *9*, 319–327.

(120)   Schett, G.; Stach, C.; Zwerina, J.; Voll, R.; Manger, B. *Arthritis Rheum.* **2008**, *58*, 2936–2948.

(121)   O'Rielly, D. D.; Rahman, P. *Pharmacogenomics Pers. Med.* **2010**, *3*, 15–31.

(122)   Smolen, J. S.; Landew, R.; Breedveld, F. C.; Dougados, M.; Emery, P.; Gaujoux-Viala, C.; Gorter, S.; Knevel, R.; Nam, J.; Schoels, M.; Aletaha, D.; Buch, M.; Gossec, L.; Huizinga, T.; Bijlsma, J. W. J. W.; Burmester, G.; Combe, B.; Cutolo, M.; Gabay, C.; Gomez-Reino, J.; Kouloumas, M.; Kvien, T. K.; Martin-Mola, E.; McInnes, I.; Pavelka, K.; van, R. P.; Scholte, M.; Scott, D. L.; Sokka, T.; Valesini, G.; van, V. R.; Winthrop, K. L.; Wong, J.; Zink, A.; van, d. H. D. *Ann. Rheum. Dis.* **2010**, *69*, 964–975.

(123)   Yasuda, H.; Shima, N.; Nakagawa, N.; Yamaguchi, K.; Kinosaki, M.; Mochizuki, S.; Tomoyasu, A.; Yano, K.; Goto, M.; Murakami, A.; Tsuda, E.; Morinaga, T.; Higashio, K.; Udagawa, N.; Takahashi, N.; Suda, T. *Proc. Natl. Acad. Sci. U. S. A.* **1998**, *95*, 3597–3602.

(124)   Cutolo, M.; Sulli, A.; Pizzorni, C.; Seriolo, B.; Straub, R. H. *Ann. Rheum. Dis.* **2001**, *60*, 729–735.

(125)   Chan, E. S. L.; Cronstein, B. N. *Nat. Rev. Rheumatol.* **2010**, *6*, 175–178.

(126)   Deaglio, S.; Dwyer, K. M.; Gao, W.; Friedman, D.; Usheva, A.; Erat, A.; Chen, J.-F.; Enjyoji, K.; Linden, J.; Oukka, M.; Kuchroo, V. K.; Strom, T. B.; Robson, S. C. *J. Exp. Med.* **2007**, *204*, 1257–1265.

(127)   Alarcon, G. S. *Nat. Clin. Pract. Rheum.* **2006**, *2*, 592–593.

(128)   Whitehead, V. M.; Perrault, M. M.; Stelcner, S. *Cancer Res.* **1975**, *35*, 2985–2990.

(129)   Gan, H.-T.; Chen, Y.-Q.; Ouyang, Q. *J. Gastroenterol. Hepatol.* **2005**, *20*, 1016–1024.

(130)   Jacob, N.; Stohl, W. *Arthritis Res. Ther.* **2011**, *13*, 228.

(131)   Kyburz, D.; Brentano, F.; Gay, S. *Nat. Clin. Pract. Rheumatol.* **2006**, *2*, 458–459.

(132)   Hennessy, E. J.; Parker, A. E.; O'Neill, L. A. *J. Nat. Rev. Drug Discov.* **2010**, *9*, 293–307.

(133)   Vogel, S. N.; Fitzgerald, K. A.; Fenton, M. J. *Mol. Interv.* **2003**, *3*, 466–477.

(134)   Davis, J. P.; Cain, G. A.; Pitts, W. J.; Magolda, R. L.; Copeland, R. A. *Biochemistry* **1996**, *35*, 1270–1273.

(135)   Strand, V.; Scott, D. L.; Simon, L. S. Eds., *Novel Therapeutic Agents for the Treatment of Autoimmune Diseases*, Decker: Hamilton, Ontario, Canada, 1997.

(136)   Makowka, L.; Sher, L. S.; Cramer, D. V. *Immunol. Rev.* **1993**, *136*, 51–70.

(137)   Gerwin, N.; Hops, C.; Lucke, A. *Adv. Drug Deliv. Rev.* **2006**, *58*, 226–242.

(138)   Chevalier, X. *Curr. Drug Targets* **2010**, *11*, 546–560.

(139)   Janssens, H. J. E. M.; Janssen, M.; van, d. L. E. H.; van, R. P. L. C. M.; van, W. C. *Lancet* **2008**, *371*, 1854–1860.

(140)   Gaffo, A. L.; Saag, K. G. *Nat. Clin. Pract. Rheumatol.* **2009**, *5*, 12–13.

(141)   Terkeltaub, R. *Nat. Rev. Rheumatol.* **2010**, *6*, 30–38.

(142)   Martinon, F.; Petrilli, V.; Mayor, A.; Tardivel, A.; Tschopp, J. *Nature* **2006**, *440*, 237–241.

(143)   Busso, N.; So, A. *Arthritis Res. Ther.* **2010**, *12*, 206.

(144)   Molad, Y. *Curr. Rheumatol. Rep.* **2002**, *4*, 252–256.

(145)   Lee, M.-H. H.; Graham, G. G.; Williams, K. M.; Day, R. O. *Drug Saf.* **2008**, *31*, 643–665.

(146)   Takano, Y.; Hase-Aoki, K.; Horiuchi, H.; Zhao, L.; Kasahara, Y.; Kondo, S.; Becker, M. A. *Life Sci.* **2005**, *76*, 1835–1847.

(147)   de Bont, J. M.; Pieters, R. *Nucleosides Nucleotides Nucleic Acids* **2004**, *23*, 1431–1440.

(148)   Barten, L. J.; Allington, D. R.; Procacci, K. A.; Rivey, M. P. *Drug Des. Dev. Ther.* **2010**, *4*, 343–366.

(149)   Rudick, R. A.; Goelz, S. E. *Exp. Cell Res.* **2011**, *317*, 1301–1311.

(150)   Lalive, P. H.; Neuhaus, O.; Benkhoucha, M.; Burger, D.; Hohlfeld, R.; Zamvil, S. S.; Weber, M. S. *CNS Drugs* **2011**, *25*, 401–414.

(151)   Nicholas, R.; Giannetti, P.; Alsanousi, A.; Friede, T.; Muraro, P. A. *Drug Des. Deve.l Ther.* **2011**, *5*, 255–274.

(152) Ozerlat, I. *Nat. Rev. Neurol.* **2011,** *7,* 246.
(153) Noseworthy, J. H.; Kirkpatrick, P. *Nat. Rev. Drug Discov.* **2005,** *4,* 101–102.
(154) Clifford, D. B.; DeLuca, A.; Simpson, D. M.; Arendt, G.; Giovannoni, G.; Nath, A. *The Lancet Neurol.* **2010,** *9,* 438–446.
(155) Vermersch, P.; de Seze, J.; Delisse, B.; Lemaire, S.; Stojkovic, T. *Multiple Sclerosis* **2002,** *8,* 377–381.
(156) Panitch, H.; Goodin, D.; Francis, G.; Chang, P.; Coyle, P.; O'Connor, P.; Li, D.; Weinshenker, B. *J. Neurol. Sci.* **2005,** *239,* 67–74.
(157) Rudick, R. A. *Arch. Neurol.* **1994,** *51,* 125–128.
(158) Scott, L. J.; Figgitt, D. P. *CNS Drugs* **2004,** *18,* 379–396.
(159) Kelchtermans, H.; Billiau, A.; Matthys, P. *Trends Immunol.* **2008,** *29,* 479–486.
(160) Hamzehloo, A.; Etemadifar, M. *Arch. Iran Med.* **2006,** *9,* 111–114.
(161) Arnon, R.; Aharoni, R. *Proc. Natl. Acad. Sci. U. S. A.* **2004,** *101,* 14593–14598.
(162) Khan, O. A.; Tselis, A. C.; Kamholz, J. A.; Garbern, J. Y.; Lewis, R. A.; Lisak, R. P. *Mult. Scler.* **2001,** *7,* 349–353.
(163) Brinkmann, V.; Billich, A.; Baumruker, T.; Heining, P.; Schmouder, R.; Francis, G.; Aradhye, S.; Burtin, P. *Nat. Rev. Drug Discov.* **2010,** *9,* 883–897.
(164) Manitpisitkul, W.; Wilson, N. S.; Haririan, A. *Expert Opin. Drug Saf.* **2010,** *9,* 959–969.
(165) Snyder, J. J.; Israni, A. K.; Peng, Y.; Zhang, L.; Simon, T. A.; Kasiske, B. L. *Kidney Int.* **2009,** *75,* 317–326.
(166) Mueller, N. J. *Transpl. Infect. Dis.* **2008,** *10,* 379–384.
(167) Maltzman, J. S.; Koretzky, G. A. *J. Clin. Invest.* **2003,** *111,* 1122–1124.
(168) Borel, J. F.; Kis, Z. L. *Transplant. Proc.* **1991,** *23,* 1867–1874.
(169) Ponticelli, C. *Ann. N. Y. Acad. Sci.* **2005,** *1051,* 551–558.
(170) Carroll, P. B.; Thomson, A. W.; McCauley, J.; Abu-Elmagd, K.; Rilo, H. R.; Irish, W.; McMichael, J.; Van, T. D. H.; Starzl, T. E.; *Landes,* **1996,** 211–220.
(171) Driggers, E. M.; Hale, S. P.; Lee, J.; Terrett, N. K. *Nat. Rev. Drug Discov.* **2008,** *7,* 608–624.
(172) Nghiem, P.; Pearson, G.; Langley, R. G. *J Am. Acad. Dermatol.* **2002,** *46,* 228–241.
(173) Brown, E. J.; Albers, M. W.; Shin, T. B.; Ichikawa, K.; Keith, C. T.; Lane, W. S.; Schreiber, S. L. *Nature* **1994,** *369,* 756–758.
(174) Ballou, L. M.; Lin, R. Z. *J. Chem. Biol.* **2008,** *1,* 27–36.
(175) Bierer, B. E.; Mattila, P. S.; Standaert, R. F.; Herzenberg, L. A.; Burakoff, S. J.; Crabtree, G.; Schreiber, S. L. *Proc. Natl. Acad. Sci. U. S. A.* **1990,** *87,* 9231–9235.
(176) Saemann, M. D.; Remuzzi, G. *Nat. Rev. Nephrol.* **2009,** *5,* 611–612.
(177) Pham, P. T.; Pham, P. C.; Danovitch, G. M.; Ross, D. J.; Gritsch, H. A.; Kendrick, E. A.; Singer, J.; Shah, T.; Wilkinson, A. H. *Transplantation* **2004,** *77,* 1215–1220.
(178) Ware, C. F. *Cytokine Growth Factor Rev.* **2008,** *19,* 183–186.
(179) Butun, B. *Anti-Inflammatory Anti-Allergy Agents Med. Chem.* **2010,** *9,* 8–23.
(180) Kopf, M.; Bachmann, M. F.; Marsland, B. J. *Nat. Rev. Drug Discov.* **2010,** *9,* 703–718.
(181) Leung, L.; Cahill, C. M. *J. Neuroinflamm.* **2010,** *7.*
(182) Bathon, J. M.; Martin, R. W.; Fleischmann, R. M.; Tesser, J. R.; Schiff, M. H.; Keystone, E. C.; Genovese, M. C.; Wasko, M. C.; Moreland, L. W.; Weaver, A. L.; Markenson, J.; Finck, B. K. *New Engl. J. Med.* **2000,** *343,* 1586–1593.
(183) Lipsky, P. E.; van der Heijde, D. M. F. M.; St. Clair, E. W.; Furst, D. E.; Breedveld, F. C.; Kalden, J. R.; Smolen, J. S.; Weisman, M.; Emery, P.; Feldmann, M.; Harriman, G. R.; Maini, R. N. *New Engl. J. Med.* **2000,** *343,* 1594–1602.
(184) Bain, B.; Brazil, M. *Nat. Rev. Drug Discov.* **2003,** *2,* 693–694.
(185) Pappas, D. A.; Bathon, J. M.; Hanicq, D.; Yasothan, U.; Kirkpatrick, P. *Nat. Rev. Drug Discov.* **2009,** *8,* 695–696.
(186) Melmed, G. Y.; Targan, S. R.; Yasothan, U.; Hanicq, D.; Kirkpatrick, P. *Nat. Rev. Drug Discov.* **2008,** *7,* 641–642.
(187) Feldmann, M.; Williams, R. O.; Paleolog, E. *Ann. Rheum. Dis.* **2010,** *69,* i97–i99.

(188)   Rubbert-Roth, A.; Finckh, A. *Arthritis Res. Ther.* **2009,** *11.*
(189)   Tak, P. P.; Kalden, J. R. *Arthritis Res. Ther.* **2011,** *13,* S5.
(190)   Gur, A.; Oktayoglu, P. *Anti-Inflammatory Anti-Allergy Agents Med. Chem.* **2010,** *9,* 24–34.
(191)   Cohen, S. B. *Best Pract. Res., Clin. Rheumatol.* **2010,** *24,* 553–563.
(192)   Beum, P. V.; Lindorfer, M. A.; Beurskens, F.; Stukenberg, P. T.; Lokhorst, H. M.; Pawluczkowycz, A. W.; Parren, P. W. H. I.; van, d. W. J. G. J.; Taylor, R. P. *J. Immunol.* **2008,** *181,* 822–832.
(193)   Kausar, F.; Mustafa, K.; Sweis, G.; Sawaqed, R.; Alawneh, K.; Salloum, R.; Badaracco, M.; Niewold, T. B.; Sweiss, N. J. *Exp. Opin. Biol. Ther.* **2009,** *9,* 889–895.
(194)   Keating, M. J.; Dritselis, A.; Yasothan, U.; Kirkpatrick, P. *Nat. Rev. Drug Discov.* **2010,** *9,* 101–102.
(195)   Anon. *Nat. Rev. Drug Discov.* **2011,** *10,* 243–245.
(196)   Calero, I.; Sanz, I. *Discov. Med.* **2010,** *10,* 416–424.
(197)   Korhonen, R.; Moilanen, E. *Basic Clin. Pharmacol. Toxicol.* **2009,** *104,* 276–284.
(198)   Larsen, C. P.; Pearson, T. C.; Adams, A. B.; Tso, P.; Shirasugi, N.; Strobert, E.; Anderson, D.; Cowan, S.; Price, K.; Naemura, J.; Emswiler, J.; Greene, J.; Turk, L. A.; Bajorath, J.; Townsend, R.; Hagerty, D.; Linsley, P. S.; Peach, R. J. *Am. J. Transplant.* **2005,** *5,* 443–453.
(199)   Scheinecker, C.; Smolen, J.; Yasothan, U.; Stoll, J.; Kirkpatrick, P. *Nat. Rev. Drug Discov.* **2009,** *8,* 273–274.
(200)   Kremer, J. M.; Blanco, R.; Brzosko, M.; Burgos-Vargas, R.; Halland, A.-M.; Vernon, E.; Ambs, P.; Fleischmann, R. *Arthritis Rheumatism* **2011,** *63,* 609–621.
(201)   Reich, K.; Yasothan, U.; Kirkpatrick, P. *Nat. Rev. Drug Discov.* **2009,** *8,* 355–356.
(202)   Melnikova, I. *Nat. Rev. Drug Discov.* **2009,** *8,* 767–768.
(203)   Drell, W.; Welch, A. D. *Pharmacol. Ther.* **1989,** *41,* 195–206.
(204)   Geyer, M.; Mueller-Ladner, U. *Curr. Opin. Rheumatol.* **2010,** *22,* 246–251.
(205)   Braddock, M.; Quinn, A. *Nat. Rev. Drug Discov.* **2004,** *3,* 330–340.
(206)   Swart, J. F.; Barug, D.; Moehlmann, M.; Wulffraat, N. M. *Expert Opin. Biol. Ther.* **2010,** *10,* 1743–1752.
(207)   Daniel E, F. *Clin. Ther.***2004,** *26,* 1960–1975.
(208)   Dhimolea, E. *MAbs* **2010,** *2,* 3–13.
(209)   Kubota, T.; Koike, R. *Mod Rheumatol* **2010,** *20,* 213–221.
(210)   Quartier, P. *Open Access Rheumatol. Res. Rev.* **2011,** *3,* 9–18.
(211)   Hoffman, H. M.; Yasothan, U.; Kirkpatrick, P. *Nat. Rev. Drug Discov.* **2008,** *7,* 385–386.
(212)   Dubois, E. A.; Rissmann, R.; Cohen, A. F. *Br. J. Clin. Pharmacol.* **2011,** *71,* 639–641.
(213)   McDermott, M. F. *Drugs Today* **2009,** *45,* 423–430.
(214)   Navarro-Compan, V.; Navarro-Sarabia, F. *Clin. Med. Insights: Ther.* **2011,** *3,* 195–203.
(215)   Fenton, M. J.; Vermeulen, M. W. *Infect. Immun.* **1996,** *64,* 683–690.
(216)   Keane, J. *Rheumatol.* **2005,** *44,* 714–720.

Antibacterial Drugs

Audrey Chan, Jason Cross, Yong He, Blaise Lippa, and Dominic Ryan

1   Introduction

Perhaps more than any other therapeutic area, the discovery of antibiotics had an immediate and enormous impact on human health and life expectancy. From the pioneering work of Fleming and Domagk, the wide-scale availability of antibiotics starting in the 1930s and 1940s had an effect on human culture that is hard to overestimate. Bacterial infections went from a constant threat to claim any life, including the young and strong, to a point where it was widely believed that bacterial infection was a solved issue. However, bacteria are literally everywhere and constantly evolving. It is estimated that there are 5×10^{30} prokaryotic cells on earth, composing 450×10^{15} g of carbon. For comparison, this weight is estimated to be 60–100% of the total carbon in all plant life on earth.[1] In addition, bacteria have been on earth for an estimated 3.5 billion years, versus ~ 200,000 years for humans.[2,3] Thus, they have a long track record of evolution to adapt to change, human-made antibiotics being a relatively recent challenge.

Despite the widespread belief that antibiotics are always effective, resistance has become an increasingly common event; it is estimated that 2 million people acquire bacterial infections in U.S. hospitals, and 90,000 die as a result. About 70% of those infections are resistant to at least one drug.[4,5] At the same time, new drug development of antibiotics has slowed substantially, with few new antibiotics discovered in the last 20 years.[6,7] This review will briefly describe the history of small-molecule antibiotics, challenges to antibacterial drug discovery, briefly summarize key representative antibiotic classes, and suggest promising future directions to develop new and effective antibiotics.

2   The Rise and Decline of Antibiotics

The modern era of antibacterial drug discovery began with the 1929 publication by Alexander Fleming on the observation that a *penicillium* mold inhibited the growth of various bacteria on an agar plate.[8] He stated, "for convenience and to avoid the repetition of the rather cumbersome phrase 'Mould broth filtrate.' The name 'penicillin' will be used." Despite this early discovery, work slowed on penicillin until mass production initiated in the early 1940s during World War II. Penicillin is an irreversible inhibitor of penicillin binding proteins (PBPs), which are necessary for bacterial cell wall formation.[9] Separately, Gerhard Domagk at I. G. Farben in Germany showed that a sulfonamide dye was effective in treating bacterial infections in mice in 1931.[10,11] A 1932 patent on the discovery was filed, which included the drug prontosil, and additional publications were made in 1935. At this time it was realized that the dye was a prodrug, and the actual active agent was *para*-aminobenzenesulfonamide, the first member of the sulfa class of antibiotics.

Drugs of this type are now known to inhibit bacterial folate synthesis and thus interfere with DNA replication. Although not as effective as penicillin, prontosil became the first commercially available antibiotic in 1935.

The discovery of novel antibiotics was central to the growth of pharmaceutical companies in the 1940s and 1950s. Many began this process through helping to create a large-scale fermentation and isolation of penicillin to aid in the war effort. Subsequently, they developed multiple new antibiotics over the course of the next 2–3 decades.[12] Most of these were isolated natural products, which, like penicillin, were isolated and scaled through fermentation. Due to the ubiquity of bacteria in nature, and in many cases their constant threat to other life forms, not surprisingly, nature had created many unique ways of killing bacteria through a variety of different mechanisms. This included targeting bacterial PBPs, which interfere with cell wall formation; the ribosome, interfering with protein biogenesis; DNA gyrase and topoisomerase IV, which causes bacterial death through DNA aberrations; and bacterial membrane disruptors, to name a few. Great advantages in the pursuit of discovering antibiotics are the highly predictive *in vitro* and *in vivo* preclinical disease models, resulting in a lower attrition rate in the clinic.[13]

By the 1970s, the U.S. Surgeon General W. H. Stewart is reported to have said: "the book on infectious diseases could now be closed."[13] However, due to the abundance of bacteria, and their rapid proliferation, evolution is constant, and invariably resistance began to develop against all classes of antibiotics. For example, methicillin-resistant *Staphylococcus aureus* (MRSA), which is resistant to most commonly used antibiotics, was first identified in 1961 and today is by far the most prevalent form of *S. aureus* infections in US hospitals.[14] Gram-negative infections have also become quite worrisome. For instance, *Pseudomonas aeruginosa* is often resistant to fluoroquinolones and many other therapies, which can cause a high mortality rate in hospitals.[15] Infections also commonly interfere with other therapies, e.g., cancer therapies, and often cause death in weakened patients.

In the last 40 years, only three new classes of antibiotics have been developed, and the overall number of new drugs approved has also dwindled.[6] Most of the newly approved drugs are analogs of classical drugs, such as a new cephalosporin, or fluoroquinolone. While these can have significant value, they also often suffer from the same resistance mechanisms as their precursors. Particularly in the case of Gram-negative bacteria, there are times when virtually every drug is ineffective due to resistance.[15] This has caused a large increase in the use of polymyxin antibiotics for these infections, a class of drugs first commercialized in the 1950s, but rarely used until now due to significant renal toxicity.[16] However, as a drug of last resort with few other options, clinicians are more and more willing to take that risk of toxicity due to the desperate need of the patient. Unfortunately, today there are relatively few new agents in the pipeline, in particular for Gram-negative infections, and most large pharmaceutical companies have exited this area of research.[17,18]

3   The Unique Challenges of Antibacterial Drug Discovery

There are a number of significant challenges involved in the discovery and development of novel antibiotics. These hurdles, many of which are largely unique to this field, have contributed to the low number of new drug approvals in the area, in particular in the last 20 years.

3.1 Lead Matter

The vast majority of antibiotics developed over the past 70 years are either direct natural products or based on a natural product and made through semisynthesis.[13] Ironically, the first commercial antibiotic class, the sulfas, were fully synthetic and derived from a dye. While there are other examples of fully synthetic antibiotics, most prominently the fluoroquinolones and oxazolidinones, most successful antibiotics are still natural product derived. Natural products as starting points tend to limit synthetically feasible options to optimize a particular lead due to their synthetic complexity, and thus a limited number of sites where the scaffold can be modified by synthetic chemistry. Often this synthetic complexity also lengthens the timeline necessary to advance a project due to the larger number of steps required to complete analogs, which lowers overall analog output. Separately, it is well established that high-throughput screens of fully synthetic molecules have rarely given leads suitable to advance as antibiotics, in contrast to other therapeutic areas.[19] The reasons for this are not clear but are likely related to bacterial targets being different than human targets, which were the basis for most compounds in a pharmaceutical company's screening file. The lack of this source of leads thus often severely limits options to begin a new antibacterial project.

3.2 Bacterial Targets

Although several bacterial genomes have been sequenced, and all essential genes are known, few bacterial targets have proven to be druggable. Some have argued that nature, through millions of years of evolution, has already found the best targets.[20] One challenge is the preference for an antibiotic to inhibit more than one target to keep resistance rates low. Depending on the target, this is often not possible due to low homology with a second target and the complexities of simultaneously optimizing two potency properties across multiple strains of bacteria. In other cases, inhibiting a target may work in one species, but other bacterial species have different pathways making that target not essential, or homology is low. Due to the difficulty in advancing a narrow-spectrum antibiotic, this often is a dead end for a target.

3.3 Resistance

Due to the rapid proliferation of bacteria, they are often able to develop resistance quickly, particularly against a single agent.[21] In addition, bacteria have developed methods to increase resistance and genetic diversity, such as plasmid-encoded resistance genes that are easier to spread, and pili, cellular appendages that connect directly to other bacterial cells, allowing them to share genetic material. All antibiotics generate resistance, sometimes quickly after commercial launch. This often limits the use and lifetime of an antibiotic and in some cases prevents them from reaching any commercial value. This also ironically makes the discovery of new antibiotics a never-ending imperative.

3.4   *Efflux and Permeability*

Bacteria have been on earth for >3 billion more years than humans and have needed to evolve considerably over that time to adapt. In addition, they typically grow in contact with a range of other fungi, bacteria, and other life forms that are antagonistic, and all trying to control the same space. Bacteria also are often in otherwise inhospitable conditions such as environments containing low oxygen levels, hypotonic conditions, or with high amounts of destructive chemicals such as bile salts in the intestinal track. Thus, bacteria have evolved a large number of defense mechanisms. These include a cell wall to survive in low salt concentrations and other structural challenges, a wide range of efflux pumps to eject harmful chemicals, and in the case of Gram-negative bacteria a second cellular membrane to restrict permeability.[22] This outer membrane creates a very difficult barrier for antibiotics to enter, largely limited to small pores in the surface most available to small, polar drugs. Because the inner membrane of Gram-negatives more closely resembles a typical mammalian membrane, it favors lipophilic molecules for passive permeability. Thus, to attack an intracellular Gram-negative target, the drug must cross both membranes, which favor different properties, while simultaneously avoiding efflux pumps. These challenges are the main reason why it has been generally more difficult to target Gram-negative organisms and why infections of that type possess a higher medical need.

3.5   *Lack of Diagnostics*

There are few rapid diagnostic tests available to type an infection when a patient visits a doctor.[23] This fact, plus the need to treat ill patients quickly before their infection spreads, causes clinicians to often prescribe antibiotics empirically, at least initially. Thus, a broad-spectrum antibiotic is the most useful since it will cover multiple different types of bacteria when the exact infection type is unknown. While the development of drugs of this type has been achievable in the past, it is becoming more difficult now, in particular to cover highly resistant species such as *P. aeruginosa*. From a drug discovery perspective, there are targets that by necessity would be specific for a single bacterial species. Here, medical need is often high, but because of how antibiotics are used, a narrow-spectrum antibiotic faces high development challenges if it does not cover a broad range of bacterial infections.

3.6   *Clinical Trials*

Due to the lack of quick and effective diagnostics and the relatively lower frequency of highly resistant bacterial infections, it is difficult to advance a narrow-spectrum antibiotic in clinical trials.[7] To do so either could involve later stage treatment after a patient has failed an initial therapy and the infection has been typed, or requires combination therapy with another drug. Since the other drug will be fairly effective, a head-to-head superiority trial must be conducted against that other drug to ensure the combination is superior. However, due to the lower frequency of highly resistant infections, clear superiority can be difficult to achieve with a reasonably sized clinical trial.

3.7 Commercial Landscape

The antibiotic market has many generic and inexpensive drugs that retain effectiveness for most common infections.[24] The highest medical need, and the most common way to improve on older medicines, is to target resistant bacteria. Because this is an area of high medical need, it is encouraged by all. However, since all drugs are prone to at least some resistance, once a new therapeutic is available, clinicians may save it as a drug of last resort and not prescribe it widely. This ensures that less resistance will develop against the drug in the community and gives a clinician a high-probability option for cure should initial therapies fail. While this may make medical sense, it also makes it more difficult for a company to achieve a profit in the limited window while the drug is on patent, which would in turn enable further antibacterial research.

4 Antibiotic Classes

4.1 Cell Wall Biosynthesis Inhibitors

Bacterial cell wall biosynthesis pathways are well established antibacterial targets.[25,26,27,28] β-Lactams and glycopeptides are two major classes of cell wall biosynthesis inhibitors widely used clinically. The less frequently used cell wall inhibitors include fosfomycin,[29] cycloserine,[30] and bacitracin.[31]

4.1.1 β-Lactam/β-Lactamase Inhibitors

Natural, semisynthetic and synthetic β-lactam derivatives,[32,33] including penicillins, cephalosporins, carbapenems, monobactams, and others, are the most well-known and widely used antibiotics in the history of modern medicine. β-Lactam antibiotics[34] inhibit bacterial cell wall biosynthesis by irreversibly binding to the transpeptidases known as penicillin-binding proteins (PBPs), which are essential for the cross-linking of peptidoglycan layers of the cell wall.[35,36] The success of these drugs is due to their rapid bactericidal (killing bacteria rather than just preventing their growth) activity, good pharmacokinetics and excellent safety profiles. The use of these drugs as the major line of defense for treating bacterial infections has resulted in the rapid emergence of resistance. The main resistance mechanisms for the β-lactam class include the mutation of PBPs, especially in Gram-positive bacteria such as MRSA and penicillin-resistant *Streptococcus pneumoniae* (PRSP), and expression of β-lactamases.[37,38,39] Gram-negative pathogens also utilize efflux pumps and reduced cell permeability to decrease the susceptibility to β-lactams.[37,38,39]

Penicillins

Penicillin antibiotics are historically significant because they are the first drugs that were effective against many previously serious diseases such as syphilis and infections caused by *staphylococci* and *streptococci*. Even though penicillin G was discovered in 1929,[40] and many types of bacteria have developed resistance to it, penicillin continues to be a first-line treatment for the bacterial infections caused by susceptible, usually Gram-positive, organisms.

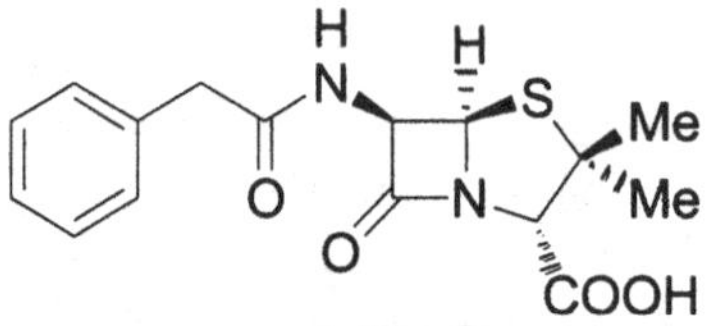

penicillin G

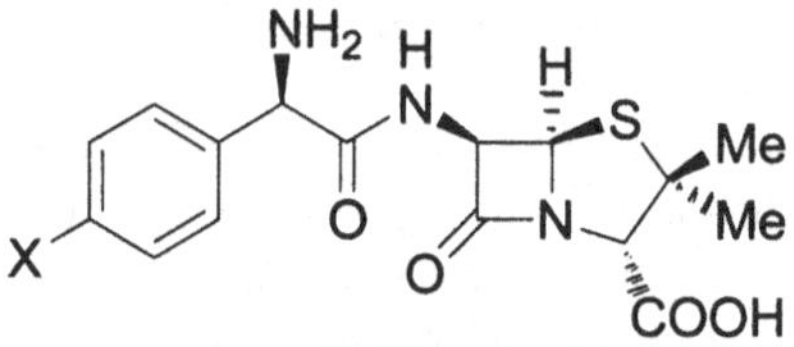

ampicillin, X = H
amoxicillin, X = OH

A major development since penicillin in this class was ampicillin,[41] launched in 1961. It offers a broader spectrum of activity than the original penicillins.[42] Like penicillin, it is effective against Gram-positive organisms such as *staphylococci* and *streptococci*. However, ampicillin also demonstrates activity against some Gram-negative organisms such as *H. influenzae* and *Proteus spp.* The enhanced Gram-negative activity is due to the introduction of the amino group on the left-hand amide side chain, which helps the drug penetrate the outer membrane of Gram-negative bacteria. Amoxicillin,[43] launched in 1972, has similar spectrum coverage[44] but improved oral bioavailability (95% vs. 40% for ampicillin).[45,46]

Further modifications of the amide side chain yielded β-lactamase-resistant penicillins, such as flucloxacillin,[47] dicloxacillin,[48] and methicillin.[49] These show increased activity against penicillinase-producing bacteria species. The steric hindrance of the bulky aromatic substitution prevents the drug from being hydrolyzed by the β-lactamase enzymes, while still allowing the drug to bind to the PBPs. However, these drugs are still ineffective against the MRSA strains that have subsequently emerged.

Another key variation of penicillins is the antipseudomonal penicillins, such as piperacillin[50] and ticarcillin.[51] They are useful for their activity against Gram-negative bacteria.[52] Due to the widespread β-lactamase resistance, the clinical use of penicillins is usually in combination with β-lactamase inhibitors to preserve their antibacterial properties.[53]

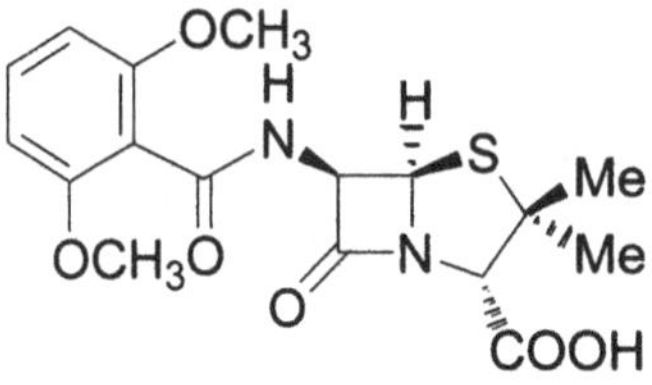

methicillin

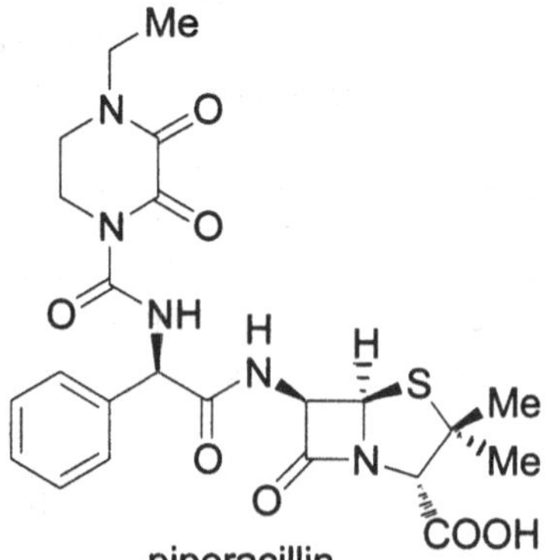

piperacillin

Cephalosporins

Like all other β-lactams, cephalosporin antibiotics[54] also have a natural origin. In 1953, Abraham and Newton[55] isolated cephalosporin C from the crude extract of *Cephalosporium acremonium*, with its chemical structure elucidated in 1959.[56] Cephalosporin C exhibited several remarkable properties, including Gram-positive and Gram-negative antibacterial activities. Following that, the development of a chemical process for the conversion of cephalosporin C to 7-aminocephalo-sporanic acid (7-ACA)[57] allows for feasible and fruitful chemical modifications to provide several generations of cephalosporins to the market. Modification of the 7-ACA side chain resulted in the development of the first semisynthetic variation, cephalothin,[58] which was launched by Eli Lilly in 1964.

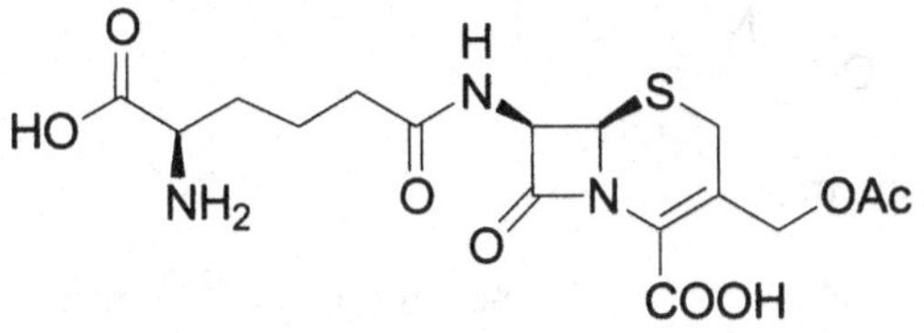

cephalosporin C

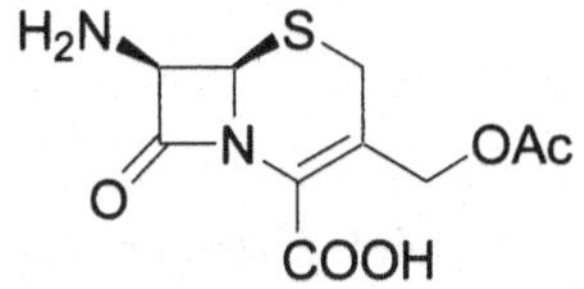

7-aminocephalosporanic acid
(7-ACA)

The antibacterial activity and stability towards β-lactamases of cephalosporins have been gradually improved through chemical modifications.[54] They are divided into four generations based on their antibacterial activity profile and spectrum coverage.[59,60] The first and second generations (cephalothin, cephalexin, cefaclor, cefoxitin, cefuroxime, etc.) are active against Gram-positive bacteria, including methicillin-susceptible *staphylococci* and *streptococci*, with very moderate activity against Gram-negative bacteria that do not produce β-lactamases.

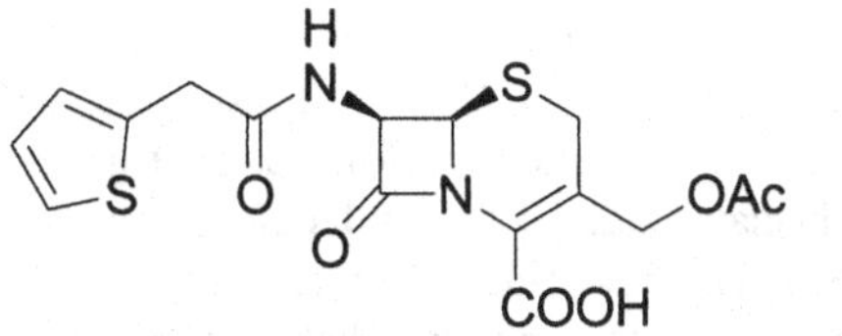

cephalothin

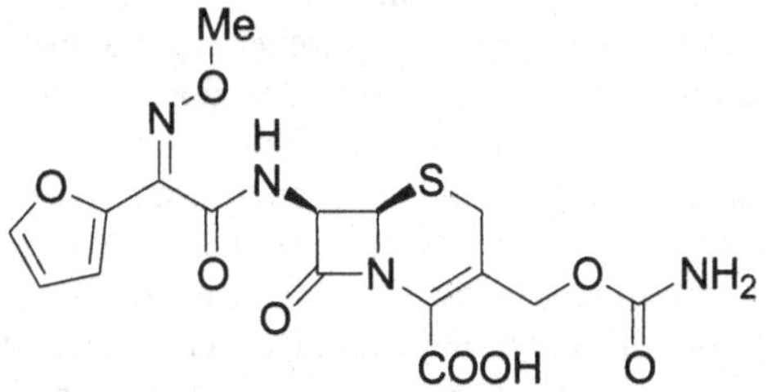

cefuroxime

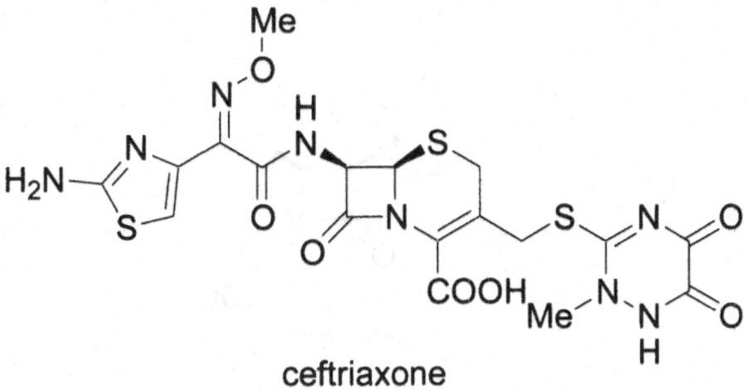

ceftriaxone

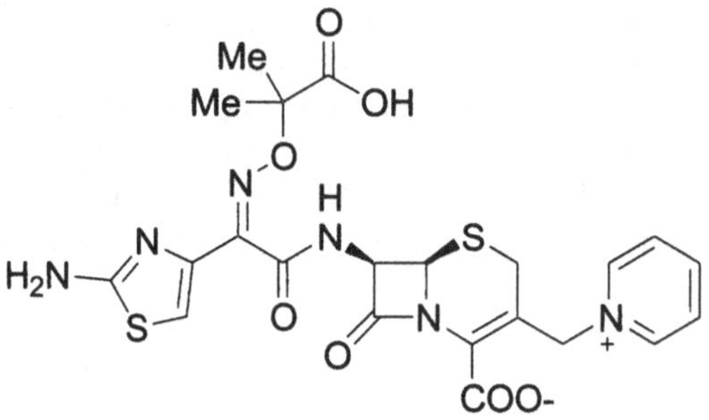

ceftazidime

The third-generation cephalosporins, such as cefixime[61] and ceftriaxone,[62,63] have a broad spectrum of activity. Compared to the first and second generations, they show decreased activity against Gram-positive bacteria, in particular members of the third generation that are orally available, and those with antipseudomonal activity. One common structural feature of many compounds in the third (and later) generations of cephalosporins is the aminothiazole oxime moiety to the left of the amide position. This moiety improves the antibacterial activity and reduces the susceptibility against β-lactamases. However, increasing levels of extended-spectrum β-lactamases (ESBLs) and other β-lactamases in many pathogens are reducing the clinical utility of this class of antibiotics as well. It is also worth noting that some third-generation cephalosporins also have anti-pseudomonal activity in addition to their broad-spectrum coverage, such as ceftazidime[64] and cefoperazone.[65] Most of the cephalosporins and other β-lactam antibiotic have a relatively short half-life (0.5–2 h in human) and require TID (three times daily) or QID (four times daily) dosing. However, ceftriaxone[62] is an exception with a notably long half-life (~6 h in human), which allowed it to be dosed once daily.[63]

Fourth-generation cephalosporins, represented by cefepime,[66] cover an extended Gram-negative spectrum, including multidrug-resistant *Enterobacter* and *Klebsiella* species, and *P. aeruginosa*, while maintaining similar activity against Gram-positive organisms as first-generation cephalosporins. The zwitterionic structure on the right-hand side of the molecule helps the drug penetrate the outer membrane of Gram-negative bacteria, resulting in enhanced antibacterial activity.[67] They also have a greater resistance to β-lactamases than third-generation cephalosporins.[67]

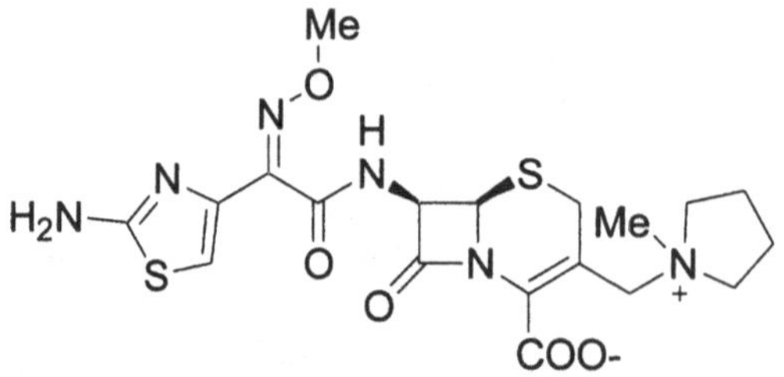

cefepime

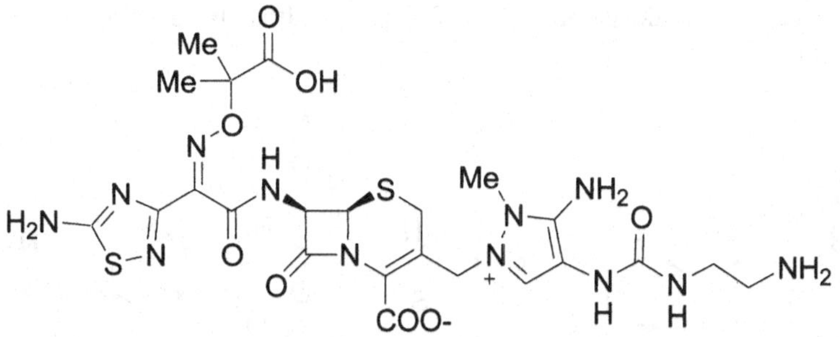

CXA-101

Ceftolozane (previously CXA-101, FR264205), a novel extended-spectrum cephalosporin in late stages of clinical development, shows excellent *in vitro* and *in vivo* activities against *P. aeruginosa*, including multidrug-resistant clinical isolates.[68] Its spectrum coverage against Gram-positive and Gram-negative bacteria is similar to third- or fourth-generation cephalosporins, such as ceftazidime, but its antipseudomonal activity is the most potent among all currently available cephalosporins and other β-lactam antibiotics, including carbapenems.[68] Compared to the structure of ceftazidime, CXA-101 incorporates an aminothiadiazole ring on the left side to replace an aminothiazole and a complex heterocycle with a urea-linked amine tail instead of a simple pyridine ring. Both structural modifications result in enhanced intrinsic potency against *P. aeruginosa*, improved stability against certain β-lactamases (such as AmpC),[69] better cell membrane penetration, and less efflux,[69] all of which attribute to the overall superior activity of CXA-101 against drug-resistant *P. aeruginosa*. Combination of CXA-101 with tazobactam (CXA-201) further improves its spectrum coverage against other β-lactamase-producing Gram-negative pathogens.[70,71] CXA-201 is under clinical development by Cubist for the first-line treatment of certain serious Gram-negative bacterial infections, including those caused by multidrug-resistant *P. aeruginosa*.

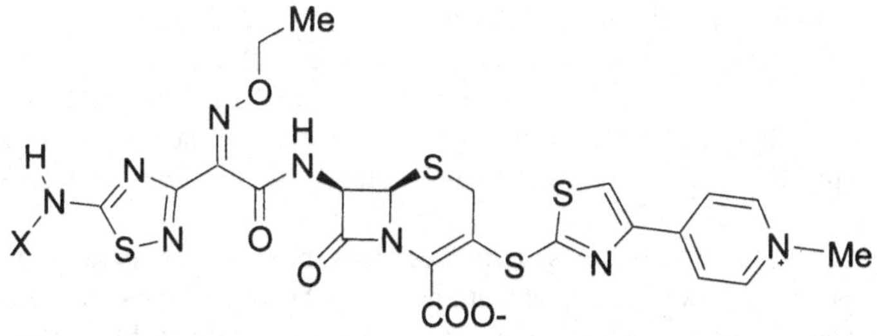

ceftaroline, X = H
ceftaroline fosamil, X = P(=O)(OH)₂

Cephalosporins generally lack activity against MRSA, but recently, anti-MRSA cephalosporins[72,73] have been developed. Ceftaroline[74] is the first anti-MRSA cephalosporin approved in the U.S. in 2010. It displays excellent activity against MRSA and other Gram-positive bacteria. It retains the activity of later generation cephalosporins coupled with broad-spectrum activity against Gram-negative bacteria. Ceftaroline fosamil is an *N*-phosphate prodrug, to improve the poor aqueous solubility of the drug for parenteral uses. Another anti-MRSA cephalosporin under clinical development is ceftobiprole.[75] It is also developed in a prodrug form, ceftobiprole medocaril.

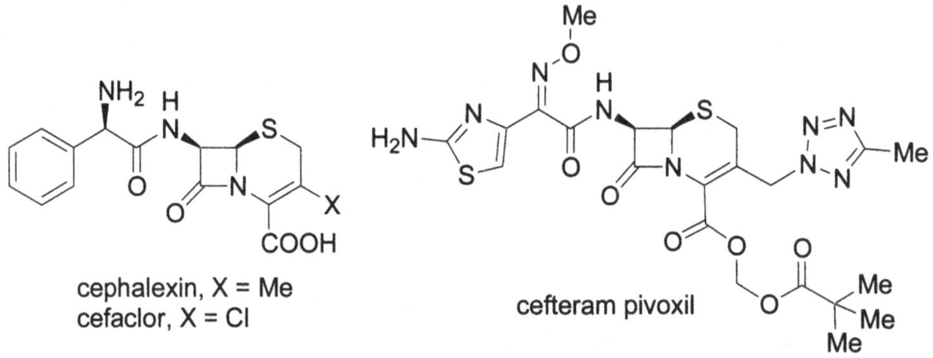

cephalexin, X = Me
cefaclor, X = Cl

cefteram pivoxil

Most of the cephalosporins have no or very low oral bioavailability and have to be administered parenterally. Only a few early-generation cephalosporins are orally available.[76] For example, cephalexin[77] and cefaclor[78] are small and lipophilic enough to achieve excellent oral bioavailability. To achieve oral bioavailability of large and polar cephalosporins, ester prodrugs have been developed,[76] such as cefuroxime axetil[79] and cefteram pivoxil.[80]

Carbapenems

The discovery of thienamycin in the late 1970s from *Streptomyces cattleya*[81] led to the development of carbapenem antibiotics.[82,83] Carbapenems are generally extremely potent against almost all clinically important aerobic or anaerobic Gram-positive and Gram-negative bacteria.[82] They usually demonstrate superior activity against Gram-negative bacteria, including *P. aeruginosa,* compared to other β-lactams. They are not susceptible to most β-lactamases, such as ESBLs and AmpCs, due to the presence of the *trans*-hydroxyethyl group at the C6 position of the β-lactam ring, which prevents hydrolysis from occurring.[84] Recently, carbapenemases such as KPCs and metallo β-lactamases have emerged. Although the incidence rate is relatively low in clinical settings, they do pose a threat to the clinical effectiveness of carbapenems.[85] Thus, carbapenems are typically used as a second line of defense by physicians for serious infections to reduce resistance generation.

    Imipenem,[86,87] launched in the U.S. in 1985, was the first carbapenem to enter the market. Thienamycin was unable to be developed commercially because of its instability in both solid state and concentrated solution.[88] This instability problem was solved by converting the primary amine group in thienamycin into an amidine group to give imipenem. Such conversion not only improved the chemical stability but also enhanced the compound's activity. Imipenem is susceptible to hydrolysis by the renal enzyme dehydropeptidase-I (DHP-I).[89] One solution to this problem was coadministration with a dipeptidase inhibitor, which led to the successful commercialization of primaxin, a 1:1 mixture of imipenem and cilastatin.[88,90]

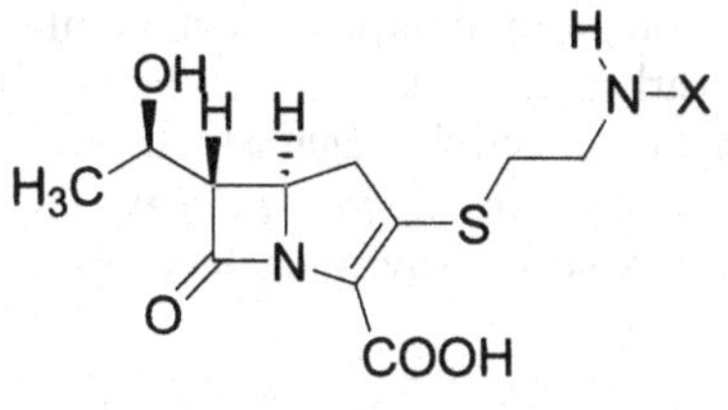

thienamycin, X = H
imipenem, X = CH=NH

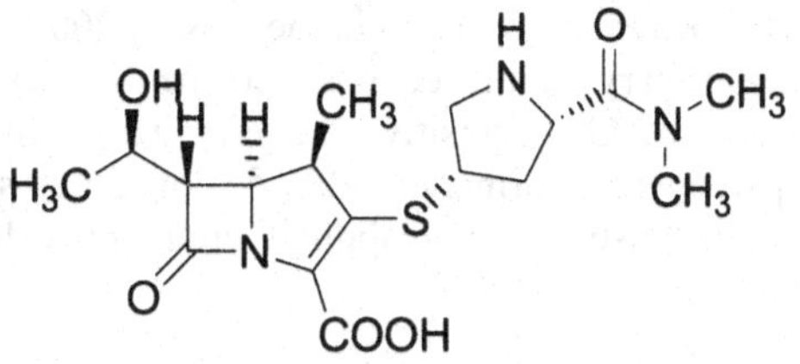

meropenem

Meropenem,[91] approved in 1996, is the second marketed carbapenem in the U.S. Introduction of a methyl group to the core ring system prevented the hydrolysis by the DHP enzyme,[92] while maintaining the antibacterial activity. The incorporation of the hydrophobic pyrrolidinyl side chain also enhanced meropenem's potency against *P. aeruginosa* and other Gram-negative pathogens, but the activity against Gram-positive pathogens slightly decreased compared to imipenem.[93]

Ertapenem,[94] approved in 2001, is the third marketed carbapenem in the US. The unique *meta*-substituted benzoic acid side chain produces a net negative charge at physiological pH that increases its binding to plasma proteins to ~90% and increases the drug's half-life in human (~ 4 h), compared to imipenem and meropenem. Ertapenem is dosed once daily (QD), while imipenem and meropenem are both dosed three times daily (TID). However, the anionic characteristics of the benzoic acid moiety combined with larger molecular weight and increased lipophilicity resulted in impaired penetration of ertapenem through the outer membrane porin and caused decreased activity against Gram-negative pathogens, such as *P. aeruginosa* and *Acinetobacter baumannii*.

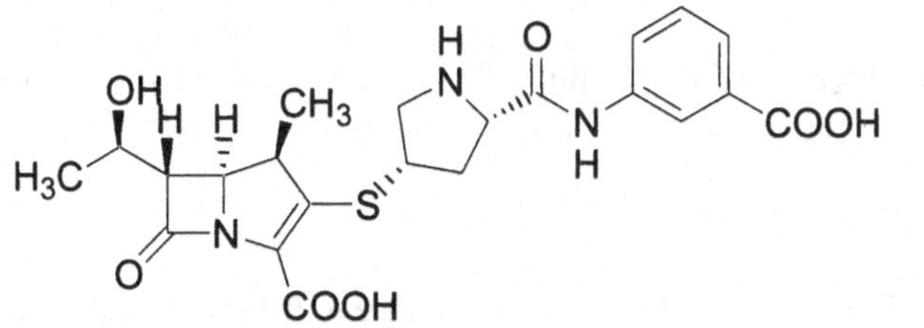

ertapenem

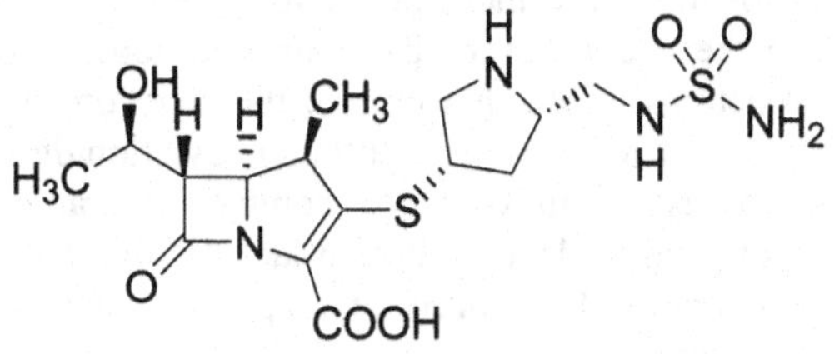

doripenem

Doripenem,[95] approved in the U.S. in 2007 for complicated intra-abdominal infection and complicated urinary tract infections, became the fourth carbapenem marketed in the U.S. It covers a wide range of Gram-positive bacteria while maintaining activity against numerous Gram-negative pathogens. Doripenem also demonstrates superior *P. aeruginosa* activity. Presumably, the sulfamide-substituted pyrrolidinyl group provides enhanced activity against Gram-negative pathogens.

Monobactams

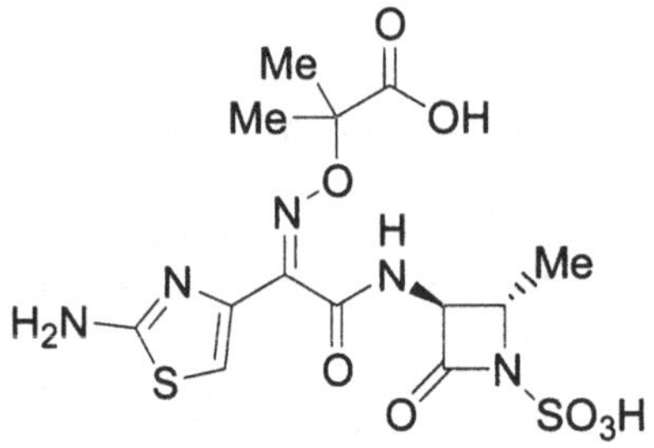

aztreonam 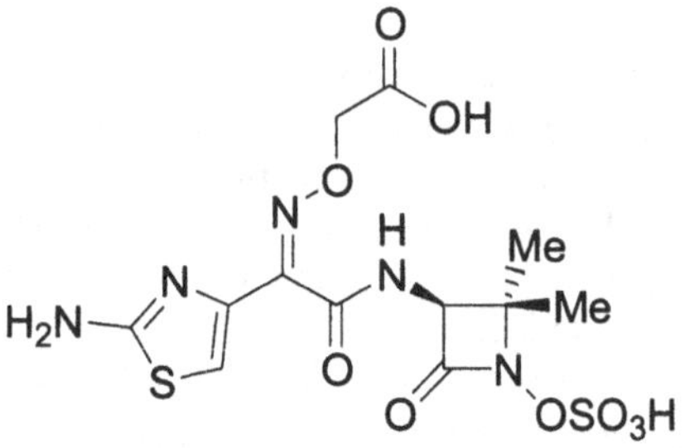   tigemonam

Monobactams, advanced in the late 1970s,[96,97] are compounds with a β-lactam ring that is not fused to a second ring (in contrast to most other β-lactams, which have a fused bicyclic β-lactam ring system).[98] Naturally occurring monobactams exhibit only weak antibacterial activity. Subsequent medicinal chemistry modifications realized their antibacterial potential and led to the development of aztreonam,[99] which is the only commercially available monobactam antibiotic. Aztreonam, approved in 1986, has a limited Gram-negative spectrum. Tigemonam[100] is another monobactam which underwent clinical development. Its antimicrobial spectrum[101] is similar to aztreonam, and it has high oral bioavailability.[102] It should be noted that monobactams are the only β-lactam antibiotics which are stable to Class B metallo β-lactamases.[103]

β-Lactamase Inhibitors

The most widespread mechanism of bacterial resistance to β-lactam antibiotics is the production of β-lactamases by bacteria.[37] β-Lactamases are enzymes that hydrolyze the β-lactam ring and they are most effective against penicillins and cephalosporins. While only a handful of β-lactamases were known in the early 1970s, there are more than 890 different β-lactamases known today.[104] They pose a major threat to the usefulness of β-lactam antibiotics. There are two types of β-lactamases based on their functions: those that require the zinc ion for their function (Class B, metallo β-lactamase) and those that require serine through acylation/deacylation (Class A, C, and D, serine-dependant β-lactamases).[105] In order to combat the rapidly evolving β-lactamase resistance, scientists have continued to develop new β-lactams that are not easily hydrolyzed by these enzymes. In addition, inhibitors have been developed to inactivate β-lactamases and thus preserve the activity of the β-lactam.

β-Lactamase inhibitors[106,107,9] on the market include clavulanic acid, sulbactam, and tazobactam, which are all derived from the β-lactam family. These inhibitors have good coverage against most of the class A enzymes, including ESBLs, TEMs, SHVs, and CTX-Ms. However, they are not active against KPCs, Class B metallo β-lactamases, and Class C enzymes (such as AmpC), and only have marginal activity for Class D enzymes.

Combinations of β-lactams and β-lactamase inhibitors (amoxicillin/clavulanic acid, ampicillin/sulbactam, and piperacillin/tazobactam) are used extensively to make the drugs more effective against organisms that express β-lactamases.

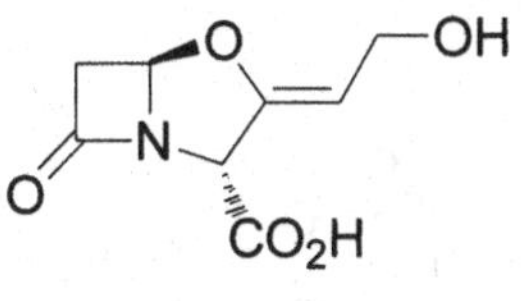

clavulanic acid 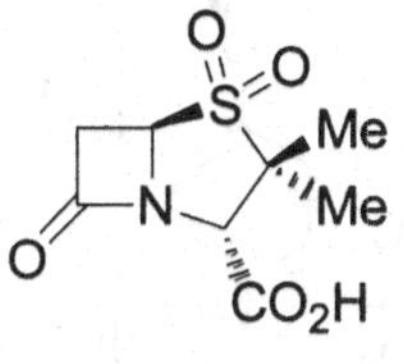   sulbactam 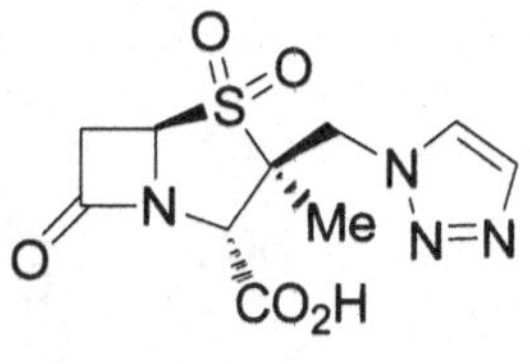   tazobactam

Most of the β-lactamase inhibitors are poor PBP binders and therefore are usually devoid of any antibacterial activity when used alone.

A new generation of β-lactamase inhibitors under clinical development is NXL-104, which is a non-β-lactam β-lactamase inhibitor. It has broader β-lactamase coverage to include KPCs and class C enzymes.[108,109] Currently it is under clinical development in combination with ceftazidime and ceftaroline.

There are no known effective inhibitors on the market for Class B enzymes (metallo β-lactamases).[103]

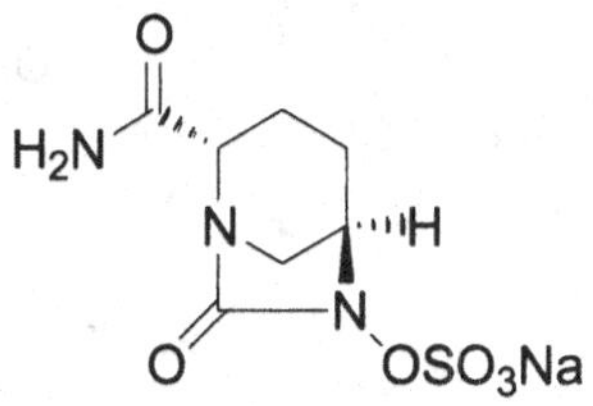

NXL-104

Glycopeptides

Glycopeptide antibiotics[110] inhibit cell wall peptidoglycan synthesis. They bind to the acyl-D-alanyl-D-alanine unit in peptidoglycan preventing the addition of new units to the peptidoglycan.[111] These antibiotics are only effective against Gram-positive bacteria, because of the permeability barrier of the intact outer membrane of Gram-negative bacteria. They are used to treat serious Gram-positive bacterial infections resistant to other antibiotics, e.g., β-lactams.

Vancomycin[112] was the first glycopeptide antibiotic used clinically. It is a natural product isolated from *Amycolatopsis orientalis* in 1953.[113] Because of the rapid development of penicillin resistance by *Staphylococci* at that time, it was quickly approved by the FDA in 1958 with the initial indication to treat penicillin-resistant *S. aureus* infections. Due to its poor oral bioavailability, it is given intravenously for most Gram-positive bacterial infections. Oral vancomycin was approved by the FDA in 1986 for the treatment of *Clostridium difficile* associated diarrhea.[114] It is not orally absorbed into the blood stream and remains in the gastrointestinal tract to eradicate *C. difficile*.

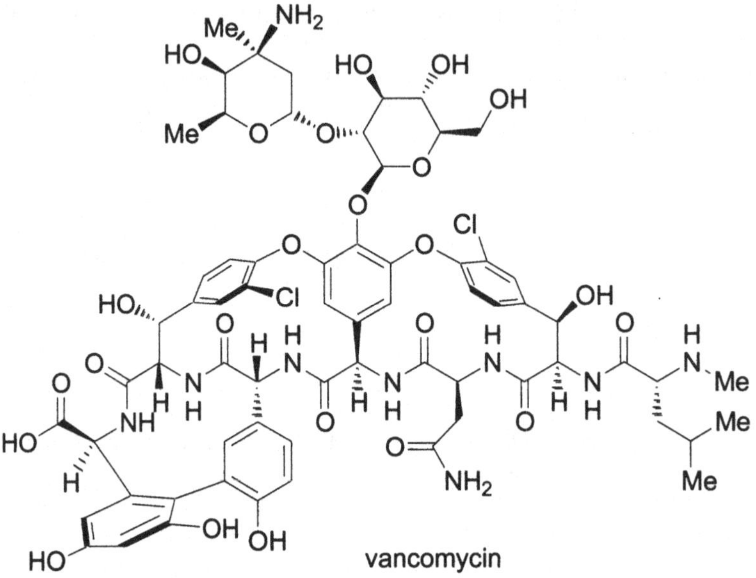

vancomycin

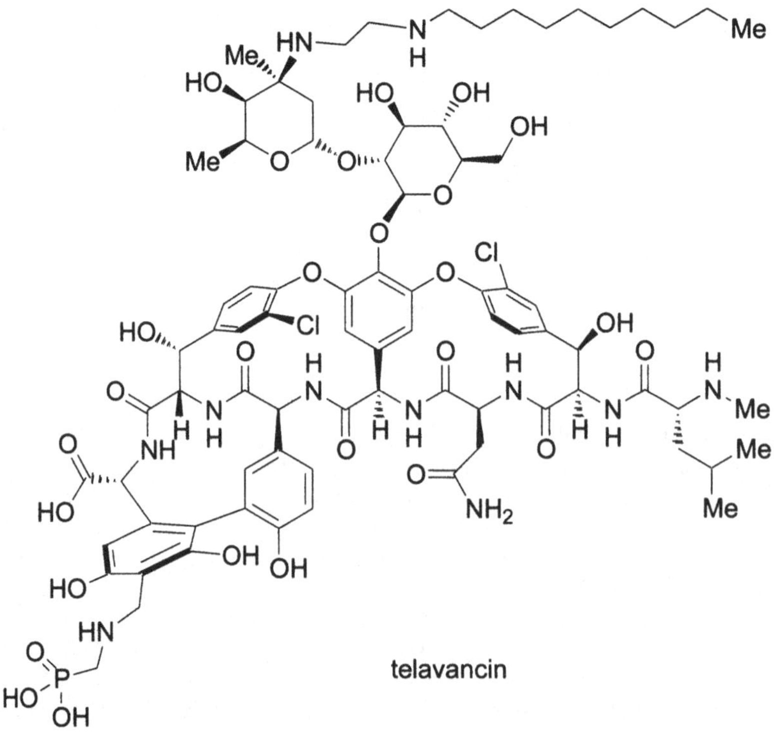

telavancin

The frequency of resistance to the glycopeptide antibiotics has increased significantly over the past decades. Vancomycin-resistant *Enterococcus* (VRE) emerged in 1987. Resistance in more common pathogens developed since 1990, including vancomycin-intermediate *S. aureus* (VISA) and vancomycin-resistant *S. aureus* (VRSA).[115,116] The mechanism of resistance to vancomycin in VRE and VRSA strains involves the alteration to the D-alanyl-D-alanine terminal amino acid residues to which vancomycin binds.[110]

Other marketed glycopeptides include the natural product teicoplanin[117] and, more recently, the semisynthetic-derived telavancin,[118] approved in 2009 for treating complicated skin and skin structure infections (cSSSI.) Telavancin has a unique dual mechanism of action: It inhibits the bacterial cell wall synthesis, like other glycopeptides, but also disrupts bacterial membranes by depolarization.[119]

4.2   *Ribosome Inhibitors*

The ribosome is one of nature's largest, oldest, and most complex biomolecular machines. The ribosome interacts with many different transfer RNAs (tRNA) and proteins in order to accomplish the impressive task of translating the genetic information encoded by messenger RNA (mRNA) to assemble amino acids into proteins.[120] Bacterial ribosomes are made up of two subunits, the small 30S and the large 50S ribosomal subunits, and together make up the 70S ribosome.[121,122] Both subunits consist of ribosomal RNA (rRNA) and many proteins (r-proteins). In a typical bacterium, the smaller 30S subunit is made up of a 16S rRNA and 21 r-proteins while the larger 50S subunit consists of two rRNA molecules, 5S and 23S rRNAs, and 34 r-proteins. Each subunit has specific functions during the synthesis of new proteins. The small 30S subunit interacts with and decodes the sequence of the mRNA, while the large 50S subunit houses the peptidyl transferase site where peptide bonds are formed between the incoming amino acids and the growing peptide chain. With the successive addition of each amino acid, the peptide chain extends from the peptidyl transferase center through the tunnel and eventually emerges at the back of the 50S subunit.

Protein synthesis is an essential process in all living cells and therefore the ribosome is an important target of many different types of antibiotics.[123–125] Most of the antibiotics that interact with the ribosome render the bacterium incapable of synthesizing new proteins and thus quickly inhibit bacterial growth. The 30S subunit is targeted by drugs that include tetracyclines, glycylcyclines, and the aminoglycosides, which hinder the subunit in carrying out its principal function of deciphering the genetic information encoded in the mRNA. A wide range of drugs binds to the 50S subunit to interfere with its main functions in controlling guanosine triphosphate (GTP) hydrolysis, the formation of peptide bonds, and channeling the peptide through the subunit tunnel. A common binding site in the 50S subunit for multiple drugs is located at the upper part of the tunnel together with the peptidyl-transferase center. This region accommodates the macrolides, lincosamides, streptogramin B (together as the MLS$_B$ compounds), and oxazolidinones.

4.2.1   *Aminoglycosides*

In contrast to the serendipitous discovery of penicillin by Alexander Fleming, the isolation of streptomycin was the result of the systematic and assiduous research by Selman A. Waksman and his graduate student Albert Schatz at Rutgers University, which later earned Professor Waksman a Nobel Prize.[126] Streptomycin is the first of the class of aminoglycoside antibiotics and was discovered in 1944 by systematic screening of natural products. Streptomycin was the first antibiotic effective against *Mycobaterium tuberculosis*; however its use against tuberculosis has steadily declined due to increased resistance.[127,128]

streptomycin

In the following two decades, many other aminoglycosides were isolated from soil bacteria, including neomycin, isolated in 1949, which exhibits poor oral bioavailability coupled with high toxicity.[129,130] Kanamycin (Kantrex) was discovered in 1957 and was effective against meningitis and pneumonia, but its use has declined due to widespread resistance.[130,131] Gentamicin (Garamycin) was isolated from the actinomycete *Micromonospora purpurea* (the "micin" spelling reflects the different genus of origin) in 1963 by Marvin Weinstein's group at Schering-Plough and was a major breakthrough in the treatment of Gram-negative infections, including *P. aeruginosa*.[130,132,133] Gentamicin continues to be one of the more commonly used aminoglycosides because of its low cost and reliable activity against Gram-negative aerobes.

gentamicin (Garamycin)   tobramycin (Nebsin)

Tobramycin (Nebsin) was isolated in 1967 from *Streptomyces tenebrarius* and it was put into clinical use in the 1970s.[134,135] Similar to gentamicin, tobramycin exhibits activity against most Gram-negative bacteria with tobramycin being generally more active than gentamicin against *P. aeruginosa*, which makes tobramycin the preferred choice for the treatment of *Pseudomonas* species infections.

As the number of aminoglycosides isolated from natural products began to dwindle along with pressure of the increasing resistance problem, scientists began a new era of the Golden Age of antibiotic medicinal chemistry, during which scientists modified natural antibiotic scaffolds to improve the potency and spectrum to include resistant strains. Amikacin (Amikin) is a semisynthetic derivative of kanamycin and was introduced in 1972.[136] Due to its different chemical structure, amikacin is particularly effective when used against aminoglycoside-resistant pathogens because it is less susceptible to inactivating enzymes.[137,138] For serious nosocomial infections caused by Gram-negative bacilli, amikacin may be the preferred agent of choice. Other semisynthetic derivatives include sisomicin (1970), debekacin (1971), netilmicin (1975), and arbekecin (1990).[130]

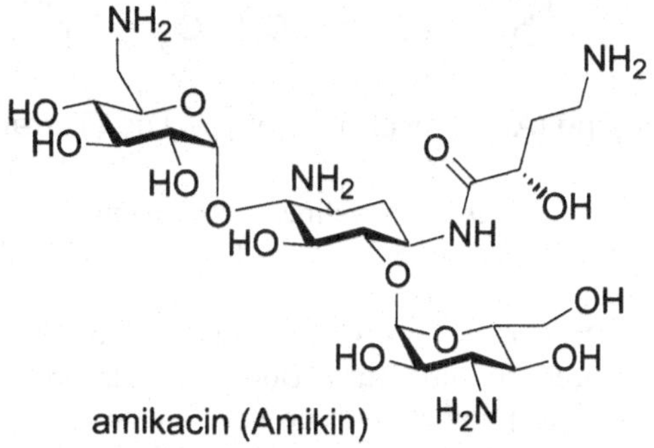

amikacin (Amikin)

Aminoglycosides are highly potent, broad-spectrum antibiotics, but their activity is most notable against a variety of Gram-negative pathogens, including *Acinetobacter* spp., *Citrobacter* spp., *Enterobacter* spp., *E. coli*, *Klebsiella* spp., *Serratia* spp., *Proteus* spp., *Morganella* spp., and *P. aeruginosa*.[139–141] For Gram-positive infections, aminoglycosides are generally used in combination with other active agents to provide synergy.[140–143] When used as a single agent, they have limited activity against *Staphylococci*, *Enterococci*, and *Viridans streptococci*.[142,144,145]

All of the aminoglycosides have essentially the same mode of action in that they selectively bind to the acceptor site (A-site) decoding region of the bacterial 16S rRNA within the 30S ribosomal subunit causing mistranslation of mRNA or premature termination of protein synthesis.[146–150] In addition, aminogylcosides damage the outer membrane of Gram-negative bacteria, either through direct interactions, or through ribosome inhibition, causing aberrant proteins to produce membranes with impaired semipermeability that increase the passage of small molecules, such as the antibiotic, into the cell. The protein's mistranslation along with the increasingly high intracellular drug concentration leads to bacterial cell death. Aminoglycosides are generally bactericidal. Resistance to aminoglycosides can be divided into three mechanisms: (a) decrease in intracellular drug concentration due to membrane permeability/ efflux mechanisms, (b) target site modification by mutational modification of 16S rRNA and/or r-proteins, and (c) enzymatic drug modification.[151–154]

Despite their apparent advantages of being broad-spectrum bactericidal antibiotics, aminoglycosides exhibit undesirable toxicities. Nephrotoxicity and ototoxicity are the most prevalent among this class of antimicrobials, while neuromuscular blockade is less common.[155,156] The exact mechanism of these toxicities is not fully understood, although it is known that the cationic form of the aminoglycosides accumulates in renal cortical cells and induces loss of intracellular ions and proteins, which results in damage to proximal tubules and leads to toxicity.[157] Nephrotoxicity is usually reversible by terminating treatment and it is dose dependent. Ototoxicity is usually irreversible and may involve vestibular and/or cochlear sensory cell damage.[158]

4.2.2   Tetracyclines and Glycylcyclines

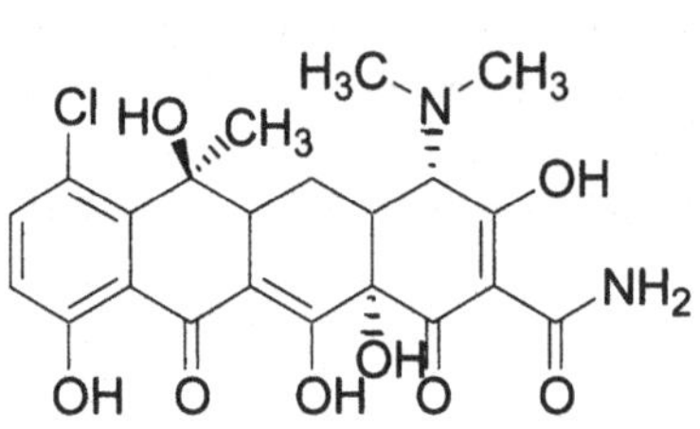

chlortetracycline (Aureomycin) 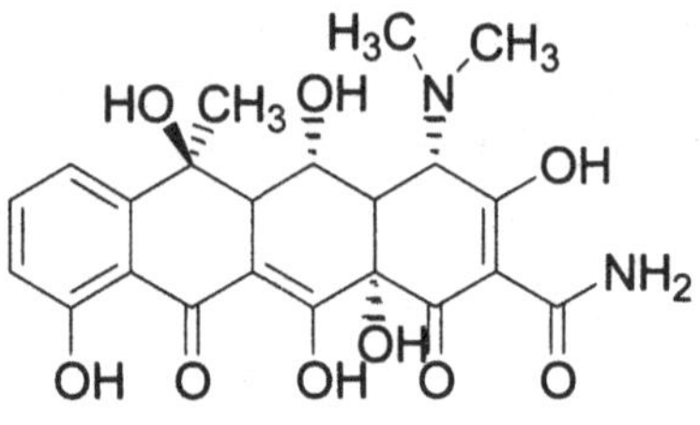   oxycycline (Terramycin)

Chlortetracycline (Aureomycin) and oxytetracycline (Terramycin) were the first members of the tetracycline class of antibiotics to be discovered in the late 1940s during the golden age of antibiotic discovery as a result of natural product screening programs.[159,160] These compounds were found to be highly effective against Gram-positive and Gram-negative bacteria, thus becoming a class of broad-spectrum antibiotics. Both chlortetracyline and oxytetracycline are relatively easy and inexpensive to produce with no severe side effects and have favorable oral bioavailability and pharmacokinetic parameters which propelled both onto the market as effective broad-spectrum antibiotics.[161] The tetracycline antimicrobial family continued to grow as more members of this group of antibiotics were isolated as fermentation products produced by soil-dwelling organisms. It was also recognized by Stokstad et al. that the fermented products from the chlortetracycline producing soil organism are growth-promoting agents in the agricultural industry.[162] Naturally, with the extensive use of tetracyclines in both clinics and agriculture, resistant isolates rapidly emerged.[163,164]

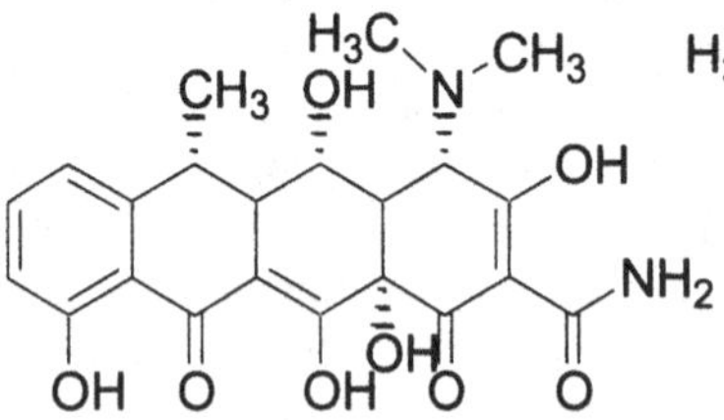

doxycycline (Vibramycin) 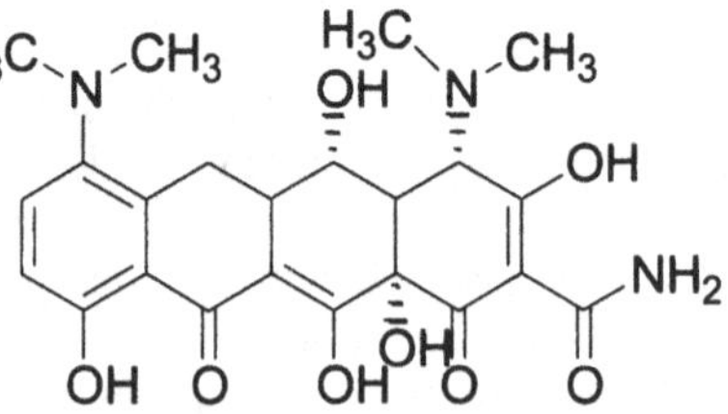   minocycline (Minocine)

To battle the rapid emergence of resistant pathogens, new tetracyclines were explored via the chemical modification approach since the traditional natural product screening efforts proved to be difficult.[165] Like the aminoglycosides, focus on the tetracyclines turned to semisynthetic efforts and is represented by the second-generation tetracyclines. Doxycycline (Vibramycin) was discovered and clinically developed by Pfizer and approved for clinical use in 1967.[160] Minocycline was synthesized by Lederle Laboratories 1967 and was clinically used under the brand name Minocycline in 1972.[160] Both of these compounds are more lipophilic than their parent compounds, and as a result, they have better absorption and are improved in some other pharmacokinetic parameters.[161]

The tetracyclines exert their antibacterial effect by binding to the small 30S subunit of the ribosome and allosterically inhibit binding of the amino acyl-tRNA at the A-site and halting protein synthesis.[166,167] Tetracyclines are generally bacteriostatic with a few exceptions that are bactericidal.[168] The use of tetracyclines has declined in recent decades due to the emergence of resistant strains of bacteria. The two more common mechanisms of resistance are mediated by increased drug efflux out of the cell by a family of Tet proteins located on the cytoplasmic surface of the cell membrane and by disruption of the tetracycline-ribosomal interaction by ribosomal protection proteins.[160,169–171] Other mechanisms of resistance include enzymatic inactivation of the drug through monohydroxylation and alteration of the target site through 16S RNA mutation.[172,173]

Research to find tetracycline analogs that circumvent the resistance mechanisms has led to the development of the glycylcyclines, which represent the third and latest generation of tetracyclines. Tigecycline (Tigacil), developed by Francis Tally, is the first clinically approved member of the glycylcyclines that gained FDA approval in 2005.[174,175] It was initially marketed by Wyeth and now by Pfizer. Tigecycline contains an N-alkyl-glycylamido group at the C9 position of the minocycline scaffold.[176] This modification results in a more efficient interaction with the ribosome (improved binding of up to five-fold compared to minocycline) and avoids the classic tetracycline-resistance mechanisms, efflux and ribosomal mutation.[177–179] Tigecycline is active against many Gram-positive and Gram-negative bacteria and anaerobes, including MRSA and A. baumannii, but it lacks potency against Pseudomonas spp.[174] Like most tetracyclines, tigecycline is bacteriostatic with the same mechanism of action against the 30S ribosomal subunit. It is in clinical use for the treatment of complicated skin and skin structure infections (cSSSIs) and complicated intra-abdominal infections.[180]

tigecycline (Tygacil)

4.2.3 Macrolides and Ketolides

In 1957, Robert B. Woodward proposed the term "macrolide" to represent macrolactone glycoside antibiotics.[181,182] Macrolides, produced largely from the actinomycete species, are characterized by a 12- to16-membered lactone ring that is attached to at least one sugar moiety. Pikromycin was the first macrolide isolated in 1950 during the golden years of natural product antibiotics.[183] In 1952, erythromycin, a 14-membered macrolactone with two sugar moieties, was discovered and quickly introduced to clinical use due to its safety and spectrum of activity.[184] Erythromycin is active against Gram-positive pathogens such as S. aureus, S. pneumoniae, and S. pyogenes and is used mainly to treat respiratory tract infections, skin and soft-tissue infections, and urogenital tract infections.[185–188] It has modest activity against Gram-negative pathogens such as E. coli and Klebsiella, and it has no activity against Pseudomonas strains.[188] Eyrthromycin suffers from short half-life, poor oral bioavailability, and gastrointestinal side effects caused by its acid instability.[189] Following the discovery of erythromycin, several other macrolides were isolated, including oleandomycin, josamycin, and tylosin, but erythromycin remained the most widely used antibiotic in this class.[190]

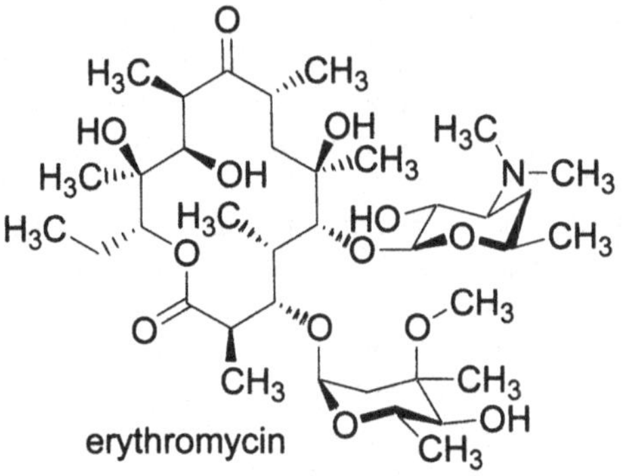

erythromycin

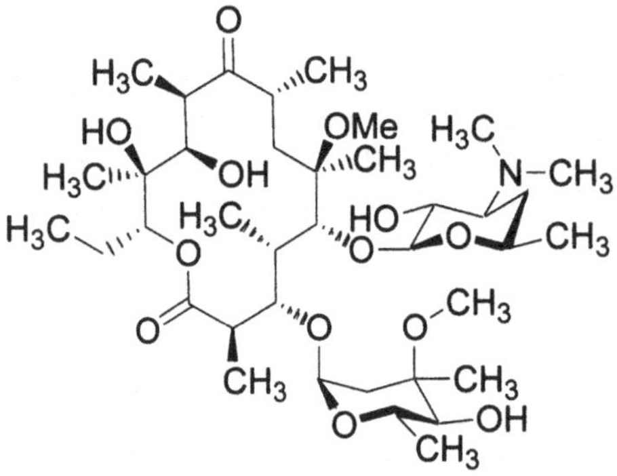

clarithromycin (Biaxin)

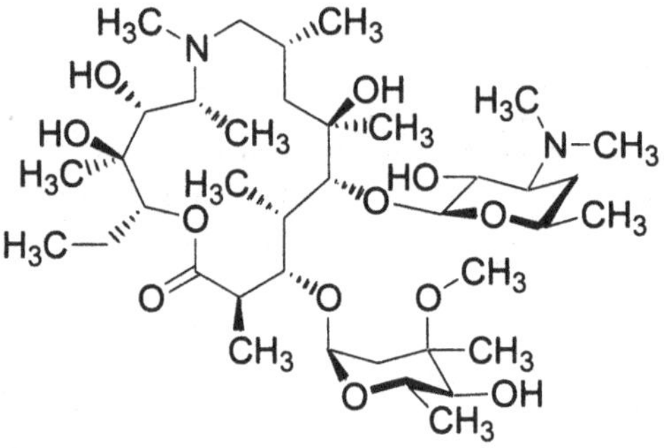

azithromycin (Zithromax)

Prompted by the need to improve the pharmacokinetic properties, enhance the activity, and expand the bacterial spectra of erythromycin, a second generation of macrolides was developed via a semisynthetic approach in the 1970s and 1980s. Five semisynthetic derivatives of erthromycin were developed and commercialized to include clarithromycin (Biaxin), dirithromycin (Dynebac), roxithromycin (Rulide), flurithromycin, and azithromycin (Zithromax).[191–195] Among the five, clarithromycin and azithromycin are marketed worldwide whereas the rest have much more limited distribution. The conversion of the C6 hydroxyl group of erythromycin to a methoxy group produces clarithromycin in a six-step synthetic sequence which significantly increases the cost of the drug relative to erythromycin.[196] Clarithromycin has excellent tissue distribution, especially in lungs.[197] Azithromycin is derived from the insertion of an amino group into erythromycin, resulting in a 15-membered ring. Azithromycin has superior antibacterial activity against *H. influenzae* compared to both erythromycin and clarithromycin.[197] In addition, azithromycin has minimal risk against drug-drug interactions and QTc prolongation.[197,198] Both azithromycin and clarithromycin have improved oral bioavailability and an extended half-life, enabling them to be taken orally once daily or twice daily, respectively.[190]

While the second generation of macrolides was driven largely by the need to improve pharmacokinetic properties, third-generation macrolide efforts began in the 1990s and focused on the development of new macrolides with activity against multidrug-resistant pathogens because resistance had arisen suddenly and rapidly in the 1980s and 1990s. The research for a third-generation macrolide led to the discovery of a new semi-synthetic series, the ketolide. Ketolides have a ketone functionality in place of the L-clandinose group at C3 of clarithromycin, and they also have an aromatic or heteroaromatic ring attached by a tether to the macrolactone ring.[199,200] The aromatic or heteroaromatic ring is hypothesized to be important because it enables additional interactions with the bacterial ribosome, thus providing improved potency and overcomes one of the mechanisms of resistance in certain pathogens.[190]

The ketolide telithromycin (Ketek) is the only third-generation macrolide in clinical use as of 2004.[190] Telithromycin possess the C3 ketone as well as a C11/C12 cyclic carbamate with an alkyl-heteroaromatic side chain at the N11a position, and this transformation is prepared from clarithromycin using an eight-step chemical sequence.[201] Telithromycin has excellent potency against both susceptible and resistant S. pneumoniae, including macrolide and penicillin-resistant strains.[202] It is also active against penicillin-susceptible S. aureus, fastidious Gram-negative bacteria such as H. influenzae and M. catarrhalis, and atypical and intracellular pathogens such as M. pneumoniae, C. pneumoniae, and L. pneumophilia.[203]

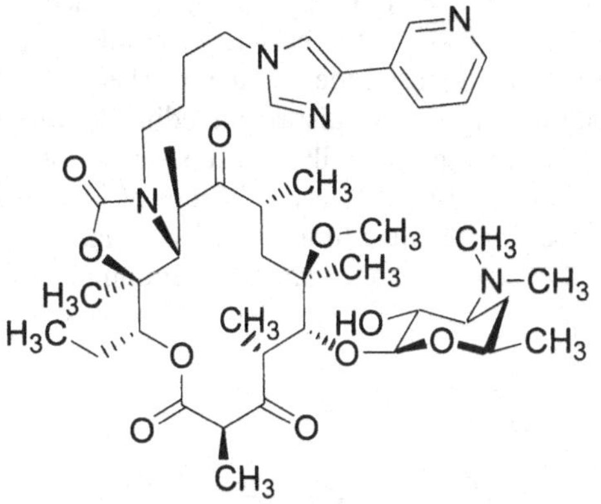

telithromycin (Ketek) 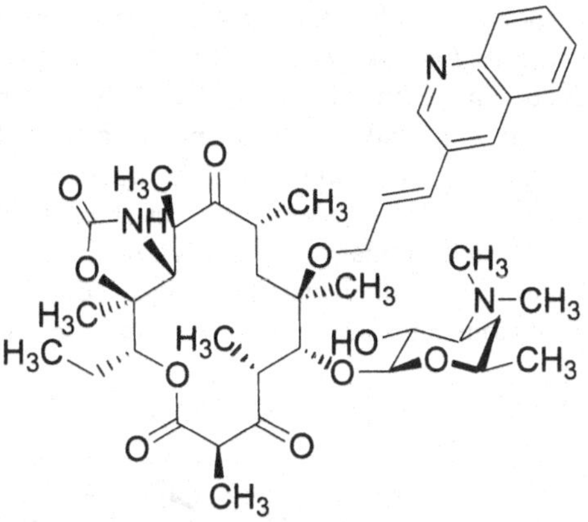   cethromycin (ABT-773)

Cethromycin (ABT-773) is a second ketolide that was initially discovered by Abbott and is now under development by Advanced Life Sciences in Phase III clinical trials. Cethromycin, like telithromycin, has the C3-keto group and the C11/C12 cyclic carbamate, but the alkyl-heteroaromatic side chain is attached in an ether linkage to the C6 hydroxyl group on the macrolactone.[193] It was determined that the C6 alkyl side chain orients the heteroaromatic group in such a way that it can interact in the same manner as the carbamate side chain in telethromycin. Cethromycin has similar activities as telithromycin against macrolide-susceptible and macrolide-resistant S. pneumoniae, but unlike telithromycin, it is also active against macrolide-resistant S. pyogenes.[190] In addition to the expanded spectrum, ketolides also have the property of bactericidal activity against S. pneumoniae, S. pyogenes, H. influenzae, and M. catarrhalis.[203]

Macrolides and ketolides reversibly bind to the 50S subunit of the bacterial ribosome and inhibit RNA-dependent protein synthesis by preventing transpeptidation and translocation reactions.[204,205] Both types of compounds bind to domain V of the 23S rRNA,but ketolides have 10- to 100-fold higher affinity to the ribosome than erythromycin.[206] The ketolides also have a higher binding affinity to domain II of the 23S rRNA allowing these compounds to gain antibacterial activity against macrolide-resistant strains.[207]

Macrolide resistance is due to either active drug efflux or alteration of the drug-binding site on the ribosome by methylation, which is common in macrolide, lincosamide, and streptogramin resistance, hence the term MLS$_B$ resistance.[204,208] The efflux mechanism is mediated by the macrolide efflux gene (*mef*). Resistance by methylation (*erm* gene) occurs on an adenine residue in domain V of the 23S rRNA and it prevents binding of the macrolides to domain V and results in high-level macrolide resistance.

Macrolides and ketolides are generally well tolerated, but the most common side effect is gastrointestinal disturbance.[209] Clarithromycin has been shown to induce prolongation of the QT interval and may lead directly to ventricular arrhythmia while telithromycin only induces modest increases in the QT interval. Telithromycin is also associated with blurred vision, exacerbation of myasthenia gravis, and liver toxicity.

4.2.4   Lincosamides

Lincosamides represent a small class of antibiotics that is chemically derived from amino acid and sugar moieties. The first natural lincosamide is lincomycin (Lincocin), produced from *Streptomyces lincolnensis* and clinically approved in 1960.[210] Nine years later, clindamycin was introduced into clinical use as the only semisynthetic lincosamide on the market.[211,212] Both lincomycin and clindamycin are potent against Gram-positive organisms, including *S. aureus, S. epidermidis, S. pneumonia,* and *S. pyogenes*.[213] Clindamycin generally has superior activity over lincomycin. However, both are resistant to Gram-negative bacteria. Currently, the use of lincomycin and clindamycin has drastically declined due to their limited antibacterial spectrum, the emergence of resistance, and severe adverse effects, including granulocytopenia, thrombocytopenia, and pseudomembranous colitis.[214]

lincomycin (Lincocin)   clindamycin (Cleocin)

Lincosamides exert their antimicrobial activity by binding to the 50S ribosomal subunit and there is evidence that it overlaps with erythromycin binding sites.[215] This binding inhibits protein synthesis by affecting the process of peptide chain initiation and stimulates dissociation of peptidyl-tRNA from the ribosome.[213] Like the macrolides, resistance (the MLS$_B$ resistance) occurs mainly by methylation at A2058 on 23S rRNA by certain genes encoding methyltransferases (*erm* genes) and thus reduces the binding affinity of lincosamides to the ribosome target.

4.2.5   Streptogramins

The streptogramins, discovered in the 1950s, are a unique class of antibacterials that are natural products produced by the *Streptomyces* spp.[216,217] Each member of the class is composed of at least two structurally unrelated compounds, group A and group B.[218] Group A streptogramins are cyclic polyunsaturated macrolactones, and group B are cyclic hexa- or heptadepsipeptides. The first generation combination of natural streptogramins is pristinamycin (Pyostacine), which is a 70:30 ratio of pristinamycin IIA and pristinamycin IA.[219] Pristinamycin is particularly active against Gram-positive pathogens such as species of *Staphylococci* and *Streptococci*. However, the use of pristinamycin in the hospital setting has been limited because of the absence of a parenteral form due to its poor solubility in water.[218]

The poor solubility of pristinamycin led to the discovery of a semisynthetic streptogramin combination, dalfopristin and quinupristine (Synercid) that was developed in the 1990s by Rhone-Poulenc and received FDA approval in 1999.[220] Synercid is a 70:30 mixture of dalfopristin and quinupristin and is intravenously administered. The spectrum of activity of the combination of dalfopristin/quinupristin includes most Gram-positive pathogens such as species of *Staphylococci*, *Streptococci*, and *Enterococci*.[221] Furthermore, the combination is also aerobic Gram-negatives including *M. catarrhalis*, *Legionella* spp., *Mycoplasma* spp., *Neisseria* spp., and to a less degree *H. influenzae*.

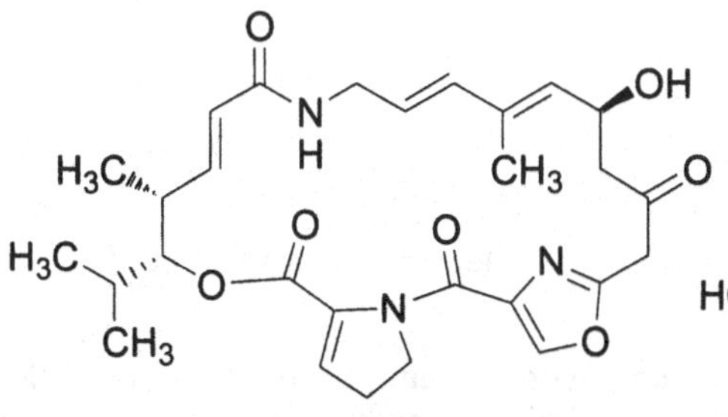

Group A
pristinamycin IIA (Pyostacin)

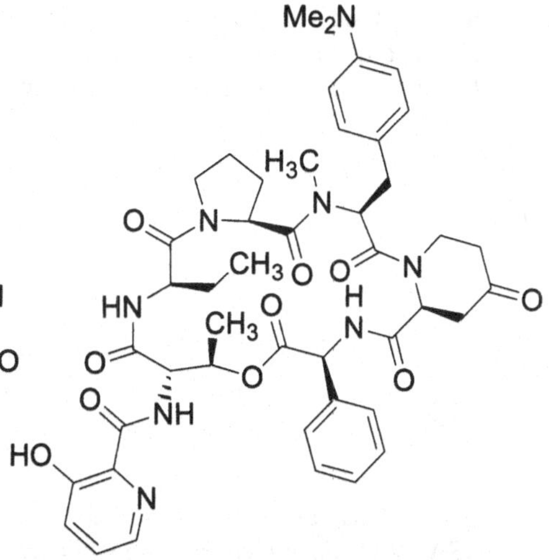

Group B
pristinamycin IA (Pyostacin)

Group A and group B streptogramins act synergistically to inhibit bacterial growth. Depending on the bacterial strain, the combination of dalfopristin/quinupristin is generally about 10-fold more potent than either drug alone.[222] Each of the components of streptogramins is bacteriostatic on its own but the combination is bactericidal against *Staphylococci* and *Streptococci*.[223] The synergism of the combination is hypothesized to be due to the conformational changes that results from the binding of the group A streptogramin on the 50S ribosome, which increases the affinity of group B streptogramins on the ribosome.[221]

Both groups A and B streptogramins bind to the bacterial 50S ribosomal subunit and interfere with the elongation cycle, thereby inhibiting protein synthesis.[224] Specifically, group A streptogramins interfere with the proper positioning of aminoacyl-tRNA and peptidyl-tRNA in the peptidyl transferase center and block peptide bond formation, while group B streptogramins cause the premature dissociation of peptidyl-tRNA from the ribosome. The three major acquired resistance mechanisms for streptogramins are active efflux, target modification (the MLS_B resistance), and enzymatic inactivation of streptogramin components.[225]

4.2.6   Oxazolidinones

Oxazolidinones are a new class of synthetic antibiotics discovered by DuPont in 1978 that possess novel structures compared to existing antibacterial agents.[226] The core pharmacophore of this class is a five-membered oxazolidin-2-one ring with an aromatic moiety and substitutions at the N3 and C5 positions of the heterocycle. The C5 stereochemistry is critical with the S-enantiomer having greater antibacterial activity than the R-enantiomer.[227] In 1987, the pioneering work of scientists at DuPont culminated in the identification of two clinical candidates of the oxazolidinone class, DuP-105 and DuP-721.[228] Unfortunately, development of both compounds was terminated because of toxicity. The oxazolidinone research continued at The Upjohn Company, and in 1996, two nontoxic derivatives of oxazolidinones were identified, of which one is a piperazine derivative, eperezolid, and the second is a morpholine derivative, linezolid (Zyvox).[229] Linezolid entered Phase I clinical trials only six months after eperezolid and it ultimately outperformed eperezolid to become the first oxazolidinone to be FDA approved in 2000. It has been shown that eperezolid requires a larger dose and has bone marrow toxicity.

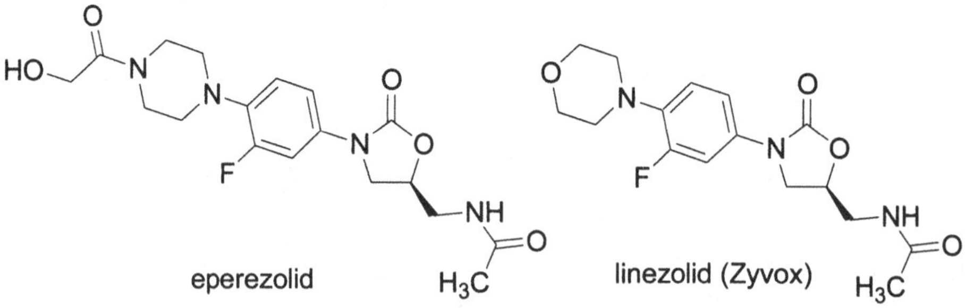

eperezolid   linezolid (Zyvox)

Linezolid has an excellent activity spectrum against Gram-positive bacteria, including VRE, MRSA, and *Streptococcus pneumoniae,* but limited activity against Gram-negative microbes.[230] Its clinical indications include hospital- and community-acquired pneumonia, uncomplicated and complicated skin and soft-tissue infections (cSSTIs), and infections caused by vancomycin-resistant enterococci.[226] Along with its impressive Gram-positive coverage, linezolid is bacteriostatic with high bioavailability and favorable ADME properties, which permit administration in either intravenous or oral form.

Linezolid's common adverse effects are diarrhea, nausea, headaches, reversible myelotoxicity, lactic acidosis, and peripheral and optic neuropathies.[231,232] Linezolid, like most oxazolidinones, is a reversible and nonselective inhibitor of monoamine oxidase (MAOI), but because it is a weak MAOI, linezolid does not produce serotonin syndrome when taken alone.[233]

However, when administered together with a serotonin reuptake inhibitor it causes serotonin syndrome.

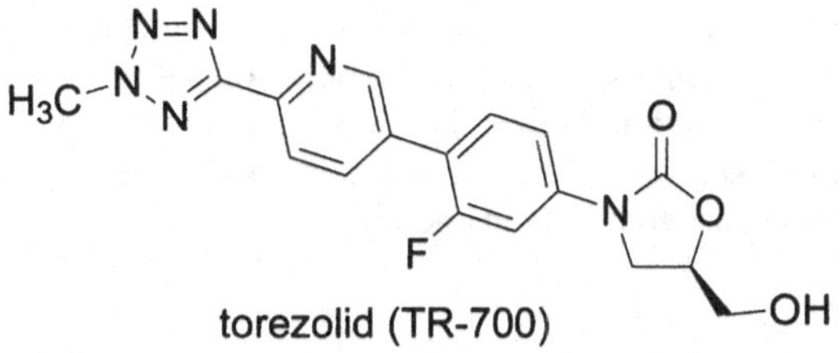

torezolid (TR-700)

Oxazolidinones have a unique mechanism of action in that the compounds bind to the peptidyltrasferase center in the ribosomal 50S subunit and prevent the formation of the N-formyl-methionyl-tRNA-ribosome-mRNA ternary complex (the 70S ribosomal initiation complex), which is an essential step in the initiation of bacterial protein synthesis.[226] This novel mode of action allows for little cross-resistance to occur between oxazolidinones and existing classes of antibacterial agents. Resistance to oxazolidinones, mainly linezolid, is limited to mutation in chromosomal genes encoding 23S rRNA and *cfr* methyltransferase (rare).[226,234]

The search for a second generation oxazolidinone is ongoing with the goal to expand the spectrum of antimicrobial activity to include Gram-negative pathogens and improve its selectivity profile, in particular to eliminate the inhibition of MAO enzymes. Torezolid (TR-700) is the only second-generation oxazolidinone in Phase III trials and was discovered by Dong-A Pharmaceuticals and licensed to Trius Therapeutics.[235] Torezolid is masked as a phosphate prodrug (TR-701) to improve its solubility and bioavailability properties. Torezolid has improved activity against most Gram-positive bacteria, including MSSA, MRSA, and VRE, compared to linezolid and is active against linezolid-resistant strains. Torezolid has favorable druglike properties and can be developed as oral and intravenous formulations with once-daily dosing.

4.3   DNA and RNA Synthesis Inhibitors

DNA and RNA syntheses are critical functions for cell survival and proliferation, and these attributes make biochemical pathways associated with these functions ideal candidates for targeting by antibacterial agents. The first antibiotics used clinically, the sulfonamides, essentially starve the cell of the nucleotides required for DNA replication to proceed by preventing their synthesis. Blocking transcription is also a viable antibacterial strategy, since this shuts down gene expression and interferes with mRNA synthesis, which is required for protein synthesis.

4.3.1   Folate Synthesis Inhibitors

Synthesis of purines, thymidine, and some amino acids requires the use of tetrahydrofolate as a cofactor. Since bacteria synthesize folate cofactors de novo, inhibition of enzymes in this pathway is an effective antibacterial strategy.[236]

The sulfa drugs inhibit dihydropterate synthase (DHPS) by competing with *p*-aminobenzoic acid. Prontosil, later found to be a prodrug of sulfanilamide, was discovered in 1932 and first used clinically in 1933 to treat a case of staphylococcal septicemia.[237] Gerhard Domagk, who recognized the therapeutic value of Prontosil, received the Nobel Prize for Physiology or

Medicine in 1939. Due to the synthetic ease of attaching other groups to the sulfonamide moiety, additional sulfa drugs were rapidly discovered. One of the most successful of these is sulfamethoxazole, which is used in combination therapy[238] (with trimethoprim) for the treatment of urinary tract infections, acute otitis media, chronic bronchitis, and traveler's diarrhea. Resistance often occurs due to single-step point mutations to DHPS.[239] Severe toxic effects can be observed after sulfonamide administration, possibly due to formation of a hydroxylamine metabolite,[240] and include skin rash, Stevens-Johnson syndrome, toxic epidermal necrolysis, hepatic necrosis, agranulocytosis, and aplastic anemia.

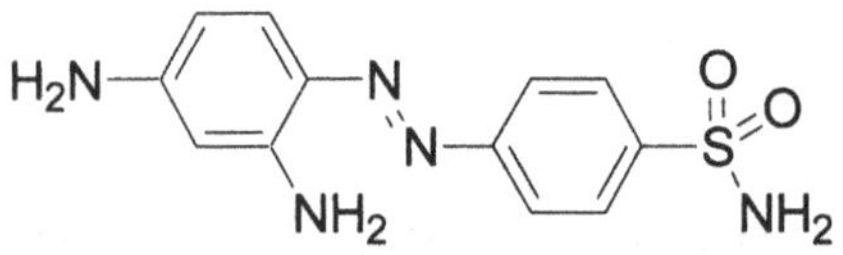

Prontosil 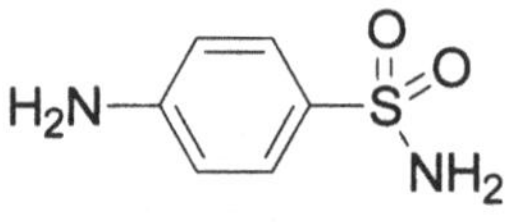   sulfanilamide

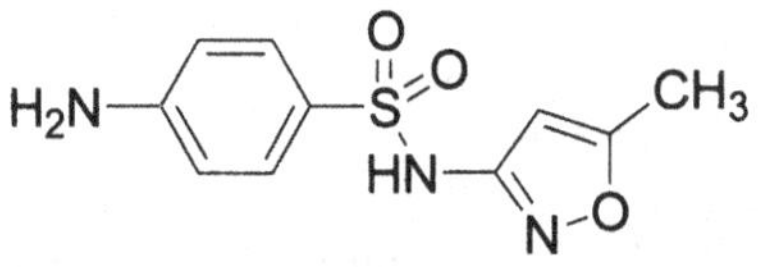

sulfamethoxazole

Dihydrofolate reductase (DHFR) inhibitors have also proven to be clinically useful in the treatment of urinary tract infections, particularly when codosed with a sulfa drug, due to increased efficacy observed with concurrent inhibition of two enzymes in the folate biosynthesis pathway.[236] Resistance to DHFR inhibitors is often due to mutation of the target protein, reducing the K_i of the drug and making it less efficacious.[239] Trimethoprim (Proloprim) is the most clinically relevant of the DHFR inhibitors and inhibits the enzyme by competing with dihydrofolate.[241] It is commonly administered as a 1:5 combination with sulfamethoxazole (Bactrim). Gertrude Elion and George Hitchings shared the 1988 Nobel Prize in Physiology or Medicine, partially for the discovery of trimethoprim.

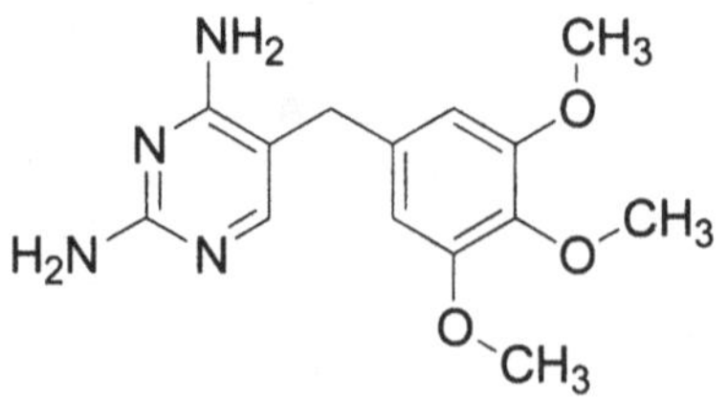

trimethoprim (Proloprim)

4.3.2   DNA Gyrase and Topoisomerase IV Inhibitors

DNA gyrase and topoisomerase IV are attractive targets for antibacterials, since these enzymes are responsible for the critical processes of DNA supercoiling and decatenation.[242] Their inhibition interferes with DNA transcription, replication, and repair, leading to DNA cleavage and cell death.

Structural similarity between DNA gyrase and topoisomerase IV presents an opportunity for inhibitors to target both enzymes, with the potential benefit of increased efficacy and reduced rates of target-based mutational resistance.

The quinolone class of antibiotics[243] has a purely synthetic origin, being related to an impurity from the manufacture of the antimalarial chloroquine that showed Gram-negative activity.[244] Their mode of action was not initially known, but it was subsequently shown that they target DNA gyrase and/or topoisomerase IV by inserting into the DNA strand breaks and preventing enzymatic turnover.[245] Quinolone's bactericidal activity tends to be concentration dependent, with lower inhibitor concentrations being bacteriostatic.[246] In general, their Gram-negative activity is due to inhibition of DNA gyrase, while Gram-positive activity is due to topoisomerase IV inhibition.[247] Known resistance mechanisms include drug efflux[248] and target protein mutation, resulting in decreased quinolone binding.[249] There are no clear examples of bacterial chemical modification of quinolones that lead to loss of antibacterial activity.[243]

The first commercialized quinolone antibiotic was nalidixic acid[250,251] (NegGram) and was marketed in 1964 by Sterling-Winthrop. While it was bactericidal and could be delivered orally, it suffered from relatively poor pharmacokinetics and a narrow spectrum (primarily *E. coli*) that limited it to treatment of urinary tract infections.

nalidixic acid (NegGram)

Norfloxacin[252] (Noroxin), introduced by Merck Sharp & Dohme in 1983, added a 7-piperizine, responsible for *P. aeruginosa* activity, as well as the characteristic 6-fluoro group that dramatically improved potency and has been retained in most subsequent analogs, and gives rise to the label "fluoroquinolones".[253] Norfloxacin showed improved potency (comparable to many fermentation-derived antibiotics) and broader Gram-negative spectrum while introducing limited Gram-positive activity; however, low blood levels still limited its use to Gram-negative urinary tract infections. Ciprofloxacin[254] (Cipro) is an important member of the fluoroquinolone antibiotic class and one that still sees significant use. Launched by Bayer in 1986, ciprofloxacin retained the overall structure of norfloxacin but replaced the *N*1-ethyl group with an *N*1-cyclopropyl. This change affected both PK (improved blood, tissue, and intracellular levels as well as IV formulation) and potency,[253] resulting in a broader scope of indications, including complicated UTIs, prostatitis, sexually transmitted diseases, and nosocomial infections. Levofloxacin[255] (Levaquin), launched in Japan by Daiichi Sankyo in 1993, is an example of a tricyclic quinolone, incorporating a third fused ring connecting the *N*1- and 8-positions. Its antibacterial profile is similar to that of ciprofloxacin,[253] but with increased activity against *Streptococcus* spp., expanding its use to include treatment of community-acquired pneumonia. Further expansion of Gram-positive coverage and introduction of modest antianaerobic activity were achieved with moxifloxacin[256] (Avelox), introduced by Bayer in 1999, although it is not approved for the latter indication. The addition of the 8-methoxy group appears to expand Gram-positive activity, while the bulkier bicyclic amine at the 7-position is suggested to reduce efflux.

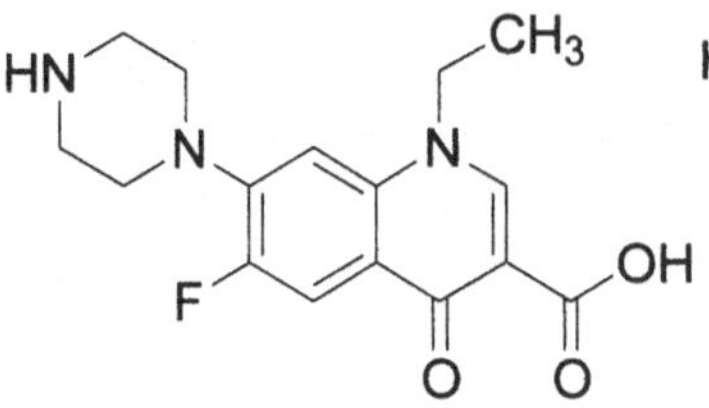

norfloxacin (Noroxin)

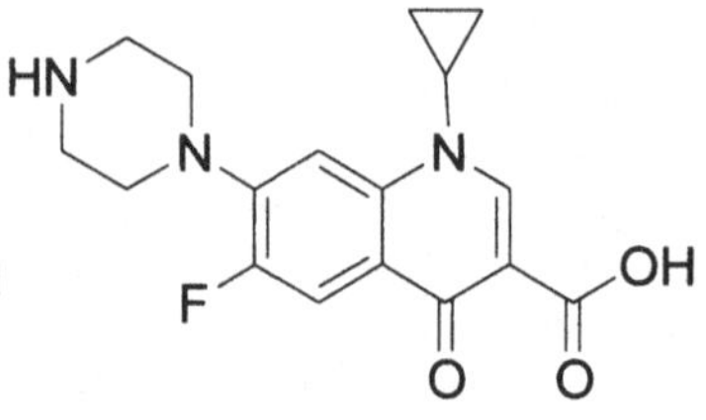

ciprofloxacin (Cipro)

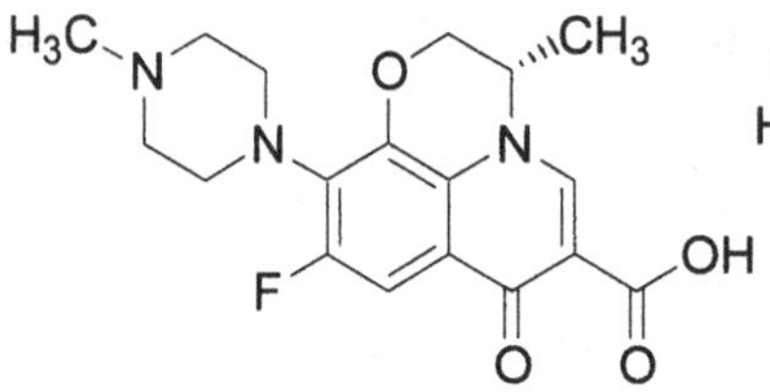

levofloxacin (Levaquin)

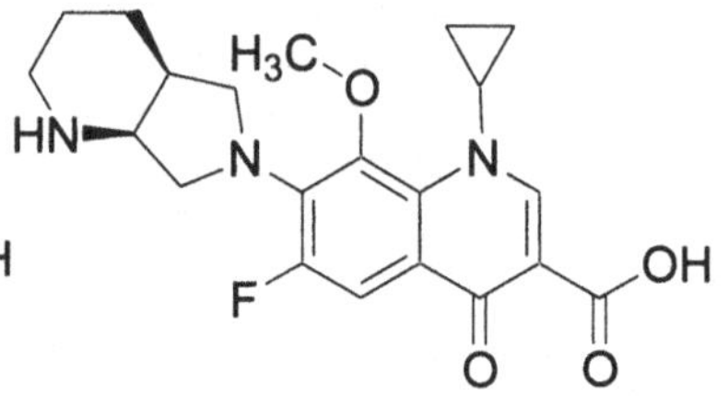

moxifloxacin (Avelox)

Many promising new fluoroquinolones have failed in recent clinical trials or been withdrawn postmarket, often due to toxicity concerns rather than efficacy. Indeed, many of these drugs looked to broaden antibacterial coverage and expand clinical indications. For example, trovafloxacin[257] (Trovan), which contains an *N*-aryl moiety, was approved in 1998 for intra-abdominal infections but was discontinued in 2001 due to hepatotoxicity concerns.[258,259] There are currently several fluoroquinolones in Phase II clinical development, including delafloxacin (Rib-X), JNJ-Q2 (Johnson & Johnson), finafloxacin (MerLion), and levonadifloxacin (Wockhardt), which seek to improve or modify antibacterial coverage and expand on approved indications for this class.

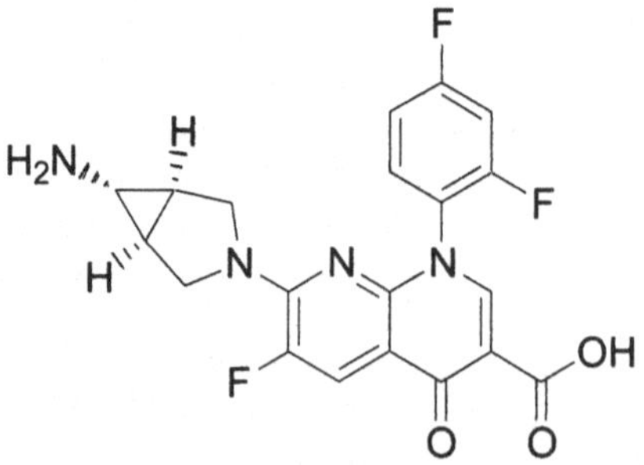

trovafloxacin (Trovan)

An alternative strategy for inhibiting DNA gyrase is by targeting the ATPase domain of gyrase B, thereby eliminating the energy source the enzyme required for proper function.

Novobiocin[260] (Albamycin), a courmermycin produced by *Streptomyces* spp. and approved in 1964, is the only ATPase inhibitor of DNA gyrase that has been successfully approved and marketed. It was used to treat *staphylococcal* infections but was withdrawn due to poor clinical efficacy and toxicity concerns.

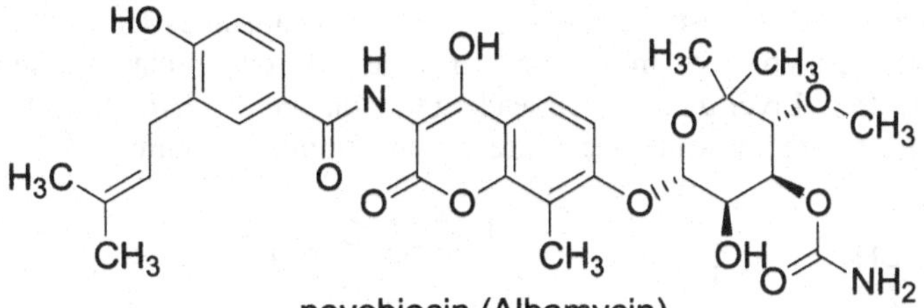

novobiocin (Albamycin)

4.3.3 Anaerobic DNA Synthesis Inhibitors

The nitroimidazole antibiotic metronidazole[261] (Flagyl) was approved for use in the U.S. in 1963 after previously being launched in 1960 by G. D. Searle in France. Metronidazole has no protein target but instead acts directly on DNA.[261] Its activity is contingent on intracellular reduction of the 5-nitro group to a radical anion or other reactive moieties by pyruvate-ferredoxin oxidoreductase. These reactive species are responsible for lethal damage to DNA, including single strand breaks. Metronidazole resistance is most often associated with lower drug activation within the cell, preventing the reactive species from forming.[262] Since a reduced environment is a requisite for conversion of the parent compound to its active form, metronidazole only shows clinically relevant efficacy against obligate anaerobic organisms, including *Bacteroides* spp., *Clostridium* spp., and *Peptostreptococcus* spp., as well as sensitive protozoa. It is often prescribed as a combination therapy along with a fluoroquinolone, cephalosporin, or tetracycline, and is used most commonly for intra-abdominal infections, skin and soft-tissue infections, bacterial septicemia caused by sensitive organisms, endocarditis, gynecological infections, and trichomoniasis. There is potential for central nervous system toxicity,[263] including peripheral neuropathy, as well as leucopenia associated with long-term use.

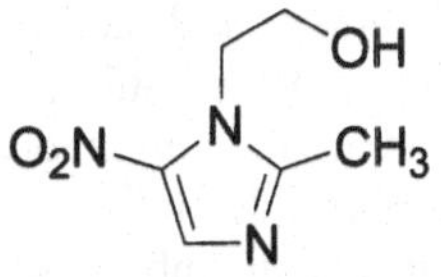

metronidazole (Flagyl)

4.3.4 RNA Polymerase Inhibitors

In addition to disruption of DNA synthesis, inhibition of DNA-dependent RNA synthesis is an effective strategy for shutting down bacterial transcription, leading to cell death. The rifamycins act by inhibiting RNA polymerase activity, in particular the initiation phase before chain elongation can occur. Isolation of rifamycins from the actinomycete *Amycolatopsis mediterranei* was first performed at Lepetit SA in 1959, leading to the chemical characterization of rifamycin B, which showed modest antibacterial activity.[264] Rifampin (Rifadin), approved in the U.S. in 1971

(1968 in Italy), was discovered via semisynthetic transformation of rifamycin B. It is indicated for treatment of tuberculosis and asymptomatic carriers of *N. meningitides*, although it is not indicated for the treatment of meningococcal infection. It also shows *in vitro* activity against other bacteria, including *S. aureus, S. epidermidis, H, influenza*, and *M. leprae*, but clinical efficacy has not been proven. Resistance against rifampin can occur rapidly and is most often conferred against the entire compound class. The most common resistance mechanism is by single-step mutation to RNA polymerase;[265] less common mechanisms include reduced cellular uptake and biochemical modification of the drug. Liver dysfunction leading to jaundice and hyperbilirubinemia, as well as immunoallergic effects, are known to occur after administration of rifampin.[266]

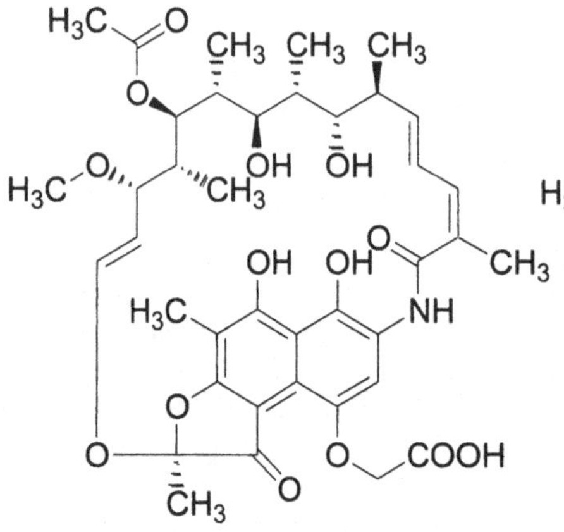

rifamycin B 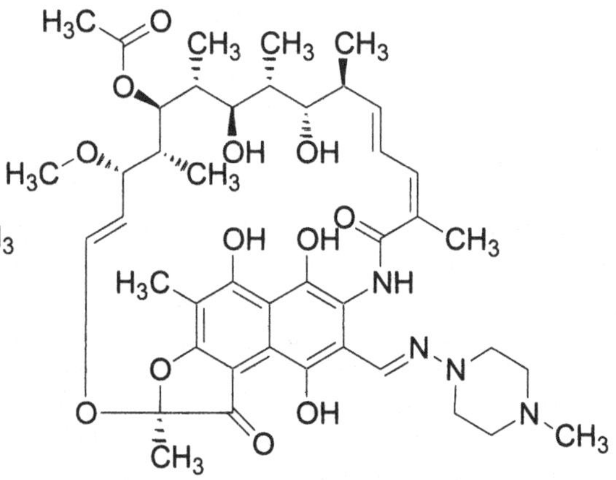   rifampin (Rifadin)

4.4   Membrane Inhibitors

The bacterial membrane is essential for bacterial survival as it provides selective permeability for cellular homeostasis and metabolic energy transduction.[267] The membrane is made up of about one-third of the proteins in the cell and is the site for crucial processes, such as active transport of nutrients and wastes, bacterial respiration, establishment of proton-motive force in association with respiratory enzymes, ATP generation, and cell-cell communication.[268] Therefore, the bacterial membrane is recognized as an important antibacterial target site as membrane-damaging agents can cause disruption of membrane architecture and functional integrity, which may lead to leakage of cytosolic contents and eventual cell death.[268]

    One class of antibiotics that are membrane-damaging agents is the lipopeptides. Lipopeptide antibiotics represent an old class of antibiotics that was discovered over 60 years ago, which includes the polymyxins and the more recently approved daptomycin. These agents generally consist of a hydrophilic cyclic peptide that is attached to a hydrophobic chain which facilitates insertion into the lipid bilayer of bacterial membranes.

4.4.1   Polymyxins

The class of polymyxins, discovered in 1947 from *Bacillus polymyxa*, consists of polymyxins A–E, of which only polymyxin B (PMB) and polymyxin E, also known as colistin, are currently on the market.[269–271] PMB and colistin share a common primary sequence, the only difference being at position 6 which is occupied by D-Phe in PMB and D-Leu in colistin. PMB is a mixture of at least four closely related components, polymyxin B_1 to B_4, with polymyxin B_1 and B_2 as the major components, which have an *N*-terminal acylated by (*S*)-6-methyloctanoic acid and (*S*)-6-methylheptanoic acid, respectively. Like PMB, colistin also has two main components, colistin A and colistin B, and are acylated by (*S*)-6-methyloctanoic acid and (*S*)-6-methylheptanoic acid, respectively.[272,273]

   Both PMB and colistin have no activity against Gram-positive bacteria and anaerobes but are active against a variety of Gram-negative pathogens.[274] The great majority of isolates of *E. coli*, *Klebsiella* spp., *Enterobacter* spp., *P. aeruginosa* (including MDR isolates), and *Acinetobacter* spp., all important nosocomial pathogens, are usually susceptible to polymyxins. Additionally, *Salmonella* spp., *Citrobacter* spp., *Shigella* spp., *Pasteurella* spp., *Haemophillus* spp., and *Stenotrophomonas maltophilia* are usually susceptible. *Proteus* spp., *Providencia* spp., and most isolates of *Serratia* spp. possess intrinsic resistance to polymyxins. Isolates of *Brucella* spp., *Neisseria* spp., *Chromobacterium* spp., and *Burkholderia* spp. are resistant.

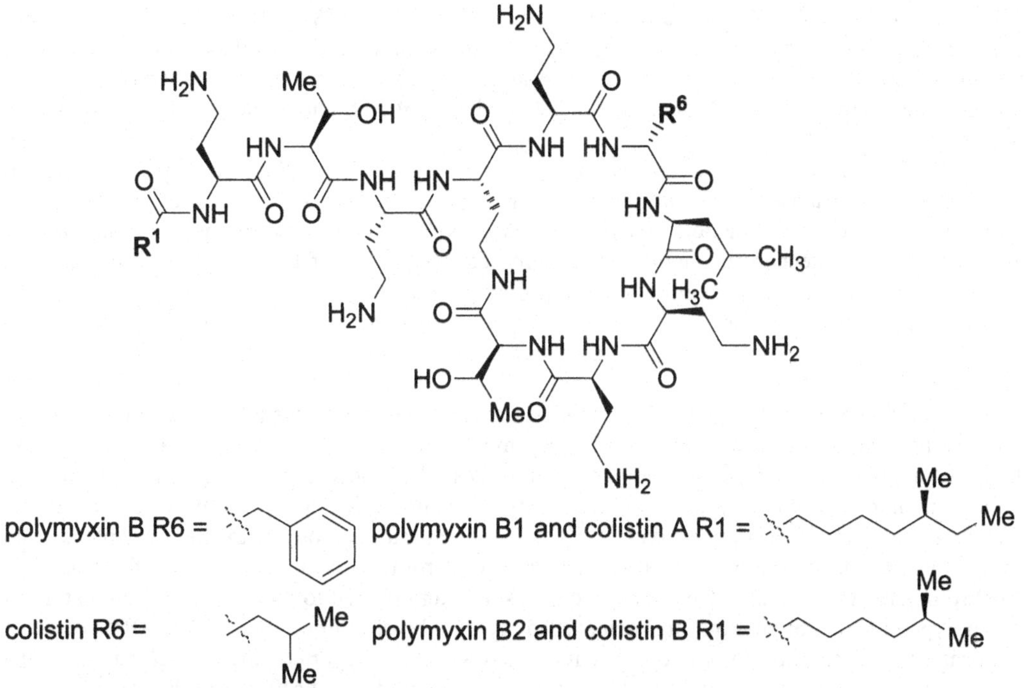

Polymyxin B and colistin are rapid-acting bactericidal agents.[16] The initial target of polymyxins is the lipopolysaccharide (LPS) component of the membrane which bears negative charges and confers integrity and stability to the bacterial outer membrane. LPS is composed of three domains, the O antigen, the core polysaccharide, and lipid A. Lipid A consists of a β-1′-6-linked D-glucosamine (GLcN) disaccharide that is phosphorylated at the 1- and 4′-positions and contains six fatty acyl chains.[75] Lipid A acts as a hydrophobic anchor with the tight packing of fatty acyl chains helping to stabilize the overall outer membrane structure.[275] The polycationic peptide ring of polymyxin binds to the negatively charged lipid A phosphates to displace the calcium and magnesium divalent cations that normally function to bridge and stabilize the LPS containing outer membrane monolayer.[276–280] Because the peptides have affinities for LPS that are at least three orders of magnitude higher than the divalent cations Ca^{2+} and Mg^{2+}, they competitively displace the ions and this initial electrostatic interaction temporarily stabilizes the complex and brings the N-terminal fatty acyl side chain of the polymyxin molecule into proximity of the outer membrane.[281,282] The fatty acid side chain of polymyxin further interacts with the LPS and inserts into the outer membrane.[278] Polymyxins produce a disruptive physiochemical effect, leading to permeability changes in the outer membrane and are subsequently taken up via a self-promoted uptake pathway leading to cell death.[283] The removal of the lipid component of the polymyxins results in polymyxin nonapeptide, which is less active. However, polymyxin nonapeptide does sensitize bacteria to the effect of other antibiotics and human neutrophiles.[284–286] Most bacterial mechanisms of resistance to polymyxins involve the modification of LPS, since the first critical step in polymyxin action on Gram-negative bacteria is the electrostatic interaction between the positively charged peptide and the negatively charged LPS.[278] Numerous species have developed mechanisms for the modification of lipid A by reducing its net negative charge which stop or reduce the initial electrostatic interaction.[287–290] In E. coli, K. pneumoniae, P. aeruginosa, Salmonella enteric, and S. typhimurium, the modification of the phosphates of lipid A with positive charged moieties, such as 4-amino-4-deoxy-L-arabinose (L-Ara4N) and/or phosphoethanolamine (Petn), reduces the negative charge of LPS and thereby increases resistance to polymyxins.[287,291,292]

Alternatively, in K. pneumoniae, the presence of capsule polysaccharide (CPS) is critical for polymyxin resistance.[290] CPS limits the interaction of polymyxins with their target sites. Thus, upregulation of CPS production confers increased polymyxin resistance. In P. aeruginosa, a genetically stable mechanism involves the increased production of the outer membrane protein H1. It is believed that this protein exerts its protective effects by functionally replacing divalent cations in the cell membrane.[293,294] In S. typhimurium, the gene mig-14 is involved in polymyxin resistance, and in Vibrio cholerae, the outer membrane porin OmpU is responsible for resistance to polymyxins.[289,295]

Studies from before the 1970s identified several adverse events attributed to polymyxins. A small percentage of patients exposed to polymyxins develop allergic manifestations including fever, eosinophilia, and macular and urticarial rashes.[296,297] However, the more prominent side effects involve neurotoxicity and nephrotoxicity.[298] Neurotoxicities of polymyxins are usually manifested as paresthesias and ataxia.[296,299,300] These symptoms along with diplopia, ptosis, and nystagmus are usually resolved after decrease or prompt discontinuation of therapy.[300,301] Nephrotoxicity, analogous to the aminoglycosides, is likely related to the highly cationic nature of the polypeptide.[98] Polymyxins are reabsorbed by the brush-border membrane (BBM) of the proximal tubuli leading to kidney damage.[302] With these reported adverse effects and the emergence of other antimicrobial agents, polymyxins quickly fell out of favor in the 1970s.

Since the mid-1990s, there has been a greatly renewed interest in their clinical use due to the emergence of multidrug-resistant Gram-negative bacilli (mainly P. aeruginosa, A. baumannii, and recently, K. pneumoniae), in parallel with the lack of new antibacterials.[303,304] Polymyxins have certainly been invaluable for the therapy of serious nosocomial pathogens. However, concerns regarding nephrotoxicity and neurotoxicity continue to persist. The substantial gap of knowledge of the polymyxin pharmacology is another problem. Therefore, further investigations on the pharmacokinetics, pharmacodynamics, and toxicodynamics of polymyxins and its efficacy alone and in combination with other antibiotics are urgently required.

4.4.2　Daptomycin

While the polymyxins are undergoing a revival, daptomycin is the new star in the lipopeptide class that has claimed recent headlines. Daptomycin is a natural product of a soil actinomycete and is produced by the microorganism *Streptomyces roseosporus*.[305,306] Daptomycin was first identified by Eli Lilly and Company in the mid-1980s and underwent clinical trials with BID dosing. The trials were suspended in 1991 because of skeletal muscle toxicity leading to elevated serum levels of creatine phosphokinase (CPK).[307] In 1997, daptomycin was out licensed to Cubist Pharmaceuticals which reevaluated its toxicokinetics, and re-introduced it into clinical trials in 1999. In 2003, the FDA approved daptomycin for the treatment of cSSSIs with a novel dosing regimen, and it is marketed under the name Cubicin. In 2006 an additional FDA approval was granted for the treatment of bloodstream infections and right-sided endocarditis caused by MSSA and MRSA. Following that, the once-daily daptomycin is now approved for the treatment of cSSSIs in North America, Europe, and several other countries.

Daptomycin is active against aerobic and anaerobic Gram-positive bacteria including multidrug-resistant strains. Its spectrum includes MRSA, MSSA, glycopeptide-intermediate *S.*

aureus (GISA), methicillin-resistant coagulase-negative *Staphylococcus* spp. (CoNS), and VRE.[308–316] Furthermore, daptomycin is also effective against a variety of *streptococcal* groups, including *S. pyogenes* and *S. agalactiae*.[308–310,316] It is also active against vancomycin-resistant strains of *Lactobacillus*, *Pediococcus*, and *Leuconostoc*.[305,317,318] It has activity against *Bacillus* species and corynebacteria, whereas *Listeria* may have elevated MIC values.[319,320] Daptomycin has activity against some anaerobic organisms such as *Peptostreptococci* and *Clostridia*, including *C. difficile,* but it has no activity against actinomyces.[321]

Daptomycin is a 13-membered amino acid cyclic lipopeptide antibacterial agent containing a hydrophilic core and a hydrophobic tail, similar to the polymyxins.[322] The anionic character of daptomycin is masked by its complexation with divalent cations such as Ca^{2+}, enabling the lipopeptide to perturb bacterial membranes.[323–325] A mechanism of action proposed for daptomycin involves two elements: the lipophilic decanoyl side chain is bound first and then inserted into the bacterial cell membrane in a Ca^{2+}-dependent manner, without entry into the cytoplasm.[122] This affects the cell membrane, triggering a rapid leakage of K^+ ions. This results in a subsequent loss of the ion concentration gradient, depolarization of the cell membrane, and a disruption in the production of macromolecules, such as RNA, DNA, and other proteins, and causes rapid cell death.[325,326] Daptomycin possesses the unique feature that it is rapidly bactericidal without being bacteriolytic, which eliminates lysis-associated side effects, and has a prolonged postantibiotic effect.[327] Furthermore, daptomycin is active in both exponential and stationary phases of bacterial growth, a property that is highly unusual.[328]

The incidence of daptomycin resistance in clinical isolates is very low. Mutations in *mprF*, *rpoB*, *rpoC,* and *yycG* have been implicated in resistance of staphylococci to daptomycin.[329–331] Other mechanisms may also be involved, as not all daptomycin nonsusceptible isolates have been shown to express one of these mutations. Daptomycin is not effective in lung infections, which may be attributable to the lipid content of pulmonary surfactants which inhibits its antibacterial activity.[332,333]

As mentioned earlier, the discontinuation of the first clinical trial of daptomycin was due to skeletal muscle toxicity when the drug was administered twice daily. Subsequent animal studies by scientists at Cubist Pharmaceuticals revealed that once-daily dosing is much less frequently associated with this adverse effect.[334] As shown in these experiments, daptomycin effects on skeletal muscle increased two- to four-fold upon fractionation of the daily dose. It is hypothesized that skeletal toxicity is reversible and its repair is more efficient when dosing is less frequent. In healthy volunteers, doses up to 12 mg/kg daily (QD) were not associated with muscle toxicity.[335] Daptomycin is generally well tolerated with the most common side effects being gastrointestinal disturbances, headache, and injection site reactions.[336]

Daptomycin represents a new class of antibacterials, a rarity in the last 30 years. Its unique mechanism of action and its low resistance profile together with its rapid bactericidal action make it a favorable alternative to vancomycin for MDR cocci. Although daptomycin proved efficacious in the treatment of cSSSIs, infective right-sided endocarditis, and bacteremia, further clinical evidence is needed to establish its effectiveness in osteomyelitis, joint infection, prosthetic material infection, and left-sided endocarditis.

A new semisynthetic derivative of daptomycin developed by Cubist Pharmaceuticals is CB-183,315, which replaces the decanoyl lipophilic tail of daptomycin with an alkyl aromatic group.[337] CB-183,315 exhibits potent bactericidal activity against clinical isolates of *C. difficile,* which is a spore-forming, Gram-positive, anaerobic bacterium that exists in the intestinal tract.[338,339] CB-183,315 is currently in Phase III clinical development as an oral treatment for *C.*

difficile infection (CDI), also commonly known as *C. difficile*-associated diarrhea (CDAD), which is the most common cause of health care-related infectious diarrhea.[337,338,340]

5   Emerging Strategies to Discover New Antibacterial Drugs

5.1   *New Compounds, In or Near the Clinic*

A very recent review[341] covers the status of antibiotics in the clinic up to April 2011, and many of the most exciting new candidates are described above. To summarize, as additional companies have left this field of research, and due to the difficulty of advancing novel antibacterials beyond the discovery stage, the pipeline for antibacterials has continued to contract. This is particularly true of the discovery of new antibiotics for Gram-negative infections and new classes of antibiotics. More recently, a few interesting preclinical reports have appeared. One example relates to the chemical modification of vancomycin, which is very challenging due to the lengthy synthesis, macromolecular construction, and number of stereocenters. This challenge was taken up by the Boger lab.[342] They made a key change to the molecule that restores activity of the scaffold to otherwise vancomycin-resistant pathogens (below).

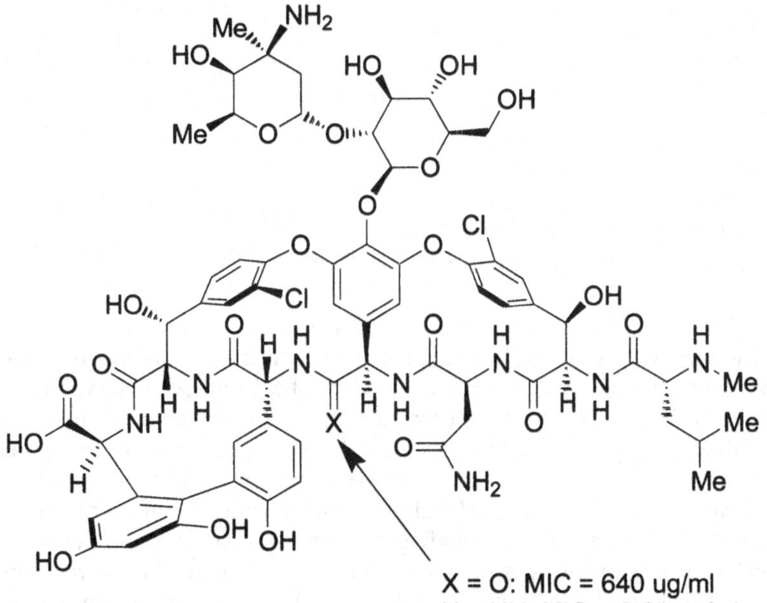

X = O: MIC = 640 ug/ml
X = NH: MIC = 0.31 ug/ml

5.2   *Tools for Discovery*

The efficiency of discovery activities in any drug discovery program depends on the relevance of the assay guidance. This is certainly true of antibacterial discovery, and one example is provided by Pfizer which has described[343] a rich cascade of assay tools to provide a high level of confidence in advancing HTS hits. This was illustrated in a specific example showing the discovery of hits from an assay for *E. coli* phosphopantetheine adenyltransferase. Because this enzyme can process in two directions, HTS follow-up consisted of two discrete biochemical assays and kinetic analysis to characterize the mode of action. They used an array of tool compounds in conjunction with these assays, supplemented by isothermal titration calorimetry to understand the binding behavior of each series. They further used ^{19}F NMR to monitor the displacement of fragment probes they previously identified in an STD NMR fragment screen.

Bioinformatics analysis of bacterial genomes was instrumental in the first wave of target screening by HTS.[19] Despite the difficulties in biochemical target screening, such analysis remains an important tool in the identification of potential targets.[344] Genomics tools are used not only in target identification but increasingly in engineering systems to validate or screen targets in new ways. For example, inhibitors of a component of the 30S ribosome RpsD were found with a whole-cell screen where the specific target was sensitized by a xylose-dependent antisense suppression of RpsD. This was instrumental in the identification of Glabramycin C.[345] A different inhibitor for this target, pleosporone 1, was found[346] by screening a different natural product source.

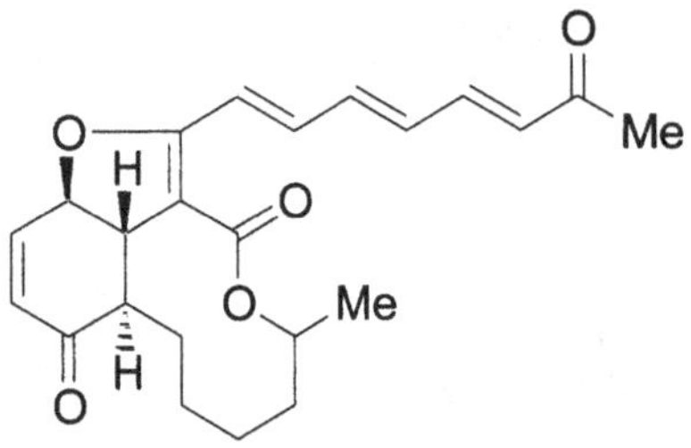

glabramycin C
MIC = 1 µg/ml *S. pneumoniae* CL2883
MIC = 16 µg/ml *S. aureus* ATCC 29213

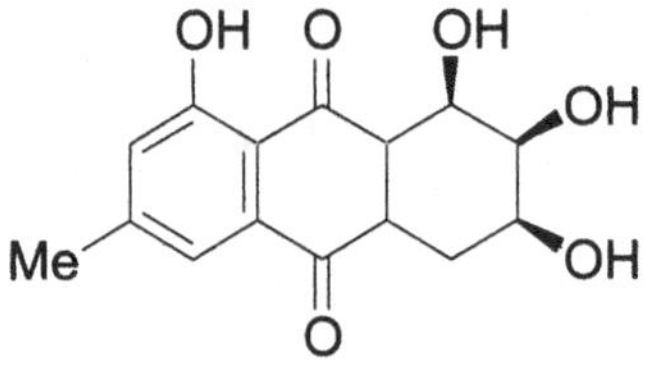

pleosporone 1
MIC = 4µM *S. pneumoniae* CLL2883

Merck has used a platform[347] of inducible suppression of 245 different *S. aureus* genes to generate specific sensitivity to individual targets or to pool them into groups for broader profiling. With this platform they recently identified a new class of a Gram-positive type II topoisomerase inhibitor, kibdelomycin A. Other approaches to target sensitization have been described and discussed in Silver's review[348] where it is pointed out that, despite the wide range of tools available, there has not yet been a significant improvement in drug discovery success rates.

The permeability of bacterial cell walls is different from that of eukaryotic cell walls and may contribute[348] to the difficulty of finding penetrating ligands when screening molecules typically directed at eukaryotic systems. This remains a largely unexplored topic from the perspective of generating tools that might help guide SAR optimization and remains an opportunity for future work.

Efflux pumps are one of the most important hurdles to overcome in antibacterial discovery. Some have explored efflux pump inhibitors as described in a recent review[349] and while there are strategies to attempt this, the multiplicity of pumps, difficulty of saturation, and species specificity present significant challenges.[22]

Antisense technologies have been proposed[350] as antibacterial therapeutic agents with significant acknowledged challenges of obtaining bacterial cell penetration but with strategies of attaching moieties known to be transported across bacterial cell membranes used to surmount such problems. AVI is active in this space in an example targeting *Burkholderia*[351] and with patent applications[352] involving RNAi. Antisense has also been proposed[353] as a mechanism to circumvent the very rapid emergence of resistance to rifamycins that target RNA polymerase.

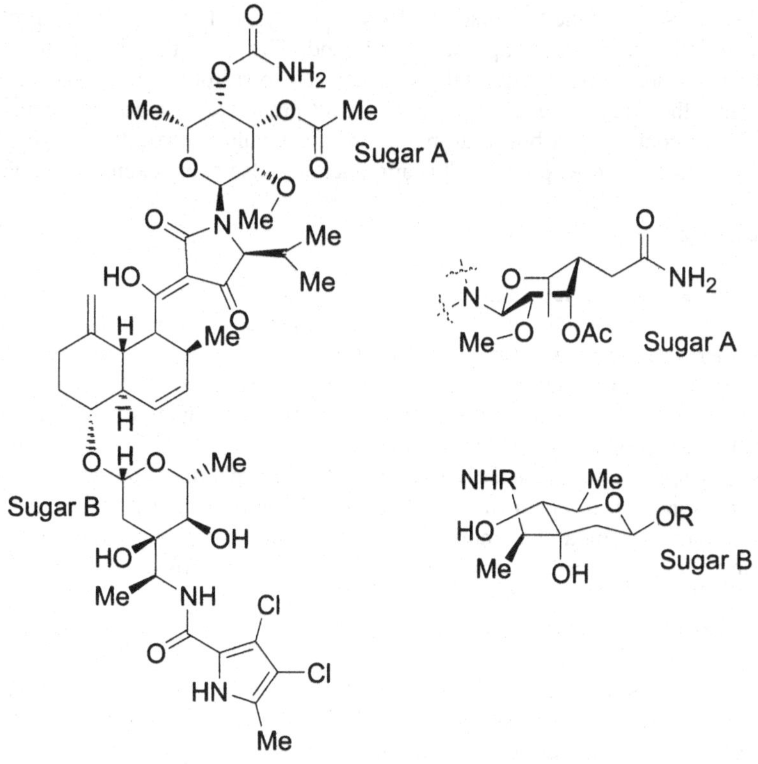

kibdelomycin

Antisense methods have also been used to identify and confirm resistance mechanisms, including for Daptomycin[353] by shutting down a gene that is upregulated in the resistant phenotype. A restoration of the sensitive phenotype upon gene silencing is taken as evidence for the role played by the silenced gene in producing resistance. The use of these technologies will likely face significant challenges when resistance is based on changes in bacterial cell permeability or in addressing bacterial species spectrum limitations.

Structure-based discovery tools such as X-ray[354] and computational methods[355] for using the target structures[356] or mapping chemical space[357] for ligands continue to be well integrated into the discovery process. Their strength lies in helping to guide the generation of analogs of a given chemical series through an efficient path and to some extent the discovery of new chemical directions. The limitations of target-based discovery are attributable not to the use of these tools but rather to the relevance of the chemical space to the antibacterial hurdles cited above. With careful attention to the challenges of permeation, efflux, and spectrum, these tools can play an increasing role in driving both novelty and efficiency.

Lastly, and not strictly a tool for discovery, covalent ligands represent an important adjunct strategy in the design of antibacterial ligands. The strategy has a parallel in other therapeutic areas. The goal is to increase residence time of a ligand on the target, and one of the best examples within the antibacterial space is the β-lactam class of antibiotics where the lactam ring is opened irreversibly by the target serine. Slow hydrolysis of the inactivated protein achieves an effect comparable to a much stronger affinity ligand. The same principle is used in DFG-out kinase inhibition[358] where loop reorganization is required to restore a functional protein and this slow step increases the effectiveness of the ligand. A different example can be found in GSK '052 (formerly AN3365) containing a boron atom, where the ligand is expected to form a reversible covalent complex with the target protein. This approach has also been taken up by Avila.[359]

5.3   Screening

Silver points out in a recent review that neither the target-based cell nor biochemical screening has produced a new class of antibacterial drugs or perhaps even any advanced clinical compounds.[349] Silver, and another recent review from Brotz-Oesterhelt[360] highlight some key reasons behind the failures. These include inhibiting targets that may only be essential in the absence of a clinical environment (fatty acid biosynthesis targets) and screening for compounds that hit with no regard for the potential to get into bacterial cells.

Screening approaches are necessarily limited by the chemical space sampled. Schreiber has advocated an approach to chemical space, termed *diversity-oriented synthesis*, populated by more complex structures containing more stereocenters and more rings, reminiscent of natural products. Spring[361] has published an example of the use of this in antibacterial discovery.

5.4   New Targets

Strong interests in targets involved in a multitude of bacterial cell processes exist. A target in the synthesis of LPS was recently reported by Nicotra.[362,363] LPS has several components that must be assembled to produce a functional outer membrane. One such pathway is the Kdo pathway, which includes several essential genes, notably arabinose 5-phosphate isomerase, which has a redundant enzyme only in *E. coli*, and Kdo 8-phosphate synthase (KdsA, also duplicated in *E. coli*) and CMP-Kdo synthesis by KdsB. This pathway could prove to contain several promising new targets.

Hopkins[364] argues that the traditional approach of identifying a single essential gene is flawed, using as examples the β-lactam targets (multiple PBPs) and fluoroquinolone targets (multiple proteins in the gyrase/DNA complex). Hopkins suggests that a network topology analysis to identify critical nodes can shape future discovery. A related direction proposes[365] targeting metabolic networks in bacteria as a way to identify sensitive points of intervention likely to produce lethal results when inhibited. This method certainly deserves further consideration. A different related concept is targeting membrane viability directly through physical perturbation of the membrane, as described above for daptomycin. This is an important contrast to target- or pathway directed drug discovery, and would in fact be missed by targeted discovery efforts. This does not guarantee that resistance cannot emerge, but clearly cannot produce single-point mutants that prevent binding of the drug as is found in many other resistance mechanisms.

Finally, the number of tractable and essential targets would increase if single species drugs were to be viable commercially. Tuberculosis is the only species where this is being done successfully, and there is much debate[366] about the merits of extending the approach to the common infectious pathogens such as *P. aeruginosa*.

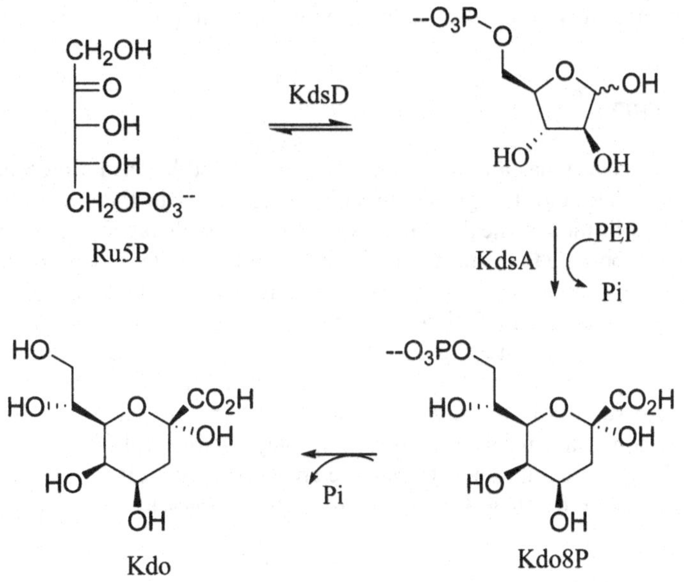

CH$_2$OH
=O
—OH
—OH
CH$_2$OPO$_3^{--}$

Ru5P

KdsD

$^{--}$O$_3$P—O
OH
HO OH

KdsA
PEP
Pi

HO
HO''' O CO$_2$H
'''OH
HO
OH

Kdo

Pi

$^{--}$O$_3$PO
HO''' O CO$_2$H
'''OH
HO
OH

Kdo8P

5.5   *Emerging Strategy*

HTS-driven discovery of antibacterial targets has met with limited success as outlined in previous reviews. While HTS has been used successfully to find hits in biochemical screens, those hits have usually proven difficult to translate into marketable drugs. The challenges have been in overcoming the array of hurdles of sufficient pan-species biochemical activity followed by achieving durable bacterial cell penetration resulting in MICs and finally having a sufficient therapeutic safety index. Phenotypic screening of compounds or natural products can be an effective tool but often at the expense of resources required for ligand deconvolution and perhaps target identification. Given that so many effective antibacterial agents derive from natural products, one can assume that the problems encountered in HTS-driven discovery are more about being in the right chemical space than not having the tools needed.

One strategy is to look in complementary areas to improve the odds of success. Hunting in a new chemical space may be possible, but simply building a large screening deck that is in a very different space from existing ones poses numerous challenges. While details differ among vendors, none have large purchasable collections that look more like natural products.

Another way to look in the new chemical space is to start with lower molecular weight compounds that implicitly sample a larger chemical space universe. Lower molecular weight means less decorated scaffolds that represent the potential of broad structural elaboration. Fragment screening has therefore become an important component of the discovery strategy. Effective use of such fragment hits is greatly enhanced by having clear understanding of the binding mode and thus access to protein crystallography and appropriate biophysical tools such as SPR or ITC followed by project-specific triaging tools and reference compounds.

Another way to enhance the exploration of novel and possibly natural product-like space is to screen compounds with greater structural complexity yet synthetic accessibility. diversity oriented synthesis has the ability to supply compounds in this space. Given the history of antibacterial drug discovery, one also cannot ignore natural products, especially from novel sources.

6   Conclusions

Antibacterial drug discovery has a rich history of discovery and has had a profound effect on life today as we know it. Through the ground-breaking discoveries in the 1920s–1950s, antibiotics became a critical tool in modern health-care. However, in this therapeutic area the need for new antibiotics can never be satisfied due to the multitude of bacteria and resistance mechanisms. Multiple established bacterial targets exist with a range of inhibitors, mainly natural product derived, but also fully synthetic, including the first antibiotic commercialized. While some modern drug discovery technologies, such as HTS, in antibacterials have not yet produced a commercialized drug, continued optimization of which technologies to use and where will increase the odds for success in the future. In the search for new antibacterials, other strategies such as covalent binding and phenotypic screening play a larger role and likely will achieve a greater degree of success than in other therapeutic areas. Although discovering new antibiotics is difficult, it is an imperative for society, and we must never go back to the pre-penicillin era.

References

(1)   Whitman, W. B.; Coleman, D. C.; Wiebe, W. J. *Proc. Natl. Acad. Sci. USA* **1998**, *95*, 6578–6583.
(2)   Schopf, J. W. *Proc. Natl. Acad. Sci. USA* **1994**, *91*, 6735–6742.
(3)   Woese, C. R.; Kandler, O.; Wheelis, M. L. *Proc. Natl. Acad. Sci. USA* **1990**, *87*, 4576–4579.
(4)   Spellberg, B.; Guidos, R.; Gilbert, D.; Bradley, J.; Boucher, H. W.; Scheld, W. M.; Bartlett, J. G.; Edwards, J. J. *Clin. Infect. Dis.* **2008**, *46*, 155–164.
(5)   Boucher, H. W.; Talbot, G. H.; Bradley, J. S.; Edwards, J. E.; Gilbert, D.; Rice, L. B.; Scheld, M.; Spellberg, B.; Bartlett, J. *Clin. Infect. Dis.* **2009**, *48*, 1–12.
(6)   Talbot, G. H.; Bradley, J.; Edwards, J. E. J.; Gilbert, D.; Scheld, M.; Bartlett, J. G. *Clin. Infect. Dis.* **2006**, *42*, 657–668.
(7)   Fischbach, M. A.; Walsh, C. T. *Science* **2009**, 1089–1093.
(8)   Fleming, A. *Brit. J. Exp. Pathol.* **1929**, *10*, 226–236.
(9)   Drawz, S. M.; Bonomo, R. A. *Clin. Microbiol. Rev.* **2010**, *23*, 160–201.
(10)   Raju, T. N. *Lancet.* **1999**, *353*, 681.
(11)   Van, M. A. S. *J. Vet. Pharmacol. Ther.* **1994**, *17*, 309–316.
(12)   Walsh, C., Ed. *Antibiotics: Actions, Origins, Resistance.* ASM Press, 2003.
(13)   Von Nussbaum, F.; Brands, M.; Hinzen, B.; Weigand, S.; Häbich, D. *Angew. Chem. Int. Ed.* **2006**, *45*, 5072–5129.
(14)   Klevens, R. M.; Morrison, M. A.; Nadle, J.; Petit, S.; Gershman, K.; Ray, S.; Harrison, L. H.; Lynfield, R.; Dumyati, G.; Townes, J. M.; Craig, A. S.; Zell, E. R.; Fosheim, G. E.; McDougal, L. K.; Carey, R. B.; Fridkin, S. K.; *JAMA* **2007**, *298*, 1763–1771.
(15)   Peleg, A. Y.; Hooper, D. C. *N. Engl. J. Med.* **2010**, *362*, 1804–1813.
(16)   Evans, M. E.; Feola, D. J.; Rapp, R. P. *Ann. Pharmacother.* **1999**, *33*, 960–967.
(17)   Taubes, G. *Science* **2008**, *321*, 356–361.
(18)   Nathan, C. *Nature* **2004**, *431*, 899–902.
(19)   Payne, D. J.; Gwynn, M. N.; Holmes, D. J.; Pompliano, D. L. *Nat. Rev. Drug Discov.* **2006**, *6*, 29–40.
(20)   Bumann, D. *Curr. Opin. Microbiol.* **2008**, *11*, 387–392.

(21) Martinez, J. L.; Baquero, F. *Antimicrob. Agents Chemother.* **2000**, *44*, 1771–1777.
(22) Weingart, H.; Petrescu, M.; Winterhalter, M. *Curr. Drug Targets* **2008**, *9*, 789–796.
(23) Bootsma, M. C. J. *Proc. Natl. Acad. Sci. USA* **2006**, *103*, 5620–5625.
(24) So, A. D.; Gupta, N.; Brahmachari, S. K.; Chopra, I.; Munos, B.; Nathan, C.; Outterson, K.; Paccaud, J. P.; Payne, D. J.; Peeling, R. W.; Spigelman, M.; Weigelt, J. *Drug Resist. Update* **2011**, *14*, 88–94.
(25) Bugg, T. D. H.; Braddick, D.; Dowson, C. G.; Roper, D. I. *Trends Biotechnol.* **2011**, *29*, 167–173.
(26) Silver, L. L. *Curr. Opin. Microbiol.* **2003**, *6*, 431–438.
(27) Bugg, T. D. H.; Walsh, C. T. *Nat. Prod. Rep.* **1992**, *9*, 199–215.
(28) Green, D. W. *Expert Opin. Ther. Targets* **2002**, *6*, 1–19.
(29) Kahan, F. M.; Kahan, J. S.; Cassidy, P. J.; Kropp, H. *Ann. N. Y. Acad. Sci.* **1974**, *235*, 364–386.
(30) Neuhaus, F. C.; Lynch, J. L. *Biochemistry* **1964**, *3*, 471–480.
(31) Siewert, G.; Strominger, J. L. *Proc. Natl. Acad. Sci. U. S. A.* **1967**, *57*, 767–773.
(32) Bruggink, A. *Synthesis of β-Lactam Antibiotics: Chemistry, Biocatalysis & Process Integration*, Springer, 2001.
(33) Morin, R. B.; Gorman, M. *Chemistry and Biology of β-Lactam Antibiotics: Nontraditional β-Lactam Antibiotics*; Academic Press, 1982.
(34) Queener, S. F.; Webber, J. A.; Queener, S. W. *β-Lactam Antibiotics for Clinical Use*, M. Dekker, 1986.
(35) Koch, A. L. *Clin. Microbiol. Rev.* **2003**, *16*, 673–687.
(36) Ghuysen, J. M. *Trends Microbiol.* **1994**, *2*, 372–380.
(37) Fisher, J. F.; Meroueh, S. O.; Mobashery, S. *Chem. Rev.* **2005**, *105*, 395–424.
(38) Arias, C. A.; Murray, B. E. *N. Engl. J. Med.* **2009**, *360*, 439–443.
(39) Thomson, J. M.; Bonomo, R. A. *Curr. Opin. Microbiol.* **2005**, *8*, 518–524.
(40) Fleming, A. *Rev. Infect. Dis.* **1980**, *2*, 129–139.
(41) Acred, P.; Brown, D. M.; Turner, D. H.; Wilson, M. J. *Br. J. Pharmacol. Chemother.* **1962**, *18*, 356–369.
(42) Sutherland, R.; Rolinson, G. N. *J. Clin. Pathol.* **1964**, *17*, 461–465.
(43) Handsfield, H. H.; Clark, H.; Wallace, J. F.; Holmes, K. K.; Turck, M. *Antimicrob. Agents Chemother.* **1973**, *3*, 262–265.
(44) Neu, H. C.; Winshell, E. B. *Antimicrob. Agents Chemother.* **1970**, *10*, 407–410.
(45) Croydon, E. A.; Sutherland, R. *Antimicrob. Agents Chemother. (Bethesda)* **1970**, *10*, 427–430.
(46) Welling, P. G.; Huang, H.; Koch, P. A.; Craig, W. A.; Madsen, P. O. *J. Pharm. Sci.* **1977**, *66*, 549–552.
(47) Sutherland, R.; Croydon, E. A. P.; Rolinson, G. N. *Br. Med. J.* **1970**, *4*, 455–460.
(48) Marcy, S. M.; Klein, J. O. *Med. Clin. North Am.* **1970**, *54*, 1127–1143.
(49) Daikos, G. K.; Kontomichalou, P.; Paradelis, A. *Dtsch. Med. Wochenschr.* **1963**, *88*, 1678–1689.
(50) Fu, K. P.; Neu, H. C. *Antimicrob. Agents Chemother.* **1978**, *13*, 358–367.
(51) Klastersky, J.; Henri, A.; Daneau, D. *J. Clin. Pharmacol.* **1974**, *14*, 172–175.
(52) Verbist, L. *Antimicrob. Agents Chemother.* **1978**, *13*, 349–357.
(53) Fass, R. J.; Prior, R. B. *Antimicrob. Agents Chemother.* **1989**, *33*, 1268–1274.
(54) Bryskier, A. *J. Antibiot.* **2000**, *53*, 1028–1037.
(55) Newton, G. G. F.; Abraham, E. P. *Nature* **1955**, *175*, 548.
(56) Abraham, E. P.; Newton, G. G. F. *Biochem. J.* **1961**, *79*, 377–393.
(57) Morin, R. B.; Jackson, B. G.; E. H. Flynn; Roeske, R. W. *J. Am. Chem. Soc.* **1962**, *84*, 3400–3401.
(58) Chauvette, R. R.; Flynn, E. H.; Jackson, B. G.; Lavagnino, E. R.; Morin, R. B.; Mueller, R. A.; Pioch, R. P.; Roeske, R. W.; Ryan, C. W.; Spencer, J. L.; Van Heyningen, E. *J. Am. Chem. Soc.* **1962**, *84*, 3401–3402.
(59) O'Callaghan, C. H. *J. Antimicrob. Chemother.* **1975**, *1*, 1–12.
(60) Bryskier, A.; Aszodi, J.; Chantot, J. F. *Expert Opin. Invest. Drugs* **1994**, *3*, 145–171.
(61) Brogden, R. N.; Campoli-Richards, D. M. *Drugs* **1989**, *38*, 524–550.
(62) Nahata, M. C.; Barson, W. J. *Drug Intell. Clin. Pharm.* **1985**, *19*, 900–906.
(63) Richards, D. M.; Heel, R. C.; Brogden, R. N.; Speight, T. M.; Avery, G. S. *Drugs* **1984**, *27*, 469–527.

(64) O'Callaghan, C. H.; Acred, P.; Harper, P. B.; Ryan, D. M.; Kirby, S. M.; Harding, S. M. *Antimicrob. Agents Chemother.* **1980,** *17,* 876–883.

(65) Brogden, R. N.; Carmine, A.; Heel, R. C.; Morley, P. A.; Speight, T. M.; Avery, G. S. *Drugs* **1981,** *22,* 423–460.

(66) Okamoto, M. P.; Nakahiro, R. K.; Chin, A.; Bedikian, A.; Gill, M. A. *Am. J. Hosp. Pharm.* **1994,** *51,* 463–477; quiz 541–542.

(67) Nikaido, H.; Liu, W.; Rosenberg, E. Y. *Antimicrob. Agents Chemother.* **1990,** *34,* 337–342.

(68) Takeda, S.; Nakai, T.; Wakai, Y.; Ikeda, F.; Hatano, K. *Antimicrob. Agents Chemother.* **2007,** *51,* 826–830.

(69) Takeda, S.; Ishii, Y.; Hatano, K.; Tateda, K.; Yamaguchi, K. *Int. J. Antimicrob. Agents* **2007,** *30,* 443–445.

(70) Livermore, D. M.; Mushtaq, S.; Ge, Y. *J. Antimicrob. Chemother.* **2010,** *65,* 1972–1974.

(71) Titelman, E.; Karlsson, I. M.; Ge, Y.; Giske, C. G. *Diagn. Microbiol. Infect. Dis.* **2011,** *70,* 137–141.

(72) Springer, D.M. *Curr. Med. Chem. Anti-Infect. Agents* **2002,** *1,* 269–279.

(73) Page, M. G. P. *Curr. Opin. Pharm.* **2006,** *6,* 480–485.

(74) Duplessis, C.; Crum-Cianflone, N. F. *Clin. Med. Rev. Ther.* **2011,** *3,* a2466.

(75) Barbour, A.; Schmidt, S.; Rand, K. H.; Derendorf, H. *Int. J. Antimicrob. Agents* **2009,** *34,* 1–7.

(76) Morrow, J. D. *Am. J. Med. Sci.* **1992,** *303,* 35–39.

(77) Wick, W. E. *Appl. Environ. Microbiol.* **1967,** *15,* 765–769.

(78) Sanders, C. C. *Antimicrob. Agents Chemother.* **1977,** *12,* 490–497.

(79) Perry, C. M.; Brogden, R. N. *Drugs* **1996,** *52,* 125–158.

(80) Okamoto, S.; Hamana, Y.; Inoue, M.; Mitsuhashi, S. *Antimicrob. Agents Chemother.* **1987,** *31,* 1111–1116.

(81) Kahan, J. S.; Kahan, F. M.; Goegelman, R.; Currie, S. A.; Jackson, M.; Stapley, E. O.; Miller, T. W.; Miller, A. K.; Hendlin, D.; Mochales, S.; Hernandez, S.; Woodruff, H. B.; Birnbaum, J. *J. Antibiot.* **1979,** *32,* 1–12.

(82) Moellering, R. C., Jr; Eliopoulos, G. M.; Sentochnik, D. E. *J. Antimicrob. Chemother.* **1989,** *24 (Suppl A),* 1–7.

(83) Bonfiglio, G.; Russo, G.; Nicoletti, G. *Expert Opin. Investig. Drugs* **2002,** *11,* 529–544.

(84) Beadle, B. M.; Shoichet, B. K. *Antimicrob. Agents Chemother.* **2002,** *46,* 3978–3980.

(85) Pfeifer, Y.; Cullik, A.; Witte, W. *Int. J. Med. Microbiol.* **2010,** *300,* 371–379.

(86) Lipman, B.; Neu, H. C. *Med. Clin. North Am.* **1988,** *72,* 567–579.

(87) Barza, M. *Ann. Intern. Med.* **1985,** *103,* 552–560.

(88) Kahan, F. M.; Kropp, H.; Sundelof, J. G.; Birnbaum, J. *J. Antimicrob. Chemother.* **1983,** *12 (Suppl D),* 1–35.

(89) Kropp, H.; Sundelof, J. G.; Hajdu, R.; Kahan, F. M. *Antimicrob. Agents Chemother.* **1982,** *22,* 62–70.

(90) Birnbaum, J.; Kahan, F. M.; Kropp, H.; Macdonald, J. S. *Am. J. Med.* **1985,** *78,* 3–21.

(91) Fish, D. N.; Singletary, T. J. *Pharmacotherapy* **1997,** *17,* 644–669.

(92) Fukasawa, M.; Sumita, Y.; Harabe, E. T.; Tanio, T.; Nouda, H.; Kohzuki, T.; Okuda, T.; Matsumura, H.; Sunagawa, M. *Antimicrob. Agents Chemother.* **1992,** *36,* 1577–1579.

(93) Edwards, J. R.; Turner, P. J.; Wannop, C.; Withnell, E. S.; Grindey, A. J.; Nairn, K. *Antimicrob. Agents Chemother.* **1989,** *33,* 215–222.

(94) Odenholt, I. *Expert Opin. Investig. Drugs* **2001,** *10,* 1157–1166.

(95) Chahine, E. B.; Ferrill, M. J.; Poulakos, M. N. *Am. J. Health-Syst. Ph.* **2010,** *67,* 2015–2024.

(96) Sykes, R. B.; Bonner, D. P.; Bush, K.; Georgopapadakou, N. H.; Wells, J. S. *J. Antimicrob. Chemother.* **1981,** *8 (Suppl E),* 1–16.

(97) Imada, A.; Kitano, K.; Kintaka, K.; Muroi, M.; Asai, M. *Nature* **1981,** *289,* 590–591.

(98) Cimarusti, C. M.; Sykes, R. B. *Med. Res. Rev.* **1984,** *4,* 1–24.

(99) Wise, R.; Andrews, J. M.; Hancox, J. *J. Antimicrob. Chemother.* **1981,** *8 (Suppl E),* 39–47.

(100) Chin, N. X.; Neu, H. C. *Antimicrob. Agents Chemother.* **1988,** *32,* 84–91.

(101) Tanaka, S. K.; Summerill, R. A.; Minassian, B. F.; Bush, K.; Visnic, D. A.; Bonner, D. P.; Sykes, R. B. *Antimicrob. Agents Chemother.* **1987,** *31,* 219–225.

(102) Clark, J. M.; Olsen, S. J.; Weinberg, D. S.; Dalvi, M.; Whitney, R. R.; Bonner, D. P.; Sykes, R. B. *Antimicrob. Agents Chemother.* **1987**, *31*, 226–229.

(103) Walsh, T. R.; Toleman, M. A.; Poirel, L.; Nordmann, P. *Clin. Microbiol. Rev.* **2005**, *18*, 306–325.

(104) Bush, K.; Jacoby, G. A. *Antimicrob. Agents Chemother.* **2010**, *54*, 969–976.

(105) Bush, K.; Jacoby, G. A.; Medeiros, A. A. *Antimicrob. Agents Chemother.* **1995**, *39*, 1211–1233.

(106) Perez-Llarena, F. J.; Bou, G. *Curr. Med. Chem.* **2009**, *16*, 3740–3765.

(107) Bebrone, C.; Lassaux, P.; Vercheval, L.; Sohier, J.-S.; Jehaes, A.; Sauvage, E.; Galleni, M. *Drugs* **2010**, *70*, 651–679.

(108) Stachyra, T.; Levasseur, P.; Péchereau, M.-C.; Girard, A.-M.; Claudon, M.; Miossec, C.; Black, M. T. *J. Antimicrob. Chemother.* **2009**, *64*, 326–329.

(109) Bonnefoy, A.; Dupuis-Hamelin, C.; Steier, V.; Delachaume, C.; Seys, C.; Stachyra, T.; Fairley, M.; Guitton, M.; Lampilas, M. *J. Antimicrob. Chemother.* **2004**, *54*, 410–417.

(110) Kahne, D.; Leimkuhler, C.; Lu, W.; Walsh, C. *Chem. Rev.* **2005**, *105*, 425–448.

(111) Barna, J. C. J.; Williams, D. H. *Annu. Rev. Microbiol.* **1984**, *38*, 339–357.

(112) Levine, D. P. *Clin. Infect. Dis.* **2006**, *42*, S5–S12.

(113) Mccormick, M. H.; Mcguire, J. M.; Pittenger, G. E.; Pitterger, R. C.; Stark, W. M. *Antibiot. Annu.* **1955**, *3*, 606–611.

(114) Silva, J.; Batts, D. H.; Fekety, R.; Plouffe, J. F.; Rifkin, G. D.; Baird, I. *Am. J. Med.* **1981**, *71*, 815–822.

(115) Smith, T. L.; Pearson, M. L.; Wilcox, K. R.; Cruz, C.; Lancaster, M. V.; Robinson-Dunn, B.; Tenover, F. C.; Zervos, M. J.; Band, J. D.; White, E.; Jarvis, W. R. *N. Engl. J. Med.* **1999**, *340*, 493–501.

(116) Hamilton-Miller, J. M. *Infection* **2002**, *30*, 118–124.

(117) Campoli-Richards, D. M.; Brogden, R. N.; Faulds, D. *Drugs* **1990**, *40*, 449–486.

(118) Smith, W. J.; Drew, R. H. *Drugs Today* **2009**, *45*, 159–173.

(119) Higgins, D. L.; Chang, R.; Debabov, D. V.; Leung, J.; Wu, T.; Krause, K. M.; Sandvik, E.; Hubbard, J. M.; Kaniga, K.; Schmidt, D. E.; Gao, Q.; Cass, R. T.; Karr, D. E.; Benton, B. M.; Humphrey, P. P. *Antimicrob. Agents Chemother.* **2005**, *49*, 1127–1134.

(120) Green, R.; Noller, H. F. *Ann. Rev. Biochem.* **1997**, *66*, 679–716.

(121) Dahlberg, A. E.; Zimmermann, R. A. *Encyclopedia of Microbiology*, Arnette Blackwell: Paris, 1992.

(122) Garrett, R. A.; Douthwaite, S.; Liljas, A.; Matheson, A. T.; Moore, P. B.; Noller, H. F. *The Ribosome: Structure, Function, Antibiotics and Cellular Interactions*, 2nd ed., American Society for Microbiology: Washington DC, 2000.

(123) Vazquez, D. *Inhibitors of Protein Synthesis*, Spinger Verlag: Berlin, 1979.

(124) Gale, E. F.; Cundliffe, E.; Reynolds, P. E.; Richmond, M. H.; Waring, M. J. *The Molecular Basis of Antibiotic Actions*, Wiley: London, 1981.

(125) Spahn, C. M.; Prescott, C. D. *J. Mol. Med.* **1996**, *74*, 423–439.

(126) Schatz, A.; Waksman, S. A. *Proc. Soc. Exp. Biol. Med.* **1944**, *57*, 244–248.

(127) Waksman, S. A. *My Life with the Micorbes*, Simon and Schuster: New York, 1954.

(128) Waksman, S. A. *The Conquest of Tuberculosis*, University of California Press: Berkeley, 1964.

(129) Waksman, S. A.; Lechevalier, H. A. *Science* **1949**, *109*, 305–307.

(130) Beaucaire, G. *J. Chemother.* **1995**, *7*, 111–123.

(131) Umezawa, H.; Ueda, M.; Maeda, K.; Yagishita, K.; Kondo, S.; Okami, Y.; Utahara, R.; Osata, Y.; Nitta, K.; Takeucki, T. *J. Antibiot.* **1957**, *10*, 181–189.

(132) Weinstein, M. J.; Luedemann, G. M.; Oden, E. M.; Wagman, G. H. *J. Med. Chem.* **1963**, *6*, 463–464.

(133) Kumar, C. G.; Himabindu, M.; Jetty, A. *Crit. Rev. Biotechnol.* **2008**, *28*, 173–212.

(134) Higgins, C. E.; Kastners, R. E. *Antimicrob. Agents Chemother.* **1967**, *7*, 324–331.

(135) Brogden, R. N.; Pinder, R. M.; Sawyer, P. R.; Speight, T. M.; Avery, G. S. *Drugs* **1967**, *12*, 166–200.

(136) Kawaguchi, H.; Naito, T.; Nakagowa, S.; Fugijawa, K. *J. Antibiot.* **1972**, *25*, 695.

(137) Price, K. E.; Pursiano, T. A.; DeFuria, M. D.; Wright, G. E. *Antimicrob. Agents Chemother.* **1974**, *5*, 143–152.

(138) Chambers, H. F.; Sande, M. A. In *The Pharmacological Basis of Therapeutics*; McGraw-Hill: New York, 1996, pp. 1103–1121.

(139) Moellering, R. C. *Rev. Infect. Dis.* **1983**, *5*, S212.
(140) Young, L. S.; Hewitt, W. L. *Antimicrob. Agents Chemother.* **1973**, *4*, 617.
(141) Vakulenko, S. B.; Mobashery, S. *Clin. Microbiol. Rev.* **2003**, *16*, 430–450.
(142) Calderwood, S. A.; Wennersten, C.; Moellering, R. C.; Kunz, L. J.; Krogstad, D. J. *Antimicrob. Agents Chemother.* **1979**, *12*, 401.
(143) Gutschik, E.; Jepsen, O. B.; Mortensen, I. *J. Infect. Dis.* **1977**, *135*, 832.
(144) Watanakunakorn, C.; Glotzbecker, C. *Antimicrob. Agents Chemother.* **1977**, *12*, 346.
(145) Weinstein, A.; Moellering, R. C. J. *JAMA* **1973**, *223*, 1030.
(146) Moazed, D.; Noller, H. F. *Nature* **1987**, *327*, 389–394.
(147) Purohit, P.; Stem, S. *Nature* **1994**, *370*, 659–662.
(148) Fourmy, D.; Recht, M. I.; Blanchard, S. C.; Puglisi, J. D. *Science* **1996**, *274*, 1367–1371.
(149) Mingeot-Leclercq, M. P.; Glupczynski, Y.; Tulkens, P. M. *Antimicrob. Agents Chemother.* **1999**, *43*, 727–737.
(150) Noller, H. F. *Science* **2005**, *309*, 1508–1514.
(151) Magnet, S.; Blanchard, J. S. *Chem. Rev.* **2005**, *105*, 477–498.
(152) Poole, K. *J. Antimicrob. Chemother.* **2005**, *56*, 20–51.
(153) Jana, S.; Deb, J. K. *Curr. Drug Targets* **2005**, *6*, 353–361.
(154) Davies, J. *Science* **1994**, *264*, 375–382.
(155) Watanabe, A.; Nagai, J.; Adachi, Y.; Katsube, T.; Kitahara, Y.; Murakami, T.; Takano, M. *J. Control. Release* **2004**, *95*, 423–433.
(156) Corrado, A. P.; Morais, I. P.; Prado, W. A. *Acta. Physiol. Pharmacol. Lactinoam.* **1989**, *39*, 419–430.
(157) Rougier, F.; Claude, D.; Maurin, M.; Sedoglavic, A.; Ducher, M.; Corvaisier, S.; Jelliffe, R.; Maire, P. *Antimicrob. Agents Chemother.* **2003**, *47*, 1010–1016.
(158) Nakashima, T.; Teranishi, M.; Hibi, T.; Kobayashi, M.; Umemura, M. *Acta Oto-Laryngol.* **2000**, *120*, 904–911.
(159) Duggar, B. M. *Ann. NY Acad. Sci.* **1948**, *51*, 177–181.
(160) Chopra, I.; Roberts, M. *Microbiol. Mol. Biol. Rev.* **2001**, *65*, 232–260.
(161) Agwuh, K. N.; MacGowan, A. *J. Antimicrob. Chemother.* **2006**, *58*, 256–265.
(162) Stokstad, E. L. R.; Jukes, T. H.; Pierce, J.; Page, A. C., Jr.; Franklin, A. L. *J. Biol. Chem.* **1949**, *180*, 647–654.
(163) Watanabe, T. *Bacteriol. Rev.* **1963**, *27*, 87–115.
(164) Roberts, M. C. *FEMS Microbiol. Rev.* **1996**, *19*, 1–24.
(165) Wright, G. D. *Nat. Rev. Microbiol.* **2007**, *5*, 175–186.
(166) Maxwell, I. H. *Biochim. Biophys. Acta.* **1967**, *138*, 337–346.
(167) Brodersen, D. E.; Clemons Jr., W. M.; Carter, A. P.; Morgan-Warren, R. J.; Wimberly, B. T.; Ramakrishnan, V. *Cell* **2000**, *103*, 1143–1154.
(168) Schnappinger, D.; Hillen, W. *Arch. Microbiol.* **1996**, *165*, 359–369.
(169) Levy, S. B.; McMurry, L. *Biochem. Biophys. Res. Commun.* **1974**, *56*, 1060–1068.
(170) Yamaguchi, A.; Udagawa, T.; Sawai, T. *J. Biol. Chem.* **1990**, *265*, 4809–4813.
(171) Burdett, V. *J. Biol. Chem.* **1991**, *266*, 2872–2877.
(172) Yang, W.; Moore, I. F.; Koteva, K. P.; Bareich, D. C.; Hughes, D. W.; Wright, G. D. *J. Biol. Chem.* **2004**, *279*, 52346–52352.
(173) Ross, J. I.; Eady, E. A.; Cove, J. H.; Cunliffe, W. J. *Antimicrob. Agents Chemother.* **1998**, *42*, 1702–1705.
(174) Rose, W. E.; Rybak, M. J. *Pharmacotherapy* **2006**, *26*, 1099–1110.
(175) Projan, S. J. *Clin. Infect. Dis.* **2010**, *50 (Suppl 1)*, S24–S25.
(176) Pankey, G. A. *J. Antimicrob. Chemother.* **2005**, *56*, 470–480.
(177) Bergeron, J.; Ammirati, M.; Danley, D.; James, L.; Norcia, M.; Retsema, J.; Strick, C. A.; Su, W.-G.; Sutcliffe, J.; Wondrack, L. *Antimicrob. Agents Chemother.* **1996**, *40*, 2226–2228.
(178) Zhanel, G. G.; Homenuik, K.; Nichol, K.; Noreddin, A.; Vercaigne, L.; Embil, J.; Gin, A.; Karlowsky, J. A.; Hoban, D. J. *Drugs* **2004**, *64*, 63–88.
(179) Chopra, I. *Drug Resist. Update* **2002**, *5*, 119–125.

(180) Doan, T..-L.; Fung, H. B.; Mehta, D.; Riska, P. F. *Clin. Ther.* **2006**, *28*, 1079–1106.
(181) Woodward, R. B. *Angew. Chem.* **1957**, *69*, 585.
(182) Woodward, R. B. *Angew. Chem.* **1957**, *69*, 50.
(183) Brockmann, H.; Henkel, W. *Chem. Ber.* **1951**, *84*, 284–288.
(184) McGuire, J. M.; Bunch, R. L.; Anderson, R. C.; Boaz, H. E.; Flynn, E. H.; Powell, H. M.; Smith, J. W. *Antibiot. Chemother.* **1952**, *2*, 281.
(185) Washington, J. A., II; Wilson, W. R. *Mayo Clin. Proc.* **1985**, *60*, 271.
(186) Washington, J. A., II; Wilson, W. R. *Mayo Clin. Proc.* **1985**, *60*, 189.
(187) Ma, Z.; Nemoto, P. *Curr. Med. Chem. Anti-Infec. Agents* **2002**, *1*, 15.
(188) Vester, B.; Douthwaite, A. *Antimicrob. Agents Chemother.* **2001**, *45*, 1.
(189) Kurath, P.; Jones, P. H.; Egan, R. S.; Perun, T. J. *Experimentia* **1971**, *27*, 362.
(190) Katz, L.; Ashley, G. W. *Chem. Rev.* **2005**, *105*, 499–528.
(191) Bonnefoy, A.; Guitton, M.; Delachaume, C.; Le Priol, P.; Girard, A. M. *Antimicrob. Agents Chemother.* **2001**, *45*, 1688.
(192) Bonnefoy, A.; Le Priol, P. *J. Antimicrob. Chemother.* **2001**, *47*, 471.
(193) Ma, Z.; Clark, R. F.; Brazzale, A.; Wang, S.; Rupp, M. J.; Li, L.; Griesgraber, G.; Zhang, S.; Yong, H.; Phan, L. T.; Nemoto, P. A.; Chu, D. T.; Plattner, J. J.; Zhang, X.; Zhong, P.; Cao, Z.; Nilius, A. M.; Shortridge, V. D.; Flamm, R.; Mitten, M.; Meulbroek, J.; Ewing, P.; Alder, J.; Or, Y. S. *J. Med. Chem.* **2001**, *44*, 4137.
(194) Shortridge, V. D.; Zhong, P.; Cao, Z.; Beyer, J. M.; Almer, L. S.; Ramer, N. C.; Doktor, S. Z.; Flamm, R. K. *Antimicrob. Agents Chemother.* **2002**, *46*, 783.
(195) Yassin, H. M.; Dever, L. L. *Expert Opin. Investig. Drugs* **2001**, *10*, 353.
(196) Morimoto, S.; Aldachi, T.; Matsunaga, T.; Kashimura, M.; Asaka, T.; Watanabe, Y.; Sota, K.; Sekiuchi, K. USP 4990602 (1991).
(197) Alvarez-Elcoro, S.; Enzler, M. J. *Mayo Clin. Proc.* **1999**, *74*, 613–634.
(198) Amsden, G. W. *Ann. Pharmacother.* **1995**, *29*, 906–917.
(199) Denis, A.; Agouridas, C. *Bioorg. Med. Chem. Lett.* **1998**, *8*, 2427.
(200) Agouridas, C.; Denis, A.; Auger, J. M.; Benedetti, Y.; Bonnefoy, A.; Bretin, F.; Chantot, J. F.; Dussarat, A.; Fromentin, C.; D'Ambrieres, S. G.; Lachaud, S.; Laurin, P.; Le Martret, O.; Loyau, V.; Tessot, N. *J. Med. Chem.* **1998**, *41*, 4080.
(201) Denis, A.; Agouridas, C.; Auger, J. M.; Benedetti, Y.; Bonnefoy, A.; Bretin, F.; Chantot, J. F.; Dussarat, A.; Fromentin, C.; D'Ambrieres, S. G.; Lachaud, S.; Laurin, P.; Le Martret, O.; Loyau, V.; Tessot, N.; Pejac, J. M.; Perron, S. *Bioorg. Med. Chem. Lett.* **1999**, *9*, 3075.
(202) Doern, G. V.; Brown, S. D. *J. Infect.* **2004**, *48*, 56–65.
(203) Zhanel, G. G.; Walters, M.; Noreddin, A.; Vercaigne, L. M.; Wierzbowski, A.; Embil, J. M.; Gin, A. S.; Douthwaite, S.; Hoban, D. J. *Drugs* **2002**, *62*, 1771.
(204) Gaynor, M.; Mankin, A. S. *Curr. Topics Med. Chem.* **2003**, *3*, 949–960.
(205) Sturgill, M. G.; Rapp, R. P. *Ann. Pharmacother.* **1992**, *26*, 1099–1108.
(206) Hansen, L. H.; Mauvais, P.; Douthwaite, S. *Mol. Microbiol.* **1999**, *31*, 623–631.
(207) Douthwaite, S.; Hansen, L. H.; Mauvais, P. *Mol. Microbiol.* **2000**, *36*, 183–193.
(208) Clark, J. P.; Langston, E. *Mayo Clin. Proc.* **2003**, *78*, 1113–1124.
(209) Zuckerman, J. *Infect. Dis. Clin. N. Am.* **2004**, *18*, 621–649.
(210) Phillips, I. *J. Antimicrob. Chemother.* **1981**, *7*, 11–18.
(211) Birkenmeyer, R. D. USP 3435025 (1969).
(212) Birkenmeyer, R. D.; Kagan, F. *J. Med. Chem.* **1970**, *13*, 616–619.
(213) Spizek, J.; Novotna, J.; Rezanka, T. *Adv. App. Microbiol.* **2004**, *56*, 121–154.
(214) O'Grady, F. *Antibiotic and Chemotherapy: Anti-Infective Agents and Their Use in Therapy*; 7th ed.; Churchill Livingstone: New York, 1997.
(215) Schlunzen, F.; Zarivach, R.; Harms, J.; Bashan, A.; Tocilj, A.; Albrecht, R.; Yonath, A.; Franceschi, F. *Nature* **2001**, *413*, 814.
(216) Vaquez, D. In *Antibiotics*; Springer-Verlag, 1967; Vol. 1, pp. 387–403.
(217) Vaquez, D. In *Antibiotics*; Springer-Verlag, 1975; Vol. 3, pp. 521–534.

(218) Pechere, J. C. *Drugs* **1996**, *51*, 13–19.

(219) Paris, J. M.; Barriere, J. C.; Smith, C.; Bost, P. E. In *Recent Progress in The Chemical Synthesis of Antibiotics*; Springer-Verlag, 1990, pp. 182–248.

(220) Barriere, J. C.; Paris, J. M. *R. Soc. Chem. Spec. Publ.* **1997**, *198*, 27–41.

(221) Abbanat, D.; Macielag, M.; Bush, K. In *Antimicrobial Pharmacodynamics in Theory and Clinical Practice*, Informa Healthcare: New York, 2007.

(222) Neu, H. C.; Chin, N. X.; Gu, J. W. *J. Antimicrob. Chemother.* **1992**, *30*, 83–94.

(223) Vannuffel, P.; Cocito, C. *Drugs* **1996**, *51*, 20.

(224) Beyer, D.; Pepper, K. *Expert Opin. Investig. Drugs* **1998**, *7*, 591–599.

(225) Mukhtar, T. A.; Wright, G. D. *Chem. Rev.* **2005**, *105*, 529–542.

(226) Bozdogan, B.; Appelbaum, P. C. *Int. J. Antimicrob. Agent.* **2004**, *23*, 113–119.

(227) Gregory, W. A.; Brittelli, D. R.; Wang, C.-L.; Wuonola, M. A.; McRipley, R. J.; Eustice, D. C.; Eberly, V. S.; Bartholomew, P. T.; Slee, A. M.; Forbes, M. *J. Med. Chem.* **1989**, *32*, 1673.

(228) Slee, A. M.; Wuonola, M. A.; McRipley, R. J.; Zajac, I.; Zawada, M. J.; Bartholomew, P. T. *Antimicrob. Agents Chemother.* **1987**, *31*, 1791–1797.

(229) Brickner, S. J.; Hutchinson, D. K.; Barbachyn, M. R.; Manninen, P. R.; Ulanowicz, D. A.; Garmon, S. A. etl. al. *J. Med. Chem.* **1996**, *39*, 673–679.

(230) Diekema, D. J.; Jones, R. N. *Lancet.* **2001**, *358*, 1975–1982.

(231) Birmingham, M. C.; Rayner, C. R.; Meagher, A. K.; Flavin, S. M.; Batts, D. H.; Schentag, J. J. *Clin. Infect. Dis.* **2003**, *36*, 159–168.

(232) Wilcox, M. H. *Expert Opin. Pharmacother.* **2005**, *6*, 2315–2326.

(233) Poce, G.; Zappia, G.; Porretta, G. C.; Botta, B.; Biava, M. *Expert Opin. Ther. Patents* **2008**, *18*, 97–121.

(234) Xu, J.; Golshani, A.; Aoki, H.; Remme, J.; Chosay, J.; Shinabarger, D. L.; Ganoza, M. C. *Biochem. Biophys. Res. Commun.* **2005**, *328*, 471–476.

(235) Im, W. B.; Choi, S. H.; Park, J.-Y.; Choi, S. H.; Finn, J.; Yoon, S.-H. *Eur. J. Med. Chem.* **2011**, *46*, 1027–1039.

(236) Kompis, I. M.; Islam, K.; Then, R. L. *Chem. Rev.* **2005**, *105*, 593–620.

(237) Hardman, J.; Limbird, L.; Gilman, A. *Goodman & Gilman's The Pharmacological Basis of Therapeutics*, 10th ed., McGraw-Hill: New York, 2001.

(238) Allison, M. E. M.; Kennedy, A. C.; McGeachie, J.; McDonald, G. A. *Scot. Med. J.* **1969**, *14*, 355–360.

(239) Huovinen, P.; Sundstrom, L.; Swedberg, G.; Skold, O. *Antimicrob. Agents Chemother.* **1995**, *39*, 279–289.

(240) Nakamura, H.; Uetrecht, J.; Cribb, A. E.; Miller, M. A.; Zahid, N.; Hill, J.; Josephy, P. D.; Grant, D. M.; Spielberg, S. P. *J. Pharmacol. Exp. Ther.* **1995**, *274*, 1099–1104.

(241) Hawser, S.; Lociuro, S.; Islam, K. *Biochem. Pharmacol.* **2006**, *71*, 941–948.

(242) Maxwell, A. *Trends Microbiol.* **1997**, *5*, 102–109.

(243) Mitscher, L. A. *Chem. Rev.* **2005**, *105*, 559–592.

(244) Chu, D. T. W.; Shen, L. L. In *Fifty Years of Antimicrobials: Past Perspectives and Future Trends*, Society for General Microbiology, 1995; pp. 111–137.

(245) Laponogov, I.; Sohi, M. K.; Veselkov, D. A.; Pan, X.-S.; Sawhney, R.; Thompson, A. W.; McAuley, K. E.; Fisher, L. M.; Sanderson, M. R. *Nat. Struct. Mol. Biol.* **2009**, *16*, 667–669.

(246) Appelbaum, P. C.; Hunter, P. A. *Int. J. Antimicrob. Ag.* **2000**, *16*, 5–15.

(247) Drlica, K.; Hooper, D. C. In *Quinolone Antimicrobial Agents*; 2003; pp. 19–40.

(248) Jalal, S.; Ciofu, O.; Hoiby, N.; Gotoh, N.; Wretlind, B. *Antimicrob. Agents Chemother.* **2000**, *44*, 710–712.

(249) Tanaka, M.; Wang, T.; Onodera, Y.; Uchida, Y.; Sato, K. *J. Infect. Chemother.* **2000**, *6*, 131–139.

(250) Lesher, G. Y.; Froelich, E. J.; Gruett, M. D.; Hays, J.; Brundage, R. P. *J. Med. Chem.* **1962**, *5*, 1063–1065.

(251) McChesney, E. W.; Froelich, E. J.; Lesher, G. Y.; Crain, A. V. R.; Rosi, D. *Toxicol. Appl. Pharm.* **1964**, *6*, 292–309.

(252) Koga, H.; Itoh, A.; Murayama, S.; Suzue, S.; Irikura, T. *J. Med. Chem.* **1980**, *23*, 1358–1363.

(253) Domagala, J. M. *J. Antimicrob. Chemother.* **1994**, *33*, 685–706.

(254) Wise, R.; Andrews, J. M.; Edwards, L. J. *Antimicrob. Agents Chemother.* **1983**, *23*, 559–564.

(255) Atarashi, S.; Yokohama, S.; Yamazaki, K.; Sakano, K.; Imamura, M.; Hayakawa, I. *Chem. Pharm. Bull.* **1987**, *35*, 1896–1902.

(256) Nightingale, C. H. *Pharmacotherapy* **2000**, *20*, 245–256.

(257) Ling, T. K. W.; Liu, E. Y. M.; Cheng, A. F. B. *Chemotherapy* **1999**, *45*, 22–27.

(258) Lazarczyk, D. A.; Goldstein, N. S.; Gordon, S. C. *Dig. Dis. Sci.* **2001**, *46*, 925–926.

(259) Sun, Q.; Zhu, R.; Foss, F. W., Jr; Macdonald, T. L. *Bioorg. Med. Chem. Lett.* **2007**, *17*, 6682–6686.

(260) Hoeksema, H.; Johnson, J. L.; Hinman, J. W. *J. Am. Chem. Soc.* **1955**, *77*, 6710–6711.

(261) Edwards, D. I. *J. Antimicrob. Chemother.* **1993**, *31*, 9–20.

(262) Land, K. M.; Johnson, P. J. *Drug Resist. Update* **1999**, *2*, 289–294.

(263) Deenadayalu, V. P.; Orinion, E. J.; Chalasani, N. P.; Yoo, H. Y. *Clin. Gastroenterol. Hepatol.* **2005**, *3*, xxix.

(264) Floss, H. G.; Yu, T.-W. *Chem. Rev.* **2005**, *105*, 621–632.

(265) Ezekiel, D. H.; Hutchins, J. E. *Nature* **1968**, *220*, 276–277.

(266) Grosset, J.; Leventis, S. *Clin. Infect. Dis.* **1983**, *5*, S440–S446.

(267) Hancock, R. E. *Trends Microbiol.* **1997**, *5*, 37–42.

(268) Zhang, Y. M.; Rock, C. O. *J. Lipid. Res.* **2009**, *50*, S115–S119.

(269) Ainsworth, G. C.; Brown, A. M.; Brownlee, G. *Nature* **1947**, *160*, 263.

(270) Stansly, P. G.; Shepherd, R. G.; White, H. J. *Johns Hopkins Hosp. Bull.* **1947**, *81*, 43–54.

(271) Koyama, Y.; Kurosasa, A.; Tsuchiya, A.; Takakuta, K. *J. Antibiot.* **1950**, *3*, 457–458.

(272) Orwa, J. A.; Govaerts, C.; Bussan, R.; Roets, E.; Van Schepdael, A.; Hoogmartens, J. *J. Chromatogr. A* **2001**, *912*, 369–373.

(273) Storm, D. R.; Rosental, K. S.; Swanson, P. E. *Annu. Rev. Biochem.* **1977**, *46*, 723–763.

(274) Kucers, A.; Crowe, S.; Grayson, M. L.; Hoy, J. *The Use of Antibiotics*, 5th ed., Butterworth-Heinemann: Oxford, England, 1997.

(275) Nikaido, H. *Microbiol. Mol. Biol. Rev.* **2003**, *67*, 593–656.

(276) Hancock, R. E. *J. Med. Microbiol.* **1997**, *46*.

(277) Hancock, R. E.; Lehrer, R. *Trends Biotechnol.* **1998**, *16*, 82–88.

(278) Clausell, A.; Garcia-Subirats, M.; Pujol, M.; Busquets, M. A.; Rabanal, F.; Cajal, Y. *J. Phys. Chem. B* **2007**, *111*, 551–563.

(279) Melo, M. N.; Ferre, R.; Castanho, M. A. *Nat. Rev. Microbiol.* **2009**, *7*, 245–250.

(280) Powers, J. P.; Hancock, R. E. *Peptides* **2003**, *24*, 1681–1691.

(281) Groisman, E. A.; Kayser, J.; Soncini, F. C. *J. Bacteriol.* **1997**, *179*, 7040–7045.

(282) Schindler, M.; Osborn, M. J. *Biochemistry* **1979**, *18*, 4425–4430.

(283) Hancock, R. E. W. *Lancet.* **1997**, *349*, 418–422.

(284) Rose, F.; Heuer, K. U.; Sibelius, U.; Hombach-Klonisch, S.; Kiss, L.; Seeger, W.; Grimminger, F. *J. Infect. Dis.* **2000**, *182*, 191–199.

(285) Tsuberry, H.; Ofek, I.; Cohen, S.; Fridkin, M. *J. Med. Chem.* **2000**, *43*, 3085–3092.

(286) Warren, H. S.; Kania, S. A.; Siber, G. R. *Antimicrob. Agents Chemother.* **1985**, *28*, 107–112.

(287) Moskowitz, S. M.; Ernst, R. K.; Miller, S. I. *J. Bacteriol.* **2004**, *186*, 575–579.

(288) Winfield, M. D.; Groisman, E. A. *Proc. Natl. Acad. Sci. USA* **2004**, *101*, 17162–17167.

(289) Brodsky, I. E.; Ernst, R. K.; Miller, S. I.. *J. Bacteriol.* **2002**, *184*, 3203–3213.

(290) Campos, M. A.; Vargas, M. A.; Regueiro, V. *Infect. Immun.* **2004**, *72*, 7107–7114.

(291) Lee, H.; Hsu, F. F.; Turk, J.; Groisman, E. A. *J. Bacteriol.* **2004**, *186*, 4124–4133.

(292) Helander, I. M.; Kilpelainen, I.; Vaara, M. *Mol. Microbiol.* **1994**, *11*, 481–487.

(293) Bell, A.; Hancock, R. E. *J. Bacteriol.* **1989**, *171*, 3211–3217.

(294) Nicas, T. I.; Hancock, R. E. *J. Gen. Microbiol.* **1983**, *129*, 509–517.

(295) Mathur, J.; Waldor, M. K. *Infect. Immun.* **2004**, *72*, 3577–3583.

(296) Kock-Weser, J.; Sidel, V. W.; Federman, P.; Kanarek, D. C. F.; Eaton, A. E. *Ann. Intern. Med.* **1970**, *72*, 857–868.

(297) Ledson, M. J.; Gallagher, M. J.; Cowperthwaite, C.; Convery, R. P.; Walshaw, M. J. *J. Eur. Respir.* **1998,** *12,* 592–594.

(298) Falagas, M. E.; Kasiakou, S. K. *Crit. Care* **2006,** *10,* R27.

(299) Hopper, J., Jr.; Jawetz, E.; Hinman, F., Jr. *Am. J. Med. Sci.* **1953,** *225,* 402–409.

(300) Lindesmith, L. A.; Baines, R. D., Jr.; Bigelo, D. B.; Petty, T. L. *Ann. Intern. Med.* **1968,** *68,* 318–327.

(301) Wolinsky, E.; Hines, J. D. *N. Engl. J. Med.* **1972,** *266,* 759–762.

(302) Moestrup, S. K.; Cui, S.; Vorum, H.; Bregengard, C.; Bjorn, S. E.; Norris, K.; Glieman, J.; Christensen, E. I. *J. Clin. Invest.* **1995,** *96,* 1404–1413.

(303) Falagas, M. E.; Kasiakou, S. K. *Clin. Infect. Dis.* **2005,** *40,* 1333–1341.

(304) Li, J.; Nation, R. L. *Clin. Infect. Dis.* **2006,** *43,* 663–664.

(305) Kings, A.; Phillips, I. *J. Antimicrob. Chemother.* **2001,** *48,* 219–223.

(306) Baltz, R. H.; Miao, V.; Wrigley, S. W. *Nat. Prod. Rep.* **2005,** *22,* 717–741.

(307) Tally, F. P.; DeBruin, M. F. *J. Antimicrob. Chemother.* **2000,** *46,* 523–526.

(308) Barry, A. L.; Fuchs, P. C.; Brown, S. D. *Antimicrob. Agents Chemother.* **2001,** *45,* 1919–1922.

(309) Critchley, I. A.; Draghi, D. C.; Sahm, D. F.; Thornsberry, C.; Jones, M. E.; Karlowsky, J. A. *J. Antimicrob. Chemother.* **2003,** *51,* 639–649.

(310) Fluit, A. C.; Schmitz, F. J.; Verhoef, J.; Milatovic, D. *Int. J. Antimicrob. Agents* **2004,** *24,* 59–66.

(311) Fluit, A. C.; Schmitz, F. J.; Verhoef, J.; Milatovic, D. *Antimicrob. Agents Chemother.* **2004,** *48,* 1007–1011.

(312) Peterson, P. J.; Bradford, P. A.; Weiss, W. J.; Murphy, T. M.; Sum, P. E.; Projan, S. J. *Antimicrob. Agents Chemother.* **2002,** *46,* 2595–2601.

(313) Richter, S. S.; Kealey, D. E.; Murray, C. T.; Heilmann, K. P.; Coffman, S. L.; Doern, G. V. *J. Antimicrob. Chemother.* **2003,** *52,* 123–127.

(314) Rybak, M. J.; Hershberger, E.; Moldovan, T.; Grucz, R. G. *Antimicrob. Agents Chemother.* **2000,** *44,* 1062–1066.

(315) Snydman, D. R.; Jacobus, N. V.; McDermott, L. A.; Supran, S. E. *Antimicrob. Agents Chemother.* **2000,** *44,* 1710-1712.

(316) Streit, J. M.; Jones, R. N.; Sader, H. S. *J. Antimicrob. Chemother.* **2004,** *53,* 669–674.

(317) Suh, B. *Diagn. Microbiol. Infect. Dis.* **2010,** *66,* 111–115.

(318) Golan, Y.; Poutsiaka, D. D.; Tozzi, S.; Tozzi, S.; Hadley, S.; Snydman, D. R. *J. Antimicrob. Chemother.* **2001,** *47,* 364–365.

(319) Citron, D. M.; Appleman, M. D. *J. Clin. Microbiol.* **2006,** *44,* 3814–3818.

(320) Spanjaard, I.; Vandenbroucke-Grauls, C. M. *Antimicrob. Agents Chemother.* **2008,** *52,* 1850–1851.

(321) Goldstein, E. J.; Citron, D. M.; Merriam, C. V.; Warren, Y. A.; Tyrrell, K. L.; Fernandex, H. T. *Antimicrob. Agents Chemother.* **2003,** *47,* 337–341.

(322) Tedesco, K. L.; Rybak, M. J. *Pharmacotherapy* **2004,** *24,* 41–57.

(323) Straus, S. K.; Hancock, R. E. W. *Biochim. Biophys. Acta* **2006,** *1758,* 1215–1223.

(324) Jung, D.; Powers, J. P.; Straus, S. K.; Hancock, R. E. W. *Chem. Phys. Lipid* **2008,** *154,* 120–128.

(325) Lakey, J. H.; Ptak, M. *Biochemistry* **1988,** *27,* 4639–4645.

(326) Silverman, J. A.; Perimutter, N. G.; Shapiro, H. M. *Antimicrob. Agents Chemother.* **2003,** *47,* 2538–2544.

(327) English, B. K.; Maryniw, E. M.; Talati, A. J.; Meals, E. A. *Antimicrob. Agents Chemother.* **2006,** *50,* 2225–2227.

(328) Kanafani, Z. A.; Corey, G. R. *Expert Rev. Anti. Infect. Ther.* **2007,** *5,* 177–184.

(329) Friedman, L.; Alder, J. D.; Silverman, J. A. *Antimicrob. Agents Chemother.* **2006,** *50,* 2137–2145.

(330) Julian, K.; Kosowska-Shick, K.; Whitener, C.; Roos, M.; Labischinski, H.; Rubio, A.; Parent, L.; Ednie, L.; Koeth, L.; Bogdanovich, T.; Applebaum, P. C. *Antimicrob. Agents Chemother.* **2007,** *51,* 3445–3448.

(331) Murthy, M. H.; Olson, M. E.; Wickert, R. W.; Fey, P. D.; Jalali, Z. *J. Med. Microbiol.* **2008,** *57,* 1036–1038.

(332) Pertel, P. E.; Bernardo, P.; Fogerty, C.; Matthews, P.; Northland, R.; Benvenuto, M.; Thorne, G. M.; Luperchio, S. A.; Arbeit, R. D.; Alder, J. D. *Clin. Infect. Dis.* **2008,** *46,* 1142–1151.

(333) Silverman, J. A.; Morton, L. I.; Vanpraagh, A. D.; Li, T.; Alder, J. D. *J. Infect. Dis.* **2005**, *191*, 2149–2152.

(334) Oleson, F. B.; Berman, C. L.; Kirkpatrick, J. B.; Regan, K. S.; Lai, J. -J.; Tally, F. P. *Antimicrob. Agents Chemother.* **2000**, *44*, 2948–2953.

(335) Benvenuto, M.; Benziger, D. P.; Yankelev, S.; Vigliani, G. *Antimicrob. Agents Chemother.* **2006**, *50*, 3245–3249.

(336) Kosmidis, C.; Levine, D. P. *Expert Opin. Pharmacother.* **2010**, *11*, 615–625.

(337) Yin, N.; He, Y.; Herradura, P.; Pearson, A.; Li, J.; Mascio, C.; Townsend-Howland, K.; Silverman, J.; Steenbergen, J.; Thorne, G.; Citron, D.; Van Praagh, A. D. G.; Mortin, L. I.; Pawlick, R.; Oleson, R.; Keith, D.; Metcalf, C. In *50th Intersci. Conf. Antimicrob. Agents Chemother. (ICAAC)*; F1-1612: Boston, 2010.

(338) Mortin, L. I.; Van Praagh, A. D. G.; Zhang, S.; Arya, A.; Chuong, L.; Kang, C.; Zhang, X.-X.; Li. T. In *50th Intersci. Conf. Antimicrob. Agents Chemother. (ICAAC)*; B-707: Boston, 2010.

(339) Snydman, D. R.; Jacobus, N. V.; McDermott, L. A. In *50th Intersci. Conf. Antimicrob. Agents Chemother. (ICAAC)*; Boston, 2010.

(340) Mascio, C.; Townsend, K.; Silverman, J. In *50th Intersci. Conf. Antimicrob. Agents Chemother. (ICAAC)*; C1-097: Boston, 2010.

(341) Butler, M. S.; Cooper, M. A. *J. Antibiot.* **2011**, *64*, 413–425.

(342) Xie, J.; Pierce, J. G.; James, R. C.; Okano, A.; Boger, D. L. *J. Am. Chem. Soc.* **2011**, *133*, 13946–13949.

(343) Miller, J. R.; Thanabal, V.; Melnick, M. M.; Lall, M.; Donovan, C.; Sarver, R. W.; Lee, D.-Y.; Ohren, J.; Emerson, D. *Chem. Biol. Drug Des.* **2010**, *75*, 444–454.

(344) In *Comprehensive Medicinal Chemistry II*, Vol. 2, Elsevier, 2007; pp. 731–748.

(345) Jayasuriya, H.; Zink, D.; Basilio, A.; Vicente, F.; Collado, J.; Bills, G.; Goldman, M. L.; Motyl, M.; Huber, J.; Dezeny, G.; Byrne, K.; Singh, S. B. *J. Antibiot.* **2009**, *62*, 265–269.

(346) Zhang, C.; Ondeyka, J. G.; Zink, D. L.; Basilio, A.; Vicente, F.; Collado, J.; Platas, G.; Huber, J.; Dorso, K.; Motyl, M.; Byrne, K.; Singh, S. B. *Bioorg. Med. Chem. Lett.* **2009**, *17*, 2162–2166.

(347) Donald, R. G. K.; Skwish, S.; Forsyth, R. A.; Anderson, J. W.; Zhong, T.; Burns, C.; Lee, S.; Meng, X.; LoCastro, L.; Jarantow, L. W.; Martin, J.; Lee, S. H.; Taylor, I.; Robbins, D.; Malone, C.; Wang, L.; Zamudio, C. S.; Youngman, P. J.; Phillips, J. W. *Chem. Biol.* **2009**, *16*, 826–836.

(348) Silver, L. *Clin. Microbiol. Rev.* **2011**, *24*, 71–109.

(349) Amaral, L.; Fanning, S. *Curr. Med. Chem.* **2011**, *18*, 2969–2980.

(350) Woodford, N.; Wareham, D. W. *J. Antimicrob. Chemother.* **2009**, *63*, 225–229.

(351) Greenberg, D. E.; Marshall-Batty, K. R.; Brinster, L. R.; Zarember, K. A.; Shaw, P. A.; Mellbye, B. L.; Iversen, P. L.; Holland, S. M.; Geller, B. L. *J. Infec. Dis.* **2010**, *201*, 1822–1830.

(352) Bai, H.; Zhou, Y.; Hou, Z.; Xue, X.; Meng, J.; Luo, X. *Infect. Disord.: Drug Targets* **2011**, *11*, 175–187.

(353) Rubio, A.; Conrad, M.; Haselbeck, R. J.; Kedar, G. C.; Brown-Driver, V.; Finn, J.; Silverman, J. A. *Antimicrob. Agents Chemother.* **2011**, *55*, 364–367.

(354) Barker, J. J. *Drug Discov. Today* **2006**, *11*, 391–404.

(355) O'Shea, R.; Moser, H. E. *J. Med. Chem.* **2008**, *51*, 2871–2878.

(356) Škedelj, V.; Tomaši , T.; Mašič, L. P.; Zega, A. *J. Med. Chem.* **2011**, *54*, 915–929.

(357) Nathan, C. *Cell Host & Microbe* **2011**, *9*, 343–348.

(358) Liu, Y.; Gray, N. S. *Nat. Chem. Biol.* **2006**, *2*, 358–364.

(359) Singh, J.; Petter, R. C.; Baillie, T. A.; Whitty, A. *Nat. Rev. Drug Discov.* **2011**, *10*, 307–317.

(360) Brötz-Oesterhelt, H.; Sass, P. *Future Microbiol.* **2010**, *5*, 1553–1579.

(361) Galloway, W. R. J. D.; Bender, A.; Welch, M.; Spring, D. R. *Chem. Commun.* **2009**, 2446.

(362) Airoldi, C.; Sommaruga, S.; Merlo, S.; Sperandeo, P.; Cipolla, L.; Polissi, A.; Nicotra, F. *ChemBioChem* **2011**, *12*, 719–727.

(363) Cipolla, L.; Polissi, A.; Airoldi, C.; Gabrielli, L.; Merlo, S.; Nicotra, F. *Curr. Med. Chem.* **2011**, *18*, 830–852.

(364) Hopkins, A. L. *Nat. Chem. Biol.* **2008**, *4*, 682–690.

(365) Shen, Y.; Liu, J.; Estiu, G.; Isin, B.; Ahn, Y.-Y.; Lee, D.-S.; Barabási, A.-L.; Kapatral, V.; Wiest, O.; Oltvai, Z. N. *P. Natl. Acad. Sci.* **2010,** *107,* 1082–1087.
(366) Then, R. L.; Sahl, H.-G. *Curr. Pharm. Des.* **2010,** *16,* 555–566.

11

Antiviral Drug Discovery

Nicholas A. Meanwell, Stanley V. D'Andrea, Christopher W. Cianci,
Ira B. Dicker, Kap-Sun Yeung, Makonen Belema, and Mark Krystal

1   Introduction

The major viruses that are known to infect human beings are compiled in Table 1, a synopsis that represents a broad range of acute and chronic infections that manifest as a variety of physiological and pathophysiological effects. The development and implementation of vaccines has been a very successful strategy to prevent many viral infections, most notably polio, smallpox, hepatitis B virus (HBV), measles, mumps and rubella, and very recently rotavirus and papilloma virus.[1] Annual influenza vaccination campaigns are a familiar part of the Fall season in the northern hemisphere, required because of genetic drift associated with this virus. However, there are many viral diseases that require therapeutic intervention because vaccines either have not been developed or lack therapeutic efficacy. For example, the HBV vaccine introduced in 1992 is not therapeutically effective, leaving the 400 million worldwide carriers of the virus to rely upon antiviral drugs to control viral replication, preserve liver function, and ultimately prevent the development of liver cirrhosis and cancer. The modern era of small-molecule antiviral drug discovery is associated with introduction of the nucleoside analog acyclovir (1) for the treatment of herpes simplex virus (HSV) infection in 1978.[2–4] Although not the first marketed antiviral agent, acyclovir (1) was the first drug to be used widely to control a viral infection, propelled by the introduction of the prodrug valacyclovir (2), which markedly improves oral bioavailability and offers a more convenient, less frequent dosing regimen.[5–8]

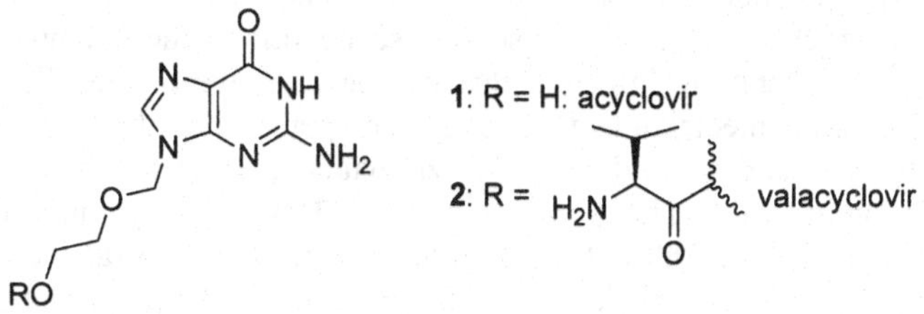

1: R = H: acyclovir

2: R =   valacyclovir

Table 1. The Major Viruses Causing Human Infection and Disease

Virus	Family	Genetic Basis of Virus	Effect
Smallpox	Vaccinia	DNA[1]	Skin abrasions
Rotavirus	Reovirus	ds RNA[2]	Diarrhea
Poliovirus	Picornavirus	+ve RNA[3]	Poliomyelitis
Hepatitis A Virus	Picornavirus	+ve RNA	Hepatitis
Rhinovirus	Rhinovirus	+ve RNA	Respiratory infection, cold
Rubella	Togavirus	+ve RNA	Respiratory infection, German measles
Yellow Fever Virus	Flavivirus	+ve RNA	Yellow fever
Hepatitis C Virus	Flavivirus	+ve RNA	Hepatitis
West Nile Virus	Flavivirus	+ve RNA	Meningitis, encephalitis
Norovirus	Calcivirus	+ve RNA	Acute gastroenteritis
Herpes Simplex Virus	Herpesevirus	DNA	Cold, genital sores
Varicella Zoster Virus	Herpesevirus	DNA	Chicken pox, shingles
Human Cytomegalovirus	Herpesevirus	DNA	Various
Epstein-Barr Virus	Herpesevirus	DNA	Mononucleosis
Adenovirus	Adenovirus	DNA	Eye infection, cold symptoms
Human Papilloma Virus	Papovirus	DNA	Genital and skin warts
Parvovirus B19	Parvovirus	DNA	Rash
Hepatitis B Virus	Hepadnavirus	ds/ss DNA[5]	Hepatitis
Coronavirus	Coronavirus	+ve RNA (segmented)	Cold symptoms
SARS-CoV	Coronavirus	+ve RNA (segmented)	Severe acute respiratory syndrome
Respiratory Syncytial Virus	Paramyxovirus	-ve RNA	Bronchitis
Measles Virus	Paramyxovirus	-ve RNA	Respiratory
Mumps Virus	Paramyxovirus	-ve RNA	Respiratory
Parainfluenza Viruses	Paramyxovirus	-ve RNA	Lower respiratory tract infections
Bunyavirus	Bunyavirus	-ve RNA[4]	Fever
Hantavirus	Bunyavirus	-ve RNA	Hemorrhagic fever
Influenza Virus	Orthomyxovirus	-ve RNA (segmented)	Respiratory
Ebola Virus	Filovirus	-ve RNA	Hemorrhagic fever
Human Immunodeficiency Virus	Retrovirus	+ve RNA	Acquired Immunodeficiency Syndrome (AIDS)

[1]Double strand DNA; [2]Double strand RNA; [3]Positive strand RNA; [4]Negative strand RNA; [5]Combination of double- and single-strand DNA

HBV and the human immunodeficiency virus-1 (HIV-1) and hepatitis C virus (HCV) are the most problematic viruses currently infecting humans. The discovery of HIV-1 infection in 1981 as the cause of acquired immunodeficiency syndrome (AIDS) and the subsequent spread of this virus triggered a tremendous effort in both basic virology and drug discovery focused on providing therapeutic agents. This enterprise was successful to the extent that HIV-1 is now largely viewed more as a chronic infection that can be effectively controlled by combinations selected from the 25 unique medications that have been approved by the Spring of 2012 since the advent of the nucleoside azidothymidine (AZT, zidovudine, **3**) as the first antiretroviral agent approved for marketing in 1987.[9] The discovery of AZT (**3**) has its origin in a seminal publication from the Prusoff group that described the antiviral properties of several iodinated, pyrimidine-based, deoxynucleoside analogs, including **4** and **5**, that inhibited herpes simplex virus replication in cell culture.[10,11] The safety and selectivity of acyclovir (**1**) for HSV set a standard for the nucleoside class of antiviral agents that were explored extensively in the 1980s and 1990s, with marketed inhibitors of hepatitis B virus (HBV) and HIV-1 emerging from those efforts.[12] However, human toxicity with this class of drugs has been unpredictable, as exemplified by the problems encountered by fialuridine (FIAU, **6**) which was explored clinically for the treatment of HBV but caused severe toxicity to the liver in several patients after 13 weeks of therapy.[13,14] This was due to the sudden onset of hepatic failure and lactic acidosis that was not reversible; 5 patients died and 2 survived after liver transplants. The toxicity was subsequently attributed to the inhibition of human mitochondrial DNA polymerase-γ by FIAU (**6**).

3, zidovudine (AZT)   **4**

5   **6**, fialuridine (FIAU)

The molecular characterization of HIV-1 in 1983, just 2 years after the first reports of the physiological effects of this retrovirus that has infected approximately 60 million people worldwide and been responsible for almost 30 million deaths, catalyzed what has become a 30-year commitment to identify and develop drugs capable of controlling virus replication.[15,16] These efforts have been highly successful, with 25 individual drugs representing 6 distinct mechanistic classes currently marketed that, when used in combinations known as highly active retroviral therapy (HAART), have rendered HIV-1 a controllable, chronic infection rather than the death sentence that characterized this disease in the first decade after its discovery.[17] It was the advent of the HIV-1 protease inhibitors in late 1995 and early 1996 that transformed the treatment of HIV-1 infection by allowing combinations with nucleoside analogs that blunted the emergence of resistance, a significant problem when treatment relied upon a single drug. Although early combination regimens necessitated the administration of multiple tablets on difficult and challenging dosing schedules, the approval in 2006 of Atripla provided the convenience of a once-a-day therapy that improves compliance.[18,19] Atripla combines the nucleoside analogs tenofovir disoproxil fumarate (**7**), the fumarate salt of a prodrug of tenofovir (**8**), and emtricitabine (**9**) with the nonnucleoside reverse transcriptase inhibitor (NNRTI) efavirenz (**10**) into a single, fixed-dose tablet. It is a consequence of these successes that the development of new antiretroviral therapeutics is focused as much on identifying agents with good safety profiles and convenient dosing regimens as it is on efficacy.

7: R = [structure] tenofovir disoproxil

8: R = H, tenofovir

9, emtricitabine (FTC)

10, efavirenz

The clinical experience with therapeutic treatment of HIV-1 infection has had a significant impact on the development of direct-acting antiviral agents that inhibit HCV, markedly compacting timelines for both the discovery and development process. A key difference between HIV-1 and HCV is that the latter is curable, responding to immune stimulation or antiviral therapy following the establishment of a chronic infection. A significant advance in the therapy of HCV infection was the introduction in 2002 of a combination of pegylated interferon-α (Peg-IFNα) and the nucleoside analog ribavirin (11), a drug regimen that effected cure rates of 40–55%.[20] However, neither of these agents are specific antiviral agents, with Peg-IFNα stimulating immune function while the actions of ribavirin (11) are still enigmatic 30 years after its marketing approval.[21,22] The molecular cloning of HCV in 1989 facilitated the elucidation of the genomic organization and characterization of the key enzymes involved in replication, setting the stage for the identification of effective inhibitors.[23,24] The advent of replicons, initially subgenomic but subsequently genomic, that recapitulated basic aspects of viral replication provided a significant boost to HCV drug discovery that was followed by the development of a replicating virus that allowed study of all aspects of the virus lifecycle.[25–27] These systems allowed convenient assessment of the effects of HCV inhibitors in a cell-based setting and facilitated the discovery of inhibitors using a chemical genomics screening approach.[28] The licensing of the HCV NS3 protease inhibitors boceprevir (12), which occurred on May 13, 2011, and telaprevir (13), approved a few days thereafter, represented a significant milestone in the development of direct-acting HCV therapeutics and these agents are currently used as add-on therapy to Peg-IFNα and ribavirin (11). However, there are multiple HCV inhibitors representing a range of mechanistic approaches currently under evaluation in clinical studies, with several trials assessing the efficacy of combinations of direct-acting antiviral agents that have the potential to replace Peg-IFNα and ribavirin (11). Combinations of small-molecule agents will likely be required to treat HCV in order to minimize the emergence of resistance.[29] The replication rates for HCV infection are estimated to be 10–100-fold higher than HIV-1 and, since the HCV NS5B polymerase has poor proofreading capacity, it has been estimated that all possible single and double mutants are synthesized multiple times every day in an infected individual.[30,31] The anticipated limitations of monotherapeutic regimens have been realized in clinical trials with telaprevir (13) while more recent studies indicate that two drug combinations can also lead to the development of resistance.[31–33]

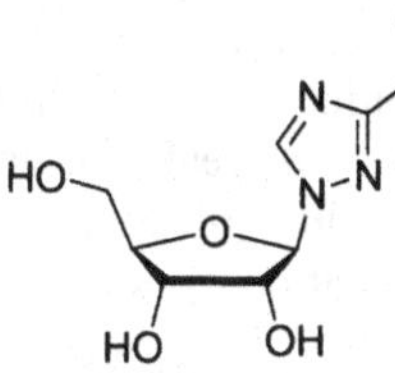

11, ribavirin 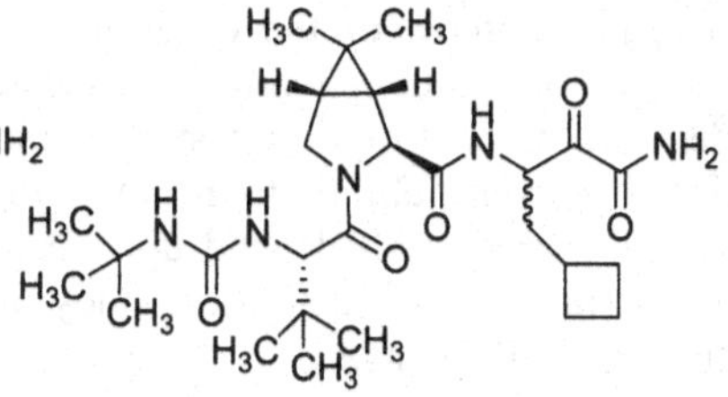   12, boceprevir

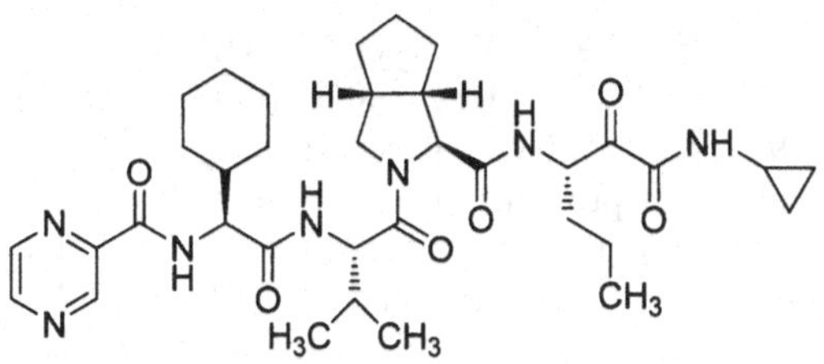

13, telaprevir

The first marketed antiviral agent was the influenza inhibitor amantadine (14) for which *in vitro* and *in vivo* activity were described in 1964,[34] a discovery that was quickly followed by clinical studies that facilitated its licensing as a prophylactic and therapeutic agent in 1966.[35–38] The homologue rimantadine (15) was approved by the FDA in 1994,[38–40] while the influenza neuraminidase inhibitors zanamavir (16) and GS4071 (17), the former administered topically while the latter is orally bioavailable as the ethyl ester prodrug oseltamivir (18), were approved in 1999.[41,42]

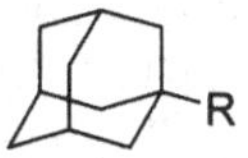

14: R = NH₂: amantadine
15: R = CH(CH₃)NH₂: rimantadine   16, zanamivir

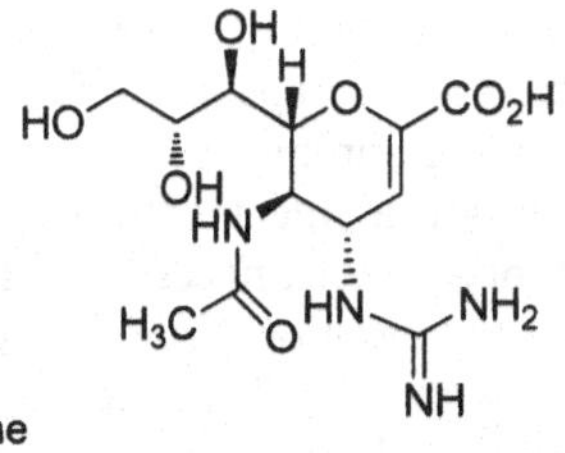

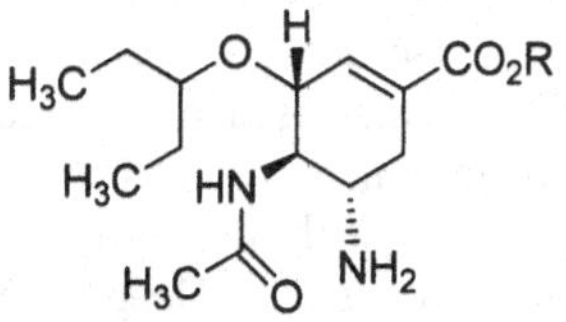

17: R = H, GS4071
18: R = C₂H₅, oseltamivir

One of the keys to the success of antiviral drug discovery and development has been the identification and characterization of the infectious agent and the development of cell-based systems that facilitate detailed study of the virus lifecycle. These model systems typically recapitulate full replication of the virus in cell culture or rely upon the replication of subgenomic elements in host cells and usually provide effective and predictive models for efficacy in humans. That HCV was not identified until 1989 suggests that there may be additional infectious agents awaiting discovery, anticipating the proposal that viruses may underlie several chronic diseases, including multiple sclerosis, arthritis and other autoimmune diseases, and several cancers, among others.[43–49] The rapid molecular characterization of the coronavirus responsible for the outbreak of severe acute respiratory syndrome (SARS) in Hong Kong between November 2002 and July

2003 provides an elegant example of the capability of contemporary molecular virology.[50–53] However, the rapid spread of this infection to 37 countries within a few months, disseminated by the facility of modern jet travel, provides a stark demonstration of our vulnerability to highly infectious viral infections.   The ability to respond to virus outbreaks of this type with the development of antiviral therapy will depend upon the capability to rapidly identify infectious agents and create the *in vitro* systems that allow study of replication cycles and the identification of potent and effective antiviral agents.

2   Human Immunodeficiency Virus-1 Inhibitors

The primate lentiviruses include the HIV types 1 (HIV-1) and 2 (HIV-2) and simian immunodeficiency viruses (SIVs) of which HIV-1 and HIV-2 infect humans, HIV-1-like viruses infect chimpanzees and SIV variants infect African monkeys.[53,54]   HIV-1 infection in humans causes AIDS, a major global health problem in which the body's immune system breaks down, leaving the victim vulnerable to opportunistic infections.[55,56]   The Joint United Nations Program on HIV/AIDS (UNAIDS) estimates that, at the end of 2010, 34 million individuals worldwide were infected with HIV-1, with the majority (~68%) living in sub-Saharan Africa, while in the United States, there were approximately 1.2 million individuals living with HIV-1 infections.[57] Since the beginning of the pandemic in 1981, AIDS is estimated to have killed nearly 25 million people worldwide, including 425,000 Americans, although the number of people dying from AIDS-related diseases fell in 2010 to 1.8 million, down from the peak of 2.2 million annually in the mid-2000s.[57]   The changing demographics are attributed to the broader use of antiretroviral therapy, particularly in lower income countries, that has resulted in an estimated 700,000 deaths being averted in 2010.   Notably, AIDS has replaced malaria and tuberculosis as the world's deadliest infectious disease and it was the sixth leading cause of death worldwide in 2008.[58]

Following the approval of AZT (3) in 1987, the 24 additional distinct antiretroviral drugs that have been approved for use in combination therapy by the Spring of 2012 are compiled in Table 2 (note that fosamprenavir is a prodrug formulation of amprenavir).[59–61]   When used in combination, these drugs represent an effective therapeutic regimen to control the replication of HIV-1, reducing the disease to a manageable, chronic infection, but they are not curative.[9,62,63] Missed drug treatments contribute to the development of drug resistance due to the archived nature of the virus within long-lived memory T cells, necessitating treatment courses that are for the life of the patient.[64–66]   Despite this clear limitation, drug therapy for HIV-1 disease has been remarkably successful at extending life, with mortality rates of those on HAART regimens approaching that of the general population.[67–69]   In this section, the history of the development of HIV-1 therapies is reviewed and the potential of novel targets and approaches beyond the currently available, direct-acting antiretrovirals (DAAs) is explored.[70]

Table 2. Drugs Marketed for the Treatment of HIV-1 Infection

Drug Generic Name	Trade Name	Launch Year
Nucleoside reverse transcriptase inhibitors (NRTIs)		
Zidovudine (AZT)	Retrovir	1987
Didanosine (ddI)	Videx	1991
Zalcitabine (ddC)	Hivid	1992*
Stavudine (d4T)	Zerit	1995
Lamivudine (3TC)	Epivir	1998
Abacavir	Ziagen	1999
Tenofovir disoproxil	Viread	2001
Emtricitabine (FTC)	Emtriva	2003
Non-nucleoside reverse transcriptase inhibitors (NNRTIs)		
Nevirapine	Viramune	1996
Efavirenz	Sustiva, Stocrin	1998
Delavirdine	Rescriptor	1999
Etravarine	Intelence	2008
Rilpirivine	Endurant	2011
Protease inhibitors		
Saquinavir	Invirase	1995
Indinavir	Crixivan	1996
Ritonavir	Norvir	1996
Nelfinavir	Viracept	1997
Amprenavir	Agenerase, Prozei	1999
Lopinavir + ritonavir	Kaletra, Aluvia	2000
Atazanavir	Reyataz, Zrivada	2003
Fosamprenavir	Lexiva, Tetzir	2005
Tipranavir	Aptivus	2006
Darunavir	Prezista	2006
Entry inhibitors		
Enfuvirtide	Fuzeon	2003
Maraviroc	Celsentri, Selzentry	2007
Integrase inhibitors		
Raltegravir	Isentress	2009

* Withdrawn from the market December 31, 2006.

Currently available drugs belong to 6 different classes that includes 8 nucleoside (nucleotide) reverse transcriptase (RT) inhibitors (NRTIs), 5 nonnucleoside RT inhibitors (NNRTIs), 9 discrete protease inhibitors (PIs), and 1 integrase inhibitor and are listed in Table 2.[60,61] These drugs target the three essential enzymes encoded by the virus: RT, protease, and integrase. Additionally, two drugs block virus entry into the host cell at different points: the fusion inhibitor enfuvirtide, which prevents the fusion of the virus envelope with the host cell membrane by compromising the function of gp41, and maraviroc, a CCR5 antagonist which interferes with the interaction of the virus with one of its key coreceptors on the host cell.[60,61] Although there are, in principle, innumerable combination drug regimens that can be used for the treatment of HIV-1, recommended regimens typically combine three drugs from at least two different classes as a means of controlling virus replication. In practice, however, only a few key combinations are typically used to treat newly diagnosed (naïve) patients based upon efficacy, convenience, and safety.[71] More complex combinations and dosing regimens must be used to treat patients who have become resistant to their current therapies or to manage difficult drug-drug interactions (DDI).[71–73]

2.1 *HIV-1 Entry Inhibitors*

HIV-1 drugs can broadly be categorized according to which mechanistic step in the life-cycle they inhibit.[74] HIV-1 initiates access to its host cell by the binding of the viral envelope protein gp120, which is expressed as trimers on the surface of the virion, to the host cell receptor CD4, the prelude to a series of carefully choreographed events that offer multiple opportunities for therapeutic intervention.[75–82] Small-molecule agents that interfere with the attachment of gp120 to CD4 have been described and are potentially attractive anti-HIV-1 drugs.[83–89] One such agent, BMS-663068 (**19**), the phosphonooxymethyl prodrug of BMS-626529 (**20**), is currently in clinical development formulated as the tris(hydroxymethyl)aminomethane salt.[90–93] The prodrug element of BMS-663068 (**19**) is designed to increase the solubility of BMS-626529 (**20**), which is designated as a BCS (Biophramaceutiocs Classification System) class 2 drug based on its poor solubility and high membrane permeability. The phosphonooxymethyl moiety increases the solubility of the compound in the gut, thereby reducing dissolution-limited bioavailability.[93,94] Cleavage of the phosphate moiety of **19** by alkaline phosphatase found on the brush border membrane of enterocytes leads to the liberation of BMS-626529 (**20**) and one molecule of formaldehyde after the spontaneous degradation of the chemically unstable hydroxymethyl intermediate, a prodrug strategy that relies upon absorption of the parent drug being faster than precipitation in the gut to be successful (Scheme 1).[94–96]

Scheme 1. Mechanism of activation of the prodrug BMS-663068 (**19**)
of the HIV-1 attachment inhibitor BMS-626529 (**20**).

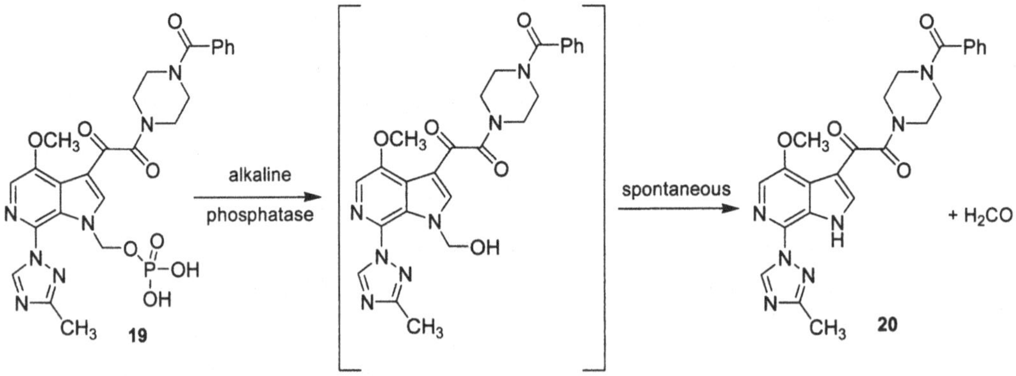

The interaction of gp120 with CD4 effects a conformational change in the viral protein that exposes binding sites for one of two cellular coreceptors, the chemokine G-protein-coupled receptors (GPCRs) CCR5 or CXCR4.[97–99] The engagement of a co-receptor by gp120 triggers the later steps in the process, the rearrangement of viral gp41 into its fusion-active conformation. Inhibitors of the interaction of gp120 with CCR5 and CXR4 have been identified, although only the CCR5 antagonist maraviroc (**21**) has been approved for therapy, while plerixafor (AMD-3100, **22**) and AMD-11070 (**23**) are representative of the kind of basic amine derivative that has been characterized as antagonizing the CXCR4 receptor.[79,100–111] Plerixafor (AMD-3100, **22**) has been approved as a hematopoietic stem cell mobilizer but was not pursued as an antiretroviral agent due to its lack of oral bioavailability and cardiac effects. CCR5 receptors are used by M-tropic viruses that are predominantly responsible for initial infection while T-tropic viruses exploit CXCR4 as the coreceptor, with these viruses usually associated with advanced infection. There has been considerable focus on CCR5 antagonists since this locus does not appear to be essential in humans, as revealed by individuals whose CCR5 receptors lack 32 amino acids from their C-terminus (CCR5-Δ32).[112] Individuals homozygous for this deletion, which creates a nonfunctional receptor, are protected from contracting HIV-1 infection, while heterozygotes can be infected but exhibit much slower disease progression.[113] However, therapy targeted to CCR5 receptors is not efficacious against CXCR4-tropic viruses and there has been concern that the use of CCR5 receptor antagonists would lead to coreceptor switching by selecting for a virus population that prefers to use the CXCR4 receptor.[82,114,115] However, in practice this does not appear to have been a major issue with the clinical use of maraviroc (21).[82,114,115] The optimization of a piperidine derivative, discovered using high-throughput screening (HTS), that led to maraviroc (21) illuminated some interesting aspects of drug design.[100,101,103,116–119] Persistent Type 1 cytochrome P450 (CYP 450) 2D6 inhibition was a liability with early piperidine derivatives, which led to an approach that sought to sterically hinder the basic nitrogen atom as a means of interfering with its postulated association with the Fe atom of the enzyme.[118,119] This tactic culminated in the adoption of the tropane bicyclic ring system as the key scaffold for optimization. A second problem involved minimizing interaction with the $K_v11.1$ cardiac potassium ion channel encoded by the human *ether-a-go-go*-related gene (hERG), inhibition of which can result in potentially fatal arrhythmias associated with long QT syndrome, in which repolarization of the heart is slowed.[118,119] The inhibition of the hERG ion channel was independent of the pKa of the central basic amine, and both the 3-isopropyl-5-methyl-1,2,4-triazole and the difluorocyclohexylamide were identified as structural elements that, when combined in maraviroc (21), conferred excellent antiviral potency (EC_{90} = 2 nM for inhibition of HIV_{BAL} replication in PM-1 cells) accompanied by modest hERG inhibition (19% at 10 μM), a low molecular weight (513), and reasonable lipophilicity (Log D = 2.1).[100,101,103,116–119]

21 (maraviroc)

22 (AMD-3100)   **23 (AMD-11070)**

HIV-1 gp41 is deployed in a trimeric form on the virion surface in a metastable state that is maintained by its close association with gp120.[120–123] The gp41 protein incorporates 4 structural elements that are critical to its function—a hydrophobic fusion peptide, heptad repeat segments in the amino (HR-N) and carboxy (HR-C) regions and a transmembrane domain, which are organized topologically as depicted in Figure 1. The conformational changes in gp120 induced by consecutive engagement with CD4 and a coreceptor liberate gp41, which undergoes a significant structural rearrangement in which the hydrophobic fusion peptide is released from a protected environment and inserts into the host cell membrane. The amino terminus heptad repeats associate into a trimeric species in a fashion that creates binding grooves that recognize the carboxy terminus heptad repeats.[120–123] The association of HR-C with HR-N creates a hairpin-shaped, 6-helix bundle structure for gp41 whose formation draws the host and viral membranes into proximity, a critical prelude to membrane fusion.[124,125] The observation that peptides derived from HR-C exhibit antiviral activity by successfully competing with native HR-C for association with HR-N was an important discovery that not only provided critical insights into the biochemical mechanism of virus-host fusion but also set the stage for the development of enfuvirtide (T-20, **24**), a 36 residue peptide comprised of residues 127–169 of gp41 HR-C that exhibits an EC_{50} of ~1 ng/mL in cell culture.[126–132] Enfuvirtide (**24**) is produced by chemical synthesis and is marketed as Fuzeon®. However, the peptidic nature of enfuvirtide (**24**) prevents oral absorption, necessitating subcutaneous administration by injection twice daily, an inconvenient dosing regimen that has relegated the drug primarily to use in salvage therapy for those with HIV-1 infections whose viral load cannot be adequately controlled by oral medications.[133–137] Small molecule inhibitors that mimic the action of enfuvirtide (**24**) have been sought since the elucidation of the mechanism of function of gp41 and the identification of a potential binding pocket for small molecules.[138–142] However, inhibitors of this type have largely proven to be elusive, with the indole **25** the most recently characterized compound that demonstrates an EC_{50} = 800 nM in a HIV-1 fusion assay and is an active inhibitor of viral replication in cell culture with EC_{50}s of 0.92–2.80 µM.[143–145]

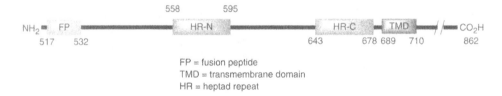

Figure 1. Topological arrangement of HIV-1 gp41.

$$Ac\text{-}_{643}YTSLIHSLIEESQNQQEKNEQELLELDKWASLWNWF_{678}\text{-}NH_2$$

24 (enfuvirtide, T-20)

25

2.2   *HIV-1 Reverse Transcriptase Inhibitors*

Once HIV-1 has gained entry into the cell, the virus sheds its protein coat, releasing its genetic information (RNA strand) and the polymerase enzyme, which is packaged in the virion, into the cytosol where it can take advantage of the availability of substrate to reverse transcribe the RNA genome into a double-stranded DNA copy.[146-148] RT is an essential enzyme that assembles as an asymmetric heterodimer, composed of one p66 subunit (560 amino acids) and a second p51 subunit (440 amino acids), both of which are encoded by the same sequence in the viral genome.[149,150] The p66 subunit adopts a topography that is similar to that of a right hand comprised of fingers (amino acids 1–85 and 118–155), palm (amino acids 86–117 and 156–237), and thumb (amino acids 238–318) domains.[150-157] The active site contains three aspartates (110, 185, and 186), with the functional catalytic pocket being formed and activated after substrate binding by the association of the fingers and palm domains.[158] The process of reverse transcription is complex, as RT acts at different stages of the process, initially as an RNA-dependent DNA polymerase and later as a DNA-dependent DNA polymerase. The replication process is not continuous across the incoming template but rather requires two separate strand transfer events. There is also an embedded RNaseH activity within the last 120 amino acids of the p66 subunit that hydrolyzes the RNA strand as it is being copied to DNA.[159] In principle, inhibition of either the polymerase or RNaseH functions will inhibit virus replication, but, at this time and despite numerous efforts, only the polymerase function has been successfully exploited for drug development.[159-161]

The first HIV-1 drugs were those that inhibited the reverse transcriptase (RT) enzyme by virtue of acting as competitive inhibitors of the normal deoxyribonucleotide substrates that interfere with elongation of the growing chain, typically by the absence of a 3-hydroxyl moiety, thereby stalling the synthesis of complementary DNA by acting as obligate chain terminators.[162-165] Nucleosides and the nucleotide analog tenofovir (**8**) are prodrugs that must be converted by host cell kinases into 5'-triphosphate nucleotides or, in the case of tenofovir (**8**), the diphosphorylated phosphonate, in order to act as competitive substrates and be incorporated in the nascent DNA (Scheme 2).[162-165] That the inhibitory activity of the nucleoside analogs is dependent on their propensity to be converted to the corresponding triphosphates in cell-based assays represents a significant complication in the interpretation of structure-activity relationships (SARs), which is often unpredictable during the discovery phase.

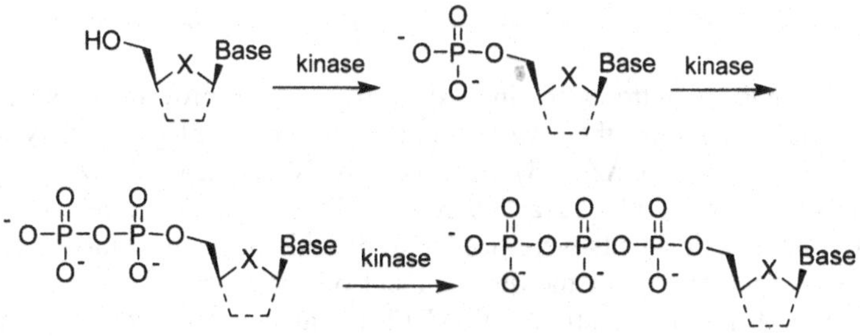

Scheme 2. Phosphorylation of nucleoside analogs

Nevertheless, this has been a particularly successful approach to HIV-1 drug discovery with 8 of the 25 strucuturally distinct approved ARVs members of this class (Table 2). Although AZT (3) was the first agent licensed for the treatment of HIV-1, just 4 years after the identification of the virus as the etiologic agent responsible for AIDS, it was originally prepared in 1964 as part of a study of nucleoside derivatives as potential anticancer agents.[166] In 1974 AZT (3) was shown to have antiretroviral activity, which justified its selection among the first compounds screened for HIV-1 inhibition in the early days following identification and characterization of the virus.[167] The HIV-1 inhibitory activity of AZT (3) was first described in 1985 using what was then a newly developed HIV-1 cell culture assay.[168] AZT (3) was shown to be a substrate analog of thymidine, with incorporation of AZT triphosphate into the growing DNA chain resulting in termination.[169] Similar to AZT (3), d4T (stavudine, 26) and ddC (zalcitabine, 27) were also first prepared as potential anticancer agents,[170–176] while the HIV-1 inhibitory activity of ddI (28) was reported contemporaneously with that of ddC (27). Ultimately all three compounds were successfully developed and approved, an important step toward both revolutionizing and rationalizing HIV-1 therapy. However, there were many problems associated with poor tolerability and the emergence of resistance with the use of these drugs.[177] In particular, ddI (28) was poorly tolerated due to an association with pancreatitis and peripheral neuropathy, restricting its use to salvage therapy.[178,179] ddC (27) also caused significant toxicity, delaying its development and limiting its dosage and efficacy, problems that ultimately led to its withdrawal from the market at the end of 2006.[180] A major issue that was not well characterized at that time but was later determined to be characteristic of the class of dideoxynucleoside analogs was that many of the side effects associated with these agents were driven by mitochondrial toxicity.[181–185] This is due to the nucleoside analog triphosphate being incorporated by mitochondrial DNA polymerase γ into mitochondrial DNA, leading to a range of toxicities, including lipodystrophy, lactic acidosis, and peripheral neuropathy, while the toxicity associated with AZT (3) was due, in part, to its bone marrow suppression.[186–189]

26 (d4T, stavudine)   27 (ddC, zalcitabine)   28 (ddI, didanosine)

The initial NRTIs introduced into therapy were approved as single agents but, unfortunately, as monotherapies they were limited in their ability to slow HIV-1 disease progression.[190,191] Resistance to AZT (3) emerged rapidly and tolerability was poor and, as a consequence, only patients with advanced-stage HIV-1 disease were administered these drugs.[192,193] Safer therapeutic agents were needed to provide for combination therapy in order to fully suppress HIV-1 and to prevent the development of resistance. Two groups independently characterized the HIV-1 inhibitory activity of d4T (26), which was selected for development after the characterization of several analogs.[194–201] This nucleoside analog had the best combination of safety, showing reduced toxicity toward both mouse and human bone marrow progenitor cells, potency, and pharmaceutics properties when compared to the other compounds of interest, and clinical development culminated in FDA approval in 1994.[202–204] d4T (26) became an important part of HIV therapy, particularly in the dawning era of combined therapies until its use in the

Western world was largely superseded by newer NRTIs with further improved safety profiles. These included the acyclic nucleotide analog tenofovir (**8**), emtricitabine (FTC, **9**), lamivudine (3TC, **29**), and abacavir (**30**), the only marketed carbocyclic nucleoside analog.

29 (lamivudine, 3TC)   **30** (abacavir)   **31** (GS-7340)

Ultimately, the more recently introduced nucleotide analog tenofovir (**8**) was shown to possess a significantly improved safety profile compared to d4T (**26**).[205] The compelling clinical data eventually propelled the use of tenofovir (**8**), marketed as the prodrug tenofovir disoproxil fumarate (**7**), to the role of a backbone in combination therapies. More recently, the phosphoramidate derivative GS-7340 (**31**) has been advanced into clinical trials as a tissue-targeting prodrug of tenofovir (**8**). This prodrug relies upon the intracellular cleavage of the alanine ester moiety as the trigger for prodrug degradation, in contrast to tenofovir disoproxil fumarate (**7**), which is released less discriminately *in vivo*.[206–211]

The second class of HIV-1 RT inhibitor is comprised of the NNRTIs, a family of allosteric inhibitors of RT function that bind to the substrate- or product-bound forms of the enzyme complex, thereby blocking elongation by either a noncompetitive or uncompetitive inhibition mode.[212–214] The first HIV-1 NNRTIs were discovered in 1989 by phenotypic screening that identified 1-(2-2-hydroxyethoxymethyl)-6-(phenylthio)thymine (HEPT, **32**) followed in 1990 by the potent benzimidazol-2-thione derivative TIBO (**33**) as a compound that potently inhibited virus replication with an EC_{50} = 1.5 nM.[215–219] This discovery catalyzed considerable additional effort to further probe this mechanistic class of HIV-1 inhibitor, an enterprise that has been highly successful with over 50 distinct chemotypes identified and characterized to date.[212–214] Despite their structural diversity, all NNRTIs bind to the same allosteric site located ~10–15 Å from the catalytic center in the palm domain of the p66 subunit of HIV-1 RT and many conform to a butterfly-shaped topography.[212–214] However, despite the broad range of compounds that have been characterized as HIV-1 NNRTIs, only 5 drugs based on this mechanism have been licensed for marketing: efavirenz (**10**), nevirapine (**34**), delavirdine (**35**), etravirine (**36**), and rilpivirine (**37**).[220–229] The two most recent marketed entries in this class, etravirine (**36**) and rilpivirine (**37**), represent second-generation inhibitors that offer improved resistance and side-effect profiles along with the convenience associated with once daily dosing for rilpivirine (**37**).[226–229] The most recent NNRTI to enter clinical trials is lersivirine (**38**), the product of an exercise in drug design focused on carefully controlling physical properties during the optimization stage and candidate selection process.[230–232] Lersivirine (**38**) is a potent inhibitor of HIV-1 in cell culture (EC_{50} = 4 nM) that satisfied targeted physical properties with a low *c*Log*P* of 2.1 and ligand efficiency (LE) = 0.43 and lipophilic ligand efficiency (LLE) = 4.92, based on enzyme inhibition data, where LE = $\Delta G_{binding}$/heavy atom count and LLE = pIC_{50}–*c*Log*P*.[230–232] The control of physical properties during lead optimization is an area of emerging interest focused upon reducing drug candidate attrition.[233]

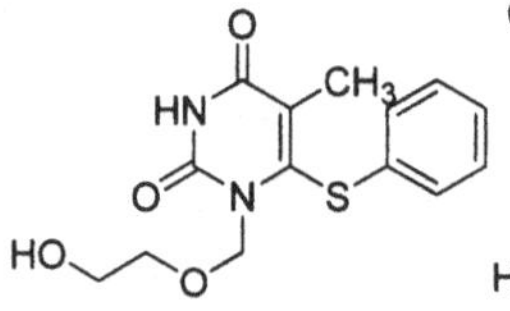

32 (HEPT) 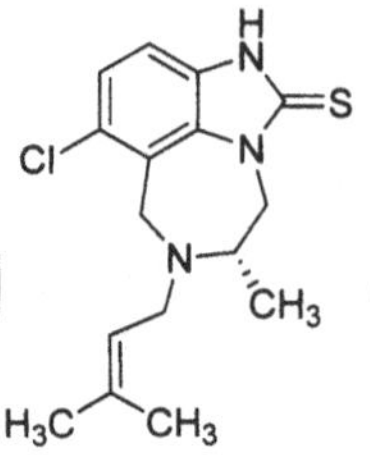   **33** (TIBO) 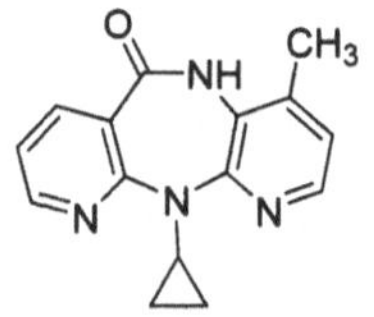   **34** (neviripine)

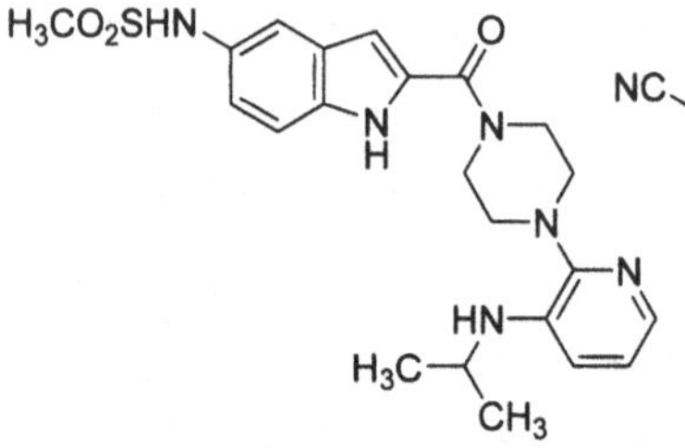

35 (delavirdine) 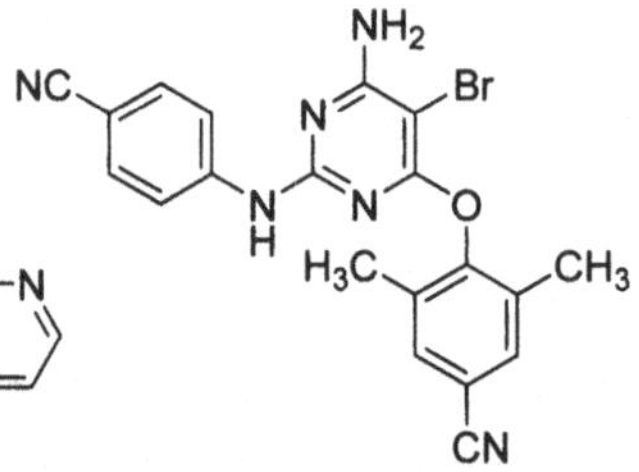   **36** (etravirine)

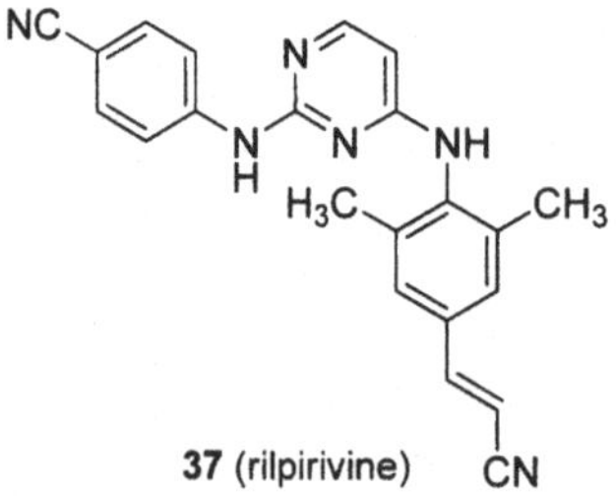

37 (rilpirivine) 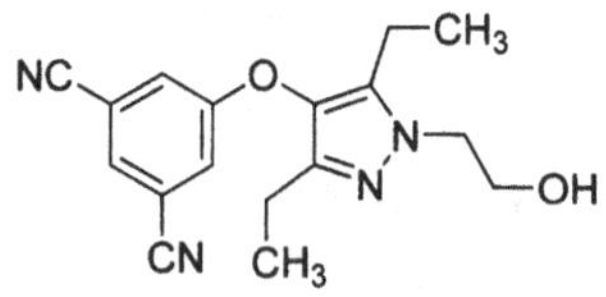   **38** (lersivirine)

Over 40 amino acid substitutions have been shown to be associated with NNRTI resistance, with most substitutions mapping to the NNRTI binding pocket in three domains (amino acids 98–108, 178–190, and 225–238) of the p66 subunit.[234–236] The two changes most frequently observed in patients failing an NNRTI-containing HARRT regimen are K103N and Y181C.[235,237] Additional substitutions outside of the NNRTI pocket which confer NNRTI resistance have also been characterized. These include Y318F, which is toward the 5′ end of the polymerase gene, and N348I, which resides in the domain connecting the polymerase and RNaseH domains.[238–240] Interestingly, the N348I substitution has been associated with resistance to both the NRTI AZT (**3**) and the NNRTI nevirapine (**34**).[241–245] Some resistance substitutions are reported to reduce the affinity of NNRTIs for the binding site while certain changes, such as V106A, P236L, and Y181C, may either reduce or increase RNaseH cleavage activities.[246–248]

The first-generation NNRTIs efavirenz (**10**), nevirapine (**34**), and delavirdine (**35**) were initially studied in what were essentially treatment-naïve patients. Subsequent studies established strong cross-resistance between these HIV-1 inhibitors, leading to treatment guidelines in which patients with emergent resistance to one agent should not be switched to the other. Today, the most common HAART backbone therapies use efavirenz (10) or nevirapine (34) in combination with two other agents. However, the first-generation NNRTIs exhibit a low barrier to the development of resistance and must therefore be combined with at least two other non-NNRTI drugs, typically two NRTIs or an NRTI and a protease inhibitor (PI). They are generally safe and well tolerated, although hepatotoxicity and severe rash are associated with the use of nevirapine (34), while efavirenz (10) causes CNS side effects early in therapy.[249,250] Because of their long plasmas half-lives, they are both amenable to once-daily regimens. Most noteworthy of these combinations is the triple-combination-medication Atripla, which combines the nucleoside analogs tenofovir disoproxil fumarate (7) and emtricitabine (9) with efavirenz (10) into a single pill.[251] However, due to the emergence of NNRTI resistance, it has been a long-standing goal to develop an NNRTI with little cross-resistance to the other members in common use. This goal was partially realized in the last few years with the approval of etravirine (36) and, most recently, rilpivirine (37). A sense for how these newer NNRTIs may be used clinically in order to avoid or overcome efavirenz (10) resistance is the substitution of rilpivirine (37) in place of the former in the new rilpivirine (37)/tenofovir disoproxil fumarate (7)/emtricitabine (9) combination marketed as Complera.[252]

2.3 HIV-1 Protease Inhibitors

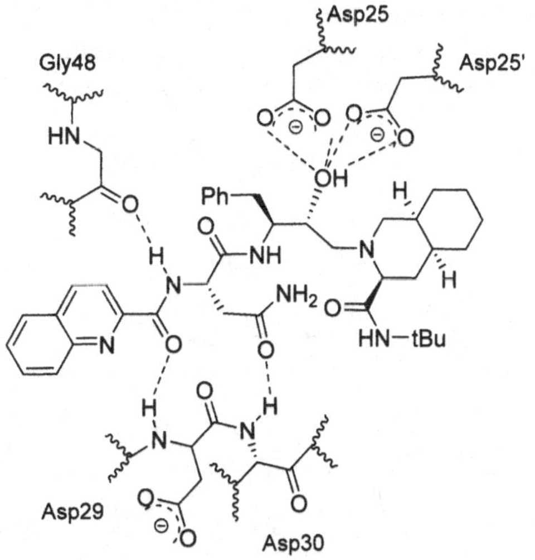

Figure 2. Key interactions between saquinavir (39) and HIV-1 protease

The era of HIV-1 protease inhibitor drugs began with the introduction of saquinavir (39), ritonavir (40), and indinavir (41) in 1996.[253] From the very beginning, these drugs were administered as part of a multidrug regimen that fully suppressed the virus, providing a more durable outcome due to a much lower emergence of resistance when compared to monotherapies.[254] Triple-drug therapies included 2 NRTIs, either AZT (3) and lamivudine (29), and a protease inhibitor, or stavudine (26), lamivudine (29), and a protease inhibitor.[254] Unfortunately, these drugs were dosed multiple times per day and had toxicities that reduced compliance, which in turn ultimately led to the development of resistance. Consequently, to spare this new drug class, the standard-of-care guidelines that were implemented indicated that therapy should still only be initiated for advanced HIV-1 disease.

       The HIV-1 protease is an aspartyl protease that is formed by the symmetrical dimerization of monomeric, 99-residue subunits, each of which contributes a catalytic aspartate residue.[253] Protease inhibitor design was markedly facilitated by the availability of X-ray crystallographic structures that allowed direct observation of inhibitors bound to the enzyme.[255,256] Saquinavir (**39**) was the first marketed inhibitor of HIV-1 protease (Table 2), a compound that relied upon the hydroxyethylene moiety as an isostere of the tetrahedral transition state, a design element originating in the extensive exploration of inhibitors of the mammalian aspartyl protease renin.[257–259] In the cocrystal structure, the critical hydroxyl moiety of saquinavir (**39**) interacts with both catalytic aspartate residues (Figure 2), a successful design feature of all of the marketed HIV-1 protease inhibitors with the exception of tipranavir (**47**), which relies upon the enol moiety as the transition state mimic and is derived from a lead discovered by high-throughput screening.[260,261] There are several interesting aspects of drug design and development embedded in these molecules.[250–257] Of note, the thioether element of nelfinavir (**42**) projects the phenyl group deeper in to the S3 pocket of the enzyme as a means of increasing affinity.[262] The phosphate moiety in fosamprenavir (**44**) acts as a prodrug element that is cleaved presystemically by alkaline phosphatase enzymes located on the brush border membrane of the gut, a modification introduced to improve the pharmaceutical properties of amprenavir (**43**) and reduce the pill burden.[263] Ritonavir (**40**) is a potent, mechanism-based inhibitor of CYP 450 3A4 (CYP 3A4) and P-glycoprotein and is now used primarily as a pharmacokinetic booster of HIV-1 inhibitors, particularly saquinavir (**39**), lopinavir (**45**), atazanavir (**46**), tipranavir (**47**), and darunavir (**48**).[264–273]

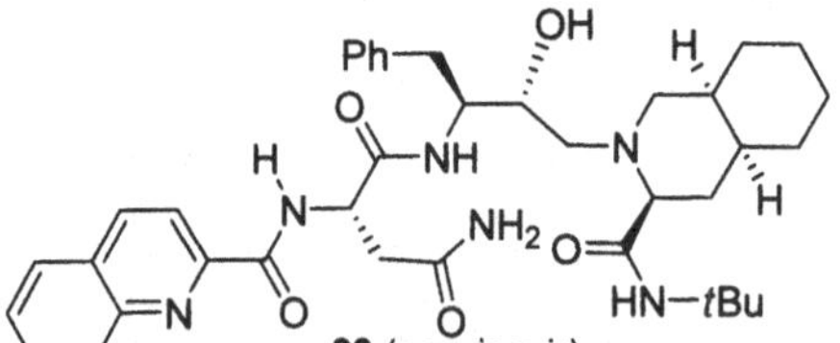

39 (saquinavir)

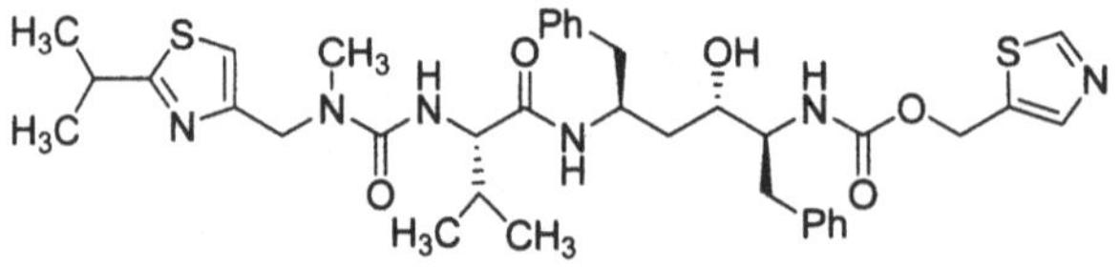

40 (ritonavir)

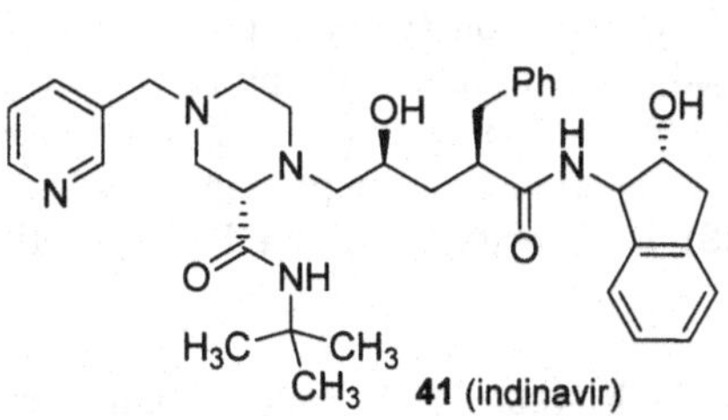

41 (indinavir)

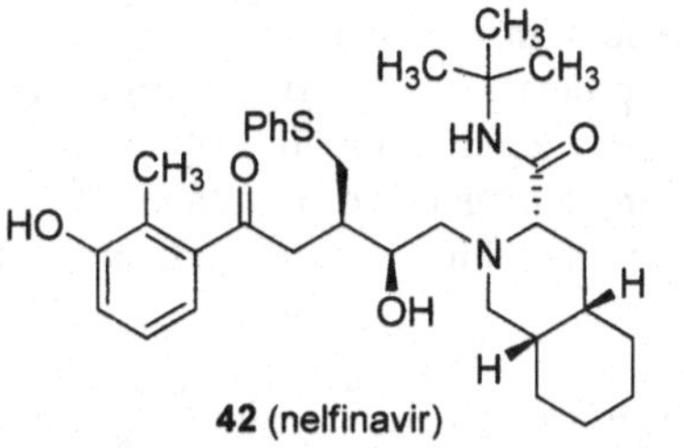

42 (nelfinavir)

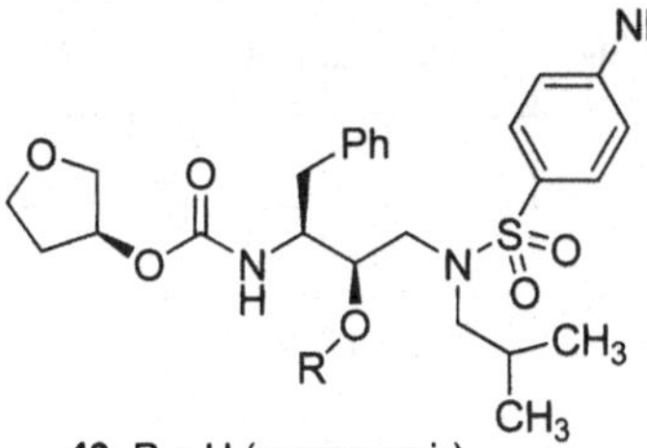

43: R = H (amprenavir)
44: R = PO(OH)$_2$ (fos-amprenavir)

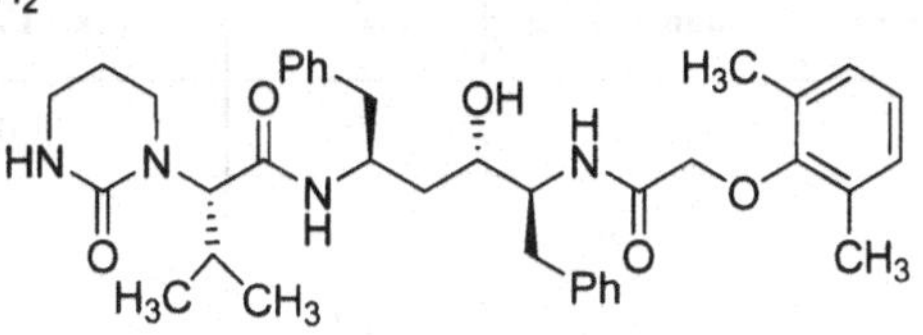

45 (lopinavir)

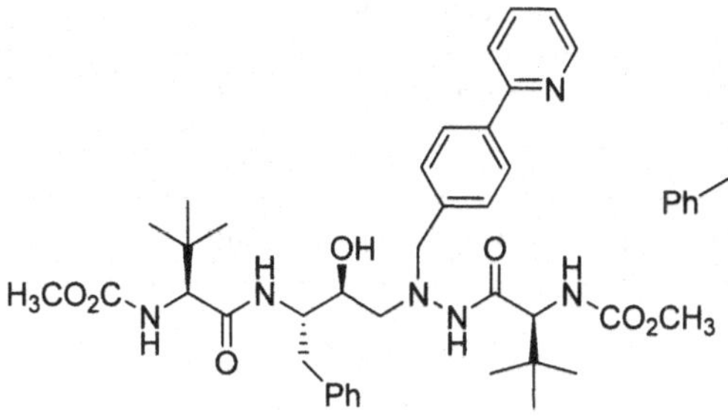

46 (atazanavir)

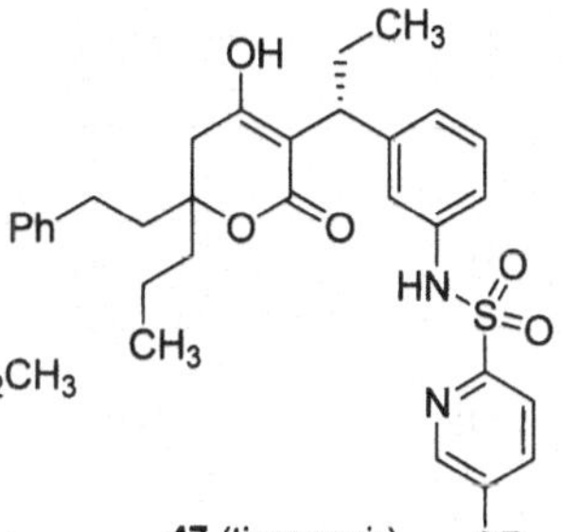

47 (tipranavir)

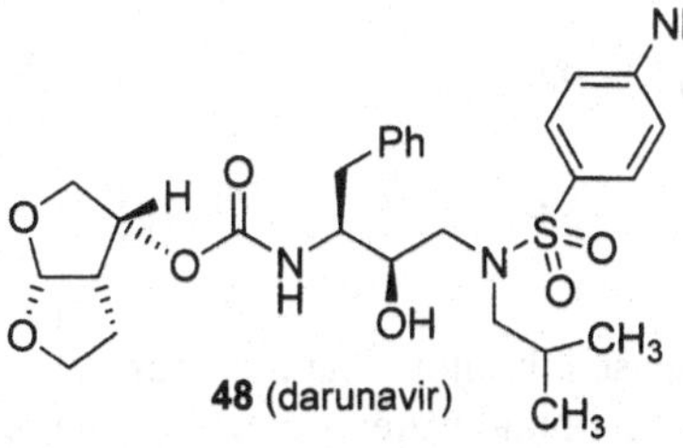

48 (darunavir)

An interesting analysis of the marketed protease inhibitors evaluated the thermodynamic signatures of the individual molecules using isothermal titration calorimetry to dissect the enthalpic and entropic components of the binding affinity.[233,274] The results are compiled in Table 3 and reveal a pattern in which the underlying energetics of the most recently introduced protease inhibitor darunavir (**48**) relies much more on an enthalpic rather than entropic component to the free energy of association.[274] The high-enthalpic component of binding of darunavir (**48**) is perhaps not surprising since this molecule is based on amprenavir (**43**), a compound with a thermodynamic signature that shows greater dependence on an enthalpic contribution than the other HIV-1 protease inhibitors. Darunavir (**48**) was designed to optimize hydrogen-bonding interactions and extend minimally beyond the substrate-binding groove as a means of reducing the potential for the emergence of resistance.[269,276,277] This drug design principle has been effectively manifested in the clinic where darunavir (**48**) is associated with a high genetic barrier to resistance.

Table 3. Thermodynamic Signatures for Marketed HIV Protease Inhibitors.

Compound	Launch Date	K_d (pM)	ΔG (kcal/mole)	ΔH (kcal/mole)	TΔS (kcal/mole)
Saquinavir	1995	400	−13.0	1.2	−14.2
Indinavir	1996	480	−12.4	1.8	−14.2
Ritonavir	1996	29	−13.7	4.3	−9.4
Nelfinavir	1997	260	−12.8	3.1	−15.9
Amprenavir Fosamprenavir	1999 2003	390	−13.2	−6.9	−6.3
Lopinavir	2000	7.7	−15.1	−3.8	−11.3
Atazanavir	2003	150	−14.3	−4.2	−10.1
Tipranavir	2005	8	−14.6	−0.7	−13.9
Darunavir	2006	4.5	−15.0	−12.7	−2.3

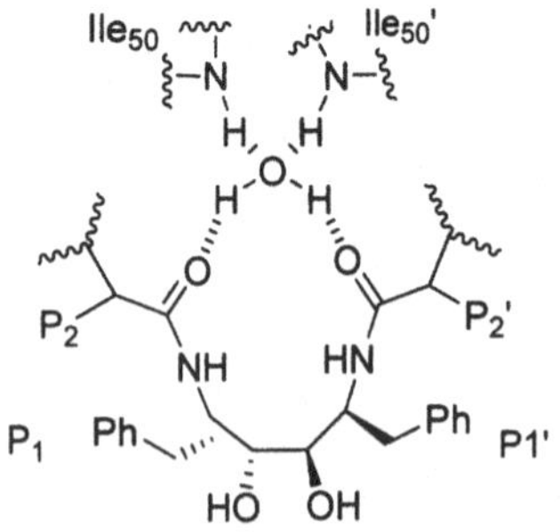

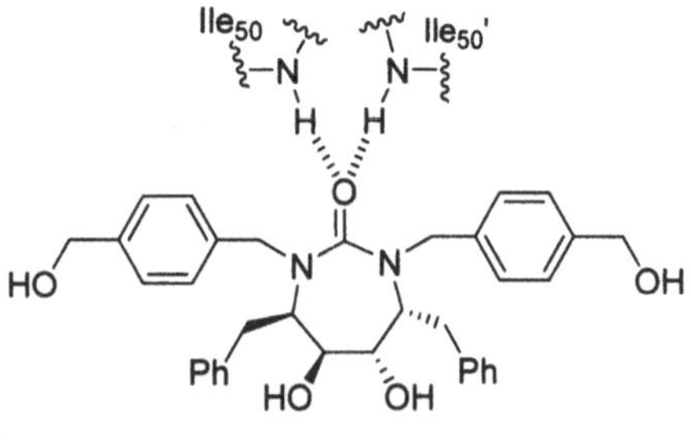

Figure 3. Design of HIV-1 protease inhibitors that incorporate structural elements that interact with the flap residues and displace a bound water molecule observed in cocrystals of the enzyme with bound peptidomimetic inhibitors.

Another interesting principle in drug design is the displacement from ligand–protein complexes of water molecules that mediate drug–target interactions, providing a potential gain in affinity based on the entropic contribution of releasing the bound water molecule.[275,278,279] This approach, in which structural elements are introduced into the inhibitor to mimic the interfacial water, is perhaps underutilized in drug design, a function of requiring X-ray cocrystal structure data and the fact that success has, on occasion, been elusive. The latter is a function of the individual circumstance associated with the protein and ligand, particularly the need to establish good H-bonding interactions, the strength of which are sensitive to geometry.[274] However, the design of of HIV-1 protease inhibitors has provided a compelling example of this tactical approach to drug design.[280–282] The water molecule that mediates the interaction between the NHs of the 2 flap residues $Ile_{50}/Ile_{50'}$ and the P2/P2' carbonyl moieties of linear, peptidomimetic inhibitors was replaced by incorporating a H-bonding accepting moiety into a cyclic inhibitor, a design concept depicted in Figure 3.[280–282] As an additional benefit, cyclic ureas of this type present a conformationally restrained and preorganized inhibitor to the protease, a potential thermodynamic advantage manifest as enhanced potency. Urea **49** that emerged from the early phase of this study is a potent HIV-1 protease inhibitor with $Ki = 0.27$ nM that inhibits HIV-1 replication in cell culture with an EC_{90} of 57 nM.[280–282]

The protease inhibitor class is perhaps the most efficacious of the antiretroviral classes, in part because of their generally high genetic barrier to resistance.[283] The early use of the first-generation protease inhibitors saquinavir (**39**), ritonavir (**40**), indinavir (**41**), and nelfinavir (**42**) did not take advantage of boosting by CYP 3A4 inhibition but did allow for dual class triple-combination therapy that heralded the HAART era.[283,284] However, the suboptimal pharmacokinetic properties of these inhibitors meant that although the genetic barriers were high, resistance did ultimately emerge to these first generation protease inhibitors. Resistance occurred through two distinct mechanisms: alteration of the enzyme active site such that the inhibitor bound less avidly or substitutions at or near the protease cleavage sites in the gag polyprotein, the substrate for the enzyme. Resistance in the protease itself usually occurs in the catalytic site through the stepwise accumulation of substitutions that result in an expansion of the binding pocket and reduced affinity for the inhibitors. A series of primary mutations at one of ~17 sites in the protease protein have been identified to date and, usually, the more substitutions the protease enzyme contains, the more resistant it is to certain inhibitors.[285] However, resistance comes at a price: The accumulation of these primary mutations also decreases viral fitness. This leads to the accumulation of secondary accessory mutations, sometimes called minor mutations.[283,285,286] These minor mutations by themselves do not affect the sensitivity of the enzyme to the various protease inhibitors. Instead, they function by compensating for the fitness defect of the primary mutations, thereby improving the replication capacity of the virus.[283,285,286]

A major advance in protease inhibitor therapy came with the realization that as a very potent CYP 3A4 inhibitor, low doses of ritonavir (**40**), below that generally used for antiretroviral effect, could be used to increase plasma levels of co-dosed protease inhibitors that are metabolized by CYP 3A4.[273] Although the initial boosting regimens were implemented with the first-generation protease inhibitors, it was the development of more potent and safer second generation inhibitors that heralded a new era in HAART. These second-generation inhibitors include amprenavir (**43**), now dosed as the prodrug fosamprenavir (**44**), lopinavir (**45**), atazanavir (**46**), tipranavir (**47**), and darunavir (**48**).[283] The pharmacokinetic boost provided by ritonavir (**40**) allowed for less frequent dosing, once a day for many drugs, and a higher genetic barrier to resistance, a consequence of the much improved daily trough plasma concentrations. The boosting effect of ritonavir (**40**) is so efficient and the resistance barrier so high that the protease inhibitors lopinavir (**45**) and darunavir (**48**) have been examined in what is, in essence, monotherapy studies in HIV-1-infected individuals.[284]

Most protease inhibitors have an overlapping cross-resistance profile, although the newer agents require a greater number of primary mutations in the protease to induce clinically relevant resistance. Thus, stepwise substitutions at positions in the binding cleft, amino acids 23, 30, 32,

47, 48, 50, and 52, the flap region, residues 46 and 54, and the interior of the enzyme, amino acids 76, 88 and 90, usually decrease susceptibility to many protease inhibitors.[285] Another four primary mutations, at residues 24, 33, 53, and 73, have differential effects on the various protease inhibitors.[287] However, in some cases a primary mutation may affect protease inhibitors in different ways. For example, the second-generation inhibitor atazanavir (**46**) induces an I50L primary mutation that partially encodes resistance to this agent but sensitizes the protease to many of the other protease inhibitors.[288] The coverage of initial protease resistance by boosted second-generation protease inhibitors allowed for the sequential use of protease inhibitors in patients who developed resistance to first generation agents. This is especially true for the more potent second-generation inhibitors tipranavir (**47**) and darunavir (**48**).[283]

Resistance to protease inhibitors can also be induced by changes at or around the substrate cleavage sites in the gag polyprotein. These substitutions are frequently secondary to primary mutations in the enzyme and are thought to allow the substrate to fit more effectively into the altered catalytic site of the enzyme.[283] Thus, changes at the P1/P6 cleavage site, L449F or P453L, have been shown to be associated with the I50V substitution in protease.[289,290] However, it has also been found that changes in a cleavage site alone can decrease susceptibility to protease inhibitors. Mutations at the nucleocapsid/p1 cleavage site, A431V, K436E, and/or I437V/T, can be selected individually *in vitro* and result in resistance to a protease inhibitor, a phenomenon that may also directly affect drug susceptibility *in vivo*.[291,292] Substitutions at non-cleavage-site positions in the gag polyprotein have also been selected *in vitro* and shown to contribute to resistance to protease inhibitors, although the mechanism by which this occurs is currently unknown.[293,294]

2.4   *HIV-1 Integrase Inhibitors*

HIV-1 integrase catalyzes the integration of proviral DNA into the host cell DNA, an essential step in the HIV-1 viral lifecycle and the key step in establishing a permanent infection. Raltegravir (**50**) was approved for marketing in 2007 as the first HIV-1 integrase inhibitor, the culmination of considerable effort that was based on clearly defining the biochemical staging of enzyme function.[295–304] Initial early lead structures were frequently based on catechols, hydrazides, or coumarins, all of which failed to show antiviral activity in cell culture by a mechanism that could be reliably attributed to inhibition of virus genome integration. The critical biochemical development was the understanding that the strand transfer step, in which the viral DNA is inserted into the host cell DNA, was the key enzymatic process susceptible to inhibition rather than assembly of the enzyme on viral substrate or the 3'-cleavage reaction in which 2 nucleotides are removed from the termini of the double-stranded viral DNA.[297] A more effective screening assay based on this mechanistic insight identified the diketoacid derivative **51** as the first specific inhibitor of HIV-1 integrase that demonstrated antiviral activity in cell culture and which was quickly optimized to the more potent analog **52**.[305] These compounds bind to a complex of HIV-1 and the viral DNA susbtrate, with the diketoacid moiety binding to the two Mg^{2+} atoms involved in catalysis, forming a ternary complex that interferes with the binding to host cell double-stranded DNA.[297–299,306–308] In this fashion, the diketoacid moieties of **51** and **52** function as a phosphate isostere, a relatively rare example of successful emulation of this moiety.[279] By adopting a strategy of incorporating elements of the diketoacid pharmacophore into a dihydroxypyrimidine ring system, more potent integrase inhibitors were developed that ultimately led to the discovery of raltegravir (**50**), marketed as Isentress.[300,309,310] Two other integrase inhibitors that are in advanced clinical trials and have demonstrated antiviral effects in HIV-1-infected patients are elvitegravir (**53**) and dolutegravir (**54**).[311–315] Elvitegravir (**53**) is being developed as part of a quadruple combination with tenofovir disoproxil fumarate (**7**), emtricitabine (**9**), and the PK enhancer cobicistat (**55**), a compound derived from ritonavir (**40**) but which is, by design, devoid of HIV-1 protease inhibition and which protects the integrase inhibitor from CYP 3A4-mediated metabolism.[315–317] Dolutegravir (**54**) is currently in Phase III clinical trials where it is being evaluated as a triple combination therapy with lamivudine (**29**) and abacavir (**30**).[318]

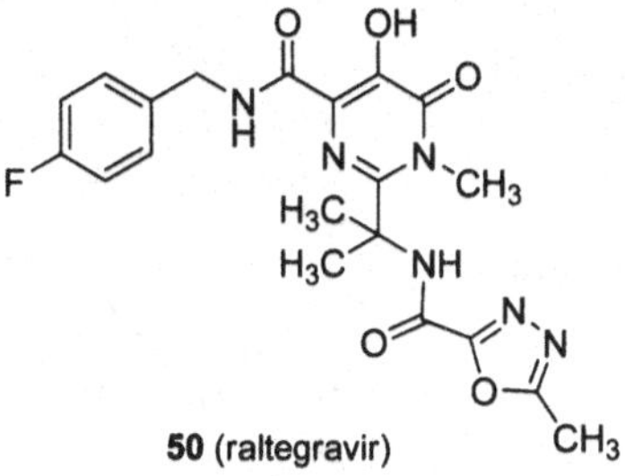

50 (raltegravir)

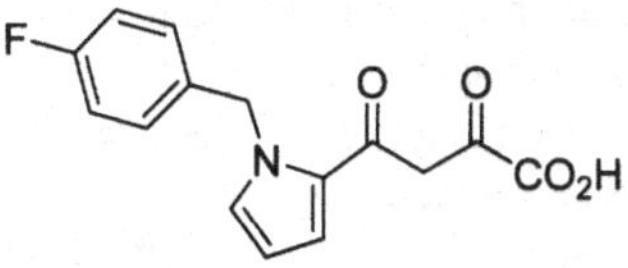

51

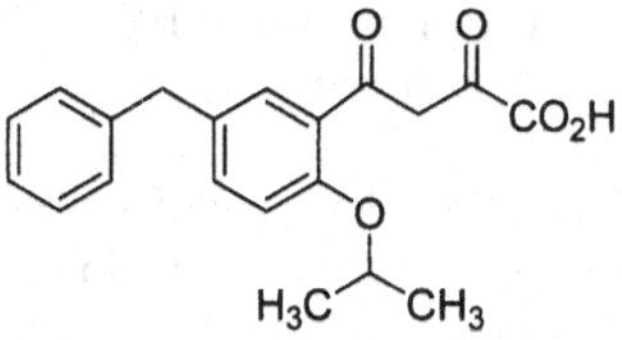

52

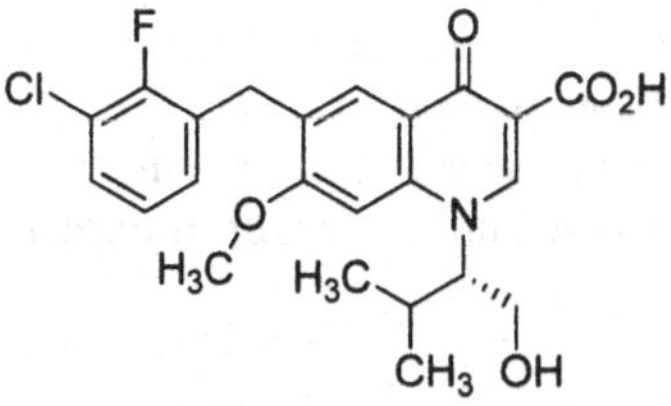

53 (elvitegravir)

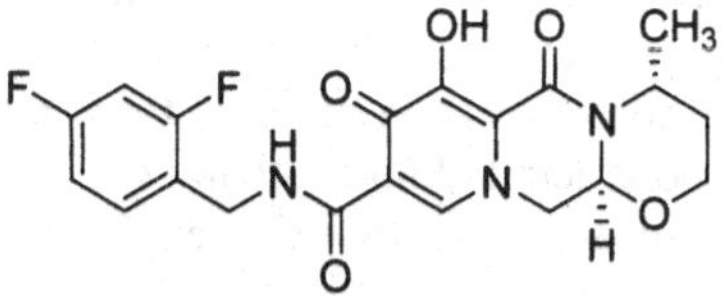

54 (dolutegravir)

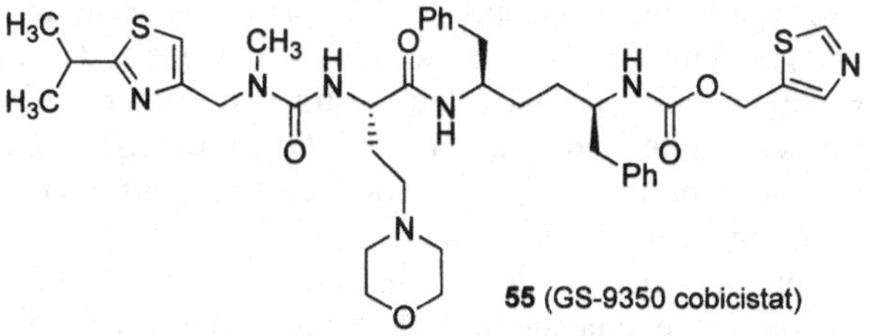

55 (GS-9350 cobicistat)

The most recent approach to interfering with the function of HIV-1 integrase depends on disrupting a critical association with the host cell transcriptional coactivator lens epithelium-derived growth factor (LEDG-F).[319–322] The LEDG-F binding site on HIV-1 integrase is well conserved and LEDG-F is believed to act as a tether that associates the enzyme with host cell chromatin, guiding the integration of viral DNA into chromosomal DNA.   The quinoline derivative **56** has been characterized as an effective inhibitor of LEDG-F binding to HIV-1 integrase and is an active antiviral agent in cell culture, EC_{50} = 2.35 μM.[322]   An X-ray crystallogaphic structure of **56** bound to the HIV-1 integrase enzyme has been solved, data that will facilitate the design of more potent compounds that are anticipated to exhibit a different resistance profile from active-site inhibitors such as raltegravir (**50**).   The first representative of this class of HIV-1 inhibitor, BI-224436, entered clinical trials in 2011 where it was evaluated in a Phase 1a dose escalation study to assess bioavailability and pharmacokinetics in healthy volunteers.

56

The development of effective HIV-1 inhibitors has provided several informative insights into the design of antiviral agents.   As a virus with a polymerase that exhibits low fidelity of reproduction of the viral genetic information, it quickly became apparent that combination therapy would be essential for the development of a durable drug regimen.   This was realized in a practical sense with the advent of the protease inhibitors that formed a critical component of early drug combinations.   With a contemporary armamentarium of 25 drugs that is still growing, many of which can be used in combination, HIV-1 infection is now considered to be a manageable, chronic infection.   However, as many patients now have been infected for extended periods of time, there is a rising appreciation for the effect of the infection on long-term health.   The design of an effective vaccine to prevent or treat HIV-1 infection has proven to be elusive while a cure is far from the current horizon.   Nevertheless, there is a rising sense of optimisim that a cure may be possible with a more detailed understanding of latent virus and a number of experiments in both a preclinical and clinical setting are being explored.

3   Hepatitis B Virus Inhibitors

Hepatitis B virus (HBV) is a small, enveloped, DNA virus that exists as eight genotypes that belong to the *Hepadnaviridae* family and causes hepatitis in humans and animals.[323-326]   The HBV genome comprises a relaxed, circular, partially double-stranded DNA of approximately 3200 base pairs that encodes the large, middle, and small HBV surface proteins, core and precore proteins, polymerase enzyme, and the X protein.[324,327-329]   After entering hepatocytes, the relaxed circular DNA is converted to a covalent closed circular DNA (cccDNA) which functions as the transcriptional template for the four viral RNAs that are translated to the HBV proteins.   The pregenomic RNA also serves as the template for HBV genome replication.[324,327-329]
   It is estimated that 2 billion people worldwide have been infected with HBV, a statistic that has resulted in more than 350 million chronic carriers.[323,330]   Chronic HBV infection leads to the development of liver cirrhosis, decompensation, and hepatocellular carcinoma, which accounts for over half million deaths per year worldwide.   Although universal HBV vaccination programs

implemented since 1986 using a recombinant, yeast-derived vaccine that targets the HBV surface antigen HBsAg have controlled the spread of HBV infection, antiviral drug therapy is required for those chronically infected with the virus to minimize disease progression toward life-threatening sequelae.[326,331,332] The first drug made available for treating chronic HBV infection is a recombinant form of interferon-α2b, which has recently been supplanted by pegylated interferon-α2a and 2b due to the more convenient, once-a-week dosing schedule and a higher response rate resulting from the improved plasma stability of the interferon by pegylation.[326,333] Treatment with interferon, which is administered by a subcutaneous injection, leads to undetectable HBV DNA levels in a significant percentage of patients with the benefit of lack of emergence of viral resistance. However, the overall cure rate is low and use is limited by the extensive appearance of multiple adverse side effects, including flulike symptoms, fatigue, anorexia, and emotional instability.[326,331,333]

In addition to interferon, there are several orally administered nucleoside analogs licensed for treating chronic HBV infection that can conveniently be categorized into three classes based on structure.[334] These are the L-nucleosides lamivudine (**29**, 3TC) and the β-L-enantiomer of the natural nucleoside thymidine, telbivudine (**57**, LdT),[335–337] the acyclic nucleoside phosphonates adefovir dipivoxil (**58**)[338,339] and tenofovir disoproxil fumarate (**7**),[340,341] which are prodrugs of adefovir (**59**) and tenofovir (**8**), respectively, and the carbocyclic deoxyguanosine analog entecavir (**60**).[342–344] These nucleoside analogs all target the HBV polymerase that is responsible for replication of the viral genome, an enzyme that functions as a reverse transcriptase for the synthesis of the negative-strand DNA from the pre-genomic RNA template as well as a DNA polymerase to synthesize the viral positive-strand DNA from negative-strand DNA. Clinical trials with nucleoside analogs have shown that suppression of viral replication, as measured by serum HBV DNA levels, resulted in enhanced treatment outcomes, including the normalization of serum alanine aminotransferase levels and improvements in liver histology. Although virologic resistance remains a concern with nucleoside analog therapy, newer agents approved subsequent to lamivudine (**29**), particularly adefovir dipivoxil (**58**), tenofovir disoproxil (**7**) and entecavir (**60**), have shown much improved resistance profiles.[345–347]

57 (telbivudine)

58: R = H₂C-O...

59: R = H (adefovir)

60 (entecavir)

Since the first phosphorylation to form a nucleoside monophosphate is often the rate-limiting step in the pathway from nucleoside analog to its triphosphate, a molecular scaffold that can adopt a conformation to position the 5'-hydroxyl group relative to the nucleobase for efficient phosphorylation by host cell kinases, in addition to the recognition of triphosphate by the target polymerase, is important to ensure inhibition activity. An interesting development in the antiviral nucleoside research is the appreciation that unnatural L-nucleosides, mainly cytidine and thymidine analogs, can be phosphorylated by deoxycytidine and thymidine kinases with similar or greater efficiency than their natural D-enantiomers.[348] In fact the L-enantiomer of β-2',3'-dideoxycytidine analogs are preferentially favored by deoxycytidine kinase, resulting in more potent antiviral activity compared to the D-enantiomer. Furthermore, L-nucleosides also exhibit more favorable metabolic stability and improved toxicity profiles. For example, while lamivudine (**29**) is inert to deamination by deoxycytidine deaminase, its D-enantiomer is deaminated to the inactive 2'-deoxy-3'-thiauridine. Moreover, the D-enantiomer of lamivudine (**29**) exhibits undesirable mitochondrial toxicity, a function of the triphosphate inhibiting host cellular DNA polymerases β and γ. The 5'-triphosphate derived from telbivudine (**57**), which inhibits HBV DNA polymerase by competing with its natural substrate thymidine-5'-triphosphate, does not significantly inhibit human DNA polymerases α, β, or γ.

A second approach to facilitating nucleoside analog metabolism to phosphate derivatives relies upon restricting the conformationally flexible 5-membered, 2'-deoxyribose ring of the natural nucleosides in a bioactive conformation by using more rigid scaffolds. This is a challenging task since the natural nucleosides can adopt several pseudorotational conformations and it is not well understood which is preferentially recognized by host cell kinases. The identification of the natural product oxetanocin A (**61**) and, subsequently, lobucavir (**62**), nucleoside analogs that are active against HIV-1, HBV and herpesviruses, however, suggests the feasibility of this approach.[349] The mobile tetrahydrofuran moiety of the natural nucleosides can be replaced by a conformationally more rigid 4-membered ring that favors a single puckered conformation, represented by the oxetane and cyclobutane moieties in oxetanocin A (**61**) and lobucavir (**62**), respectively. Based on structural studies and molecular modeling analyses, the hypothesis was formulated that the distance between the 5'-carbon atom and the C2 carbon of the nucleobase as well as the dihedral angle between these two groups are important determinants for recognition by kinases and, hence, antiviral activity. These design criteria were used to design lobucavir (**62**) and ultimately led to the discovery of the extremely potent carbopentacyclic analog entecavir (**60**), which is a clinical effective treatment of HBV infection at an oral dose as low as 0.5 mg per day.

61 (oxetanocin A)   **62** (lobucavir)

In contrast to the rigidification approach, deannulation of the ribose ring to an acyclic moiety confers these molecules with the flexibility to present many conformations, one of which is presumably readily recognized by the appropriate host cell kinase and readily phosphorylated. The phosphonate derivatives tenofovir (**8**) and adefovir (**59**) are representatives of this tactical approach to drug design, with the phosphonate acting as a mimic of a nucleoside monophosphate that effectively circumvents the first phosphorylation step and confers stability towards hydrolysis by phosphomonoesterases.[164] The molecular flexibility associated with these two compounds has also been postulated to be a key factor contributing to the suppression of the development of virologic resistance during antiviral treatment. This idea is supported by the observation that the diphosphate derivative of tenofovir (**8**) adopts multiple conformations in crystal structures of complexes of HIV-1 reverse transcriptase with template primer.[350]

The polar phosphonate group of acyclic nucleotide analogs is frequently associated with nephrotoxicity resulting from accumulation of the drug in the renal proximal tubules, a toxicity profile that limits the dose of adefovir dipivoxil (**58**) that can be used in the clinic for treating chronically HBV-infected patients. The polar nature of the phosphonate group found in tenofovir (**8**) and adefovir (**59**) also precludes efficient absorption across the lipid bilayer of the intestine that is essential for oral exposure. For example, the oral bioavailability of tenofovir (**8**) is between 5–6% in rats and monkeys and 17% in dogs.[351] This circumstance is markedly improved by masking the phosphonate group with the prodrug elements isopropyloxycarbonyloxymethyl and pivaloyloxymethyl for tenofovir (**8**) and adefovir (**59**), respectively. As depicted in Scheme 3 for tenofovir disoproxil (**7**), the cleavage is initiated by the hydrolysis of the isopropyloxycarbonyl group by carboxyesterases to afford isopropyl alcohol and an unstable carbonate intermediate which undergoes rapid decomposition in two steps to generate tenofovir monoester accompanied by the spontaneous release of carbon dioxide and formaldehyde. The monoester subsequently enters a similar degradation pathway to ultimately deliver tenofovir (**8**). Tenofovir disoproxil fumarate (**7**) shows chemical stability at pH 7.4 at 37 °C and in dogs it is more stable in the intestinal homogenates than in plasma but of negligible stability in the liver, where esterase activity is high.[351] These properties prevent premature hydrolysis of the prodrug before or during absorption through the intestine but allow rapid conversion thereafter to form the parent drug. The bioavailability of the parent tenofovir (**8**) in dogs is improved to 30% after oral administration of the disoproxil prodrug **7** and appears to correlate with the intestinal stability of the prodrugs rather than their lipophilicity. For adefovir (**59**), membrane permeability is increased from <10 nm/s to 80 nm/s in a Caco-2 permeability assay by masking as adefovir dipivoxil (**58**), which contributes to an approximate 3-fold improvement in oral bioavailability in humans.[352]

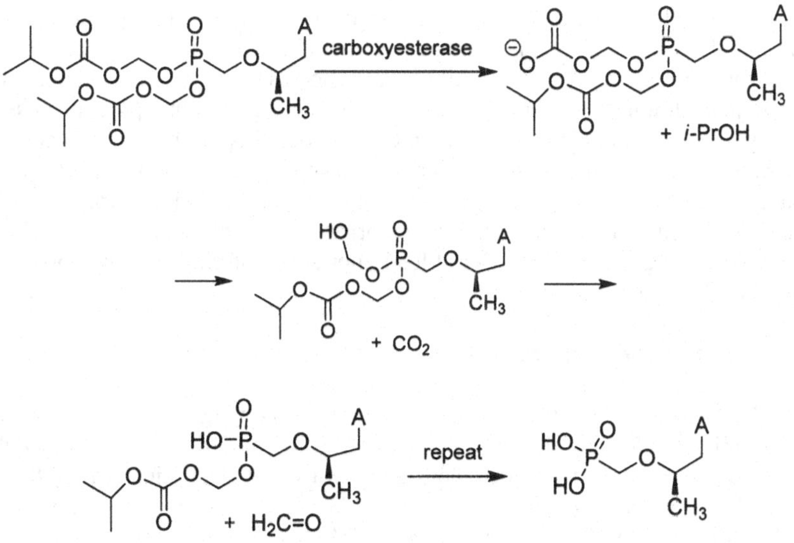

Scheme 3. Unmasking of the nucleoside phosphonate prodrug of tenofovir (**8**)

4   Hepatitis C Virus Inhibitors

The worldwide incidence of HCV infections is approaching 200 million, with over 5 million in the U.S., representing a chronic disease that is largely asymptomatic and progresses slowly over several decades to cause severe liver damage to the host.[353–355] HCV therapy initially relied upon treatment with the immune stimulant interferon-α (IFN-α), subsequently improved by the addition of daily doses of the nucleoside analog ribavirin (11) and the introduction of pegylated interferon-α (PEG-IFN-α), which reduced the dosing frequency of the protein to once-weekly subcutaneous injections. This combination improved response rates, defined as a sustained virological response (SVR) in which individuals are virus free 24 weeks after the cessation of therapy, from < 20% with IFN-α to 40–55% with the optimized combination therapy.[356] However, the isolation and detailed characterization of the positive-strand RNA virus responsible for hepatitis C infection in 1989 set the stage for a drug discovery campaign focused on discovering and developing direct-acting antiviral agents (DAAs) that specifically target viral proteins and nucleic acid.[357–359] These HCV inhibitors were anticipated to offer considerable potential for improved response rates with therapy more tolerable than PEG-IFN-α and ribavirin (11), which are associated with a range of side effects that challenge patient compliance.[360] However, a significant challenge for the clinical use of DAAs is the replication rate of HCV *in vivo*, estimated at 10^{12} virions per day, which is 10–100-fold higher than for HIV, that, when combined with a HCV RNA-dependent RNA polymerase (RdRp) that demonstrates poor fidelity of replication of the small, ~10,000-base genome, predicts for the rapid emergence of resistant virus.[361–363] Indeed, under conditions of monotherapy with a DAA, resistance emerges within several days, an anticipated outcome that mandates the use of combinations of DAAs that possess orthogonal resistance profiles, most effectively accomplished by combining agents with complementary viral targets.[364,365] The initial approaches to identifying DAAs to treat HCV infection focused on the NS3 protease and NS5B RdRp since the activity of these enzymes could be readily recapitulated *in vitro*, an approach accelerated by the development of replicating subgenomic cell culture systems and virus, initially a genotype 2a strain but subsequently the more prevalent genotype 1, that allowed assessment of DAAs in cell-based systems.[366–377] The chimpanzee model of HCV infection provides the closest approximation for the human condition while mouse models incorporating chimeric livers have also been developed, although neither have been extensively adopted in a preclinical setting.[378–380] Most commonly, the development of HCV inhibitors has relied upon the determination of activity of candidate compounds in cell culture systems followed by the demonstration of appropriate pharmacokinetic properties in preclinical species, typically after oral administration, and with liver exposure assessed to confirm delivery to the target organ. In the clinic, efficacy can conveniently be assessed by monitoring the decline in viral load after administering a single dose of drug to an infected individual.

4.1   HCV NS3 Protease Inhibitors

The HCV NS3 protease emerged as an antiviral target following characterization as a serine protease responsible for cleaving the viral polyprotein at the NS3-4A, NS4A-4B, NS4B-5A and NS5B-5B junctions, an effort facilitated considerably by the determination of the solid-state structure using X-ray crystallography.[381–386] The NS3 protease inhibitors boceprevir (12) and telaprevir (13), which in May 2011 were the first HCV DAAs to be approved for marketing, are tetrapeptide derivatives that rely upon extensive contacts between the inhibitor and protease in addition to the formation of a covalent bond between the catalytic serine-139 and the activated carbonyl moiety that is embedded in the α-keto amide element.[387–393] These agents are slow binding, mechanism-based inhibitors of the enzyme that form stable but reversible complexes with the protein, as depicted in Figure 4.[387–393]

Figure 4. Mechanistic basis of serine protease inhibition by α-keto amide derivatives

However, the first NS3 protease inhibitor to enter clinical evaluation was BILN-2061 (**63**), which established clinical proof of concept for the DAA class following 2 days of dosing to HCV-infected subjects.[394–397] BILN-2061 (**63**) is a potent inhibitor of HCV NS3 protease, the design of which originates with the observation that the enzyme is subject to product inhibition by the carboxy terminal moiety of substrate and substrate-derived peptides.[398–400] The discovery of the quinoline P2* moiety projecting from the P2 proline was the result of a classical SAR exploration while the macrocyclic element that links the P1 and P3 Cα substituents provides conformational preorganization to the molecule that is based on the proximity of these moieties in X-ray crystal structures of the enzyme bound to inhibitors, a common topographical phenomenon in many protease–substrate interactions.[401–405] The follow-on compound to BILN-2061 (**63**) is BI-201335 (**64**), a tripeptide-based NS3 inhibitor that is an acyclic molecule that removes the challenge of conducting ring-closing metathesis-mediated macrocycle construction at scale.[406–411]

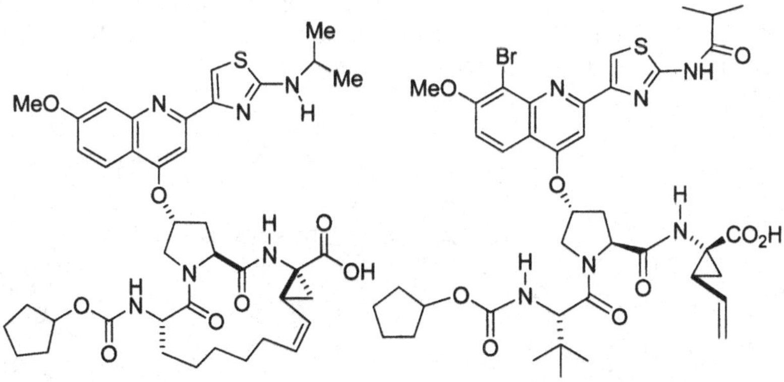

63 (BILN-2061)   **64 (BI-201335)**

Two additional NS3 inhibitors that rely on an overt acid moiety at P1 are GS-9451 (**65**), related to BI-201335 (**64**), and the macrocyclic phosphinic acid GS-9256 (**66**), both of which have entered clinical trials.[412,413]

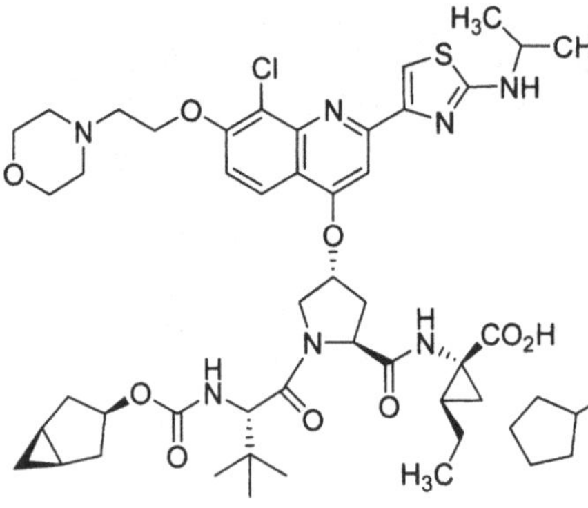

65 (GS-9451) 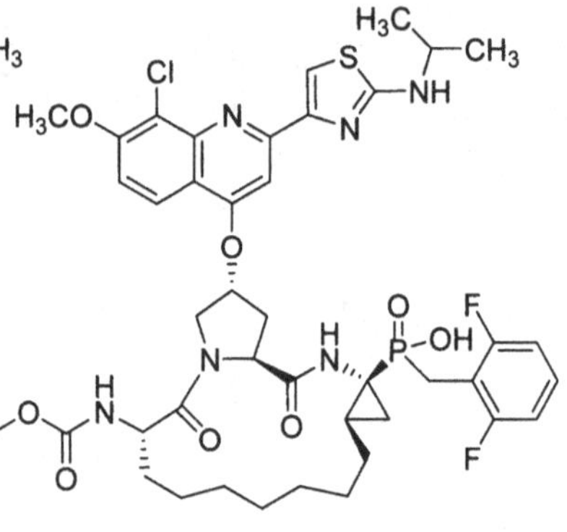   66 (GS-9256)

An advance in the design of HCV NS3 protease inhibitors came with the identification of the cyclopropyl acylsulfonamide moiety as a structural motif that preserved the acidic element at P1 while allowing access to the small P1′ pocket, conferring potency enhancements of up to 1000-fold compared to simple carboxylic acid analogs.[279] BMS-605339 (67) was the first of this chemotype advanced into clinical trials but was subsequently replaced by BMS-650032 (68). Several other protease inhibitors that rely upon the cyclopropyl acylsulfonamide acid isostere and which have demonstrated efficacy in HCV-infected subjects include simiprevir (69, TMC-435),[414–418] danoprevir (70, ITMN-191),[419,420] vaniprevir (71, MK-7009),[421–423] and MK-5172 (72), the first of the second-generation protease inhibitors that exhibits activity towards viruses resistant to first-generation compounds.[424,425] These inhibitors are all based on macrocyclic structures, with simiprevir (69) and danoprevir (70) uniting the P1–P3 elements in a fashion similar to BILN-2061 (63), while vaniprevir (71) and MK-5172 (72) take advantage of the proximity of the P4 moiety and P2* elements in the bound state.[424–427] These molecules offer considerable potential in the treatment of HCV infection as part of combination therapies with either PEG-IFN/ribavirin or direct-acting antiviral agents with complementary mechanisms.

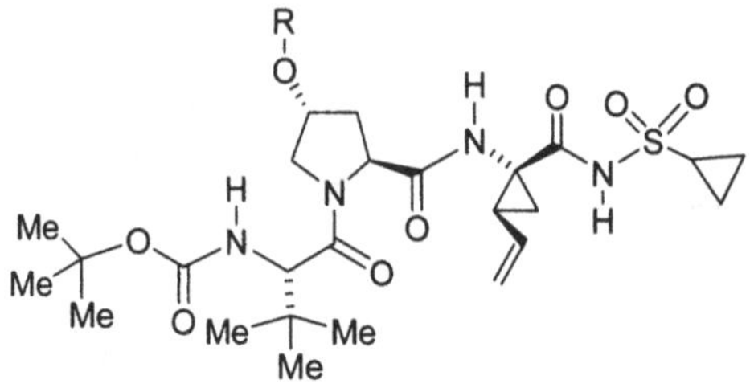

67: R 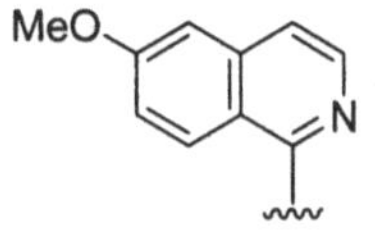 =   BMS-605339

68: R =

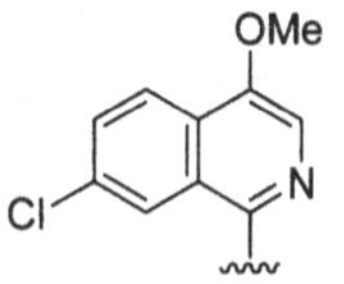

BMS-650032
(asunaprevir)

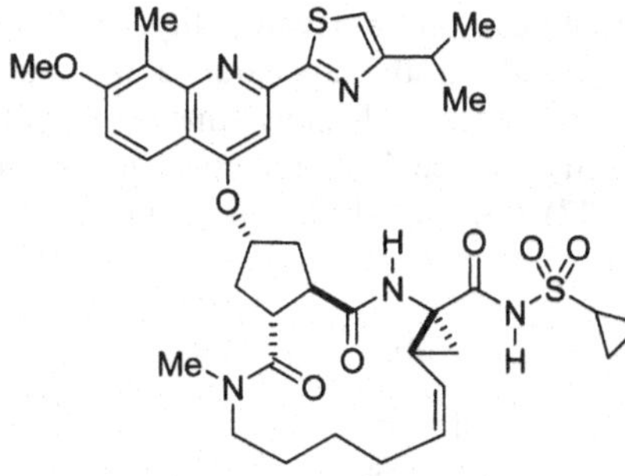

69 (TMC-435, simeprevir)

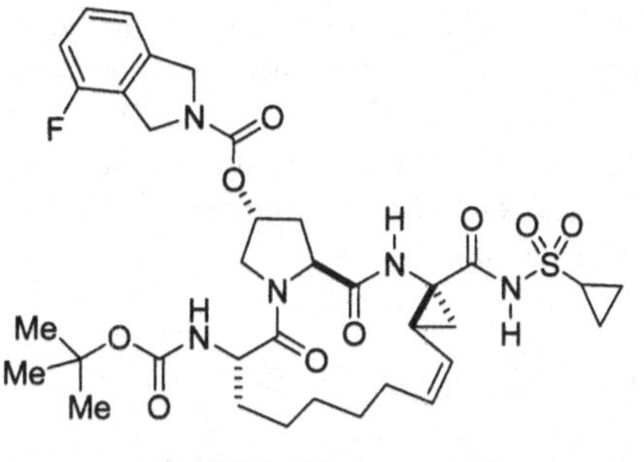

70 (ITMN-191, danoprevir)

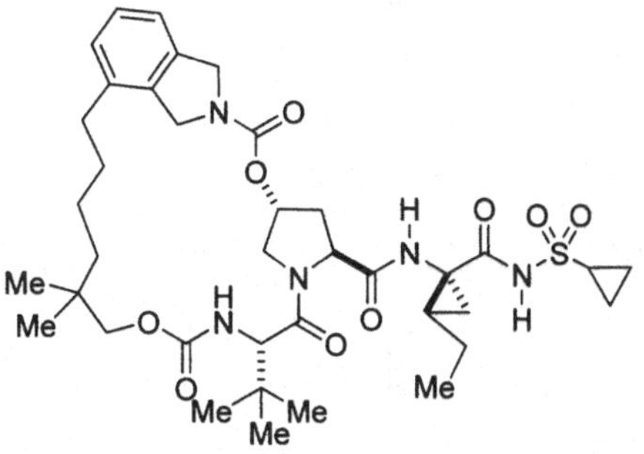

71 (MK-7009)

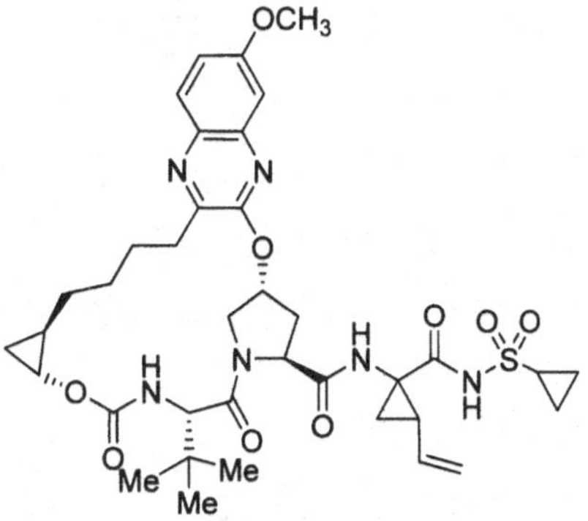

72 (MK-5172)

4.2   HCV NS5B RNA-Dependent RNA Polymerase Inhibitors

The HCV NS5B protein is an RNA-dependent RNA polymerase that is responsible for the replication of the viral genome that represents an important and tractable target for drug design and development.[428–431] X-ray crystallography studies revealed that the protein adopts a right-handed three-dimensional structure that is composed of fingers, palm, and thumb subdomains, features that are common among viral polymerases.[432–434] The active site of NS5B where catalysis occurs resides in the palm domain and, in the unliganded state, this is fully enclosed as a result of contacts made between the fingers and thumb domains.[432–434] HCV NS5B inhibitors can be categorized into two classes: nucleoside inhibitors, compounds that interact with the active site of the polymerase as substrate mimetics and effect the termination of the replication process, and nonnucleoside inhibitors, compounds that inhibit the enzyme's function by interacting with one of several allosteric sites.[428–430,435] In general, the former inhibitor class exhibits a superior resistance barrier and pan-genotype inhibitor activity, properties that are sought after in the emerging optimal standard of care for HCV.[428,436,437]

4.2.1   Nucleoside/tide Inhibitors (NIs)

The first nucleoside inhibitors to be evaluated clinically for the treatment of HCV were the 2'-methyl cytidine derivative NM-107 (**73**), advanced as the valine prodrug valopicitabine (**74**, NM-283),[438–440] and R-1479 (**75**), a 4'-azido ribonucleotide advanced as the triple-ester prodrug balapiravir (**76**, R-1626), a modification designed to improve oral bioavailability.[441-444] However, development of both compounds was terminated based on the appearance of unacceptable toxicological effects in clinical trials.   More recent successes with the 2'-methyl ribonucleotide series have focused on the guanine derivative 2'-methyl-G, the 5'-monophosphate of which has been configured as different prodrugs in IDX-184 (**77**) and INX-189 (**78**), both of which offer promise as therapeutic agents.[445-447]

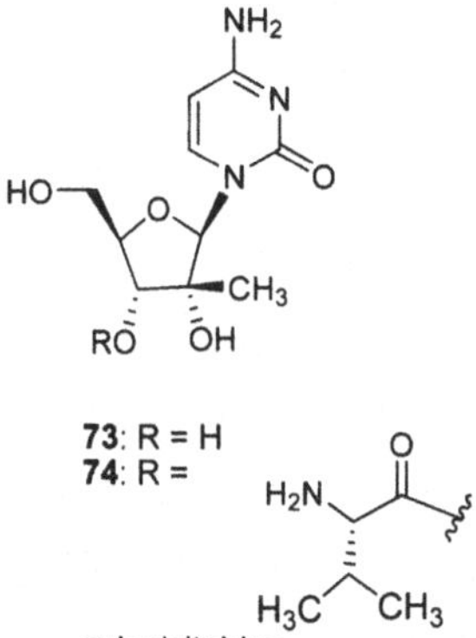

73: R = H
74: R =

valopicitabine

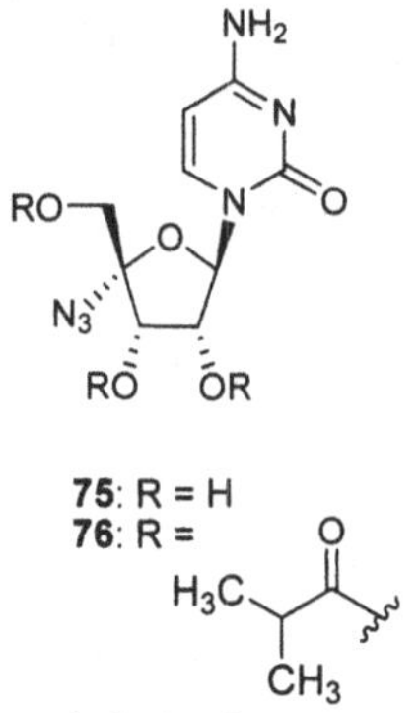

75: R = H
76: R =

balapiravir

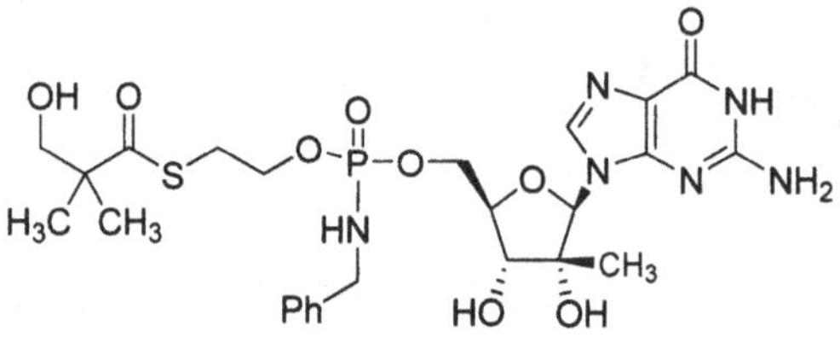

77 (IDX-184)

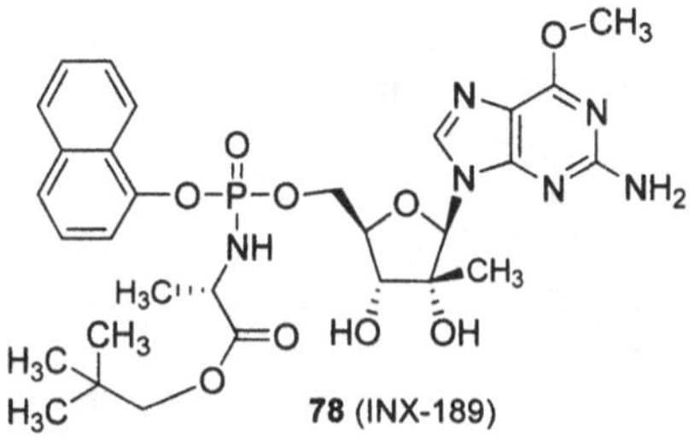

78 (INX-189)

2'-α-Fluoro-2'-β-methyl nucleosides are potent and selective inhibitors of HCV NS5B polymerase that exhibit antiviral activity in cell culture systems and are clinically efficacious.[428,436] The first of this class advanced into clinical trials was the nucleoside PSI-6130 (**79**), which exhibits an $EC_{90} = 4.6$ μM in an HCV replicon assay, has a high genetic barrier to resistance, and is not cross-resistant with R-1479 (**75**).[436,448-451]   However, PSI-6130 (**79**) demonstrated only modest oral bioavailability in preclinical studies, a problem resolved with the synthesis of the diester prodrug mericitabine (**80**), a compound that demonstrated pronounced antiviral effects in HCV-infected subjects after either 2-weeks of monotherapy or 4 weeks of combination therapy with pegylated-IFN/ribavirin at doses of 500–1500 mg BID.[452,453] In addition, in a pioneering DAA combination study, coadministration of mericitabine (**80**) with the protease inhibitor danoprevir (**70**) for 14-days to either treatment-naïve subjects or those who had failed a Peg-IFN/RBV regimen, resulted in a potent antiviral effect, with viral titers reduced by a mean ≥ 4.9log$_{10}$ at day 14.[454] Moreover, 25% of subjects had HCV RNA levels below the limit of detection at the end of therapy, anticipating the potential of interferon-free drug regimens.[454] Commensurate with the high genetic barrier to resistance seen *in vitro*, clinically significant resistance has not been observed in early clinical trials with mericitabine (**80**).

PSI-6206 (**81**), a metabolite of PSI-6130 (**79**) that arises by deamination of the cytosine and was observed in both hepatocytes *in vitro* and rhesus monkeys, is inactive in HCV replicon assays at concentrations of up to 100 μM.[455,456] However, its triphosphate form, PSI-7409 (**82**), is a potent inhibitor of HCV polymerase in an enzymatic assay. Moreover, the mean half-life of PSI-7409 (**82**) in primary human hepatocytes is 38 h, considerably longer than that of the triphosphate derivative of PSI-6130 (**79**), which is 4.7 h, suggesting the potential of PSI-6206 (**81**) if the triphosphate form could be effectively delivered to cells. Detailed metabolism studies conducted in replicons and hepatocytes established that the poor antiviral activity associated PSI-6206 (**81**) is due to the molecule being a poor substrate for 5'-phosphorylation, the initial metabolic activation step preceding triphosphate formation.[456,457] These observations prompted adoption of a kinase bypass strategy that relies upon delivering the monophosphate derivative to cells, a tactic that necessitates the use of prodrug elements to mask the highly polar nature of the phosphate moiety, physical properties that limit membrane permeability.[428,436,458] This strategy was successful with the identification of PSI-7977 (**83**) as the preferred, single (*S*)-p diastereomer derived from the initially examined 1:1 mixture, designated as PSI-7851.[457,459,460] PSI-7977 (**83**) is a potent inhibitor of HCV replication in a genotype 1b replicon, EC$_{90}$ = 420 nM, that has been advanced into clinical trials. Unmasking of the prodrug moiety of PSI-7977 (**83**) is thought to be initiated by cleavage of the alanine ester by either cathepsin A (*CatA*) or carboxylesterase 1 (*CES1*) to afford the acid **84** which nucleophilically attacks phosphorus, expelling phenol as a leaving group to give the highly labile **85**, a highly reactive species subject to spontaneous and rapid hydrolysis to the alaninyl phosphate **86** (Scheme 4).[461] The final metabolic step in prodrug unmasking is performed by histidine triad nucleotide-binding protein 1 (Hint1) that provides the monophosphate **87**, which is phosphorylated sequentially by UMP-CMP kinase and nucleoside diphosphate kinase to afford the active triphosphate derivatrive.[461]

79: R = H (PSI-6130)
80: R = COCH(CH$_3$)$_2$ (mericitabine)

81: R= H (PSI-6206)
82: R = triphosphate (PSI-7409)

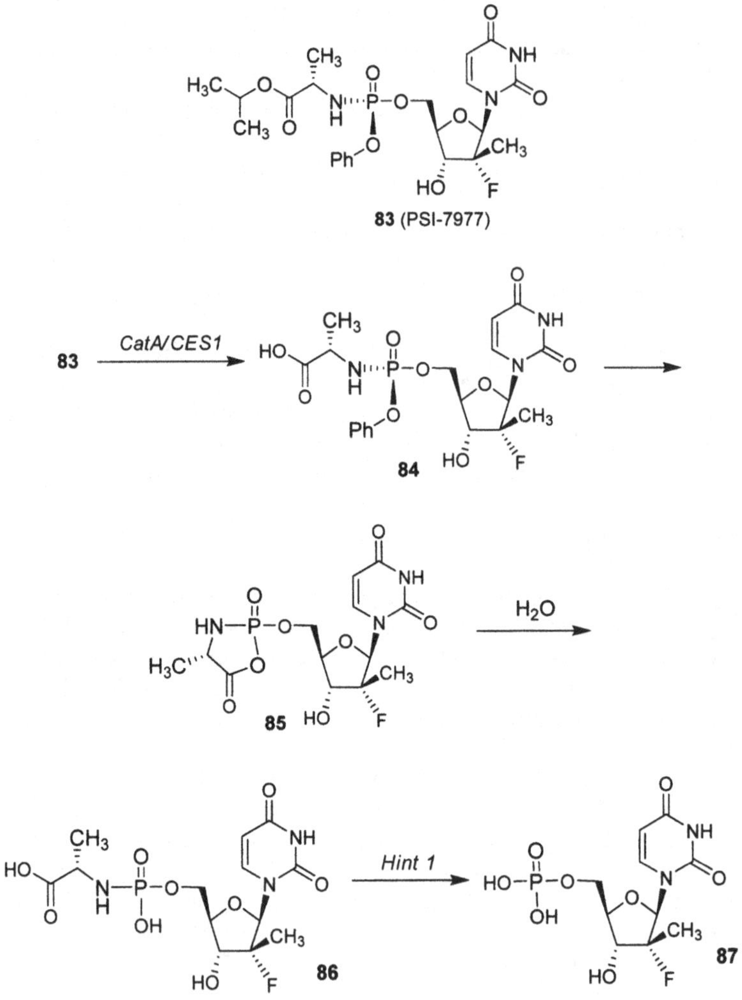

Scheme 4. Metabolic Processing of PSI-7977 (**83**)

Early clinical trials with PSI-7977 (**83**) in treatment-naïve genotype-1 subjects who were administered daily 400-mg doses for 12 weeks in conjunction with Peg-IFN/RBV (24 weeks) suggest considerable promise, with 98% of those completing the treatment regimen achieving a sustained virological response defined as undetectable viral load 12 weeks after the end of therapy.[462]

Another advanced 2'-α-fluoro-2'-β-methyl nucleotide derivative is PSI-352938 (**88**), a double prodrug of 2'-α-fluoro-2'-β-methyl-guanosine that contains a cyclic monophosphate moiety and an *O*-ethyl-derivatized purine moiety.[463–465] An alternative prodrug of 2'-α-fluoro-2'-β-methyl-guanosine is PSI-353661 (**89**), which uses the same phosphoramidate prodrug developed with PSI-7977 (**83**) but an *O*-methoxy derivative of the purine.[466,467] This nucleoside analog is a potent antiviral agent in cell culture with an EC$_{90}$ = 8 nM in a replicon assay.[466,467] PSI-7977 (**83**) and PSI-352938 (**88**) were designed for their complementarity, based on pyrimidine and purine nucleosides, respectively, that exhibit nonoverlapping resistant profiles and are dependent on different metabolic pathways to unmask the prodrug elements.[468]

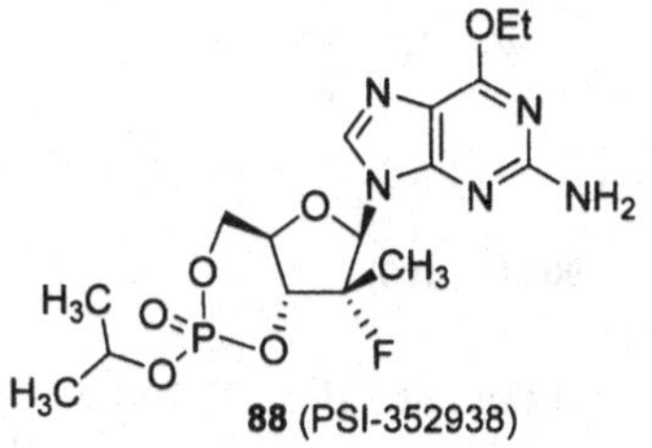

88 (PSI-352938)

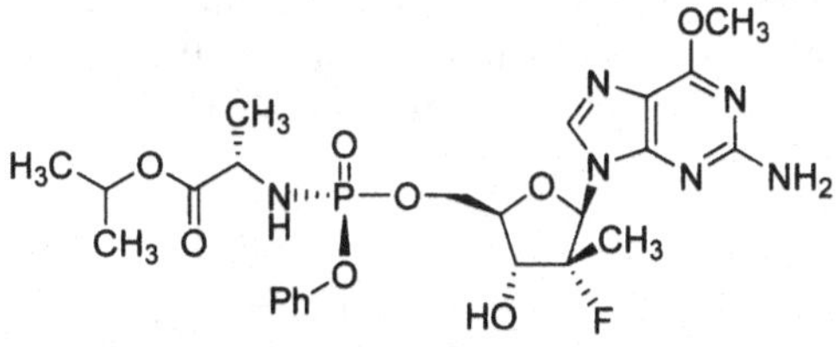

89 (PSI-353661)

4.2.2   *Non-nucleoside HCV NS5B Inhibitors (NNIs)*

NNIs are categorized into four classes depending on the region of the NS5B protein that they interact with and are, consequently, referred to as thumb-I, thumb-II, palm-I, and palm-II inhibitors, or alternatively as site I, II, III, and IV inhibitors, respectively.[428,469–473] The multiplicity of the biochemical sites for intervention has sustained considerable effort directed towards optimizing a structurally diverse array of inhibitors suitable for development, and several candidates have advanced into clinical trials.

Benzimidazole-based inhibitors that bind to thumb site I and interfere with the association of the fingers loop were the first NNIs to be discovered, with leads originating from high throughput screens.[474–480] Several of these inhibitors were advanced into clinical trials where they exhibited a pronounced effect on reducing viral load but have subsequently been abandoned, including JTK-109 (**90**) and MK-3281 (**91**).[479,481] BI-207127 (structure not disclosed) has been evaluated in early clinical trials at doses of 400–800 mg TID in conjunction with Peg-IFN/RBV while a more recent study has assessed this compound in combination with the NS3 protease inhibitor BI-201335 (**64**) and ribavirin (**11**) dosed over 4 weeks with good response rates to therapy.[482]

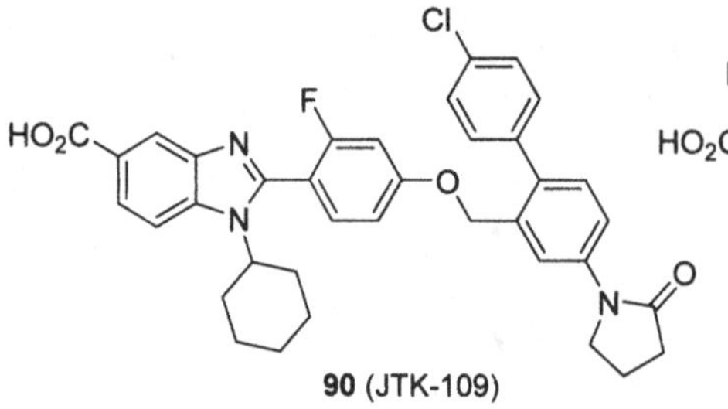

90 (JTK-109) 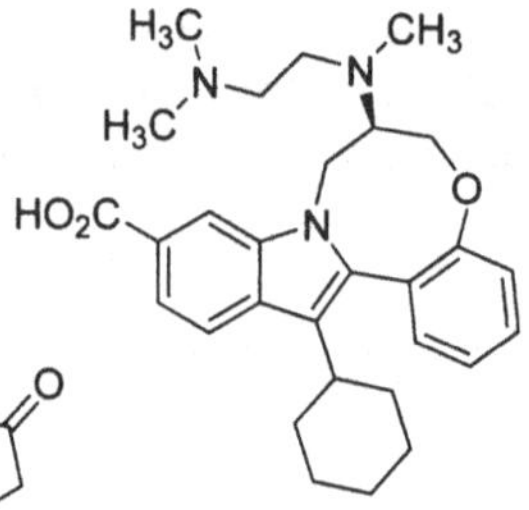   91 (MK-3281)

Filibuvir (**92**) is a site II NNI that, at a dose of 700 mg BID for 3–10 days, reduced viral load by $2.30\log_{10}$ in HCV-infected subjects naïve to PegIFN/RBV therapy and is similarly effective in treatment-experienced patients.[483–485] VX-222 (**93**) is the culmination of the study of a series of thiophene derivatives that also bind to thumb site II and, interestingly, this compound retains activity toward the M423T mutant that reduces sensitivity to filibuvir (**92**) by 250-fold.[486–488]

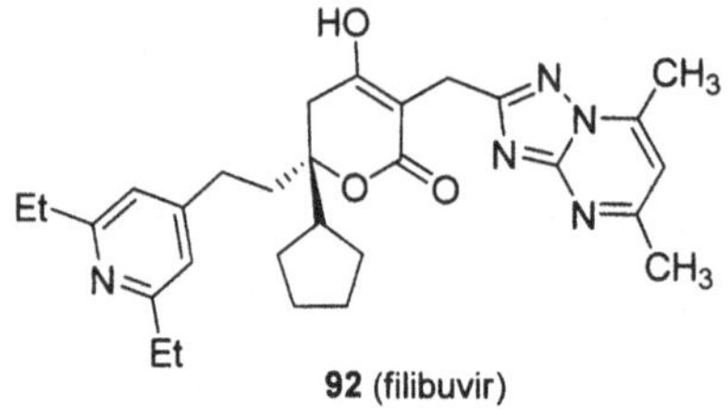

92 (filibuvir) 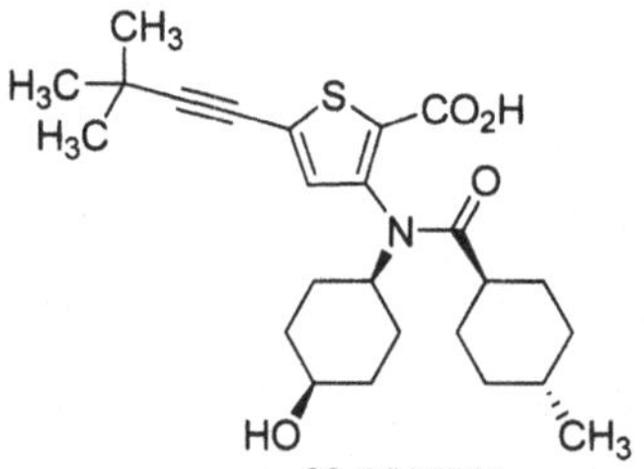   93 (VX-222)

Setrobuvir (**94**) is a site III NNI that belongs to the benzothiazidine chemotype class that originated as a high-throughput screening (HTS) lead and was the basis for the optimization to **95**, a compound considered for clinical evaluation.[489–494] Setrobuvir (**94**) has been examined in Phase 2 clinical trials at doses of 200–400 mg BID as add-on therapy to Peg-IFN/RBV. The structurally related benzothiadiazine **96** (A-837093) is a potent HCV NS5B inhibitor sensitive to several mutations in site III, including S368A, Y448H, G554D, Y555C and D559G, that demonstrates antiviral activity following oral dosing to HCV-infected chimpanzees, reducing viral load by $>1.4\log_{10}$.[495–497]

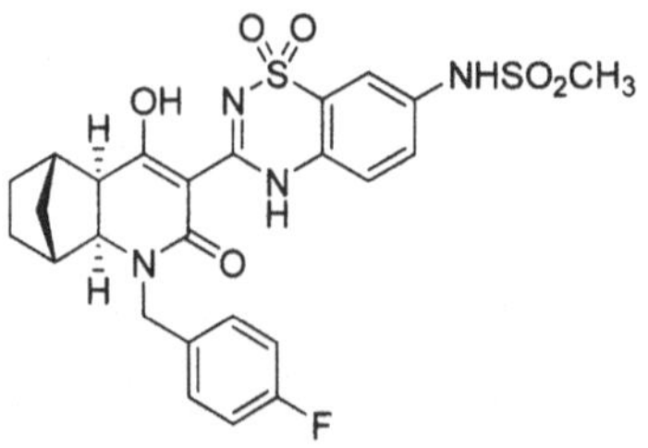

94 (setrobuvir) 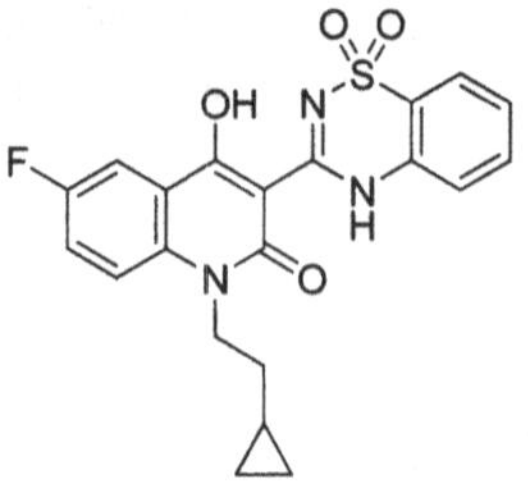   95

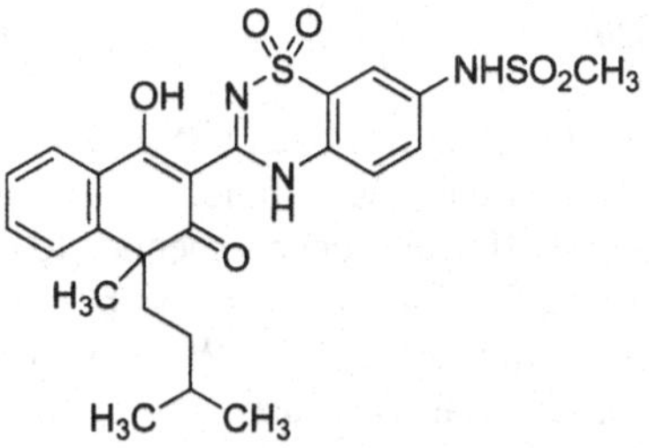

96 (A-837093)

Nesbuvir (HCV-796, **97**) is a site IV NNI that binds proximal to the catalytic elements of the enzyme and was the first NNI to demonstrate proof of concept (POC) in clinical trials.[498,499] Nesbuvir (**97**) exhibits the broadest genotype coverage among the first-generation NS5B allosteric site inhibitors that entered clinical trials and exhibited encouraging antiviral efficacy at doses ranging from 100–1000 mg BID in combination with Peg-IFN/RBV. However, development was terminated due to the appearance of severe hepatotoxicity several weeks into the study.[500]

Tegobuvir (**98**) is an NNI discovered using a replicon screen, but this compound is inactive in HCV polymerase biochemical assays.[501,502] Resistance to tegobuvir (**98**) maps to a region that partially overlaps with that of site III and site IV inhibitors. In studies directed at illuminating the reasons underlying the discordance between the biochemical and the replicon activities, it was discovered that the replicon potency of tegobuvir (**98**) erodes by 50–90-fold when tested in the presence of a CYP 450 inhibitor, implicating a metabolite in the expression of biological activity.[502] In clinical studies, tegobuvir (**98**) has been examined in combination with Peg-IFN/RBV and the NS3 protease inhibitor GS-9256 (**66**) with or without ribavirin (**11**) for 28 days.[503] In the latter study, the combination of tegobuvir (**98**) 40 mg BID and GS-9256 (**66**) 75 mg BID produced a median $4.1 \log_{10}$ reduction in viral load which was enhanced to a $5.1 \log_{10}$ reduction in the ribavirin arm.[503]

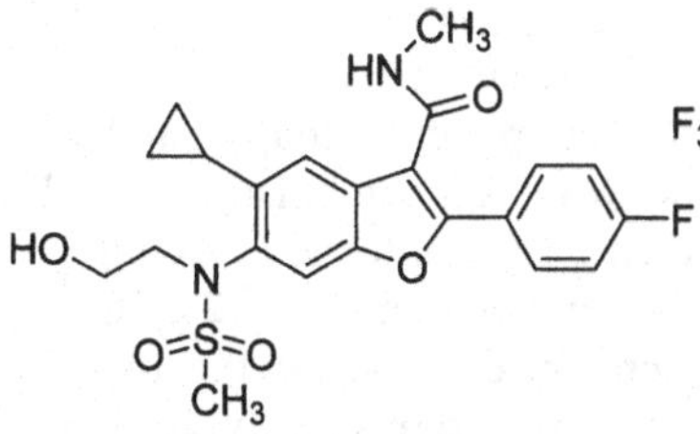

97 (nesbuvir, HCV-796) 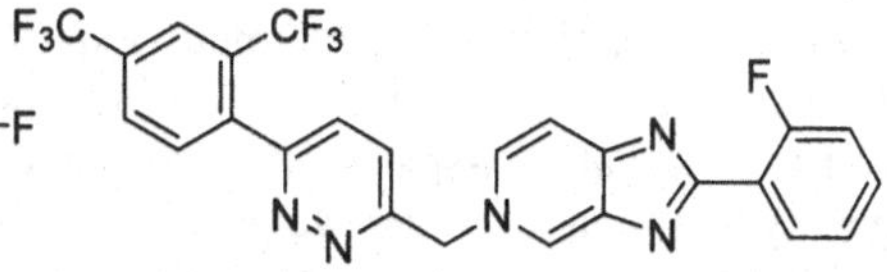   **98** (tegobuvir)

The long-held belief that some aspect of interferon-based therapy is required to effect a durable HCV cure has been challenged by early results with combinations of direct acting agents against multiple targets, resulting in a significant paradigm shift in the design of current HCV clinical studies. It is anticipated that nucleoside or non-nucleoside inhibitors of HCV NS5B will become an integral component of optimal therapy that is likely to emerge based on DAA combinations.

4.3   *Inhibitors of HCV NS5A*

HCV NS5A is a multifunctional protein that has been characterized as the master regulator of the HCV life cycle and has neither known enzymatic functions nor a counterpart in the eukaryotic genome.[504–507] Although a substantial effort has been expended to develop an understanding of the role and function of NS5A in virus replication, progress has been incremental and the picture that has emerged is one of considerable complexity.[504–507] With the heavy focus on the enzymes NS3 protease and NS5B polymerase as drug targets, early interest in pursuing NS5A inhibitors was slow to evolve, a circumstance that changed dramatically with the clinical validation established by daclatasvir (**99**, BMS-790052).[28,506–508] In a classic application of forward chemical genetics, a unique replicon assay system was used to screen over a million compounds and identify the iminothiazolidinone BMS-858 (**100**) as a modestly potent inhibitor, EC_{50} of 0.58 µM, of HCV genotype-1b replication for which resistance was mapped to the first 100 residues of the NS5A protein.[509–511] However, BMS-858 (**100**) was found to be chemically unstable and a related compound was shown to undergo a radical-mediated dimerization under the assay conditions to afford **101** as a mixture of 2 components that demonstrated potent genotype 1b replicon inhibition.[509–511] The pharmacophore for HCV NS5A inhibition was reduced to the stilbene **102**, the symmetry of which complements the solid-state structure of the amino terminus domain of HCV NS5A.[511–513] Daclatasvir (**99**) emerged from a significant iterative campaign of structural refinement focused on enhancing potency, expanding the virology profile to encompass a broad range of HCV genotypes, and optimizing pharmacokinetic properties to afford a compound that is active in cell-based assays at picomolar concentrations.[28] The high inherent potency translated into a significant clinical effect where mean declines in viral load of 1.8–3.3$\log_{10}$ were measured following single doses ranging from 1 to 100 mg.[28,514] However, in a 14-day monotherapy trial, the majority of patients experienced a rebound in viral load due to the selection of resistant virus with mutations correlating with those selected for *in vitro*.[515,516] In a Phase 2a study conducted in treatment-naïve subjects, treatment with daclatasvir (**99**) (10 or 60 mg QD) in combination with Peg-IFN/RBV for 48 weeks resulted in 83% SVR-24 compared to 25% for a Peg-IFN/Rbv control group.[33] In addition, daclatasvir (**99**) (60 mg QD) in combination with the protease inhibitor asunaprevir (**68**) (600 mg BID) with or without Peg-IFN/Rbv effected 36 and 90% SVR-24, respectively, in null-responders, a treatment experienced patient group that has failed Peg-IFN/RBV therapy.[33] Although the SVR-24 rate for the two DAA combinations was low, this was the first demonstration that IFN-sparing therapy could cure HCV infection. Acting on the finding that all the genotype-1b subjects that were in the DAA combination study (2 of the 11) achieved SVR-24, a follow-up 24-week therapy study was conducted with just the two DAAs in ten genotype-1b null responders.[517] The results have been very promising, with nine subjects that completed the study and one subject that discontinued treatment after two weeks of therapy achieving undetectable levels of viremia 24 weeks after the cessation of therapy (SVR-24).[517]

99 (daclatasvir, BMS-790052)

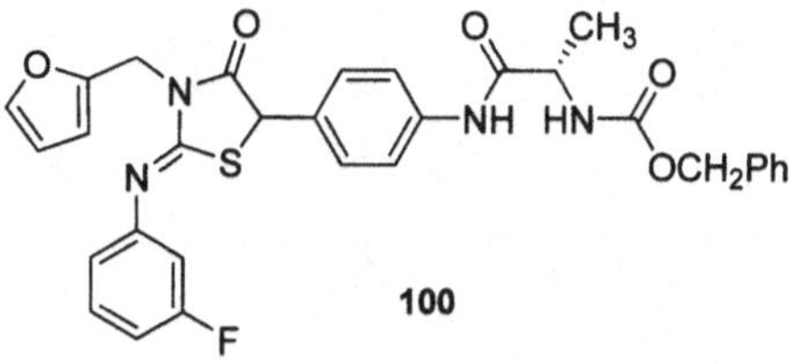

100

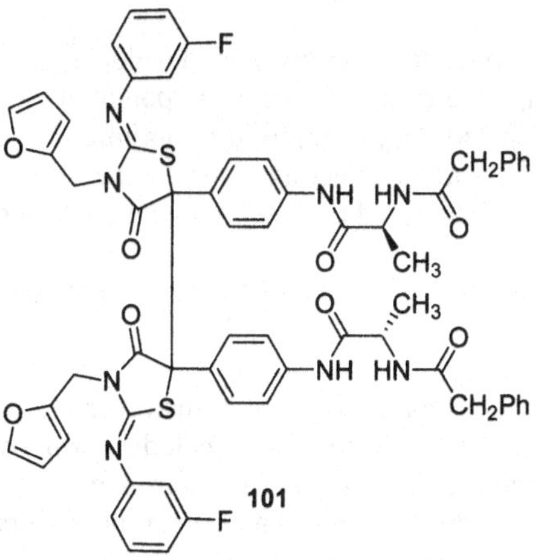

101

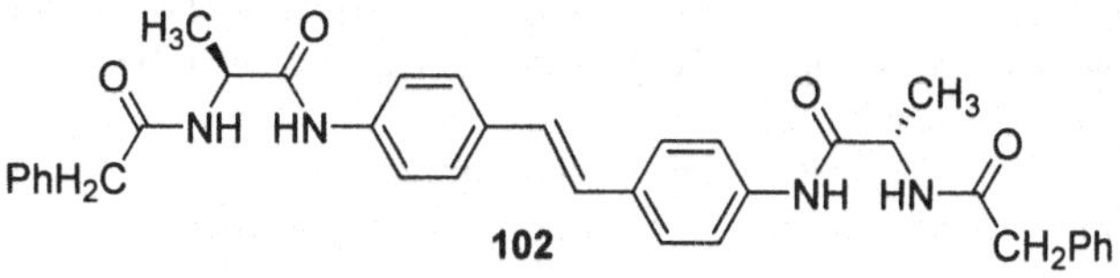

102

An alternate NS5A-targeting chemotype is represented by A-831 (**103**, AZD-2836), which was the first HCV NS5A inhibitor to enter clinical trials, although development was discontinued due to inadequate exposure and formulation issues.[507] Further optimization of this series afforded AZD-7295 (structure not disclosed), a potent genotype 1b replicon inhibitor, EC_{50} = 0.007 μM, but which is significantly weaker towards a genotype 1a replicon, EC_{50} = 1.24 μM. In a 5-day multiple ascending-dose study, AZD7295 exhibited a maximum mean viral decline of $2.1\log_{10}$ at a dose of 233 mg TID in genotype 1b-infected subjects but was inactive in individuals infected with HCV genotype 1a, presumably a reflection on the poor *in vitro* potency combined with pharmacokinetic properties inadequate to provide concentrations that surpass the EC_{50}.[507] Despite the significant structural differences that exist between the chemotypes represented by daclatasvir (**99**) and A-831 (**103**), they may share similar modes of interaction with the NS5A protein since they express a partially overlapping resistance profile.[506,507]

103 (A-831, AZD-2836)

The intensity of the effort to discover and develop direct-acting HCV inhibitors has contributed to a rich clinical pipeline comprised of many potential therapies that address a range of mechanistic targets. It is anticipated that, when used in optimized combinations, these agents will offer an all oral drug regimen capable of effecting a cure in the majority of patients and HCV may be the first chronic viral disease to be cured by small-molecule inhibitors.

5   Inhibitors of Respiratory Viruses—Influenza and Respiratory Syncytial Virus

Respiratory virus infections are a principal cause of human morbidity and mortality among all age groups.[518–520]   Several of these virus infections, including rhinovirus, influenza, respiratory syncytial virus (RSV), and parainfluenza viruses, are confined primarily to mucosal airway cells and result in an acute infection. Adenoviruses induce a similar respiratory illness but can be more persistent, while infections that begin initially in the upper respiratory tract and spread systemically are caused by pathogens such as measles and hantavirus. Additionally, emerging viruses, including coronavirus, metapneumovirus and bocavirus, have been identified as etiologic agents in upper respiratory disease.[521]

The large number of viruses that cause similar illnesses are problematic to patient management due to the lack of rapid and accurate identification methods for these agents and the insufficient battery of antiviral agents to either treat or prevent infection.[521–524]   Nevertheless, effective prophylactic or therapeutic agents have been developed for some of the more common respiratory virus infections such as influenza and RSV while the search for clinical candidates to treat other respiratory viruses is ongoing.

5.1   Influenza Virus

The first account of an influenza epidemic is believed to have occurred in 412 BC, as recorded by the father of modern medicine, Hippocrates. However, the term influenza was first coined in 1357 AD, derived from the Italian word for influence, stemming from the belief that annual disease epidemics targeting the lungs arose due to an influence of the stars that were present in the sky during outbreaks.[525]   We now know that influenza is as a single-strand, negative sense, segmented RNA virus that is a member of the orthomyxovirus family. Spreading through aerosol droplets, fomites or contact with contaminated surfaces, influenza causes acute, contagious viral infections of the respiratory tract mucosa during winter seasonal epidemics.[526]   The poor replication fidelity associated with the influenza virus polymerase promotes antigenic drift, which leads to the global seasonal influenza epidemics that are estimated to be responsible for 250,000–500,000 deaths annually.[527]   Cyclical appearances of new strains caused by antigenic shift result in seasonal pandemics with a higher potential for human morbidity and mortality. These strains are usually introduced as the result of the recombination of a human strain with an avian or swine virus during coinfection, whereby the human strain acquires surface protein genes (among others) from the nonhuman virus.[528] These pandemics can be exceptionally severe, as evidenced most prominently by the 1918 Spanish influenza pandemic with exhibited fatality rates exceeding 2.5% and worldwide total death estimates exceeding 50 million.[529,530]

Amantadine (14) was the first anti-influenza agent approved in 1966 and was followed many years later by the second-generation compound rimantadine (15), which was approved in 1994.[37,531–533] Mechanistically, the adamantanes are unique in that they interfere with the ion channel activity of the viral M2 envelope protein, which performs an essential function during the viral uncoating process.[534,535] The M2 channel mediates the influx of protons into the virion, thereby destabilizing the association of the influenza M1 matrix protein with ribonucleoprotein (RNP) complexes prior to release into the cytoplasm and RNP translocation to the infected cell nucleus.[534,536,537] Adamantanes as a therapy for influenza infection can be effective, but there are a number of shortcomings, including lack of activity against influenza B viruses, central nervous system side effects, and a low barrier to the development of resistance to these agents.[538–541] Since the 2007 influenza season, more than 90% of all circulating H3N2 viruses as well as the current swine-origin H1N1 subtype have exhibited resistance to the adamantine drugs, although 30–50% of other H1N1 isolates remain susceptible.[542,543] In addition, a majority of the H5N1 avian viruses, which can cause considerable morbidity and mortality in humans, are now resistant to the adamantanes, potentially due to the extensive overuse of these agents in animals by chicken farmers in northern China.[544–546]

5.1.1   Influenza Neuraminidase Inhibitors

The influenza neuraminidase (NA) is a tetrameric envelope protein with sialidase activity that promotes nascent virus release (budding) from infected cells *via* the cleavage of sialic acid residues at the termini of host cell membrane glyocoproteins.[547–549] The enzyme may also aid in the mobility of influenza within the airway mucosa.[550,551] Early studies demonstrating that synthetic sialic acid (*N*-acetylneuraminic acid) analogs were modestly potent inhibitors of NA enzyme activity *in vitro* and virus growth in cell culture provided the basis for the development of zanamavir (16) and oseltamivir (17) as clinically useful influenza drugs. 2-Deoxy-2-3-dehydro-*N*-acetylneuraminic acid (DANA, 104) and its *N*-trifluoroacetyl homologue 105 inhibit NA enzyme activity *in vitro*, with the latter compound demonstrating an IC_{50} 2 μM and interfering with virus replication in tissue culture with an EC_{50} of 2 μM.[552,553] The successful development of neuraminidase inhibitors represented a compelling example of the application of rational drug design that exploited X-ray crystal structures of the influenza NA apoprotein.[40,41,548,554] The structure of influenza NA revealed that the active site consisted of three lobes lined with a highly conserved array of amino acid residues among multiple influenza A and B strains, indicating that the development of a broad-spectrum agent was feasible.[40,41,549,555,556]

Rational drug design was a viable approach to the identification of potent influenza neuraminidase inhibitors once the structure determination of the enzyme in complex with its sialic acid receptor was achieved.[40,549,556] A comprehensive analysis of the cocrystal structure of influenza neuraminidase with the unsaturated sialic acid transition state analog DANA (104) identified a region of negative charge in the enzyme proximal to the 4-hydroxyl of the compound.[557,558] This led to the concept of introducing basic moieties at the 4-position of DANA

(104) with the anticipation that formation of a salt bridge with the carboxylic acid moiety of Glu-119 would yield a more avidly binding species, a prediction confirmed with the synthesis and evaluation of the 4-amino analog 106, which was 100-fold more potent than the prototype.[40,556] Further optimization focused on exploiting the potential to engage a second proximal glutamate residue (Glu-227), accomplished with the larger and more basic guanidino group in the context of zanamivir (16), which exhibited sub-nanomolar potency and a high degree of selectivity for influenza neuraminidases over those of bacterial or mammalian origin.[40,556,559-561] The planarity associated with the α,β-unsaturated carboxylic acid moiety of zanamivir (16) mimics the topography of the oxonium ion transition state intermediate thought to form during the hydrolysis of sialic acid as it is cleaved from glycoconjugates by influenza neuraminidase (Scheme 5).[562]

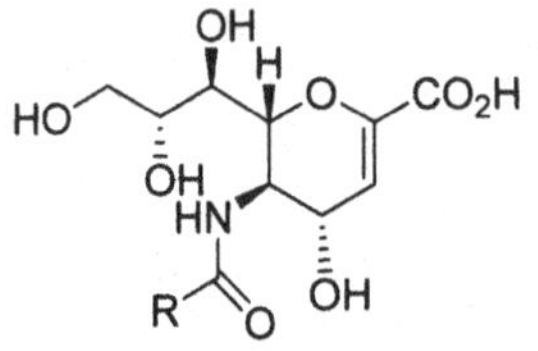

104: R = CH$_3$ (DANA)
105: R = CF$_3$ (FANA)

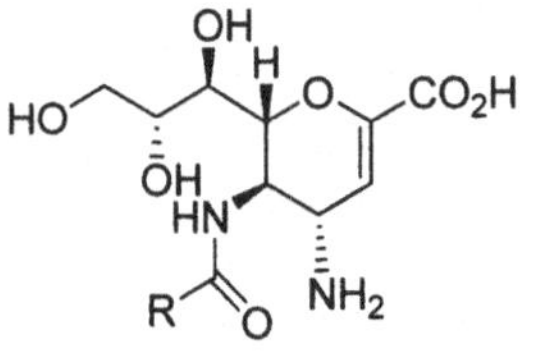

106

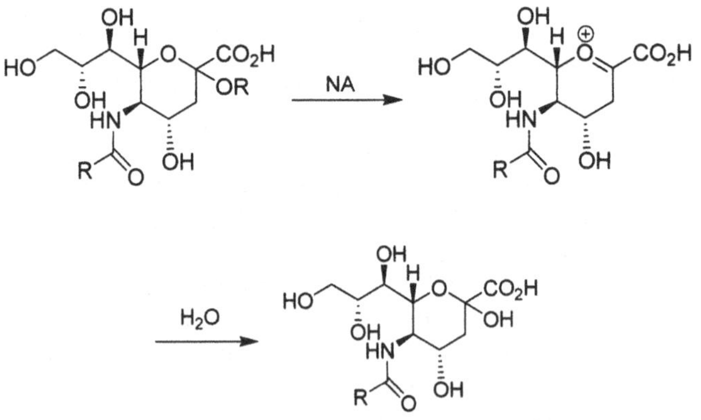

Scheme 5. Mechanism of the neuraminidase-mediated cleavage of neuraminic acid

The highly basic guanidine moiety of zanamivir (16) limits its oral bioavailability and restricts therapy to topical administration, a delivery mode that is efficacious in both murine and ferret models of influenza infection when administered intranasally or by aerosol.[40,563,564] Following successful human clinical trials, zanamivir (16) was approved for marketing within the United States in July of 1999 in the form of an inhaled powder under the trade name of Relenza®.[565,566]

In a successful effort to identify an orally bioavailable influenza NA inhibitor, GS-4071 (17) was designed as a transition state analog in which the cyclohexene scaffold facilitates more precise isosterism of the oxonium ion generated during the sialidase cleavage step.[41,562,567] GS-

4071 (**17**) relies upon a less basic primary amine to interact with the region of negative charge in the enzyme, a moiety associated with reduced potency compared to the guanidine found in zanamivir (**16**). To further enhance potency and introduce more druglike elements anticipated to be compatible with oral bioavailability, a clearly visible hydrophobic pocket in the crystal structure was exploited by the introduction of a lipophilic 3-pentyl moiety in place of the hydrophilic glycerol side chain of sialic acid and other inhibitor analogs [41,567] Although considerably less polar than zanamivir (**16**), the oral bioavailability of GS-4071 (**17**) in the rat was only 5%, prompting evaluation of prodrugs with the ethyl ester oseltamivir (**18**) successfully delivering GS-4071 (**17**) to plasma following oral administration.[569,570] In October 1999, oseltamivir (**18**) was approved, just 3 months after zanamivir (**16**), and is marketed under the trade name of Tamiflu®.

Since the approval of these neuraminidase inhibitors, Tamiflu® has consistently been the more widely prescribed influenza treatment, attributed to the convenience of oral dosing when compared to the more unwieldly inhalation treatment required with Relenza®. However, sales of these neuraminidase inhibitors fluctuate on a yearly basis, dependent on the severity of the influenza season and the predictions for the emergence of a new pandemic strain of virus. Sales of Tamiflu® soared with the rise in the number of highly pathogenic H5N1 avian (HPAI) influenza cases reported in Asia during 2003 and the subsequent spread to Africa and Asia in 2006, along with the possible threat of an impending H5N1 pandemic.[571,572] The anticipation of a severe avian influenza H5N1 pandemic created a wave of government and institutional stockpiling.[573–575] With human mortality rates observed in H5N1 infections exceeding 60%, a mutation or reassortment event resulting in efficient human-to-human H5N1 transmission would potentially be a devastating development.[572,576]

The emergence of resistance is always a problem with influenza viruses due to the high mutation rate of the viral polymerase, the selective advantage of resistant viruses in the face of drug treatment, and the nature of the influenza season, where virus is transmitted through human-human contact. For example, during the single influenza season in 2008, the circulating strain of H1N1 influenza evolved to become resistant to oseltamivir (**18**), a result of the acquisition of an H274Y mutation. In addition, a growing number of H5N1 cases resistant to this drug have been reported, ostensibly due to resistance in the avian population through overuse on farms.[577,578] In 2009, a new pandemic strain of flu (H1N1pdm09) emerged that was initially sensitive to oseltamivir (**18**), but resistance is once again developing.[579,580] As a consequence, there remains an unmet medical need for new influenza antivirals and the suggestion that new influenza inhibitors should be combined with the current neuraminidase inhibitors as a means of combating the development of resistance.[581]

Peramivir (**107**) is the third marketed neuraminidase inhibitor that provides an intravenous option for hospitalized patients, although at this point it is only approved in Japan. Similar to its predecessors, peramivir (**107**) was developed using a rational design, structure-based approach.[582,583] However, the scaffold for peramivir (**107**) is a cyclopentane ring system, smaller than the six-membered templates found in zanamivir (**16**) and GS-4071 (**17**).[582,583] From an analysis of the cocrystal structure of DANA (**104**) complexed with influenza A neuraminidase, it was noted that the four active-site binding pockets each interacted with the four different functional groups extending from the six-membered heterocycle but there was little direct interaction with the ring itself.[558,585] An acidic pocket containing the Arg-118, Arg-292, and Arg-

371 triad interacts with the carboxylic acid moiety of DANA (**104**) while a hydrophobic pocket binds the glycerol side chain and a second hydrophobic pocket is engaged by the acetamido group. A fourth negatively charged pocket, encompassing Glu-119 and Glu-227, defines the region where the C4-hydroxyl moiety resides. Evaluation of a furan-based analog of DANA (**104**), compound **108**, illustrated that the precise position of the coordinating central ring was not as important as the positions of the interacting functional groups themselves.[582–584] Replacing the furan core of **108** with a cyclopentane moiety and simplification of the substituents afforded **109**, a compound with potency comparable to DANA (**104**) that bound to the enzyme in a mode that projected the guanidino group into the negatively charged pocked in a fashion analogous to that observed for zanamivir (**16**).[582,583]   However, when the glycerol side chain was exchanged for a more hydrophobic *n*-butyl or *iso*-butyl moiety, designed to achieve a better interaction with the hydrophobic pocket, the guanidino group positioning was slightly altered. Intriguingly, this was advantageous, as the zanamivir escape mutations in which Gln119 is replaced by glycine or alanine remained susceptible to peramivir (**107**).[585–587] The *iso*-butyl side chain was ultimately selected because it bound to both hydrophobic pockets and demonstrated fewer conformational differences between the influenza A and B neuraminidase active sites.[582,583]

    The initial efficacy data in animal models of influenza and preliminary clinical studies suggested oral efficacy with peramivir (**107**).[587] However, during phase III clinical prophylaxis trials, peramivir (**107**) failed to reduce influenza shedding, most likely due to the relatively low blood levels achieved.[588] A parenteral formulation was developed and intravenously administered peramivir (**107**) has been studied in 4 clinical trials but has yet to be approved for use in the United Sates. Nevertheless, the 2009 H1N1 swine flu pandemic prompted the FDA to issue an emergency use authorization in the United Sates for peramivir IV use.[589]

107 (peramivir)   **108**

109   **110**: R = H (laninamivir)
          111: R = CH$_3$(CH$_2$)$_6$CO

    The most recent neuraminidase entrant to be approved is laninamivir (**111**), the octanoate prodrug of R-125489 (**110**), which is approved for use in Japan.[590] R-125489 (**110**) is the 7-*O*-

methyl derivative of zanamivir (16) and exhibits a similar broad range of activity against most influenza strains.[591–593] The physical properties of laninamivir (111) are not compatible with oral dosing, so in order to improve on zanamivir (16), a prodrug was developed that was designed to promote an extended retention time in the lung, providing a long-acting neuraminidase inhibitor.[593] Preclinical studies have shown that the prodrug is cleaved to the active agent in the lung, where it resides for extended periods of time, and clinical studies have shown that the active agent exhibits a half-life of 3 days in man after a single dose.[593,594] In 2010, a single dose of laninamivir (111), which is marketed as Inavir®, was approved as a therapeutic agent for the treatment of influenza infection in Japan.

5.2 Respiratory Syncytial Virus (RSV)

In 1956, during a respiratory illness outbreak, an unusual virus was isolated from a symptomatic laboratory chimpanzee and designated as chimpanzee coryza agent.[595] The following year, identical viruses were isolated from a child with pneumonia and another with croup.[596,597] Due to the characteristic cytopathic effect of this virus which induced the formation of syncytia throughout cellular monolayers during infection, it was designated respiratory syncytial virus (RSV). RSV is a member of the pneumovirus genus within the paramyxovirus family of viruses. Among infants and young children, RSV is the chief causative agent of virus-induced lower respiratory tract infection and is especially pathogenic in premature and immunocompromised infants.[598] Serology studies demonstrated that nearly every child contracts RSV by 2 years of age.[598,599] In addition, RSV has been shown to be responsible for serious illness in select adult populations, including the elderly and those with cardiopulmonary disease and bone marrow transplant patients.[600–602]

The guanosine nucleoside analog ribavirin (11) is a broad-spectrum antiviral agent active against a number of DNA and RNA viruses, including RSV.[603–605] Multiple mechanisms of action have been proposed to explain the antiviral properties of ribavirin (11), including inhibition of inosine monophosphate dehydrogenase and viral polymerase inhibition by incorporation into the growing nucleic acid chain, as well as immunomodulation.[606] An aerosolized form of ribavirin (11), marketed as Virazole®, was approved for marketing in 1986 for the treatment of RSV infection in children. Although it remains the only antiviral available for the treatment of RSV infection, ribavirin (11) is not currently favored for use for this indication due to the mode of delivery, lack of cost effectiveness, questionable efficacy in studies subsequent to its approval, toxicity, and teratogenicity.[607–611] In this light, there is a severe unmet medical need for effective RSV antivirals, although ribavirin (11) has proven to be a useful agent when used in combination with interferon-based therapies for the treatment of hepatitis C virus infections.[612]

5.2.1 RSV Immunoprophylaxis

In view of the fact that severe RSV disease is restricted primarily to infants, the infirmed elderly and the immune compromised, it was speculated that the serum of healthy adults may have sufficient titers of neutralizing antibodies to prevent or reduce the severity of RSV infection, while the high risk groups do not. Evidence supporting this hypothesis was reported in 1976, when it was observed that the titer of maternal circulating RSV-neutralizing antibodies was inversely proportional to the severity of RSV-induced pneumonia suffered by their infants.[613] Later studies revealed that the levels of IgG directed against the RSV fusion glycoprotein (F protein), in particular, were highly correlated with the prevention of a second RSV infection.[614] Moreover, multiple experiments revealed that topical or intraperitoneal administration of human immunoglobulins containing RSV neutralizing antibodies were efficacious in cotton rat and owl monkey models of RSV infection.[615–617] These findings prompted clinical trials using intravenous IgG (IGIV) for the prevention of RSV disease. IGIV diminished RSV nasal shedding and the reduction of nasopharyngeal RSV titers was noted, but no improvement in symptom severity or reduction in either hospitalization stay or oxygen requirement was demonstrated.[618] Further trials suggested that insufficient antibody titers called for an RSV-enriched IgGIV formulation (RSV-IGIV). RSV-IGIV was obtained from pooled human plasma that contained high-titer neutralizing antibodies by screening donors using ELISAs and microneutralization assays.[619] After moderate success in clinical trials, RSV-IGIV was approved in 1996 for the prevention of serious RSV disease in preterm infants under the brand name RespiGam®.[620,621] Unfortunately, there were issues associated with RSV-IGIV involving intravenous delivery of high volumes over long periods (15 mL/kg over 2–4 h).[622] Furthermore, safety was a significant concern, with the potential for transmission of blood-borne pathogens using a pooled human blood product.[623] Consequently, RespiGam® was withdrawn from the market once a more appropriate substitute became available.

In order to surmount the shortfalls of RSV-IGIV polyclonal immunotherapy, attention was focused on development of monoclonal antibody RSV therapy. The viral envelope G and F glycoproteins play the greatest role in the induction of RSV neutralizing antibodies.[624] The function of the G glycoprotein is to promote host cell adsorption while the F glycoprotein mediates fusion of viral and cellular membranes.[625,626] Certain monoclonal antibodies (MAbs) that immunoprecipitate the F protein also prevented cell-to-cell fusion, thereby inhibiting syncytia formation.[626] That the F protein is more highly conserved across strains than the G glycoprotein, along with the ability of the F protein to induce neutralizing antibodies, made it an attractive viral target for development of an immunoprophylaxis agent.[627,628] Characterization of the RSV F protein with panels of MAbs identified three distinct neutralizing antigenic sites, denoted A, B, and C.[629,630] Moreover, eight of these particular epitopes directed against antigenic site A were conserved in 22 of 23 RSV subgroup A and B clinical isolates that were collected over a 30-year period between 1956 and 1986.[630] These findings served as the impetus for the development of the first antiviral biologic agent. Palivizumab was approved in 1998 and marketed under the brand name of Synagis® as a recombinant humanized monoclonal antibody recommended for the prevention of serious lower respiratory tract disease in pediatric patients at high risk for RSV infection. It is delivered as a once-a-month IM injection, usually for the first 2 years of life, during the winter months. The antibody is a chimeric IgG with an antigen recognition site derived from a mouse monoclonal (MAb 1129).[630] Palivizumab has a human immunoglobulin backbone to improve patient immune tolerance and maintain favorable pharmacokinetic properties.[631,632] Since its approval, palivizumab has been often scrutinized due to its cost-benefit ratio, as it exhibits moderate efficacy in the face of high expense.[633–637] Hence, there remains a significant unmet medical need for the development of more efficacious anti-RSV agents for use in prevention and therapy.

5.2.2 Small-Molecule RSV Inhibitors

Small-molecule inhibitors of RSV have been sought as part of an attempt to identify compounds that would act prophylactically or therapeutically and offer the convenience of oral dosing.[638-644] RSV inhibitors have typically been discovered using virus infectivity assays that are mechanistically agnostic and rely upon a chemical genetics strategy to deduce the mode of action. The most common mechanism of inhibition of RSV discovered by this process has been molecules that interfere with virus-host cell membrane fusion and several advanced molecules have been described.[643-645] The bis-benzimidazole derivative BABIM (112) was the first RSV fusion inhibitor to be described that was ultimately determined to share structural elements and a silhouette with a series of benzimidazole derivatives that appeared to be mechanistically identical.[644,646-653] BMS-433771 (115), which was derived from optimization of the screening leads 113 and 114, has been characterized as an orally effective inhibitor of RSV fusion that acts by interfering with the function of the viral fusion protein by binding into a pocket formed by association of the N-terminus heptad repeats that subsequently recognize structural elements in the carboxy terminus heptad repeats.[651-658] The site of action was determined with the aid of the photoaffinity label 116, which contains a diazirine moiety that under the influence of UV light ejects N_2 to afford a highly reactive carbene that inserts into proximal elements of a protein and is an established and effective technique for identifying protein targets of small-molecule drugs.[653,659-662] Structural coincidence between BABIM (112) and BMS-433771 (115) was accomplished with the amidine 117, which retained potent inhibitory activity towards virus developed to be resistant to BMS-433771 (115).[658]

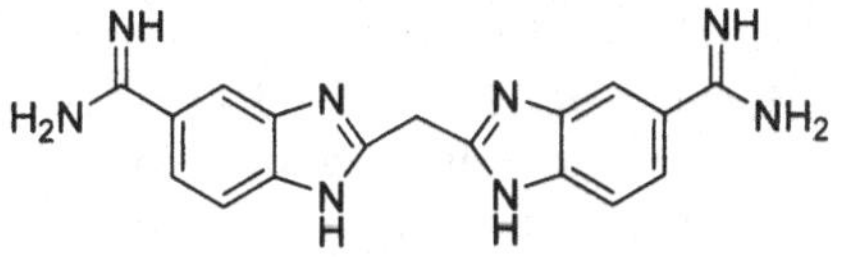

112 (BABIM)

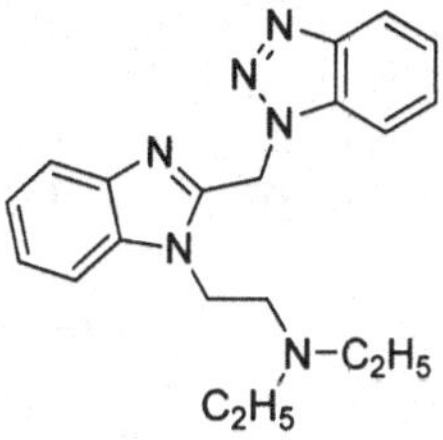

113

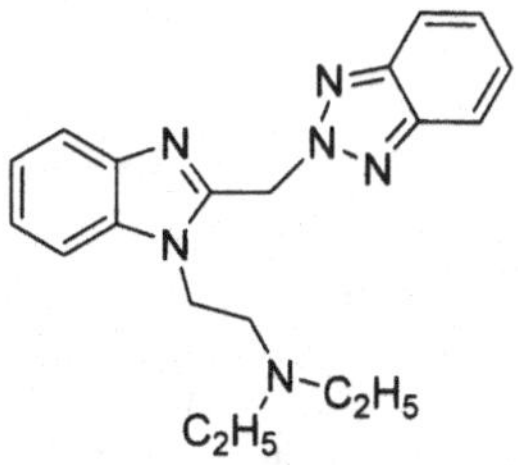

114

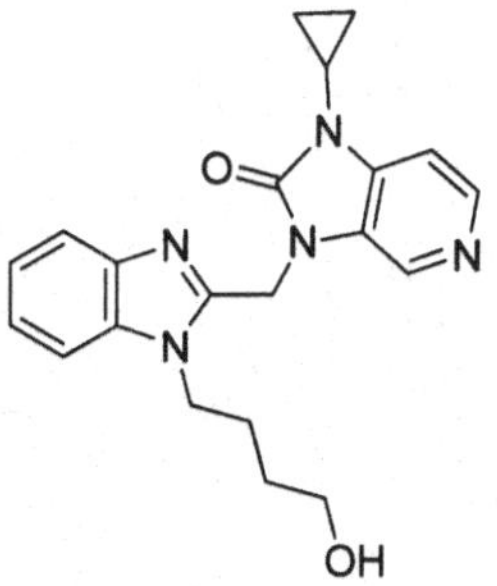

115 (BMS-433771)

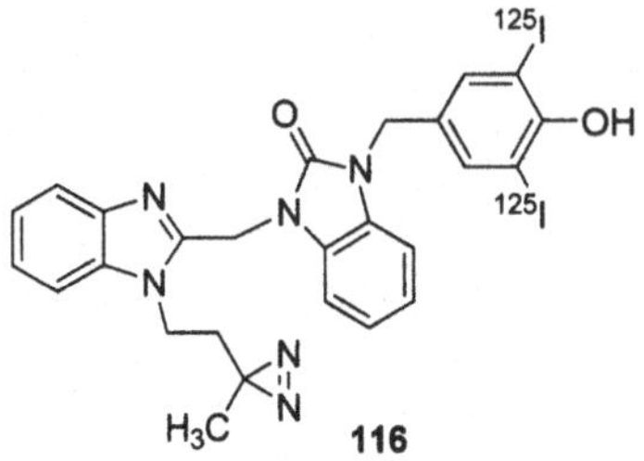

116

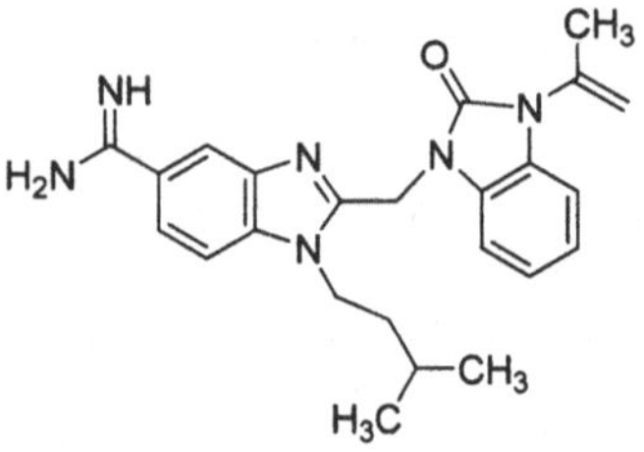

117

Several other RSV fusion inhibitors have been described that act in a mechanistically similar fashion, including the dendrimer-like triazine derivatives CL-387626 (**118**) and RFI-641 (**119**) and the phenol derivative VP-14637 (**120**).[663–674] None of these compounds are orally bioavailable and are delivered topically, either by aerosolization or intranasally, and VP-14637 (**120**), which has been show to be effective in the cotton rat model of RSV infection, has been advanced into clinical trials.[671–674] TMC-353121 (**121**) is an orally bioavailable RSV fusion inhibitor that is evolved from the screening lead JNJ-1789008 (**122**), a compound subjected to careful structural optimization designed to reduce the extended lung retention times seen with early leads.[675–679] Lung retention time correlated with basicity, leading to a focus on compounds with a pKa of <9 that led to the identification of TMC-353121 (**121**) as a potent RSV inhibitor that exhibits a $T_{1/2}$ in lung tissue of 25.1 h, markedly shorter than for earlier compounds, some of which demonstrated a $T_{1/2}$ of >150 h. TMC-353121 (**121**) significantly reduced viral titers in RSV-infected cotton rats when administered orally, intravenously, or by aerosol and the compound is also active in a BALB/c mouse model of infection.[675–679] The mode of action of TMC-353121 (**121**) was elucidated by a series of experiments that involved the characterization of resistant virus and photoaffinity labeling using the diazirine derivative **123** which labeled the RSV F protein.[680] An X-ray crystallographic structure of TMC-353121 (**121**) bound to the RSV F protein 6 helix bundle in its fusogenic conformation revealed detailed molecular interactions that indicated that the compound bound to a hydrophobic pocket formed in the amino terminus trimer of heptad repeats but, interestingly, identified stabilizing interactions with elements of the carboxy terminal heptad repeats.[680]

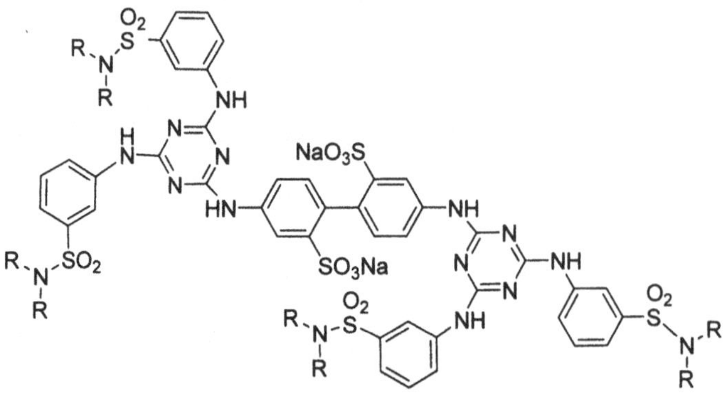

118: R = CH$_2$CH$_2$CONH$_2$ (CL-387626)
119: R = CH$_2$CONH$_2$ (RFI-641)

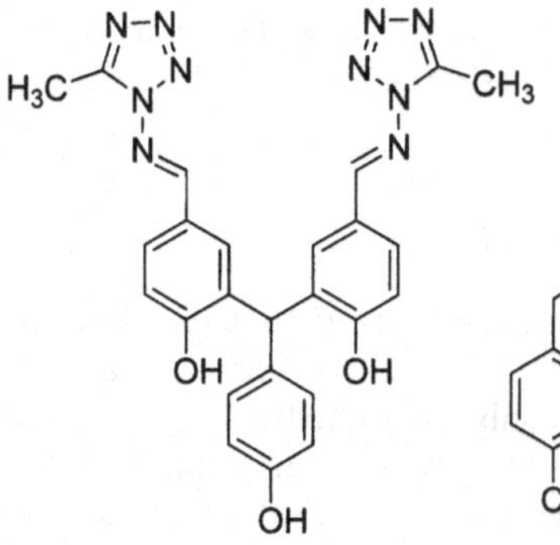

120 (VP-14637) 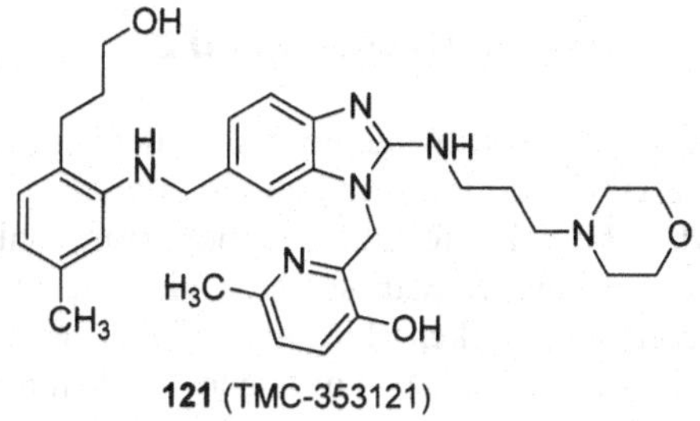 121 (TMC-353121)

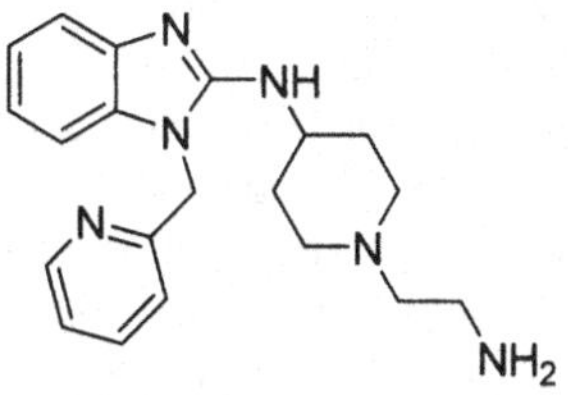

122 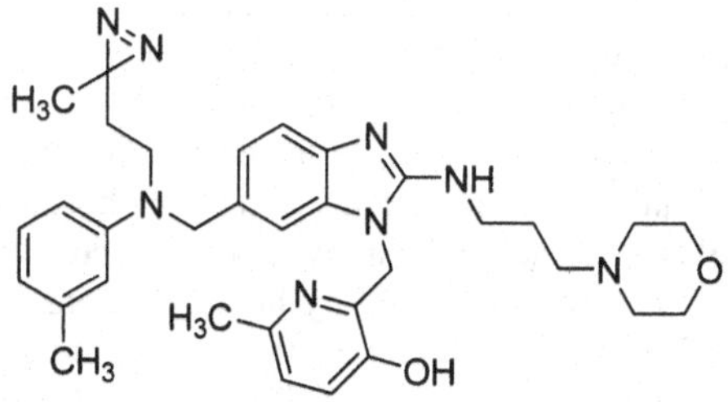 123

Another important inhibitor of RSV is the benzodiazepine RSV-604 (**124**), which has been advanced into clinical trials.[681,682] RSV-604 (**124**) was optimized from a screening lead and resistance was found to map to the viral nucleocapsid protein, which binds to and protects the viral RNA and is an important element in the viral replication complex. RSV-604 (**124**) exhibits an EC_{50} = 600 nM in cell culture, is active towards both RSV-A and RSV-B strains and a range of clinical isolates, and is orally bioavailable in the rat.[681,682] Although the mode of action of RSV-604 (**124**) is not fully defined, a crystalline form of the RSV nucleocapsid protein has been identified that might facilitate cocrystallization.[683]

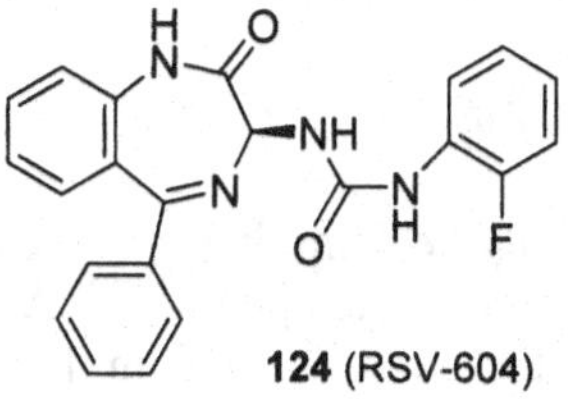

124 (RSV-604)

6   *Herpesviridae* Inhibitors

The *Herpesviridae* family of double-stranded DNA viruses that are known to infect humans comprises herpes simplex viruses type 1 and type 2 (HSV-1 and HSV-2), varicella zoster virus (VZV), the cause of chicken pox and shingles, human cytomegalovirus (HCMV), Epstein-Barr virus (EBV, human herpesvirus 4), human herpesviruses 6 type-A and type B (HHV-6-A and

HHV-6-B) and human herpesviruses type-7 (HHV-7) and type-8 (HHV-8), the latter known more commonly as Kaposi's sarcoma virus that was a hallmark of advanced HIV infections prior to the advent of HAART.[684,685]

6.1   Herpes Simplex Viruses 1 and 2

HSV infections have a worldwide presence, infecting an estimated 60–95% of the population, wth both subtypes establishing latent infections in the host. HSV-1 is more typically associated with oral disease while HSV-2 infections are more commonly associated with the genitals.[686–688] HSV is spread by direct contact with secreted virus at sites of infection, entering the host through mucous membranes or broken skin, after which it is transported along sensory neurons to the ganglia where it establishes a state of latency. When replication is reactivated, the virus can travel along the peripheral sensory neurons to mucosal sites or the skin.[686–689]

    Therapeutic options for the treatment of HSV infections are orally active nucleoside analogs acyclovir (6), valacyclovir (7), and famciclovir (125), a double prodrug of penciclovir (127) that is metabolized through the intermediacy of 126 that was approved for marketing in 2006.[689–693] The specificity of acyclovir (6) and famciclovir (125) for HSV and VZV is a function of the viral thymidine kinase which efficiently phosphorylates the 5'-hydroxyl of these agents as the first critical step in metabolic activation, with conversion to the active triphosphate, a substrate of the viral polymerase, completed by host cell kinases. The intracellular half-life of penciclovir triphosphate in HSV-infected cells is 10–20 h, considerably longer than the 1 h or less reported for acyclovir triphosphate.[693] Acyclovir (6) is typically dosed at 200 mg 5 times a day while the dosing regimen for both valacyclovir (7) and famciclovir (125) is less frequent, either BID or TID.[689]

125: R = OAc: famciclovir
126: R = H

127: penciclovir

6.2   Varicella Zoster Virus (VZV)

In children, VZV infection is typically self-limiting and presents as a rash with vesicular skin lesions that are commonly referred to as chicken pox.[696–698] A live, attenuated vaccine is available for the prevention of chicken pox that was licensed in 1999 but has been a source of some controversy.[698,699] In adults primary VZV infection can be associated with a more severe disease while the virus can cause significant problems to the fetus of pregnant women. Reactivation of VZV in adults in later life presents as a vesicular rash, typically appearing on the thorax and accompanied by considerable neurological pain that can persist for longer than a month after the initial appearance of rash, referred to as postherpetic neuralgia. Therapy with the nucleoside analogs acyclovir (6), valacyclovir (7), and famciclovir (125) can relieve the symptoms of shingles but must be initiated within 3 days of the onset of symptoms to be most successful.

The bicyclic furanonucleoside derivative **128** is the most potent VZV inhibitor yet reported, exhibiting sub-nanomolar EC$_{50}$s in cell culture.[700–704] This compound is poorly active toward thymidine kinase-deficient VZV strains, indicative of phosphorylation as a critical step in metabolic activation. Indeed, this class of nucleoside analog is a substrate for VZV thymidine kinase but not for the HSV enzymes, providing an explanation for the observed selectivity towards VZV.[705] The lipophilicity of **128** leads to limited solubility, a problem addressed by the valine prodrug FV-100 (**129**), a compound that has recently entered clinical trials for the treatment of shingles.[700–704]

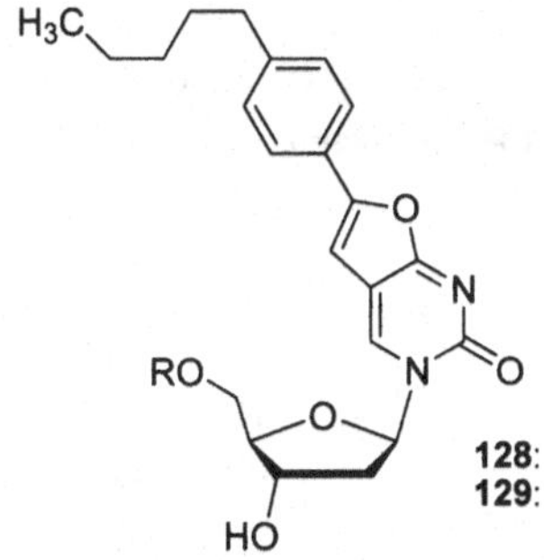

128: R = H
129: R = Val (FV-100)

6.3 Human Cytomegalovirus (HCMV) Inhibitors

HCMV is a widespread virus that infects the majority of the world's population but is a limited problem in healthy individuals.[706–708] However, HCMV infection presents a significant challenge in the immunocompromised patient population, particularly those who have undergone bone marrow or solid organ transplants, and is an opportunistic problem in those infected with HIV-1. The small-molecule HCMV DNA polymerase inhibitors ganciclovir (**130**) preferably administered as the valine ester prodrug valganciclovir (**131**), cidofovir (**132**), and foscarnet (**133**) are approved for the prevention and treatment of HCMV infection.[709–717] Ganciclovir (**130**) and cidofovir (**132**) act as enzyme substrates after metabolic activation by phosphorylation while the pyrophosphate analog foscarnet (**133**) inhibits all human herpes virus polymerases by a mechanism that remains to be definitively elucidated.[709–717] The phosphothiorate antisense oligonucleotide (ASO) fomivirsen (**134**), which targets an immediate early gene locus, is the first of this drug chemotype to be licensed for marketing approved for intra-ocular application in the treatment of HCMV-associated retinitis in 1998.[718–720] Fomivirsen (**134**) is applied topically by intravitreal injection at doses of 165 or 330 μg once weekly for the treatment of HCMV-induced retinitis in AIDS patients, a delivery mode from which systemic exposure is limited. High-dose valaciclovir (**7**) has also been used prophylactically to prevent HCMV in bone marrow and renal transplant recipients.[721]

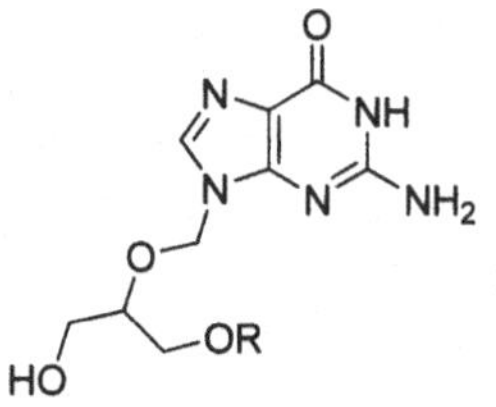

130: R = H: ganciclovir
131: R = Val: valganciclovir   **132**: cidofovir 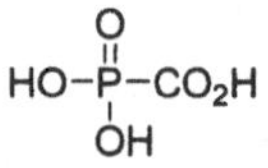   **133**: foscarnet

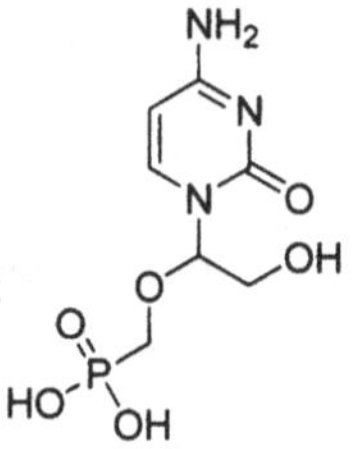

5'-GCGTTTGCTCTTCTTCTTGCG-3'
134: fomivirsen

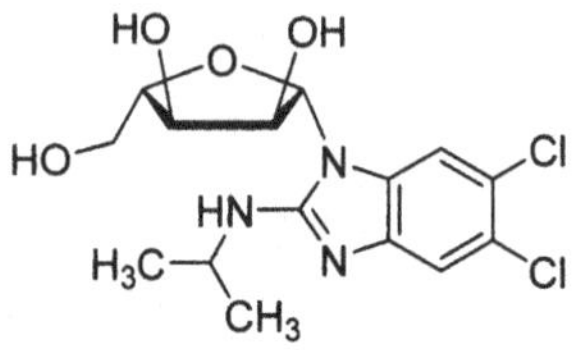

135: maribavir 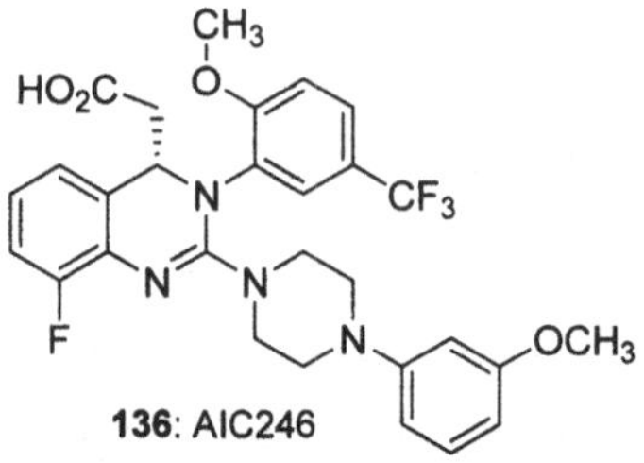   **136**: AIC246

6.4   Epstein-Barr Virus (EBV, Human Herpesvirus 4) Inhibitors

EBV is an ubiquitous virus, discovered in 1964, that has established a persistent, latent infection in 90% of the world's population and has been shown to be associated with infectious mononucleosis, the most frequent clinical symptom of this disease.[722] Infection can occur in early childhood via the oral compartment as the site of entry with transmission due to exposure to oral secretions with shared toys or utensils a common vector. Infection is largely asymptomatic in young children but presents as infectious mononucleosis in adolescents and young adults typically after a 6-week incubation period following exposure. Symptoms typically appear initially as malaise followed by fever, sore throat, and fatigue; hepatitis is observed in 80% of cases but is usually subclinical in the vast majority of cases. EBV can also be acquired from blood transfusions, attributed to infection of circulating memory B-cells in the peripheral circulation, and from organ transplants. The median duration of infectious mononucleosis is 16 days, but recovery is typically slow, with fatigue persisting for several months before resolving. However, reinfection is not common. Therapy is largely palliative in nature, with analgesics prescribed for the management of fever and pain.[722] Several nucleoside analogs have been explored as therapy for EBV with valacyclovir (**7**) exhibiting a significant effect on reducing viral load and resolving symptoms compared to placebo in a small study.[722,723] Ganciclovir (**130**) inhibits EBV replication *in vitro* but the application of its prodrug valganciclovir (**131**) has not been explored in a clinical setting.

7   Epilogue

The emergence of HIV-1 infection provided an important stimulus to the design and development of antiviral therapies which was a largely a fledgling enterprise in 1981 when the virus was first encountered. In the intervening years, antiviral drug design has come of age, providing a cadre of potent and effective HIV-1 inhibitors that, when combined as triple therapy, provide excellent control over virus replication, reducing what was once a death sentence to a manageable, chronic disease. The learnings from HIV-1 science and drug discovery have been applied very effectively to the problem presented by HCV infection, with a heavy focus on combinations of direct-acting antiviral agents that are capable of curing this infection in the absence of pegylated-interferon α and ribavirin. The pace of change resulting from innovative clinical trials of combinations of HCV inhibitors has been remarkable with evidence that this will continue as the optimal combinations are identified. The current longer term focus appears to be fixed firmly on identifying a fixed-dose regimen capable of treating all genotypes with a convenient, single tablet ingested on a daily basis. Indeed, combinations of direct-acting HCV inhibitors offer the promise of curing a chronic viral infection, an objective that would be unique in the history of antiviral therapy.

References

1.   Plotkin, S. A. *Nat. Med.* **2005**, *11*, S5.
2.   Field, H. J.; De Clrecq, E. *Microbiol. Today* **2004**, 31, 58.
3.   Elion, G. B.; Furman, P. A.; Fyfe, J. A.; De Miranda, P.; Beauchamp, L.; Schaeffer, H. J. *Proc. Natl. Acad. Sci. U.S.A.* **1977**, *74*, 5716.
4.   Schaeffer, H. J.; Beauchamp, L.; De Miranda, P.; Elion, G. B.; Bauer, D. J.; Collins, P. *Nature* **1978**, *272*, 583.
5.   Beauchamp, L. M.; Orr, G. F.; De Miranda, P.; Burnette, T.; Krenitsky, T. A. *Antiviral Chem. Chemother.* **1992**, *3*, 157.
6.   Purifoy, D. J.; Beauchamp, L. M.; de Miranda, P.; Ertl, P.; Lacey, S.; Roberts, G.; Rahim, S. G.; Darby, G.; Krenitsky, T. A.; Powell, K. L. *J. Med. Virol.* **1993**, *Suppl 1*, 139.
7.   Ormrod, D.; Scott, L. J.; Perry, C. M. *Drugs* **2000**, *59*, 839.
8.   Ormrod, D.; Goa, K. *Drugs* **2000**, *59*, 1317.
9.   Broder, S. *Antiviral Res.* **2010**, *85*, 1.
10.   Prusoff, W. H. *Biochim. Biophys. Acta* **1959**, *32*, 295.
11.   Perkins, E. S.; Wood, R. M.; Sears, M. L.; Prusoff, W. H.; Welch, A. D. *Nature* **1962**, *194*, 985.
12.   Darby, G. *Antiviral Chem. Chemother.* **1995**, *6*, 54.
13.   McKenzie, R.; Fried, M. W.; Sallie, R.; Conjeevaram, H.; Di Bisceglie, A. M.; Park, Y.; Savarese, B.; Kleiner, D.; Tsokos, M.;. Luciano, C.; Pruett, T., Stotka, J. L.; Straus, S. E.; Hoofnagle, J. H. *New Engl. J. Med.* **1995**, *333*, 1099.
14.   Colacino, J. M. *Antiviral Res.* **1996**, *29*, 125.
15.   Gallo, R. C.; Sarin, P. S.; Gelmann, E. P.; Robert-Guroff, M.; Richardson, E.; Kalyanaraman, V. S.; Mann, D.; Sidhu, G. D.; Stahl, R. E.; Zolla-Pazner, S.; Leibowitch, J.; Popovic, M. *Science* **1983**, *220*, 865.
16.   Barre-Sinoussi, F.; Chermann, J. C.; Rey, F.; Nugeyre, M. T.; Chamaret, S.; Gruest, J.; Dauguet, C.; Axler-Blin, C.; Vezinet-Brun, F.; Rouzioux, C.; Rozenbaum, W.; Montagnier, L. *Science* **1983**, *220*, 868.
17.   De Clercq, E. *Int. J. Antimicrob. Agents* **2009**, *33*, 307.
18.   Gallant, J. E.; DeJesus, E.; Arribas, J. R..; Pozniak, A. L.; Gazzard, B.; Campo, R. E.; Lu, B.; McColl, D.; Chuck, S.; Enejosa, J.; Toole, J. J.; Cheng, A. K.; for the Study 934 Group. *N. Engl. J. Med.* **2006**, *354*, 251.

19.   Frampton, J. E.; Croom, K. F. *Drugs* **2006**, *66*, 1501.
20.   Scott, L. J.; Perry, C. M. *Drugs* **2002**, *62*, 507.
21.   Aghemo, A.; Grazia Rumi, M.; Colombo, M. *Nat. Rev Gastroenterol. Hepatol.* **2010**, *7*, 485.
22.   Jain, M. K.; Zoellner, C. *Exp. Opin. Pharmacother.* **2010**, *11*, 673.
23.   Choo, Q. L.; Kuo, G.; Weiner, A. J.; Overby, L. R.; Bradley, D. W.; Houghton, M. *Science* **1989**, *244*, 359.
24.   Choo, Q.-L. Richman, K. H.; Han, J. H.; Berger, K.; Lee, C.; Dong, C.; Gallegos, C.; Coit, D.; Medina-Selby, A.; Barr, P. J.; Weiner, A. J.; Bradley, D. W.; Kuo, G.; Houghton, M. *Proc. Natl. Acad. Sci. U.S.A.* **1991**, *88*, 2451.
25.   Lohmann, V.; Körner, F.; Koch, J. O.; Herian, U.; Theilmann, L.; Bartenschlager, R. *Science* **1999**, *285*, 110.
26.   Bartenschlager, R. *J. Hepatol.* **2005**, *43*, 210.
27.   Buch, J.; Purcell, R. H. *Proc. Natl. Acad. Sci. U.S.A.* **2006**, *103*, 3500.
28.   Gao, M.; Nettles, R. E.; Belema, M.; Snyder, L. B.; Nguyen, V. N.; Fridell, R. A.; Serrano-Wu, M. H.; Langley, D. R.; Sun, J.-H.; O'Boyle II, D. R.; Lemm, J. A.; Wang, C.; Knipe, J. O.; Chien, C.; Colonno, R. J.; Grasela, D. M.; Meanwell, N. A.; Hamann, L. G. *Nature* **2010**, *465*, 96.
29.   Koev, G.; Kati, W. *Exp. Opin. Investig. Drugs* **2008**, *17*, 303.
30.   Neumann, A. U.; Lam, N. P.; Dahari, H.; Gretch, D. R.; Wiley, T. E.; Layden, T. J.; Perelson, A. S. *Science* **1998**, *282*, 103.
31.   Guedj, J.; Rong, L.; Dahari, H.; Perelson, A. S. *J. Viral Hepat.* **2010**, *17*, 825.
32.   Rong, L.; Dahari, H.; Ribeiro, R. M.; Perelson, A. S. *Sci. Transl. Med.* **2010**, *2*, 1.
33.   Lok, A. S.; Gardiner, D. F.; Lawitz, E.; Martorell, C.; Everson, G. T.; Ghalib, R.; Reindollar, R.; Rustgi, V.; McPhee, F.; Wind-Rotolo, M.; Persson, A.; Zhu, K.; Dimitrova, D. I.; Eley, T.; Guo, T.; Grasela, D. M.; Pasquinelli, C. *New Engl. J. Med.* **2012**, *366*, 216.
34.   Davies, W. L.; Grunert, R. R.; Haff, R. F.; McGahen, J. W.; Neumayer, E. M.; Paulshock, M.; Watts, J. C.; Wood, T. R.; Hermann, E. C.; Hoffmann, C. E. *Science* **1964**, *144*, 862.
35.   Wendel, H. A.; Snyder, M. T.; Pell, S. *Clin. Pharmacol. Ther.* **1966**, *7*, 38.
36.   Quilligan, J. J., Jr.; Hirayama, M.; Baernstein, H. D., Jr. *J. Pediatr.* **1966**, *69*, 572.
37.   Finklea, J. F.; Hennessy, A. V.; Davenport, F. M. *Am. J. Epidemiol.* **1967**, *85*, 403.
38.   Wingfield, W. L.; Pollack, D.; Grunert, R. R. *New Engl. J. Med.* **1969**, *281*, 579.
39.   Dolin, R.; Reichman, R. C.; Madore, H. P.; Maynard, R.; Linton, P. N.; Webber-Jones, J. *New Engl. J. Med.* **1982**, *307*, 580.
40.   Tominack, R. L.; Hayden, F. G. *Infectious Dis. Clin. North Am.* **1987**, *1*, 459.
41.   von Itzstein, M.; Mu, W.-Y.; Kok, G. B.; Pegg, M. S.; Dyasan, J. C.; Jim, B.; Phan, T. V.; Smythe, M. L.; White, H. F.; Ollver, S. W.; Colman, P. M.; Varghese, J. N.; Ryan, D. M.; Woods, J. M.; Bethell, R. C.; Hotham, V. J.; Cameron, J. M.; Penn, C. R. *Nature* **1993**, *363*, 418.
42.   Kim, C. U.; Lew, W.; Williams, M. A.; Liu, H.; Zhang, L.; Swaminathan, S.; Bischofberger, N.; Chen, M. S.; Mendel, D. B.; Tai, C. Y.; Laver, W. G.; Stevens, R. C. *J. Am. Chem. Soc.* **1997**, *119*, 681.
43.   Maghzi, A.-H., Marta, M., Bosca, I., Etemadifar, M., Dobson, R., Maggiore, C., Giovannoni, G., Meier, U.-C. *Pathophysiology* **2011**, *18*, 13.
44.   Kakalacheva, K., Muenz, C., Luenemann, J. D. *Biochim. Biophys. Acta* **2011**, *1812*, 132.
45.   Milo, R.; Kahana, E. *Autoimmunity Rev.* **2010**, *9*, A387.
46.   Franssila, R.; Hedman, K. *Clin. Rheumatol.* **2006**, *20*, 1139.
47.   Kadow, J. F.; Regueiro-Ren, A.; Weinheimer, S. P. *Curr. Opin. Investig. Drugs* **2002**, *3*, 1574.
48.   Boccardo, E.; Villa, L. L. *Curr. Med. Chem.* **2007**, *14*, 2526.
49.   Tong, T. R. *Perspectives Med. Virol.* **2007**, *16*, 43.
50.   Cheng, V. C. C.; Lau, S. K. P.; Woo, P. C. Y.; Yuen, K. Y. *Clin. Microbiol. Rev.* **2007**, *20*, 660.
51.   Yeung, K.-S., Meanwell, N. A. *Infectious Disorders: Drug Targets* **2007**, *7*, 29.
52.   Wenzel, R. P.; Bearman G., Edmond, M. B. *Arch. Med. Res.* **2005**, *36*, 610.
53.   Apetrei, C.; Robertson, D. L.; Marx, P. A. *Frontiers Biosci.* **2004**, *9*, 225.

54. Chahroudi, A.; Bosinger, S. E.; Vanderford, T. H.; Paiardini, M.; Silvestri, G. *Science* **2012**, *335*, 1188.

55. Powderly, W. G. *Clin. Infect. Dis.* **2000**, *31*, 597.

56. Willemot, P.; Klein, M. B. *Exp. Rev. Anti-Infective Ther.* **2004**, *2*, 521.

57. UNAIDS World AIDS Day Report 2011 available at: http://www.unaids.org/en/media/unaids/contentassets/documents/unaidspublication/2011/JC2216_WorldAIDSday_report_2011_en.pdf

58. World Health Organization data fact sheet no. 310, updated June, 2011 available at: http://www.who.int/mediacentre/factsheets/fs310/en/index.html

59. Kolata, G. *Science* **1987**, *235*, 1570.

60. de Clercq, E. *Int. J. Antimicrob. Agents* **2009**, *33*, 307.

61. de Clercq, E. *Rev. Med. Virol.*, **2009**, *19*, 287.

62. Choudhary, S. K.; Margolis, D. M. *Ann. Rev. Pharmacol. Toxicol.* **2011**, *51* 397.

63. Levy, J. A. *New Engl. J. Med.* **2009**, *360*, 724.

64. Finzi, D.; Blankson, J.; Siliciano, J. D.; Margolick, J. B.; Chadwick, K.; Pierson, T.; Smith, K.; Lisziewicz, J.; Lori, F.; Flexner, C.; Quinn, T. C.; Chaisson, R. E.; Rosenberg, E.; Walker, B.; Gange, S.; Gallant, J.; Siliciano, R. F. *Nat. Med.* **1999**, *5*, 512.

65. Pierson, T.; McArthur, J.; Siliciano, R. F. *Ann. Rev. Immunol.* **2000**, *18*, 665.

66. Siliciano, J. D.; Siliciano, R. F. *J. Antimicrob. Chemother.* **2004**, *54*, 6.

67. Palella, F. J.; Delaney, K. M.; Moorman, A. C.; Loveless, M. O.; Fuhrer, J.; Satten, G. A.; Aschman, D. J.; Holmberg, S. D. *N. Engl. J. Med.*, **1998**, *38*, 853.

68. Lima, V. D., Harrigan, R., Bangsberg, D. R., Hogg, R. S., Gross, R., Yip, B., Montaner, J. S. G., *J. Acq. Immune Defic. Syndr.* **2009**, *50*, 529.

69. Harrison, K. M.; Song, R.; Zhang X. *J. Acq. Immune Defic. Syndr.* **2010**, *53*, 124.

70. Bhaskaran, K.; Hamouda, O.; Sannes, M.; Boufassa, F.; Johnson, A. M.; Lambert, P. C.; Porter, K. *JAMA* **2008**, *300*, 51.

71. Thompson, M. A.; Aberg, J. A.; Cahn, P.; Montaner, J. S. G.; Rizzardini, G.; Telenti, A.; Gatell, J. M.; Guenthard, H. F.; Hammer, S. M.; Hirsch, M. S.; Jacobsen, D. M.; Reiss, P.; Richman, D. D.; Volberding, P. A.; Yeni, P.; Schooley, R. T. *JAMA* **2010**, *304*, 321.

72. Deeks, S. G. *Lancet* **2003**, *362*, 2002.

73. Stephan, C.; Dauer, B.; Khaykin, P.; Stuermer, M.; Gute, P.; Klauke, S.; Staszewski, S. *Curr. HIV Res.* **2009**, *7*, 320.

74. Nozza, S.; Galli, L.; Visco, F.; Soria, A.; Canducci, F.; Salpietro, S.; Gianotti, N.; Bigoloni, A.; Della Torre, L.; Tambussi, G.; Lazzarin, A.; Castagna, A. *AIDS* **2010**, *24*, 924.

75. Qian, K.; Morris-Natschke, S. L.; Lee, K.-H. *Med. Res. Rev.* **2009**, *29*, 369.

76. Tilton, J. C.; Doms, R. W. *Antiviral Res.* **2010**, *85*, 91.

77. Hertje, M.; Zhou, M.; Dietrich, U. *ChemMedChem* **2010**, *5*, 1825.

78. Caffrey, M. *Trends Microbiol.* **2011**, *19*, 191.

79. Singh, I. P.; Chauthe, S. K. *Exp. Opin. Therapeutic Patents* **2011**, *21*, 227.

80. Singh, I. P.; Chauthe, S. K. *Exp. Opin. Therapeutic Patents* **2011**, *21*, 399.

81. Teissier, E.; Penin, F.; Pecheur, E.-I. *Molecules* **2011**, *16*, 221.

82. Lobritz, M. A.; Ratcliff, A. N.; Arts, E. J. *Viruses* **2010**, *2*, 1069.

83. Kadow, J. F.; Bender, J.; Regueiro-Ren, A.; Ueda, Y.; Wang, T.; Yeung, K.-S.; Meanwell, N. A. In *Antiviral Drugs: From Basic Discovery Through Clinical Trials,* Kazmierski, W. M., ed., Wiley: Hoboken, NJ., **2011**, pp149–162.

84. Lin, P. F.; Blair, W.; Wang, T.; Spicer, T.; Guo, Q.; Zhou, N.; Gong, Y.-F.; Wang, H. G.; Rose, R.; Yamanaka, G.; Robinson, B.; Li, C. B.; Fridell, R.; Deminie, C.; Demers, G.; Yang, Z.; Zadjura, L.; Meanwell, N.; Colonno, R. *Proc. Natl. Acad. Sci. USA* **2003**, *100*, 11013.

85. Wang, T.; Zhang, Z.; Wallace, O. B.; Deshpande, M.; Fang, H.; Yang, Z.; Zadjura, L. M.; Tweedie, D. L.; Huang, S.; Zhao, F.; Ranadive, S.; Robinson, B. S.; Gong, Y.-F.; Ricarrdi, K.; Spicer, T. P.; Deminie, C.; Rose, R.; Wang, H.-G. H.; Blair, W. S.; Shi, P.-Y.; Lin, P.-F.; Colonno, R. J.; Meanwell, N. A. *J. Med. Chem.* **2003**, *46*, 4236.

86.   Meanwell, N. A.; Wallace, O. B.; Fang, H.; Wang, H.; Deshpande, M.; Wang, T.; Yin, Z.; Zhang, Z.; Pearce, B. C.; James, J.; Yeung, K.-S.; Qiu, Z.; Wright, J. J. K.; Yang, Z.; Zadjura, L.; Tweedie, D. L.; Yeola, S.; Zhao, F.; Ranadive, S.; Robinson, B. A.; Gong, Y.-F.; Wang, H.-G. H.; Spicer, T. P.; Blair, W. S.; Shi, P.-Y.; Colonno, R. J.; Lin, P.-f. *Bioorg. Med. Chem. Lett.* **2009**, *19*, 1977.

87.   Meanwell, N. A.; Wallace, O. B.; Wang, H.; Deshpande, M.; Pearce, B. C.; Trehan, A.; Yeung, K.-S.; Qiu, Z.; Wright, J. J. K.; Robinson, B. A.; Gong, Y.-F.; Wang, H.-G. H.; Spicer, T. P.; Blair, W. S.; Shi, P.-Y.; Lin, P.-f. *Bioorg. Med. Chem. Lett.* **2009**, *19*, 5136.

88.   Wang, T.; Kadow, J. F.; Zhang, Z.; Yin, Z.; Gao, Q.; Wu, D.; Parker, D. D.; Yang, Z.; Zadjura, L.; Robinson, B. A.; Gong, Y.-F.; Blair, W. S.; Shi, P.-Y.; Spicer, T. P.; Yamanaka, G.; Lin, P.-F.; Meanwell, N. A. *Bioorg. Med. Chem. Lett.* **2009**, *19*, 5140.

89.   Wang, T.; Yin, Z.; Zhang, Z.; Bender, J. A.; Yang, Z.; Johnson, G.; Yang, Z.; Zadjura, L. M.; D'Arienzo, C. J.; Parker, D. D.; Gesenberg, C.; Yamanaka, G. A.; Gong, Y.-F.; Ho, H.-T.; Fang, H.; Zhou, N.; McAuliffe, B. V.; Eggers, B. J.; Fan, L.; Nowicka-Sans, B.; Dicker, I. B.; Gao, Q.; Colonno, R. J.; Lin, P.-F.; Meanwell, N. A.; Kadow, J. F. *J. Med. Chem.*, **2009**, *52*, 7778.

90.   Nettles, R.E.; Schürmann, D.; Zhu, L.; Stonier, M.; Huang, S.-P.; Chang, I.; Chien, C.; Krystal, M.; Wind-Rotolo, M.; Ray, N.; Hanna, G.J.; Bertz, R.; Grasela, D. *J. Infect. Dis.*, **2012**, *206*, 1002.

91.   Nowicka-Sans, B.; Gong, Y.-F.; McAuliffe,B.; Dicker, I.; Ho, H.-T.; Zhou, N.; Eggers, B.; Lin, P.-F.; Ray, N.; Wind-Rotolo, M.; Zhu, L.; Majumdar, A.; Stock, D.; Lataillade, M.; Hanna, G.J.; Matiskella, J.D.; Ueda, Y.; Wang, T.; Kadow, J.F.; Meanwell, N.A.; Krystal, M. *Antimicrob. Agents Chemother.*, **2012**, *56*, 3498.

92.   Kadow, J. F.; Ueda, Y.; Connolly, T. P.; Wang, T.; Chen, C.-P.; Yeung, K.-S.; Bender, J.; Yang, Z.; Zhu, J.; Mattiskella, J.; Regueiro-Ren, A.; Yin, Z.; Zhang, Z.; Farkas, M.; Yang, X.; Wong, H.; Smith, D.; Raghaven, K. S.; Pendri, Y.; Staab, A.; Soundararajan, N.; Meanwell, N.; Zheng, M.; Parker, D. D.; Adams, S.; Ho, H.-T.; Yamanaka, G.; Nowicka-Sans, B.; Eggers, B.; McAuliffe, B.; Fang, H.; Fan, L.; Zhou, N.; Gong, Y.-f; Colonno, R. J.; Lin, P.-F.; Brown, J.; Grasela, D. M.; Chen, C.; Nettles, R. E. Abstracts of Papers, 241st ACS National Meeting & Exposition, Anaheim, CA, United States, March 27–31, 2011, MEDI-29.

93.   Kadow, J. F.; Ueda, Y.; Meanwell, N. A.; Connolly, T. P.; Wang, T.; Chen, C.-P.; Yeung, K.-S.; Zhu, J.; Bender, J. A.; Yang, Z.; Parker, D.; Lin, P.-F.; Colonno, R. J.; Mathew, M.; Morgan, D.; Zheng, M.; Chen, C.; Grasela, D. *J. Med. Chem.* **2012**, *55*, 2048.

94.   Amidon, G. L.; Lennernaes, H.; Shah, V. P.; Crison, J. R. *Pharm. Res.* **1995**, *12*, 413.

95.   Heimbach, T.; Oh, D.-M.; Li, L. Y.; Forsberg, M.; Savolainen, J.; Leppaenen, J.; Matsunaga, Y.; Flynn, G.; Fleisher, D. *Pharm. Res.* **2003**, *20*, 848.

96.   Heimbach, T.; Oh, D.-M.; Li, L. Y.; Rodriguez-Hornedo, N.; Garcia, G.; Fleisher, D. *Int. J. Pharm.* **2003**, *261*, 81.

97.   Princen, K.; Schols, D. *Cytokine Growth Factor Rev.* **2005**, *16*, 659.

98.   Kuhmann, S. E.; Hartley, O. *Ann. Rev. Pharmacol. Toxicol.* **2008**, *48*, 425.

99.   Lederman, M. M.; Penn-Nicholson, A.; Cho, M.; Mosier, D. *JAMA* **2006**, *296*, 815.

100.   Palani, A.; Tagat, J. R. *J. Med. Chem.* **2006**, *49*, 2851.

101.   Wood, A.; Armour, D. *Prog. Med. Chem.* **2005**, *43*, 239.

102.   Dorr, P.; Westby, M.; Dobbs, S.; Griffin, P.; Irvine, B.; Macartney, M.; Mori, J.; Rickett, G.; Smith-Burchnell, C.; Napier, C.; Webster, R.; Armour, D.; Price, D.; Stammen, B.; Wood, A.; Perros, M. *Antimicrob. Agents Chemother.* **2005**, *49*, 4721.

103.   Meanwell, N. A.; Kadow, J. F. *Curr. Opin. Investig. Drugs* **2007**, *8*, 669.

104.   Perry, C. M. *Drugs* **2010**, *70*, 1189.

105.   Bridger, G. J.; Skerlj, R. T.; Padmanabhan, S.; Martellucci, S. A.; Henson, G. W.; Struyf, S.; Witvrouw, M.; Schols, D.; De Clercq, E. *J. Med. Chem.* **1999**, *42*, 3971.

106.   Este, J. A.; Cabrera, C.; De Clercq, E.; Struyf, S.; Van Damme, J.; Bridger, G.; Skerlj, R. T.; Abrams, M. J.; Henson, G.; Gutierrez, A.; Clotet, B.; Schols, D. *Molec. Pharmacol.* **1999**, *55*, 67.

107.   Bridger, G. J.; Skerlj, R. T. *Adv. Antiviral Drug Design* **1999**, *3*, 161.

108.   Vartanian, J. P. *Curr. Opin. Anti-Infective Investig. Drugs* **2000**, *2*, 310.

109.   Choi, W.-T.; Duggineni, S.; Xu, Y.; Huang, Z.; An, J. *J. Med. Chem.*, **2012**, *55*, 977.

110. Skerlj, R. T.; Bridger, G. J.; Kaller, A.; McEachern, E. J.; Crawford, J. B.; Zhou, Y.; Atsma, B.; Langille, J.; Nan, S.; Veale, D.; Wilson, T.; Harwig, C.; Hatse, S.; Princen, K.; De Clercq, E.; Schols, D. *J. Med. Chem.* **2010**, *53*, 3376.

111. Moyle, G.; DeJesus, E.; Boffito, M.; Wong, R. S.; Gibney, C.; Badel, K.; MacFarland, R.; Calandra, G.; Bridger, G.; Becker, S. *Clin. Infect. Dis.* **2009**, *48*, 798.

112. Carrington, M.; Dean, M.; Martin, M. P.; O'Brien, S. J. *Hum. Mol. Genet.* **1999**, *8*, 1939.

113. Samson, M; Libert, F.; Doranz, B. J., Rucker, J.; Liesnard, C.; Farber, C. M.; Saragosti, S.; Lapoumeroulie, C.; Cognaux, J.; Forceille, C.; Muyldermans, G.; Verhofstede, C.; Burtonboy, G.; Georges, M.; Imai, T.; Rana, S.; Yi, Y.; Smyth, R. J.; Collman, R. G.; Doms, R. W.; Vassart, G.; Parmentier, M. *Nature* **1996**, 382, 722.

114. Westby, M.; van der Ryst, E. *Antiviral Chem. Chemother.* **2005**, *16*, 339.

115. Westby, M.; van der Ryst, E. *Antiviral Chem. Chemother.* **2010**, *20*, 179.

116. Dorr, P.; Perros, M. *Exp. Opin. Drug Discov.* **2008**, *3*, 1345.

117. Dorr, P.; Stupple, P. *Antiviral Drugs* **2011**, 117.

118. Price, D. A.; Armour, D.; De Groot, M.; Leishman, D.; Napier, C.; Perros, M.; Stammen, B. L.; Wood, A. *Bioorg. Med. Chem. Lett.* **2006**, *16*, 4633.

119. Price, D. A.; Armour, D.; de Groot, M.; Leishman, D.; Napier, C.; Perros, M.; Stammen, B. L.; Wood, A. *Curr. Topics Med. Chem.* **2008**, *8*, 1140.

120. Colman, P. M.; Lawrence, M. C. *Nat. Rev. Molec. Cell Biol.* **2003**, *4*, 309.

121. Harrison, S. C., *Nat. Struct. Molec. Biol.* **2008**, *15*, 690.

122. Melikyan, G. B. *Retrovirology* **2008**, *5*, 111. DOI: 10.1186/1742-4690-5-111.

123. Melikyan, G. B. *Curr. Topics Membranes* **2011**, *68*, 81.

124. Lu, M.; Blacklow, S. C.; Kim, P. S. *Nat. Struct. Biol.* **1995**, *2*, 1075.

125. Chan, D. C.; Fass, D.; Berger, J. M.; Kim, P. S. *Cell* **1997**, *89*, 263.

126. Wild, C.; Oas, T.; McDanal, C.; Bolognesi, D.; Matthews, T. *Proc. Natl. Acad. Sci. USA* **1992**, *89*, 10537.

127. Jiang, S.; Lin, K.; Strick, N. *Nature* **1993**, *365*, 113.

128. Jiang, S.; Lin, K.; Strick, N.; Neurath, A. R. *Biochem. Biophys. Res. Commun.* **1993**, *195*, 533.

129. Wild, C. T.; Shugars, D. C.; Greenwell, T. K.; McDanal, C. B.; Matthews, T. J. *Proc. Natl. Acad. Sci. USA* **1994**, *91*, 9770.

130. Furuta, R. A.; Wild, C. T.; Weng, Y.; Weiss, C. D. *Nat. Struct. Biol.* **1998**, *5*, 276.

131. Matthews, T.; Salgo, M.; Greenberg, M.; Chung, J.; DeMasi, R.; Bolognesi, D. *Nature Rev. Drug Discov.* **2004**, *3*, 215.

132. Schneider, S. E.; Bray, B. L.; Mader, C. J.; Friedrich, P. E.; Anderson, M. W.; Taylor, T. S.; Boshernitzan, N.; Niemi, T. E.; Fulcher, B. C.; Whight, S. R.; White, J. M.; Greene, R. J.; Stoltenberg, L. E.; Lichty, M. *J. Peptide Sci.* **2005**, *11*, 744.

133. Bray, B. L. *Nat. Rev. Drug Discov.* **2003**, *2*, 587.

134. Joly, V.; Jidar, K.; Tatay, M.; Yeni, P. *Exp. Opin. Pharmacother.* **2010**, *11*, 2701.

135. Oldfield, V.; Keating, G. M.; Plosker, G. *Drugs* **2005**, *65*, 1139.

136. Fung, H. B.; Guo, Y. *Clin. Ther.* **2004**, *26*, 352.

137. McKinnell, J. A.; Saag, M. S. *Curr. Opin. HIV AIDS* **2009**, *4*, 513.

138. Chan, D. C.; Chutkowski, C. T.; Kim, P. S. *Proc. Natl. Acad. Sci. USA* **1998**, *95*, 15613.

139. Debnath, A. K. *Exp. Opin. Investig. Drugs* **2006**, *15*, 465.

140. Liu, S.; Wu, S.; Jiang, S. *Curr. Pharm. Des.* **2007**, *13*, 143.

141. Cai, L.; Jiang, S. *ChemMedChem* **2010**, *5*, 1813.

142. Teixeira, C.; Gomes, J. R. B.; Gomes, P.; Maurel, F. *Eur. J. Med. Chem.* **2011**, *46*, 979.

143. Zhou, G.; Wu, D.; Hermel, E.; Balogh, E.; Gochin, M. *Bioorg. Med. Chem. Lett.* **2010**, *20*, 1500.

144. Gochin, M.; Zhou, G.-Y.; Phillips, A. H. *ACS Chem. Biol.* **2011**, *6*, 267.

145. Zhou, G.; Wu, D.; Snyder, B.; Ptak, R. G.; Kaur, H.; Gochin, M. *J. Med. Chem.* **2011**, *54*, 7220.

146. Dvorin, J. D.; Malim, M. H. *Curr. Topics Microbiol. Immunol.* **2003**, *281*, 179.

147. Smith, A. E.; Helenius, A. *Science* **2004**, *304*, 237.

148. Hulme, A. E.; Perez, O.; Hope, T. J. *Proc. Natl. Acad. Sci. USA* **2011**, *108*, 9975.

149. Sarafianos, S. G.; Marchand, B.; Das, K.; Himmel, D. M.; Parniak, M. A.; Hughes, S. H.; Arnold, E. *J. Mol. Biol.* **2009**, *385*, 693.
150. Kohlstaedt, L. A.; Wang, J.; Friedman, J. M.; Rice, P. A.; Steitz, T. A. *Science* **1992**, *256*, 1783.
151. Ren, J.; Esnouf, R.; Garman, E.; Somers, D.; Ross, C.; Kirby, I.; Keeling, J.; Darby, G.; Jones, Y.; Stuart, D.; Stammers, D. *Nat. Struct. Biol.* **1995**, *2*, 293.
152. Esnouf, R.; Ren, J.; Ross, C.; Jones, Y.; Stammers, D.; Stuart, D. *Nat. Struct. Biol.* **1995**, *2*, 303.
153. Hsiou, Y.; Ding, J.; Das, K.; Clark, A. D., Jr.; Hughes, S. H.; Arnold, E. *Structure* **1996**, *4*, 853.
154. Rodgers, D. W.; Gamblin, S. J.; Harris, B. A.; Culp, J. S.; Hellmig, B.; Woolf, D. J.; Debouck, C.; Harrison, S. C.; Ray, S. *Proc. Natl. Acad. Sci. USA* **1995**, *92*, 1222.
155. Jacobo-Molina, A.; Ding, J.; Nanni, R. G.; Clark, A. D. Jr.; Lu, X.; Tantillo, C.; Williams, R. L.; Kamer, G.; Ferris, A. L.; Clark, P.; Hizi, A.; Hughes, S. H.; Arnold, E. *Proc. Natl. Acad. Sci. USA* **1993**, *90*, 6320.
156. Smerdon, S. J.; Jager, J.; Wang, J.; Kohlstaedt, L. A.; Chirino, A. J.; Friedman, J. M.; Rice, P. A.; Steitz, T. A. *Proc. Natl. Acad. Sci. USA* **1994**, *91*, 3911.
157. Huang, H.; Chopra, R.; Verdine, G. L.; Harrison, S. C. *Science* **1998**, *282*, 1669.
158. de Béthune, M.-P. *Antiviral Res.* **2010**, *85*, 75.
159. Beilhartz, G. L.; Gotte, M. *Viruses* **2010**, *2*, 900.
160. Yu, F.; Liu, X.; Zhan, P.; De Clercq, E. *Mini-Rev. Med. Chem.* **2008**, *8*, 1243.
161. Wendeler, M.; Beilhartz, G. L.; Beutler, J. A.; Gotte, M.; Le Grice, S. F. J. *HIV Therapy* **2009**, *3*, 39.
162. De Clercq, E. *Infect. Dis. Ther.* **2003**, *30*, 485.
163. Painter G. R; Almond, M. R; Mao S.; Liotta D. C. *Curr. Topics Med. Chem.* **2004**, *4*, 1035.
164. De Clercq, E.; Holy, A. *Nat. Rev. Drug Discov.* **2005**, *4*, 928.
165. De Clercq, E. *Antiviral Res.* **2007**, *75*, 1.
166. Horwitz, J. P.; Chua, J.; Noel, M. *J. Org. Chem.* **1964**, *29*, 2076.
167. Ostertag, W.; Roesler, G.; Krieg, C. J.; Kind, J.; Cole, T.; Crozier, T.; Gaedicke, G.; Steinheider, G.; Kluge, N.; Dube, S. *Proc. Natl. Acad. Sci. USA* **1974**, *71*, 4980.
168. Mitsuya, H.; Weinhold, K. J.; Furman, P. A.; St. Clair, M. H.; Lehrman, S. N.; Gallo, R. C.; Bolognesi, D.; Barry, D. W.; Broder, S. *Proc. Natl. Acad. Sci. USA* **1985**, *82*, 7096.
169. Furman, P. A.; Fyfe, J. A.; St. Clair, M. H.; Weinhold, K.; Rideout, J. L.; Freeman, G. A.; Lehrman, S. N.; Bolognesi, D. P.; Broder, S.; Mitsuya, H.; Barry, D. W. *Proc. Natl. Acad. Sci. USA* **1986**, *83*, 8333.
170. Horwitz, J. P.; Chua, J.; Da Rooge, M. A.; Noel, M. *Tet. Lett.* **1964**, *5*, 2725.
171. Martin, J. C.; Hitchcock, M. J. M.; De Clercq, E.; Prusoff, W. H. *Antiviral Res.* **2010**, *85*, 34.
172. Horwitz, J. P.; Chua, J.; Noel, M.; Donatti, J. T. *J. Org. Chem.* **1967**, *32*, 817.
173. Jeffries, D. J. *J. Antimicrob. Chemother.* **1989**, *23 (Suppl. A)* 29.
174. Devineni, D.; Gallo, J. M. *Clin. Pharmacokinet.* **1995**, *28*, 351.
175. Moyle, G. *Exp. Opin. Investig. Drugs* **1998**, *7*, 451.
176. Mitsuya, H.; Broder, S. *Proc. Natl. Acad. Sci. USA* **1986**, *83*, 1911.
177. Morris-Jones, S.; Moyle, G.; Easterbrook, P. J. *Exp. Opin. Investig Drugs* **1997**, *6*, 1049.
178. Lambert, J. S.; Seidlin, M.; Reichman, R. C.; Plank, C. S.; Laverty, M.; Morse, G. D.; Knupp, C.; McLaren, C.; Pettinelli, C.; Valentine, F. T.; Dolin, R. *New Engl. J. Med.* **1990**, *322*, 1333.
179. Cooley, T. P.; Kunches, L. M.; Saunders, C. A.; Ritter, J. K.; Perkins, C. J.; McLaren, C.; McCaffrey, R. P.; Liebman, H. A. *New Engl. J. Med.* **1990**, *322*, 1340.
180. Berger, A. R.; Arezzo, J. C.; Schaumburg, H. H.; Skowron, G.; Merigan, T.; Bozzette, S.; Richman, D.; Soo, W. *Neurology* **1993**, *43*, 358.
181. Moyle, G. *Antiviral Ther.* **2005**, *10 (Suppl. 2)*, M47.
182. Fokunang, C. N.; Hitchcock, J.; Spence, F.; Tembe-Fokunang, E. A.; Burkhardt, J.; Levy, L.; George, C. *Int. J. Pharmacol.* **2006**, *2*, 152.
183. Kohler, J. J.; Lewis, W. *Environ. Molec. Mutagen.* **2007**, *48*, 166.
184. Koczor, C. A.; Lewis, W. *Exp. Opin. Drug Metab. Toxicol.* **2010**, *6*, 1493.
185. Apostolova, N.; Blas-Garcia, A.; Esplugues, J. V. *Trends Pharmacol. Sci.* **2011**, *32*, 715.

186.   Chen, C.-H.; Vazquez-Padua, M..; Cheng, Y.-C. *Mol. Pharmacol.* **1990**, *39*, 625.
187.   Brinkman, K.; Smeitink, J. A.; Romijn, J. A.; Reiss, P. *Lancet* **1999**, *354*, 1112.
188.   Kontorinis, N.; Dieterich, D. *Semin. Liver Dis.* **2003**, *23*, 173.
189.   Richman, D. D.; Fischl, M. A.; Grieco, M. H.; Gottlieb, M. S.; Volberding, P. A.; Laskin, O. L.; Leedom, J. M.; Groopman, J. E.; Mildvan, D.; Hirsch, M S; Jackson, G. G.; Durack, D. T.; Nusinoff-Lehrmann, S.; the AZT Collaborative Working Group. *New Engl. J. Med.* **1987**, *317*, 192.
190.   Shulman, N.; Winters, M. *Curr. Drug Targets Infect. Disord.* **2003**, *3*, 273.
191.   Volberding, P. A., Graham, N. M. H. *JAIDS* **1994**, *7* (*Suppl. 2.*), S12.
192.   Kozal, M. J.; Shafer, R. W.; Winters, M. A.; Katzenstein, D. A.; Merigan, T. C. *J. Infect. Dis.* **1993**, *167*, 526.
193.   Sande, M. A.; Carpenter, C. C.; Cobbs, C. G.; Holmes, K. K.; Sanford, J. P. *JAMA* **1993**, *270*, 2583.
194.   August, E. M.; Marongiu, M. E.; Lin, T. S.; Prusoff, W. H. *Biochem. Pharmacol.* **1988**, *37*, 4419.
195.   Lin, T. S.; Chen, M. S.; McLaren, C.; Gao, Y. S.; Ghazzouli, I.; Prusoff, W. H. *J. Med. Chem.* **1987**, *30*, 440.
196.   Lin, T. S.; Schinazi, R. F.; Prusoff, W. H. *Biochem. Pharmacol.* **1987**, *36*, 2713.
197.   Lin, T. S.; Schinazi, R. F.; Chen, M. S.; Kinney-Thomas, E.; Prusoff, W. H. *Biochem. Pharmacol.* **1987**, *36*, 311.
198.   Baba, M.; Pauwels, R.; Herdewijn, P.; De Clercq, E.; Desmyter, J.; Vandeputte, M. *Biochem. Biophys. Res. Commun.* **1987**, *142*, 128.
199.   Balzarini, J.; Kang, G. J.; Dalal, M.; Herdewijn, P.; De Clercq, E.; Broder, S.; Johns, D. G. *Molec. Pharmacol.* **1987**, *32*, 162.
200.   Mansuri, M. M.; Starrett, J. E., Jr.; Ghazzouli, I.; Hitchcock, M. J. M.; Sterzycki, R. Z.; Brankovan, V.; Lin, T. S.; August, E. M.; Prusoff, W. H. *J. Med. Chem.* **1989**, *32*, 461.
201.   Mansuri, M. M.; Starrett, J. E., Jr.; Wos, J. A.; Tortolani, D. R.; Brodfuehrer, P. R.; Howell, H. G.; Martin, J. C. *J. Org. Chem.* **1989**, *54*, 4780.
202.   Sommadossi, J.-P.; Carlisle, R. *Antimicrob. Agents Chemother.* **1987**, *31*, 452.
203.   Kawaguchi, T.; Fukushima, S.; Ohmura, M.; Mishima, M.; Nakano M. *Chem. Pharm. Bull.* **1989**, *37*, 1944.
204.   Martin, J. C.; Hitchcock, M. J. M.; Fridland, A.; Ghazzouli, I.; Kaul, S.; Dunkle, L. M.; Sterzycki, R. Z.; Mansuri, M. M. *Ann. New York Acad. Sci.* **1990**, *616*, 22.
205.   Gallant, J. E.; Staszewski, S.; Pozniak, A. L.; DeJesus, E.; Suleiman, J. M. A. H.; Miller, M. D.; Coakley, D. F.; Lu, B.; Toole, J. J.; Cheng, A. K.; Myers, R. A.; Wolfe, P.; Stryker, R.; Schneider, S.; Kooshian, G. S.; Ruane, P.; Letendre, S.; Lampiris, H.; Beall, G.; Witt, M.; Simon, G.; Timpone, J.; Sension, M.; Juba, P.; Hernandez, J.; Campo, R.; Yangco, B.; Pierone, G., Jr.; Stephens, J.; Kessler, H. A.; Berger, D.; Wheat, J.; Greenberg, R. N.; Hellinger, J.; Tashima, K.; Morris, A. B.; Clay, P. G.; Tebas, P.; Markowitz, M.; Wohl, D.; Jemsek, J. G.; Pegram, S.; Slater, L.; Santana, J. L.; Sepulveda-Arzola, G.; Morales, J. O.; West, T.; Brand, J. D.; Bellos, N. C.; Borucki, M.; Barnett, B. J.; Green, S. L.; Craven, P. C.; Casiro, A.; Cassetti, I.; Cahn, P.; Benetucci, J. A.; Pedro, R.; Hayden, R. L.; Madruga, J. V. R.; Uip, D. E.; Timerman, A.; Mendonca, J. S.; Lewi, D. S.; Schechter, M.; Koenig, E.; Vittecoq, D.; Troisvallets, D.; Livrozet, J. M.; Bouvet, E.; Salmon-Ceron, D.; Sereni, D.; Arasteh, K.; Plettenberg, A.; Weitner, L.; Jager, H.; Lazzarin, A.; Esposito, R.; Guaraldi, G.; Concia, E.; Clotet, B.; Gonzalez-Lahoz, J.; Pulido, F.; Rubio, R.; Lopez-Aldeguer, J.; Friedl, A.; Opravil, M.; De Ruiter, A.; Easterbrook, P.; Williams, I.; Chen, S.-S.; Isaacson, E.; Jaffe, H. S.; Lu, B.; Margot, N.; Rooney, J. F.; Sayre, J.; Tran, S.; Fliederbaum, P.; James, J.; Schmidt, A.; Uffelman, K.; Capone, P.; Mingione, C.; Sidi, A.; Holmstrom, T.; Rodriguez-Amaya, K.; Sandholdt, I. *JAMA* **2004**, *292*, 191.
206.   Chapman, H.; Kernan, M.; Prisbe, E.; Rohloff, J.; Sparacino, M.; Terhorst, T.; Yu, R. *Nucleosides, Nucleotides Nucleic Acids* **2001**, *20*, 621.
207.   Lee, W. A.; He, G.-X.; Eisenberg, E.; Cihlar, T.; Swaminathan, S.; Mulato, A.; Cundy, K. C. *Antimicrob. Agents Chemother.* **2005**, *49*, 1898.
208.   Birkus, G.; Wang, R.; Liu, X.; Kutty, N.; MacArthur, H.; Cihlar, T.; Gibbs, C.; Swaminathan, S.; Lee, W.; McDermott, M. *Antimicrob. Agents Chemother.* **2007**, *51*, 543.

209.   Birkus, G.; Kutty, N.; He, G.-X.; Mulato, A.; Lee, W.; McDermott, M.; Cihlar, T. *Mol. Pharmacol.* **2008,** *74,* 92.

210.   Cahard, D; McGuigan, C.; Balzarini, J. *Mini-Rev. Med. Chem.* **2004,** *4,* 371.

211.   Mehellou, Y.; Balzarini, J.; McGuigan, C. *ChemMedChem* **2009,** *4,* 1779.

212.   de Béthune, M.-P. *Antiviral Res.* **2010,** *85,* 75.

213.   Zhan, P.; Liu, X. *Exp. Opin. Therapeutic Patents* **2011,** *21,* 717.

214.   Zhan, P.; Chen, X.; Li, D.; Fang, Z.; De Clercq, E.; Liu, X. *Med. Res. Rev.* **2012,** DOI: 10.1002/med.20241.

215.   Baba, M.; Tanaka, H.; De Clercq, E.; Pauwels, R.; Balzarini, J.; Schols, D.; Nakashima, H.; Perno, C. F.; Walker, R. T.; Miyasaka, T. *Biochem. Biophys. Res. Commun.* **1989,** *165,* 1375.

216.   Miyasaka, T.; Tanaka, H.; Baba, M.; Hayakawa, H.; Walker, R. T.; Balzarini, J.; De Clercq, E. *J. Med. Chem.* **1989,** *32,* 2507.

217.   Tanaka, H.; Baba, M.; Hayakawa, H.; Sakamaki, T.; Miyasaka, T.; Ubasawa, M.; Takashima, H.; Sekiya, K.; Nitta, I.; Shigeta, S.; Walker, R. T.; Balzarini, J.; De Clercq, E. *J. Med. Chem.* **1991,** *34,* 349.

218.   Debyser, Z.; Pauwels, R.; Andries, K.; Desmyter, J.; Kukla, M.; Janssen, P. A. J.; De Clercq, E. *Proc. Natl. Acad. Sci. U.S.A.* **1991,** *88,* 1451.

219.   Pauwels, R.; Andries, K.; Desmyter, J.; Schols, D.; Kukla, M. J.; Breslin, H. J.; Raeymaeckers, A.; Van Gelder, J.; Woestenborghs, R.; Heykants, J.; Schellekens, K.; Janssen, M. A. C.; De Clercq, E.; Janssen, P. A. J. *Nature* **1990,** *343,* 470.

220.   Milinkovic, A.; Martinez, E. *Exp. Rev. Anti-Infect. Ther.* **2004,** *2,* 367.

221.   Grozinger, K.; Proudfoot, J.; Hargrave, K. *Drug Discov. Devel.* **2006,** *1,* 353.

222.   Adams, W. J.; Aristoff, P. A.; Jensen, R. K.; Morozowich, W.; Romero, D. L.; Schinzer, W. C.; Tarpley, W. G.; Thomas, R. C. *Pharm. Biotechnol.* **1998,** *11,* 285.

223.   Scott, L. J.; Perry, C. M. *Drugs* **2000,** *60,* 1411.

224.   Young, S. D.; Britcher, S. F.; Tran, L. O.; Payne, L. S.; Lumma, W. C.; Lyle, T. A.; Huff, J. R.; Anderson, P. S.; Olsen, D. B.; Carroll, S. S. *Antimicrob. Agents Chemother.* **1995,** *39,* 2602.

225.   Maggiolo, F. *J. Antimicrob. Chemother.* **2009,** *64,* 910.

226.   Ludovici, D. W.; De Corte, B. L.; Kukla, M. J.; Ye, H.; Ho, C. Y.; Lichtenstein, M. A.; Kavash, R. W.; Andries, K.; de Béthune, M.-P.; Azijn, H.; Pauwels, R.; Lewi, P. J.; Heeres, J.; Koymans, L. M. H.; de Jonge, M. R.; Van Aken, K. J. A.; Daeyaert, F. F. D.; Das, K.; Arnold, E.; Janssen, P. A. J. *Bioorg. Med. Chem. Lett.* **2001,** *11,* 2235.

227.   Deeks, E. D.; Keating, G. M. *Drugs* **2008,** *68,* 2357.

228.   Janssen, P. A. J.; Lewi, P. J.; Arnold, E.; Daeyaert, F.; de Jonge, M.; Heeres, J.; Koymans, L.; Vinkers, M.; Guillemont, J.; Pasquier, E.; Kukla, M.; Ludovici, D.; Andries, K.; de Béthune, M.-P.; Pauwels, R.; Das, K.; Clark, A. D., Jr.; Frenkel, Y. V.; Hughes, S. H.; Medaer, B.; De Knaep, F.; Bohets, H.; De Clerck, F.; Lampo, A.; Williams, P.; Stoffels, P. *J. Med. Chem.* **2005,** *48,* 1901.

229.   Garvey, L.; Winston, A. *Exp. Opin. Investig. Drugs* **2009,** *18,* 1035.

230.   Mowbray, C. E.; Burt, C.; Corbau, R.; Gayton, S.; Hawes, M.; Perros, M.; Tran, I.; Price, D. A.; Quinton, F. J.; Selby, M. D.; Stupple, P. A.; Webster, R.; Wood, A. *Bioorg. Med. Chem. Lett.* **2009,** *19,* 5857.

231.   Corbau, R.; Mori, J.; Phillips, C.; Fishburn, L.; Martin, A.; Mowbray, C.; Panton, W.; Smith-Burchnell, C.; Thornberry, A.; Ringrose, H.; Knochel, T.; Irving, S.; Westby, M.; Wood, A.; Perros, M. *Antimicrob. Agents Chemother.* **2010,** *54,* 4451.

232.   Faetkenheuer, G.; Staszewski, S.; Plettenburg, A.; Hackman, F.; Layton, G.; McFadyen, L.; Davis, J.; Jenkins, T. M. *AIDS* **2009,** *23,* 2115.

233.   Meanwell, N. A. *Chem. Res. Toxicol.* **2011,** *24,* 1420.

234.   Menendez-Arias, L.; Betancor, G.; Matamoros, T. *Antiviral Res.* **2011,** *92,* 139.

235.   Ceccherini-Silberstein, F.; Svicher, V.; Sing, T.; Artese, A.; Santoro, Maria, M.; Forbici, F.; Bertoli, A.; Alcaro, S.; Palamara, G.; Monforte, A. d'A.; Balzarini, J.; Antinori, A.; Lengauer, T.; Perno, C. F. *J. Virol.* **2007,** *81,* 11507.

236.   Tambuyzer, L.; Azijn, H.; Rimsky, L. T; Vingerhoets, J.; Lecocq, P.; Kraus, G.; Picchio, G.; de Béthune, M.-P. *Antiviral Ther.* **2009,** *14,* 103.
237.   Soriano, V.; de Mendoza, C. *HIV Clin. Trials* **2002,** *3,* 237.
238.   Harrigan, P. R.; Salim, M.; Stammers, D. K.; Wynhoven, B.; Brumme, Z. L.; McKenna, P.; Larder, B.; Kemp, S. D. *J. Virol.* **2002,** *76,* 6836.
239.   Cheung, P. K; Wynhoven, B.; Harrigan P. R. *AIDS Rev.* **2004,** *6,* 107.
240.   Nikolenko, G. N.; Delviks-Frankenberry, K. A.; Palmer, S.; Maldarelli, F.; Fivash, M. J.; Coffin, J. M.; Pathak, V. K. *Proc. Natl. Acad. Sci. U.S.A.* **2007,** *104,* 317.
241.   Waters, J. M.; O'Neal, W.; White, K. L.; Wakeford, C.; Lansdon, E. B.; Harris, J.; Svarovskaia, E. S.; Miller, M. D.; Borroto-Esoda, K. *Antiviral Ther.* **2009,** *14,* 231.
242.   Yap, S.-H.; Sheen, C.-W.; Fahey, J.; Zanin, M.; Tyssen, D.; Lima, V. D.; Wynhoven, B.; Kuiper, M.; Sluis-Cremer, N.; Harrigan, P. R.; Tachedjian, G. *PLoS Med.* **2007,** *4,* e335.
243.   Gupta, S.; Fransen, S.; Paxinos, E. E.; Stawiski, E.; Huang, W.; Petropoulos, C. J. *Antimicrob. Agents Chemother.* **2010,** *54,* 1973.
244.   Delviks-Frankenberry, K. A.; Nikolenko, G. N.; Maldarelli, F.; Hase, S.; Takebe, Y.; Pathak, V. K. *J. Virol.* **2009,** *83,* 8502.
245.   Hachiya, A.; Kodama, E. N.; Sarafianos, S. G.; Schuckmann, M. M.; Sakagami, Y.; Matsuoka, M.; Takiguchi, M.; Gatanaga, H.; Oka, S. *J. Virol.* **2008,** *82,* 3261.
246.   D'Cruz, O. J.; Uckun, F. M. *J. Antimicrob. Chemother.* **2006,** *57,* 411.
247.   Archer, R. H.; Dykes, C.; Gerondelis, P.; Lloyd, A.; Fay, P.; Reichman, R. C.; Bambara, R. A.; Demeter, L. M. *J. Virol.* **2000,** *74,* 8390.
248.   Gerondelis, P.; Archer, R. H.; Palaniappan, C.; Reichman, R. C.; Fay, P. J.; Bambara, R. A.; Demeter, L. M. *J. Virol.* **1999,** *73,* 5803.
249.   Podzamczer, D.; Fumero, E. *Exp. Opin. Pharmacother.* **2001,** *2,* 2065.
250.   Best, B. M.; Goicoechea, M. *Exp. Opin. Drug Metab. Toxicol.* **2008,** *4,* 965.
251.   Deeks, E. D.; Perry, C. M. *Drugs* **2010,** *70,* 2315.
252.   De Clercq, E. *Biochem. Pharmacol.* **2012,** *83,* DOI: dx.doi.org/10.1016/j.bcp.2012.03.024.
253.   Ghosh, A. K.; Anderson, D. D.; Mitsuya, M. In *Burger's Medicinal Chemistry, Drug Discovery and Development,* *7th Ed.,* **2010,** Abraham, D. J.; Rotella, D. P., Eds., Vol. 7, Chapter 1, pp 1–74.
254.   Moyle, G. *Exp. Opin. Investig. Drugs* **1998,** *7,* 413.
255.   Navia, M.; Fitzgerald, P.; McKeever, B.; Leu, C.; Heimbach, J.; Herber, W.; Sial, I.; Darke, P.; Springer, J. *Nature* **1989,** *337,* 615.
256.   Wlodawer, A.; Miller, M.; Jaskolski, M.; Sathyanarayana, B.; Baldwin, E.; Weber, I.; Selk, L.; Clawson, L,; Schneider, J.; Kent, B. *Science,* **1989,** *245,* 616.
257.   Roberts, N. A.; Martin, J. A.; Kinchington, D.; Broadhurst, A. V.; Craig, J. C.; Duncan, I. B.; Galpin, S. A.; Handa, B. K.; Kay, J; Krohn, A.; Lambert, R.; Merrett, J.; Mills, J.; Parkes, K; Redshaw, S; Ritchie, A.; Taylor, D.; Thomas, G.; Machin, P. *Science* **1990,** *248,* 358.
258.   Blundell, T. L.; Cooper, J.; Foundling, S. I.; Jones, D. M.; Atrash, B.; Szelke, M. *Biochemistry* **1987,** *26,* 5585.
259.   Harbeson, S. L.; Rich, D. H. *J. Med. Chem.* **1989,** *32,* 1378.
260.   Turner, S. R.; Strohbach, J. W.; Tommasi, R. A.; Aristoff, P. A.; Johnson, P. D.; Skulnick, H. I.; Dolak, L. A.; Seest, E. P.; Tomich, P. K.; Bohanon, M. J.; Horng, M.-M.; Lynn, J. C.; Chong, K.-T.; Hinshaw, R. R.; Watenpaugh, K. D.; Janakiraman, M. N.; Thaisrivongs, S. *J. Med. Chem.* **1998,** *41,* 3467.
261.   Thaisrivongs, S.; Strohbach, J. W. *Biopolymers* **1999,** *51,* 51.
262.   Kaldor, S. W.; Kalish, V. J.; Davies II, J. F.; Shetty, B. V.; Fritz, J. E.; Appelt, K.; Burgess, J. A.; Campanale, K. M.; Chirgadze, N. Y.; Clawson, D. K.; Dressman, B. A.; Hatch, S. D.; Khalil, D. A.; Kosa, M. B.; Lubbehusen, P. P.; Muesing, M. A.; Patick, A. K.; Reich, S. H.; Su, K. S.; Tatlock, J. H. *J. Med. Chem.* **1997,** *40,* 3979.
263.   Spaltenstein, A.; Kazmierski, W. M.; Miller, J. F.; Samano, V. *Curr. Topics Med. Chem.* **2005,** *5,* 1589.

264.   Dorsey, B.; Levin, R.; McDaniel, S.; Vacca, J.; Guare, J.; Darke, P.; Zugay, J.; Emini, E.; Schleif, W.; Quintero, J.; Lin, J.; Chen, I.; Holloway, M.; Fitzgerald, P.; Axel, M.; Ostovic, D.; Anderson, P.; Huff, J. *J. Med. Chem.* **1994**, *37*, 3443.

265.   Kempf, D. J.; Sham, H. L.; Marsh, K. C.; Flentge, C. A.; Betebenner, D.; Green, B. E.; McDonald, E.; Vasavanonda, S.; Saldivar, A.; Wideburg, N. E.; Kati, W. M.; Ruiz, L.; Zhao, C.; Fino, L.; Patterson, J.; Molla, A.; Plattner, J. J.; Norbeck, D. W. *J. Med. Chem.* **1998**, *41*, 602.

266.   Kempf, D. J. *Infectious Disease Ther.* **2002**, *25*, 49.

267.   Kim, E.; Baker, C.; Dwyer, M.; Murcko, M; Rao, B.; Tung, R.; Navia, M. *J. Am. Chem. Soc.* **1995**, *117*, 1181.

268.   Bold, G.; Faessler, A.; Capraro, H.-G.; Cozens, R.; Klimkait, T.; Lazdins, J.; Mestan, J.; Poncioni, B.; Roesel, J.; Stover, D.; Tintelnot-Blomley, M.; Acemoglu, F.; Beck, W.; Boss, E.; Eschbach, M.; Huerlimann, T.; Masso, E.; Roussel, S.; Ucci-Stoll, K.; Wyss, D.; Lang, M. *J. Med. Chem.* **1998**, *41*, 3387.

269.   Surleraux, D. L. N. G.; Tahri, A.; Verschueren, W. G.; Pille, G. M. E.; de Kock, H. A.; Jonckers, T. H. M.; Peeters, A.; De Meyer, S.; Azijn, H.; Pauwels, R.; de Béthune, M.-P.; King, N. M.; Prabu-Jeyabalan, M.; Schiffer, C. A.; Wigerinck, P. B. T. P. *J. Med. Chem.* **2005**, *48*, 1813.

270.   Kempf, D. J.; Marsh, K. C.; Kumar, G.; Rodrigues, A. D.; Denissen, J. F.; McDonald, E.; Kukulka, M. J.; Hsu, A.; Granneman, G. R.; Baroldi, P. A.; Sun, E.; Pizzuti, D.; Plattner, J. J.; Norbeck, D. W.; Leonard, J. M. *Antimicrob. Agents Chemother.* **1997**, *41*, 654.

271.   Busse, K. H.; Penzak, S. R. *Exp. Rev. Clin. Pharmacol.* **2008**, *1*, 533.

272.   Xu, L.; Desai, M. C. *Curr. Opin. Investig. Drugs* **2009**, *10*, 775.

273.   Hull, M. W.; Montaner, J. S. G. *Ann. Med.* **2011**, *43*, 375.

274.   Freire, E. *Drug Discov. Today* **2008**, *13*, 869.

275.   Ladbury, J. E.; Klebe, G.; Freire, E. *Nat. Rev. Drug Discov.* **2010**, *9*, 23.

276.   Ghosh, A. K. *J. Med. Chem.* **2008**, *52*, 2163.

277.   Ghosh, A. K.; Chapsal, B. D.; Weber, I. T.; Mitsuya, H. *Acc. Chem. Res.* **2008**, *41*, 78.

278.   Michel, J.; Tirado-Rives, J.; Jorgensen, W. L. *J. Am. Chem. Soc.* **2009**, *131*, 15403.

279.   Meanwell, N. A. *J. Med. Chem.* **2011**, *54*, 2529.

280.   Lam, P. Y. S.; Jadhav, P. K.; Eyermann, C. J.; Hodge, C. N.; Ru, Y.; Bacheler, L. T.; Meek, J. L.; Otto, M. J.; Rayner, M. M.; Wong, Y. N.; Chang, C.-H.; Weber, P. C.; Jackson, D. A.; Sharpe, T. R.; Erikson-Viitanen, S. *Science* **1994**, *263*, 380.

281.   Lam, P. Y. S.; Ru, Y.; Jadhav, P. K.; Aldrich, P. E.; DeLucca, G. V.; Eyermann, C. J.; Chang, C.-H.; Emmett, G.; Holler, E. R.; Daneker, W. F.; Li, L.; Confalone, P. N. McHugh, R. J.; Han, Q.; Li, R.; Markwalder, J. A.; Seitz, S. P.; Sharpe, T. R.; Bacheler, L. T.; Rayner, M. M.; Klabe, R. M. Shum, L.; Winslow, D. L.; Kornhauser, D. M.; Jackson, D. A.; Erickson-Viitanen, S.; Hodge C. N. *J. Med. Chem.* **1996**, *39*, 3514.

282.   De Lucca, G. V.; Erickson-Viitanen, S.; Lam, P. Y. S. *Drug Discov. Today* **1997**, *2*, 6.

283.   Wensing, A. M. J.; van Maarseveen, N. M.; Nijhuis, M. *Antiviral Res.* **2010**, *85*, 59.

284.   Perez-Valero, I.; Bayon, C.; Cambron, I.; Gonzalez, A.; Arribas, J. R. *J. Antimicrob.Chemother.* **2011**, *66*, 1954.

285.   Shafer, R. W.; Schapiro, J. M. *AIDS Rev.* **2008**, *10*, 67.

286.   Shafer, R. W.; Rhee, S. Y.; Pillay, D.; Miller, V.; Sandstrom, P.; Schapiro, J. M.; Kuritzkes, D. R.; Bennett, D. *AIDS* **2007**, *21*, 215.

287.   Rhee, S. Y.; Taylor, J.; Wadhera, G.; Ben-Hur, A.; Brutlag, D. L.; Shafer, R. W. *Proc. Natl. Acad. Sci. USA* **2006**, *103*, 17355.

288.   Weinheimer, S.; Discotto, L.; Friborg, J.; Yang, H.; Colonno, R. *Antimicrob. Agents Chemother.* **2005**, *49*, 3816.

289.   Maguire, M. F.; Guinea, R.; Griffin, P.; Macmanus, S.; Elston, R. C.; Wolfram, J.; Richards, N.; Hanlon, M. H.; Porter, D. J.; Wrin, T.; Parkin, N.; Tisdale, M.; Furfine, E.; Petropoulos, C.; Snowden, B. W.; Kleim, J. P. *J. Virol.* **2002**, *76*, 7398.

290.   Prado, J. G.; Wrin, T.; Beauchaine, J.; Ruiz, L.; Petropoulos, C. J.; Frost, S. D.; Clotet, B.; D'Aquila, R. T.; Martinez-Picado, J. *AIDS* **2002**, *16*, 1009.

291.   Nijhuis, M.; van Maarseveen, N. M.; Lastere, S.; Schipper, P.; Coakley, E.; Glass, B.; Rovenska, M.; de Jong, D.; Chappey, C.; Goedegebuure, I. W.; Heilek-Snyder, G.; Dulude, D.; Cammack, N.; Brakier-Gingras, L.; Konvalinka, J.; Parkin, N.; Krausslich, H. G.; Brun-Vezinet, F.; Boucher, C. A. *PLoS Med.* **2007**, *4*, e36.

292.   Dam, E.; Quercia, R.; Glass, B.; Descamps, D.; Launay, O.; Duval, X.; Krausslich, H. G.; Hance, A. J.; Clavel, F. *PLoS Pathog.* **2009**, *5*, e1000345.

293.   Gatanaga, H.; Suzuki, Y.; Tsang, H.; Yoshimura, K.; Kavlick, M. F.; Nagashima, K.; Gorelick, R. J.; Mardy, S.; Tang, C.; Summers, M. F.; Mitsuya, H. *J. Biol. Chem.* **2002**, *277*, 5952.

294.   Callebaut, C.; Stray, K.; Tsai, L.; Williams, M.; Yang, Z. Y.; Cannizzaro, C.; Leavitt, S. A.; Liu, X.; Wang, K.; Murray, B. P.; Mulato, A.; Hatada, M.; Priskich, T.; Parkin, N.; Swaminathan, S.; Lee, W.; He, G. X.; Xu, L.; Cihlar, T. *Antimicrob. Agents Chemother.* **2011**, *55*, 1366.

295.   Jaskolski, M.; Alexandratos, J. N.; Bujacz, G.: Wlodawer, A. *FEBS J.* **2009**, *276*, 2926.

296.   Cabrera, C. *Curr Opin Investig Drugs* **2008**, *9*, 885.

297.   Hazuda, D.; Iwamoto, M.; Wenning, L. *Annu. Rev. Pharmacol. Toxicol.* **2009**, *49*, 377.

298.   Nowotny, M. *EMBO Reports* **2009**, *10*, 144.

299.   McColl, D. J.; Chen, X. *Antiviral Res.* **2010**, *85*, 101.

300.   Summa, V.; Petrocchi, A.; Bonelli, F.; Crescenzi, B.; Donghi, M.; Ferrara, M.; Fiore, F.; Gardelli, C.; Gonzalez Paz, O.; Hazuda, D. J.; Jones, P.; Kinzel, O.; Laufer, R.; Monteagudo, E.; Muraglia, E.; Nizi, E.; Orvieto, F.; Pace, P.; Pescatore, G.; Scarpelli, R.; Stillmock, K.; Witmer, M. V.; Rowley, M. *J. Med. Chem.* **2008**, *51*, 5843.

301.   Steigbigel, R. T.; Cooper, D. A.; Kumar, P. N.; Eron, J. E.; Schechter, M.; Markowitz, M.; Loutfy, M. R.; Lennox J. L.; Gatell, J. M.; Rockstroh, J. K.; Katlama, C.; Yeni, P.; Lazzarin, A.; Clotet, B.; Zhao, J.; Chen, J.; Ryan, D. M.; Rhodes, R. R.; Killar, J. A.; Gilde, L. R.; Strohmaier, K. M.; Meibohm, A. R.; Miller, M. D.; Hazuda, D. J.; Nessly, M. L.; DiNubile, M. J.; Isaacs, R. D.; Nguyen, B.-Y.; Teppler, H. *New Engl. J. Med.* **2008**, *359*, 339.

302.   Cooper, D. A.; Steigbigel, R. T.; Gatell, J. M.; Rockstroh, J. K.; Katlama, C.; Yeni, P.; Lazzarin, A.; Clotet, B.; Kumar, P. N.; Eron, J. E.; Schechter, M.; Markowitz, M.; Loutfy, M. R.; Lennox, J. L.; Zhao, J.; Chen, J.; Ryan, D. M.; Rhodes, R. R.; Killar, J. A.; Gilde, L. R.; Strohmaier, K. M.; Meibohm, A. R.; Miller, M. D.; Hazuda, D. J.; Nessly, M. L.; DiNubile, M. J.; Isaacs, R. D.; Teppler, H.; Nguyen, B.-Y. *New Engl. J. Med.* **2008**, *359*, 355.

303.   Croxtall, J. D.; Lyseng-Williamson, K. A.; Perry, C. M. *Drugs* **2008**, *68*, 131.

304.   Croxtall, J. D.; Keam, S. J. *Drugs* **2009**, *69*, 1059.

305.   Wai, J. S.; Egbertson, M. S.; Payne, L. S.; Fisher, T. E.; Embrey, M. W.; Tran, L. O.; Melamed, J. Y.; Langford, H. M.; Guare, J. P., Jr.; Zhuang, L.; Grey, V. E.; Vacca, J. P.; Holloway, M. K.; Naylor-Olsen, A. M.; Hazuda, D. J.; Felock, P. J.; Wolfe, A. L.; Stillmock, K. A.; Schleif, W. A.; Gabryelski, L. J.; Young, S. D. *J. Med. Chem,* **2000**, *43*, 4923.

306.   Hare, S.; Gupta, S. S.; Valkov, E.; Engelman, A.; Cherepanov, P. *Nature* **2010**, *464*, 232.

307.   Li, X.; Krishnan, L.; Cherepanov, P.; Engelman, A. *Virology* **2011**, *411*, 194.

308.   Maertens, G. N.; Hare, S.; Cherepanov, P. *Nature* **2010**, *468*, 326.

309.   Summa, V.; Petrocchi, A.; Matassa, V. G.; Gardelli, C.; Muraglia, E.; Rowley, M.; Paz, O. G.; Laufer, R.; Monteagudo, E.; Pace, P. *J. Med. Chem.* **2006**, *49*, 6646.

310.   Petrocchi, A.; Koch, U.; Matassa, V. G.; Pacini, B.; Stillmock, K. A.; Summa, V. *Bioorg. Med. Chem. Lett.* **2007**, *17*, 350.

311.   Sato, M.; Kawakami, H.; Motomura, T.; Aramaki, H.; Matsuda, T.; Yamashita, M.; Ito, Y.; Matsuzaki, Y.; Yamataka, K.; Ikeda, S.; Shinkai, H. *J. Med. Chem.* **2009**, *52*, 4869.

312.   Shimura, K.; Kodama, E.; Sakagami, Y.; Matsuzaki, Y.; Watanabe, W.; Yamataka, K.; Watanabe, Y.; Ohata, Y.; Doi, S.; Sato, M.; Kano, M.; Ikeda, S.; Matsuoka, M. *J. Virol.* **2008**, *82*, 764.

313.   Klibanov, O. M. *Curr. Opin. Investig. Drugs* **2009**, *10*, 190.

314.   Kobayashi, M.; Yoshinaga, T.; Seki, T.; Wakasa-Morimoto, C.; Brown, K. W.; Ferris, R.; Foster, S. A.; Hazen, R. J.; Miki, S.; Suyama-Kagitani, A.; Kawauchi-Miki, S.; Taishi, T.; Kawasuji, T.; Johns, B. A.; Underwood, M. R.; Garvey, E. P.; Sato, A.; Fujiwara, T. *Antimicrob. Agents Chemother.* **2011**, *55*, 813.

315. Carl, J.; Lenz, C.; Rockstroh, J. K. *Expert Opin. Investig. Drugs* **2011,** *20,* 537.
316. Xu, L.; Liu, H.; Murray, B. P.; Callebaut, C.; Lee, M. S.; Hong, A.; Strickley, R. G.; Tsai, L. K.; Stray, K. M.; Wang, Y.; Rhodes, G. R.; Desai, M. C. *ACS Med. Chem. Lett.* **2010,** *1,* 209.
317. Cohen, C.; Elion, R.; Ruane, P.; Shamblaw, D.; DeJesus, E.; Rashbaum, B.; Chuck, S. L.; Yale, K.; Liu, H. C.; Warren, D. R.; Ramanathan, S.; Kearney, B. P. *AIDS* **2011,** *25,* F7.
318. Min, S.; Sloan, L.; DeJesus, E.; Hawkins, T.; McCurdy, L.; Song, I.; Stroder, R.; Chen, S.; Underwood, M.; Fujiwara, T.; Piscitelli, S.; Lalezari, J. *AIDS* **2011,** *25,* 1737.
319. Walker, M. A. *Curr. Opin. Investig. Drugs* **2009,** *10,* 129.
320. De Luca, L.; Ferro, S.; Morreale, F.; Chimirri, A. *Mini-Rev. Med. Chem.* **2011,** *11,* 714.
321. De Luca, L.; Ferro, S.; Morreale, F.; De Grazia, S.; Chimirri, A. *ChemMedChem* **2011,** *6,* 1184.
322. Christ, F.; Voet, A.; Marchand, A.; Nicolet, S.; Desimmie, B. A.; Marchand, D.; Bardiot, D.; Van der Veken, N. J.; Van Remoortel, B.; Strelkov, S. V.; De Maeyer, M.; Chaltin, P.; Debyser, Z. *Nat. Chem. Biol.* **2010,** *6,* 442.
323. Ganem, D.; Prince, A. M. *New Engl. J. Med.* **2004,** *350,* 1118.
324. Locarnini, S.; Zoulim, F. *Antiviral Ther.* **2010,** *15 (Suppl.* 3), 3.
325. Rehermann, B.; Nascimbeni, M. *Nat. Rev. Immunol.* **2005,** *5,* 215.
326. Kwon, H.; Lok, A. S. *Nat. Rev. Gastroenterol. Hepatol.* **2011,** *8,* 275.
327. Block, T. M.; Guo, H.; Guo, J.-T. *Clinics Liver Dis.* **2007,** *11,* 685.
328. Lucifora, J.; Zoulim, F. *Future Virol.ogy* **2011,** *6,* 599.
329. Schaedler, S.; Hildt, E. *Viruses* **2009,** *1,* 185.
330. Pumpens, P.; Grens, E.; Nassal, M. *Intervirology* **2003,** *45,* 218.
331. Delaney IV, W. E.; Borroto-Esoda K. *Curr. Opin. Pharmacol.* **2008,** *8,* 1.
332. Poland, G. A.; Jacobson, R. M. *New Engl. J. Med.* **2004,** *35,* 2832.
333. Keating, G. M. *Drugs* **2009,** *69,* 2633.
334. Fung, J.; Lai, C.-L.; Seto, W.-K.; Yuen, M.-F. *J. Antimicrob. Chemother.* **2011,** *66,* 2715.
335. Jarvis, B.; Faulds, D. *Drugs* **1999,** *58,* 101.
336. Standring, D. N.; Bridges, E. G.; Placidi, L.; Faraj, A.; Loi, A. G.; Pierra, C.; Dukhan, D.; Gosselin, G.; Imbach, J.-L.; Hernandez, B.; Juodawlkis, A.; Tennant, B.; Korba, B.; Cote, P.; Cretton-Scott, E.; Schinazi, R. F.; Myers, M.; Bryant, M. L.; Sommadossi, J.-P. *Antiviral Chem. Chemother.* **2001,** *12 (Suppl. 1),* 119.
337. McKeage, K.; Keam, S. J. *Drugs* **2010,** *70,* 1857.
338. Starrett, J. E. Jr.; Tortolani, D. R.; Russell, J.; Hitchcock, M. J. M.; W., Valerie; Martin, J. C.; Mansuri, M. M. *J. Med. Chem.* **1994,** *37,* 1857.
339. Dando, T. M.; Plosker, G. L. *Drugs* **2003,** *63,* 2215.
340. Adusumilli, S. *Drugs Today* **2009,** *45,* 679.
341. Jenh, A. M.; Pham, P. A. *Exp. Rev. Anti-Infect. Ther.* **2010,** *8,* 1079.
342. Scott, L. J.; Keating, G. M. *Drugs* **2009,** *69,* 1003.
343. Bisacchi, G. S.; Chao, S. T.; Bachard, C.; Daris, J. P.; Innaimo, S.; Jacobs, G. A.; Kocy, O.; Lapointe, P.; Martel, A.; Merchant, Z.; Slusarchyk, W. A.; Sundeen, J. E.; Young, M. G.; Colonno, R. Zahler, R. *Bioorg. Med. Chem. Lett.* **1997,** *7,* 127.
344. Innaimo, S. F.; Seifer, M.; Bisacchi, G. S.; Standring, D. N.; Zahler, R.; Colonno, R. J. *Antimicrob. Agents Chemother.* **1997,** *41,* 1444.
345. Zoulim, F.; Locarnini, S. *Gastroenterology* **2009,** *137,* 1593.
346. Holness, G.; Carriero, D. C.; Dieterich, D. T. *Exp. Rev. Gastroenterol. Hepatol.* **2009,** *3,* 693.
347. Locarnini, S. A.; Yuen, L. *Antiviral Ther.* **2010,** *15,* 451.
348. Gumina, G.; Chong, Y.; Choo, H.; Song, G.-Y.; Chu C. K. *Curr. Topics Med. Chem.* **2002,** *2,* 1065.
349. Wilber, R.; Kreter, B.; Bifano, M.; Danetz, S.; Lehman-McKeeman, L.; Tenney, D. J.; Meanwell, N.; Zahler, R.; Brett-Smith, H. In *Antiviral Drugs: From Basic Discovery Through Clinical Trials,* Kazmierski, W. M., ed., Wiley: Hoboken, NJ, **2011,** pp 401–416.
350. Tuske, S; Sarafianos, S. G.; Clark, A. D.; Ding, J.; Naeger, L. K.; White, K. L.; Miller, M. D.; Gibbs, C. S.; Boyer, P. L.; Clark, P.; Wang, G.; Gaffney, B. L.; Jones, R. A.; Jerina, D. M.; Hughes, S. H.; Arnold, E. *Nat. Struct. Molec. Biol.* **2004,** *11,* 469.

351. Shaw, J.-P.; Sueoka, C. M.; Oliyai, R.; Lee, W. A.; Arimilli, M. N.; Kim, C. U.; Cundy, K. C. *Pharma. Res.* **1997,** *14,* 1824.
352. Li, F.; Maag, H.; Alfredson T. *J. Pharm. Sci.* **2008,** *97,* 1109.
353. Lavanchy D. *Clin. Microbiol. Infect.* **2011,** *17,* 107–115.
354. Chak, E.; Talal, A. H.; Sherman, K. E.; Schiff, E. R.; Saab, S. *Liver Int.* **2011,** *31,* 1090.
355. Pawlotsky, J.-M. *Trends Microbiol.* **2004,** *12,* 96.
356. Hoofnagle, J. H., Seeff, L. B. *New Engl. J. Med.* **2006,** *355,* 2444.
357. Choo, Q.-L.; Kuo, G.; Weiner, A. J.; Overby, L. R.; Bradley, D. W.; Houghton, M. *Science,* **1989,** *244,* 359.
358. Kuo, G.; Choo, Q.-L.; Alter, H. J.; Gitnick, G. I.; Redeker, A. G.; Purcell, R. H.; Miyamura, T.; Dienstag, J. L.; Alter, M. J.; Stevens, C. E.; Tegtmeier, G. E.; Bonino, F.; Colombo, M.; Lee, W.-S.; Kuo, C.; Berger, K.; Shuster, J. R.; Overby, L. R.; Bradley, D. W.; Houghton, M. *Science* **1989,** *244,* 362.
359. Houghton, M. *J. Hepatol.* **2009,** *51,* 939.
360. Keam, S. J.; Cvetković, R. S. *Drugs* **2008,** *68,* 1273.
361. Neumann, A. U.; Lam, N. P.; Dahari, H.; Gretch, D. R.; Wiley, T. E.; Layden, T. J.; Perelson, A. S. *Science* **1998,** *282,* 103.
362. Guedj, J.; Rong, L.; Dahari, H.; Perelson, A. S. *J. Viral Hepatitis* **2010,** *17,* 825.
363. Rong, L.; Perelson, A. S. *Crit. Rev. Immunol.* **2010,** *30,* 131.
364. Rong, L.; Dahari, H.; Ribeiro R. M; Perelson, A. S. *Sci. Transl. Med.* **2010,** *2,* 30ra32.
365. Gelman, M. A.; Glenn, J. S. *Trends Molec. Med.* **2011,** *17,* 34.
366. De Francesco, R.; Carfi, A. *Adv. Drug Deliv. Rev.* **2007,** *59,* 1242.
367. Kwong, A. D.; McNair, L.; Jacobson, I.; George, S. *Curr. Opin. Pharmacol.* **2008,** *8,* 522.
368. Bartenschlager, R. *Nature Rev. Drug Discov.* **2002,** *1,* 911.
369. Appel, N.; Schaller, T.; Penin, F.; Bartenschlager, R. *J. Biol. Chem.* **2006,** *281,* 9833.
370. Bartenschlager, R.; Sparacio, S. *Virus Res.* **2007,** *127,* 195.
371. Bartenschlager, R. *Arzneim. Forschung* **2010,** *60,* 695.
372. Lohmann, V.; Korner, F.; Koch, J.-O.; Herian, U.; Theilmann, L.; Bartenschlager, R. *Science* **1999,** *285,* 110.
373. Bartenschlager, R.; Pietschmann, T. *Proc. Natl. Acad. Sci. USA* **2005,** *102,* 9739.
374. Kato, T.; Date, T.; Miyamoto, M.; Furusaka, A.; Tokushige, K.; Mizokami, M.; Wakita, T. *Gastroenterology* **2003,** *125,* 1808.
375. Wakita, T.; Pietschmann, T.; Kato, T.; Date, T.; Miyamoto, M.; Zhao, Z.; Murthy, K.; Habermann, A.; Kraeusslich, H.-G.; Mizokami, M.; Bartenschlager, R.; Liang, T. J. *Nat. Med.* **2005,** *11,* 791.
376. Lindenbach, B. D.; Evans, M. J.; Syder, A. J.; Woelk, B.; Tellinghuisen, T. L.; Liu, C. C.; Maruyama, T.; Hynes, R. O.; Burton, D. R.; McKeating, J. A.; Rice, C. M. *Science* **2005,** *309,* 623.
377. Zhong, J; Gastaminza, P.; Cheng, G.; Kapadia, S.; Kato, T.; Burton, D. R.; Wieland, S. F.; Uprichard, S. L.; Wakita, T.; Chisari, F. V. *Proc. Natl. Acad. Sci. USA* **2005,** *102,* 9294.
378. Koike, K.; Moriya, K.; Matsuura, Y. *Hepatol. Res.* **2010,** *40,* 69.
379. Akari, H.; Iwasaki, Y.; Yoshida, T.; Iijima, S. *Microbiol. Immunol.* **2009,** *53,* 53.
380. Boonstra, A.; van der Laan, L. J. W.; Vanwolleghem, T.; Janssen, H. L. A. *Hepatology* **2009,** *50,* 1646.
381. Bartenschlager, R.; Ahlborn-Laake, L.; Mous, J.; Jacobsen, H. *J. Virol.* **1993,** *67,* 3835.
382. Hijikata, M.; Mizushima, H.; Tanji, Y.; Komoda, Y.; Hirowatari, Y.; Akagi, T.; Kato, N.; Kimura, K.; Shimotohno, K. *Proc. Natl. Acad. Sci. USA* **1993,** *90,* 10773.
383. Love, R. A.; Parge, H. E.; Wickersham, J. A.; Hostomsky, Z.; Habuka, N.; Moomaw, E. W.; Adachi, T.; Hostomska, Z. *Cell* **1996,** *87,* 331.
384. Kim, J. L.; Morgenstern, K. A.; Lin, C.; Fox, T.; Dwyer, M. D.; Landro, J. A.; Chambers, S. P.; Markland, W.; Lepre, C. A.; O'Malley, E. T.; Harbeson, S. L.; Rice, C. M.; Murcko, M. A.; Caron, P. R.; Thomson, J. A. *Cell* **1996,** *87,* 343.
385. Raney, K. D.; Sharma, S. D.; Moustafa, I. M.; Cameron, C. E. *J. Biol. Chem.* **2010,** *285,* 22725.
386. Chen, K. X.; Njoroge, F. G. *Curr. Opin. Investig. Drugs* **2009,** *10,* 821.

387. Perni, R. B.; Almquist, S. J.; Byrn, R. A.; Chandorkar, G.; Chaturvedi, P. R.; Courtney, L. F.; Decker, C. J.; Dinehart, K.; Gates, C. A.; Harbeson, S. L.; Heiser, A.; Kalkeri, G.; Kolaczkowski, E.; Lin, K.; Luong, Y.-P.; Rao, B. G.; Taylor, W. P.; Thomson, J. A.; Tung, R. D.; Wei, Y.; Kwong, A. D.; Lin, C. *Antimicrob. Agents Chemother.* **2006,** *50,* 899.

388. Lin, C.; Kwong, A. D.; Perni, R. B. *Infectious Disorders: Drug Targets* **2006,** *6,* 3.

389. Kwong, A. D.; Kauffman, R. S.; Hurter, P.; Mueller, P. *Nat. Biotechnol.* **2011,** *29,* 993.

390. Venkatraman, S.; Bogen, S. L.; Arasappan, A.; Bennett, F.; Chen, K.; Jao, E.; Liu, Y.-T.; Lovey, R.; Hendrata, S.; Huang, Y.; Pan, W.; Parekh, T.; Pinto, P.; Popov, V.; Pike, R.; Ruan, S.; Santhanam, B.; Vibulbhan, B.; Wu, W.; Yang, W.; Kong, J.; Liang, X.; Wong, J.; Liu, R.; Butkiewicz, N.; Chase, R.; Hart, A.; Agrawal, S.; Ingravallo, P.; Pichardo, J.; Kong, R.; Baroudy, B.; Malcolm, B.; Guo, Z.; Prongay, A.; Madison, V.; Broske, L.; Cui, X.; Cheng, K.-C.; Hsieh, T. Y.; Brisson, J.-M.; Prelusky, D.; Korfmacher, W.; White, R.; Bogdanowich-Knipp, S.; Pavlovsky, A.; Bradley, P.; Saksena, A. K.; Ganguly, A.; Piwinski, J.; Girijavallabhan, V.; Njoroge, F. G. *J. Med. Chem.* **2006,** 49, 6074.

391. Prongay, A. J.; Guo, Z.; Yao, N.; Pichardo, J.; Fischmann, T.; Strickland, C.; Myers, Jr., J.; Weber, P. C.; Beyer, B. M.; Ingram, R.; Hong, Z.; Prosise, W. W.; Ramanathan, L.; Taremi, S. S.; Yarosh-Tomaine, T.; Zhang, R.; Senior, M.; Yang, R.-S.; Malcolm, B.; Arasappan, A.; Bennett, F.; Bogen, S. L.; Chen, K.; Jao, E.; Liu, Y.-T.; Lovey, R. G.; Saksena, A. K.; Venkatraman, S.; Girijavallabhan, V.; Njoroge F. G.; Madison, V. *J. Med. Chem.* **2007,** *50,* 2310.

392. Njoroge, F. G.; Chen, K. X.; Shih N.-Y.; Piwinski, J. J. *Acc. Chem. Res.* **2008,** *41,* 50.

393. Malcolm, B. A.; Liu, R.; Lahser, F.; Agrawal, S.; Belanger, B.; Butkiewicz, N.; Chase, R.; Gheyas, F.; Hart, A.; Hesk, D.; Ingravallo, P.; Jiang, C.; Kong, R.; Lu, J.; Pichardo, J.; Prongay, A.; Skelton, A.; Tong, X.; Venkatraman, S.; Xia, E.; Girijavallabhan, V.; Njoroge, F. G. *Antimicrob. Agents Chemother.* **2006,** *50,* 1013.

394. Lamarre, D.; Andreson, P. C.; Bailey, M.; Bealieu, P.; Bolger, G.; Bonneau, P.; Bös, M.; Cameron, D. R.; Cartier, M.; Cordingley, M. G.; Faucher, A. M.; Goudreau, N.; Kawai, S. H.; Kukolj, G.; Legacé, L.; LaPlante, S. R.; Narjes, H.; Poupart, M. A.; Rancourt, J.; Sentjens, R. E.; St. George, R.; Simoneau, B.; Steinmann, G.; Thibeault, D.; Tsantrizos, Y. S.; Weldon, S. M.; Yong, C. L.; Llinàs-Brunet, M. *Nature* **2003,** *426,* 186.

395. Hinrichsen, H.; Benhamou, Y.; Wedemeyer, H.; Reiser, M.; Sentjens, R. E.; Calleja, J. L.; Forns, X.; Erhardt, A.; Croenlein, J.; Chaves, R. L.; Yong, C.-L.; Nehmiz, G.; Steinmann, G. G. *Gastroenteroloy* **2004,** *127,* 1347.

396. Llinas-Brunet, M.; Bailey, M. D.; Bolger, G.; Brochu, C.; Faucher, A.-M.; Ferland, J. M.; Garneau, M.; Ghiro, E.; Gorys, V.; Grand-Maitre, C.; Halmos, T.; Lapeyre-Paquette, N.; Liard, F.; Poirier, M.; Rheaume, M.; Tsantrizos, Y. S.; Lamarre, D. *J. Med. Chem.* **2004,** *47,* 1605.

397. Tsantrizos, Y. *Acc. Chem. Res.* **2008,** *41,* 1252.

398. Steinkuehler, C.; Biasiol, G.; Brunetti, M.; Urbani, A.; Koch, U.; Cortese, R.; Pessi, A.; De Francesco, R. *Biochemistry* **1998,** *37,* 8899.

399. Ingallinella, P.; Altamura, S.; Bianchi, E.; Taliani, M.; Ingenito, R.; Cortese, R.; De Francesco, R.; Steinkuehler, C.; Pessi, A. *Biochemistry* **1998,** *37,* 8906.

400. Llinas-Brunet, M.; Bailey, M.; Ddziel, R.; Fazal, G.; Gorys, V.; Goulet, S.; Halmos, T.; Maurice, R.; Poirier, M.; Poupart, M.-A.; Rancourt, J.; Thibeault, D.; Wernic, D.; Lamarre, D. *Bioorg. Med. Chem. Lett.* **1998,** *8,* 2719.

401. Di Marco, S.; Rizzi, M.; Volpari, C.; Walsh, M. A.; Narjes, F.; Colarusso, S.; De Francesco, R.; Matassa, V. G.; Sollazzo, M. *J. Biol. Chem.* **2000,** *275,* 7152.

402. Fairlie, D. P.; Abbenante, G.; March, D. R. *Curr. Med. Chem.* **1995,** *2,* 654.

403. McGeary, R. P.; Fairlie, D. P. *Curr. Opin. Drug Discov. Develop.* **1998,** *1,* 208.

404. Tyndall, J. D. A.; Nall, T.; Fairlie, D. P. *Chem. Rev.* **2005,** *105,* 973.

405. Madala, P. K; Tyndall, J. D. A; Nall, T.; Fairlie, D. P. *Chem. Rev.* **2010,** *110,* PR1.

406. Llinas-Brunet, M.; Bailey, M. D.; Goudreau, N.; Bhardwaj, P. K.; Bordeleau, J.; Bos, M.; Bousquet, Y.; Cordingley, M. G.; Duan, J.; Forgione, P.; Garneau, M.; Ghiro, E.; Gorys, V.; Goulet, S.; Halmos, T.; Kawai, S. H.; Naud, J.; Poupart, M.-A.; White, P. W. *J. Med. Chem.* **2010,** *53,* 6466.

407. White, P. W.; Llinas-Brunet, M.; Amad, M.; Bethell, R. C.; Bolger, G.; Cordingley, M. G.; Duan, J.; Garneau, M.; Lagace, L.; Thibeault, D.; Kukolj, G. *Antimicrob. Agents Chemother.* **2010**, *54*, 4611.
408. Lemke, C. T.; Goudreau, N.; Zhao, S.; Hucke, O.; Thibeault, D.; Llinas-Brunet, M.; White, P. W. *J. Biol. Chem.* **2011**, *286*, 11434.
409. Manns, M. P.; Bourliere, M.; Benhamou, Y.; Pol, S.; Bonacini, M.; Trepo, C.; Wright, D.; Berg, T.; Calleja, J. L.; White, P. W.; Stern, J. O.; Steinmann, G.; Yong, C.-L.; Kukolj, G.; Scherer, J.; Boecher, W. O. *J. Hepatol.* **2011**, *54*, 1114.
410. Yee, N. K.; Farina, V.; Houpis, I. N.; Haddad, N.; Frutos, R. P.; Gallou, F.; Wang, X.-J.; Wei, X.; Simpson, R. D.; Feng, X.; Fuchs, V.; Xu, Y.; Tan, J.; Zhang, L.; Xu, J.; Smith-Keenan, L. L.; Vitous, J.; Ridges, M. D.; Spinelli, E. M.; Johnson, M.; Donsbach, K.; Nicola, T.; Brenner, M.; Winter, E.; Kreye, P.; Samstag, W. *J. Org. Chem.* **2006**, *71*, 7133.
411. Shu, C.; Zeng, X.; Hao, M.-H.; Wei, X.; Yee, N. K.; Busacca, C. A.; Han, Z.; Farina, V.; Senanayake, C. H. *Org. Lett.* **2008**, *10*, 1303.
412. Sheng, X. C.; Appleby, T.; Butler, T.; Cai, R.; Chen, X.; Cho, A.; Clarke, M. O.; Cottell, J.; Delaney IV, W. E. Doerffler, E.; Link, J.; Ji, M.; Pakdaman, R.; Pyun, H.-J.; Wu, Q.; Xu, J.; Kim, C. U. *Bioorg. Med. Chem. Lett.* **2012**, *22*, 2629.
413. Sheng, X. C.; Casarez, A.; Cai, R.; Clarke, M. O.; Chen, X.; Cho, A.; Delaney, IV, W.; Doerffler, E. E.; Ji, M.; Mertzman, M.; Pakdaman, R.; Pyun, H.-J. Rowe, T.; Wu, Q.; Xu, J.; Kim, C. U. *Bioorg. Med. Chem. Lett.* **2012**, *22*, 1394.
414. Raboisson, P.; de Kock, H.; Rosenquist, A.; Nilsson, M.; Salvador-Oden, L.; Lin, T.-I.; Roue, N.; Ivanov, V.; Wahling, H.; Wickstrom, K.; Hamelink, E.; Edlund, M.; Vrang, L.; Vendeville, S.; Van de Vreken, W.; McGowan, D.; Tahri, A.; Hu, L.; Boutton, C.; Lenz, O.; Delouvroy, F.; Pille, G.; Surleraux, D.; Wigerinck, P.; Samuelsson, B.; Simmen, K. *Bioorg. Med. Chem. Lett.* **2008**, *18*, 4853.
415. Cummings, M. D.; Lindberg, J.; Lin, T.-I.; de Kock, H.; Lenz, O.; Lilja, E.; Fellander, S.; Baraznenok, V.; Nystrom, S.; Nilsson, M.; Vrang, L.; Edlund, M.; Rosenquist, A.; Samuelsson, B.; Raboisson, P.; Simmen, K. *Angew. Chem. Intl. Ed.* **2010**, *49*, 1652.
416. Lin, T.-I.; Lenz, O.; Fanning, G.; Verbinnen, T.; Delouvroy, F.; Scholliers, A.; Vermeiren, K.; Rosenquist, A.; Edlund, M.; Samuelsson, B.; Vrang, L.; de Kock, H.; Wigerinck, P.; Raboisson, P.; Simmen, K. *Antimicrob. Agents Chemother.* **2009**, *53*, 1377.
417. Davies, S. *Drugs Future* **2009**, *34*, 545.
418. Tsantrizos, Y. S. *Curr. Opin. Investig. Drugs* **2009**, *10*, 871.
419. Seiwert, S. D.; Andrews, S. W.; Jiang, Y.; Serebryany, V.; Tan, H.; Kossen, K.; Rajagopalan, P. T. R.; Misialek, S.; Stevens, S. K.; Stoycheva, A.; Hong, J.; Lim, S. R.; Qin, X.; Rieger, R.; Condroski, K. R.; Zhang, H.; Do, M. G.; Lemieux, C.; Hingorani, G. P.; Hartley, D. P.; Josey, J. A.; Pan, L.; Beigelman, L.; Blatt, L. M. *Antimicrob. Agents Chemother.* **2008**, *52*, 4432.
420. Deutsch, M.; Papatheodoridis, G. V. *Curr. Opin. Investig. Drugs* **2010**, *11*, 951.
421. McCauley, J. A.; McIntyre, C. J.; Rudd, M. T.; Nguyen, K. T.; Romano, J. J.; Butcher, J. W.; Gilbert, K. F.; Bush, K. J.; Holloway, M. K.; Swestock, J.; Wan, B.-L.; Carroll, S. S.; DiMuzio, J. M.; Graham, D. J.; Ludmerer, S. W.; Mao, S.-S.; Stahlhut, M. W.; Fandozzi, C. M.; Trainor, N.; Olsen, D. B.; Vacca, J. P.; Liverton, N. J. *J. Med. Chem.* **2010**, *53*, 2443.
422. Liverton, N. J.; Carroll, S. S.; Di Muzio, J.; Fandozzi, C.; Graham, D. J.; Hazuda, D.; Holloway, M. K.; Ludmerer, S. W.; McCauley, J. A.; McIntyre, C. J.; Olsen, D. B.; Rudd, M. T.; Stahlhut, M.; Vacca, J. P. *Antimicrob. Agents Chemother.* **2009**, *54*, 305.
423. Hammond, E.; Lucas, A.; Lucas, M.; Phillips, E.; Gaudieri, S. *Drugs Future* **2010**, *35*, 803.
424. Harper, S.; McCauley, J. A.; Rudd, M. T.; Ferrara, M.; DiFilippo, M.; Crescenzi, B.; Koch, U.; Petrocchi, A.; Holloway, M. K.; Butcher, J. W.; Romano, J. J.; Bush, K. J.; Gilbert, K. F.; McIntyre, C. J.; Nguyen, K. T.; Nizi, E.; Carroll, S. S.; Ludmerer, S. W.; Burlein, C.; DiMuzio, J. M.; Graham, D. J.; McHale, C. M.; Stahlhut, M. W.; Olsen, D. B.; Monteagudo, E.; Cianetti, S.; Giuliano, C.; Pucci, V.; Trainor, N.; Fandozzi, C. M.; Rowley, M.; Coleman, P. J.; Vacca, J. P.; Summa, V.; Liverton, N. J. *ACS Med. Chem. Lett.* **2012**, *3*, 332.

425.   Morikawa, K.; Lange, C. M.; Gouttenoire, J.; Meylan, E.; Brass, V.; Penin, F.; Moradpour, D. *J. Viral Hepatitis* **2011**, *18*, 305.

426.   Venkatraman, S.; Njoroge, F. G. *Exp. Opin. Therapeutic Patents* **2009**, *19*, 1277.

427.   Avolio, S.; Summa, V. *Curr. Topics Med. Chem.* **2010**, *10*, 1403.

428.   Sofia, M. J.; Chang, W.; Furman, P. A.; Mosley, R. T.; Ross, B S. *J. Med. Chem.* **2012**, *55*, 2481

429.   Walker, M. P.; Hong, Z. *Curr. Opin. Pharmacol.* **2002**, *2*, 534.

430.   Beaulieu, P. L.; Tsantrizos, Y. S. *Curr. Opin. Investig. Drugs* **2004**, *5*, 838.

431.   Lohmann, V.; Roos, A.; Korner, F.; Koch, J. O.; Bartenschlager, R. *J. Viral Hepatitis* **2000**, *7*, 167.

432.   Ago, H.; Adachi, T.; Yoshida, A.; Yamamoto, M.; Habuka, N.; Yatsunami, K.; Miyano, M. *Structure* **1999**, *7*, 1417.

433.   Lesburg, C. A.; Cable, M. B.; Ferrari, E.; Hong, Z.; Mannarino, A. F.; Weber, P. C. *Nat. Struct. Biol.* **1999**, *6*, 937.

434.   Bressanelli, S.; Tomei, L.; Roussel, A.; Incitti, I.; Vitale, R. L.; Mathieu, M.; De Francesco, R.; Rey, F. A. *Proc. Natl. Acad. Sci. U. S. A.* **1999**, *96*, 13034.

435.   Legrand-Abravanel, F.; Nicot, F.; Izopet, J. *Expert Opin. Investig. Drugs* **2010**, *19*, 963.

436.   Sofia, M. J.; *Antiviral Chem. Chemother.* **2011**, *22*, 23.

437.   McCown, M. F.; Rajyaguru, S.; Le Pogam, S.; Ali, S.; Jiang, W.-R.; Kang, H.; Symons, J.; Cammack, N.; Najera, I. *Antimicrob. Agents Chemother.* **2008**, *52*, 1604.

438.   Benzaria, S.; Bardiot, D.; Bouisset, T.; Counor, C.; Rabeson, C.; Pierra, C.; Storer, R.; Loi, A. G.; Cadeddu, A.; Mura, M.; Musiu, C.; Liuzzi, M.; Loddo, R.; Bergelson, S.; Bichko, V.; Bridges, E.; Cretton-Scott, E.; Mao, J.; Sommadossi, J.-P.; Seifer, M.; Standring, D.; Tausek, M.; Gosselin, G.; La Colla, P. *Antiviral Chem. Chemother.* **2007**, *18*, 225.

439.   Pierra, C.; Benzaria, S.; Amador, A.; Moussa, A.; Mathieu, S.; Storer, R.; Gosselin, G. *Nucleosides, Nucleotides Nucleic Acids* **2005**, *24*, 767.

440.   Pierra, C.; Amador, A.; Benzaria, S.; Cretton-Scott, E.; D'Amours, M.; Mao, J.; Mathieu, S.; Moussa, A.; Bridges, E. G.; Standring, D. N.; Sommadossi, J.-P.; Storer, R.; Gosselin, G. *J. Med. Chem.* **2006**, *49*, 6614.

441.   Smith, D. B.; Martin, J. A.; Klumpp, K.; Baker, S. J.; Blomgren, P. A.; Devos, R.; Granycome, C.; Hang, J.; Hobbs, C. J.; Jiang, W.-R.; Laxton, C.; Le Pogam, S.; Leveque, V.; Ma, H.; Maile, G.; Merrett, J. H.; Pichota, A.; Sarma, K.; Smith, M.; Swallow, S.; Symons, J.; Vesey, D.; Najera, I.; Cammack, N. *Bioorg. Med. Chem. Lett.* **2007**, *17*, 2570.

442.   Klumpp, K.; Leveque, V.; Le Pogam, S.; Ma, H.; Jiang, W.-R.; Kang, H.; Granycome, C.; Singer, M.; Laxton, C.; Hang, J. Q,; Sarma, K.; Smith, D. B.; Heindl, D.; Hobbs, C. J.; Merrett, J. H.; Symons, J.; Cammack, N.; Martin, J. A.; Devos, R.; Najera, I. *J. Biol. Chem.* **2006**, *281*, 3793.

443.   Li, F.; Wu, X.; Hadig, X.; Huang, S.; Hong, L.; Tran, T.; Brandl, M.; Alfredson, T. *Drug Develop. Indus. Pharmacy* **2010**, *36*, 413.

444.   Smith, D. B.; Kalayanov, G.; Sund, C.; Winqvist, A.; Pinho, P.; Maltseva, T.; Morisson, V.; Leveque, V.; Rajyaguru, S.; Le Pogam, S.; Najera, I.; Benkestock, K.; Zhou, X.-X.; Maag, H.; Cammack, N.; Martin, J. A.; Swallow, S.; Johansson, N. G.; Klumpp, K.; Smith, M. *J. Med. Chem.* **2009**, *52*, 219.

445.   Zhou, X.-J.; Pietropaolo, K.; Chen, J.; Khan, S.; Sullivan-Bolyai, J.; Mayers, D. *Antimicrob. Agents Chemother.* **2011**, *55*, 76.

446.   McGuigan, C.; Madela, K.; Aljarah, M.; Gilles, A.; Brancale, A.; Zonta, N.; Chamberlain, S.; Vernachio, J.; Hutchins, J.; Hall, A.; Ames, B.; Gorovits, E.; Ganguly, B.; Kolykhalov, A.; Wang, J.; Muhammad, J.; Patti, J. M.; Henson, G. *Bioorg. Med. Chem. Lett.* **2010**, *20*, 4850.

447.   Vernachio, J. H.; Bleiman, B.; Bryant, K. D.; Chamberlain, S.; Hunley, D.; Hutchins, J.; Ames, B.; Gorovits, E.; Ganguly, B.; Hall, A.; Kolykhalov, A.; Liu, Y.; Muhammad, J.; Raja, N.; Walters, C. R.; Wang, J.; Williams, K.; Patti, J. M.; Henson, G.; Madela, K.; Aljarah, M.; Gilles, A.; McGuigan, C. *Antimicrob. Agents Chemother.* **2011**, *55*, 1843.

448.   Cole, P.; Castaner, R.; Bolos, J. *Drugs Future* **2009**, *34*, 282.

449. Stuyver, L. J.; McBrayer, T. R.; Tharnish, P. M.; Clark, J.; Hollecker, L.; Lostia, S.; Nachman, T.; Grier, J.; Bennett, M. A.; Xie, M.-Y.; Schinazi, R. F.; Morrey, J. D.; Julander, J. L.; Furman, P. A.; Otto, M. J. *Antiviral Chem. Chemother.* **2006**, *17*, 79.

450. McCown, M. F.; Rajyaguru, S.; Le Pogam, S.; Ali, S.; Jiang, W.-R.; Kang H.; Symons, J.; Cammack, N.; Najera, I. *Antimicrob. Agents Chemother.* **2008**, *52*, 1604.

451. Ali, S.; Leveque, V.; Le Pogam, S.; Ma, H.; Philipp, F.; Inocencio, N.; Smith, M.; Alker, A.; Kang, H.; Najera, I.; Klumpp, K.; Symons, J.; Cammack, N.; Jiang, W.-R. *Antimicrob. Agents Chemother.* **2008**, *52*, 4356.

452. Asif, G.; Hurwitz, S. J.; Shi, J.; Hernandez-Santiago, B. I.; Schinazi, R. F. *Antimicrob. Agents Chemother.* **2007**, *51*, 2877.

453. Le Pogam, S.; Seshaadri, A.; Ewing, A.; Kang, H.; Kosaka, A.; Yan, J.-M.; Berrey, M.; Symonds, B.; de la Rosa, A.; Cammack, N.; Najera, I. *J. Infect. Dis.* **2010**, *202*, 1510.

454. Gane, E. J.; Roberts, S. K.; Stedman, C. A. M.; Angus, P. W.; Ritchie, B.; Elston, R.; Ipe, D.; Morcos, P. N.; Baher, L.; Najera, I.; Chu, T.; Lopatin, U.; Berrey, M. M.; Bradford, W.; Laughlin, M.; Shulman, N. S.; Smith, P. F. *Lancet* **2010**, *376*, 1467.

455. Murakami, E.; Niu, C.; Bao, H.; Micolochick Steuer, H. M.; Whitaker, T.; Nachman, T.; Sofia, M. A.; Wang, P.; Otto, M. J.; Furman, P. A. *Antimicrob. Agents Chemother.* **2008**, *52*, 458.

456. Ma, H.; Jiang, W.-R.; Robledo, N.; Leveque, V.; Ali, S.; Lara-Jaime, T.; Masjedizadeh, M.; Smith, D, B.; Cammack, N; Klumpp, K.; Symons, J. *J. Biol. Chem.* **2007**, *282*, 29812.

457. Lam, A. M.; Murakami, E.; Espiritu, C.; Micolochick Steuer, H. M.; Niu, C.; Keilman, M.; Bao, H.; Zennou, V.; Bourne, N.; Julander, J. G.; Morrey, J. D.; Smee, D. F.; Frick, D. N.; Heck, J. A.; Wang, P.; Nagarathnam, D.; Ross, B. S.; Sofia, M. J.; Otto, M. J.; Furman, P. A. *Antimicrob. Agents Chemother.* **2010**, *54*, 3187.

458. Ray, A. S.; Hostetler, K. Y. *Antiviral Res.* **2011**, *92*, 277.

459. Sofia, M. J.; Bao, D.; Chang, W.; Du, J.; Nagarathnam, D.; Rachakonda, S.; Reddy, P. G.; Ross, B. S.; Wang, P.; Zhang, H.-R.; Bansal, S.; Espiritu, C.; Keilman, M.; Lam, A. M.; Steuer, H. M. M.; Niu, C.; Otto, M. J.; Furman, P. A. *J. Med. Chem.* **2010**, *53*, 7202.

460. Ross, B. S.; Ganapati R. P.; Zhang, H.-R.; Rachakonda, S.; Sofia, M. J. *J. Org. Chem,* **2011**, *76*, 8311.

461. Murakami, E.; Tolstykh, T.; Bao, H.; Niu, C.; Micolochick Steuer, H. M.; Bao, D.; Chang, W.; Espiritu, C.; Bansal, S.; Lam, A. M.; Otto, M. J.; Sofia, M. J.; Furman, P. A. *J. Biol. Chem.* **2010**, *285*, 34337.

462. Reviriego, C. *Drugs Future* **2012**, *37*, 175.

463. Reddy, P.; Bao, D.; Chang, W.; Chun, B.-K.; Du, J.; Nagarathnam, D.; Rachakonda, S.; Ross, B. S.; Zhang, H.-R.; Bansal, S.; Espiritu, C. L.; Keilman, M.; Lam, A. M.; Niu, C.; Micolochick Steuer, H.; Furman, P. A.; Otto, M. J.; Sofia, M. J. *Bioorg. Med. Chem. Lett.* **2010**, *20*, 7376.

464. Reddy, P. G.; Chun, B.-K.; Zhang, H.-R.; Rachakonda, S.; Ross, B. S.; Sofia, M. J. *J. Org. Chem.* **2011**, *76*, 3782.

465. Lam, A. M.; Espiritu, C.; Murakami, E.; Zennou, V.; Bansal, S.; Micolochick Steuer, H. M.; Niu, C.; Keilman, M.; Bao, H.; Bourne, N.; Veselenak, R. L.; Reddy, P. G.; Chang, W.; Du, J.; Nagarathnam, D.; Sofia, M. J.; Otto, M. J.; Furman, P. A. *Antimicrob. Agents Chemother.* **2011**, *55*, 2566.

466. Chang, W.; Bao, D.; Chun, B.-K.; Naduthambi, D.; Nagarathnam, D.; Rachakonda, S.; Reddy, P. G.; Ross, B. S.; Zhang, H.-R.; Bansal, S.; Espiritu, C. L.; Keilman, M.; Lam, A. M.; Niu, C.; Micolochick Steuer, H.; Furman, P. A.; Otto, M. J.; Sofia, M. J. *ACS Med. Chem. Lett.* **2011**, *2*, 130.

467. Furman, P. A.; Murakami, E.; Niu, C.; Lam, A. M.; Espiritu, C.; Bansal, S.; Bao, H.; Tolstykh, T.; Micolochick Steuer, H.; Keilman, M.; Zennou, V.; Bourne, N.; Veselenak, R. L.; Chang, W.; Ross, B. S.; Du, J.; Otto, M. J.; Sofia, M. J *Antiviral Res.* **2011**, *91*, 120.

468. Lam, A. M.; Espiritu, C.; Bansal, S.; Micolochick Steuer, H. M.; Zennou, V.; Otto, M. J.; Furman, P. A. *J. Virol.* **2011**, *85*, 12334.

469. Tramontano, E. *Mini-Rev. Med. Chem.* **2008**, *8*, 1298.

470. Beaulieu, P. L. *Exp. Opin. Therapeutic Patents* **2009**, *19*, 145.

471.   Watkins, W. J.; Ray, A. S.; Chong, L. S. *Curr. Opin. Drug Discov. Devel.* **2010**, *13*, 441.

472.   Powdrill, M. H.; Bernatchez, J. A.; Gotte, M. *Viruses* **2010**, *2*, 2169.

473.   Legrand-Abravanel, F.; Nicot, F.; Izopet, J. *Exp. Opin. Investig. Drugs* **2010**, *19*, 963.

474.   Beaulieu, P. L.; Bos, M.; Bousquet, Y.; Fazal, G.; Gauthier, J.; Gillard, J.; Goulet, S.; LaPlante, S.; Poupart, M.-A.; Lefebvre, S.; McKercher, G.; Pellerin, C.; Austel, V.; Kukolj, G. *Bioorg. Med. Chem. Lett.* **2004**, *14*, 119.

475.   Beaulieu, P. L.; Boes, M.; Bousquet, Y.; DeRoy, P.; Fazal, G.; Gauthier, J.; Gillard, J.; Goulet, S.; McKercher, G.; Poupart, M.-A.; Valois, S.; Kukolj, G. *Bioorg. Med. Chem. Lett.* **2004**, *14*, 967.

476.   Beaulieu, P. L.; Bousquet, Y.; Gauthier, J.; Gillard, J.; Marquis, M.; McKercher, G.; Pellerin, C.; Valois, S.; Kukolj, G. *J. Med. Chem.* **2004**, *47*, 6884.

477.   LaPlante, S. R.; Jakalian, A.; Aubry, N.; Bousquet, Y.; Ferland, J.-M.; Gillard, J.; Lefebvre, S.; Poirier, M.; Tsantrizos, Y. S.; Kukolj, G.; Beaulieu, P. L. *Angew. Chem. Int. Ed.* **2004**, *43*, 4306.

478.   Ishida, T.; Suzuki, T.; Hirashima, S.; Mizutani, K.; Yoshida, A.; Ando, I.; Ikeda, S.; Adachi, T.; Hashimoto, H. *Bioorg. Med. Chem. Lett.* **2006**, *16*, 1859.

479.   Hirashima, S.; Suzuki, T.; Ishida, T.; Noji, S.; Yata, S.; Ando, I.; Komatsu, M.; Ikeda, S.; Hashimoto, H. *J. Med. Chem.* **2006**, *49*, 4721.

480.   Beaulieu, P. L. *Curr. Opin. Drug Discov. Devel.* **2006**, *9*, 618.

481.   Narjes, F.; Crescenzi, B.; Ferrara, M.; Habermann, J.; Colarusso, S.; del Rosario Rico Ferreira, M.; Stansfield, I.; Mackay, A. C.; Conte, I.; Ercolani, C.; Zaramella, S.; Palumbi, M.-C.; Meuleman, P.; Leroux-Roels, G.; Giuliano, C.; Fiore, F.; Di Marco, S.; Baiocco, P.; Koch, U.; Migliaccio, G.; Altamura, S.; Laufer, R.; De Francesco, R.; Rowley, M. *J. Med. Chem.* **2011**, *54*, 289.

482.   Zeuzem, S.; Asselah, T.; Angus, P.; Zarski, J.-P.; Larrey, D.; Muellhaupt, B.; Gane, E.; Schuchmann, M.; Lohse, A.; Pol, S.; Bronowicki, J.-P.; Roberts, S.; Arasteh, K.; Zoulim, F.; Heim, M.; Stern, J. O.; Kukolj, G.; Nehmiz, G.; Haefner, C.; Boecher, W. O. *Gastroenterology* **2011**, *141*, 2047.

483.   Li, H.; Tatlock, J.; Linton, A.; Gonzalez, J.; Jewell, T.; Patel, L.; Ludlum, S.; Drowns, M.; Rahavendran, S. V.; Skor, H.; Hunter, R.; Shi, S. T.; Herlihy, K. J.; Parge, H.; Hickey, M.; Yu, X.; Chau, F.; Nonomiya, J.; Lewis, C. *J. Med. Chem.* **2009**, *52*, 1255.

484.   Shi, S. T.; Herlihy, K. J.; Graham, J. P.; Nonomiya, J.; Rahavendran, S. V.; Skor, H.; Irvine, R.; Binford, S.; Tatlock, J.; Li, H.; Gonzalez, J.; Linton, A.; Patick, A. K.; Lewis, C. *Antimicrob. Agents Chemother.* **2009**, *53*, 2544.

485.   Wagner, F.; Thompson, R.; Kantaridis, C.; Simpson, P.; Troke, P. J. F.; Jagannatha, S.; Neelakantan, S.; Purohit, V. S.; Hammond, J. L. *Hepatology* **2011**, *54*, 50.

486.   Chan, L.; Das, S. K.; Reddy, T. J.; Poisson, C.; Proulx, M.; Pereira, O.; Courchesne, M.; Roy, C.; Wang, W.; Siddiqui, A.; Yannopoulos, C. G.; Nguyen-Ba, N.; Labrecque, D.; Bethell, R.; Hamel, M.; Courtemanche-Asselin, P.; L'Heureux, L.; David, M.; Nicolas, O.; Brunette, S.; Bilimoria, D.; Bedard, J. *Bioorg. Med. Chem. Lett.* **2004**, *14*, 793.

487.   Chan, L.; Pereira, O.; Reddy, T. J.; Das, S. K.; Poisson, C.; Courchesne, M.; Proulx, M.; Siddiqui, A.; Yannopoulos, C. G.; Nguyen-Ba, N.; Roy, C.; Nasturica, D.; Moinet, C.; Bethell, R.; Hamel, M.; L'Heureux, L.; David, M.; Nicolas, O.; Courtemanche-Asselin, P.; Brunette, S.; Bilimoria, D.; Bedard, J. *Bioorg. Med. Chem. Lett.* **2004**, *14*, 797.

488.   Yi, G.; Deval, J.; Fan, B.; Cai, H.; Soulard, C.; Ranjith-Kumar, C. T.; Smith, D. B.; Blatt, L.; Beigelman, L.; Kao, C. C. *Antimicrob. Agents Chemother.* **2012**, *56*, 830.

489.   Ruebsam, F.; Tran, C. V.; Li, L.-S.; Kim, S. H.; Xiang, A. X.; Zhou, Y.; Blazel, J. K.; Sun, Z.; Dragovich, P. S.; Zhao, J.; McGuire, H. M.; Murphy, D. E.; Tran, M. T.; Stankovic, N.; Ellis, D. A.; Gobbi, A.; Showalter, R. E.; Webber, S. E.; Shah, A. M.; Tsan, M.; Patel, R. A.; LeBrun, L. A.; Hou, H. J.; Kamran, R.; Sergeeva, M. V.; Bartkowski, D. M.; Nolan, T. G.; Norris, D. A.; Kirkovsky, L. *Bioorg. Med. Chem. Lett.* **2009**, *19*, 451.

490.   Ellis, D. A.; Blazel, J. K.; Tran, C. V.; Ruebsam, F.; Murphy, D. E.; Li, L.-S.; Zhao, J.; Zhou, Y.; McGuire, H. M.; Xiang, A. X.; Webber, S. E.; Zhao, Q.; Han, Q.; Kissinger, C. R.; Lardy, M.; Gobbi, A.; Showalter, R. E.; Shah, A. M.; Tsan, M.; Patel, R. A.; LeBrun, L. A.; Kamran, R.;

Bartkowski, D. M.; Nolan, T. G.; Norris, D. A.; Sergeeva, M. V.; Kirkovsky, L. *Bioorg. Med. Chem. Lett.* **2009**, *19*, 6047.

491.   Ruebsam, F.; Murphy, D. E.; Tran, C. V.; Li, L.-S.; Zhao, J.; Dragovich, P. S.; McGuire, H. M.; Xiang, A. X.; Sun, Z.; Ayida, B. K.; Blazel, J. K.; Kim, S. H.; Zhou, Y.; Han, Q.; Kissinger, C. R.; Webber, S. E.; Showalter, R. E.; Shah, A. M.; Tsan, M.; Patel, R. A.; Thompson, P. A.; LeBrun, L. A.; Hou, H. J.; Kamran, R.; Sergeeva, M. V.; Bartkowski, D. M.; Nolan, T. G.; Norris, D. A.; Khandurina, J.; Brooks, J.; Okamoto, E.; Kirkovsky, L. *Bioorg. Med. Chem. Lett.* **2009**, *19*, 6404.

492.   Dragovich, P. S.; Thompson, P. A.; Ruebsam, F. *World Patent Appl.* WO-2010/042834 A1 April 15th, 2010.

493.   Dhanak, D.; Duffy, K. J.; Johnston, V. K.; Lin-Goerke, J.; Darcy, M.; Shaw, A. N.; Gu, B.; Silverman, C.; Gates, A. T.; Nonnemacher, M. R.; Earnshaw, D. L.; Casper, D. J.; Kaura, A.; Baker, A.; Greenwood, C.; Gutshall, L. L.; Maley, D.; DelVecchio, A.; Macarron, R.; Hofmann, G. A.; Alnoah, Z.; Cheng, H.-Y.; Chan, G.; Khandekar, S.; Keenan, R. M.; Sarisky, R. T. *J. Biol. Chem.* **2002**, *277*, 38322.

494.   Tedesco, R.; Shaw, A. N.; Bambal, R.; Chai, D.; Concha, N. O.; Darcy, M. G.; Dhanak, D.; Fitch, D. M.; Gates, A.; Gerhardt, W. G.; Halegoua, D. L.; Han, C.; Hofmann, G. A.; Johnston, V. K.; Kaura, A. C.; Liu, N.; Keenan, R. M.; Lin-Goerke, J.; Sarisky, R. T.; Wiggall, K. J.; Zimmerman, M. N.; Duffy, K. J. *J. Med. Chem.* **2006**, *49*, 971.

495.   Wagner, R.; Larson, D. P.; Beno, D. W. A.; Bosse, T. D.; Darbyshire, J. F.; Gao, Y.; Gates, B. D.; He, W.; Henry, R. F.; Hernandez, L. E.; Hutchinson, D. K.; Jiang, W. W.; Kati, W. M.; Klein, L. L.; Koev, G.; Kohlbrenner, W.; Krueger, A. C.; Liu, J.; Liu, Y.; Long, M. A.; Maring, C. J.; Masse, S. V.; Middleton, T.; Montgomery, D. A.; Pratt, J. K.; Stuart, P.; Molla, A.; Kempf, D. J. *J. Med. Chem.* **2009**, *52*, 1659.

496.   Lu, L.; Dekhtyar, T.; Masse, S.; Pithawalla, R.; Krishnan, P.; He, W.; Ng, T.; Koev, G.; Stewart, K.; Larson, D.; Bosse, T.; Wagner, R.; Pilot-Matias, T.; Mo, H.; Molla, A. *Antiviral Res.* **2007**, *76*, 93.

497.   Chen, C.-M.; He, Y.; Lu, L.; Lim, H. B.; Tripathi, R. L.; Middleton, T.; Hernandez, L. E.; Beno, D. W. A.; Long, M. A.; Kati, W. M.; Bosse, T. D.; Larson, D. P.; Wagner, R.; Lanford, R. E.; Kohlbrenner, W. E.; Kempf, D. J.; Pilot-Matias, T. J.; Molla, A. *Antimicrob. Agents Chemother.* **2007**, *51*, 4290.

498.   Howe, A. Y. M.; Cheng, H.; Johann, S.; Mullen, S.; Chunduru, S. K.; Young, D. C.; Bard, J.; Chopra, R.; Krishnamurthy, G.; Mansour, T.; O'Connell, J. *Antimicrob. Agents Chemother.* **2008**, *52*, 3327.

499.   Kneteman, N. M.; Howe, A. Y. M.; Gao, T.; Lewis, J.; Pevear, D.; Lund, G.; Douglas, D.; Mercer, D. F.; Tyrrell, D. L. J.; Immermann, F.; Chaudhary, I.; Speth, J.; Villano, S. A.; O'Connell, J.; Collett, M. *Hepatology* **2009**, *49*, 745.

500.   Feldstein, A.; Kleiner, D.; Kravetz, D.; Buck, M. *J. Clin. Gastroenterol.* **2009**, *43*, 374.

501.   Vliegen, I.; Paeshuyse, J.; De Burghgraeve, T.; Lehman, L. S.; Paulson, M.; Shih, I.-H.; Mabery, E.; Boddeker, N.; De Clercq, E.; Reiser, H.; Oare, D.; Lee, W. A.; Zhong, W.; Bondy, S.; Puerstinger, G.; Neyts, J. *J. Hepatol.* **2009**, *50*, 999.

502.   Shih, I.-H.; Vliegen, I.; Peng, B.; Yang, H.; Hebner, C.; Paeshuyse, J.; Purstinger, G.; Penaux, M.; Tian, Y.; Mabery, E.; Qi, X.; Bahador, G.; Paulson, M.; Lehman, L. S.; Bondy, S.; Tse, W.; Reiser, H.; Lee, W. A.; Schmitz, U.; Neyts, J.; Zhong, W. *Antimicrob. Agents Chemother.* **2011**, *55*, 4196.

503.   Zeuzem, S.; Buggisch, P.; Agarwal, K.; Marcellin, P.; Sereni, D.; Klinker, H.; Moreno, C.; Zarski, J.-P.; Horsmans, Y.; Mo, H.; Arterburn, S.; Knox, S.; Oldach, D.; McHutchison, J, G.; Manns, M. P., Foster, G. R. *Hepatology*, **2012**, *55*, 749.

504.   Szabo, G. *Gastroenterol.* **2006**, *130*, 995.

505.   Schmitz, U.; Tan, S.-L., *Recent Patents Anti-Infect. Drug Discov.* **2008**, *3*, 77.

506.   Cordek, D. G.; Bechtel, J. T.; Maynard, A. T.; Kazmierski, W. M. Cameron, C. E. *Drugs Future* **2011**, *36*, 691.

507.   Najarro, P.; Mathews, N.; Cockerill, S. *Hepatitis C*. Tan, S.-L.; He, Y. eds., Caister Academic Press: Norwich, UK, **2011**, pp271–292.

508.   Gish, R. G.; Meanwell, N. A. *Clin. Liver Dis.* **2011,** *15,* 627.

509.   Lemm, J. A.; O'Boyle, D., II; Liu, M.; Nower, P. T.; Colonno, R.; Deshpande, M. S.; Snyder, L. B.; Martin, S. W.; St. Laurent, D. R.; Serrano-Wu, M. H.; Romine, J. L.; Meanwell, N. A.; Gao, M. *J. Virol.* **2010,** *84,* 482.

510.   Lemm, J. A.; Leet, J. E.; O'Boyle, D. R., II; Romine, J. L.; Huang, X. S.; Schroeder, D. R.; Alberts, J.; Cantone, J. L.; Sun, J.-H.; Nower, P. T.; Martin, S. W.; Serrano-Wu, M. H.; Meanwell, N. A.; Snyder, L. B.; Gao, M. *Antimicrob. Agents Chemother.* **2011,** *55,* 3795.

511.   Romine, J. L.; St. Laurent, D. R.; Leet, J. E.; Martin, S. W.; Serrano-Wu, M. H.; Yang, F.; Gao, M.; O'Boyle, D. R.; Lemm, J. A.; Sun, J.-H.; Nower, P. T.; Huang, X.; Deshpande, M. S.; Meanwell, N. A.; Snyder, L. B. *ACS Med. Chem. Lett.* **2011,** *2,* 224.

512.   Tellinghuisen, T. L; Marcotrigiano, J.; Rice, C. M. *Nature* **2005,** *435,* 374.

513.   Love, R. A; Brodsky, O.; Hickey, M. J; Wells, P. A; Cronin, C. N. *J. Virol.* **2009,** *83,* 4395.

514.   Nettles, R. E.; Gao, M.; Bifano, M.; Chung, E.; Persson, A.; Marbury, T. C.; Goldwater, R.; DeMicco, M. P.; Rodriguez-Torres, M.; Vutikullird, A.; Fuentes, E.; Lawitz, E.; Lopez-Talavera, J. C.; Grasela, D. M. *Hepatology* **2011,** *54,* 1956.

515.   Fridell, R. A.; Wang, C.; Sun, J.-H.; O'Boyle, D. R., II; Nower, P.; Valera, L.; Qiu, D.; Roberts, S.; Huang, X.; Kienzle, B.; Bifano, M.; Nettles, R. E.; Gao, M. *Hepatology* **2011,** *54,* 1924.

516.   Fridell, R. A.; Qiu, D.; Wang, C.; Valera, L.; Gao, M. *Antimicrob. Agents Chemother.* **2010,** *54,* 3641.

517.   Chayama, K.; Takahashi, S.; Toyoata, J.; Karino, Y.; Ikeda, K.; Ishikawa, H.; Watanabe, H.; McPhee, F.; Hughes, E.; Kumada, H. *Hepatology,* **2012,** *55,* 742.

518.   Zhang, S.; Zhang, W.; Tang, Y. W. *Curr. Infect. Disease Reports* **2011,** *13,* 149.

519.   Hayden, F.; de Jong, M. D. *Antiviral Ther.* **2007,** *12,* 579.

520.   Marcos, M. A.; Camps, M.; Pumarola, T.; Marinex, J. A.; Martinez, E.; Mensa, J.; Garcia, E.; Penarroja, G.; Dambrava, P.; Casas, I.; de Anta, M. T. J.; Torres, A. *Antiviral Ther.* **2006,** *11,* 351.

521.   Pavia, A. T. *Clin. Infect. Dis.* **2011,** *52, Suppl 4,* S284.

522.   Ruuskanen, O.; Lahti, E.; Jennings, L. C.; Murdoch, D. R. *Lancet* **2011,** *377,* 1264.

523.   Burke, R. L.; Vest, K. G.; Eick, A. A.; Sanchez, J. L.; Johns, M. C.; Pavlin, J. A.; Jarman, R. G.; Mothershead, J. L.; Quintana, M.; Palys, T.; Cooper, M. J.; Guan, J.; Schnabel, D.; Waitumbi, J.; Wilma, A.; Daniels, C.; Brown, M. L.; Tobias, S.; Kasper, M. R.; Williams, M.; Tjaden, J. A.; Oyofo, B.; Styles, T.; Blair, P. J.; Hawksworth, A.; Montgomery, J. M.; Razuri, H.; Laguna-Torres, A.; Schoepp, R. J.; Norwood, D. A.; Macintosh, V. H.; Gibbons, T.; Gray, G. C.; Blazes, D. L.; Russell, K. L.; Rubenstein, J.; Hathaway, K.; Gibbons, R.; Yoon, I. K.; Saunders, D.; Gaywee, J.; Stoner, M.; Timmermans, A.; Shrestha, S. K.; Velasco, J. M.; Alera, M. T.; Tannitisupawong, D.; Myint, K. S.; Pichyangkul, S.; Woods, B.; Jerke, K. H.; Koenig, M. G.; Byarugaba, D. K.; Mangen, F. W.; Assefa, B.; Williams, M.; Brice, G.; Mansour, M.; Pimentel, G.; Sebeny, P.; Talaat, M.; Saeed, T.; Espinosa, B.; Faix, D.; Maves, R.; Kochel, T.; Smith, J.; Guerrero, A.; Maupin, G.; Sjoberg, P.; Duffy, M.; Garner, J.; Canas, L.; Macias, E.; Kuschner, R. A.; Shanks, D.; Lewis, S.; Nowak, G.; Ndip, L. M.; Wolfe, N.; Saylors, K. *BMC Public Health* **2011,** *11, Suppl 2,* S6.

524.   Casiano-Colon, A. E.; Hulbert, B. B.; Mayer, T. K.; Walsh, E. E.; Falsey, A. R. *J. Clin. Virol.* **2003,** *28,* 169.

525.   Hoehling, A. A. *The Great Epidemic*; Little, Brown: Boston, MA, 1961.

526.   Griffin, M. R.; Coffey, C. S.; Neuzil, K. M.; Mitchel, E. F., Jr.; Wright, P. F.; Edwards, K. M. *Arch. Intern. Med.* **2002,** *162,* 1229.

527.   Hedlund, M.; Larson, J. L.; Fang, F. *Viruses* **2010,** *2,* 1766.

528.   Palese, P.; Shaw, M. L. *Fields Virology 5th Ed.*, Knipe, D. M.; Howley, P. M., ed., Lippincott Williams and Wilkins, **2007,** Vol. II., Chapter 47, p 1647.

529.   Taubenberger, J. K.; Reid, A. H. *Perspect. Med. Virol.* **2002,** *7,* 101.

530.   Taubenberger, J. K.; Palese, P. *Influenza Virol.* **2006,** 299.

531.   Maugh, T. H. *Science* **1976,** *192,* 128.

532.   Koff, W. C.; Knight, V. *J. Virol.* **1979,** *31,* 261.

533. Monto, A. S.; Ohmit, S. E.; Hornbuckle, K.; Pearce, C. L. *Antimicrob. Agents Chemother.* **1995**, *39*, 2224.
534. Pinto, L. H.; Holsinger, L. J.; Lamb, R. A. *Cell* **1992**, *69*, 517.
535. Hay, A. J. *Semin. Virol.* **1992**, *3*, 21.
536. Helenius, A. *Cell* **1992**, *69*, 577.
537. Zhirnov, O. P. *Virology* **1992**, *186*, 324.
538. Couch, R. B. *New Engl. J. Med.* **1997**, *337*, 927.
539. Hayden, F. G.; Hoffman, H. E.; Spyker, D. A. *Antimicrob. Agents Chemother.* **1983**, *23*, 458.
540. Betts, R. F. *Curr. Opin. Infect. Dis.* **1991**, *4*, 804.
541. Hayden, F. G.; Sperber, S. J.; Belshe, R. B.; Clover, R. D.; Hay, A. J.; Pyke, S. *Antimicrob. Agents Chemother.* **1991**, *35*, 1741.
542. Pabbaraju, K.; Wong, S.; Kits, D. K; Fox, J. D. *Can. J. Infect. Dis. Med. Microbiol.* **2010**, *21*, e87.
543. Lan, Y.; Zhang, Y.; Dong, L.; Wang, D.; Huang, W.; Xin, L.; Yang, L.; Zhao, X.; Li, Z.; Wang, W.; Li, X.; Xu, C.; Yang, L.; Guo, J.; Wang, M.; Peng, Y.; Gao, Y.; Guo, Y.; Wen, L.; Jiang, T.; Shu, Y. *Antiviral Ther.* **2010**, *15*, 853.
544. Poland, G. A.; Jacobson, R. M.; Targonski, P. V. *Vaccine* **2007**, *25*, 3057.
545. He, G.; Qiao, J.; Dong, C.; He, C.; Zhao, L.; Tian, Y. *Antiviral Res.* **2008**, *77*, 72.
546. Cyranoski, D. *Nature Med.* **2005**, *11*, 909.
547. Gottschalk, A. *Arch. Biochem. Biophys.* **1957**, *69*, 37.
548. Varghese, J. N.; Laver, W. G.; Colman, P. M. *Nature* **1983**, *303*, 35.
549. Varghese, J. N.; McKimm-Breschkin, J. L.; Caldwell, J. B.; Kortt, A. A.; Colman, P. M. *Proteins* **1992**, *14*, 327.
550. Burnet, F. M. *Austral. J Exp. Biol. Med. Sci.* **1948**, 26, 389.
551. Klenk, H. D.; Rott, R. *Adv. Virus Res.* **1988**, *34*, 247.
552. Meindl, P.; Bodo, G.; Palese, P.; Schulman, J.; Tuppy, H. *Virology* **1974**, *58*, 457.
553. Palese, P.; Schulman, J. L.; Bodo, G.; Meindl, P. *Virology* **1974**, *59*, 490.
554. Colman, P. M.; Hoyne, P. A.; Lawrence, M. C. *J. Virol.* **1993**, *67*, 2972.
555. Burmeister, W. P.; Ruigrok, R. W. H.; Cusack, S. *EMBO J.* **1992**, *11*, 49.
556. Dyason, J. C.; von Itzstein, M. *Australian J. Chem.* **2001**, *54*, 663.
557. Burmeister, W. P.; Henrissat, B.; Bosso, C.; Cusack, S.; Ruigrok, R. W. *Structure* **1993**, *1*, 19.
558. Bossart-Whitaker, P.; Carson, M.; Babu, Y. S.; Smith, C. D.; Laver, W. G.; Air, G. M. *J. Mol. Biol.* **1993**, *232*, 1069.
559. Holzer, C. T.; von Itzstein, M.; Jin, B.; Pegg, M. S.; Stewart, W. P.; Wu, W. Y. *Glycoconjugate J.* **1993**, *10*, 40.
560. Woods, J. M.; Bethell, R. C.; Coates, J. A.; Healy, N.; Hiscox, S. A.; Pearson, B. A.; Ryan, D. M.; Ticehurst, J.; Tilling, J. *Antimicrob. Agents Chemother.* **1993**, *37*, 1473.
561. Thomas, G. P.; Forsyth, M.; Penn, C. R.; McCauley, J. W. *Antiviral Res.* **1994**, *24*, 351.
562. Taylor, N. R.; von Itzstein, M. *J. Med. Chem.* **1994**, *37*, 616.
563. Ryan, D. M.; Ticehurst, J.; Dempsey, M. H. *Antimicrob. Agents Chemother.* **1995**, *39*, 2583.
564. Ryan, D. M.; Ticehurst, J.; Dempsey, M. H.; Penn, C. R. *Antimicrob. Agents Chemother.* **1994**, *38*, 2270.
565. Hayden, F. G.; Osterhaus, A. D.; Treanor, J. J.; Fleming, D. M.; Aoki, F. Y.; Nicholson, K. G.; Bohnen, A. M.; Hirst, H. M.; Keene, O.; Wightman, K.; Adam, G.; Van Pottelbergy, H.; Godefroid, M.; Ruuskanen, O.; Makela, M.; Simmons, A.; Luciani, J.; Richir, J.; Cozic, J. A.; Behar, M.; Lehm, R.; Pregliasco, F.; Crovari, P.; De Groot, R.; Rothbarth, P.; Claas, E.; Rimmelzwaan, G.; van der Woude, J. C.; Hauge, H.; Hercz, I. Innvik, K.; Haegde Naess, M.; Sand, O.; Tomassen T.; Jane, C.; Bordas, J. M.; Palomo, M.; Alonso, M.; Vellaneuva, M. A.; Sandberg, T.; Ahlm, C.; Pauksens, K.; Glimaker, M.; Duffy, M. F.; Khan, D. S. A.; Orr, P.; Clecner, B.; Dylewski, J.; Forward, K.; Miedzinski, L. J.; Walmsley S.; Williams, K.; Zoutman, D. E.; Adelglass, J.; Andruczk, R. C.; Becker, S.; Campbell, S.; Elinoff, V. A.; Fischer, L. A.; Gelderman, L. I.; Graff, A.; Lobo, M.; Henry, D.; Howard, T. M.; Kerzner, B.; Williams, K. H.; Kobayashi, R.; Mikolich, D. J.; O'Rourke, J.; Ribino, J.; Pace, S. A.; Payne, L. E.; Post, G. L.; Puopolo, A.; Rhudy, J.; Ruoff, G.

E.; Ryder-Benz, J.; Schoenberger, J.; Serfer, H. M.; Sklar, B. M.; Sun, W.; Whitlock, W. *New Engl. J. Med.* **1997,** *337,* 874.

566.   Hayden, F. G.; Gubareva, L. V.; Monto, A. S.; Klein, T. C.; Elliot, M. J.; Hammond, J. M.; Sharp, S. J.; Ossi, M. J.; Bremner, A. D.; Broker, R.; Fleming, D. M.; Henry, D.; Herlocher, L.; Keeney, R.; Perry, J.; Pichichero, M.; Ruuskanen, O.; Stoltz, R.; Turner, R.; VanHook, C. *New Engl. J. Med.* **2000,** *343,* 1282.

567.   Lew, W.; Chen, X.; Kim, C. U. *Curr. Med. Chem.* **2000,** *7,* 663.

568.   Mendel, D. B.; Tai, C. Y.; Escarpe, P. A.; Li, W.; Sidwell, R. W.; Huffman, J. H.; Sweet, C.; Jakeman, K. J.; Merson, J.; Lacy, S. A.; Lew, W.; Williams, M. A.; Zhang, L.; Chen, M. S.; Bischofberger, N.; Kim, C. U. *Antimicrob. Agents Chemother.* **1998,** *42,* 640.

569.   Li, W.; Escarpe, P. A.; Eisenberg, E. J.; Cundy, K. C.; Sweet, C.; Jakeman, K. J.; Merson, J.; Lew, W.; Williams, M.; Zhang, L.; Kim, C. U.; Bischofberger, N.; Chen, M. S.; Mendel, D. B. *Antimicrob. Agents Chemother.* **1998,** *42,* 647.

570.   Eisenberg, E. J.; Bidgood, A.; Cundy, K. C. *Antimicrb. Agents Chemother.* **1997,** *41,* 1949.

571.   Li, K. S.; Guan, Y.; Wang, J.; Smith, G. J.; Xu, K. M.; Duan, L.; Rahardjo, A. P.; Puthavathana, P.; Buranathai, C.; Nguyen, T. D.; Estoepangestie, A. T.; Chaisingh, A.; Auewarakul, P.; Long, H. T.; Hanh, N. T.; Webby, R. J.; Poon, L. L.; Chen, H.; Shortridge, K. F.; Yuen, K. Y.; Webster, R. G.; Peiris, J. S. *Nature* **2004,** *430,* 209.

572.   Webby, R. J.; Webster, R. G. *Science* **2003,** *302,* 1519.

573.   Cinti, S.; Chenoweth, C.; Monto, A. S. *Infect. Control Hos. Epidemiol.* **2005,** *26,* 852.

574.   Ward, P.; Small, I.; Smith, J.; Suter, P.; Dutkowski, R. *J. Antimicrob. Chemother.* **2005,** *55 Suppl 1,* i5.

575.   Li, Y.; Hsu, E. B.; Links, J. M. *Biosecurity and Bioterrorism: Biodefense Strategy, Practice, and Science* **2010,** 8, 119.

576.   Trampuz, A.; Prabhu, R. M.; Smith, T. F.; Baddour, L. M. *Mayo Clinic Proc.* **2004,** *79,* 523.

577.   Myers, J. L.; Hensley, S. E. *Exp. Rev. Anti-infect. Ther.* **2011,** *9,* 385.

578.   Normile, D.; Enserink, M. *Science* **2007,** *315,* 448.

579.   Leung, Y. H.; Lim, W. L.; Wong, M. H.; Chuang, S. K. *Epidemiol. Infect.* **2011,** *139,* 41.

580.   Meijer, A.; Jonges, M.; Abbink, F.; Ang, W.; van Beek, J.; Beersma, M.; Bloembergen, P.; Boucher, C.; Claas, E.; Donker, G.; van Gageldonk-Lafeber, R.; Isken, L.; de Jong, A.; Kroes, A.; Leenders, S.; van der Lubben, M.; Mascini, E.; Niesters, B.; Oosterheert, J. J.; Osterhaus, A.; Riesmeijer, R.; Riezebos-Brilman, A.; Schutten, M.; Sebens, F.; Stelma, F.; Swaan, C.; Timen, A.; van 't Veen, A.; van der Vries, E.; te Wierik, M.; Koopmans, M. *Antiviral Res.* **2011,** *92,* 81.

581.   Krystal, M. R.; Meanwell, N. A. *Curr. Opin. Investig. Drugs* **2009,** *10,* 746.

582.   Babu, Y. S.; Chand, P.; Bantia, S.; Kotian, P.; Dehghani, A.; El-Kattan, Y.; Lin, T. H.; Hutchison, T. L.; Elliott, A. J.; Parker, C. D.; Ananth, S. L.; Horn, L. L.; Laver, G. W.; Montgomery, J. A. *J. Med. Chem.* **2000,** *43,* 3482.

583.   Chand, P.; Kotian, P. L.; Dehghani, A.; El-Kattan, Y.; Lin, T.-H.; Hutchison, T. L.; Babu, Y. S.; Bantia, S.; Elliott, A. J.; Montgomery, J. A. *J. Med. Chem.* **2001,** *44,* 4379.

584.   Yamamoto, T.; Kumazawa, H.; Inami, K.; Teshima, T.; Shiba, T. *Tetrahedron Lett.* **1992,** *33,* 5791.

585.   Burmeister, W. P.; Ruigrok, R. W.; Cusack, S. *EMBO J.* **1992,** *11,* 49.

586.   Gubareva, L. V.; Kaiser, L.; Hayden, F. G. *Lancet* **2000,** *355,* 827.

587.   Sidwell, R. W.; Smee, D. F. *Exp. Opin. Investig. Drugs* **2002,** *11,* 859.

588.   Barroso, L.; Treanor, J.; Gubareva, L.; Hayden, F. G. *Antiviral Ther.* **2005,** *10,* 901.

589.   Mancuso, C. E.; Gabay, M. P.; Steinke, L. M.; Vanosdol, S. J. *Annals Pharmacother.* **2010,** *44,* 1240.

590.   Ikematsu, H.; Kawai, N. *Exp. Rev. Anti-infect. Ther.* **2011,** *9,* 851.

591.   Yamashita, M.; Tomozawa, T.; Kakuta, M.; Tokumitsu, A.; Nasu, H.; Kubo, S. *Antimicrob. Agents Chemother.* **2009,** *53,* 186.

592.   Vavricka, C. J.; Li, Q.; Wu, Y.; Qi, J.; Wang, M.; Liu, Y.; Gao, F.; Liu, J.; Feng, E.; He, J.; Wang, J.; Liu, H.; Jiang, H.; Gao, G. F. *PLoS Pathogens* **2011,** *7,* e1002249.

593. Kubo, S.; Tomozawa, T.; Kakuta, M.; Tokumitsu, A.; Yamashita, M. *Antimicrob. Agents Chemother,* **2010,** *54,* 1256.
594. Ishizuka, H.; Yoshiba, S.; Okabe, H.; Yoshihara, K. *J. Clin. Pharmacol.* **2010,** *50,* 1319.
595. Blount, R. E., Jr.; Morris, J. A.; Savage, R. E. *Proc. Soc. Experimental Biol. Med.* **1956,** *92,* 544.
596. Chanock, R.; Finberg, L. *Am. J. Hyg.* **1957,** *66,* 291.
597. Chanock, R.; Roizman, B.; Myers, R. *Am. J. Hyg.* **1957,** *66,* 281.
598. Chanock, R. M.; Parrott, R. H. *Pediatrics* **1965,** *36,* 21.
599. Chanock, R. M.; Kim, H. W.; Vargosko, A. J.; Deleva, A.; Johnson, K. M.; Cumming, C.; Parrott, R. H. *JAMA* **1961,** *176,* 647.
600. Englund, J. A.; Sullivan, C. J.; Jordan, M. C.; Dehner, L. P.; Vercellotti, G. M.; Balfour, H. H. J. *Ann. Intern. Med.* **1988,** *109,* 203.
601. Falsey, A. R.; Cunningham, C. K.; Barker, W. H.; Kouides, R. W.; Yuen, J. B.; Menegus, M.; Weiner, L. B.; Bonville, C. A.; Betts, R. F. *J. Infect. Dis.* **1995,** *172,* 389.
602. Garcia, R.; Raad, I.; Abi-Said, D.; Bodey, G.; Champlin, R.; Tarrand, J.; Hill, L. A.; Umphrey, J.; Neumann, J.; Englund, J.; Whimbey, E. *Infect. Control Hosp. Epidemiol.* **1997,** *18,* 412.
603. Witkowski, J. T.; Robins, R. K.; Sidwell, R. W.; Simon, L. N. *J. Med. Chem.* **1972,** *15,* 1150.
604. Sidwell, R. W.; Huffman, J. H.; Khare, G. P.; Allen, L. B.; Witkowski, J. T.; Robins, R. K. *Science* **1972,** *177,* 705.
605. Prusiner, P.; Sundaralingam, M. *Nature: New Biol.* **1973,** *244,* 116.
606. Graci, J. D.; Cameron, C. E. *Rev. Med. Virol.* **2006,** *16,* 37.
607. Krilov, L. *Exp. Opin. Ther. Patents* **2002,** *12,* 441.
608. Levin, M. J. *J. Pediatrics* **1994,** *124,* S22.
609. Randolph, A. G.; Wang, E. E. *Arch. Pediatrics Adolescent Med.* **1996,** *150,* 942.
610. Ferm, V. H.; Willhite, C.; Kilham, L. *Teratology* **1978,** *17,* 93.
611. Kochhar, D. M.; Penner, J. D.; Knudsen, T. B. *Toxicol. Appl. Pharmacol.* **1980,** *52,* 99.
612. Jain, M. K.; Zoellner, C. *Exp. Opin. Pharmacother.* **2010,** *11,* 673.
613. Lamprecht, C. L.; Krause, H. E.; Mufson, M. A. *J. Infect. Dis.* **1976,** *134,* 211.
614. Kasel, J. A.; Walsh, E. E.; Frank, A. L.; Baxter, B. D.; Taber, L. H.; Glezen, W. P. *Viral Immunol.* **1987,** *1,* 199.
615. Prince, G. A.; Hemming, V. G.; Horswood, R. L.; Baron, P. A.; Chanock, R. M. *J. Virol.* **1987,** *61,* 1851.
616. Prince, G. A.; Hemming, V. G.; Horswood, R. L.; Chanock, R. M. *Virus Res.* **1985,** *3,* 193.
617. Hemming, V. G.; Prince, G. A.; London, W. T.; Baron, P. A.; Brown, R.; Chanock, R. M. *Antimicrob. Agents Chemother.* **1988,** *32,* 1269.
618. Hemming, V. G.; Rodriguez, W.; Kim, H. W.; Brandt, C. D.; Parrott, R. H.; Burch, B.; Prince, G. A.; Baron, P. A.; Fink, R. J.; Reaman, G. *Antimicrob. Agents Chemother.* **1987,** *31,* 1882.
619. Siber, G. R.; Leszcynski, J.; Pena-Cruz, V.; Ferren-Gardner, C.; Anderson, R.; Hemming, V. G.; Walsh, E. E.; Burns, J.; McIntosh, K.; Gonin, R.; Anderson, L. J. *J. Infect. Dis.* **1992,** *165,* 456.
620. Groothuis, J. R.; Simoes, E. A.; Levin, M. J.; Hall, C. B.; Long, C. E.; Rodriguez, W. J.; Arrobio, J.; Meissner, H. C.; Fulton, D. R.; Welliver, R. C.; Tristram, D. A., Siber, G. R.; Prince, G. A.; Van Raden, M.; Hemming, V. G.; the Respiratory Syncytial Virus Immune Globulin Study Group. *New Engl. J. Med.* **1993,** *329,* 1524.
621. Conner, E.; and the PREVENT Study Group. *Pediatrics* **1997,** *99,* 93.
622. Groothuis, J. R.; Hoopes, J. M.; Hemming, V. G. *Adv. Ther.* **2011,** *28,* 110.
623. Wright, M.; Piedimonte, G. *Pediat. Pulmonol.* **2010,** *46,* 324.
624. Walsh, E. E.; Brandriss, M. W.; Schlesinger, J. J. *J. Gen. Virol.* **1987,** *68,* 2169.
625. Levine, S.; Klaiber-Franco, R.; Paradisco, P. R. *J. Gen. Virol.* **1987,** *68,* 2521.
626. Walsh, E. E.; Hruska, J. *J. Virol.* **1983,** *47,* 171.
627. Johnson, P. R., Jr.; Olmsted, R. A.; Prince, G. A.; Murphy, B. R.; Alling, D. W.; Walsh, E. E.; Collins, P. L. *J. Virol.* **1987,** *61,* 3163.
628. Johnson, P. R.; Collins, P. L. *J. Gen. Virol.* **1988,** *69,* 2901.
629. Fernie, B. F.; Cote, P. J., Jr.; Gerin, J. L. *Proc. Soc. Exp. Biol. Med.* **1982,** *171,* 266.

630. Beeler, J. A.; Coelingh, K. *J. Virol.* **1989**, *63*, 2941.
631. Mufson, M. A.; Belshe, R. B.; Orvell, C.; Norrby, E. *J. Clin. Microbiol.* **1987**, *25*, 1535.
632. Johnson, S.; Oliver, C.; Prince, G. A.; Hemming, V. G.; Pfarr, D. S.; Wang, S. C.; Dormitzer, M.; O'Grady, J.; Koenig, S.; Tamura, J. K.; Woods, R.; Bansal, G.; Couchenour, D.; Tsao, E.; Hall, W. C.; Young, J. F. *J. Infect. Dis.* **1997**, *176*, 1215.
633. Storch, G. A. *Pediatrics* **1998**, *102*, 648.
634. Hegele, R. G. *Eur. Respir. J.* **2011**, *38*, 246.
635. Moler, F. W.; Brown, R. W.; Faix, R. G.; Gilsdorf, J. R. *Pediatrics* **1999**, *103*, 495.
636. Eppes, S. C. *Pediatrics* **1999**, *103*, 534.
637. Hu, J.; Robinson, J. L. *World J. Pediatr.* **2010**, *6*, 296.
638. Wyde, P. R. *Antiviral Res.* **1998**, *39*, 63.
639. Meanwell, N. A.; Krystal, M. *Drug Discov. Today* **2000**, *5*, 241.
640. Prince, G. A. *Exp. Opin. Investig. Drugs* **2001**, *10*, 297.
641. Douglas, J. L. *Exp. Rev. Anti-Infect. Ther.* **2004**, *2*, 625.
642. Sidwell, R. W.; Barnard, D. L. *Antiviral Res.* **2006**, *71*, 379.
643. Carter, M.; Cockerill, G. S. *Annual Rep. Med. Chem.* **2008**, *43*, 229.
644. Meanwell, N. A.; Krystal, M. *Drugs Future* **2007**, *32*, 441.
645. Bonfanti, J.-F.; Roymans, D. *Curr. Opin. Drug Discov. Develop.* **2009**, *12*, 479.
646. Dubovi, E. J.; Geratz, J. D.; Tidwell, R. R. *Virology* **1980**, *103*, 502.
647. Dubovi, E. J.; Geratz, J. D.; Shaver, S. R.; Tidwell, R. R. *Antimicrob. Agents Chemother.* **1981**, *19*, 649.
648. Tidwell, R. R.; Geratz, J. D.; Dubovi, E. J. *J. Med. Chem.* **1983**, *26*, 294.
649. Tidwell, R. R.; Geratz, J. D.; Clyde, W. A., Jr.; Rosenthal, K. U.; Dubovi, E. J. *Antimicrob. Agents Chemother* **1984**, *26*, 591.
650. Cianci, C.; Meanwell, N.; Krystal, M. *J. Antimicrob. Chemother.* **2005**, *55*, 289.
651. Cianci, C.; Yu, K.-L.; Combrink, K.; Sin, N.; Pearce, B.; Wang, A.; Civiello, R.; Voss, S.; Luo, G.; Kadow, K.; Genovesi, E. V.; Venables, B.; Gulgeze, H.; Trehan, A.; James, J.; Lamb, L.; Medina, I.; Roach, J.; Yang, Z.; Zadjura, L.; Colonno, R.; Clark, J.; Meanwell, N.; Krystal, M. *Antimicrob. Agents Chemother.* **2004**, *48*, 413.
652. Cianci, C.; Genovesi, E. V.; Lamb, L.; Medina, I.; Yang, Z.; Zadjura, L.; Yang, H.; D'Arienzo, C.; Sin, N.; Yu, K.-L.; Combrink, K.; Li, Z.; Colonno, R.; Meanwell, N.; Clark, J.; Krystal, M. *Antimicrob. Agents Chemother.* **2004**, *48*, 2448.
653. Cianci, C.; Langley, D. R.; Dischino, D. D.; Sun, Y.; Yu, K.-L.; Stanley, A.; Roach, J.; Li, Z.; Dalterio, R.; Colonno, R.; Meanwell, N. A.; Krystal, M. *Proc. Natl. Acad. Sci. USA* **2004**, *101*, 15046.
654. Yu, K.-L.; Zhang, Y.; Civiello, R. L.; Kadow, K. F.; Cianci, C.; Krystal, M.; Meanwell, N. A. *Bioorg. Med. Chem. Lett.* **2003**, *13*, 2141.
655. Yu, K.-L.; Zhang, Y.; Civiello, R. L.; Trehan, A. K.; Pearce, B. C.; Yin, Z.; Combrink, K. D.; Gulgeze, H. B.; Wang, X. A.; Kadow, K. F.; Cianci, C. W.; Krystal, M.; Meanwell, N. A. *Bioorg. Med. Chem. Lett.* **2004**, *14*, 1133.
656. Yu, K.-L,; Wang, X. A.; Civiello, R. L.; Trehan, A. K.; Pearce, B. C.; Yin, Z.; Combrink, K. D.; Gulgeze, H. B.; Zhang, Y.; Kadow, K. F.; Cianci, C. W.; Clarke, J.; Genovesi, E. V.; Medina, I.; Lamb, L.; Wyde, P. R.; Krystal, M.; Meanwell, N. A. *Bioorg. Med. Chem. Lett.* **2006**, *16*, 1115.
657. Yu, K.-L.; Sin, N.; Civiello, R. L.; Wang, X. A.; Combrink, K. D.; Gulgeze, H. B.; Venables, B. L.; Wright, J. J. K.; Dalterio, R. A.; Zadjura, L.; Marino, A.; Dando, S.; D'Arienzo, C.; Kadow, K. F.; Cianci, C. W.; Li, Z.; Clarke, J.; Genovesi, E. V.; Medina, I.; Lamb, L.; Colonno, R. J.; Yang, Z.; Krystal, M.; Meanwell, N. A. *Bioorg. Med. Chem. Lett.* **2007**, *17*, 895.
658. Wang, X. A.; Cianci, C. W.; Yu, K.-L; Combrink, K. D.; Thuring, J. W.; Zhang, Y.; Civiello, R. L.; Kadow, K. F.; Roach, J.; Li, Z.; Langley, D. R.; Krystal, M.; Meanwell, N. A. *Bioorg. Med. Chem. Lett.* **2007**, *17*, 4592.
659. Moss, R. A. *Acc. Chem. Res.* **2006**, *39*, 267.
660. Dubinsky, L.; Krom, B. P.; Meijler, M. M. *Bioorg. Med. Chem.* **2012**, *20*, 554.

661. Das, J. *Chem. Rev.* **2011,** *111,* 4405.
662. Hashimoto, M.; Hatanaka, Y. *Eur. J. Org. Chem.* **2008,** 2513.
663. Ding, W.-D.; Mitsner, B.; Krishnamurthy, G.; Aulabaugh, A.; Hess, C. D.; Zaccardi, J.; Cutler, M.; Feld, B.; Gazumyan, A.; Raifeld, Y.; Nikitenko, A.; Lang, S. A.; Gluzman, Y.; O'Hara, B.; Ellestad, G. A. *J. Med. Chem.* **1998,** *41,* 2671.
664. Aulabaugh, A.; Ding, W.-D.; Ellestad, G. A.; Gazumyan, A.; Hess, C. D.; Krishnamurthy, G.; Mitsner, B.; Zaccardi, J. *Drugs Future* **2000,** *25,* 287.
665. Gazumyan, A.; Mitsner, B.; Ellestad, G. A. *Curr. Pharm. Des.* **2000,** *6,* 525.
666. Nikitenko, A. A.; Raifeld, Y. E.; Wang, T. Z. *Bioorg. Med. Chem. Lett.* **2001,** *11,* 1041.
667. Razinkov, V.; Gazumyan, A.; Nikitenko, A.; Ellestad, G.; Krishnamurthy, G. *Chem. Biol.* **2001,** *8,* 645.
668. Huntley, C. C.; Weiss, W. J.; Gazumyan, A.; Buklan, A.; Feld, B.; Hu, W.; Jones, T. R.; Murphy, T.; Nikitenko, A. A.; O'Hara, B.; Prince, G.; Quartuccio, S.; Raifeld, Y. E.; Wyde, P.; O'Connell, J. F. *Antimicrob. Agents Chemother.* **2002,** *46,* 841.
669. Weiss, W. J.; Murphy, T.; Lynch, M. E.; Frye, J.; Buklan, A.; Gray, B.; Lenoy, E.; Mitelman, S.; O'Connell, J.; Quartuccio, S.; Huntley, C. *J. Med. Primatol.* **2003,** *32,* 82.
670. Razinkov, V.; Huntley, C.; Ellestad, G.; Krishnamurthy, G. *Antiviral Res.* **2002,** 55, 189.
671. McKimm-Breschkin, J. *Curr. Opin. Investig. Drugs* **2000,** *1,* 425.
672. Douglas, J. L.; Panis, M. L.; Ho, E.; Lin, K.-Y.; Krawczyk, S. H.; Grant, D. M.; Cai, R.; Swaminathan, S.; Cihlar, T. *J. Virol.* **2003,** *77,* 5054.
673. Douglas, J. L.; Panis, M. L.; Ho, E.; Lin, K.-Y.; Krawczyk, S. H.; Grant, D. M.; Cai, R.; Swaminathan, S.; Chen, X.; Cihlar, T. *Antimicrob. Agents Chemother.* **2005,** *49,* 2460.
674. Wyde, P. R.; Laquerre, S.; Chetty, S. N.; Gilbert, B. E.; Nitz, T. J.; Pevear, D. C. *Antiviral Res.* **2005,** *68,* 18.
675. Bonfanti, J.-F.; Ispas, G.; Van Velsen, F.; Olszewska, W.; Gevers, T.; Roymans, D. In *Antiviral Drugs: From Basic Discovery Through Clinical Trials,* Kazmierski, W. M., Ed. Wiley: Hoboken, NJ, **2011,** pp341–352.
676. Bonfanti, J.-F.; Doublet, F.; Fortin, J.; Lacrampe, J.; Guillemont, J.; Muller, P.; Queguiner, L.; Arnoult, E.; Gevers, T.; Janssens, P.; Szel, H.; Willebrords, R.; Timmerman, P.; Wuyts, K.; Janssens, F.; Sommen, C.; Wigerinck, P.; Andries, K. *J. Med. Chem.* **2007,** *50,* 4572.
677. Bonfanti, J.-F.; Meyer, C.; Doublet, F.; Fortin, J.; Muller, P.; Queguiner, L.; Gevers, T.; Janssens, P.; Szel, H.; Willebrords, R.; Timmerman, P.; Wuyts, K.; van Remoortere, P.; Janssens, F.; Wigerinck, P.; Andries, K. *J. Med. Chem.* **2008,** *51,* 875.
678. Rouan, M.-C.; Gevers, T.; Roymans, D.; de Zwart, L.; Nauwelaers, D.; De Meulder, M.; van Remoortere, P.; Vanstockem, M.; Koul, A.; Simmen, K.; Andries, K. *Antimicrob. Agents Chemother.* **2010,** *54,* 4534.
679. Olszewska, W.; Ispas, G.; Schnoeller, C.; Sawant, D.; Van de Casteele, T.; Nauwelaers, D.; Van Kerckhove, B.; Roymans, D.; De Meulder, M.; Rouan, M. C.; Van Remoortere, P.; Bonfanti, J. F.; Van Velsen, F.; Koul, A.; Vanstockem, M.; Andries, K.; Sowinski, P.; Wang, B.; Openshaw, P.; Verloes, R. *Eur. Respiratory J.* **2011,** *38,* 401.
680. Roymans, D.; De Bondt, H. L.; Arnoult, E.; Geluykens, P.; Gevers, T.; Van Ginderen, M.; Verheyen, N.; Kim, H.; Willebrords, R.; Bonfanti, J.-F.; Bruinzeel, W.; Cummings, M. D.; Van Vlijmen, H.; Andries, K. *Proc. Natl. Acad. Sci. USA* **2010,** *107,* 308.
681. Henderson, E. A.; Alber, D. G.; Baxter, R. C.; Bithell, S. K.; Budworth, J.; Carter, M. C.; Chubb, A.; Cockerill, G. S.; Dowdell, V. C. L.; Fraser, I. J.; Harris, R. A.; Keegan, S. J.; Kelsey, R. D.; Lumley, J. A.; Stables, J. N.; Weerasekera, N.; Wilson, L. J.; Powell, K. L. *J. Med. Chem.* **2007,** *50,* 1685.
682. Chapman, J.; Abbott, E.; Alber, D. G.; Baxter, R. C.; Bithell, S. K.; Henderson, E. A.; Carter, M. C.; Chambers, P.; Chubb, A.; Cockerill, G. S.; Collins, P. L.; Dowdell, V. C. L.; Keegan, S. J.; Kelsey, R. D.; Lockyer, M. J.; Luongo, C.; Najarro, P.; Pickles, R. J.; Simmonds, M.; Taylor, D.; Tyms, S.; Wilson, L. J.; Powell, K. L. *Antimicrob. Agents Chemother.* **2007,** *51,* 3346.

683. El Omari, K.; Scott, K.; Dhaliwal, B.; Ren, J.; Abrescia, N. G. A.; Budworth, J.; Lockyer, M.; Powell, K. L.; Hawkins, A. R.; Stammers, D. K. *Acta Crystallographica, Sec. F: Struct. Biol. Crystallization Commun.* **2008,** *F64,* 1019.

684. Siakallis, G.; Spandidos, D. A.; Sourvinos, G. *Antiviral Ther.* **2009,** *14,* 1051.

685. Dropulic, L. K.; Cohen, J. I. *Clin. Pharmacol. Ther.* **2010,** *88,* 610.

686. Forsgren, M.; Klapper, P. E. *Principles and Practice of Clinical Virology* 6th Ed., **2009,** 95–131, Wiley: Chichester, UK, 2009, pp. 95-131.

687. Roizman, B.; Knipe, D. M.; Whitley, R. J. *Fields Virology 5th Ed.*, Knipe, D. M.; Howley, P. M., ed., Lippincott Williams and Wilkins, **2007,** Vol. II., Chapter 67, pp 2501.

688. Xu, F.; Sternberg, M. R.; Kottiri, B. J.; McQuillan, G. M.; Lee, F. K.; Nahmias, A. J.; Berman, S. M.; Markowitz, L. E. *JAMA* **2006,** *296,* 964.

689. Brady, R. C.; Bernstein, D. I. *Antiviral Res.* **2004,** *61,* 73.

690. Kleymann, G. *Exp. Opin. Investig. Drugs* **2005,** *14,* 135.

691. Simpson, D.; Lyseng-Williamson, K. A. *Drugs* **2006,** *66,* 2397.

692. Vere Hodge, R. A. *Antiviral Chem. Chemother.* **1993,** *4,* 67.

693. Chakrabarty, A.; Tyring, S. K.; Beutner, K.; Rauser, M. *Antiviral Chem. Chemother.* **2004,** *15,* 251.

694. Coen, D. M.; Schaffer, P. A. *Nature Rev. Drug Discov.* **2003,** *2,* 278.

695. Mamidyala, S. K.; Firestine, S. M. *Expert Opin. Ther. Patents* **2006,** *16,* 1463.

696. Quinlivan, M.; Breuer, J. *Rev. Med. Virol.* **2006,** *16,* 225.

697. Partridge, D. G.; McKendrick, M. W. *Exp. Opin. Pharmacother.* **2009,** *10,* 797.

698. Visalli, R. J. *Expert Opin. Ther. Patents* **2004,** *14,* 355.

699. Breuer, J. *J. Clin. Pathol.* **2001,** *54,* 743.

700. McGuigan, C; Barucki, H.; Blewett, S.; Carangio, A.; Erichsen, J. T.; Andrei, G.; Snoeck, R.; De Clercq, E.; Balzarini, J. *J. Med. Chem.* **2000,** *43,* 4993.

701. McGuigan, C.; Brancale, A.; Barucki, H.; Srinivasan, S.; Jones, G.; Pathirana, R.; Carangio, A.; Blewett, S.; Luoni, G.; Bidet, O.; Jukes, A.; Jarvis, C. Andrei, G.; Snoeck, R.; De Clercq, E.; Balzarini, J. *Antiviral Chem. Chemother.* **2001,** *12,* 77.

702. Balzarini, J.; McGuigan, C. *J. Antimicrob. Chemother.* **2002,** *50,* 5.

703. McGuigan, C.; Balzarini, J. *Antiviral Res.* **2006,** *71,* 149.

704. McGuigan, C.; Pathirana, R. N.; Migliore, M.; Adak, R.; Luoni, G.; Jones, A. T.; Díez-Torrubia, A.; Camarasa, M.-J.; Velázquez, S.; Henson, G.; Verbeken, E.; Sienaert, R.; Naesens, L.; Snoeck, R.; Andrei, G. Balzarini, J. *J. Antimicrob. Chemother.* **2007,** *60,* 1316.

705. Sienaert, R.; Naessens, L.; Brancale, A.; De Clercq, E.; McGuigan, C.; Balzarini, J. *Mol. Pharm.* **2002,** *61,* 249.

706. De Clercq, E. *J. Antimicrob. Chemother.* **2003,** *51,* 1079.

707. Boeckh, M.; Geballe, A. P. *J. Clin. Investig.* **2011,** *121,* 1673.

708. Emery, V.C.; Hassan-Walker, A. F. *Drugs* **2002,** *62,* 1853.

709. Matthews, T.; Boehme, R. *Rev. Infect. Dis.* **1988,** *10* (Suppl. 3), S490.

710. McGavin, J. K.; Goa, K. L. *Drugs* **2001,** *61,* 1153.

711. Crumpacker, C. S. *New Engl. J. Med.* **1996,** *335,* 721.

712. Fromtling, R. A.; Castaner, J. *Drugs Future* **1996,** *21,* 1003.

713. De Clercq, E. *Antiviral Res.* **2002,** *55,* 1.

714. Plosker, G. L.; Noble, S. *Drugs* **1999,** *58,* 325.

715. Chrisp, P.; Clissold S. P. *Drugs* **1991,** *41,* 104.

716. Schreiber, A.; Haerter, G.; Schubert, A.; Bunjes, D.; Mertens, T.; Michel, D. *Exp. Opin. Pharmacother.* **2009,** *10,* 191.

717. Gerard, L.; Salmon-Ceron, D. *Int. J. Antimicrob. Agents* **1995,** *5,* 209.

718. Sorbera, L. A.; Rabasseda, X.; Castaner, J. *Drugs Future* **1998,** *23,* 1168.

719. Orr, R. M. *Curr. Opin. Molec. Therapeutics* **2001,** *3,* 288.

720. Perry, C. M.; Balfour, J. A. B. *Drugs* **1999,** *57,* 375.

721. Lowance, D.; Neumayer, H. H.; Legendre, C.; Squiflet, J.-P.; Kovarik, J.; Brennan, P. J.; Norman, D.; Mendez, R.; Keating, M. R.; Coggon, G. L.; Crips, A.; Lee, I. C. and the International

Valacyclovir Cytomegalovirus Prophylaxis Transplantation Study Group. *New Engl. J. Med.* **1999,** *340*, 1462.

722.   Odumade, O. A.; Kristin A. Hogquist, K. A.; Balfour, H. H., Jr. *Clin. Microbiol Rev.* **2011,** *24*, 193.

723.   Balfour, H. H. *New Engl. J. Med.* **1999,** *340*, 1255.

724.   Lin, J. C., Smith, M. C. Pagano, J. S. *J. Virol.* **1984,** *50*, 50.

Index

Printed in the United States
By Bookmasters